Electromagnetic Vortices

IEEE Press
445 Hoes Lane
Piscataway, NJ 08854

Electromagnetic Vortices: Wave Phenomena and Engineering Applications

Edited by

Zhi Hao Jiang
Southeast University

Douglas H. Werner
The Pennsylvania State University

IEEE Antennas and Propagation Society

IEEE Press Series on Electromagnetic Wave Theory

IEEE PRESS
WILEY

Published by John Wiley & Sons, Inc., Hoboken, New Jersey.
Published simultaneously in Canada.

Library of Congress Cataloging-in-Publication Data

Names: Jiang, Zhi Hao, editor. | Werner, Douglas H., 1960– editor.
Title: Electromagnetic vortices : wave phenomena and engineering applications / edited by Zhi Hao Jiang, Douglas H. Werner.
Description: Hoboken, New Jersey : Wiley-IEEE Press, [2021] | Includes bibliographical references and index.
Identifiers: LCCN 2021030660 (print) | LCCN 2021030661 (ebook) | ISBN 9781119662822 (cloth) | ISBN 9781119662969 (adobe pdf) | ISBN 9781119662877 (epub)
Subjects: LCSH: Electromagnetic waves. | Spheromaks. | Angular momentum. | Nanostructures.
Classification: LCC QC718.5.E45 .E44 2021 (print) | LCC QC718.5.E45 (ebook) | DDC 530.14/1–dc23
LC record available at https://lccn.loc.gov/2021030660
LC ebook record available at https://lccn.loc.gov/2021030661

Cover Design: Wiley

Set in 9.5/12.5pt STIXTwoText by Straive, Pondicherry, India

10 9 8 7 6 5 4 3 2 1

M 03M000030000004705 180122

To my wife Pingjuan Li Werner, with grateful appreciation
of her loving support

"It was many and many a year ago,
In a kingdom by the sea (Monterey, CA, 1988),
Where I lost my heart to one you may know,
By the name of Pingjuan Li..."
Douglas H. Werner
(Poem modified from *'Annabel Lee'* by E. A. Poe)

To my parents for their eternal love and understanding
and
to the city of Nanjing where I always belong
Zhi Hao Jiang

Contents

About the Editors

Zhi Hao Jiang is a professor with the State Key Laboratory of Millimeter Waves, School of Information Science and Engineering, Southeast University, Nanjing, P.R. China. He received the BS degree in radio engineering from the Southeast University, Nanjing, in 2008, and the PhD degree in electrical engineering from The Pennsylvania State University, University Park, PA, USA, in 2013. From 2013 to 2016, he was a postdoctoral fellow with the Computational Electromagnetics and Antennas Research Laboratory (CEARL), Department of Electrical Engineering, The Pennsylvania State University. His current research interests include microwave/millimeter-wave/terahertz antennas, millimeter-wave passive circuits, impedance surfaces, metamaterials, and analytical methods in the electromagnetics.

Douglas H Werner holds the John L and Genevieve H McCain Chair Professorship in Electrical Engineering and is the director of the Computational Electromagnetics and Antennas Research Laboratory (CEARL), and also a faculty member of the Materials Research Institute (MRI), at The Pennsylvania State University. Prof. Werner has received numerous awards and recognitions for his work in the areas of electromagnetics and optics. He holds 20 patents, has published over 950 technical papers and proceedings articles, and is the author of 6 books and 35 book chapters. He is a fellow of the Institute of Electrical and Electronics Engineers (IEEE), the Institute of Engineering and Technology (IET), the Optical Society (OSA), the Applied Computational Electromagnetics Society (ACES), and the Progress in Electromagnetics Research (PIER) Electromagnetics Academy. He is also a Senior Member of the National Academy of Inventors (NAI), International Union of Radio Science (URSI), and the International Society for Optics and Photonics (SPIE).

List of Contributors

Shah Nawaz Burokur
LEME, UPL, Univ Paris Nanterre
Ville d'Avray, France

Filippo Capolino
Department of Electrical Engineering and Computer Science
University of California
Irvine, CA, USA

Menglin L. N. Chen
Department of Electrical and Electronic Engineering
The University of Hong Kong
Hong Kong, P.R. China

Yuqi Chen
College of Information Science and Electronic Engineering
Zhejiang University
Hangzhou, P.R. China

Xiaoming Chen
School of Electronic and Information Engineering, Xi'an Jiaotong University, Xi'an, P.R. China

Tie Jun Cui
State Key Laboratory of Millimeter-Waves, School of Information Science and Engineering, Southeast University
Nanjing, P.R. China

Alessio D'Errico
Department of Physics
University of Ottawa
Ottawa, ON, Canada

Mina Hanifeh
Department of Electrical Engineering and Computer Science
University of California
Irvine, CA, USA

Wei Hong
State Key Laboratory of Millimeter-Waves
School of Information Science and Engineering
Southeast University
Nanjing, P.R. China

Naoki Honma
Faculty of Science and Engineering
Iwate University
Morioka, Iwate, Japan

Nanzhe Hu
Department of Electrical and Computer Engineering, University of Southern California
Los Angeles, CA, USA

Hao Huang
Department of Electrical and Computer Engineering, University of Southern California
Los Angeles, CA, USA

Zhi Hao Jiang
State Key Laboratory of Millimeter-Waves, School of Information Science and Engineering, Southeast University
Nanjing, P.R. China

Li Jun Jiang
Department of Electrical and Electronic Engineering, The University of Hong Kong
Hong Kong, P.R. China

Lei Kang
Department of Electrical Engineering
The Pennsylvania State University
University Park, PA, USA

Ebrahim Karimi
Department of Physics
University of Ottawa
Ottawa, ON, Canada

Natalia M. Litchinitser
Department of Electrical and Computer Engineering, Duke University
Durham NC, USA

Cong Liu
Department of Electrical and Computer Engineering, University of Southern California
Los Angeles, CA, USA

Kentaro Murata
Faculty of Science and Engineering
Iwate University
Morioka, Iwate, Japan

Kai Pang
Department of Electrical and Computer Engineering, University of Southern California
Los Angeles, CA, USA

Anastasios Papathanasopoulos
Department of Electrical and Computer Engineering, University of California
Los Angeles, CA, USA

Cheng-Wei Qiu
Department of Electrical and Computer Engineering, National University of Singapore
Singapore

Yahya Rahmat-Samii
Department of Electrical and Computer Engineering, University of California
Los Angeles, CA, USA

Yongxiong Ren
Department of Electrical and Computer Engineering, University of Southern California
Los Angeles, CA, USA

Wei E. I. Sha
Key Laboratory of Micro-nano Electronic Devices and Smart Systems of Zhejiang Province, College of Information Science and Electronic Engineering, Zhejiang University
Hangzhou, P.R. China

Haoqian Song
Department of Electrical and Computer Engineering, University of Southern California
Los Angeles, CA, USA

Hao Song
Department of Electrical and Computer Engineering, University of Southern California
Los Angeles, CA, USA

Xinzhou Su
Department of Electrical and Computer Engineering, University of Southern California
Los Angeles, CA, USA

Fabrizio Tamburini
ZKM Karlsruhe, Lorenzstraße, Berlin, Germany

Bo Thidé
Swedish Institute of Space Physics
Ångström Laboratory, Uppsala, Sweden

Moshe Tur
School of Electrical Engineering
Tel Aviv University
Ramat Aviv, Israel

Zhixia Wang
College of Information Science and Electronic Engineering
Zhejiang University, Hangzhou
P.R. China

Xinyue Wang
College of Information Science and
Electronic Engineering, Zhejiang University
Hangzhou, P.R. China

Douglas H. Werner
Department of Electrical Engineering
The Pennsylvania State University
University Park, PA, USA

Alan E. Willner
Department of Electrical and Computer
Engineering, University of Southern California
Los Angeles, CA, USA

Guodong Xie
Department of Electrical and Computer
Engineering, University of Southern California
Los Angeles, CA, USA

Xiaoyan Y. Z. Xiong
Department of Electrical and Electronic
Engineering, The University of Hong Kong
Hong Kong, P.R. China

Xiaowen Xiong
College of Information Science and
Electronic Engineering, Zhejiang University
Hangzhou, P.R. China

Wei Xue
School of Electronic and Information
Engineering, Xi'an Jiaotong University
Xi'an, P.R. China

Yuanjie Yang
School of Physics, University of Electronic
Science and Technology of China
Chengdu, P.R. China

Jianjia Yi
School of Information and Communications
Engineering, Xi'an Jiaotong University
Xi'an, P.R. China

Jia Yuan Yin
School of Physics and Optoelectronic
Engineering, Xidian University
Xi'an, P.R. China

Runzhou Zhang
Department of Electrical and Computer
Engineering, University of Southern California
Los Angeles, CA, USA

Xianmin Zhang
College of Information Science and
Electronic Engineering, Zhejiang University
Hangzhou, P.R. China
Ningbo Research Institute, Zhejiang University
Ningbo, P.R. China

Zhe Zhao
Department of Electrical and Computer
Engineering, University of Southern California
Los Angeles, CA, USA

Jiayu Zheng
College of Information Science and
Electronic Engineering
Zhejiang University
Hangzhou, P.R. China

Shilie Zheng
College of Information Science and
Electronic Engineering, Zhejiang University
Hangzhou, P.R. China
Ningbo Research Institute
Zhejiang University, Ningbo, P.R. China

Huibin Zhou
Department of Electrical and Computer
Engineering, University of Southern California
Los Angeles, CA, USA

Zelin Zhu
College of Information Science and Electronic
Engineering, Zhejiang University, Hangzhou,
P.R. China

Preface

The content of this book is focused on the cutting-edge research area of electromagnetic vortices including fundamental theory, wave phenomena, and also the applications of vortex waves in a variety of disciplines. Akin to vortices that exist widely throughout nature, ranging from a fluid vortex with a central flow singularity to the countless spiral galaxies in the universe, vortices also exist in electromagnetic and optical fields which are manifested in what are called vortex waves. Such vortex waves refer to a class of electromagnetic waves carrying nonvanishing orbital angular momentum (OAM), which are characterized by their helical phase-front and annular-shaped intensity distribution, indicating a singularity at the beam center. This family of waves encompasses unconventional waveforms with complex structures such as vector vortex beams, higher-order Bessel beams, etc. Their structured wavefronts endow them with unprecedented wave behavior that breaks conventional bounds enabling new physical properties, revealing a subtle connection between macroscopic physical optics and microscopic quantum optics. These unique properties have further led to the creation of revolutionary new ways of exploiting wave-matter interactions, especially when coupled with interdisciplinary research areas such as nanophotonics, metamaterials, communications, nonlinear optics, quantum information, etc. Ever since the introduction of the concept of electromagnetic vortices in 1989 by Coullet et al. and the first demonstration of OAM in paraxially propagating vortex beams by Allen et al., there has been a dramatically increasing number of scientists and engineers enthusiastically devoting their efforts to this field of exploration. In particular, over the past few years, the field of electromagnetic vortices has witnessed tremendous advancements in many of its associated subareas, which have jointly enabled devices and even systems for a wide variety of applications including astrophysical observation, particle manipulation, nonlinear optics, microlasers, fiber optics, wireless communications, and even quantum cryptography, just to name a few. As such, electromagnetic vortices hold great promise for revolutionizing developments in many research areas and triggering key disruptive technologies that have the potential to change the daily lives of human beings.

The main goal of this book is to provide an all-inclusive and in-depth overview of many of the recent and important accomplishments regarding the various topics associated with the field of electromagnetic vortices. In this book, a comprehensive treatment of the field has been compiled from a group of leading researchers throughout the world on subjects ranging from the fundamental theoretical principles and discoveries of new physical wave behaviors to state-of-the-art device designs and illustrative application examples encompassing a wide range of related sub-areas. In particular, topics that will be covered include the basics of electromagnetic vortices, the generation and detection of various forms of vortex waves with homogeneous and inhomogeneous wavefronts from the microwave regime all the way down to optical wavelengths, the multiplexing and demultiplexing of vortex beams, nonlinear and active vortex beams, communication and sensing that exploits vortex beams, manipulation of structured photon states for quantum information, etc. Multidisciplinary research outcomes linking electromagnetic vortices to nanophotonics, quantum, wireless communications, astronomy, and optical forces are presented, which may serve to inspire interest from scientists and engineers currently outside this field. Throughout the book, a plethora of transformational application examples have also been included, encompassing OAM-based wireless communication systems, radar detection, lasers, chirality detection, magnetic and chirality force microscopy, quantum security, as well as several others. Moreover, these examples were carefully selected for their transformational impact on pushing the research envelope in the field of electromagnetic vortices from investigations of physical phenomena to real-world engineering applications. Importantly, each chapter is contributed by an internationally recognized author or group of authors covering the latest research results. We hope that this book will be an indispensable resource for researchers, faculty, and graduate students in academia as well as professionals working in the telecommunication, optical engineering, and even future quantum information industries.

This book contains 14 invited chapters contributed by the leading experts in the field of electromagnetic vortices, which are divided into four parts. Part I includes 2 chapters that are focused on providing an introduction to the fundamentals and basics of electromagnetic vortices, Part II contains a collection of 6 chapters emphasizing the physical wave phenomena associated with electromagnetic vortices, Part III has 4 chapters that highlight several of their disruptive engineering applications, and the last two chapters in Part IV cover some important multidisciplinary explorations in the field. A brief summary of each chapter is furnished as follows. Chapter 1 provides an introduction to the fundamentals of vortex beams and discusses their relationship with conventional pencil beams. It also includes a brief overview of the generation methods as well as potential applications of vortex waves in communications from the antenna standpoint. Chapter 2 presents a detailed step-by-step derivation of the physical foundations of OAM radios from the perspective of classical electrodynamic theory, followed by a short discussion regarding their physical implementation and astrophysics applications. Chapter 3 provides an overview of the generation of scalar vortex beams at microwave frequencies using coding metasurfaces, including both reflective and transmissive designs, and spoof surface plasmon polariton waveguiding circuits. Chapter 4 illustrates the design methodology and experimental characterization of vortex beam generators by utilizing a combination of transformation optics and additive manufacturing. By employing 3-D printing enabled all-dielectric platforms that are nearly non-dispersive with respect to their frequency response, the resulting devices possess a very broad bandwidth. Chapter 5 considers the basic properties and generation methods for vector vortex beams, which represent a more generalized family of waveforms under the umbrella of electromagnetic vortices. By coupling various high-efficiency transmit-array cells with a globally coherent phase-only synthesis strategy, high-capacity generation and multiplexing of vector vortex beams are demonstrated (both numerically and experimentally) in the millimeter-wave regime. Chapter 6 reviews the metamaterial approaches for achieving electromagnetic vortices at optical wavelengths using nanofabrication techniques and provides a brief overview of their optical applications, with a particular focus on active optical vortex emitters, i.e. OAM lasers. Chapter 7 continues the discussion of the generation of optical vortices, emphasizing the methodologies for achieving special types of vortices with intriguing physical properties, which are different from conventional scalar vortex beams, including perfect vortices, fractional vortices, anomalous vortices, and vortices with varying topological charges. Chapter 8 investigates both the generation and detection of electromagnetic vortices by employing metasurfaces and photonic crystals with defects. In addition, the nonlinear effects associated with vortex beams are also described, including the proposal of a precise numerical boundary element method and the formulation of a general angular momentum conservation law for studying the second harmonic generation of vortex waves. Chapter 9 introduces the application of vortex beams for radar detection, multiple-in multiple-out (MIMO) wireless systems, and spatial field digital modulation. In order to fulfill the system requirements with a simplified solution, several OAM antenna designs are proposed including a partial arc transmitting scheme for constructing higher-order OAM modes. Chapter 10 focuses on the theoretical analysis of vortex beam multiplexing for short-range microwave communications from the perspective of the antenna array, channel matrix, and beamforming circuits, arriving at an optimal solution for the array radius based on the geometric channel model. Chapter 11 consists of the application of vortex waves for wireless communications in more sophisticated and realistic multipath environments, where strong inter-channel and intra-channel interference can take place. By combining with spatial equalization and orthogonal frequency division multiplexing techniques, it is shown that OAM communication can be useful in not only the line-of-sight environment but also in rich multipath environments. Chapter 12 presents the design consideration, application challenges, system implementation, and experimental characterization of free-space OAM optical communication links. In particular, an inter-channel crosstalk mitigation scheme is proposed and the application of optical OAM communications for unmanned aerial vehicles as well as in underwater environments is discussed and demonstrated. Chapter 13 focuses on two particular vector vortex beams, i.e., radially- and azimuthally-polarized vortex beams, and provides the theoretical treatment for how they can be applied to chirality detection of subwavelength-size particles and magnetic force microscopy. Chapter 14 investigates the usage of structured photons in quantum platforms and a method for generating/detecting single-photons as well as entangled photon pairs carrying well-defined OAM states. The chapter also includes illustrative examples showcasing their applications in quantum key distribution for quantum security and quantum simulation with quantum walks.

We are indebted to all the coauthors of each of the 14 chapters for their excellent contributions to this book, which provides the readers with an invaluable resource documenting the cutting-edge research outcomes in the field. The cooperation of the contributing authors during the course of this book preparation as well as their dedication to delivering a high-quality product, especially under such globally difficult circumstances arising from the pandemic, is sincerely and deeply appreciated. Finally, we also gratefully acknowledge the editorial assistance provided by the highly professional staff of Wiley/IEEE Press. It has truly been a pleasure working with them on this book project.

Zhi Hao Jiang
Nanjing, P.R. China

Douglas H. Werner
University Park, PA, USA

Part I

Fundamentals and Basics of Electromagnetic Vortices

1

Fundamentals of Orbital Angular Momentum Beams

Concepts, Antenna Analogies, and Applications

Anastasios Papathanasopoulos and Yahya Rahmat-Samii

Department of Electrical and Computer Engineering, University of California, Los Angeles, CA, USA

1.1 Electromagnetic Fields Carry Orbital Angular Momentum

Angular momentum is the rotational equivalent of linear momentum. An object moving in a straight line carries linear momentum. Along with its linear motion, an object can also rotate in two distinct ways. A spinning object carries spin angular momentum (SAM) and an orbiting object carries orbital angular momentum (OAM), with Earth being an interesting example for consideration. The analogy between the angular momentum of a moving object and the electromagnetic wave's angular momentum is shown in Figure 1.1. Earth's spinning is associated with SAM; Earth's orbital motion around the sun is linked to OAM. Electromagnetic beams may also carry these two types of momenta. If the electric field vector rotates along the beam axis, the beam has SAM; if the electromagnetic field's wavefront is twisting, with the wave vector forming a helix, the beam has OAM.

The history of angular momentum dates back to the early twentieth century. The SAM of light was theoretically studied for the first time by Poynting in 1909 [1] and experimentally studied by Beth in 1936 [2]. SAM is intrinsic, since it does not depend on the choice of an axis, and it is only polarization dependent. If s is the SAM mode number, then $s = \pm 1$ corresponds to right- and left-hand polarized waves, and $s = 0$ corresponds to linearly polarized waves. Although the experimental demonstration of the exchange of angular momentum between circularly polarized beams and matter was performed more than 80 years ago, the work associated with light's angular momentum has been almost exclusively concerned with SAM.

It was not until 1992 that Allen et al. [3] showed that helically phased beams with a phase term $e^{jl\phi}$ (where $j = \sqrt{-1}$ is the imaginary unit, l is the OAM mode number, and ϕ is the azimuthal angle) carry OAM. The wavefront of an OAM beam is a spiral; the phase twists around the beam axis and changes $2\pi l$ after a full turn. Unlike SAM, OAM is linked to spatial distribution rather than polarization. OAM is extrinsic, since it depends on the choice of the calculation axis. The angular momentum is the composition of SAM and OAM, such that the angular momentum mode number is $j = s + l$.

Electromagnetic Vortices: Wave Phenomena and Engineering Applications, First Edition.
Edited by Zhi Hao Jiang and Douglas H. Werner.

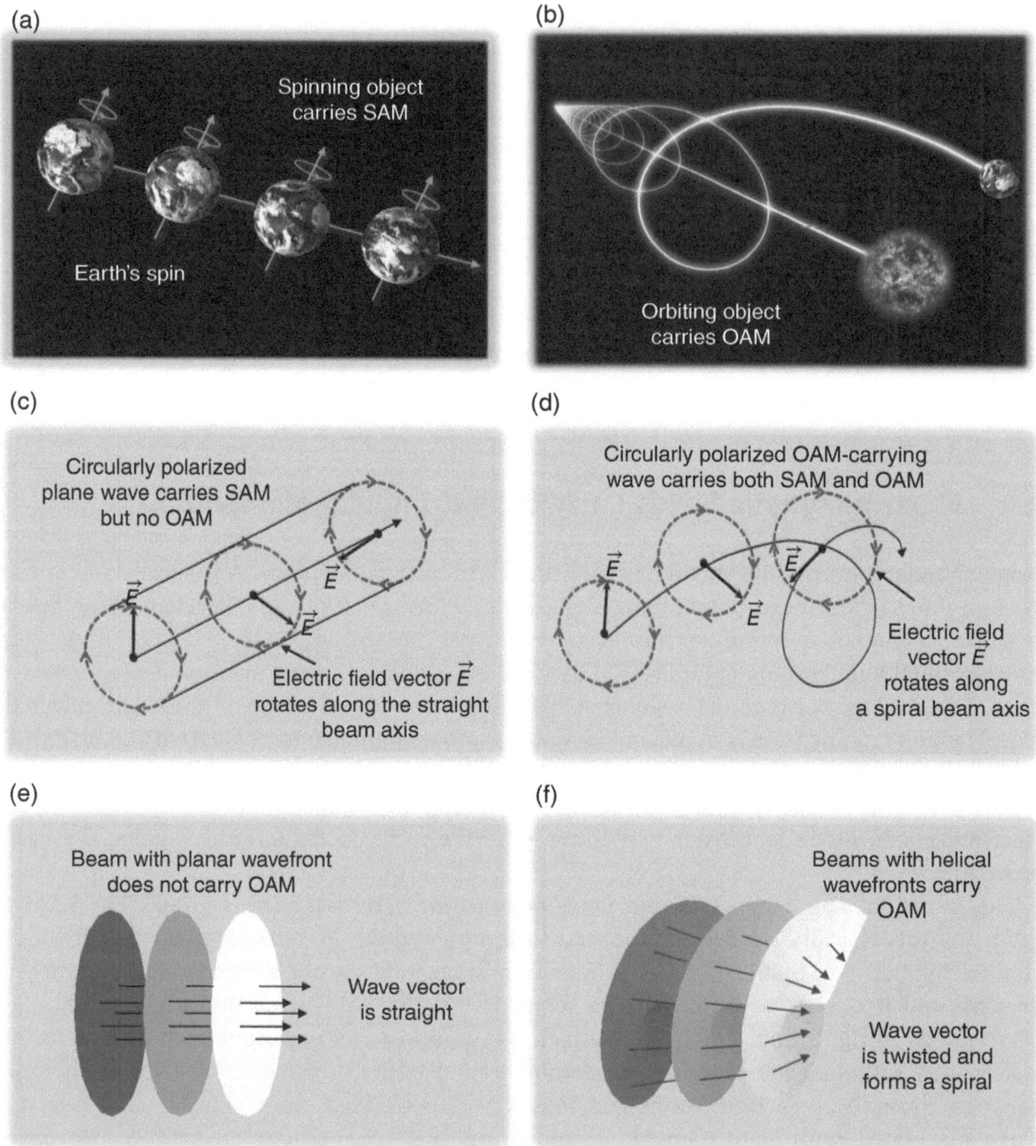

Figure 1.1 Analogy between moving object's and electromagnetic wave's angular momentum: (a) Earth spins and carries SAM; (b) Earth orbits around the sun and carries OAM; (c) a circularly polarized wave carries SAM; the electric field vector rotates along a straight beam axis; (d) a circularly polarized OAM-carrying wave carries both SAM and OAM; the electric field vector rotates along a spiral beam axis; (e) beam with planar wavefront; the wavevector is straight; (f) OAM-carrying beams with spiral wavefront; the wavevector is twisted.

1.2 OAM Beams; Properties and Analogies with Conventional Beams

Helically phased beams exhibit two unique properties: (i) the *orthogonality* of different OAM modes and (ii) the OAM beam *divergence*. As we will discuss in Section 1.3, the infinite number of orthogonal OAM modes provides an additional set of data carriers, thus OAM has the potential to increase the capacity and spectral efficiency of wireless communication links [4]. The beam divergence can be advantageous for applications that require cone-shaped patterns, such as satellite-based navigation and guidance systems that serve moving vehicles [5] but may also pose a challenge for long-distance communication links [6]. In the subsequent paragraphs, these two peculiar features are explicated, and a comparative study between conventional and OAM beams is performed to underline their differences.

The first characteristic property of OAM beams is the *orthogonality* of distinct OAM modes. In general, the electric field of an OAM beam can be written as:

$$E(\rho, \phi, z) = A(\rho, z)e^{jl\phi}, \tag{1.1}$$

where ρ is the radial distance, ϕ is the azimuthal angle, and z is the propagation distance in the cylindrical coordinate system. There exist various subsets of OAM-carrying beams, for example Laguerre–Gaussian [3], Bessel–Gaussian [7], hypergeometric–Gaussian [8], vector vortex [9], and other beams [10]. The radial distribution $A(\rho, z)$ in Eq. (1.1) defines the special subset among all OAM-carrying beams. All the previous beams carry the $e^{jl\phi}$ term, which gives rise to an OAM mode of l-order. If we consider the electric fields $E_1(\rho, \phi, z)$, $E_2(\rho, \phi, z)$ of two OAM beams with modes l_1, l_2, and radial distributions $A_1(\rho, z)$, $A_2(\rho, z)$, the following orthogonality relation is satisfied [11]:

$$\int_0^{2\pi} E_1(\rho, \phi, z)E_2^*(\rho, \phi, z)d\phi = \begin{cases} 0 & \text{if } l_1 \neq l_2 \\ A_1(\rho, z)A_2^*(\rho, z) & \text{if } l_1 = l_2 \end{cases}, \tag{1.2}$$

where the asterisk (*) denotes the complex conjugate. It follows that: (i) there is an infinite number of OAM modes, with each mode identified by the mode number l, and (ii) the infinite set of OAM states forms an orthogonal basis.

The second special feature of OAM beams is the beam *divergence*. The far-field signature of the helical wavefront is an amplitude null at the phase vortex center. Accordingly, the null size can be described in terms of a divergence angle, which represents the angle from the null to the maximum gain [12]. As the OAM beam travels through space, the radius of the 'dark zone' around the amplitude null in the center of the beam increases.

1.2.1 Laguerre–Gaussian Modes

In general, an OAM-carrying beam could refer to any beam that carries the $e^{jl\phi}$ term, regardless of the radial distribution $A(\rho, z)$ in Eq. (1.1). The Laguerre–Gaussian modes are a special subset among all OAM-carrying beams that are cylindrically symmetric solutions to the paraxial wave equation in the cylindrical coordinate system [3]. The Laguerre–Gaussian modes are chosen to be presented because they are one of the most popular examples of OAM-carrying beams (see, for example [13–18]), and a general OAM-carrying beam can be expanded in a complete basis of Laguerre–Gaussian modes [11, 19, 20]. The electric field of a linearly polarized Laguerre-Gaussian beam at $z = 0$ can be written as [3, 5]:

$$\vec{E}^{LG}(\rho, \phi, z = 0) = \sqrt{\frac{2p!}{\pi(p + |l|)!}} \frac{E_0^{LG}}{w_g} \left(\frac{\rho\sqrt{2}}{w_g}\right)^{|l|} e^{-\rho^2/w_g^2} L_p^{|l|}\left(\frac{2\rho^2}{w_g^2}\right) e^{-jl\phi}\hat{x}, \tag{1.3}$$

where ρ and ϕ are the radial and azimuthal coordinates in the cylindrical coordinate system; E_0^{LG} is a complex amplitude coefficient, l and p are integers known as azimuthal and radial mode numbers, w_g is the equivalent beam waist that can be related to the antenna aperture diameter D (refer to [5] and Appendix 1.A for more details) and is equal to the half-width of the normalized aperture field amplitude at $1/e$ controlling the transverse extent of the beam, $L_p^l(\cdot)$ is the associated Laguerre polynomial [21]:

$$L_p^l(x) = \frac{1}{p!}e^x x^{-l}\frac{d^p}{dx^p}(e^{-x}x^{p+l}) = \sum_{m=0}^{p}(-1)^m\binom{p+l}{p-m}\frac{x^m}{m!}, \tag{1.4}$$

where the binomial coefficient is [21]:

$$\binom{n}{k} = \frac{n!}{k!(n-k)!} \tag{1.5}$$

when $k \leq n$ and is zero when $k > n$. For $l = 0$, the Laguerre–Gaussian beam carries no OAM since the phase term $e^{-jl\phi}$ vanishes. For any other l, the field carries the phase term $e^{-jl\phi}$, which gives rise to an OAM state of $-l$-order. The normalized electric field intensity distributions of Laguerre–Gaussian beams with different azimuthal and radial modes l and p are shown in Figures 1.2 and 1.3. It can be observed that the number of side lobe intensity rings is equal to the integer p. For the same p, the null size (i.e. the divergence angle) increases as the azimuthal mode number l increases.

The far-field features of Laguerre–Gaussian beams were studied in [5]. The far-field expression can be found from Eq. (1.3) using the aperture field method [5] (see Appendix 1.A for the proof):

$$\begin{aligned}\vec{E}_{ff}^{LG}(r,\theta,\phi) = {} & \frac{jk_0E_0^{LG}e^{-jk_0r}}{4\pi r}(\hat{\theta}\cos\phi - \hat{\phi}\cos\theta\sin\phi)w_g(-1)^p(-j)^l \\ & \times\sqrt{\frac{2\pi p!}{(p+|l|)!}}\left(\mathrm{sgn}(l)\frac{\Psi}{\sqrt{2}}\right)^{|l|}e^{-\Psi^2/4}L_p^{|l|}\left(\frac{\Psi^2}{2}\right)e^{-jl\phi}\end{aligned} \tag{1.6}$$

and $\Psi = k_0w_g\sin\theta$ ($k_0 = 2\pi/\lambda$ is the free-space wavenumber). Equation (1.6) is a cone-shaped pattern with azimuthal symmetry. Note that the electric field maintains the phase term $e^{-jl\phi}$ in the far-field. This is a general characteristic of OAM fields (for example, the same feature is observed for the case of Bessel–Gaussian beams [5]) and a proof can be found in Appendix 1.A. The far-field expression for the Laguerre–Gaussian mode with $p = 0$ can be simplified as:

$$\vec{E}_{ff}^{LG}(r,\theta,\phi) = \frac{jk_0E_0^{LG}e^{-jk_0r}}{4\pi r}(\hat{\theta}\cos\phi - \hat{\phi}\cos\theta\sin\phi)w_g(-j)^l\sqrt{\frac{2\pi}{|l|!}}\left(\frac{\mathrm{sgn}(l)\Psi}{\sqrt{2}}\right)^{|l|}e^{-\Psi^2/4}e^{-jl\phi}.$$

For the dominant radial mode $p = 0$, the far-field expression Eq. (1.7) peaks at the elevation angle of:

$$\theta_c^{LG} = \frac{1}{\sin}\left(\frac{\sqrt{2|l|}}{k_0w_g}\right). \tag{1.7}$$

Equation (1.7) shows that the cone angle depends both on the azimuthal mode number l and the beam waist (i.e. aperture diameter, as was shown in [5]). For constant l, the cone angle decreases as we increase the beam waist w_g, i.e. the aperture diameter. For constant w_g, the cone angle increases as we increase the mode number l.

In what follows, a comparison between two classes of beams is carried out. The first is the conventional Airy disk, and the second is the OAM-carrying Laguerre–Gaussian beam. The Airy disk pattern is produced by a circular aperture with uniform amplitude and phase-field distributions and it is a common and useful model in the design of conventional aperture-type antennas, such as

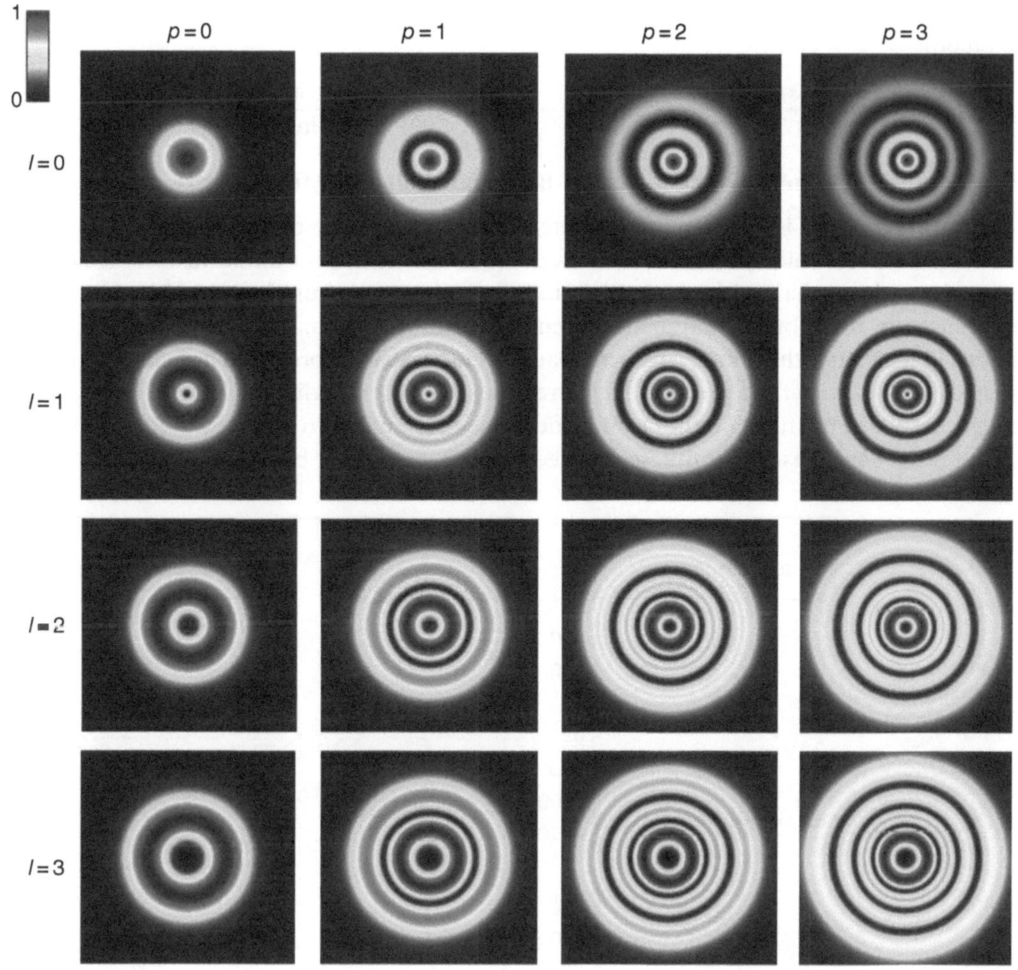

Figure 1.2 Normalized aperture field intensity distributions versus ρ/w_g of Laguerre–Gaussian beams with different azimuthal and radial modes l and p. The number of side lobe intensity rings is equal to the integer p. For the same p, the null size (i.e. the divergence angle) increases as the azimuthal mode number l increases.

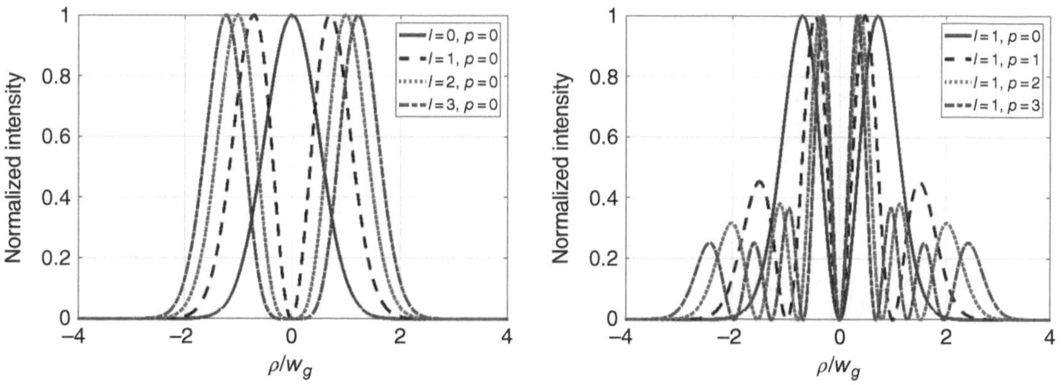

Figure 1.3 Normalized aperture field intensity line cuts of Laguerre–Gaussian beams with different azimuthal and radial modes l and p.

reflectors. The far-field electric field of the Airy disk pattern can be written as (see Appendix 1.A for the proof):

$$\vec{E}_{ff}^{AD}(r,\theta,\phi) = \frac{jk_0E_0^{AD}a^2e^{-jk_0r}}{2r}(\hat{\theta}\cos\phi - \hat{\phi}\cos\theta\sin\phi)\frac{J_1(k_0a\sin\theta)}{k_0a\sin\theta}, \quad (1.8)$$

where E_0^{AD} is a constant; J_1 is the first-order Bessel function of the first kind; $k_0 = \frac{2\pi}{\lambda}$ is the free-space wavenumber; D is the aperture diameter, and $a = D/2$ is the radius of the aperture.

What are the fundamental differences between conventional and OAM beams? In the near-field region, the aperture phase of the Airy disk is uniform and the wavefront is planar. The Airy disk can be produced by a uniformly illuminated circular aperture antenna, such as a parabolic reflector antenna [22]. On the other hand, Laguerre–Gaussian modes can be produced by helicoidal reflector antennas [23]. The aperture phase of Laguerre–Gaussian modes twirls around the beam axis and changes $2\pi l$ after a full turn (l is the OAM mode number), resulting in a spiral wavefront. Figure 1.4 illustrates the analogies and antitheses between these two types of beams.

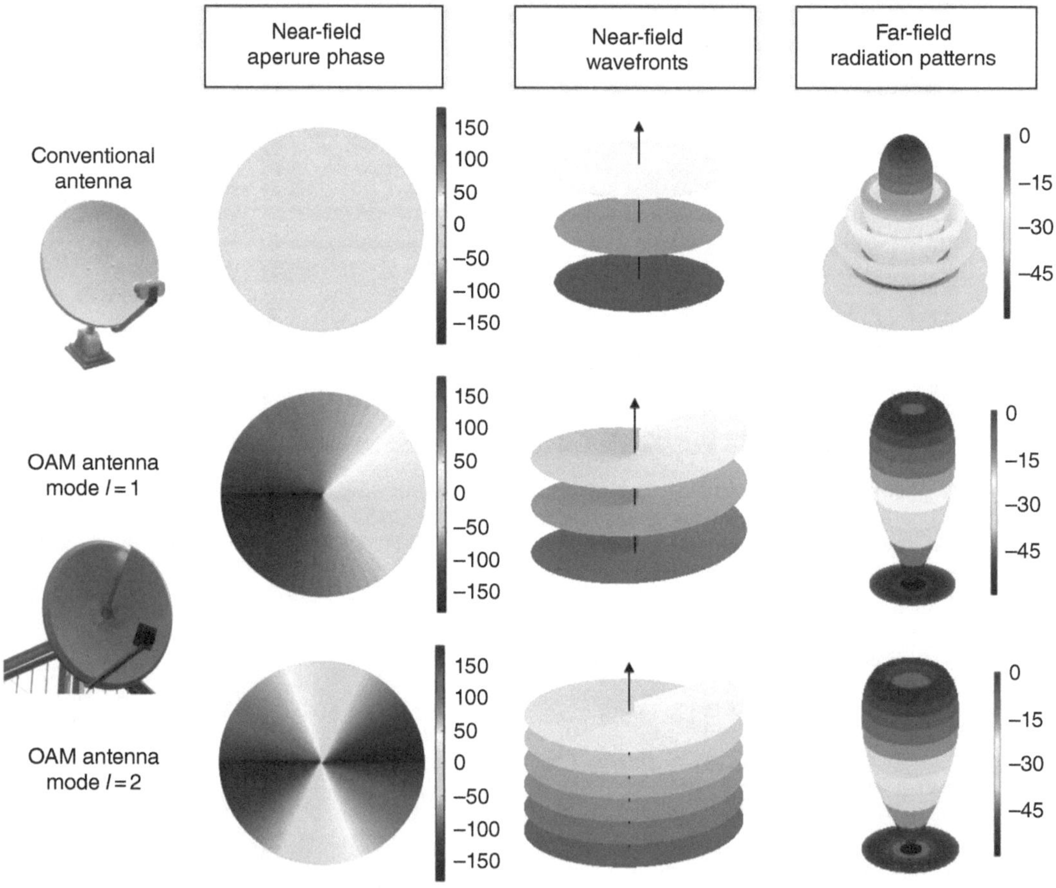

Figure 1.4 Comparison between conventional and OAM beams. A uniformly illuminated circular aperture produces a planar wavefront in the near-field and a highly directive far-field radiation pattern. An OAM-carrying Laguerre–Gaussian beam with mode number l produces a spiral wavefront in the near-field and a cone-shaped far-field pattern with an amplitude null at the phase vortex center. The OAM beam divergence increases for larger l.

The far-field characteristics of the Airy disk and the Laguerre–Gaussian beam are remarkably different. For the Airy disk case, the manifestation of the uniform aperture phase and the planar wavefront is a highly directive far-field pattern with the maximum gain at the axis of symmetry of the antenna. The locus of the points with constant phase in the far-field, i.e. the far-field wavefront S of the Airy disk, can be found from Eq. (1.8):

$$S_{AD} = k_0 r, \tag{1.9}$$

which describes a spherical wavefront. The gradient of the wavefront gives the direction of the wavevector (i.e. the geometrical optics ray direction):

$$\nabla S_{AD} = k_0 \hat{r} \tag{1.10}$$

For the Laguerre–Gaussian beam, the far-field signature of the vortex phase is a cone-shaped pattern with an amplitude null at the center. The locus of the points with constant phase in the far-field can be found from Eq. (1.6) and is described by the following equation:

$$S_{LG} = k_0 r + l\phi, \tag{1.11}$$

which describes a twisted wavefront, as shown in Figure 1.5. Note that the first term is frequency dependent similar to Eq. (1.10). The gradient of the wavefront (i.e. the wavevector direction) is:

$$\nabla S_{LG} = k_0 \hat{r} + \hat{\phi}\frac{l}{r\sin\theta} = k_0\left[\hat{r} + \hat{\phi}\frac{l}{k_0 r\sin\theta}\right]. \tag{1.12}$$

Unlike the wavevector of the Airy disk, the wavevector of a Laguerre–Gaussian beam has a radial and azimuthal component. Note that for very large distances compared to the wavelength ($k_0 r \rightarrow \infty$),

$$\nabla S_{LG} \approx k_0 \hat{r}, \quad \text{for} \quad k_0 r \rightarrow \infty. \tag{1.13}$$

In other words, the wavefront of a Laguerre–Gaussian modes at very large distances compared to the wavelength, i.e. in the very far-field, resembles a spherical wavefront, as shown in Figure 1.5.

An immediately remarkable trait of the cone-shaped far-field pattern is the reduced directivity compared to the conventional beam counterpart. The Airy disk pattern is a highly directive far-field pattern with 100% aperture efficiency. A Laguerre–Gaussian mode with $l = 1$, aperture diameter D, and beam waist $w_g = 0.415D$ has an aperture efficiency of 50%. As a result, an antenna with an aperture that is twice as large compared to the conventional counterpart is required to maintain the same directivity level. Figure 1.6 shows the directivity of the Airy disk pattern and the first-order

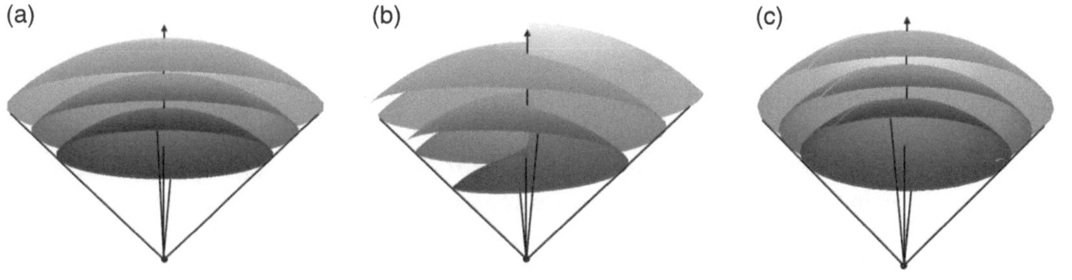

Figure 1.5 Far-field wavefront of (a) Airy disk and (b, c) Laguerre–Gaussian modes. The wavefront of the Airy disk is spherical and the wavefront of Laguerre–Gaussian modes is twisted; (c) at very large distances compared to the wavelength ($k_0 r \rightarrow \infty$), the wavefront of Laguerre–Gaussian approaches a spherical wavefront.

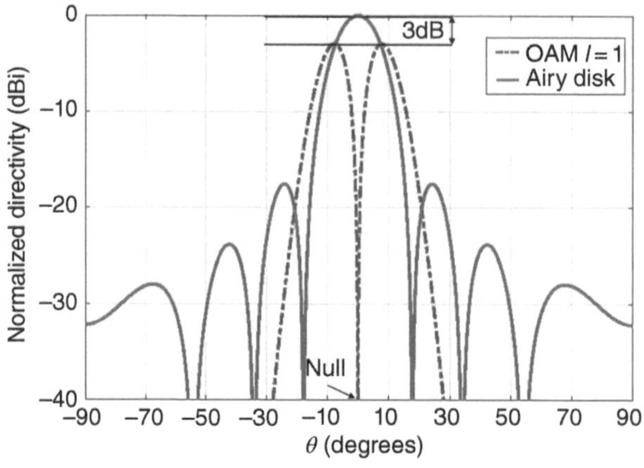

Figure 1.6 Normalized directivities of the Airy disk pattern with aperture diameter $D = 4\lambda$ and the Laguerre–Gaussian mode with mode number $l = 1$, same aperture diameter D, and beam waist $w_g = 0.415D$. The directivity of the Laguerre–Gaussian beam is 3 dB less than the Airy disk pattern for this choice of D and w_g. The curves are normalized to the maximum directivity of the Airy disk pattern.

Laguerre–Gaussian beam for $D = 4\lambda$ and $w_g = 0.415D$, normalized to the maximum directivity of the Airy disk pattern. The directivity of the Laguerre–Gaussian beam is 3 dB less than the Airy disk pattern of the same aperture size. Another distinctive characteristic is the beam divergence; the larger the OAM mode number, the larger the cone angle of the beam. That is, higher-order OAM modes diverge more rapidly with propagating distance.

To provide further insight in the OAM field distribution at various distances far from the antenna, we studied the changes of amplitude pattern shape from the reactive near-field toward the far-field. The first case of study is a conventional reflector antenna with a uniform aperture phase and −10 dB edge taper. The aperture field distribution is modeled using the two-parameter (2P) model [22, eq. (16)] (for more details see Appendix 1.A), and the field at various distances is calculated using the Fresnel–Kirchhoff diffraction integral [24]. The second case of study is a helicoidal reflector [23]. The aperture field is modeled using the [22, eq. (16)] multiplied by the phase term $e^{-jl\phi}$, for $l = 1$. The aperture diameter for both cases is $D = 10\lambda$. The changes of amplitude pattern shape from the reactive near-field toward the far-field for the two cases is shown in Figure 1.7. The pattern is calculated at $r = 4.9\lambda$, 24λ, 8000λ corresponding to the reactive near-field, radiating near-field and far-field regions [25]. It can be observed that the amplitude null at the vortex center is maintained at all distances.

1.3 Communicating Using OAM: Potentials and Challenges

The observation that OAM-carrying beams exhibit two unique properties compared to conventional beams, namely, the orthogonality and divergence, gives rise to the following question: Can OAM beams' unique characteristics potentially benefit communication links? We demonstrate that the answer is 'yes,' albeit there are still many challenges to overcome before OAM is employed in a practical application. Inspired by this question, we examine possible communication scenarios that involve OAM antennas, review potential applications of OAM beams in communication systems, and discuss technical challenges.

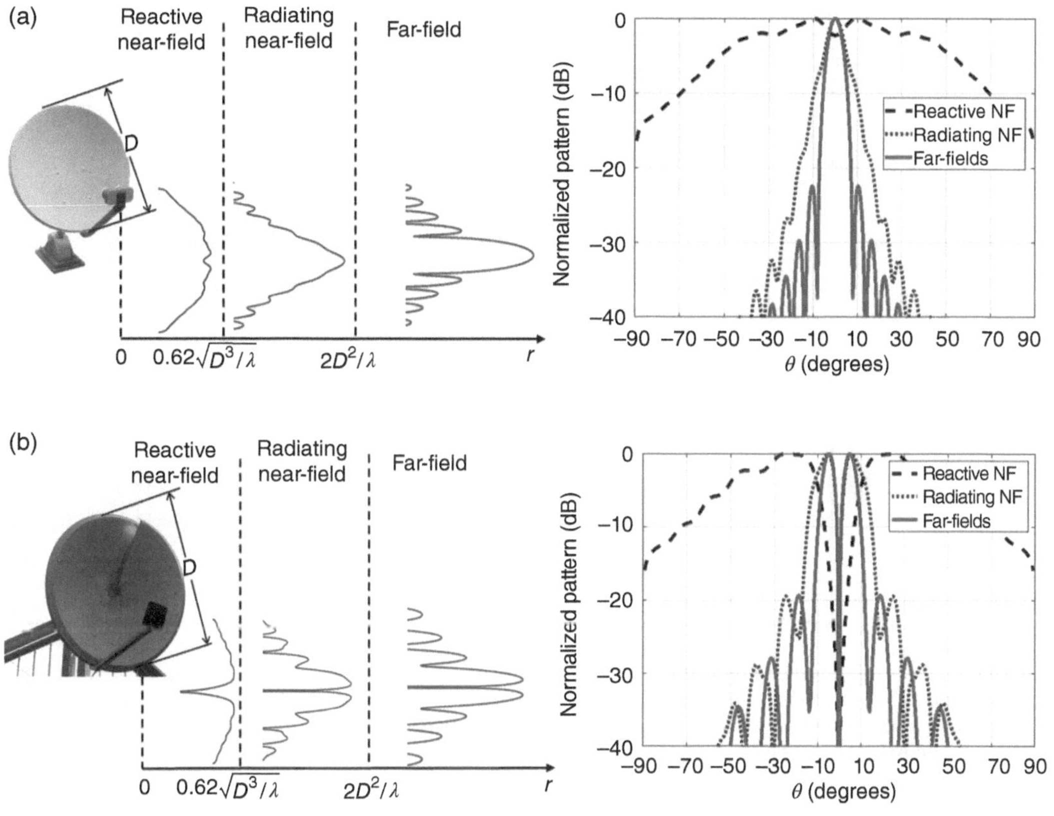

Figure 1.7 Changes of amplitude pattern shape from the reactive near field toward the far-field for (a) a conventional antenna and (b) an OAM antenna. The dimensions are $D = 10\lambda$ and the pattern is calculated at $r = 4.9\lambda$, 24λ, 8000λ corresponding to the reactive near-field, radiating near-field and far-field regions.

1.3.1 OAM Communication Link Scenarios and Technical Barriers

In general, there are two communication links that involve OAM antennas, as shown in Figure 1.8. In the first scenario, an OAM antenna establishes a communication channel with a conventional antenna (the OAM antenna can be transmitting or receiving signals). This case is no different than the classical counterpart, where two conventional antennas are utilized. The link budget is calculated using the classical Friis transmission formula based on the directional properties, distance in terms of wavelength, and orientation of the antennas [26]:

$$\frac{P_R}{P_T} = \left(\frac{\lambda}{4\pi d}\right)^2 G_T(\theta_t, \phi_t) G_R(\theta_r, \phi_r), \tag{1.14}$$

where P_R and P_T are the received and transmitted powers, respectively, λ is the operating wavelength, d represents the link distance, $G_T(\theta_t, \phi_t)$, $G_R(\theta_r, \phi_r)$ correspond to the gains along the transmitting T and receiving R antennas direction; antennas polarization mismatch is neglected for simplicity. The figure of merit of this communication link is the received power (i.e., signal-to-noise ratio) at the antenna terminals.

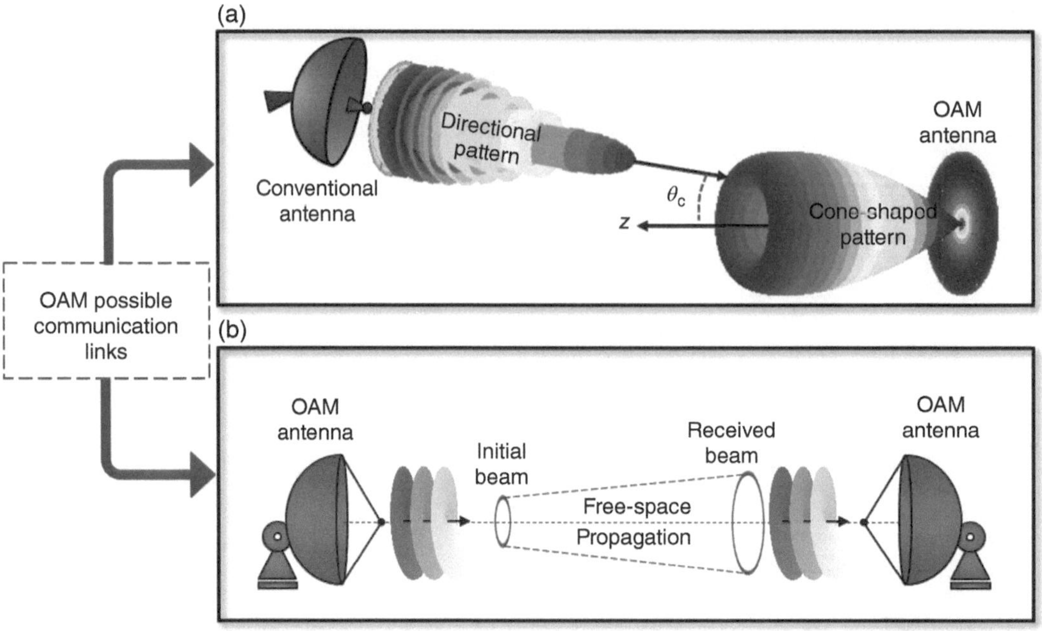

Figure 1.8 OAM possible communication links: (a) OAM to the conventional antenna. The figure of merit is the received power, which is calculated using the classical Friis transmission formula [26]. *Source:* Based on Friis [26]; (b) OAM to OAM antenna. The figures of merit are the received power and the detected OAM mode purity, which are calculated using the generalized Friis transmission equation [27]. The OAM beam divergence poses a challenge for far-field communication distances. A large receiving aperture is required. *Source:* Based on Cho and Byun [27].

In the second scenario, two OAM antennas that are facing each other constitute the communication link. The transmitting antenna generates an OAM beam of mode number l, which propagates in free space and is detected by another OAM antenna. The performance of this particular link is characterized by the received power (i.e., signal-to-noise ratio) as well as the detected OAM mode purity; a generalized Friis transmission equation that accounts for both the received power and the mode purity has been developed in [27]. A large receiving aperture is required for two reasons. First, a limited aperture would lead to power loss due to the central intensity drop of nonzero OAM modes. A series of recent studies has shown that the power loss penalty increases for larger OAM modes (larger OAM modes diverge more rapidly with propagating distance) [28–31]. Second, a substantial portion of the transmitted wavefront should be captured at the receiver endpoint to detect the OAM phase information and accurately estimate the mode number l.

An OAM-based communication link presents unique challenges that are related to the OAM beams' inherent characteristics, such as the beam divergence. Ideally, a large and perfectly aligned receiving aperture would capture the full OAM beam, as shown in Figure 1.9. The requirement to collect enough power of the divergent OAM beam with a very large aperture (compared to wavelength) significantly limits the achievable distance of OAM communication links [32], especially for the radio frequencies, where the wavelengths are longer compared to optical frequencies. In particular, Refs. [33–35] question the possibility of using OAM antennas in long-distance communication links. They highlight that to achieve a realistic signal-to-noise ratio at the receiver endpoint, the physical size of the receiving antenna becomes impractical. The impact of the finite receiving

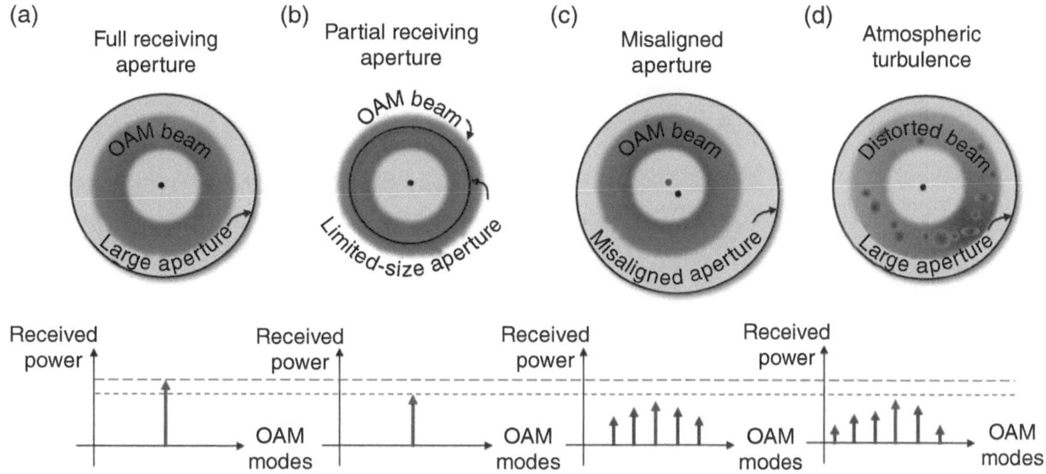

Figure 1.9 OAM communication challenges: (a) receiving with a perfectly aligned and full receiving aperture, (b) receiving with a limited-size aperture causes power loss of the detected mode, (c) receiving with a misaligned aperture, and (d) atmospheric turbulence causes power leakage in adjacent modes.

aperture size and the misalignment (lateral and angular) between the transmitter and the receiver on the performance of an optical OAM communication link was investigated in [36]. In principle, when a finite-size receiving aperture is used, power loss of the detected mode occurs (see Figure 1.9). Receiving an OAM beam with misalignment would cause power leakage in the neighboring modes and inter-mode crosstalk [37].

Another critical challenge for a practical OAM-communication link is the atmospheric turbulence. Temperature and pressure fluctuations in the atmosphere result in spatial changes in the refractive index. The propagation of optical OAM beams in the turbulent atmosphere has been numerically modeled in [38–40] and experimentally investigated in [41–45]. The main conclusion is that atmospheric turbulence leads to wavefront distortion and significant system performance degradation. Atmospheric turbulence impacts more OAM beams in optical frequencies than radio frequencies [32]. As the turbulence strength increases, the power of the transmitted OAM mode starts to leak to neighboring modes and tends to be equally distributed among modes for strong turbulence [46].

The multipath propagation phenomenon in wireless communication links using OAM is another factor that affects the system performance. The analysis, simulations, and measurements of multipath effects in a millimeter-wave communication link using OAM multiplexing at 28 GHz were presented in [47]. A schematic of multipath effects of an OAM channel caused by specular reflection from a parallel ideal reflector (ground plane) that is placed at distance h below the transmitting antenna is shown in Figure 1.10. From image theory, the reflected beam can be seen as an OAM beam of an opposite mode number (the reflection changes the sign of the OAM mode) that is placed at a distance h below the ground plane. As a result, the receiver antenna will not only receive an OAM beam with an OAM number of l from the original transmitter but also a reflected beam with an OAM number of $-l$ from an offset link. Ref. [47] showed that an OAM channel with a higher OAM number l tends to suffer more from both strong intra-channel crosstalk and strong inter-channel crosstalk with other OAM channels.

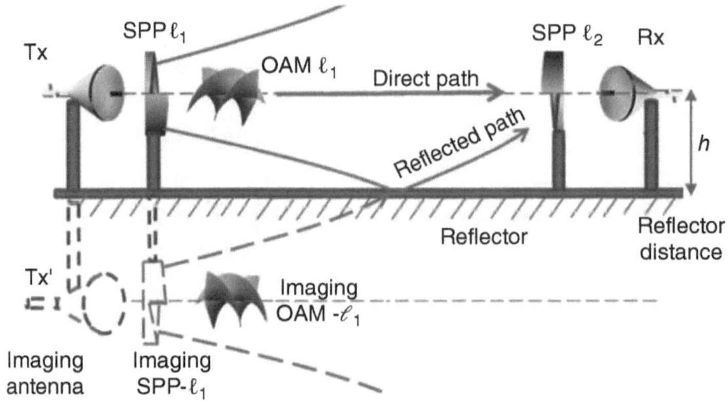

Figure 1.10 A schematic of multipath effects of an OAM channel caused by specular reflection from a parallel ideal reflector (ground plane) that is placed at distance *h* below the transmitting antenna. The receiver antenna will not only receive an OAM beam with an OAM number of *l* from the original transmitting antenna but also a reflected beam with an OAM number of −*l* from an offset link. *Source:* Yan et al. [47]; © 2016 Springer Nature. Licensed under CC BY-4.0.

1.3.2 OAM Emerging Applications and Perspectives

1.3.2.1 Free-space Communications

One application that may take advantage of the OAM cone-shaped pattern is the local satellite-based navigation and guidance system that serves moving vehicles. For such an application, the satellite is usually placed in a geostationary orbit and employs a contour beam that covers a certain geographical area [5], as shown in Figure 1.11. An antenna is mounted on the moving vehicle and is circularly polarized to alleviate multipath propagation problems caused by reflections from building walls and the ground surface. For the vehicle to continuously communicate with the satellite, a directivity of at least 3 dB in the direction of the satellite should be maintained regardless of the orientation of the vehicle [48].

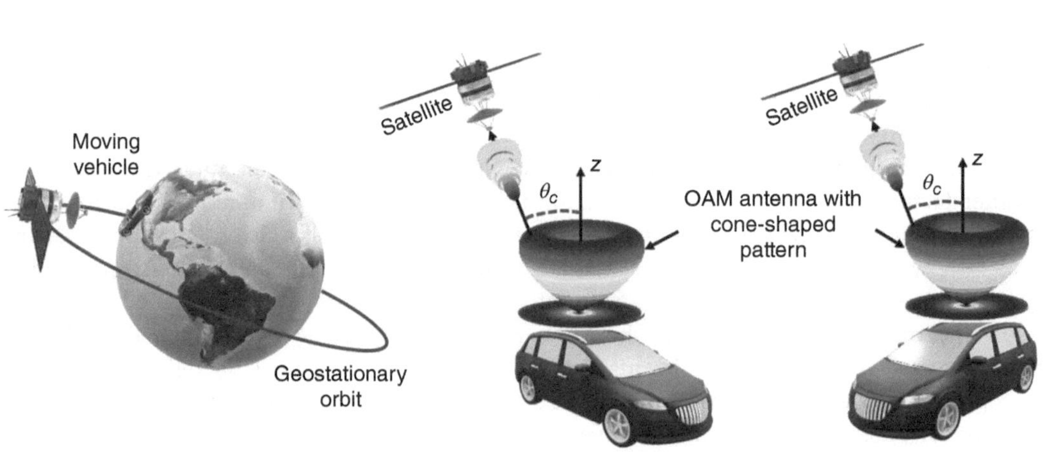

Figure 1.11 A satellite-based, geostationary navigation and guidance system that serves moving vehicles. OAM antennas with cone-shaped patterns are proposed to be mounted on the vehicle [5]. The same signal-to-noise ratio is achieved regardless of the orientation of the vehicle because of the horizontally omnidirectional radiation pattern. *Source:* Modified from Veysi et al. [5].

Several antenna designs have been suggested for communications between moving vehicles on the earth and geostationary satellites. One possible solution is an antenna with a pencil-beam pattern that sweeps out a cone in space. The beam pattern scans to continuously point toward the satellite so that uninterrupted vehicle-to-satellite communication is achieved. This solution increases the cost of the antenna system since the tracking of the satellite is needed. An alternative approach is an antenna with a fixed cone-shaped pattern, with the peak of the cone (denoted by the cone angle θ_c in Figure 1.11) pointing toward the satellite. The same signal-to-noise ratio is achieved regardless of the orientation of the vehicle because of the horizontally omnidirectional radiation pattern. The latter approach does not require the tracking of the satellite, thus the antenna cost and complexity are reduced. Among various antennas with conical patterns that have been proposed for vehicle-to-satellite communications [48–51], an OAM antenna is a powerful apparatus that systematically generates a cone-shaped pattern with high-azimuthal symmetry [5].

Besides the OAM beam divergence and the cone-shaped pattern, the second fundamental property of OAM beams, namely, the orthogonality of different OAM modes, can be employed to transfer information. OAM shift keying (OAM-SK) is one communication technology that takes advantage of the distinction of the different orthogonal OAM modes [11, 32, 52, 53]. In this communication technology, the OAM modes are used as a communication alphabet, with N data symbols encoded in the OAM mode number l of N OAM states. Figure 1.12 shows an example where OAM modes $l = 0,\ 1,\ 2$ represent the data symbols 0, 1, 2. The sequence of the OAM states is sent by the transmitter. The data are decoded at the receiver by checking the received OAM mode number. OAM-SK is a direct modulation technology that enables transferring information by encoding information in the value l of the OAM mode number.

A series of proof-of-concept experiments has elevated the scientific interest in the potential of harnessing OAM-SK for free-space information transfer. The first experiment to use OAM-SK for free-space information transfer was carried out by Gibson in 2004 [17]. In Gibson's seminal work, eight different OAM modes in the set $\ell = \{-16,\ -12,\ -8,\ -4,\ +4,\ +8,\ +12,\ +16\}$, each representing a data symbol, were used as communication alphabets. The transmitted OAM states and data symbols were accurately measured up to a transmission range of 15 m using a HeNe laser transmitter at a wavelength of 632.8 nm. Since the pioneering work of Gibson, several OAM-SK experiments have been carried out. In [54], 16 superimposed OAM modes were used to transfer information over a 3 km link of strong turbulence over the city of Vienna. The transmitter was a laser with a spatial light modulator at a wavelength of 532 nm. An incoherent detection scheme aided by an artificial neural network was employed to identify the characteristic mode patterns and distinguish among the 16 OAM modes. The latter experimental scheme has been extended to a distance of 143 km between two Canary Islands [55]. To the authors' best knowledge, this is the longest distance that an OAM-SK experiment has been reported. Therein, the OAM mode

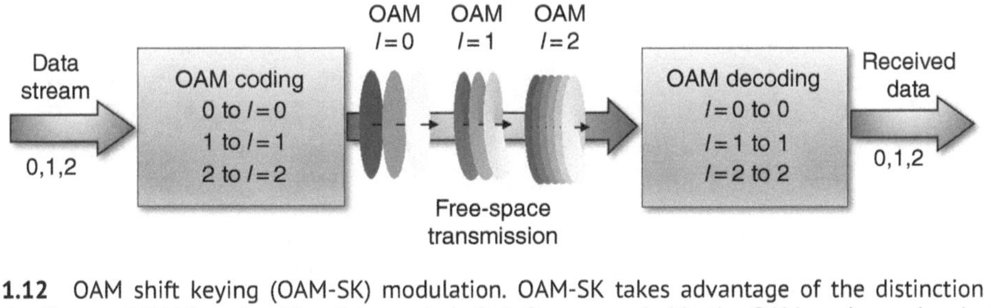

Figure 1.12 OAM shift keying (OAM-SK) modulation. OAM-SK takes advantage of the distinction of the different orthogonal OAM modes. The data symbols 0, 1, 2 are encoded in the OAM mode number l = 0, 1, 2.

patterns were observed on a receiving aperture with a diameter of roughly $D = 11$ m, and a green laser with a wavelength of 532 nm was used, corresponding to a far-field distance of $2D^2/\lambda = 4.55 \times 10^5$ km. Note that the 143 km link distance is large in absolute terms, but it cannot be considered a far-field distance. As discussed in Section 1.3.1, objections regarding the possibility of using OAM in far-field communication links have surfaced in [33–35], and the adoption of OAM in far-field wireless communications is still an open problem.

As the available wireless spectrum becomes more and more crowded under the incessant strive for high data-transfer capacities, several methods have been developed to increase the spectral efficiency. Various properties of the electromagnetic wave, such as time, polarization, and wavelength, have been employed to develop multiplexing schemes [56, 57]. Using the spatial property of electromagnetic waves as a new dimension (i.e. space division multiplexing (SDM)) has recently aroused interest as a potential method to enhance wireless communications. A special case of SDM is the mode division multiplexing (MDM), where spatially orthogonal modes act as independent data channels that can be efficiently multiplexed and de-multiplexed. Thereby, the spectral efficiency is increased by a factor equal to the number of the independent spatial modes.

One particular subset of SDM is the OAM division multiplexing (OAM-DM) [32]. In an attempt to further increase the spectral efficiency of wireless communications, OAM-DM technique exploits the orthogonality of OAM modes [58, 59]. An independent data stream (i.e. data 1, 2, ..., N in Figure 1.13) is encoded on the value l of N carrier OAM modes. The different OAM beams are multiplexed at the transmitter along with other modes, propagate through a common medium, and are de-multiplexed at the receiver. The obtained field at the receiver can be written as [32]:

$$U_{\mathrm{MUX}}(\rho, \phi) = \sum_{p=1}^{N} S_p A_p(\rho) e^{jl_p\phi}, \tag{1.15}$$

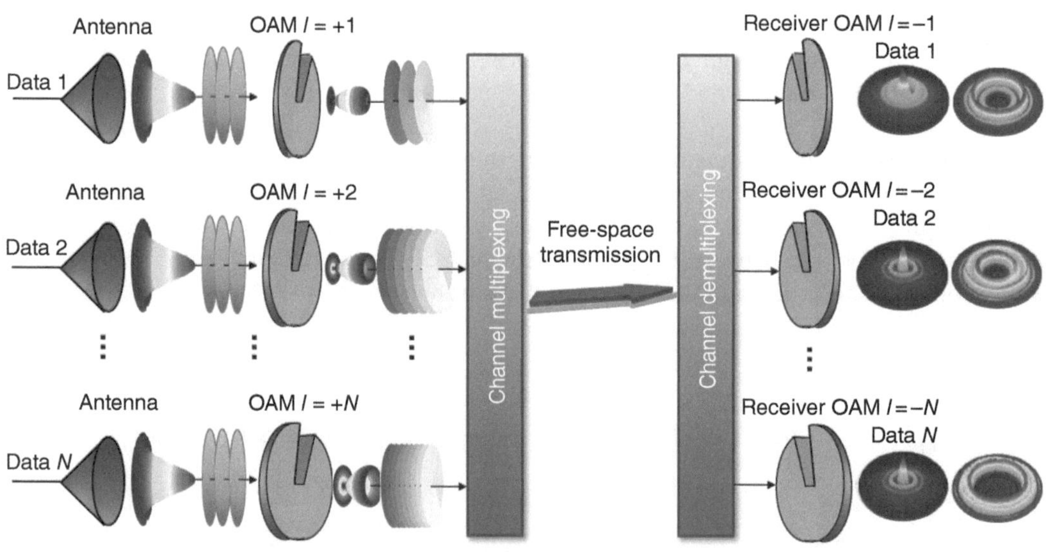

Figure 1.13 OAM division multiplexing (OAM-DM) modulation. An independent data stream (i.e. data 1, 2, ..., N) is encoded on the value l of N carrier OAM modes. The different OAM beams are multiplexed at the transmitter, propagate through a common medium, and are de-multiplexed at the receiver. Thereby, the spectral efficiency is increased by a factor N equal to the number of the carrier OAM modes.

where N is the number of OAM modes, S_p is the modulated data signal on the pth OAM mode, $A_p(\rho)$ is the complex electric field amplitude of the pth OAM mode, ρ is the radial distance, and ϕ is the azimuthal angle in the cylindrical coordinate system. At the receiver end, the multiplexed signal is multiplied by a phase factor $e^{-jl_q\phi}$, where $l_q = 1, \ 2, \ \ldots, \ N$. The OAM beam with $l = l_q$ has a central intensity peak and can be easily distinguished from other modes by means of their different spatial signatures. Finally, the data information carried by the OAM mode $l = l_p$ can be retrieved. Consequently, the spectral efficiency is increased by a factor N equal to the number of the carrier OAM modes. A schematic diagram illustrating the principles of OAM-DM is shown in Figure 1.13.

Several recent experiments have demonstrated the potential of optical OAM multiplexing to increase the spectral efficiency of wireless communication systems. In [61], a transmission capacity of 2.56 Tbit s^{-1} and a spectral efficiency of 95.7 bit s^{-1} Hz^{-1} were achieved using OAM-DM and polarization multiplexing (8 OAM modes and 2 orthogonal polarization states at a wavelength of 1550.12 nm). In [60], an aggregate capacity of 100.8 Tbit s^{-1} was achieved using 1008 data channels. Each data channel was carried on 12 OAM beams, 2 polarizations, and 42 wavelengths (1536.34–1568.5 nm), and was encoded with 100 Gbit s^{-1} quadrature phase-shift keying data. A schematic illustrating the concept of three-dimensional multiplexing to increase the multiplexed data channels by combing OAM-DM, polarization-division multiplexing, and wavelength-division multiplexing is shown in Figure 1.14. OAM-DM can be combined with other multiplexing schemes, such as time, polarization, and wavelength multiplexing, to satisfy the ever-growing need for transmission capacity.

The possibility of using OAM-DM extends beyond optical frequencies. Thidé et al. [62] were the first to numerically show that antenna arrays can generate OAM beams in radio frequencies and highlight the potential of OAM communications in the lower frequencies. The first experimental test of encoding multiple channels on the same radio frequency using OAM was performed by Tamburini et al. [23]. Subsequently, several OAM-DM experiments have been performed in the radio frequency domain. In [63], an OAM-DM 10 m microwave link operated at 10 GHz with four OAM modes was experimentally demonstrated to quadruple the spectral efficiency while keeping a low-receiver computational complexity. The antenna aperture size was 0.6 m, which corresponds to a far-field distance of $2D^2/\lambda = 24$ m and the 10 m link cannot be considered a far-field link. In [64], a 32 Gbit s^{-1} mm-wave link was demonstrated over 2.5 m at a carrier frequency of 28 GHz with a spectral efficiency of 16 bit s^{-1} Hz^{-1} using four independent OAM beams on each of the two orthogonal polarizations. The receiving and transmitting antenna aperture had circular apertures with diameters of 30 cm, which correspond to a far-field distance of $2D^2/\lambda = 16.8$ m and the 2.5 m link cannot be considered a far-field link. As discussed in Section 1.3, the adoption of OAM in far-field wireless radio frequency (RF) communications is still an open problem.

1.3.2.2 Optical Fiber Communications

Optical fibers have been successfully integrated into communication systems, offering the advantages of large bandwidth, low loss and cost, and immunity to electromagnetic interference [65]. The conventional optical fiber is known as single-mode fiber (SMF) and includes a core surrounded by a transparent cladding material with a lower index of refraction; the attenuation is typically 0.2 dB/km at 1550 nm and the core radius does not exceed 10 μm [53]. Up to now, the increasing demand for high-speed data transfer in optical networks has been covered by the conventional SMF. The growing demand for network capacity is expected to continue in the future. According to Ref. [66], the annual global data center traffic will reach 20.6 Zettabytes (zB) – or 20.6×10^{21} bytes – (1.7 zB per month) by the end of 2021, up from 6.8 zB per year (568 exabytes per month) in 2016.

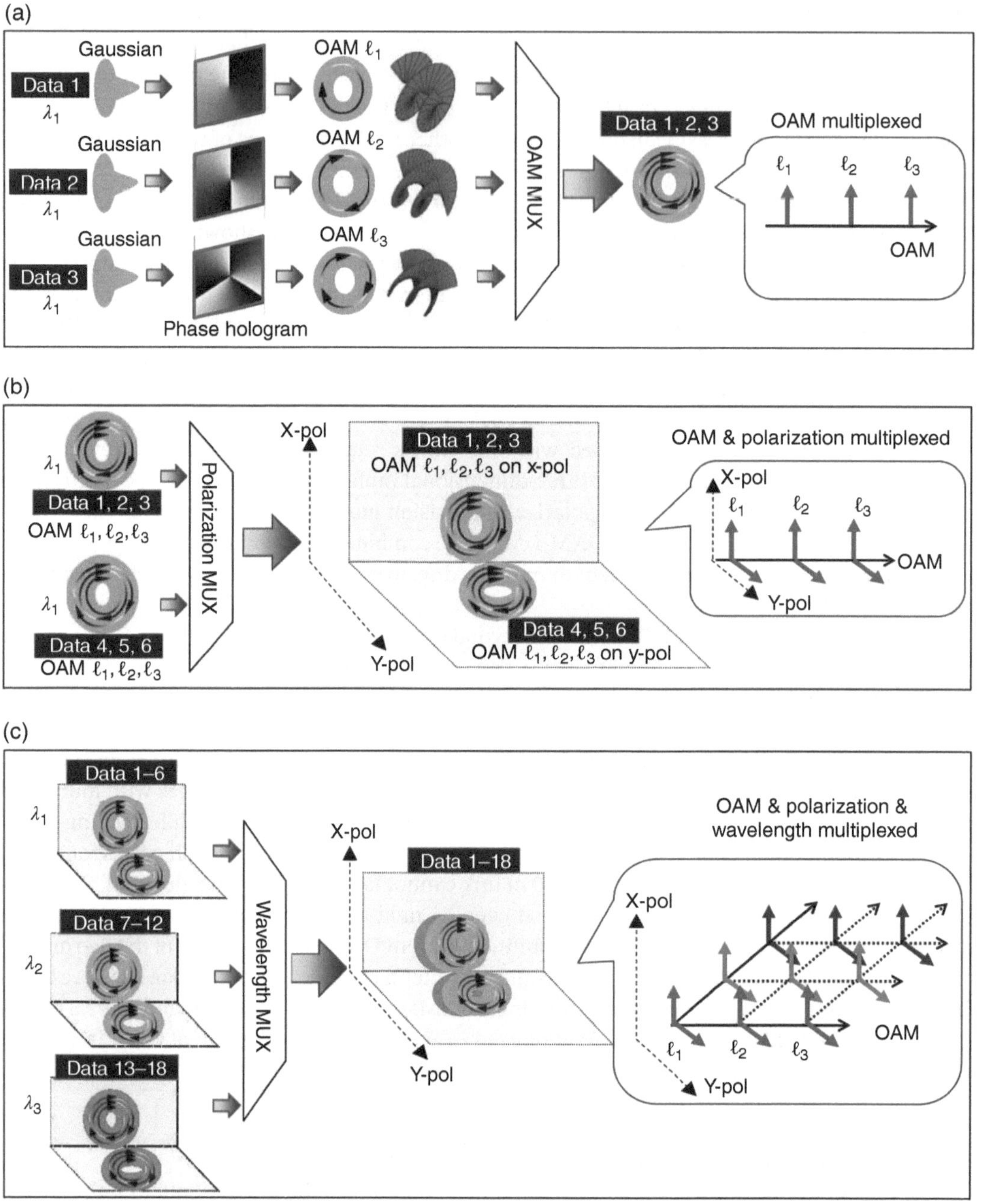

Figure 1.14 A schematic illustrating the concept of three-dimensional multiplexing to increase the multiplexed data channels by combing (a) OAM-DM, (b) polarization-division multiplexing, and (c) wavelength-division multiplexing. *Source:* Huang et al. [60] © 2014 Optical Society of America.

In an attempt to meet the everlasting demand for network capacity, alternative methods have been developed to increase the spectral efficiency of optical networks. The problem is that the conventional SMF has its inherent capacity limits; therefore, researchers have focused on innovative ways to increase the data-carrying capacity of a single optical fiber. SDM using multi-core fibers

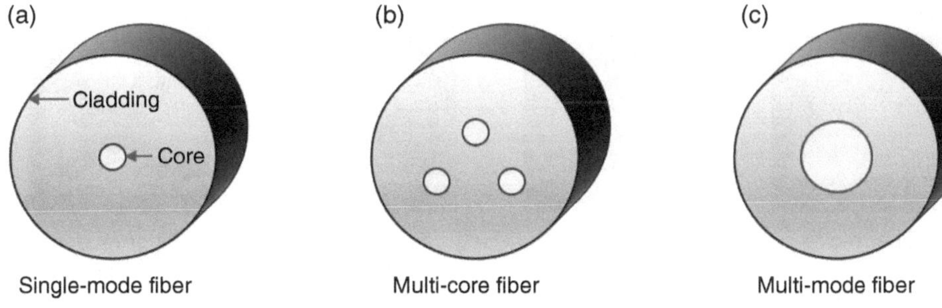

Figure 1.15 A schematic representation of different types of optical fibers.

(MCFs), multi-mode fibers (MMFs), and their combination has been considered to overcome the capacity crunch of conventional SMFs [67, 68]. A schematic of different types of optical fibers is shown in Figure 1.15 [65]. In an MCF, multiple cores share the same cladding and act as an independent data carriers; thereby, the spectral efficiency scales by the number of the cores. Unlike MCFs, MMFs have a single core that is larger than the conventional SMF, usually around 50 μm in radius. MMFs can support multiple spatial orthogonal modes each carrying distinct data and acting as a separate communication channel, thus increasing the spectral efficiency of optical fibers. The latter special case of SDM is known as MDM [67].

The most popular spatial modes for fiber MDM are the linearly polarized (LP) modes. LP modes are the superposition of the fiber eigenmodes transverse electric (TE), transverse magnetic (TM), and hybrid modes of HE and EH type [65]. LP modes are the solution of the scalar Maxwell's equations and propagate under the weakly guiding approximation, which is valid when the refractive indices of the core and the cladding are very close (within 1%) [69]. The intensity profiles of LP modes together with the constituent eigenmodes and the corresponding propagation constants β are shown in Figure 1.16a–d. The constituent eigenmodes of each LP mode have different propagation constants leading to walk-off and mode distortion. LP modes also share common eigenmodes and thus are inherently coupled during propagation leading to crosstalk. To overcome the problem of mode-coupling, MMFs rely on digital signal processing (DSP) algorithms based either on adaptive optics feedback or complex multiple-input multiple-output (MIMO) methodologies [32, 70].

To avoid inter-modal coupling and MIMO DSP complexity in MMFs, OAM modes have been proposed as an alternative modal basis set [69]. The electric field of OAM modes in fibers can be expressed in terms of HE and EH hybrid modes as [71, 72]:

$$\text{OAM}_{\pm l,m} = \begin{cases} \text{HE}^{\text{even}}_{l+1,m} \pm j\text{HE}^{\text{odd}}_{l+1,m} \\ \text{EH}^{\text{even}}_{l-1,m} \pm j\text{EH}^{\text{odd}}_{l-1,m} \end{cases} \tag{1.16}$$

The intensity profiles of OAM modes, the constituent eigenmodes, and the corresponding propagation constants β are shown in Figure 1.16e–h. Unlike LP modes, OAM modes are formed by the superposition of a single eigenmode type with even and odd symmetry [73]. That is, OAM modes are inherently de-coupled resulting in low crosstalk, and the constituent eigenmodes with even and odd symmetry share the same propagation constant leading to a reduced walk-off effect compared to LP modes [74]. As a result, OAM has the potential to increase the spectral efficiency of optical fiber communications while obviating complex DSP algorithms.

OAM-based fiber communications is a new field of research with a short history. The first paper to report the generation of an OAM-carrying optical vortex in a fiber was published in 2006 [76].

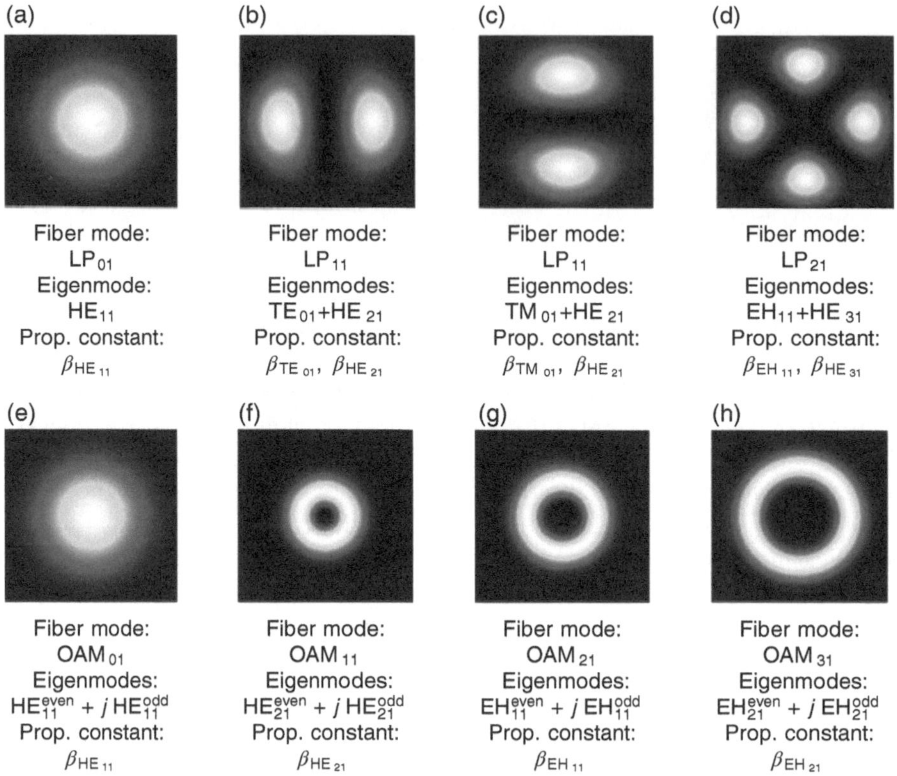

Figure 1.16 Intensity profiles of LP modes and OAM modes. The constituent eigenmodes and the propagation constants β of each mode are noted below each LP and OAM mode. Unlike OAM modes, LP modes (i) have distinct propagation constants leading to walk-off and (ii) share common eigenmodes and are inherently coupled resulting in crosstalk.

Bozinovic et al. reported the stable propagation of OAM modes in a new class of optical fibers, called *vortex fibers* [77, 78]. The vortex fiber of [78] maintained a mode purity of 97% after a propagating distance of 20 m at a wavelength of 1550 nm under bends and twists. The same group demonstrated the successful transmission of OAM states through a 0.9 km vortex fiber. In 2012, a 1.1 km long vortex fiber was presented [79]. A year later, the same 1.1 km vortex fiber was used to achieve 400 Gbit s^{-1} data transmission using four OAM modes at a single wavelength of 1550 nm, and 1.6 Tbit s^{-1} using two OAM modes over 10 wavelengths around 1550 nm with a low-complexity DSP coherent detection method [80]. The experiment of [80] suggested that OAM could provide an additional degree of freedom for data multiplexing in future fiber networks using OAM-MDM that does not require complex DSP [81]. At present, research efforts focus on the design of fibers that can support multiple OAM modes, both MMFs (see, for example Refs. [74, 82–85]) and MCFs (see Figure 1.17 and Refs. [75, 86, 87]).

1.4 OAM Generation Methods

The history of OAM dates back in 1992, when Allen et al. showed that helically phased beams with a phase term $e^{jl\phi}$ carry OAM [3]. Since then, research efforts have focused on the generation of OAM beams. Figure 1.18 shows the analogy between the generation principle of OAM and conventional

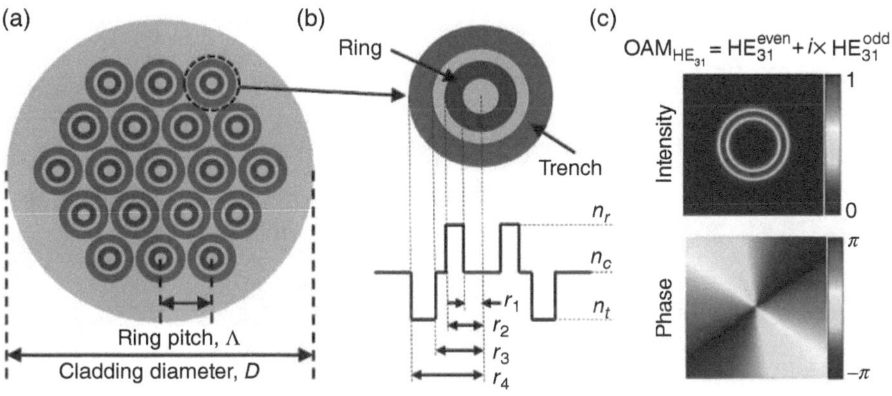

Figure 1.17 (a) Cross-section of a 19-ring multi-OAM multi-ring fiber. (b) Refractive index profile of a single ring. (c) Intensity and phase distributions of HE_{31} related OAM mode in a single ring. *Source:* Li and Wang [75] © 2014 Springer Nature.

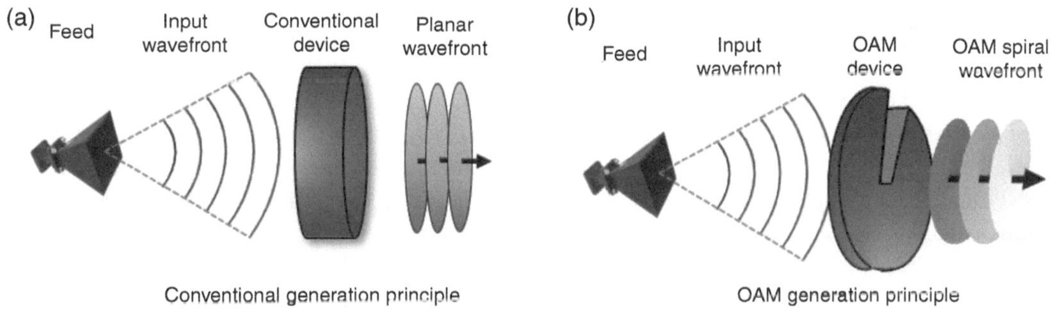

Figure 1.18 Analogy between conventional and OAM generation principle.

beams. In a conventional antenna system, a device (such as reflectors [22] and lenses [88]) is used to transform an input wavefront (for example, a spherical wavefront in Figure 1.18) into a planar wavefront, to achieve a highly directive beam in the far-field. In analogy with the conventional antenna system, an OAM system employs an OAM device to transform the input wavefront into a spiral wavefront. Figure 1.19 shows a representative example where the wavefront-transformation devices are a parabolic reflector [22] and a helicoidal reflector antenna [23], corresponding to the conventional and OAM cases, respectively. The conventional reflector antenna attempts to achieve a uniform phase in the exit aperture, whereas the helicoidal reflector aims to achieve the OAM vortex phase.

The generation of OAM-carrying beams necessitates the design of antennas that can generate the vortex aperture phase and spiral wavefront that are associated with OAM beams. Some representative designs that have been developed for this purpose are shown in Table 1.1. The spiral phase plate (SPP) has one planar and one spiral surface with a step discontinuity, and the thickness increases as the azimuth angle increases [13, 19, 89–91]. The original shape of the conventional parabolic reflector is modified to form a 'helicoidal reflector' [23] that converts the field generated by a feed into an OAM-carrying field [92–94]. A stepped spiral reflecting surface is a discrete approximation of the helicoidal reflector [101]. Planar SPPs, transmitarray [95–100], and reflectarray antennas [5, 114–117] employ elements that are distributed in planar surfaces; these elements

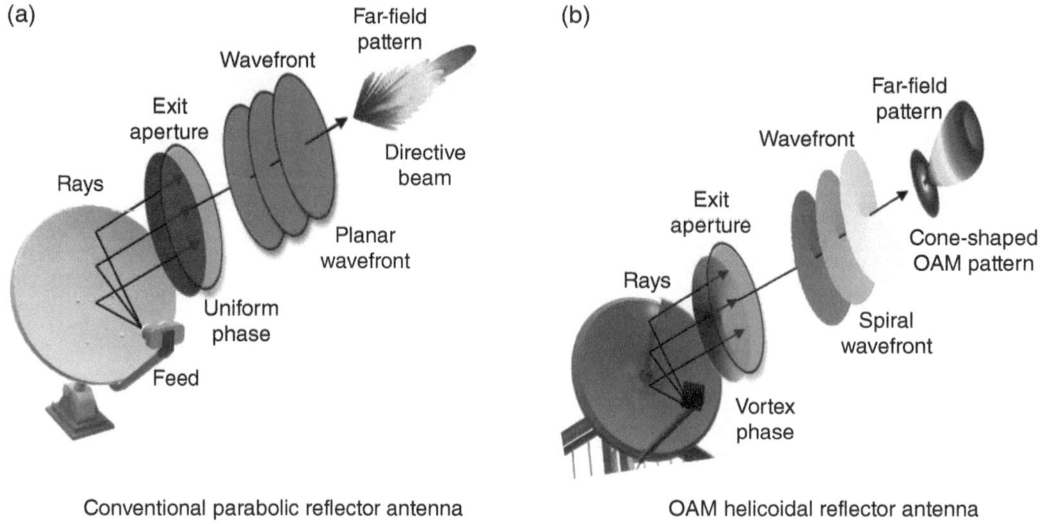

Figure 1.19 Comparison between the generation method of conventional antennas with directive far-field patterns and OAM antennas: the case of reflector antennas.

compensate the phase delay associated with incident waves from different paths emitted from the feed and generate the desired vortex phase distribution. The uniform circular array (UCA) consists of elements that are uniformly distributed on the circumference of a circle. The array elements are fed with an input signal with the same amplitude but with a successive phase delay from element to element such that after a full turn the phase has been incremented by $2\pi l$, where l is the OAM mode number [33, 62, 102–109]. The transmittance functions of holographic gratings can also be designed to generate OAM beams [110–113]. Other generation methods include cylindrical lenses [3], spatial light modulators [9], q plates [118], dielectric resonators [119], traveling-wave circular loops [120], and metasurfaces [121]. The previous designs can generate single-mode OAM beams, superposition of OAM modes, i.e., mixed-mode OAM beams [5, 95], and mode-reconfigurable OAM beams [102, 121].

1.5 Summary and Perspectives

This chapter has provided a comprehensive overview of OAM beams. We analyzed the fundamental properties of OAM beams and contrasted them with conventional beams. In addition, an appendix is provided to summarize the fundamental mathematical details when the OAM concept is used for potential antenna applications. We discussed emerging applications of OAM in free-space optical and RF communications as well as in fiber communications, highlighting potentials, and technical challenges. A comprehensive list of relevant references is included. We also summarized the generation methods of optical and RF OAM. Analogies regarding the fundamental properties, antenna communication links, and generation methods between OAM and conventional beams have been provided. More details on various aspects of OAM are discussed in various chapters of this book. We conclude by providing Table 1.2, which summarizes the theoretical and experimental milestones regarding OAM in chronological order. OAM is a field with short history of less than 30 years that has come a long way so far, and we are excited to see the future advancements of OAM.

Table 1.1 OAM generation methods and fabricated prototypes.

Generation method	Prototype photo	
Spiral phase plate (SPP) Refs.: [13, 19, 89–91] Optical and RF OAM	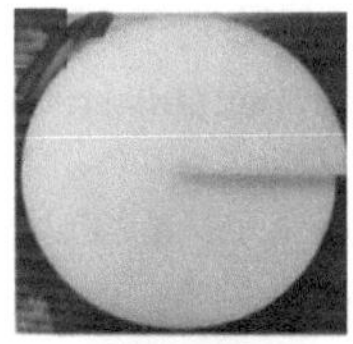	Photo credit: [90]
Helicoidal reflector antenna and reflector antenna systems Refs. [23, 92–94] RF OAM		Photo credit: [23]
Planar spiral phase plate and transmitarray antenna Refs. [95–100] RF OAM		Photo credit: [95]
Stepped spiral reflecting surface Ref. [101] RF OAM		Photo credit: [101]
Uniform circular array (UCA) Refs. [33, 62, 102–109] RF OAM	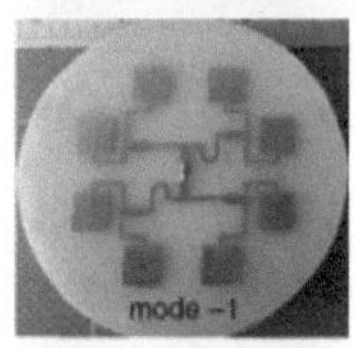	Photo credit: [102]
Holographic gratings Refs. [110–113] Optical and RF OAM		Photo credit: [110]
Reflectarray antennas Refs. [5, 114–117] RF OAM		Photo credit: [114]

Appendix 1.A OAM Far-field Calculation

The aperture field of an OAM-carrying linearly polarized field with a cylindrically symmetric distribution $E(\rho')$ can be written as:

$$\vec{E}(\rho', \phi') = E(\rho')e^{-jl\phi'}\hat{x}, \quad 0 < \rho' < a, \tag{1.A.1}$$

Table 1.2 Chronicle of milestones regarding OAM.

Reference	Year	Main contribution
[1]	1909	Theoretically studied angular momentum of circularly polarized waves
[2]	1936	Experimentally studied the SAM of light and demonstrated that SAM can cause the rotation of a mechanical system
[3]	1992	Recognized that light beams with an azimuthal phase dependence of $e^{jl\phi}$ carrying OAM
[17]	2004	Conducted the first experiment on OAM free-space optical communications
[76]	2006	Reported the generation of an OAM-carrying optical vortex in optical fibers
[62]	2007	Numerically showed that OAM can be used in the radio frequency domain
[23]	2012	Performed the first experimental test of encoding multiple channels on the same radio frequency through OAM
[80]	2013	Conducted the first OAM-MDM experiment suggesting that OAM could provide an additional degree of freedom for data multiplexing in future fiber networks
[5]	2018	Suggested a potential application that takes advantage of the OAM cone-shaped pattern in the far-field

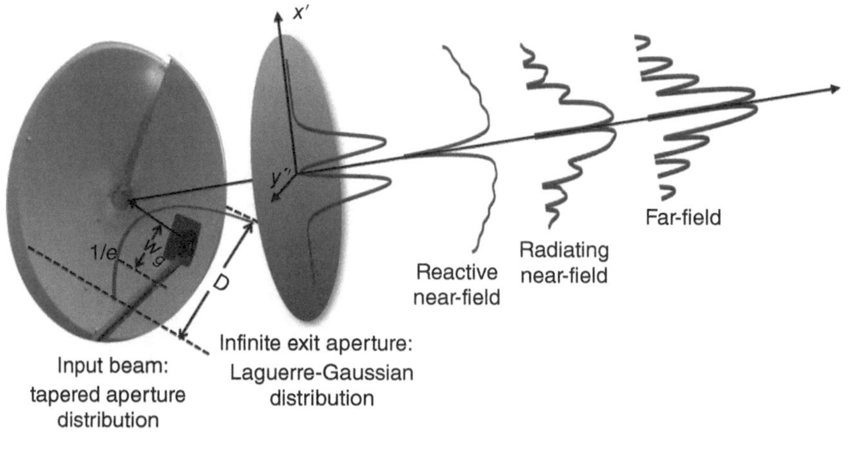

Figure 1.A.1 Schematic of the generation of OAM aperture field.

where ρ' and ϕ' are the radial and azimuthal coordinates in the cylindrical coordinate system; a is the transverse extend of the aperture field of the beam; l is the OAM order. A schematic of the generation of the OAM aperture field is shown in Figure 1.A.1. Typically, the input beam from the feed is a Gaussian-type beam with a tapered amplitude distribution. The beam waist w_g, i.e. the half-width of the normalized aperture field amplitude at $1/e$, is directly related to the OAM antenna aperture diameter D, as shown in Figure 1.A.1. For example, the equivalent beam waist in Ref. [5] was $w_g = 0.415D$ for a -12 dB taper illumination. Higher taper illumination would lead to a smaller equivalent beam waist. The role of the OAM antenna (helicoidal reflector in Figure 1.A.1) is to create the desired exit-aperture amplitude and phase distribution at the infinite exit-aperture plane. A common model for the aperture field of an OAM antenna is the Laguerre–Gaussian distribution Eq. (1.3).

The equivalent magnetic current density is calculated from [122, eq. 6-129b]:

$$\vec{M}_s = -\hat{z} \times \vec{E}(\rho', \phi') = -\hat{y}E(\rho')e^{-jl\phi'}, \quad 0 < \rho' < a. \tag{1.A.2}$$

The radiation integrals can be written as [122, eqs. 6-125c, 6-125d]:

$$L_\theta = -\cos\theta \sin\phi \int_0^a E(\rho') \left[\int_0^{2\pi} e^{-jl\phi'} e^{jk_0\rho' \sin\theta \cos(\phi - \phi')} d\phi' \right] \rho' d\rho' \tag{1.A.3}$$

$$L_\phi = -\cos\phi \int_0^a E(\rho') \left[\int_0^{2\pi} e^{-jl\phi'} e^{jk_0\rho' \sin\theta \cos(\phi - \phi')} d\phi' \right] \rho' d\rho'. \tag{1.A.4}$$

Using the integral identity [5, eq. (5)]:

$$\int_0^{2\pi} e^{-jl\phi'} e^{jk_0\rho' \sin\theta \cos(\phi - \phi')} d\phi' = 2\pi(-j)^l J_l(k_0 \sin\theta \rho') e^{-jl\phi}, \tag{1.A.5}$$

we find the expression of the far-field integrals:

$$L_\theta = -2\pi \cos\theta \sin\phi (-j)^l e^{-jl\phi} I \tag{1.A.6}$$

$$L_\phi = -2\pi \cos\phi (-j)^l e^{-jl\phi} I, \tag{1.A.7}$$

where

$$I = \int_0^a E(\rho') J_l(k_0 \sin\theta \rho') \rho' d\rho' \tag{1.A.8}$$

and $J_l(\cdot)$ is the lth order Bessel function of the first kind [21]. The expression of the far-field electric field can be written as [122, eqs. 6-122b, 6-122c]:

$$\vec{E}_{ff}(r, \theta, \phi) = \frac{jk_0 e^{-jk_0 r}}{4\pi r} \left(-\hat{\theta} L_\phi + \hat{\phi} L_\theta \right) = \frac{jk_0 e^{-jk_0 r}}{2r} - je^{-jl\phi} (\hat{\theta} \cos\phi - \hat{\phi} \cos\theta \sin\phi) I. \tag{1.A.9}$$

Note the far-field Eq. (1.A.9) maintains the $e^{-jl\phi}$ phase term.

Special Case 1: Airy Disk. The aperture field of a uniform amplitude and phase distribution is the special case of Eq. (1.A.1), where $l = 0$ (no OAM) and $E(\rho') = E_0^{AD}$. Using the integral identity [21, eq. (6.561–5)]:

$$\int_0^\delta z J_0(z) dz = \delta J_1(\delta), \tag{1.A.10}$$

we find I from Eq. (1.A.8) and the far-field expression from Eq. (1.A.9):

$$\vec{E}_{ff}^{AD}(r, \theta, \phi) = \frac{jk_0 E_0^{AD} a^2 e^{-jk_0 r}}{2r} (\hat{\theta} \cos\phi - \hat{\phi} \cos\theta \sin\phi) \frac{J_1(k_0 a \sin\theta)}{k_0 a \sin\theta}. \tag{1.A.11}$$

Special Case 2: Tapered-aperture Distribution. A physically meaningful and mathematically simple model for aperture-like antennas with uniform phase distribution is the two-parameter (2P) model [22, eq. (16)]:

$$\vec{E}^{2P}(\rho') = C + (1 - C)\left[1 - (\rho'/a)^2\right]^P, \quad 0 < \rho' < a, \tag{1.A.12}$$

where P and C are parameters that control the shape and amplitude distribution of the circular aperture. In particular, C can be related to the edge taper by ET = 20 log C. A generalized three-parameter aperture distribution model for elliptical aperture has also been developed in [123]. The far-field can be evaluated in the closed form [22, eq. (18–20)]:

$$\vec{E}_{ff}^{2P}(r,\theta,\phi) = \frac{jk_0a^2e^{-jk_0r}}{4r}(\hat{\theta}\cos\phi - \hat{\phi}\cos\theta\sin\phi)[C\Lambda_1(ka\sin\theta) + (1-C)\Lambda_{P+1}(ka\sin\theta)], \tag{1.A.13}$$

where

$$\Lambda_{P+1}(\zeta) = 2^{P+1}\Gamma(P+1)\frac{J_{P+1}(\zeta)}{\zeta^{P+1}}, \tag{1.A.14}$$

in which $\Gamma(\cdot)$ is the gamma function [21]. The far-field of the tapered-aperture distribution was studied in Section 1.2 and is shown in Figure 1.7a. The OAM tapered-aperture distribution counterpart was modeled based on Eq. (1.A.12) multiplied by the phase term $e^{-jl\phi}$:

$$\vec{E}_{OAM}^{2P}(\rho') = \left\{C + (1-C)\left[1-(\rho'/a)^2\right]^P\right\}e^{-jl\phi}, \quad 0 < \rho' < a. \tag{1.A.15}$$

The changes of amplitude pattern shape based on the aperture field of Eq. (1.A.15) from the reactive near-field toward the far-field were studied in Section 1.2 and are shown in Figure 1.7b.

Special Case 3: Laguerre–Gaussian beam. The aperture field of the Laguerre–Gaussian beams is given by Eq. (1.3). Using the integral identity [21, eq. (7.421-4)]

$$\int_0^\infty x^{\nu+1}e^{-\beta x^2}L_n^\nu(ax^2)J_\nu(xy) = 2^{-\nu-1}\beta^{-\nu-n-1}(\beta-a)^n y^\nu e^{-\frac{y^2}{4\beta}}L_n^\nu\left[\frac{ay^2}{4\beta(a-\beta)}\right], \tag{1.A.16}$$

we find I from Eq. (1.A.8) and the far-field expression from Eq. (1.A.9):

$$\begin{aligned}\vec{E}_{ff}^{LG}(r,\theta,\phi) &= \frac{jk_0E_0^{LG}e^{-jk_0r}}{4\pi r}(\hat{\theta}\cos\phi - \hat{\phi}\cos\theta\sin\phi)w_g(-1)^p(-j)^l \\ &\times\sqrt{\frac{2\pi p!}{(p+|l|)!}}\left(\frac{sgn(l)\Psi}{\sqrt{2}}\right)^{|l|}e^{-\frac{\Psi^2}{4}}L_p^{|l|}\left(\frac{\Psi^2}{2}\right)e^{-jl\phi},\end{aligned} \tag{1.A.17}$$

where $\Psi = k_0w_g\sin\theta$. For the definition of w_g refer to the first paragraph of the appendix and Figure 1.A.1. The previous discussion refers to the far-field where the radiation integral can be found in closed form. The near-field calculation using the Fresnel–Kirchhoff diffraction integral [24] was carried out numerically in Section 1.2 and the results are shown in Figure 1.7.

References

1 Poynting, J.H. (1909). The wave motion of a revolving shaft, and a suggestion as to the angular momentum in a beam of circularly polarised light. *Proceedings of the Royal Society of London* **82** (557): 560–567.

2 Beth, R.A. (1936). Mechanical detection and measurement of the angular momentum of light. *Physical Review* **50**: 115–125.

3 Allen, L., Beijersbergen, M.W., Spreeuw, R., and Woerdman, J. (1992). Orbital angular momentum of light and the transformation of Laguerre-Gaussian laser modes. *Physical Review A* **45** (11): 8185.

4 Drysdale, T.D., Allen, B., Stevens, C. et al. (2018). How orbital angular momentum modes are boosting the performance of radio links. *IET Microwaves, Antennas & Propagation* **12** (10): 1625–1632.

5 Veysi, M., Guclu, C., Capolino, F., and Rahmat-Samii, Y. (2018). Revisiting orbital angular momentum beams: Fundamentals, reflectarray generation, and novel antenna applications. *IEEE Antennas and Propagation Magazine* **60** (2): 68–81.

6 Cheng, W., Zhang, W., Jing, H. et al. (2019). Orbital angular momentum for wireless communications. *IEEE Wireless Communications* **26** (1): 100–107.

7 Gori, F., Guattari, G., and Padovani, C. (1987). Bessel-Gauss beams. *Optics Communications* **64** (6): 491–495.

8 Karimi, E., Zito, G., Piccirillo, B. et al. (2007). Hypergeometric-Gaussian modes. *Optics Letters* **32** (21): 3053–3055.

9 Maurer, C., Jesacher, A., Fürhapter, S. et al. (2007). Tailoring of arbitrary optical vector beams. *New Journal of Physics* **9** (3): 78.

10 Allen, L., Padgett, M., and Babiker IV, M. (1999). The orbital angular momentum of light, Progress in Optics. *Elsevier* **39**: 291–372.

11 Willner, A.E., Huang, H., Yan, Y. et al. (2015). Optical communications using orbital angular momentum beams. *Advances in Optics and Photonics* **7** (1): 66–106.

12 Padgett, M.J., Miatto, F.M., Lavery, M.P. et al. (2015). Divergence of an orbital-angular-momentum-carrying beam upon propagation. *New Journal of Physics* **17** (2): 023011.

13 Beijersbergen, M., Coerwinkel, R., Kristensen, M., and Woerdman, J. (1994). Helical-wavefront laser beams produced with a spiral phaseplate. *Optics Communications* **112** (5-6): 321–327.

14 Trichili, A., Rosales-Guzmán, C., Dudley, A. et al. (2016). Optical communication beyond orbital angular momentum. *Scientific Reports* **6**: 27674.

15 Oldoni, M., Spinello, F., Mari, E. et al. (2015). Space-division demultiplexing in orbital-angular-momentum-based mimo radio systems. *IEEE Transactions on Antennas and Propagation* **63** (10): 4582–4587.

16 Djordjevic, I.B. (2011). Deep-space and near-earth optical communications by coded orbital angular momentum (OAM) modulation. *Optics Express* **19** (15): 14277–14289.

17 Gibson, G., Courtial, J., Padgett, M.J. et al. (2004). Free-space information transfer using light beams carrying orbital angular momentum. *Optics Express* **12** (22): 5448–5456.

18 Ge, X., Zi, R., Xiong, X. et al. (2017). Millimeter wave communications with OAM-SM scheme for future mobile networks. *IEEE Journal on Selected Areas in Communications* **35** (9): 2163–2177.

19 Turnbull, G., Robertson, D., Smith, G. et al. (1996). The generation of free-space Laguerre-Gaussian modes at millimetre-wave frequencies by use of a spiral phaseplate. *Optics Communications* **127** (4-6): 183–188.

20 Yao, A.M. and Padgett, M.J. (2011). Orbital angular momentum: origins, behavior and applications. *Advances in Optics and Photonics* **3** (2): 161–204.

21 Gradshteyn, I.S. and Ryzhik, I.M. (2014). *Table of Integrals, Series, and Products*, 7e. Academic press.

22 Rahmat-Samii, Y. (1988). Reflector antennas. In: *Antenna Handbook: Theory, Applications, and Design* (eds. S.W. Lee and Y.T. Lo), 949–1072. Boston, MA: Springer US.

23 Tamburini, F., Mari, E., Sponselli, A. et al. (2012). Encoding many channels on the same frequency through radio vorticity: first experimental test. *New Journal of Physics* **14** (3): 033001.

24 Born, M. and Wolf, E. (2013). *Principles of Optics: Electromagnetic Theory of Propagation, Interference and Diffraction of Light*, 6e. Elsevier.

25 Balanis, C.A. (2016). *Antenna Theory: Analysis and Design*, 4e. John Wiley & Sons.

26 Friis, H.T. (1946). A note on a simple transmission formula. *Proceedings of the IRE* **34** (5): 254–256.

27 Cho, Y.H. and Byun, W.J. (2019). Generalized friis transmission equation for orbital angular momentum radios. *IEEE Transactions on Antennas and Propagation* **67** (4): 2423–2429.

28 Nguyen, D.K., Pascal, O., Sokoloff, J. et al. (2015). Antenna gain and link budget for waves carrying orbital angular momentum. *Radio Science* **50** (11): 1165–1175.

29 D. K. Nguyen, O. Pascal, J. Sokoloff, et al. (2014). Discussion about the link budget for electromagnetic wave with orbital angular momentum, *The 8th European Conference on Antennas and Propagation (EuCAP 2014)*. IEEE, pp. 1117–1121.

30 Cagliero, A., De Vita, A., Gaffoglio, R., and Sacco, B. (2015). A new approach to the link budget concept for an OAM communication link. *IEEE Antennas and Wireless Propagation Letters* **15**: 568–571.

31 Craeye, C. (2015). On the transmittance between OAM antennas. *IEEE Transactions on Antennas and Propagation* **64** (1): 336–339.

32 C. Rui, Z. Hong, M. Marco, et al. (2019). Orbital angular momentum waves: Generation, detection and emerging applications, *arXiv preprint arXiv:1903.07818*.

33 Edfors, O. and Johansson, A.J. (2011). Is orbital angular momentum (OAM) based radio communication an unexploited area? *IEEE Transactions on Antennas and Propagation* **60** (2): 1126–1131.

34 Tamagnone, M., Craeye, C., and Perruisseau-Carrier, J. (2012). Comment on 'Encoding many channels on the same frequency through radio vorticity: first experimental test'. *New Journal of Physics* **14** (11): 118001.

35 Morabito, A.F., Di Donato, L., and Isernia, T. (2018). Orbital angular momentum antennas: Understanding actual possibilities through the aperture antennas theory. *IEEE Antennas and Propagation Magazine* **60** (2): 59–67.

36 Xie, G., Li, L., Ren, Y. et al. (2015). Performance metrics and design considerations for a free-space optical orbital-angular-momentum multiplexed communication link. *Optica* **2** (4): 357–365.

37 Gao, X., Song, X., Zheng, Z. et al. (2020). Misalignment measurement of orbital angular momentum signal based on spectrum analysis and image processing. *IEEE Transactions on Antennas and Propagation* **68** (1): 521–526.

38 Anguita, J.A., Neifeld, M.A., and Vasic, B.V. (2008). Turbulence-induced channel crosstalk in an orbital angular momentum-multiplexed free-space optical link. *Applied Optics* **47** (13): 2414–2429.

39 Tyler, G.A. and Boyd, R.W. (2009). Influence of atmospheric turbulence on the propagation of quantum states of light carrying orbital angular momentum. *Optics Letters* **34** (2): 142–144.

40 Paterson, C. (2005). Atmospheric turbulence and orbital angular momentum of single photons for optical communication. *Physical Review Letters* **94** (15): 153901.

41 Rodenburg, B., Lavery, M.P., Malik, M. et al. (2012). Influence of atmospheric turbulence on states of light carrying orbital angular momentum. *Optics Letters* **37** (17): 3735–3737.

42 Ren, Y., Huang, H., Xie, G. et al. (2013). Atmospheric turbulence effects on the performance of a free space optical link employing orbital angular momentum multiplexing. *Optics Letters* **38** (20): 4062–4065.

43 Malik, M., O'Sullivan, M., Rodenburg, B. et al. (2012). Influence of atmospheric turbulence on optical communications using orbital angular momentum for encoding. *Optics Express* **20** (12): 13195–13200.

44 Chaibi, A., Mafusire, C., and Forbes, A. (2013). Propagation of orbital angular momentum carrying beams through a perturbing medium. *Journal of Optics* **15** (10): 105706.

45 Trichili, A., Salem, A.B., Dudley, A. et al. (2016). Encoding information using Laguerre Gaussian modes over free space turbulence media. *Optics Letters* **41** (13): 3086–3089.

46 Willner, A.E., Ren, Y., Xie, G. et al. (2017). Recent advances in high-capacity free-space optical and radio-frequency communications using orbital angular momentum multiplexing. *Philosophical Transactions of the Royal Society A: Mathematical, Physical and Engineering Sciences* **375** (2087): 20150439.

47 Yan, Y., Li, L., Xie, G. et al. (2016). Multipath effects in millimetre-wave wireless communication using orbital angular momentum multiplexing. *Scientific Reports* **6**: 33482.

48 Vicente-Lozano, M., Franceschetti, G., Ares-Pena, F.J., and Moreno-Piquero, E. (2002). Analysis and synthesis of a printed array for satellite communication with moving vehicles. *IEEE Transactions on Antennas and Propagation* **50** (11): 1555–1559.
49 Pan, Y.-M. and Leung, K.W. (2012). Wideband circularly polarized dielectric bird-nest antenna with conical radiation pattern. *IEEE Transactions on Antennas and Propagation* **61** (2): 563–570.
50 Lau, K. and Luk, K. (2006). A wideband circularly polarized conical-beam patch antenna. *IEEE Transactions on Antennas and Propagation* **54** (5): 1591–1594.
51 Lin, W. and Wong, H. (2014). Circularly polarized conical-beam antenna with wide bandwidth and low profile. *IEEE Transactions on Antennas and Propagation* **62** (12): 5974–5982.
52 Kai, C., Huang, P., Shen, F. et al. (2017). Orbital angular momentum shift keying based optical communication system. *IEEE Photonics Journal* **9** (2): 1–10.
53 Trichili, A., Park, K.-H., Zghal, M. et al. (2019). Communicating using spatial mode multiplexing: Potentials, challenges and perspectives. *IEEE Communications Surveys & Tutorials* **21** (4): 3175–3203.
54 Krenn, M., Fickler, R., Fink, M. et al. (2014). Communication with spatially modulated light through turbulent air across Vienna. *New Journal of Physics* **16** (11): 113028.
55 Krenn, M., Handsteiner, J., Fink, M. et al. (2016). Twisted light transmission over 143 km. *Proceedings of the National Academy of Sciences* **113** (48): 13648–13653.
56 Molisch, A. (2005). *Wireless Communications*. Wiley-IEEE Press.
57 Goldsmith, A. (2005). *Wireless Communications*. USA: Cambridge University Press.
58 Lee, D., Sasaki, H., Fukumoto, H. et al. (2017). Orbital angular momentum (OAM) multiplexing: An enabler of a new era of wireless communications. *IEICE Transactions on Communications* **100** (7): 1044–1063.
59 Cheng, W., Zhang, W., Jing, H. et al. (2018). Orbital angular momentum for wireless communications. *IEEE Wireless Communications* **26** (1): 100–107.
60 Huang, H., Xie, G., Yan, Y. et al. (2014). 100 Tbit s^{-1} free-space data link enabled by three-dimensional multiplexing of orbital angular momentum, polarization, and wavelength. *Optics Letters* **39** (2): 197–200.
61 Wang, J., Yang, J.-Y., Fazal, I.M. et al. (2012). Terabit free-space data transmission employing orbital angular momentum multiplexing. *Nature Photonics* **6** (7): 488.
62 Thidé, B., Then, H., Sjöholm, J. et al. (2007). Utilization of photon orbital angular momentum in the low-frequency radio domain. *Physical Review Letters* **99** (8): 087701.
63 Zhang, W., Zheng, S., Hui, X. et al. (2016). Mode division multiplexing communication using microwave orbital angular momentum: An experimental study. *IEEE Transactions on Wireless Communications* **16** (2): 1308–1318.
64 Yan, Y., Xie, G., Lavery, M.P. et al. (2014). High-capacity millimetre-wave communications with orbital angular momentum multiplexing. *Nature Communications* **5**: 4876.
65 Willner, A. (2019). *Optical Fiber Telecommunications*, vol. **11**. Academic Press.
66 Cisco Global Cloud Index: Forecast and Methodology, 2016–2021 White Paper - Cisco, https://www.cisco.com/c/en/us/solutions/collateral/service-provider/global-cloud-index-gci/white-paper-c11-738085.html#_Toc503317520 (accessed 2 March 2020).
67 Richardson, D., Fini, J., and Nelson, L.E. (2013). Space-division multiplexing in optical fibres. *Nature Photonics* **7** (5): 354.
68 Saridis, G.M., Alexandropoulos, D., Zervas, G., and Simeonidou, D. (2015). Survey and evaluation of space division multiplexing: From technologies to optical networks. *IEEE Communications Surveys & Tutorials* **17** (4): 2136–2156.
69 Rusch, L.A., Rad, M., Allahverdyan, K. et al. (2018). Carrying data on the orbital angular momentum of light. *IEEE Communications Magazine* **56** (2): 219–224.

70 Li, G., Bai, N., Zhao, N., and Xia, C. (2014). Space-division multiplexing: the next frontier in optical communication. *Advances in Optics and Photonics* **6** (4): 413–487.

71 Ramachandran, S. and Kristensen, P. (2013). Optical vortices in fiber. *Nanophotonics* **2** (5-6): 455–474.

72 Li, Y., Jin, L., Wu, H. et al. (2017). Superposing multiple LP modes with microphase difference distributed along fiber to generate oam mode. *IEEE Photonics Journal* **9** (2): 1–9.

73 Ramachandran, S., Gregg, P., Kristensen, P., and Golowich, S. (2015). On the scalability of ring fiber designs for OAM multiplexing. *Optics Express* **23** (3): 3721–3730.

74 Chen, S. and Wang, J. (2017). Theoretical analyses on orbital angular momentum modes in conventional graded-index multimode fibre. *Scientific Reports* **7** (1): 1–15.

75 Li, S. and Wang, J. (2014). A compact trench-assisted multi-orbital-angular-momentum multi-ring fiber for ultrahigh-density space-division multiplexing (19 rings × 22 modes). *Scientific Reports* **4**: 3853.

76 Dashti, P.Z., Alhassen, F., and Lee, H.P. (2006). Observation of orbital angular momentum transfer between acoustic and optical vortices in optical fiber. *Physical Review Letters* **96** (4): 043604.

77 Bozinovic, N., Kristensen, P., and Ramachandran, S. (2011). Long-range fiber-transmission of photons with orbital angular momentum. In: *CLEO: Science and Innovations*, CTuB1. Optical Society of America.

78 Bozinovic, N., Golowich, S., Kristensen, P., and Ramachandran, S. (2012). Control of orbital angular momentum of light with optical fibers. *Optics Letters* **37** (13): 2451–2453.

79 Ramachandran, S., Bozinovic, N., Gregg, P. et al. (2012). Optical vortices in fibres: A new degree of freedom for mode multiplexing. In: *2012 38th European Conference and Exhibition on Optical Communications*, 1–3. IEEE.

80 Bozinovic, N., Yue, Y., Ren, Y. et al. (2013). Terabit-scale orbital angular momentum mode division multiplexing in fibers. *Science* **340** (6140): 1545–1548.

81 Ingerslev, K., Gregg, P., Galili, M. et al. (2018). 12 mode, WDM, MIMO-free orbital angular momentum transmission. *Optics Express* **26** (16): 20225–20232.

82 Huang, H., Milione, G., Lavery, M.P. et al. (2015). Mode division multiplexing using an orbital angular momentum mode sorter and MIMO-DSP over a graded-index few-mode optical fibre. *Scientific Reports* **5** (1): 1–7.

83 Zhu, L., Wang, A., Chen, S. et al. (2017). Orbital angular momentum mode groups multiplexing transmission over 2.6-km conventional multi-mode fiber. *Optics Express* **25** (21): 25637–25645.

84 Wang, A., Zhu, L., Wang, L. et al. (2018). Directly using 8.8-km conventional multi-mode fiber for 6-mode orbital angular momentum multiplexing transmission. *Optics Express* **26** (8): 10038–10047.

85 Zhu, L., Wang, A., Chen, S. et al. (2018). Orbital angular momentum mode multiplexed transmission in heterogeneous few-mode and multi-mode fiber network. *Optics Letters* **43** (8): 1894–1897.

86 Li, S. and Wang, J. (2013). Multi-orbital-angular-momentum multi-ring fiber for high-density space-division multiplexing. *IEEE Photonics Journal* **5** (5): 7101007–7101007.

87 Li, S. and Wang, J. (2015). Supermode fiber for orbital angular momentum (OAM) transmission. *Optics Express* **23** (14): 18736–18745.

88 Papathanasopoulos, A., Rahmat-Samii, Y., Garcia, N., and Chisum, J.D. (2020). A novel collapsible flat-layered metamaterial gradient-refractive-index (GRIN) lens antenna. *IEEE Transactions on Antennas and Propagation* **68** (3): 1312–1321.

89 Wei, X., Liu, C., Niu, L. et al. (2015). Generation of arbitrary order Bessel beams via 3D printed axicons at the terahertz frequency range. *Applied Optics* **54** (36): 10641–10649.

90 Zhang, C. and Ma, L. (2016). Millimetre wave with rotational orbital angular momentum. *Scientific Reports* **6** (1): 1–8.

91 L. Zhu, X. Wei, J. Wang, et al. (2014). Experimental demonstration of basic functionalities for 0.1-THz orbital angular momentum (OAM) communications, *Optical Fiber Communication Conference*, OSA Technical Digest (online) (Optical Society of America, 2014), paper M3K.

92 Byun, W.-J., Lee, Y.-S., Kim, B.S. et al. (2015). Simple generation of orbital angular momentum modes with azimuthally deformed Cassegrain subreflector. *Electronics Letters* **51** (19): 1480–1482.

93 Mari, E., Spinello, F., Oldoni, M. et al. (2014). Near-field experimental verification of separation of OAM channels. *IEEE Antennas and Wireless Propagation Letters* **14**: 556–558.

94 Byun, W.J., Kim, K.S., Kim, B.S. et al. (2016). Multiplexed Cassegrain reflector antenna for simultaneous generation of three orbital angular momentum (OAM) modes. *Scientific Reports* **6**: 27339.

95 Cheng, L., Hong, W., and Hao, Z.-C. (2014). Generation of electromagnetic waves with arbitrary orbital angular momentum modes. *Scientific Reports* **4** (1): 1–5.

96 Qin, F., Wan, L., Li, L. et al. (2018). A transmission metasurface for generating OAM beams. *IEEE Antennas and Wireless Propagation Letters* **17** (10): 1793–1796.

97 Hui, X., Zheng, S., Hu, Y. et al. (2015). Ultralow reflectivity spiral phase plate for generation of millimeter-wave OAM beam. *IEEE Antennas and Wireless Propagation Letters* **14**: 966–969.

98 Chen, Y., Zheng, S., Li, Y. et al. (2015). A flat-lensed spiral phase plate based on phase-shifting surface for generation of millimeter-wave OAM beam. *IEEE Antennas and Wireless Propagation Letters* **15**: 1156 1158.

99 A. Bennis, R. Niemiec, C. Brousseau, et al. (2013). Flat plate for OAM generation in the millimeter band, *2013 7th European Conference on Antennas and Propagation (EuCAP)*. IEEE, pp. 3203–3207.

100 Niemiec, R., Brousseau, C., Mahdjoubi, K. et al. (2014). Characterization of an OAM flat-plate antenna in the millimeter frequency band. *IEEE Antennas and Wireless Propagation Letters* **13**: 1011–1014.

101 Tamburini, F., Mari, E., Thidé, B. et al. (2011). Experimental verification of photon angular momentum and vorticity with radio techniques. *Applied Physics Letters* **99** (20): 204102.

102 Bai, Q., Tennant, A., and Allen, B. (2014). Experimental circular phased array for generating OAM radio beams. *Electronics Letters* **50** (20): 1414–1415.

103 Liu, K., Liu, H., Qin, Y. et al. (2016). Generation of OAM beams using phased array in the microwave band. *IEEE Transactions on Antennas and Propagation* **64** (9): 3850–3857.

104 Kang, L., Li, H., Zhou, J. et al. (2019). A mode-reconfigurable orbital angular momentum antenna with simplified feeding scheme. *IEEE Transactions on Antennas and Propagation* **67** (7): 4866–4871.

105 Liu, Q., Chen, Z.N., Liu, Y. et al. (2018). Circular polarization and mode reconfigurable wideband orbital angular momentum patch array antenna. *IEEE Transactions on Antennas and Propagation* **66** (4): 1796–1804.

106 Zhao, M., Gao, X., Xie, M. et al. (2018). Generation of coupled radio frequency orbital angular momentum beam with an optical-controlled circular antenna array. *Optics Communications* **426**: 126–129.

107 Gong, Y., Wang, R., Deng, Y. et al. (2017). Generation and transmission of OAM-carrying vortex beams using circular antenna array. *IEEE Transactions on Antennas and Propagation* **65** (6): 2940–2949.

108 Yuan, T., Cheng, Y., Wang, H.-Q., and Qin, Y. (2016). Generation of OAM radio beams with modified uniform circular array antenna. *Electronics Letters* **52** (11): 896–898.

109 Lin, M., Gao, Y., Liu, P., and Liu, J. (2017). Theoretical analyses and design of circular array to generate orbital angular momentum. *IEEE Transactions on Antennas and Propagation* **65** (7): 3510–3519.

110 F. E. Mahmouli and S. Walker. (2012). Orbital angular momentum generation in a 60GHz wireless radio channel, *2012 20th Telecommunications Forum (TELFOR)*, Belgrade, Serbia: IEEE, (20–22 November 2012).

111 Bazhenov, V.Y., Vasnetsov, M., and Soskin, M. (1990). Laser beams with screw dislocations in their wavefronts. *JETP Letter* **52** (8): 429–431.

112 Heckenberg, N., McDuff, R., Smith, C., and White, A. (1992). Generation of optical phase singularities by computer-generated holograms. *Optics Letters* **17** (3): 221–223.

113 Arlt, J., Dholakia, K., Allen, L., and Padgett, M. (1998). The production of multiringed Laguerre–Gaussian modes by computer-generated holograms. *Journal of Modern Optics* **45** (6): 1231–1237.

114 Meng, X.-S., Wu, J.-J., Wu, Z.-S. et al. (2018). Design of multiple-polarization reflectarray for orbital angular momentum wave in radio frequency. *IEEE Antennas and Wireless Propagation Letters* **17** (12): 2269–2273.

115 Yu, S., Li, L., Shi, G. et al. (2016). Generating multiple orbital angular momentum vortex beams using a metasurface in radio frequency domain. *Applied Physics Letters* **108** (24): 241901.

116 Chen, G.-T., Jiao, Y.-C., and Zhao, G. (2018). A reflectarray for generating wideband circularly polarized orbital angular momentum vortex wave. *IEEE Antennas and Wireless Propagation Letters* **18** (1): 182–186.

117 Huang, H.-F. and Li, S.-N. (2019). High-efficiency planar reflectarray with small-size for OAM generation at microwave range. *IEEE Antennas and Wireless Propagation Letters* **18** (3): 432–436.

118 Marrucci, L., Karimi, E., Slussarenko, S. et al. (2011). Spin-to-orbital conversion of the angular momentum of light and its classical and quantum applications. *Journal of Optics* **13** (6): 064001.

119 Pan, Y., Zheng, S., Zheng, J. et al. (2016). Generation of orbital angular momentum radio waves based on dielectric resonator antenna. *IEEE Antennas and Wireless Propagation Letters* **16**: 385–388.

120 Zheng, S., Hui, X., Jin, X. et al. (2015). Transmission characteristics of a twisted radio wave based on circular traveling-wave antenna. *IEEE Transactions on Antennas and Propagation* **63** (4): 1530–1536.

121 Wu, J., Zhang, Z., Ren, X. et al. (2019). A broadband electronically mode-reconfigurable orbital angular momentum metasurface antenna. *IEEE Antennas and Wireless Propagation Letters* **18** (7): 1482–1486.

122 Balanis, C.A. (2012). *Advanced Engineering Electromagnetics*, 2e. John Wiley & Sons.

123 Duan, D.-W. and Rahmat-Samii, Y. (1992). A generalized three-parameter (3-P) aperture distribution for antenna applications. *IEEE Transactions on Antennas and Propagation* **40** (6): 697–713.

2

OAM Radio – Physical Foundations and Applications of Electromagnetic Orbital Angular Momentum in Radio Science and Technology

Bo Thidé[1] *and Fabrizio Tamburini*[2]

[1] *Swedish Institute of Space Physics, Ångström Laboratory, Uppsala, Sweden*
[2] *ZKM Karlsruhe, Lorenzstraße, Berlin, Germany*

2.1 Introduction

The purpose of this chapter is to introduce, explain, and elaborate recent ideas about experimentally substantiated and innovative physical wireless information transfer methods that hold great promise for a more efficient utilization of the frequency spectrum in both scientific and technological implementations. These pioneering ideas are put on a solid physical basis by deriving all theoretical results from first principles within the realm of microscopic classical electrodynamics (Maxwell–Lorentz theory).

The technique of transferring information by means of classical electromagnetic fields amounts to using such fields generated within a spatial volume of finite extent, encoding information onto physical observables carried by these fields, radiating the observables as volumetric densities into the surrounding space, measuring them remotely by using volume-integrating sensors, thus converting them into observables, and feeding these observables into information-decoding receivers. Of all physical observables carried by the electromagnetic field, only the linear momentum (i.e., the Poynting vector) is fully exploited in present-day radio science and technology, be it fundamental physical or information-theoretical research, implementations such as radio astronomy and radar probing, or wireless communication and navigation applications.

A fundamental physical limitation of the linear momentum vectorial observable, which exploits only the *translational* symmetry and the pertinent degrees of freedom of the electromagnetic field, is that it is single-mode in the sense that a linear momentum radio communication link comprising one single transmitting and one single receiving transducer of a conventional type (e.g., a linear half-wave dipole antenna), known as a single-input-single-output (SISO) link, provides only one physical channel per carrier frequency.

In contrast, electromagnetic angular momentum, the pseudovectorial physical observable that manifests the *rotational* symmetry of the electromagnetic field, is multimode, being a superposition of discrete, mutually independent electromagnetic angular momentum eigenmodes, each uniquely identified by the number of 2π that the phase of the field varies azimuthally during one oscillation period, i.e., by its topological charge. This allows a frequency reuse in the sense that several independent angular momentum channels can coexist simultaneously on the same carrier frequency.

Electromagnetic Vortices: Wave Phenomena and Engineering Applications, First Edition.
Edited by Zhi Hao Jiang and Douglas H. Werner.

Hence, a wireless communication system designed in such a way that each of the transmitting and receiving ends are equipped with a transducer that allows the dynamics of the charges to be two- or three-dimensional, thus enabling the direct generation and detection of angular momentum modes, would constitute an angular momentum SISO link system. Such systems have multimode capability due to the fact that different angular momentum eigenmodes are topologically distinct and linearly independent by virtue of their mutual functional orthogonality in azimuthal angular space. Even in situations when the different eigenmodes overlap in time, frequency, and physical space, the individual angular momentum eigenmodes are physically independent and separable, and they can be extracted by using suitable angular momentum mode filtering techniques. Classically, the actual number of physical angular momentum channels that can be realized in a given application and/or setup is essentially limited only by engineering and implementation issues.

In this chapter, we discuss concepts and issues related to the exploitation of the angular momentum physical layer in practical radio implementations, both theoretically and by way of examples. To supplement and support this theoretical approach, we present results from electromagnetic angular momentum experiments and observations. We also delineate the possibilities for completely new concepts in transducer design and technology that are offered by recent advances in nano- and optomechanics, spintronics, orbitronics, and skyrmionics.

In Appendix, we make a detailed, systematic derivation of the theory for electromagnetic angular momentum and put it into its proper fundamental physical context and on a firm basis by first reviewing the conservation/symmetry properties of the electromagnetic field that make wireless information transfer possible. Then we continue with a theoretical presentation of the physics itself, touching upon first quantization quantum-like aspects of classical electrodynamics and its observables. Finally, we derive generic expressions for the angular momentum density and show that for any system described by the Maxwell–Lorentz equations, the magnitude of the electromagnetic angular momentum always tends to a constant far away from an arbitrary source.

2.2 Physics

In this section we analyse the fundamental physics on which the wireless information transfer mechanisms rest and begin with the electromagnetic field. We then discuss the energy, linear momentum, and angular momentum observables and derive formulas for them in the paraxial and far-zone approximations.

2.2.1 The Classical Electromagnetic Field

We will work in microscopic classical electrodynamics where only the fields $\mathbf{E}(t, \mathbf{x})$ and $\mathbf{B}(t, \mathbf{x})$ have meaning [1, chapter VI, 2, section 3.1] and the interplay between the fields themselves as well as between the fields and the (free and bound) microscopic electric charge and current densities $\rho^{\mathrm{e}}(t, \mathbf{x})$ and $\mathbf{j}^{\mathrm{e}}(t, \mathbf{x})$ is postulated by the Maxwell–Lorentz equations.

To ensure generality, we allow for hypothetical magnetic charge and current densities, represented by the pseudoscalar $\rho^{\mathrm{m}}(t, \mathbf{x})$ and the pseudovector $\mathbf{j}^{\mathrm{m}}(t, \mathbf{x})$, respectively, that reflect the dual symmetry of the Maxwell–Lorentz theory [3–5]. Doing so, the Maxwell–Lorentz equations can be put in the symmetrical form [6, 7]:

$$\nabla\cdot\mathbf{E} = \frac{\rho^{\mathrm{e}}}{\varepsilon_0} \tag{2.1}$$

$$\nabla \cdot \mathbf{B} = \frac{\rho^{\mathrm{m}}}{\varepsilon_0 c^2} \tag{2.2}$$

$$\nabla \times \mathbf{E} + \frac{\partial \mathbf{B}}{\partial t} = -\frac{1}{\varepsilon_0 c^2}\mathbf{j}^{\mathrm{m}} \tag{2.3}$$

$$\nabla \times \mathbf{B} - \frac{1}{c^2}\frac{\partial \mathbf{E}}{\partial t} = \frac{1}{\varepsilon_0 c^2}\mathbf{j}^{\mathrm{e}} \tag{2.4}$$

where c is the speed of light in free space and ε_0 is the dielectric permittivity of free space. This set of linear, inhomogeneous, coupled, partial differential equations are the postulates of microscopic classical electrodynamics from which all properties of the classical electromagnetic field, its observables, and interactions can be derived; the usual no-magnetic-charges Maxwell theory is recovered by setting $\rho^{\mathrm{m}} = 0$ and $\mathbf{j}^{\mathrm{m}} = \mathbf{0}$. The postulates (2.1) are valid for radio and optical wavelengths alike inasmuch as they are scale invariant. Therefore they are valid for classical wireless information transfer at all frequencies.

A fundamental property of the electromagnetic field $(\mathbf{E}, \mathbf{B})$ is that it can transfer information encoded in physical observables, i.e. directly measurable intrinsic and extrinsic electromagnetic quantities, over long distances through free space as well as through waveguides and various gaseous, fluid, and solid media. However, the electrodynamic fields $\mathbf{E}$ and $\mathbf{B}$ themselves are not observables, i.e. cannot in general be measured directly. To quote [8, 9]:

> 'The modern view of the world that emerged from Maxwell's theory is a world with two layers. The first layer, the layer of the fundamental constituents of the world, consists of fields satisfying simple linear equations. The second layer, the layer of the things that we can directly touch and measure, consists of mechanical stresses and energies and forces. The two layers are connected, because the quantities in the second layer are quadratic or bilinear combinations of the quantities in the first layer. To calculate energies or stresses, you take the square of the electric field-strength or multiply one component of the field by another. The two-layer structure of the world is the basic reason why Maxwell's theory seemed mysterious and difficult. The objects on the first layer, the objects that are truly fundamental, are abstractions not directly accessible to our senses. The objects that we can feel and touch are on the second layer, and their behavior is only determined indirectly by the equations that operate on the first layer. The two-layer structure of the world implies that the basic processes of nature are hidden from our view.'

Simple radio communication setups based on the linear momentum density (Poynting vector) typically use linear antennas, often dipoles. In this setup, which is a special case of the more general setup depicted in Figure 2.A.1, the transmitting antenna is a source volume V' in the form of a very thin conducting cylinder with its reference point/midpoint $\mathbf{x}'_0$ chosen as the origin. The observation (sensor) volume V (the receiving dipole antenna) is another very thin conducting cylinder with its midpoint/reference point $\mathbf{x}_0$. The fields set up at time t in a small volume (i.e., length) element around $\mathbf{x}$ in V is the result of a weighted integration of all field contributions emitted at retarded times $t' = t - |\mathbf{x} - \mathbf{x}'|/c$ from all length elements around all points $\mathbf{x}'$ in V', taking into account that the propagation direction varies from $\hat{\mathbf{k}}'_- = \hat{\mathbf{k}}'(\mathbf{x}, \mathbf{x}'_-)$ to $\hat{\mathbf{k}}'_+ = \hat{\mathbf{k}}'(\mathbf{x}, \mathbf{x}'_+)$, corresponding to a polar angle range of $\vartheta'_- \leq \vartheta' \leq \vartheta'_+$, and that the propagation distances vary between $|\mathbf{x} - \mathbf{x}'_-|$ and $|\mathbf{x} - \mathbf{x}'_+|$. In the paraxial approximation the propagation direction is assumed not to vary ($\vartheta' = \vartheta'_0$ for all $\mathbf{x}'$) so that one can set $\hat{\mathbf{k}}'(\mathbf{x}, \mathbf{x}') \approx \hat{\mathbf{k}}'(\mathbf{x}, \mathbf{x}_0) \equiv \hat{\mathbf{n}}'(\mathbf{x})$. In the far-zone approximation, the dependence on the field contribution amplitudes on the difference in propagation distance from the different $\mathbf{x}'$ is neglected, i.e., $|\,\mathbf{x}-\mathbf{x}'\,| \approx |\,\mathbf{x}-\mathbf{x}'_0\,| = \left[\mathbf{x}'_0 = \mathbf{0}\right] = |\,\mathbf{x}\,|$.

From the above discussion it should be clear that in order to satisfactorily assess the information-carrying capability of the electromagnetic field and optimize its exploitation in information transfer and extraction implementations such as, e.g. radio communications and radio astronomy, it is necessary to analyze not only the properties of the first-order fields **E** and **B** themselves but also the properties of the second-order electromagnetic observables that comprise products of these fields and/or their derivatives. For any conceivable distribution of sources of the electromagnetic field, the physical observables (the volume-integrated densities) tend to a constant value when they propagate in free space, and the distance from the radiation source tends to infinity. This means that there does not exist a finite distance from the source at which any of these observables tends to zero or beyond, which it vanishes globally. Consequently, they can all carry information over arbitrary, large distances in free space, albeit subject to engineering limitations (spatial and temporal distribution, transducer geometry, detector sensitivity, integration time, etc.).

2.2.2 Electrodynamic Observables

Electrodynamic observables are typically generated within a volume of finite spatial extent (the source volume V', e.g. a transmitting antenna). Transmitted in the form of volumetric densities, they propagate in the surrounding space to a remotely located volume of likewise finite spatial extent (the observation/sensor volume V, e.g. a receiving antenna) over which the densities are volume integrated into observables. In classical physics terms, the observables and their volumetric densities are viewed as the result of dynamical changes in the charged particle distribution in the source volume V' and, causally, in the observation volume V. Figure 2.1 illustrates this for the case when these volumes are linear dipole antennas; see Figure 2.A.1 for a more generic setup. The information carried by the electrodynamic observables is subsequently extracted from V and decoded.

As shown in Section 2.A.2, the physical observables fulfill local as well as global equations of continuity, and are therefore conserved (constants of motion). Neither [10] showed that this conservation is a manifestation of a symmetry of the Maxwell–Lorentz theory [11–13]. In classical physics terms, the observables and their volumetric densities are viewed as the result of dynamical changes in the charged particle distribution in the source volume V and, causally, in the observation volume V' as illustrated in Figure 2.A.1.

The physical observables are carried by the electrodynamic field in the form of volumetric densities that can be expressed as linear combinations of quadratic and/or bilinear, i.e. second-order, field quantities. The simplest and most basic information-carrying EM quantity is the field energy. At any given instant, t, at any given spatial point, $\mathbf{x} = (x_1, x_2, x_3)$, the local field energy volumetric density carried by the EM field $[\mathbf{E}(t,\mathbf{x}), \mathbf{B}(t,\mathbf{x})]$ is the scalar [see Eqs. (2.A.99) and (2.A.100)]

$$u^{\mathrm{EM}}(t,\mathbf{x}) = \frac{\varepsilon_0}{2}\left(\mathbf{E}\cdot\mathbf{E} + c^2\mathbf{B}\cdot\mathbf{B}\right) = \frac{\varepsilon_0}{2}\left(E^2 + c^2B^2\right) \qquad [\mathrm{J/m^3}] \tag{2.5}$$

In a physically infinitesimally small spatial volume element [14, chapter II, p. 14][1] $\mathrm{d}V \equiv \mathrm{d}^3x$ around $\mathbf{x}$ at time t, the corresponding physically infinitesimal instantaneous field energy is

$$\mathrm{d}U^{\mathrm{EM}}(t,\mathbf{x}) \;= u^{\mathrm{EM}}\,\mathrm{d}V \quad [J] \tag{2.6}$$

1 "Such space–time regions, although containing many atoms and extending in time over many atomic periods may still be sufficiently small with respect to the macroscopic scale to be regarded infinitesimal..."

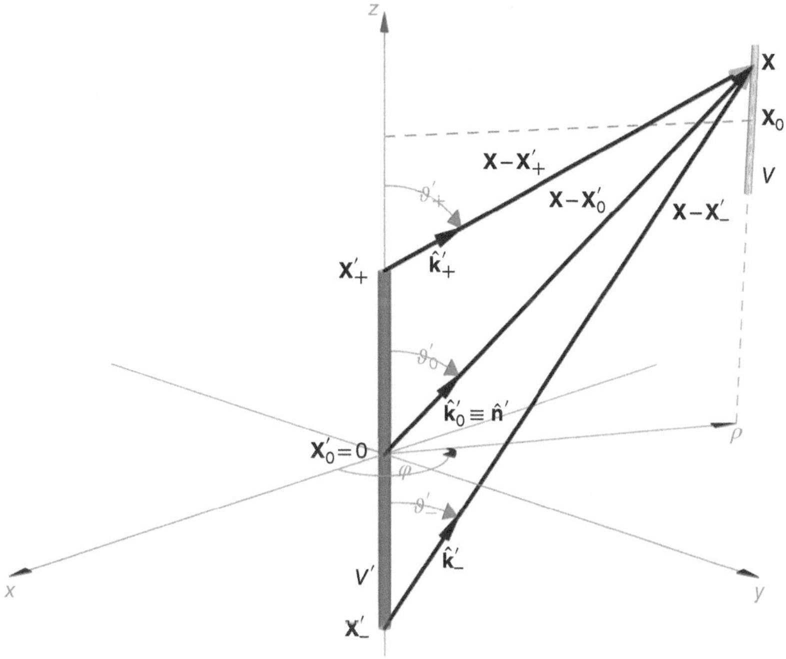

Figure 2.1 Linear antennas.

implying that at a given time t the field energy in a finite volume V is

$$U^{\mathrm{EM}}(t) = \int \mathrm{d}U^{\mathrm{EM}} = \frac{\varepsilon_0}{2}\int_V \mathrm{d}^3x\left(E^2 + c^2B^2\right) \quad [J] \tag{2.7}$$

showing that it is the square of the field, not the field itself, that is observable, similar to the situation in quantum mechanics. This is corroborated by the following further quotes from [8, 9]:

> 'We now take it for granted that electric and magnetic fields are abstractions not reducible to mechanical models. To see that this is true, we need only look at the units in which the electric and magnetic fields are supposed to be measured. The conventional unit of electric field-strength is the square-root of a joule per cubic meter. A joule is a unit of energy and a meter is a unit of length, but a square-root of a joule is not a unit of anything tangible. There is no way we can imagine measuring directly the square-root of a joule. The unit of electric field-strength is a mathematical abstraction, chosen so that the square of a field-strength is equal to an energy-density that can be measured with real instruments. The unit of energy-density is a joule per cubic meter, and therefore we say that the unit of field-strength is the square-root of a joule per cubic meter. This does not mean that an electric field-strength can be measured with the square-root of a calorimeter. It means that an electric field-strength is an abstract quantity, incommensurable with any quantities that we can measure directly.'
>
> …

'Like the Maxwell theory, quantum mechanics divides the universe into two layers. The first layer contains the wave-functions of Schrodinger, the matrices of Heisenberg and the state-vectors of Dirac. Quantities in the first layer obey simple linear equations. Their behaviour can be accurately calculated. But they cannot be directly observed. The second layer contains probabilities of particle collisions and transmutations, intensities and polarisations of radiation, expectation-values of particle energies and spins. Quantities in the second layer can be directly observed but cannot be directly calculated. They do not obey simple equations. They are either squares of first-layer quantities or products of one first-layer quantity by another.'

In Dicke radiometers [15] such as the ones used on meteorological satellites and on spacecraft measuring the cosmic microwave background (CMB) radiation [16, 17], the volume V over which the field energy is measured, i.e. integrated, is typically the efficient volume of a resistor in which the received field energy is detected as an increase in the resistor's temperature.

Early-day radio communications were based on the on–off keying of the broadband field energy produced by spark-gap transmitters and received by coherers or other square-law detectors [18]. In order to reduce the radio interference problems that were caused by these broadband, noisy signals, channelization of the available radio frequency spectrum into a raster of many channels of limited bandwidths was introduced. In today's radio contexts and scenarios, this prevalent technique is used. It amounts to a wavelength division and relies on the utilization of another conserved physical quantity, the EM field linear momentum (translational momentum).

The local volumetric density of the EM field linear momentum at $(t, \mathbf{x})$ is the ordinary vector (polar vector) field [see Eq. (2.A.112)]

$$\mathbf{g}^{\mathrm{EM}}(t,\mathbf{x}) = \varepsilon_0(\mathbf{E} \times \mathbf{B}) = \frac{\mathbf{S}}{c^2} \quad [\mathrm{N\,s/m^3}] \tag{2.8}$$

where $\mathbf{S}(t, \mathbf{x})$ is the Poynting vector, and the last step follows from the Planck relation. At any given instant t the physically infinitesimal volume element $\mathrm{d}V$ contains the corresponding physically infinitesimal EM field linear momentum

$$\mathrm{d}\mathbf{p}^{\mathrm{EM}}(t,\mathbf{x}) = \mathbf{g}^{\mathrm{EM}}\ \mathrm{d}V \quad [\mathrm{N\,s}] \tag{2.9}$$

The field linear momentum in a finite volume V at time t is therefore

$$\mathbf{p}^{\mathrm{EM}}(t) = \int \mathrm{d}\mathbf{p}^{\mathrm{EM}} = \varepsilon_0 \int_V \mathrm{d}^3x(\mathbf{E} \times \mathbf{B}) \quad [\mathrm{N\,s}] \tag{2.10}$$

Of all physical observables that may potentially be used in wireless information transfer scenarios, present-day radio science and engineering implementations almost exclusively make full use of the linear momentum vectorial observable only. For instance, as illustrated in Figure 2.1, a conventional transmitting and receiving antenna is typically wire like, i.e. engineered so that both the radiating volume V' and the observation volume V make up a very thin straight cylinder where current-carrying charges are constrained to oscillate in only one direction, along the cylinder (the z direction) [19].

In classical statistical mechanical terms, the antenna's current density is

$$\mathbf{j}^{\mathrm{e}}(t,\mathbf{x}) = \sum_s q_s \int \mathrm{d}^3v\,\mathbf{v}\,f_s(t,\mathbf{x},\mathbf{v}) = \sum_s q_s\langle \mathbf{v}_s\rangle \quad [\mathrm{Am^{-2}}] \tag{2.11}$$

where s denotes particle species, q_s their electric charge, and f_s is the (classical) distribution function that, at equilibrium, is Maxwell–Boltzmannian. The electrons (s = el) can be viewed as moving relative to a fixed neutralizing background of heavy ions (s = ion) with negligible recoil and hence

$$\mathbf{j}^{\mathrm{e}} = \mathbf{j}^{\mathrm{e}}_{\mathrm{el}} + \mathbf{j}^{\mathrm{e}}_{\mathrm{ion}} \approx \mathbf{j}^{\mathrm{e}}_{\mathrm{el}}, \quad m_{\mathrm{el}} \ll m_{\mathrm{ion}} \tag{2.12}$$

where m_s denotes mass.

Such antennas, typically with a length L of the order half a wavelength, i.e. $L \sim \lambda/2 = \pi/k$, where k is the wavenumber, emits and senses a linear momentum density vector $\mathbf{g}^{\mathrm{EM}}$ and, at the receiving end, integrates the component of the induced electric conduction current density vector $\mathbf{j}_{\mathrm{el}}$ over L along the axis (chosen as the z) of the thin, straight electrically conducting cylindrical antenna into a time-varying, spatially averaged scalar conduction current $j_{\mathrm{el}z} \propto \exp\,[\mathrm{i}(k_z z - \omega t)]$, corresponding to the z component of the photon linear momentum $p_{\mathrm{photon}_z} = \hbar k_z$ as given by the de Broglie relation [20]. This oscillating scalar current is fed into the receiver equipment (usually an electronics device) where it is processed and the wirelessly transferred information is extracted. However, in this process, the vector property of the current density is not fully exploited and information transfer capacity may therefore be wasted.

Linear momentum wireless information transfer techniques have the advantage of being simple, robust, and easy to implement, but they have a drawback that they can be wasteful when it comes to the use of the frequency spectrum. This is because electromagnetic linear momentum is single mode, permitting only one independent linear-momentum physical channel of information transfer per carrier frequency (per wave polarization (spin angular momentum [SAM]) mode, per antenna). The ensuing overcrowding of the radio frequency bands, often referred to as the 'spectrum crunch,' has become a serious problem in radio communications. It also imposes limitations on radio astronomy and other radio science applications as well as on fiber optics communications. It is doubtful whether further evolutionary refinements of conventional techniques and methods built on the Poynting vector (linear momentum density) physics layer alone will, in the long term, suffice to remedy these problems.

This situation has led to the idea of exploiting so far underutilized EM field observables that do not suffer from the same inherent fundamental physical limitations of the information-carrying capacity as those used in present-day radio techniques. Clearly, the exploitation of a new physical layer that provides additional information-carrying capabilities at the fundamental physics level, already before any multitransducer or other spectral density enhancing techniques are applied at the engineering application layer, will set the stage for new radio paradigms that may enable a wiser exploitation of the radio frequency spectrum resources as well as for the development of entirely new types of transducers.

To achieve this, Thidé et al. [21] proposed the exploration of underutilized conserved observables of the physical EM field to augment the number of information transfer degrees of freedom in the radio domain. Specifically, it was proposed that the orbital part of the electromagnetic/photonic angular momentum (rotational momentum, moment of momentum) be utilized in radio science and technology [6, 21–27]. The rationale being that it has been demonstrated in experiments at microwave and optical wavelengths that by using the electromagnetic orbital angular momentum (OAM) in addition to the SAM (wave polarization), one may enhance the information-carrying capability of wireless communications substantially [25, 26, 28–38].

The odd-parity pseudovectorial observable electromagnetic angular momentum was discussed already by Maxwell in his writings in the 1860s [39], studied by other authors since over a century [40–43], and nowadays treated in standard physics textbooks [5, 44–54] and – in more detail – in numerous monographs, e.g. Ribarič and Šušteršič [55], Fushchich and Nikitin [12], Mandel and Wolf [50], Allen et al. [56], Andrews [57], and Torres and Torner [58].

The local volumetric density of the EM field angular momentum at $(t, \mathbf{x})$ about a point $\mathbf{x}_\mathrm{m}$ is the pseudovector (axial vector) field

$$\mathbf{h}^{\mathrm{EM}}(t, \mathbf{x}; \mathbf{x}_\mathrm{m}) = \varepsilon_0(\mathbf{x} - \mathbf{x}_\mathrm{m}) \times (\mathbf{E} \times \mathbf{B}) \quad \left[\mathrm{Ns/m^2/rad^1}\right] \tag{2.13}$$

Consequently, the physically infinitesimal EM angular momentum in $\mathrm{d}V$ is

$$\mathrm{d}\mathbf{J}^{\mathrm{EM}}(t; \mathbf{x}_\mathrm{m}) = \mathbf{h}^{\mathrm{EM}}(\mathbf{x}_\mathrm{m})\mathrm{d}V \quad \left[\mathrm{Ns\,m/rad^1}\right] \tag{2.14}$$

which means that the EM field in V carries the total field angular momentum, i.e., the vectorial sum of SAM (wave polarization), SAM, and OAM about a given, fixed moment point $\mathbf{x}_\mathrm{m}$

$$\begin{aligned} \mathbf{J}^{\mathrm{EM}}(t; \mathbf{x}_\mathrm{m}) &= \int \mathrm{d}\mathbf{J}^{\mathrm{EM}}(\mathbf{x}_\mathrm{m}) = \varepsilon_0 \int_V \mathrm{d}^3x(\mathbf{x} - \mathbf{x}_\mathrm{m}) \times (\mathbf{E} \times \mathbf{B}) \\ &= \mathbf{J}^{\mathrm{EM}}(t; \mathbf{0}) - \mathbf{x}_\mathrm{m} \times \mathbf{p}^{\mathrm{EM}}(t) \quad \left[\mathrm{Ns\,m/rad^1}\right] \end{aligned} \tag{2.15}$$

Clearly, if $\mathbf{p}^{\mathrm{EM}} \neq 0$, then it is always possible to find a moment point $\mathbf{x}_\mathrm{m}$ that makes the EM field angular momentum vanish. If, on the other hand, $\mathbf{p}^{\mathrm{EM}} = \mathbf{0}$, then the field angular momentum $\mathbf{J}^{\mathrm{EM}}$ is independent of the moment point $\mathbf{x}_\mathrm{m}$. The similarities and differences in the fundamental physical properties between the EM linear and angular momenta are summarized in Table 2.1.

The capacity enhancement obtained by exploiting angular momentum is due to the fact that this quantity is already discretized (quantized) at the classical level. A beam that carries a certain amount of electromagnetic angular momentum is namely a superposition of discrete angular momentum eigenmodes, i.e. vortices [59, 60]. The z component of an OAM eigenmode is characterized by a unique integer that denotes its azimuthal quantum number (mode index), also known as the topological charge [61]. That this enhancement technique indeed works in practice has been demonstrated in an anechoic chamber experiment by Tamburini et al. [24] and soon thereafter in experiments in a real-world setting [25] as well as in numerous later experiments.

From a more fundamental physical point of view, the enhancements made possible by invoking the angular momentum physics layer compared to systems that are based solely on the linear momentum physics layer, is consistent with the fact that all classical fields carry angular momentum [51, 62, 63]. As emphasized by [64], angular momentum (also known as moment of momentum) is namely a unique physical observable in its own right, independent of and in general not derivable from linear momentum.[2] Therefore the knowledge of one of these two observables does not automatically imply the knowledge of the other or *vice versa*. That linear momentum and angular momentum are in general two independent physical observables also follows directly from Noether's theorem [10, 65]. Consequently, the dynamics of a physical system is not fully specified unless both its total linear momentum and total angular momentum are specified at all instants. Specifying only one of them is not sufficient. Exploiting only the linear momentum (Poynting) vector and not the angular momentum pseudovector of a system of charges, currents, and pertinent electromagnetic fields is like not exploiting such a system to its full capacity.

The discussion above clearly shows that in order to make optimal use of an electromechanical system for radio science and technology, including radio communications, one must, in addition to the translational degrees of freedom represented by the system's linear momentum, take also the rotational degrees of freedom, i.e. the mechanical and electromagnetic angular momenta of the

2 See in particular Section V.5: "Daniel Bernoulli and Euler on the Dependence or Independence of the Law of Moment of Momentum in 1744."

Table 2.1 Characteristics of the linear and angular momenta.

Similarities and differences between the EM linear momentum density, $\mathbf{g}^{\mathrm{EM}}(t, \mathbf{x})$, the Poynting vector, $\mathbf{S}(t, \mathbf{x})$, and the EM angular momentum density around a moment point $\mathbf{x}_{\mathrm{m}}$, $\mathbf{h}^{\mathrm{EM}}(t, \mathbf{x}; \mathbf{x}_{\mathrm{m}})$, carried by a classical electromagnetic fields ($\mathbf{E}(t, \mathbf{x})$ and $\mathbf{B}(t, \mathbf{x})$.

Property	Linear momentum density	Angular momentum density
Field point of view	$\mathbf{g}^{\mathrm{EM}} = \varepsilon_0 \mathbf{E} \times \mathbf{B} \equiv \mathbf{S}/c^2$	$\mathbf{h}^{\mathrm{EM}}(\mathbf{x}_{\mathrm{m}}) = (\mathbf{x} - \mathbf{x}_{\mathrm{m}}) \times (\varepsilon_0 \mathbf{E} \times \mathbf{B})$
Field point of view ($\mathbf{x} = r\hat{\mathbf{r}}$)	$\mathbf{g}^{\mathrm{EM}^{\mathrm{far}}} = \varepsilon_0 \mathbf{E}_{\perp}^{\mathrm{far}} \times \mathbf{B}_{\perp}^{\mathrm{far}} \equiv \frac{\mathbf{S}^{\mathrm{far}}}{c^2}$	$\mathbf{h}^{\mathrm{EM}^{\mathrm{far}}}(\mathbf{0}) = \varepsilon_0 \hat{\mathbf{r}} \times \left(r\mathbf{E}_{\parallel}^{\mathrm{far}} \times \mathbf{B}_{\perp}^{\mathrm{far}} \right) = -\varepsilon_0 r E_r^{\mathrm{far}} \mathbf{B}_{\perp}^{\mathrm{far}}$
Particle point of view ($\mathbf{x} = r\hat{\mathbf{r}}$)	$\mathbf{g}^{\mathrm{EM}^{\mathrm{far}}} \approx \frac{1}{16\pi^2 \varepsilon_0 c^4 r^2} \left\lvert \int_{V'} \mathrm{d}^3x' \left(\frac{\partial \mathbf{j}}{\partial t'} \right)_{\perp} \right\rvert^2 \hat{\mathbf{r}} \equiv \frac{\mathbf{S}^{\mathrm{far}}}{c^2}$	$\mathbf{h}^{\mathrm{EM}^{\mathrm{far}}}(\mathbf{0}) \approx \frac{1}{16\pi^2 \varepsilon_0 c^4 r^2} \int_V \mathrm{d}^3x' \left(c\rho + j_{\parallel}^{\mathrm{e}} \right) \hat{\mathbf{r}} \times \int_V \mathrm{d}^3x' \dot{\mathbf{j}}_{\perp} \hat{A}§$
SI unit	N.s.m^{-3} (kg.m^{-2}.s^{-1})	N.s.m^{-2}.rad^{-1} (kg.m^{-1}.s^{-1}.rad^{-1})
Fall-off at large distances r	$\sim r^{-2} + \mathcal{O}(r^{-3})$	$\sim r^{-2} + \mathcal{O}(r^{-3})$
Typical eigenmode phase factor	$\exp\{\mathrm{i}(kz - \omega t)\}$	$\exp[\mathrm{i}(k_z z + m\varphi - \omega t)]$, $0 \leq \varphi < 2\pi$
Quantal operator (z component)	$\hat{\mathrm{p}}_z = -\mathrm{i}\hbar \frac{\partial}{\partial z}$, $(p_z = \hbar k, k \in \mathbb{R})$	$\hat{\mathrm{L}}_z = -\mathrm{i}\hbar \frac{\partial}{\partial \varphi}$, $(L_z = \hbar m, m \in \mathbb{Z})$
Local conservation law	$\frac{\partial \mathbf{g}^{\mathrm{mech}}}{\partial t} + \frac{\partial \mathbf{g}^{\mathrm{EM}}}{\partial t} + \nabla \cdot \mathrm{T} = \mathbf{0}$	$\frac{\partial \mathbf{h}^{\mathrm{mech}}}{\partial t} + \frac{\partial \mathbf{h}^{\mathrm{EM}}}{\partial t} + \nabla \cdot \mathrm{M} = \mathbf{0}$
C symmetry (charge conjugation)	Even	Even
P symmetry (spatial inversion)	Even (polar vector/ordinary vector)	Odd (axial vector/pseudovector)
T symmetry (time reversal)	Odd	Odd

system, into account [5, 7, 44–53, 55–58, 66]. All of them are required in order for a system's two- and three-dimensional dynamics to be completely and correctly specified [67–83].

As is the case for electromagnetic linear momentum, electromagnetic angular momentum can propagate not only in free space but also in – and interact with – various propagation media [84, 85], including plasmas [6, 86–91]. This includes waveguides [92] and optical fibers [31, 93, 94], where the decay rate, due to absorption, is the same for the propagation of angular momentum as for linear momentum, since they are both bilinear in the same electric and magnetic fields.

2.2.2.1 Behavior at Very Long Distances

As shown in Section 2.A.2.1.1 for the assumed purely electric sources, the generalized and symmetrized SSPPJ formulas are

$$\mathbf{E}(t, \mathbf{x}) = \int_V \mathrm{d}^3x' \boldsymbol{\Theta}^{\mathrm{e}}(t, \mathbf{x}, \mathbf{x}') \tag{2.16}$$

$$\mathbf{B}(t,\mathbf{x}) = \frac{1}{c}\int_{V'} d^3x'\hat{\mathbf{k}}'(\mathbf{x}') \times \boldsymbol{\Theta}^{e}(t,\mathbf{x},\mathbf{x}'), \tag{2.17}$$

where the field vector density $\boldsymbol{\Theta}^{e}$ is given by Eqs. (2.A.20). The generalized and symmetrized SSPPJ formulas (2.A.45) and (2.A.46), that are exact for any combination of electric sources located in a finite volume V' and yield results that are identical to those obtained by solving Maxwell's equations from first principles, are particularly convenient for our purposes.

Paraxial approximation. In the paraxial approximation, all relative position vectors, ranging from a point $\mathbf{x}'$ the source volume V' to a point $\mathbf{x}$ in the measurement volume V, as illustrated by the range from $\mathbf{x}-\mathbf{x}'_{-}$ to $\mathbf{x}-\mathbf{x}'_{+}$ in Figure 2.1, are assumed to be parallel to the radius vector $\mathbf{r} = r\hat{\mathbf{r}}$ from the center of V', located at the origin in a spherical polar coordinate system. In this approximation, the propagation unit vector formula (2.A.29) can be approximated by:

$$\hat{\mathbf{k}}'(\mathbf{x}') \approx \hat{\mathbf{k}}'(\mathbf{0}) = \hat{\mathbf{n}}' \tag{2.18}$$

This allows the vector field $\boldsymbol{\Theta}^{e}$ to be decomposed into one part parallel to $\hat{\mathbf{r}}$ and one perpendicular to it:

$$\boldsymbol{\Theta}^{e}\phi^{e} = \boldsymbol{\Theta}^{e}_{\|}\phi^{e} + \boldsymbol{\Theta}^{e}_{\perp}\phi^{e} \tag{2.19}$$

where $\boldsymbol{\Theta}_{\|}^{\text{eparax}}$ and $\boldsymbol{\Theta}_{\perp}^{\text{e parax}}$ are given by formulas (2.A.53) and (2.A.54). From now on we shall assume that the charges and currents are electric, and therefore, for simplicity, and set $\boldsymbol{\Theta}^{e} = \boldsymbol{\Theta}$. Doing so, we can, in the paraxial approximation, write the EM fields as follows:

$$\mathbf{E}^{\text{parax}} = \mathbf{E}_{\perp}{}^{\text{parax}} + \mathbf{E}_{\|}{}^{\text{parax}} \tag{2.20}$$

$$= \int_{V'} d^3x'\boldsymbol{\Theta}_{\|}{}^{\text{parax}} + \int_{V'} d^3x'\boldsymbol{\Theta}_{\perp}{}^{\text{parax}} \tag{2.21}$$

$$\mathbf{B}^{\text{parax}} = \mathbf{B}_{\perp}^{\text{parax}} = \frac{1}{c}\hat{\mathbf{r}} \times \int_{V'} d^3x'\boldsymbol{\Theta}_{\perp}{}^{\text{parax}} \tag{2.22}$$

$$= \frac{1}{c}\hat{\mathbf{r}} \times \mathbf{E}_{\perp}^{\text{parax}} = \frac{1}{c}\hat{\mathbf{r}} \times \mathbf{E}^{\text{parax}} \tag{2.23}$$

We note that $\mathbf{B}^{\text{parax}}$ has only one vector component, perpendicular to $\hat{\mathbf{r}}$, whereas $\mathbf{E}^{\text{parax}}$ has two mutually perpendicular vector components, one perpendicular to $\hat{\mathbf{r}}$ and to $\mathbf{B}^{\text{parax}}$, and one parallel to $\hat{\mathbf{r}}$ (i.e. a radial component). It is important not to discard this radial component, since it is *only* $\mathbf{E}_{\|} = E_r\hat{\mathbf{r}}$ that contributes to the angular momentum in the far zone; the perpendicular electric field $\mathbf{E}_{\perp}$ does *not* contribute directly. These geometrical properties for the vectors always hold in the paraxial approximation, even if the far-zone approximation is not applicable.

A consistently truncated paraxial approximation expression for the electromagnetic linear momentum density is obtained by inserting the paraxially valid approximate expressions (2.A.56) and (2.A.57) for the fields into the exact expression (2.A.58). The result is

$$\mathbf{g}^{\text{EM parax}} \approx \mathbf{g}_{\|}^{\text{EM parax}} + \mathbf{g}_{\perp}^{\text{EM parax}} \tag{2.24}$$

where

$$\mathbf{g}_{\|}^{\text{EM parax}} = \varepsilon_0 E_{\perp}^{\text{EM parax}} B_{\perp}^{\text{EM parax}}\hat{\mathbf{r}} \tag{2.25}$$

$$\mathbf{g}_{\perp}^{\text{EMparax}} \approx \varepsilon_0 \mathbf{E}_{\|}^{\text{parax}}\phi^{e} \times \mathbf{B}_{\perp}^{\text{parax}} \tag{2.26}$$

In terms of given EM field sources in the form of arbitrary, prescribed electric charge and current densities, ρ and $\mathbf{j}$, respectively, the linear momentum density in the *paraxial* approximation is:

$$\mathbf{g}^{\text{EMparax}}(t, r, \vartheta, \varphi) \approx \frac{1}{16\pi^2\varepsilon_0 c^4 r^2}\left(\left|\int_{V'} d^3x' \dot{\mathbf{j}}_\perp\right|^2\right)\hat{\mathbf{r}} + \frac{1}{16\pi^2\varepsilon_0 c^4 r^3}\int_{V'} d^3x' \left(c\rho + j^{\text{e}}_{\parallel}\right)\int_{V'} d^3x' \dot{\mathbf{j}}_\perp$$

where we introduced the overdot ($\dot{}$) notation

$$\dot{\mathbf{j}} \equiv \frac{\partial \mathbf{j}\left(t - \frac{r}{c}, \mathbf{x}'\right)}{\partial t} \tag{2.27}$$

and also suppressed obvious arguments.

Hence, in terms of arbitrary, prescribed electric charge density and current density sources, ρ and $\mathbf{j}$, the linear momentum density in the *far zone*, where the paraxial approximation condition is trivially fulfilled and the radial part of $\mathbf{g}^{\text{EM}}$ is negligible, we find that

$$\mathbf{g}^{\text{EMfar}} \approx \frac{1}{16\pi^2\varepsilon_0 c^4 r^2}\left|\int_{V'} d^3x' \dot{\mathbf{j}}_\perp\right|^2 \hat{\mathbf{r}} \tag{2.28}$$

This standard formula can be found in most textbooks treating EM radiation. Figure 2.2 plots the familiar $\mathbf{g}^{\text{EMfar}}$ pattern for a short single dipole.

The angular momentum density carried by $(\mathbf{E}, \mathbf{B})$ about the origin $\mathbf{0}$ is given by the exact expression (2.A.144). Inserting the paraxially valid approximate expressions (2.A.56) and (2.A.57) for the fields into (2.A.144), we obtain

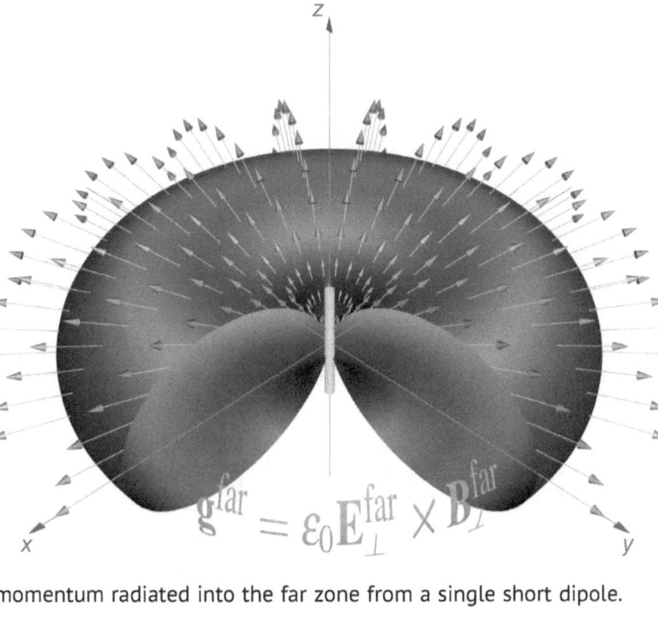

Figure 2.2 Linear momentum radiated into the far zone from a single short dipole.

$$\begin{aligned}\mathbf{h}^{\text{EMparax}}(\mathbf{0}) &= \varepsilon_0 \mathbf{r} \times \left(\mathbf{g}_{\perp}^{\text{EMparax}} + \mathbf{g}_{\parallel}^{\text{EMparax}}\right) \\ &= \varepsilon_0 (\mathbf{B}_{\perp}^{\text{parax}} \cdot \mathbf{r}) \mathbf{E}_{\parallel}^{\text{parax}} - \varepsilon_0 \left(\mathbf{E}_{\parallel}^{\text{parax}} \cdot \mathbf{r}\right) \mathbf{B}_{\perp}^{\text{parax}} + \varepsilon_0 E_{\perp}^{\text{parax}} B_{\perp}^{\text{parax}} \mathbf{r} \times \hat{\mathbf{r}}\end{aligned} \tag{2.29}$$

as follows from Eqs. (2.A.60)–(2.A.62) and the 'bac-cab' vector identity. Recalling that $\mathbf{B}_{\perp} \cdot \mathbf{r} = 0$ and that $\mathbf{r} \times \hat{\mathbf{r}} = \mathbf{0}$, Eq. (2.A.66) simplifies to

$$\mathbf{h}^{\text{EMparax}} \phi^{\text{e}}(\mathbf{0}) = -\varepsilon_0 r E_r^{\text{EMparax}} \mathbf{B}_{\perp}^{\text{parax}} \tag{2.30}$$

This shows that in cases when the paraxial approximation can be applied, the angular momentum density with respect to the origin is always proportional to the product of the distance r from the source, the amplitude of the near-zone longitudinal (radial) component of the electric field vector, which goes like r^{-2}, and the amplitude of the transverse component of the magnetic field vector, which goes like r^{-1}, so that the whole expression goes as r^{-2}, as it should. For a single frequency component ω, the angular momentum density (pseudo)vector has one DC component and one component that oscillates along the magnetic field (pseudo)vector at twice the oscillation frequency ω. The average cycle of the latter component vanishes.

Far-zone approximation. Sufficiently far away from a source volume of finite extent, such as an antenna array, the paraxial approximation condition is trivially fulfilled. There the consistently truncated approximations for the parallel and perpendicular parts of the vector field $\boldsymbol{\Theta}$ are:

$$\boldsymbol{\Theta}_{\parallel}^{\text{far}} \approx \frac{1}{4\pi\varepsilon_0 c r^2} \left(c\rho + j_{\parallel}\right) \hat{\mathbf{r}} \tag{2.31}$$

$$\boldsymbol{\Theta}_{\perp}^{\text{far}} \approx \frac{1}{4\pi\varepsilon_0 c^2 r} \hat{\mathbf{r}} \times \left(\hat{\mathbf{r}} \times \frac{\partial \mathbf{j}}{\partial t}\right) \tag{2.32}$$

From this and Eqs. (2.A.45) and (2.A.46) follows that the consistently truncated expressions for the electric and magnetic fields emitted from V' and received by V, a long distance $|\mathbf{x} - \mathbf{x}'| \approx r$ apart and expressed in the electric charge density ρ and the electric current density $\mathbf{j}$ in V', are

$$\mathbf{E}_{\parallel}^{\text{far}}(t, \mathbf{x}) \approx \frac{1}{4\pi\varepsilon_0 c r^2} \int_{V'} \mathrm{d}^3x' \left(\rho(t', \mathbf{x}')\hat{\mathbf{r}} + \mathbf{j}_{\parallel}(t', \mathbf{x}')\right) \tag{2.33}$$

$$\mathbf{E}_{\perp}^{\text{far}}(t, \mathbf{x}) \approx -\frac{1}{4\pi\varepsilon_0 c^2 r} \int_{V} \mathrm{d}^3x' \left(\frac{\partial \mathbf{j}(t', \mathbf{x}')}{\partial t}\right)_{\perp} \tag{2.34}$$

$$\mathbf{B}_{\perp}^{\text{far}}(t, \mathbf{x}) \approx \hat{\mathbf{r}} \times \mathbf{E}_{\perp}^{\text{far}}(t, \mathbf{x}) \tag{2.35}$$

where

$$\mathbf{j}_{\parallel} = (\mathbf{j} \cdot \hat{\mathbf{r}})\hat{\mathbf{r}} = j_r \hat{\mathbf{r}} = j_{\parallel} \hat{\mathbf{r}} \tag{2.36}$$

$$\mathbf{j}_{\perp} = \mathbf{j} - \mathbf{j}_{\parallel} \tag{2.37}$$

From Eqs. (2.33)–(2.35) and (2.A.68) we conclude that when the distance r between the source volume V' and the observing volume V is long enough, the angular momentum density about the moment point $\mathbf{0}$ can be approximated by

$$\mathbf{h}^{\text{EMfar}}(\mathbf{0}) \approx \frac{1}{16\pi^2\varepsilon_0 c^4 r^2} \int_{V'} \mathrm{d}^3x' \left(c\rho + j_{\parallel}^{\text{e}}\right) \hat{\mathbf{r}} \times \int_{V} \mathrm{d}^3x' \dot{\mathbf{j}}_{\perp} \tag{2.38}$$

It seems that this antenna design formula for $\mathbf{h}^{\text{EM}}$, and related formulas given above, has not been published in the standard literature. Figure 2.2 shows the well-known far-zone $\mathbf{g}^{\text{EM}}$ for a short dipole. Figure 2.3 plots the $\mathbf{h}^{\text{EMfar}}(\mathbf{0})$ for the very same short single dipole.

Figure 2.3 Angular momentum radiated into the far zone from a single short dipole.

Linear momentum density $\mathbf{g}^{\text{far}}(r, \vartheta, \varphi) = \varepsilon\left(\mathbf{E}_{\perp}^{\text{far}} \times \mathbf{B}_{\perp}^{\text{far}} \equiv \mathbf{S}^{\text{far}}/c^2\right)$ radiation pattern for a single small electric dipole (Hertzian dipole) placed at the origin and oscillating along the z axis. In the far zone the local linear momentum density field vectors (gray arrows) are directed in the radial (r) direction.

Angular momentum density $\mathbf{h}^{\text{far}}(r, \vartheta, \varphi; \mathbf{0}) = \varepsilon\hat{\mathbf{r}} \times \left(r\mathbf{E}_{\parallel}^{\text{far}} \times \mathbf{B}_{\perp}^{\text{far}}\right)$ radiation pattern for a single small electric dipole (Hertzian dipole) placed at the origin and oscillating along the z axis. In the far zone the local angular momentum density field vectors (gray arrows) are directed in the azimuthal (φ) direction.

Equations (2.A.97) and (2.A.98) can be used for designing transducers that can generate beams with a coherent superposition of angular momentum eigenmodes that can be sensed, resolved, and analyzed far away from the source volume V'. It is important that the volume V used for integrating the angular momentum density is chosen appropriately and that the angular momentum measurements are made in an optimum way, so that the r^{-2} fall-off is judiciously exploited.

2.3 Implementation

For many reasons, not least to mitigate the problems of growing frequency congestion and increasing energy consumption in radio science, radar, and wireless and fiber optic communication applications, it is desirable to strive for a more resource-conserving, efficient, and versatile use of the electromagnetic field than offered by linear-momentum (Poynting vector) techniques. In this section we discuss physical issues that are essential to consider when developing and implementing angular momentum techniques in the radio domain. These results are of interest not only for radio but also for other wavelengths, since, as was pointed out by Thidé et al. [21], it is often more convenient to perform fundamental studies of electromagnetic radiation, including those addressed in this chapter, in radio than in optics. One reason being that it is relatively easier to vary the frequency and to control the phase of radio waves and beams made from them. They can therefore be produced with a very high degree of coherence and spectral purity. Furthermore, the fact that it is possible to use digitally controlled transmitters, receivers, and antenna arrays that enable experiments to be completely software defined adds considerable flexibility and versatility compared to the typical experiment at optical wavelengths, something that increases the serendipity factor.

Let us consider a spatially limited source volume V' from which an electromagnetic beam is emitted into free space where there is no creation or annihilation of neither linear nor angular field momentum. The emission takes place during a finite time interval Δt_0, thus forming a signal 'pulse' that a sufficiently long time t_0 after the end of the emission will be propagating radially outward with speed c into the surrounding free space as discussed in Section 2.2. This signal 'pulse' will then be located in a finite volume V_0 between two concentric spherical shells, one with radius $r_0 = ct_0$ relative to the reference point $\mathbf{x}_{0'}$ in V' (see Figure 2.A.1), and another with radius $r_0 + \Delta r_0$ where $\Delta r_0 = c\Delta t_0$ [5]. According to the conservation laws (2.A.128) and (2.A.158), the cycle averaged field momentum $\mathbf{p}^{\mathrm{EM}}$ and angular momentum $\mathbf{J}^{\mathrm{EM}}$ contained in this spatially limited volume do not fall off asymptotically at large distances from the source, but tend to constant values.

Consequently, both linear and angular electromagnetic field momenta can propagate and hence be used for information transfer over in principle arbitrarily long distances. Of course, the magnitude and angular distribution of the respective momentum densities depend on the specific spatiotemporal properties of the actual radiating and receiving transducers ('antennas') used. Some transducers, such as one-dimensional linear dipole antennas used in radio today, are effective radiators and sensors of linear momentum, whereas well-defined coherent superpositions of angular momentum eigenmodes are more optimally radiated and sensed by transducers that make full use of two- or three-dimensional current distributions.

2.3.1 Wireless Information Transfer with Linear Momentum

If we identify the individual terms in the global law of conservation of linear momentum in a closed volume, Eq. (2.A.128), we can rewrite this equation as a balance equation between the time rate change of mechanical linear momentum, i.e. the force on the matter (the charged mechanical particles) in V, the time rate change of field linear momentum in V, and the flow of field linear momentum into V, across the surface S bounding V, as follows:

$$\underbrace{\int_V \mathrm{d}^3x\mathbf{f}}_{\text{Force }\mathbf{F}\text{ on the matter}} + \underbrace{\frac{\mathrm{d}}{\mathrm{d}t}\int_V \mathrm{d}^3x\varepsilon_0(\mathbf{E}\times\mathbf{B})}_{\text{Field momentum}} + \underbrace{\oint_S \mathrm{d}^2x\hat{\mathbf{n}}\cdot\mathsf{T}}_{\text{Linear momentum flow}} = \mathbf{0} \tag{2.39}$$

which, of course, is nothing but a generalized version of Newton's second law, or rather Euler's first law of motion.

In classical particle terms, Eq. (2.39) shows that electric charges subjected to an oscillating electromotive force, e.g., a current fed into an antenna from a transmitter via a transmission line, experience a mechanical force that sets the charges into translational oscillatory motion and therefore give rise to a time-varying linear conduction current density $\mathbf{j}^{\mathrm{e}}(t', \mathbf{x}')$ in V'. The conservation law (2.A.128) also shows that $\mathbf{j}^{\mathrm{e}}(t', \mathbf{x}')$ is accompanied by the simultaneous emission of a time-varying electromagnetic field linear momentum $\mathbf{p}^{\mathrm{EM}}(t)$ that propagates in free space away from V' in the form of the linear momentum density $\mathbf{g}^{\mathrm{EM}}(t, \mathbf{x}')$, as described in Section 2.2 and in Section 2.A.2.

For illustrative purposes, let us consider an electromechanical system comprising an electromagnetic field that interacts with point charges q_i, $i = 1, 2, ..., j, ..., n$, located at $\mathbf{x}_i$ at time t_i. All point charges have the same electric charge q and same mass m. The charge density ρ_j^{e} representing the jth point charge q_j at $\mathbf{x}_j$ can be described in the form of the Dirac delta distribution [*cf.* Eq. (2.11)]

$$\rho_j^{\mathrm{e}} = q\delta(\mathbf{x} - \mathbf{x}_j) \tag{2.40}$$

According to the balance Eq. (2.39), the Lorentz force acting on q_j due to the imposed linear momentum flow (the Maxwell stresses) is:

$$\begin{aligned} \mathbf{F}_j \equiv \mathbf{F}(t_j, \mathbf{x}_j) &= \int_V \mathrm{d}^3x\mathbf{f} = \int_V \mathrm{d}^3x\rho_j^{\mathrm{e}}(\mathbf{E} + \mathbf{v}^{\mathrm{mech}} \times \mathbf{B}) \\ &= q\mathbf{E}(t_j, \mathbf{x}_j) + q\mathbf{v}^{\mathrm{mech}}(t_j, \mathbf{x}_j) \times \mathbf{B}(t_j, \mathbf{x}_j) \\ &\equiv q\mathbf{E}_j + q\mathbf{v}_j^{\mathrm{mech}} \times \mathbf{B}_j = q\mathbf{E}_j + \mathbf{j}_j^{\mathrm{e}} \times \mathbf{B}_j \end{aligned} \tag{2.41}$$

Expressed in the mechanical linear momentum $\mathbf{p}_j^{\mathrm{mech}}$ of particle j, this can be written as

$$\frac{\mathrm{d}\mathbf{p}_j^{\mathrm{mech}}}{\mathrm{d}t} - \frac{q}{m}\mathbf{p}_j^{\mathrm{mech}} \times \mathbf{B}_j = q\mathbf{E}_j \tag{2.42}$$

This illustrates that the process of estimating the fields from the measured current in a linear antenna of finite one-dimensional extent where the charges experience a holonomic constraint due to their cylinder symmetric axial confinement, is difficult and can, at most, only be projective. The situation gets worse if we also include self-interaction effects, reradiation, preacceleration, and related complications. For a detailed discussion, see Rohrlich [193, chapter 6 and supplement] where the Lorentz–Abraham–Dirac equation, which describes the classical dynamics of charged particles but violates Newton's first law (Galileo's law of inertia), is discussed.

While each (conduction) electron q_j in a conventional conductive antenna, clamped mechanically to its mechanical support, can be considered to move freely and to react immediately to the imposed force $\mathbf{F}_j$, the force on the corresponding ion, which is typically five orders of magnitude and more massive, is negligible in comparison; see Eq. (2.11) in Section 2.2.2 and the associated discussion. Therefore the ions can be assumed to only provide a fixed neutralizing background, keeping the antenna electrically neutral globally, standard assumptions made in classical electron theory [14, 95]. If the antenna would be free to move in space, the ions would follow the electrons and the whole antenna would experience a force and acquire a translational motion in such a way that the system's total linear momentum is conserved. This is the classical model of the mechanical effects caused by radiation pressure as postulated in 1873 by Maxwell [96] and identified experimentally for the first time in 1901.

The conservation law (2.A.128) shows that in a remotely located sensing volume V, the reverse process takes place in that (a part of) the flow of field linear momentum density $\mathbf{g}^{\mathrm{EM}}(t, \mathbf{x})$ emitted from V' reaches V and is integrated there into a field linear momentum $\mathbf{p}^{\mathrm{EM}}(t)$ that is accompanied by a mechanical linear momentum $\mathbf{p}^{\mathrm{mech}}(t)$; see Eq. (2.A.128). This gives rise to a translational motion of charged particles and, hence, a translational (non-rotational) conduction current. If V is a very thin cylindrical conductor, as in typical radio communications scenarios where linear dipole antennas are used, this is a one-dimensional, scalar, (Ohmic) dissipative current – the receiving antenna current. Clearly, a single translational symmetric receiving antenna can only sense one-dimensional dynamics and not the two- or three-dimensional dynamics of electromagnetic angular momentum carried by the same beam.

Let us calculate the amount of linear momentum $\mathbf{p}^{\mathrm{EM}}$ carried by the electromagnetic field generated by an arbitrary source that is localized in volume V', e.g., a typical transmitting antenna. This is most readily done by evaluating the integral in the right-hand member of Eq. (2.A.125) that expresses the field linear momentum in a volume away from V' in a spherical polar coordinate system (r, ϑ, φ) with its origin at the center of the source region, e.g., the transmitting antenna phase center.

Within a solid angle cone $\Delta\Omega = \Omega_2 - \Omega_1$, defined by the polar angle range $(\vartheta_1, \vartheta_2)$ and the azimuth angle range (φ_1, φ_2), the total instantaneous linear momentum of an electromagnetic 'pulse' of finite duration $\Delta t = \Delta r_0/c$ propagating in free space is [5]

$$\mathbf{p}^{\mathrm{EM}}(t) = \int_{\Omega_1}^{\Omega_2} \mathrm{d}\Omega \int_{r_0}^{r_0 + \Delta r_0} \mathrm{d}r\, r^2\, \mathbf{g}^{\mathrm{EM}}(t, r, \vartheta, \varphi) \tag{2.43}$$

where

$$\int_{\Omega_1}^{\Omega_2} \mathrm{d}\Omega \equiv \int_{\vartheta_1}^{\vartheta_2} \int_{\varphi_1}^{\varphi_2} \sin\vartheta \mathrm{d}\vartheta \mathrm{d}\varphi \tag{2.44}$$

For very large distances $r_0 = ct_0$ from the source, we can use the expression (2.A.87) for $\mathbf{g}^{\mathrm{EM}}$. Hence we obtain

$$\mathbf{p}^{\mathrm{EMfar}}(t) = \frac{1}{16\pi^2\varepsilon_0 c^4} \left(\int_{\Omega_1}^{\Omega_2} \mathrm{d}\Omega \int_{r_0}^{r_0 + \Delta r_0} \mathrm{d}r \left| \int_{V'} \mathrm{d}^3x' \dot{\mathbf{j}}_{\perp}^{\mathrm{e}}(t', \mathbf{x}' \right| + \mathcal{O}(r^{-1}) \right) \hat{\mathbf{n}}' \tag{2.45}$$

$$\mathbf{p}^{\mathrm{EMfar}}(t) = \frac{\Delta\Omega\, \Delta r_0}{16\pi^2\varepsilon_0 c^4} \left(\left| \int_{V} \mathrm{d}^3x' \dot{\mathbf{j}}_{\perp}^{\mathrm{e}}(t', \mathbf{x}') \right| + \mathcal{O}(r^{-1}) \right) \hat{\mathbf{n}}' \tag{2.46}$$

This shows explicitly that the linear momentum carried by the field tends to a constant vector (possibly $\mathbf{0}$) when the distance that the pulse has propagated from the source, r_0, tends to infinity. This asymptotic independence of distance from a localized source, allowing the field linear momentum generated by this source to be transported all the way to infinity without radial fall-off and therefore to be irreversibly lost there is the celebrated arrow of radiation asymmetry (see Eddington [97, pp. 328–329], Jackson [52, chapter 6, 98], Zeh [99], Rohrlich [53, chapter 9, 100], and Wheeler and Zurek [101]).

The angular distribution in the far zone of the linear momentum density, $\mathbf{g}^{\mathrm{EMfar}}$ and, consequently, of the far-zone Poynting vector $\mathbf{S}^{\mathrm{far}}$ [see relation (2.A.114)], is commonly referred to as 'the radiation pattern' or, in the case of antenna engineering, 'the antenna pattern', 'beam pattern', or 'antenna diagram', particularly in the antenna engineering literature [102].

2.3.2 Wireless Information Transfer with Angular Momentum

As the global law of conservation of angular momentum in a closed volume Eq. (2.A.158) shows, the electric charges in a source volume V' that are subjected to a time-varying surface integrated angular momentum flux tensor M suffer a change in their total mechanical (matter) angular momentum $\mathbf{J}^{\mathrm{mech}}(\mathbf{x}_{\mathrm{m}})$ around a moment point $\mathbf{x}_{\mathrm{m}}$. That is, they experience a mechanical torque $\boldsymbol{\tau}(\mathbf{x}_{\mathrm{m}})$ that sets the charges into two- or three-dimensional rotational motion. In turn, this causes the emission of a superposition of electromagnetic angular momentum eigenmodes that propagate away from V' in the form of the angular momentum density $\mathbf{h}^{\mathrm{EM}}(\mathbf{x}_{\mathrm{m}})$ as described in earlier sections and in Section 2.A.2.

The angular momentum conservation law (2.A.158) can be written in the form of a balance equation between the time rate change of mechanical torque on the matter (the charged mechanical particles) in V, the time rate change of EM field angular momentum, and the flow of EM field angular momentum across the surface S, which encloses V, as follows:

$$\underbrace{\int_V \mathrm{d}^3x(\mathbf{x}-\mathbf{x}_{\mathrm{m}}) \times \mathbf{f}}_{\text{Torque } \boldsymbol{\tau}(\mathbf{x}_{\mathrm{m}}) \text{ on the matter}} + \underbrace{\frac{\mathrm{d}}{\mathrm{d}t}\int_V \mathrm{d}^3x(\mathbf{x}-\mathbf{x}_{\mathrm{m}}) \times \varepsilon_0(\mathbf{E}\times\mathbf{B})}_{\text{Field angular momentum}} + \underbrace{\oint_S \mathrm{d}^2x\hat{\mathbf{n}}\cdot\mathbf{M}(\mathbf{x}_{\mathrm{m}})}_{\text{Angular momentum flow}} = \mathbf{0} \tag{2.47}$$

This is nothing but a generalized version of Euler's second law of motion.

For illustrative purposes, let us consider the jth charge q_j in a system of n point charges q_i, $i = 1, 2, \ldots, j, \ldots, n$, each carrying the charge q. The charge density can be represented by the Dirac delta distribution as given by Eq. (2.40). Then we find that the mechanical torque on q_j is

$$\begin{aligned}
\boldsymbol{\tau}_j(\mathbf{x}_{\mathrm{m}}) \equiv \boldsymbol{\tau}(t_j, \mathbf{x}_j; \mathbf{x}_{\mathrm{m}}) &= \int_V \mathrm{d}^3x(\mathbf{x}_j - \mathbf{x}_{\mathrm{m}}) \times \mathbf{f}_j \\
&= \int_V \mathrm{d}^3x(\mathbf{x}_j - \mathbf{x}_{\mathrm{m}}) \times \left(\rho_j^{\mathrm{e}}\mathbf{E} + \mathbf{j}^{\mathrm{e}} \times \mathbf{B}\right) \\
&= \left(\mathbf{x}_j - \mathbf{x}_{\mathrm{m}}\right) \times \mathbf{F}_j
\end{aligned} \tag{2.48}$$

where $\mathbf{F}_j$ is the Lorentz force on q_j, given by Eq. (2.41).

In a remote observation, volume V where the charges are not constrained to move only along a line but are able to move freely in two or three dimensions, the reverse process takes place in that (a sufficient amount of) the flow of field angular momentum density emitted from V' is intercepted and integrated into a mechanical angular momentum $\mathbf{J}^{\mathrm{mech}}$. This gives rise to a rotational motion of the charges and, hence, to a two- or three-dimensional rotational (nontranslational) current density in V. Applications that exploit the rotational symmetry of the fields, i.e. make use of the angular momentum degrees of freedom, must therefore be implemented in techniques that are based on the angular momentum physical layer.

What was discussed above in terms of the linear momentum and the force action on each electron q_j and its associated (very massive) ion is applicable to angular momentum, except that the electrons feel a torque that gives rise to a rotational motion to a conductor that can move freely in space. This is the classical model of the mechanical effects caused by radiation torque.

Similarly to Eqs. (2.43) and (2.44), the total instantaneous angular momentum with respect to the origin ($\mathbf{x}_{\mathrm{m}} = \mathbf{0}$), within a solid angle cone $\Delta\Omega$ of an electromagnetic 'pulse' of finite duration $\Delta t = \Delta r_0/c$ propagating in free space is

$$\mathbf{J}^{\mathrm{EM}}(t; \mathbf{0}) = \int_{\Omega_1}^{\Omega_2} \mathrm{d}\Omega \int_{r_0}^{r_0 + \Delta r_0} \mathrm{d}r r^2 \; \mathbf{h}^{\mathrm{EM}}(t, r, \vartheta, \varphi; \mathbf{0}) \tag{2.49}$$

The field angular momentum about the origin $\mathbf{x}_{\mathrm{m}} = \mathbf{0}$, i.e., $\mathbf{J}^{\mathrm{EM}}(\mathbf{0})$, contained in the 'pulse' emitted from the same localized source and volume as described by Eqs. (2.43) and (2.44), is

$$\mathbf{J}^{\mathrm{EM}}(t, \mathbf{0}) = \int_{r_0}^{r_0 + \Delta r_0} \mathrm{d}r r^2 \int_{\Omega_1}^{\Omega_1} \mathrm{d}\Omega \; \mathbf{h}^{\mathrm{EM}}(t, r, \vartheta, \varphi) \tag{2.50}$$

The expressions (2.A.89) and (2.A.90) explicitly show that the angular momentum density $\mathbf{h}^{\mathrm{EM}}$ emitted from any electromagnetic radiation source has precisely the same asymptotic r^{-2} radial fall-off in all directions as the linear momentum density $\mathbf{g}^{\mathrm{EM}}$. Hence, it is also clear that $\mathbf{J}^{\mathrm{EM}}$ tends asymptotically to a constant (pseudo)vector and that it is irreversibly lost when the 'pulse' approaches infinity. This is the angular momentum analog of the linear momentum arrow of radiation.

Comparing Figures 2.2 and 2.3, it is clear that the angular distribution of the radiated *angular* momentum density $\mathbf{h}^{\mathrm{EM}}$ is very different from the angular distribution of the radiated *linear* momentum density, $\mathbf{g}^{\mathrm{EM}}$, i.e. the 'antenna pattern' for the one and the same radiating device ('antenna'); see also Eqs. (2.A.84)–(2.A.86) and (2.A.89), Thidé et al. [6], and Then and Thidé [78].

2.3.2.1 Spin Angular Momentum vs. Orbital Angular Momentum

It is shown in Section 2.A.2.3 that if the vector potential $\mathbf{A}$ fulfills the criteria for Helmholtz decomposition, it is possible to separate the analytic expression for the total field angular momentum, $\mathbf{J}^{\mathrm{EM}}$, into two exact, gauge-independent expressions. One that is not dependent on the moment point, $\mathbf{\Sigma}^{\mathrm{EM}}$, *viz.* expression (2.A.160), and one that is, $\mathbf{L}^{\mathrm{EM}}$, *viz.*, expression (2.A.161). In radio communication scenarios these expressions are identified as the spin and OAM, respectively, of the radio beam.

The *intrinsic* (coordinate independent) spin part of the angular momentum (SAM), $\mathbf{\Sigma}^{\mathrm{EM}}$, manifests itself as left- or right-hand circular polarization, i.e. as a unison rotation in time of the field vectors such that they all have one and the same direction across the phase front of the beam at given time t, and all have rotated into another direction (modulo 2π in oscillation frequency phase ωt) in each point at time $t + \Delta t$. Hence, the measurement of SAM (polarization) requires *temporal coherence.*

A common technique to observe and exploit the SAM (wave polarization) in radio is to use two colocated linear dipole antennas arranged perpendicularly to each other and to determine the polarization state by comparing the phases of the antenna currents induced in the two antennas by using interferometry. In antenna engineering, such antennas are known as turnstile antennas [74, 83, 103]. Another technique is to use longitudinal mode helical antennas; see [104, example 12.7.1, p. 461].

In contrast to SAM, the orbital part of the electrodynamic angular momentum, the OAM $\mathbf{L}^{\mathrm{EM}}$, is an *extrinsic* (coordinate dependent) property. It manifests itself as a rotation in space of the field vectors such that at any given instant they have different directions (modulo 2π in oscillation frequency phase ωt) for different azimuth angles positions φ and $\varphi + \Delta\varphi$ around the z (beam) axis in the phase plane. Hence, the measurement of OAM requires *spatial coherence.*

We emphasize that whereas the SAM is an intrinsic property of the photon, both its linear and orbital angular momenta are extrinsic properties.

2.3.2.2 Angular Momentum Transducers

As the conservation law (2.A.158) shows, angular momentum cannot be measured by single, monolithic transducers that work solely on the principle of force, e.g. are based on conduction currents flowing along a line, as is the case for linear dipole-type antennas commonly used in radio science and communications. Instead, angular momentum is optimally transmitted and received by transducers (actuators, sensors, 'antennas') that work on the principle of rotational degrees of freedom and torque, thus being capable of detecting angular momentum directly [67–83, 105–107].

In general, angular momentum transducers must support – and be able to control and sense – the three-dimensional dynamics of the charges. If, however, the rotation is confined to a single plane, it suffices that the transducers are effectively two-dimensional. Such transducers can be realized by equipping single linear-momentum antennas with two-dimensional azimuthally dependent reflectors, lenses, or phase plates [24, 37, 108–114] or by employing helicoidal parabolic antennas [25, 115]. Other examples are the flat-plate transducer developed by Bennis et al. [116] and the screen detector used by Krenn et al. [117].

Interesting possibilities for innovative monolithic transducers are offered by developments in nano- and optomechanics [83, 100, 118–126], and by atomic Bose–Einstein condensates [127]. Recent research into spintronics and orbitronics herald the arrival of transducers that are based on the interplay between charge, spin, and orbital degrees of freedom, and that may be realized with dissipationless noncharge currents [128–142], including antennas realized with condensed-matter skyrmions (magnetic vortex structures) [59, 143–146] and quantum rings [147–149].

Monolithic multidimensional antennas that make direct use of the angular momentum degree of freedom for the radio frequency range can easily be designed and manufactured with the help of the formulas given in this chapter, but they are not yet readily available. Until they are, one may use existing phased arrays of linear-momentum antennas of the type that are used in today's radio astronomy and wireless communications. This is because these multitransducer arrays, including multiple-in multiple-out (MIMO) arrays, sample a radio beam with transducers, essentially in the form of a finite set of integrating cylindrical volumes V_n, $n = 1, 2, \ldots, M$ deployed at different discrete points distributed over a certain region in space, thereby approximating, within the limitations and constraints set by the sampling theorem, [150–152], the properties of a single, monolithic two- and three-dimensional transducer that samples the field in a continuous manner [6, 21, 78, 153–171]. The fact that it is *possible* to use phased arrays consisting of N conventional linear-momentum antennas (e.g., dipole antennas) to exploit (at most) $N/2$ unaliased angular momentum modes, does of course not imply that it is *necessary* to use such arrays. Maxwell's equations allow any number and combination of sources ρ and $\mathbf{j}$ with any spatiotemporal and topological properties in the source and sensor volumes V' and V. And the currents do not have to be conduction currents in the usual sense [125].

Within the limitations just mentioned, it is possible to use arrays of linear antennas and already available radio equipment and techniques to carry out certain experimental investigations of the physical properties of a finite subset of (approximate) OAM eigenmodes and coherent superpositions thereof. This will provide useful insights into the properties of OAM information transfer and therefore will pave the way for the development of future, innovative angular momentum radio concepts based on single transducers.

The experimental investigations mentioned typically amount to measuring the phase of the currents sampled at the feedpoints of spatially distributed individual 1D antenna elements, i.e. a spatial sampling of one component of the linear momentum of the charges, and then performing a phase gradient analysis [6, 66, 116, 172–177] and postprocessing the data so that the approximate (possibly aliased) spectrum of the z component of the angular momentum, L_z, can be estimated. This way, approximations of individual OAM eigenmodes may be extracted and an approximation of the correct spectrum of individual OAM eigenmodes and the information they carry may be decoded. It should be emphasized that this way of estimating L_z is based on the application of the operator defined by formula (2.A.166) on the electric (or magnetic) field. However, as was discussed in Section 2.2, fields are, strictly speaking, not directly observable.

Proof-of-concept experiments where existing conventional radio technology was employed have shown that it is indeed possible to use the total angular momentum, i.e., the SAM + OAM physical observable, as a new physical layer for radio science and technology exploitation. These studies include numerical experiments [21] showing that it is feasible to utilize OAM in radio; controlled laboratory experiments [24, 29, 37, 178], verifying that it is possible to generate and transmit radio beams carrying noninteger OAM and to measure their OAM spectra in the form of weighted superpositions of different integer OAM eigenmodes each with its own quantum number (mode index), also known as topological charge; and outdoor experiments [25, 27, 179], verifying that in real-world settings different signals, physically encoded in different OAM modes and overlapping each other in space and time and subjected to reflection, can be transmitted wirelessly to a receiver located in the far zone and be individually extracted and decoded there.

We stress that for an electromagnetic beam of the kind used in radio astronomy, radar investigations, radio communications, and similar applications, OAM is distinctively different from – and independent of – SAM (wave polarization). If such a beam is already N-fold OAM encoded, utilizing SAM also will double the information transfer capacity by virtue of the fact that the dimension of the state space then doubles from N to $2N$.

2.3.2.3 Electric Hertzian Dipoles

A commonly used antenna in today's radio is the half-wavelength dipole. Many qualitative as well as quantitative characteristics of this antenna are fairly accurately approximated by an electric Hertzian dipole, i.e. an antenna whose length L is much shorter than the wavelength λ of its emitted (nearly monochromatic) field, $L \ll \lambda$, so that the phase variations over L of the antenna current can be neglected – a condition that simplifies the theoretical analysis considerably. Below we present expressions for the linear and angular momentum densities emitted by a single Hertzian dipole and combinations of such dipoles. The properties of Hertzian dipoles are described in many textbooks [7, 51, 52]. For convenience, we here recapitulate some of these properties.

The electric dipole moment vector with respect to an arbitrary moment point $\mathbf{x}'_0$ is defined as

$$\mathbf{d}(t',\mathbf{x}'_0) = \int_V \mathrm{d}^3x' \,(\mathbf{x}'-\mathbf{x}'_0)\,\rho^{\mathrm{e}}(t',\mathbf{x}') \tag{2.51}$$

$$= \int_V \mathrm{d}^3x' \mathbf{x}' \rho^{\mathrm{e}}(t',\mathbf{x}') - \mathbf{x}'_0 \int_V \mathrm{d}^3x' \rho^{\mathrm{e}}(t',\mathbf{x}') \tag{2.52}$$

$$= \mathbf{d}(t',\mathbf{0}) - \mathbf{x}'_0 q^{\mathrm{e}}(t') \tag{2.53}$$

where

$$q^{\mathrm{e}}(t') = \int_V \mathrm{d}^3x' \rho^{\mathrm{e}}(t',\mathbf{x}') \tag{2.54}$$

is the total net electric charge in V'.

There are two situations to consider: (i) If $q^{\mathrm{e}}(t') \neq 0$, it is always possible to choose a moment point $\mathbf{x}'_0$ so that $\mathbf{d}(t',\mathbf{x}'_0) = \mathbf{0}$. (ii) If $q^{\mathrm{e}}(t') = 0$, the dipole moment is independent of the choice of the moment point $\mathbf{x}'_0$.

Furthermore, since $\mathbf{x}'_0$ is a fixed point,

$$\frac{\partial[\mathbf{d}(t',\mathbf{x}'_0)]}{\partial t'} = \frac{\mathrm{d}[\mathbf{d}(t',\mathbf{x}'_0)]}{\mathrm{d}t'} = \frac{\mathrm{d}[\mathbf{d}(t',\mathbf{0})]}{\mathrm{d}t'} - \mathbf{x}'_0 \frac{\mathrm{d}q^{\mathrm{e}}(t')}{\mathrm{d}t'} \tag{2.55}$$

and, according to the equation of continuity for electric charge,

$$\frac{\mathrm{d}q^{\mathrm{e}}(t')}{\mathrm{d}t'} = \int_{V} \mathrm{d}^3x' \frac{\partial \rho^{\mathrm{e}}(t',\mathbf{x}')}{\partial t'} = -\oint_{S'} \mathrm{d}^2x' \hat{\mathbf{n}}' \cdot \mathbf{j}^{\mathrm{e}}(t',\mathbf{x}') \tag{2.56}$$

If there is no net flow of electric current across the surface S' that encloses V', it follows that

$$\frac{\partial [\mathbf{d}(t',\mathbf{x}'_0)]}{\partial t'} = \frac{\partial [\mathbf{d}(t',\mathbf{0})]}{\partial t'} \tag{2.57}$$

$$\frac{\partial^2 [\mathbf{d}(t',\mathbf{x}'_0)]}{\partial t'^2} = \frac{\partial^2 [\mathbf{d}(t',\mathbf{0})]}{\partial t'^2} \tag{2.58}$$

Let us consider a Hertzian electric dipole oscillating at the single angular frequency $\omega = ck$, where $k = \lambda/(2\pi)$ is the wave number, λ being the wavelength in free space, and located at the origin $\mathbf{x}' = \mathbf{0}$, which is also the moment point chosen,

$$\mathbf{d}(t',\mathbf{0}) = \mathbf{d}_{\omega}(\mathbf{0}) \cos(\omega t' + \delta) \tag{2.59}$$

If we apply the decomposition (2.A.37)–(2.A.39) on $\mathbf{d}_{\omega}(\mathbf{0})$, resulting in a parallel ($\|$) and a perpendicular ($\perp$) component

$$\mathbf{d}_{\|}(\mathbf{0}) = [\mathbf{d}_{\omega}(\mathbf{0}) \cdot \hat{\mathbf{n}}'] \hat{\mathbf{n}}' \tag{2.60}$$

$$\mathbf{d}_{\perp}(\mathbf{0}) = \mathbf{d}_{\omega}(\mathbf{0}) - \mathbf{d}_{\|} = \mathbf{d}_{\omega} - [\mathbf{d}_{\omega}(\mathbf{0}) \cdot \hat{\mathbf{n}}'] \hat{\mathbf{n}}' \tag{2.61}$$

then the electric and magnetic fields generated by the dipole can be written as

$$\begin{aligned} \mathbf{E}(t,\mathbf{x}) = {} & \frac{\mathbf{d}_{\perp}(\mathbf{0}) k^2}{4\pi\varepsilon_0 r} \left(\cos(\omega t' + \delta) - \frac{1}{kr} \sin(\omega t' + \delta) - \frac{1}{k^2 r^2} \cos(\omega t' + \delta) \right) \\ & + \frac{\mathbf{d}_{\|}(\mathbf{0}) k^2}{2\pi\varepsilon_0 r} \left(\frac{1}{kr} \sin(\omega t' + \delta) + \frac{1}{k^2 r^2} \cos(\omega t' + \delta) \right) \end{aligned} \tag{2.62}$$

$$\mathbf{B}(t,\mathbf{x}) = \frac{\hat{\mathbf{r}} \times \mathbf{d}_{\perp}(\mathbf{0}) k^2}{4\pi\varepsilon_0 c r} \left(\cos(\omega t' + \delta) - \frac{1}{kr} \sin(\omega t' + \delta) \right) \tag{2.63}$$

A single electric Hertzian dipole along the polar axis. Let us choose a Cartesian coordinate system (x_1, x_2, x_3) where the x_3 axis coincides with the polar axis of a spherical polar coordinate system (r, ϑ, φ) and let the two systems have the same origin $\mathbf{x}' = \mathbf{0}$. For a single dipole we can, without loss of generality, set the phase constant $\delta = 0$ in Eqs. (2.62) and (2.63). Straightforward calculation then yields the following exact expression for the instantaneous linear momentum density:

$$\mathbf{g}^{\mathrm{EM}}(t,\mathbf{x}) = \mathbf{g}^{\mathrm{EM}}(t,r,\vartheta,\varphi) = \frac{d^2k^4}{32\pi^2\varepsilon_0 cr^2}$$
$$\left[\left(1+\cos 2\omega t' - \frac{2}{kr}\sin 2\omega t' - \frac{2}{k^2r^2}\cos 2\omega t' + \frac{1}{k^3r^3}\sin 2\omega t'\right)\sin^2\vartheta\,\hat{\mathbf{r}}\right.$$
$$\left. + \frac{1}{kr}\left(\sin 2\omega t' + \frac{2}{kr}\cos 2\omega t' - \frac{1}{k^2r^2}\sin 2\omega t'\right)\sin 2\vartheta\,\hat{\boldsymbol{\vartheta}}\right] \tag{2.64}$$

The cycle average, i.e., the integration over a full temporal cycle, for which

$$\langle \sin 2\omega t'\rangle_t = \langle \cos 2\omega t'\rangle_t = 0 \tag{2.65}$$

of the exact formula (2.64), yields the cycle average of the linear momentum density

$$\langle \mathbf{g}^{\mathrm{EM}}\rangle_t = \frac{d^2k^4}{32\pi^2\varepsilon_0 c}\frac{\sin^2\vartheta}{r^2}\hat{\mathbf{r}} \tag{2.66}$$

The angular distribution of the linear momentum density, which is identical to the angular distribution of the Poynting vector, is commonly referred to as the 'radiation pattern,' and this well-known pattern for the Hertzian dipole is plotted in Figure 2.2. Clearly, the transverse (nonradial) part of the linear momentum density vector (and hence the Poynting vector) averages out over one oscillation cycle, which means that the cycle that averaged Poynting vector $\langle\mathbf{S}\rangle_t = c^2\langle\mathbf{g}^{\mathrm{EM}}\rangle_t$ is always purely radial all the way from the source to infinity. However, this is not the case for the Poynting vector $\mathbf{S} = c^2\mathbf{g}^{\mathrm{EM}}$ itself. As Eq. (2.64) shows, it has a transverse component that, relative to the radial component, falls off as $1/r$ with distance r and therefore is present very near the source (typically a few wavelengths) but vanishes far away from the source so that in the far zone the Poynting vector is predominantly radially directed.

To calculate the electromagnetic angular momentum density around $\mathbf{x}_{\mathrm{m}} = \mathbf{0}$ we use Eqs. (2.64) and (2.A.144) and the fact that $\mathbf{x} = r\hat{\mathbf{r}}$ and $\hat{\mathbf{r}} \times \hat{\boldsymbol{\vartheta}} = \hat{\varphi}$ to obtain the exact result (see also [42])

$$\mathbf{h}^{\mathrm{EM}}(t,r,\vartheta,\varphi;\mathbf{0}) = \frac{d^2k^3}{32\pi^2\varepsilon_0 cr^2}\left(\sin 2\omega t' + \frac{2}{kr}\cos 2\omega t' - \frac{1}{k^2r^2}\sin 2\omega t'\right)\sin 2\vartheta\,\hat{\varphi} \tag{2.67}$$

This shows that $\mathbf{h}^{\mathrm{EM}}$ has the expected r^{-2} asymptotic fall off at large distances r, has no z component, and oscillates at 2ω along $\hat{\varphi}$. This transverse angular momentum [180] would cause an electric charge, that is not located at $\vartheta = n\pi/2$, $n \in \mathbb{Z}$, to rotate around its own center of mass in the osculating plane normal to $\hat{\varphi}$. It is tempting to relate the double-frequency terms in Eqs. (2.64) and (2.67) to photon *Zitterbewegung* [181–184]. As follows from Eqs. (2.67) and (2.65), the temporal (cycle) average of the angular momentum density radiated by the single Hertzian dipole is

$$\langle \mathbf{h}^{\mathrm{EM}}\rangle_t = \mathbf{0} \tag{2.68}$$

which makes it nontrivial to measure this rotational motion.

Because of vectorial geometry, only the transverse part ($\perp\hat{\mathbf{r}}$) of the linear momentum density, where the dominating part is $\mathcal{O}(r^{-3})$, contributes to the angular momentum density in the far zone. Clearly, the radial component ($\parallel \hat{\mathbf{r}}$) of the linear momentum density (Poynting vector), which is $\mathcal{O}(r^{-2})$, does not contribute to the angular momentum density. If it did, then expression (2.50) for the total angular momentum at very large distances of a 'pulse' would grow without bounds as the 'pulse' propagates away from the source, thus violating fundamental laws of physics. See the detailed, clarifying discussion about this by Abraham [42]; an English translation of the essential part of this discussion is included in Tamburini et al. [26]. This peculiarity is also discussed by Low [185].

The fact that the angular distribution of the angular momentum density in the far zone is determined by the angular distribution of the linear momentum in the near zone and therefore is a near-zone interference phenomenon might at first seem paradoxical. However, it is the consequence of the fact that the far-zone angular momentum density, which is proportional to the vector product $\mathbf{E} \times \mathbf{B} \neq \mathbf{0}$, cannot be approximated by an expression containing the vector product of only the far-zone fields since $\mathbf{E}^{\text{far}} \times \mathbf{B}^{\text{far}} = \mathbf{0}$.

The exact expression (2.67) for the angular momentum density radiated by a single Hertzian dipole implies that, to leading order in r, this density is

$$\mathbf{h}^{\text{EM}}(t, r, \vartheta, \varphi; \mathbf{0}) = \frac{d^2k^3}{32\pi^2\varepsilon_0 cr^2}\left(\sin 2(kr - \omega t) + \mathcal{O}\left(r^{-1}\right)\right)\sin 2\vartheta\,\hat{\boldsymbol{\varphi}} \tag{2.69}$$

which shows, once again, that the asymptotic fall-off of the angular momentum density is r^{-2}. The angular distribution of $\mathbf{h}^{\text{EM}}(t, r, \vartheta, \varphi; \mathbf{0})$ is plotted in Figure 2.3.

Two colocated crossed electric Hertzian dipoles oscillating in quadrature in the azimuthal plane. The volumetric density of an observable carried by the fields generated by a plurality of sources, i.e. transmitting antennas, is a second-order product of the superposition of the individual fields. Therefore, it is not just the sum of the observable densities emitted from each source, but it also contains cross terms representing interference between the fields. This has far-reaching consequences when it comes to measuring multimode observables and using them for information transfer and other purposes.

If we specialize to two sources 1, 2, emitting individual fields $(\mathbf{E}_1, \mathbf{B}_1)$ and $(\mathbf{E}_2, \mathbf{B}_2)$, respectively, the total linear momentum density radiated is

$$\begin{aligned}\mathbf{g}^{\text{EM}} &= \varepsilon_0(\mathbf{E}_1 + \mathbf{E}_2) \times (\mathbf{B}_1 + \mathbf{B}_2)\\ &= \varepsilon_0\mathbf{E}_1 \times \mathbf{B}_1 + \varepsilon_0\mathbf{E}_2 \times \mathbf{B}_2 + \varepsilon_0\mathbf{E}_1 \times \mathbf{B}_2 + \varepsilon_0\mathbf{E}_2 \times \mathbf{B}_1\\ &= \mathbf{g}_1^{\text{EM}} + \mathbf{g}_2^{\text{EM}} + \mathbf{g}_{12}^{\text{EM}} + \mathbf{g}_{21}^{\text{EM}}\end{aligned} \tag{2.70}$$

Hence, the contributions to the total linear momentum density, and therefore to the linear current induced in a receiving antenna, from the two sources is not just the vector sum of the linear momentum densities $\mathbf{g}_1^{\text{EM}}$ and $\mathbf{g}_2^{\text{EM}}$ from the two individual sources. There is also a contribution from the two interference terms $\mathbf{g}_{12}^{\text{EM}}$ and $\mathbf{g}_{21}^{\text{EM}}$. This well-known fact is used in interferometry and beamforming.

If we calculate $\mathbf{g}^{\text{EM}}$, and from that $\mathbf{h}^{\text{EM}}$ by using Eq. (2.A.144), we find that the exact expression for the total angular momentum density carried by the superimposed fields $(\mathbf{E}_1, \mathbf{B}_1)$ and $(\mathbf{E}_2, \mathbf{B}_2)$, respectively, is

$$\begin{aligned}\mathbf{h}^{\mathrm{EM}}(\mathbf{0}) &= \varepsilon_0 \mathbf{x} \times [(\mathbf{E}_1 + \mathbf{E}_2) \times (\mathbf{B}_1 + \mathbf{B}_2)] \\ &= \mathbf{x} \times \mathbf{g}_1^{\mathrm{EM}} + \mathbf{x} \times \mathbf{g}_2^{\mathrm{EM}} + \mathbf{x} \times \mathbf{g}_{12}^{\mathrm{EM}} + \mathbf{x} \times \mathbf{g}_{21}^{\mathrm{EM}} \\ &= \mathbf{h}_1^{\mathrm{EM}} + \mathbf{h}_2^{\mathrm{EM}} + \mathbf{h}_{12}^{\mathrm{EM}} + \mathbf{h}_{21}^{\mathrm{EM}}\end{aligned} \tag{2.71}$$

i.e., a vector sum of the two independent angular momentum densities radiated from the individual sources, $\mathbf{h}_1^{\mathrm{EM}}$ and $\mathbf{h}_2^{\mathrm{EM}}$, plus the two interference terms $\mathbf{h}_{12}^{\mathrm{EM}}$ and $\mathbf{h}_{21}^{\mathrm{EM}}$. It is therefore not possible to identify and separate, for instance, the individual OAM eigenmodes in a superposition of two (or more) such modes unless the mixing (interference) is taken properly into account. As a comparison of Eqs. (2.70) and (2.71) shows, it is just as easy to engineer beams of angular momentum density as it is for linear momentum density, and then it is possible to do angular momentum interferometry.

We now study the electromagnetic linear and angular momenta emitted from a system of two colocated monochromatic Hertzian dipoles, one spatially rotated from the other by $\pi/2$, fed at the same frequency but with different temporal phase shifts. For this purpose we chose the two Hertzian dipoles $\mathbf{d}_j, j = 1, 2$ to be located at the origin and directed along the $x_1 = x$ and $x_2 = y$ axes, respectively, with their generated EM fields calculated from Eqs. (2.62) and (2.63). Doing so, by using Eq. (2.A.144), the exact expression for the total angular momentum density $\mathbf{h}^{\mathrm{EM}}$ is obtained.

When the two dipole moments have equal amplitudes, $d_1 = d_2 = d$, and are in phase quadrature relative to each other, $\delta_1 = 0$ and $\delta_2 = \pi/2$, the radiated linear momentum density is obtained. Cycle averaging the resulting formula, one finds that

$$\left\langle \mathbf{g}^{\mathrm{EM}} \right\rangle_t = \frac{d^2 k^4}{32\pi^2 \varepsilon_0 c r^2} \left[\left(1 - \cos^2\varphi \sin^2\vartheta - \frac{1}{k^3 r^3} \right) \hat{\mathbf{r}} \frac{1}{2} \left(\frac{1}{kr} + \frac{1}{kr^3 r^3} \right) \sin 2\varphi \sin 2\vartheta \, \hat{\boldsymbol{\vartheta}} \hat{\varphi} \right] \tag{2.72}$$

The cycle average of the total linear momentum radiated by two perpendicular crossed, quadrature-phased electric Hertzian dipoles during a time interval Δt_0 can be calculated by inserting expression (2.72) into formula (2.43) and (2.44) and performing the integration. The result is

$$\left\langle \mathbf{p}^{\mathrm{EM}} \right\rangle_t = \frac{d^2 k^4}{6\pi\varepsilon_0} \Delta t_0 \, \hat{\mathbf{r}} \tag{2.73}$$

The angular momentum density $\mathbf{h}^{\mathrm{EM}}(\mathbf{0})$ radiated by two perpendicular electric Hertzian dipoles located in the xy plane with their centers at the origin and fed in phase quadrature. The angular distribution is identical to that of a single electric Hertzian dipole along the z axis (see Fig. 1), the local direction of the angular and linear momentum density vector fields are totally different. Notice that the $\mathbf{h}^{\mathrm{EM}}$ vectors does not diverge with distance along the polar axis as the $\mathbf{g}^{\mathrm{EM}}$ vectors do.

For an identical configuration of crossed dipoles, the exact formula for the z component of the radiated angular momentum density $\mathbf{h}^{\mathrm{EM}}(\mathbf{0})$, valid in all of space outside the dipoles, is

$$\begin{aligned} h_z^{\mathrm{EM}}(t, r, \vartheta, \varphi; \mathbf{0}) &= \left(r\hat{\mathbf{r}} \times \mathbf{g}^{\mathrm{EM}}(t, r, \vartheta, \varphi) \right) \cdot \hat{\mathbf{z}} \\ &= \frac{d^2(\mathbf{0}) k^3}{16\pi^2 \varepsilon_0 c r^2} [(1 + \sin 2\varphi \sin 2\omega t' - \cos 2\varphi \cos 2\omega t') \\ &+ \frac{2}{kr} (\sin 2\varphi \cos 2\omega t' - \cos 2\varphi \sin 2\omega t') \\ &+ \left. \frac{1}{k^2 r^2} (1 - \sin 2\varphi \sin 2\omega t' + \cos 2\varphi \cos 2\omega t') \right] \sin^2\vartheta \end{aligned} \tag{2.74}$$

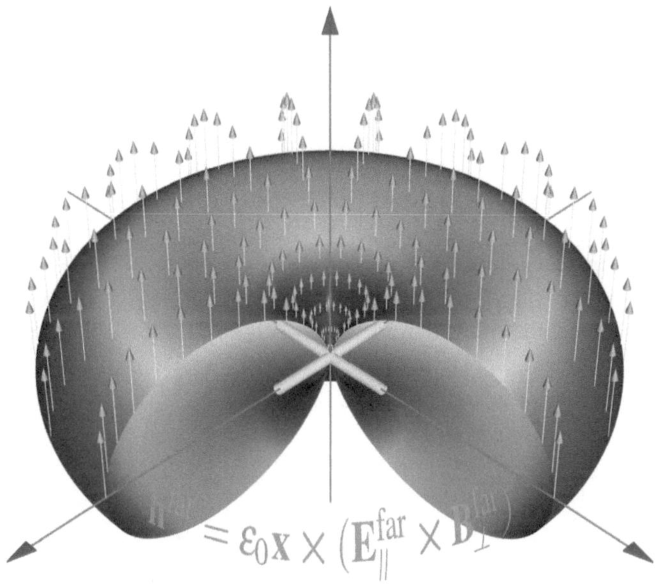

Figure 2.4 Angular momentum radiated from two crossed dipoles.

The angular distribution of $h_z^{\mathrm{EM}}(t, r, \vartheta, \varphi; \mathbf{0})$ is plotted in Figure 2.4. It is interesting to note that this angular distribution is identical to the angular distribution of the dominating part of the linear momentum density radiated by a single dipole, plotted in Figure 2.2, whereas the local directions of the two momentum density vectors are different.

To turn the z component of the angular momentum density, h_z^{EM}, radiated from two colocated Hertzian dipoles, arranged perpendicular to each other, and oscillating in quadrature into an observable z component of the total field angular momentum, $J_z^{\mathrm{EM}}(\mathbf{0})$, we must integrate the density (2.74) over an observation volume V. Choosing V in Eq. (2.50), i.e. as the region between two spherical shells with radii r_0 and $r_0 + \Delta r_0$ representing the radial extent of the 'pulse' emitted, we obtain

$$J_z^{\mathrm{EM}}(t, \mathbf{0}) = \int_{r_0}^{r_0 + \Delta r_0} \mathrm{d}r\, r^2 \int_{\vartheta_1}^{\vartheta_2} \mathrm{d}\vartheta \sin \vartheta \int_{\varphi_1}^{\varphi_2} \mathrm{d}\varphi\, h_z^{\mathrm{EM}}(\mathbf{0}) \tag{2.75}$$

where $h_z^{\mathrm{EM}}(\mathbf{0})$ is given by Eq. (2.74).

If we integrate over the azimuthal angle φ from $\varphi_1 = 0$ to $\varphi_1 = n\pi$, $n = 1,2,3, \ldots$, all temporal terms vanish and we obtain

$$J_z^{\mathrm{EM}}(\mathbf{0}) = \frac{d^2(\mathbf{0})k^3}{8\pi\varepsilon_0 c} \int_{r_0}^{r_0 + \Delta r_0} \mathrm{d}r \left(1 + \frac{1}{k^2 r^2}\right) \int_{\vartheta_1}^{\vartheta_2} \mathrm{d}\vartheta \sin^3 \vartheta \tag{2.76}$$

We point out that this time-independent result is obtained even if no cycle average is taken.

If we now integrate over all polar angles ϑ from $\vartheta_1 = 0$ to $\vartheta_2 = \pi$, we obtain the explicit, closed-form expression

$$J_z^{\mathrm{EM}}(\mathbf{0}) = \frac{d^2(\mathbf{0})k^3 \Delta r_0}{6\pi\varepsilon_0 c} \left(1 + \frac{2}{k^2 r_0^2 (1 + \Delta r_0 / r_0)}\right) \tag{2.77}$$

where the only time dependence comes from the duration of the emitted 'pulse' $\Delta t_0 = \Delta r_0/c$. When the distance r_0 that the 'pulse' has propagated is very large so that the $1/r_0^2$ contribution can be neglected, the z component of the field angular momentum can be approximated by

$$J_z^{\text{field far}}(\mathbf{0}) = \frac{d^2(\mathbf{0})k^3}{6\pi\varepsilon_0}\Delta t_0 \tag{2.78}$$

which, as expected, propagates without radial fall-off.

If the observation volume V, i.e., the angular momentum sensing transducer, is a very thin circular ring with radius r_0, located in the azimuthal (xy) plane ($\vartheta = \pi/2$), centered on the dipoles, and subtending the small symmetric polar angular interval $\vartheta_2 - \vartheta_2$ around $\vartheta = \pi/2$, the integrals in Eq. (2.76) are trivial, and the result is

$$J_z^{\text{EM}}(\mathbf{0}) = \frac{d^2(\mathbf{0})k^3}{8\pi\varepsilon_0 c}\left(1 + \frac{1}{k^2r_0^2}\right)(\vartheta_2 - \vartheta_1) \tag{2.79}$$

Expression (2.73) shows that the linear momentum is radially directed over the entire ring, thus showing that there is no net force but only net torque acting on the (charges in) the ring. This is therefore an example where only the angular momentum and not the linear momentum (Poynting vector) contributes to the motion of the charges, i.e. the current in the sensor.

As a radio analog of the celebrated Einstein-de Haas experiment [186], a microwave angular momentum torsion pendulum experiment of a type first suggested by Vul'fson [74] was performed [83]. Interpreting the results in terms of classical electron theory [14, 95] as discussed above, we found that they were in good agreement with predictions made from our analytic formula (2.79). However, comparing this formula with Eq. (2.7) in Emile et al. [83], shows that the latter has a numerical error: there is a factor of 2π missing. This explains the discrepancy of a factor of about 6 mentioned in Emile et al. [83].

2.3.3 Astronomy Applications

Electromagnetic/photonic OAM as well as optical vortices (OVs) have recently come to the fore in astrophysics, astronomy, and other space sciences [6, 187]. As an example, OVs obtained by artificially imposing OAM to the light coming from celestial sources can enable the detection and imaging of extrasolar planets [188–192] and improve the resolving power of any optical instrument by up to one order of magnitude [193].

Other recent applications in this field have been proposed for the detection of natural OAM states emitted from astrophysical sources both in the optical and in the radio domain [194]. In fact there is more than trivial evidence that OAM could be naturally imposed to the light of distant sources by different astrophysical phenomena when it traverses peculiar regions in space containing, e.g. inhomogeneities due to the presence of turbulent plasma or particular regions of spacetime such as those characterized by the distorted geometry of rotating Kerr black holes (KBHs) [195].

Light propagating in a structured astrophysical plasma is thought to generate OAM-carrying beams [90] as photons acquire an effective Proca mass through the Anderson–Higgs mechanism [98, 196]. Photons there obey a relationship between the acquired virtual mass term and the total angular momentum **J** similarly to what occurs between the SAM **Σ** and mass of a Majorana particle that can be found in the original mathematical formulation of the so-called 'Majorana tower' [197]. In this case, the OAM acquired from the inhomogeneities in the plasma resonance frequency acts as a term that reduces the total Proca photon mass [89, 90]. What actually gives a Proca mass to photons in a structured plasma is the breaking of spatial symmetry and of spatial homogeneity. In addition, there exists a characteristic scale length introduced by the plasma structure at the frequencies where the plasma is resonant with strong analogies with the models of space–time characterized by a modified action of the Lorentz group. An example is the Magueijo–Smolin model [198], and this mechanism demonstrates clear and deep analogies with the dynamics described by the Dirac equation when it introduced a characteristic scale length. Unavoidably, Berry phase effects are introduced too [199].

Gravitationally generated OAM can be detected if generated by supermassive black hole (BH) as those located at the centre of most galaxies. Their mass range spans from 10^6 to 10^{10} solar masses and beyond. For a review, see Ref. [200].

In 2011, a theoretical and numerical study by Tamburini et al. [23], showed that EM waves emitted from an accretion disk (AD) around a massive rotating BH should be endowed with specific fingerprints of OAM that depend on the model of the AD together with general relativistic (GR) effects due to the Kerr space–time dragging and gravitational lensing. The gravitationally lensed beams experience an anamorphic transformation [201], which is also accompanied by a polarization rotation [202] due to the gravitational Faraday effect [203]. The gravitational lensing also generate image deformation and rotation due to the curvature of space–time as well as other effects that affect the phase wavefront, such as wave-front warping and the gravitational Berry phase effects [204–206].

Kerr and the more general Kerr–Newman metrics not only impose gravitational Faraday rotation to a linearly polarized light but also can induce spin-to-orbital angular momentum conversion of the scattering light, as observed in inhomogeneous and anisotropic media. The detection of OAM due to gravitational lensing reveals the spacetime metrics, and the angular momentum a of the rotating gravitational lens.

After the observations of Event Horizon Telescope (EHT) of the M87$*$ BH [207, 208, 211–214] presented the first observational evidence that confirms predictions of general relativity caused OAM [210]: light propagating near a rotating BH becomes twisted and carries OAM. One can estimate the rotation parameter a from the characterization of the vorticity, namely, from the spiral spectrum [58, 84] of the electromagnetic waves emitted from the accreting matter in the surroundings of the BH, where the strong gravitational lensing from the rotating compact object changed their wavefront as hypothesized in [23]. Moreover, the characterization of the BH rotation with twisted light can be also used to improve the black-hole shadow analysis and test general relativity (see Figure 2.5).

Exploiting the properties of the Kerr metric, it was possible to measure the rotation of the famous M87$*$ BH by analysing a sequence of images acquired by the EHT [210] through the OAM imparted by the gravitational lensing [6, 187, 192]. Instead of measuring the OAM from the intensity and spatial phase distribution as in [24], the OAM content and the electromagnetic vorticity were mathematically reconstructed with a well-known technique based on the TIE that permits the reconstruction of the phase wavefront and of the OAM content of radio beams from two or more consecutive acquisitions of the same source taken at different distances. The data was extracted from the images of the brightness temperature of the Einstein ring of the M87$*$ BH, taken during four different days [215] (see Table 2.2).

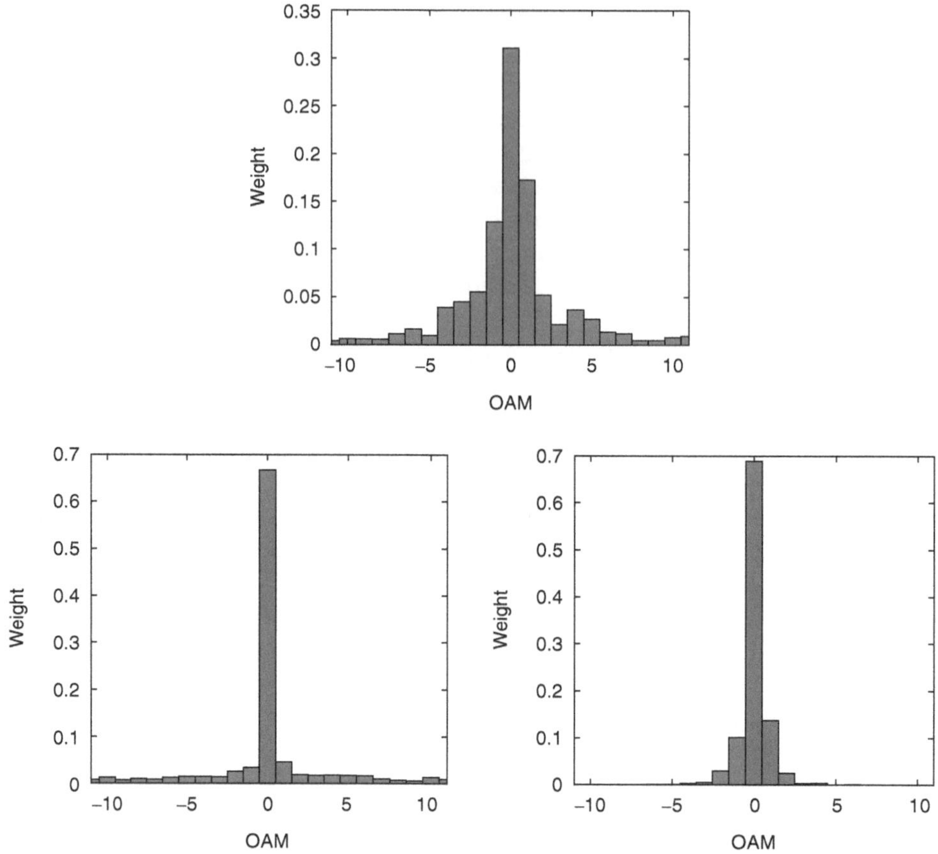

Figure 2.5 **Top:** OAM spectrum from the results of Event Horizon Telescope (EHT) visibility amplitude and phase distributions as a function of the vector baseline, taken on April 11, 2017 [207–209]. The data considered here represent the visibility amplitude and phase, provided by EHT, as a function of the vector baseline. As the spatial phase distribution in this case is present in the data, one does not need to apply the transport of intensity equation (TIE) method to recover it. To show the BH rotation, the asymmetry parameter, and the ratio q between the $m = 1$ and $m = -1$ were calculated for all data sets analyzed. Clearly, all qs in the spiral spectra are $q > 1$, indicating clockwise rotation by assuming an inclination of the BH axis of $i = 17°$ (if the angular momentum of the accretion flow and that of the BH are antialigned). This is equivalent to a scenario where instead $i = 163°$, and the angular momentum of the accretion flow and that of the BH result aligned. **Bottom (Left):** The spiral spectrum obtained with the TIE method. **Bottom (Right):** The spiral spectrum obtained from the numerical simulations agrees with experimental data and the TIE method. *Source:* Tamburini et al. [210].

Table 2.2 OAM radio from near the M87∗ BH.

Asymmetry parameter q *vs.* the rotation parameter a obtained from numerical simulations. The asymmetry parameter is obtained by dividing the height of the $l = 1$ with that of $l = -1$ bars in the spiral spectra of hypothetical Kerr black holes neighborhoods having different rotation parameters $0.5 < a < 0.9$. The values of two sample inclinations describe the rotation of the M87∗ $i = 17°$ and that with another inclination, $i = 46°$. The rotation is clockwise and averaged over different accretion disk-emission mechanisms characterized by $0.1 < \Gamma < 2$ power spectra.

Black-hole rotation parameter, a	0.50	0.60	0.80	0.90
Asymmetry parameter, q				
$i = 17°$	1.295	1.320	1.356	1.391
$i = 46°$	1.392	1.412	1.424	1.438

Appendix A

2.A.1 Theory

In classical electrodynamics, the standard methods of calculating electromagnetic fields once the sources are prescribed often amount to solving, one way or another, the Maxwell–Lorentz differential Eqs. (2.1)–(2.4) directly, or to first calculate the explicit potentials from the particular sources at hand, and only then calculate the pertinent fields. See, e.g. Jackson [52] and Balanis [102].

However, over the past decades, generic integral formulas that allow the immediate calculation of the fields directly from arbitrary given charge and current distributions, without the need for an intermediate step of first calculating the explicit scalar and vector products, have gained increasing attraction; see de Melo e Souza et al. [216] and references cited therein. In Chapter VIII of his textbook, Stratton [217] calculates, without the intervention of potentials, temporal Fourier transform expressions for the retarded electric and magnetic fields, $\mathbf{E}(t, \mathbf{x})$ and $\mathbf{B}(t, \mathbf{x})$, respectively, generated by arbitrary distributions of charge and current densities at rest relative to the observer. These formulas seem to have been first derived by Schott [218]. In chapter 14 of the second edition of their textbook on electrodynamics, Panofsky and Phillips [219] present a variant form of these Fourier expressions, and give them also in ordinary configuration space–time coordinates. Four years later, in chapter 15 of his electrodynamics textbook, Jefimenko [220] gave the Panofsky and Phillips [219] expressions for the retarded $\mathbf{E}$ and $\mathbf{B}$ fields. These expressions are sometimes referred to as the Jefimenko formulas [52, 221], but we propose that, on historical grounds, they should rather be called the Schott–Stratton–Panofsky–Phillips–Jefimenko or the SSPPJ formulas, a name we will use throughout.

As we will show below, it is possible to symmetrize the Jefimenko form of the SSPPJ formulas [220] so that they exhibit the geometric relationship between the electric and the magnetic fields in a transparent way. Allowing for magnetic charges and currents, they can be generalized to also exhibit the duality of the electromagnetic field. Furthermore, we introduce a field vector density that makes the new integral formulas attain a simple, intuitive form.

2.A.1.1 Classical Majorana-Oppenheimer Formalism and Its Affinity to First Quantization Formalism

For any prescribed set of charge and current densities, every electromagnetic observable can be calculated with, in principle, arbitrary accuracy from the first-order fields obtained as solutions of the Maxwell–Lorentz Eqs. (2.1)–(2.4). As was pointed out by Silberstein [222, 223], it is very convenient to calculate these second-order quantities by using the Majorana–Oppenheimer formalism. This formalism is also convenient for canonical (second) quantization of the electromagnetic field [2, 224, 225].

It was found by Riemann [226] that it is often convenient to combine the real-valued vector fields $\mathbf{E}$ and $\mathbf{B}$ into the complex-valued Riemann–Silberstein bivector field [222, 223, 227, 228]

$$\mathbf{G}(t, \mathbf{x}) \equiv \mathbf{E}(t, \mathbf{x}) + \mathrm{i}c\mathbf{B}(t, \mathbf{x}) \ , \ \ \mathrm{i}^2 = -1 \tag{2.A.1}$$

some properties of which are listed in Table 2.A.1. Expressing the charge and current densities in complex forms

$$\rho(t, \mathbf{x}) = \rho^{\mathrm{e}}(t, \mathbf{x}) + \frac{\mathrm{i}}{c}\rho^{\mathrm{m}}(t, \mathbf{x}) \tag{2.A.2}$$

Table 2.A.1 Various vectorial products of the Riemann–Silberstein bivector $\mathbf{G}(t, \mathbf{x}) \equiv \mathbf{E}(t, \mathbf{x}) + ic\mathbf{B}(t, \mathbf{x})$ with itself and their meaning in terms of quadratic and bilinear products of the fields **E** and **B**. An asterisk (*) denotes complex conjugate and ⊗ denotes the tensor (Kronecker, dyadic) product operator, defined in formula (2.A.117).

Complex notation	Real notation
$\mathbf{G} \cdot \mathbf{G}$	$\mathbf{E} \cdot \mathbf{E} - c^2\mathbf{B} \cdot \mathbf{B} + 2ic\mathbf{E} \cdot \mathbf{B}$
$\mathbf{G}{\cdot}\mathbf{G}^* = \sqrt{(\mathbf{G}{\cdot}\mathbf{G})(\mathbf{G}{\cdot}\mathbf{G})^*}$	$\mathbf{E} \cdot \mathbf{E} + c^2\mathbf{B} \cdot \mathbf{B} \equiv E^2 + c^2B^2$
$\mathbf{G} \times \mathbf{G}$	$\mathbf{0}$
$\mathbf{G}^* \times \mathbf{G} = -\mathbf{G} \times \mathbf{G}^*$	$2ic(\mathbf{E} \times \mathbf{B})$
$\mathbf{G} \otimes \mathbf{G}$	$\mathbf{E} \otimes \mathbf{E} - c^2\mathbf{B} \otimes \mathbf{B} + ic(\mathbf{E} \otimes \mathbf{B} + \mathbf{B} \otimes \mathbf{E})$
$\mathbf{G}^* \otimes \mathbf{G}^* = (\mathbf{G} \otimes \mathbf{G})^*$	$\mathbf{E} \otimes \mathbf{E} - c^2\mathbf{B} \otimes \mathbf{B} - ic(\mathbf{E} \otimes \mathbf{B} + \mathbf{B} \otimes \mathbf{E})$
$\mathbf{G} \otimes \mathbf{G}^*$	$\mathbf{E} \otimes \mathbf{E} + c^2\mathbf{B} \otimes \mathbf{B} - ic(\mathbf{E} \otimes \mathbf{B} - \mathbf{B} \otimes \mathbf{E})$
$\mathbf{G}^* \otimes \mathbf{G} = (\mathbf{G} \otimes \mathbf{G}^*)^*$	$\mathbf{E} \otimes \mathbf{E} + c^2\mathbf{B} \otimes \mathbf{B} + ic(\mathbf{E} \otimes \mathbf{B} - \mathbf{B} \otimes \mathbf{E})$

$$\mathbf{j}(t, \mathbf{x}) = \mathbf{j}^{\mathrm{e}}(t, \mathbf{x}) + \frac{\mathrm{i}}{c}\mathbf{j}^{\mathrm{m}}(t, \mathbf{x}) \tag{2.A.3}$$

we see that Dirac's symmetrized Maxwell–Lorentz Eqs. (2.1)–(2.4) become

$$\nabla{\cdot}\mathbf{G} = \frac{\rho}{\varepsilon_0} \tag{2.A.4}$$

$$\nabla \times \mathbf{G} - \frac{\mathrm{i}}{c}\frac{\partial \mathbf{G}}{\partial t} = \frac{\mathrm{i}}{\varepsilon_0 c}\mathbf{j} \tag{2.A.5}$$

known as the Majorana–Oppenheimer formalism.

Following Białynicki-Birula and Białynicka-Birula [225, 229], who instead of **G** use a differently normalized complex vector, *viz.* $\mathbf{F} = \mathbf{G}\sqrt{\varepsilon_0/2}$, first, we introduce the linear momentum operator for the electromagnetic field,

$$\hat{\mathrm{p}}^{\mathrm{EM}} = \sum_{i=1}^{3} \hat{\mathbf{x}}_i \hat{\mathrm{p}}_i^{\mathrm{EM}} = \sum_{i=1}^{3} \hat{\mathbf{x}}_i\left(-\mathrm{i}\hbar\frac{\partial}{\partial x_i}\right) = -\mathrm{i}\hbar\nabla \tag{2.A.6}$$

where $\hbar$ is the normalized Planck constant. Second, we introduce the vector-like symbolic construct

$$\mathrm{S} \equiv \sum_{i=1}^{3} \mathrm{S}_i\hat{\mathbf{x}}_i \tag{2.A.7}$$

where the

$$\mathrm{S}_1 = \begin{pmatrix} 0 & 0 & 0 \\ 0 & 0 & -\mathrm{i} \\ 0 & \mathrm{i} & 0 \end{pmatrix} \quad \mathrm{S}_2 = \begin{pmatrix} 0 & 0 & \mathrm{i} \\ 0 & 0 & 0 \\ -\mathrm{i} & 0 & 0 \end{pmatrix} \quad \mathrm{S}_3 = \begin{pmatrix} 0 & -\mathrm{i} & 0 \\ \mathrm{i} & 0 & 0 \\ 0 & 0 & 0 \end{pmatrix} \tag{2.A.8}$$

i.e. spin-1 matrices fulfilling the angular momentum commutation rule

$$[\mathrm{S}_i, \mathrm{S}_j] = -\mathrm{i}\varepsilon_{ijk}\mathrm{S}_k, \tag{2.A.9}$$

where

$$\varepsilon_{ijk} = \begin{cases} 1 & \text{if } i,j,k \text{ is an even permutation of } 1,2,3 \\ 0 & \text{if at least two of } i,j,k \text{ are equal} \\ -1 & \text{if } i,j,k \text{ is an odd permutation of } 1,2,3 \end{cases} \tag{2.A.10}$$

is the totally antisymmetric rank-3 pseudotensor (the Levi–Civita symbol). Third, we introduce the field helicity operator

$$\hat{\Lambda} = \frac{\mathbf{S}\cdot\hat{\mathrm{p}}^{\mathrm{EM}}}{|\,\hat{\mathbf{p}}^{\mathrm{EM}}\,|} \equiv \frac{\sum_{i=1}^{3} \mathrm{S}_i \hat{\mathrm{p}}_i^{\mathrm{EM}}}{\sqrt{\sum_{i=1}^{3} \left(\hat{\mathrm{p}}_i^{\mathrm{EM}}\right)^2}} \tag{2.A.11}$$

Then, using Eqs. (2.A.6) and (2.A.7)–(2.A.11), we can write the Maxwell–Lorentz–Dirac Eqs. (2.1)–(2.4) in free space and other domains where $\rho^{\mathrm{e}} = \rho^{\mathrm{m}} = 0$ and $\mathbf{j}^{\mathrm{e}} = \mathbf{j}^{\mathrm{m}} = \mathbf{0}$ in the form of

$$\hat{\mathrm{p}}^{\mathrm{EM}}\cdot\mathbf{G} = 0 \tag{2.A.12}$$

$$\mathrm{i}\hbar\frac{\partial \mathbf{G}}{\partial t} = \hat{\mathrm{H}}^{\mathrm{EM}}\mathbf{G} \tag{2.A.13}$$

where

$$\hat{\mathrm{H}}^{\mathrm{EM}} = c\,|\,\hat{\mathbf{p}}^{\mathrm{EM}}\,|\,\hat{\Lambda} = c\mathbf{S}\cdot\hat{\mathrm{p}}^{\mathrm{EM}} \equiv c\sum_{i=1}^{3}\mathbf{S}_i\hat{\mathrm{p}}_i^{\mathrm{EM}} \tag{2.A.14}$$

Recalling the de Broglie quantal relation [20] $\mathbf{p}^{\mathrm{EM}} = \hbar\mathbf{k}$, where $\mathbf{k} = k\hat{\mathbf{k}}$ is the wave vector, $k = 2\pi/\lambda$, and λ is the wavelength, we see that the first equation, Eq. (2.A.12), is the transversality condition, $\mathbf{k} \perp \mathbf{G}$, for electromagnetic waves propagating in free space, whereas the second equation, Eq. (2.A.13), describes the dynamics of the field and takes the form of a Schrödinger–Pauli–Dirac type equation where $\hat{\mathrm{H}}^{\mathrm{EM}}$ behaves as a Hamilton operator for the free electromagnetic field. The electromagnetic theory that corresponds to classical mechanics is the geometrical optics (a.k.a. eikonal or 'ray optics') approximation of the Maxwell–Lorentz–Dirac equations [230].

In certain aspects, the solutions of Eq. (2.A.13) can, at least formally, be viewed as first quantization photon wave functions that obey the superposition principle [1, 2, 5, 90, 183, 230–258]. It should be noted, however, that Cohen-Tannoudji et al. [51, p. 204] argue against '...suggesting (in spite of warnings) the false idea that the Maxwell waves are the wave functions of the photon.' See also the caveats put forward by Tamburini and Vicino [250] and by Andrews [259] about the nonlocalization problem of the photon wavefunction.

2.A.1.1.1 Riemann–Silberstein Electromagnetic Potentials and Fields

Given the electric and magnetic charge and current distributions, it is common practice to calculate the electromagnetic fields by solving the Maxwell–Lorentz Eqs. (2.1)–(2.4) from first principles. However, as an alternative, exact, closed-form expressions for the electromagnetic fields can be obtained directly from their scalar and vector potentials for electric charges, $(\phi^{\mathrm{e}}, \mathbf{A}^{\mathrm{e}})$, and for magnetic charges, $(\phi^{\mathrm{m}}, \mathbf{A}^{\mathrm{m}})$, as follows [52, 217, 219–221]:

$$\mathbf{E}(t,\mathbf{x}) = -\nabla\phi^{\mathrm{e}}(t,\mathbf{x}) - \frac{\partial \mathbf{A}^{\mathrm{e}}(t,\mathbf{x})}{\partial t} - \nabla\times\mathbf{A}^{\mathrm{m}}(t,\mathbf{x}) \tag{2.A.15}$$

$$\mathbf{B}(t,\mathbf{x}) = -\frac{1}{c^2}\nabla\phi^{\mathrm{m}}(t,\mathbf{x}) - \frac{1}{c^2}\frac{\partial \mathbf{A}^{\mathrm{m}}(t,\mathbf{x})}{\partial t} + \nabla\times\mathbf{A}^{\mathrm{e}}(t,\mathbf{x}) \tag{2.A.16}$$

If we introduce the complex valued scalar and vector potentials

$$\phi(t,\mathbf{x}) = \phi^{\mathrm{e}}(t,\mathbf{x}) + \frac{\mathrm{i}}{c}\phi^{\mathrm{m}}(t,\mathbf{x}) \tag{2.A.17}$$

$$\mathbf{A}(t,\mathbf{x}) = \mathbf{A}^{\mathrm{e}}(t,\mathbf{x}) + \frac{\mathrm{i}}{c}\mathbf{A}^{\mathrm{m}}(t,\mathbf{x}) \tag{2.A.18}$$

then the Riemann–Silberstein bivector (2.A.1) can be written as

$$\mathbf{G}(t,\mathbf{x}) = -\nabla\phi(t,\mathbf{x}) - \frac{\partial\mathbf{A}(t,\mathbf{x})}{\partial t} + \mathrm{i}c\nabla\times\mathbf{A}(t,\mathbf{x}) \tag{2.A.19}$$

that is more compact than Eqs. (2.A.15) and (2.A.16). The pertaining elemental/differential form is

$$\mathrm{d}\mathbf{G}(t,\mathbf{x},\mathbf{x}') = -\nabla\mathrm{d}\phi - \frac{\partial\mathrm{d}\mathbf{A}}{\partial t} + \mathrm{i}c\nabla\times\mathrm{d}\mathbf{A} \tag{2.A.20}$$

By combining Eqs. (2.A.19), (2.A.4) and (2.A.5), we obtain the following inhomogeneous wave equations for the complex potentials in Eqs. (2.A.17) and (2.A.18):

$$\frac{1}{c^2}\frac{\partial^2\Phi}{\partial t^2} - \nabla^2\Phi = \frac{\rho(t,\mathbf{x})}{\varepsilon_0} + \frac{\partial}{\partial t}\left(\nabla\cdot\mathbf{A} + \frac{1}{c^2}\frac{\partial\Phi}{\partial t}\right) \tag{2.A.21}$$

$$\frac{1}{c^2}\frac{\partial^2\mathbf{A}}{\partial t^2} - \nabla^2\mathbf{A} = \frac{1}{\varepsilon_0 c^2}\mathbf{j}(t,\mathbf{x}) - \nabla\left(\nabla\cdot\mathbf{A} + \frac{1}{c^2}\frac{\partial\Phi}{\partial t}\right) \tag{2.A.22}$$

If we utilize the gauge freedom by choosing $\nabla\cdot\mathbf{A}$ such that

$$\nabla\cdot\mathbf{A} + \frac{1}{c^2}\frac{\partial\Phi}{\partial t} = Const \in \mathbb{C} \tag{2.A.23}$$

then the last terms in the right-hand members of Eqs. (2.A.21) and (2.A.22) vanish. Equation (2.A.23) is a complex generalization of the usual real Lorenz–Lorentz gauge [7, 260, 261], not to be mistaken for the Lorentz–Lorenz polarizability relation [14].

Particular solutions for the inhomogeneous wave Eqs. (2.A.21) and (2.A.22) for the complex potentials can be obtained by the same standard methods as for the real potentials and are

$$\phi(t,\mathbf{x}) = \int_{V'}\mathrm{d}^3x'\frac{\rho(t',\mathbf{x}')}{4\pi\varepsilon_0\,|\,\mathbf{x}-\mathbf{x}'\,|} \tag{2.A.24}$$

$$\mathbf{A}(t,\mathbf{x}) = \int_{V'}\mathrm{d}^3x'\frac{\mathbf{j}(t',\mathbf{x}')}{4\pi\varepsilon_0 c^2\,|\,\mathbf{x}-\mathbf{x}'\,|} \tag{2.A.25}$$

The EM field $\mathbf{G}(t,\mathbf{x}) = \mathbf{E}(t,\mathbf{x}) + \mathrm{i}c\mathbf{B}(t,\mathbf{x})$ at time t in a volume element $\mathrm{d}V = \mathrm{d}^3x$ around a field point $\mathbf{x}$ in the sensor volume V is the sum (integral, weighted average) of the elemental E fields $\mathrm{d}\mathbf{G}(t,\mathbf{x},\mathbf{x}') = \mathrm{d}\mathbf{E}(t,\mathbf{x},\mathbf{x}') + \mathrm{i}c\mathrm{d}\mathbf{B}(t,\mathbf{x},\mathbf{x}')$ emitted at (retarded) times $t' = t - |\,\mathbf{x}-\mathbf{x}'\,|/c$ from all source volume elements $\mathrm{d}V' = \mathrm{d}^3x'$ around the source point $\mathbf{x}'$ in V'.
or, in elemental/differential form,

$$\mathrm{d}\phi(t,\mathbf{x}) = \frac{\rho(t',\mathbf{x}')}{4\pi\varepsilon_0\,|\,\mathbf{x}-\mathbf{x}'\,|}\mathrm{d}V' \tag{2.A.26}$$

$$\mathrm{d}\mathbf{A}(t,\mathbf{x}) = \frac{\mathbf{j}(t',\mathbf{x}')}{4\pi\varepsilon_0 c^2\,|\,\mathbf{x}-\mathbf{x}'\,|}\mathrm{d}V' \tag{2.A.27}$$

where the complex ρ and $\mathbf{j}$ are defined in Eqs. (2.A.2) and (2.A.3) and

$$t' = t'_{\mathrm{ret}}(t,\mathbf{x},\mathbf{x}') = t - \frac{|\,\mathbf{x}(t) - \mathbf{x}'(t')\,|}{c} \tag{2.A.28}$$

is the retarded time at which the fields were generated in the volume element dV' around $\mathbf{x}'$ and propagated causally to $\mathbf{x}$ along the local unit vector

$$\hat{\mathbf{k}}'(\mathbf{x}, \mathbf{x}') = \frac{\mathbf{x} - \mathbf{x}'}{|\mathbf{x} - \mathbf{x}'|} \tag{2.A.29}$$

arriving there at t as depicted in Figure 2.A.1; see also Figure 2.1. Identifying the real and complex parts of Eqs. (2.A.24) and (2.A.25), we obtain the corresponding solutions for the real-valued electric-charge potentials and the magnetic-charge potentials.

In order to achieve our goal to express the elemental EM field d$\mathbf{G}(t, \mathbf{x}, \mathbf{x}')$ in a source point of view, i.e. in terms of prescribed but otherwise generic charge and current sources, we insert Eqs. (2.A.26) and (2.A.27) into Eq. (2.A.20) and then integrate that expression over the source volume V' to obtain $\mathbf{G}(t, \mathbf{x})$.

After some algebra and the introduction of the noncovariant complex field vector density

$$\boldsymbol{\Theta}(t, \mathbf{x}, \mathbf{x}') = \boldsymbol{\Theta}^{\mathrm{e}}(t, \mathbf{x}, \mathbf{x}') + \frac{\mathrm{i}}{c}\boldsymbol{\Theta}^{\mathrm{m}}(t, \mathbf{x}, \mathbf{x}') \tag{2.A.30}$$

where the electric-charge $\boldsymbol{\Theta}^{\mathrm{e}}$ and its magnetic-charge dual $\boldsymbol{\Theta}^{\mathrm{m}}$ can be written as vector sums [7, 262, 263]

$$\boldsymbol{\Theta}^{\mathrm{e,m}}(t, \mathbf{x}, \mathbf{x}') = \sum_{j=1}^{4} \mathbf{b}_j^{\mathrm{e,m}}(t, \mathbf{x}, \mathbf{x}') \tag{2.A.31}$$

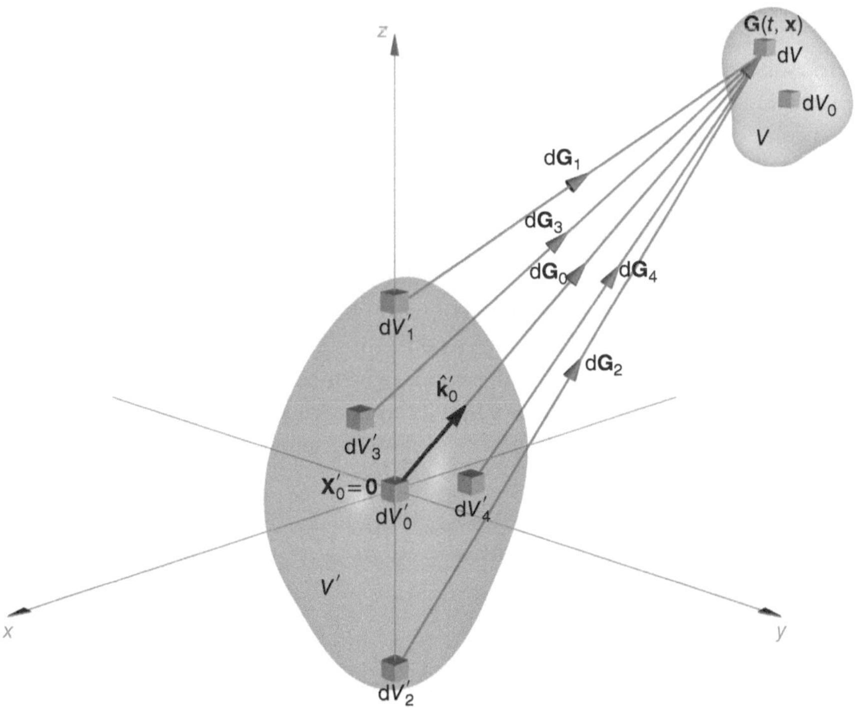

Figure 2.A.1 Source and sensor volumes.

with

$$\mathbf{b}_1^{\mathrm{e,m}}(t,\mathbf{x},\mathbf{x}') = \frac{\rho^{\mathrm{e,m}}\left(t - \frac{|\mathbf{x}-\mathbf{x}'|}{c}, \mathbf{x}'\right)}{4\pi\varepsilon_0|\mathbf{x}-\mathbf{x}'|^2}\hat{\mathbf{k}}'(\mathbf{x},\mathbf{x}') \tag{2.A.32}$$

$$\mathbf{b}_2^{\mathrm{e,m}}(t,\mathbf{x},\mathbf{x}') = \frac{\mathbf{j}^{\mathrm{e,m}}\left(t - \frac{|\mathbf{x}-\mathbf{x}'|}{c}, \mathbf{x}'\right)\cdot\hat{\mathbf{k}}'(\mathbf{x},\mathbf{x}')}{4\pi\varepsilon_0 c|\mathbf{x}-\mathbf{x}'|^2}\hat{\mathbf{k}}'(\mathbf{x},\mathbf{x}') \tag{2.A.33}$$

$$\mathbf{b}_3^{\mathrm{e,m}}(t,\mathbf{x},\mathbf{x}') = \frac{\mathbf{j}^{\mathrm{e,m}}\left(t - \frac{|\mathbf{x}-\mathbf{x}'|}{c}, \mathbf{x}'\right)\times\hat{\mathbf{k}}'(\mathbf{x},\mathbf{x}')}{4\pi\varepsilon_0 c|\mathbf{x}-\mathbf{x}'|^2}\times\hat{\mathbf{k}}'(\mathbf{x},\mathbf{x}') \tag{2.A.34}$$

$$\mathbf{b}_4^{\mathrm{e,m}}(t,\mathbf{x},\mathbf{x}') = \frac{\frac{\partial \mathbf{j}^{\mathrm{e,m}}\left(t - \frac{|\mathbf{x}-\mathbf{x}'|}{c}, \mathbf{x}'\right)}{\partial t}\times\hat{\mathbf{k}}'(\mathbf{x},\mathbf{x}')}{4\pi\varepsilon_0 c^2 \mid \mathbf{x}-\mathbf{x}' \mid}\times\hat{\mathbf{k}}'(\mathbf{x},\mathbf{x}') \tag{2.A.35}$$

we can express the elemental Riemann–Silberstein vector (2.A.20) in a source point of view, as follows:

$$\mathrm{d}\mathbf{G}(t,\mathbf{x},\mathbf{x}') = \left(\mathbf{\Theta}(t,\mathbf{x},\mathbf{x}') + \frac{\mathrm{i}}{c}\hat{\mathbf{k}}'(\mathbf{x},\mathbf{x}')\times\mathbf{\Theta}^*(t,\mathbf{x},\mathbf{x}')\right)\mathrm{d}V' \tag{2.A.36}$$

An appealing feature of this representation is that the expressions for the four vectors $\mathbf{b}_j^{\mathrm{e,m}}$ do not involve any spatial derivatives. As pointed out by Heras [264], this is an advantage as this avoids some mathematical subtleties associated with the retardation that mixes space and time [see Eq. (2.A.28)].

Formula (2.A.36), is a symmetrized (and generalized) elemental field version of the SSPPJ formulas [52, 217, 219–221] that, seemingly, have not been published before. As Eq. (2.A.36) clearly shows, there exist interesting geometric relations between the elemental fields. For instance, for purely electric sources and for purely magnetic sources, $\mathbf{dE}$ is always and everywhere perpendicular to $\mathbf{dB}$. The versatility and efficacy of these improved formulas (2.A.36) make them convenient to use in numerical computations as well as in theoretical analyses.

A.1.1.1 Purely Electric Sources

From now on, we consider only electric charges and currents, i.e., we set $\mathbf{\Theta}^{\mathrm{m}} = \mathbf{0}$. However, as is well known, magnetic charges and currents that manifest the duality of electrodynamics [265] are useful concepts in practical applications such as modeling electric slot antennas. Formulas for this case can be readily derived by following the recipes given below.

We shall frequently decompose a vector $\mathbf{v}$ into its component $\mathbf{v}_{\|}$ parallel to a unit vector $\hat{\mathbf{e}}$, and its component $\mathbf{v}_{\perp}$ transverse to $\hat{\mathbf{e}}$, according to the following scheme:

$$\mathbf{v} = \mathbf{v}_{\|} + \mathbf{v}_{\perp} \tag{2.A.37}$$

where

$$\mathbf{v}_{\|} = v_{\|}\hat{\mathbf{e}} = (\mathbf{v}\cdot\hat{\mathbf{e}})\hat{\mathbf{e}} \tag{2.A.38}$$

$$\mathbf{v}_{\perp} = \mathbf{v} - \mathbf{v}_{\|} = \mathbf{v} - (\mathbf{v}\cdot\hat{\mathbf{e}})\hat{\mathbf{e}} \tag{2.A.39}$$

We shall use this decomposition several times in the following and start with $\mathbf{\Theta}^{\mathrm{e}}$ that we write as a sum of one longitudinal component, $\mathbf{\Theta}_{\|}^{\mathrm{e}}$, parallel to the unit vector $\hat{\mathbf{k}}'(\mathbf{x},\mathbf{x}')$, and one transverse component, $\mathbf{\Theta}_{\perp}^{\mathrm{e}}$, perpendicular to $\hat{\mathbf{k}}'(\mathbf{x},\mathbf{x}')$, i.e.:

$$\boldsymbol{\Theta}^{\mathrm{e}} = \boldsymbol{\Theta}^{\mathrm{e}}_{\parallel} + \boldsymbol{\Theta}^{\mathrm{e}}_{\perp} \tag{2.A.40}$$

where

$$\boldsymbol{\Theta}^{\mathrm{e}}_{\parallel} = \mathbf{b}^{\mathrm{e}}_{1} + \mathbf{b}^{\mathrm{e}}_{2} \tag{2.A.41}$$

$$\boldsymbol{\Theta}^{\mathrm{e}}_{\perp} = \mathbf{b}^{\mathrm{e}}_{3} + \mathbf{b}^{\mathrm{e}}_{4} \tag{2.A.42}$$

This representation of $\boldsymbol{\Theta}^{\mathrm{e}}$ is very convenient when one wishes to calculate and analyze electromagnetic field quantities that are second order in the field vectors.

The local field vectors at a field point $\mathbf{x}$ in the observation volume V, $\mathbf{E}(t, \mathbf{x})$ and $\mathbf{B}(t, \mathbf{x})$, are obtained by integrating the contributions from the entire source volume V' to the field vector density $\boldsymbol{\Theta}^{\mathrm{e}}$, i.e. from Eqs. (2.A.31)–(2.A.36) for purely electric charges and currents, taking vector perpendicularity into account. The resulting elemental fields are

$$\mathrm{d}\mathbf{E}(t, \mathbf{x}, \mathbf{x}') = \boldsymbol{\Theta}^{\mathrm{e}}\mathrm{d}V' = \boldsymbol{\Theta}^{\mathrm{e}}_{\parallel}\mathrm{d}V' + \boldsymbol{\Theta}^{\mathrm{e}}_{\perp}\mathrm{d}V' \tag{2.A.43}$$

$$\mathrm{d}\mathbf{B}(t, \mathbf{x}, \mathbf{x}') = \frac{1}{c}\hat{\mathbf{k}}' \times \boldsymbol{\Theta}^{\mathrm{e}}\mathrm{d}V' = \frac{1}{c}\hat{\mathbf{k}}' \times \boldsymbol{\Theta}^{\mathrm{e}}_{\perp}\mathrm{d}V' \tag{2.A.44}$$

The fields themselves are obtained by integrating (averaging) over all field-contributing volume elements in the source volume V':

$$\mathbf{E}(t, \mathbf{x}) = \int \mathrm{d}\mathbf{E} = \int_{V'} \mathrm{d}^3x' \boldsymbol{\Theta}^{\mathrm{e}} \tag{2.A.45}$$

$$\mathbf{B}(t, \mathbf{x}) = \int \mathrm{d}\mathbf{B} = \frac{1}{c}\int_{V'} \mathrm{d}^3x' \hat{\mathbf{k}}' \times \boldsymbol{\Theta}^{\mathrm{e}} \tag{2.A.46}$$

These simple, elegant expressions for the electric and magnetic fields are exact as long as the sources are prescribed.

Once expressions for the local values of $\mathbf{E}$ and $\mathbf{B}$ at a point $\mathbf{x}$ are obtained, the volumetric densities of the observables carried by the fields, which are second order (quadratic or bilinear) in $\mathbf{E}$ and/or $\mathbf{B}$, can be written down. By integrating (averaging) these local densities, this time over all measurement-contributing points in the sensing volume V, the proper physical observables are obtained. Since this involves two volume integrations, one over V' followed by one over V, we conclude that $\boldsymbol{\Theta}^{\mathrm{e}}$ behaves as a volumetric superdensity (a density of densities). In the case of a conventional radio communication link, we realize that the fields $\mathbf{E}$ and $\mathbf{B}$ at a single instant t at a single point $\mathbf{x}$ in the receiving antenna are the sums (integrals) of many elemental fields $\mathrm{d}\mathbf{E}$ and $\mathrm{d}\mathbf{B}$ emitted at different instants t' from different points $\mathbf{x}'$ in the transmitting antenna and therefore is a weighted average of the elemental fields emitted from all points on this antenna.

A.1.1.2 Useful Approximations

The vector density $\boldsymbol{\Theta}^{\mathrm{e}}$ as given by Eqs. (2.A.40)–(2.A.42) has one component, represented by the vectors $\mathbf{b}^{\mathrm{e}}_{1}$ and $\mathbf{b}^{\mathrm{e}}_{2}$ defined in Eqs. (2.A.32) and (2.A.33), respectively. This component is parallel to the unit vector $\hat{\mathbf{k}}'(\mathbf{x}, \mathbf{x}')$ from a point $\mathbf{x}$ in the source volume V' to a point $\mathbf{x}'$ in the observation volume V over which the observables are integrated (the longitudinal field component). The other component, represented by the vectors $\mathbf{b}^{\mathrm{e}}_{3}$ and $\mathbf{b}^{\mathrm{e}}_{4}$ in Eqs. (2.A.34) and (2.A.35), is perpendicular to $\hat{\mathbf{k}}'(\mathbf{x}, \mathbf{x}')$ (the transverse field component).

A.1.2.1 The Paraxial Approximation

In a configuration where all the source points $\mathbf{x}'$ in V' are viewed from $\mathbf{x}$ in V in such a way that $\hat{\mathbf{k}}'(\mathbf{x},\mathbf{x}')$ are all nearly parallel, the paraxial approximation, illustrated in Figure 2.A.1, can be applied. This amounts to setting

$$\hat{\mathbf{k}}'(\mathbf{x},\mathbf{x}') \approx \hat{\mathbf{k}}'(\mathbf{x},\mathbf{x}'_0) \equiv \hat{\mathbf{n}}'(\mathbf{x}) \tag{2.A.47}$$

i.e., approximating the unit vector from $\mathbf{x}'$ to $\mathbf{x}$ with the $\mathbf{x}$-ward pointing normal unit vector of a sphere centered on a fixed reference point $\mathbf{x}'_0$ in V'. When we introduce this approximation into the exact expressions for the vectors $\mathbf{b}^{\mathrm{e}}_j$ in Eqs. (2.A.30)–(2.A.35), we obtain the paraxially approximate vectors

$$\mathbf{b}_1^{\mathrm{e\,parax}}(t,\mathbf{x},\mathbf{x}') = \frac{\rho^{\mathrm{e}}\left(t-\frac{|\mathbf{x}-\mathbf{x}'|}{c},\mathbf{x}'\right)}{4\pi\varepsilon_0|\mathbf{x}-\mathbf{x}'|^2}\,\hat{\mathbf{n}}'(\mathbf{x}) \tag{2.A.48}$$

$$\mathbf{b}_2^{\mathrm{e\,parax}}(t,\mathbf{x},\mathbf{x}') = \frac{\mathbf{j}^{\mathrm{e}}\left(t-\frac{|\mathbf{x}-\mathbf{x}'|}{c},\mathbf{x}'\right)\cdot\hat{\mathbf{n}}'(\mathbf{x})}{4\pi\varepsilon_0 c|\mathbf{x}-\mathbf{x}'|^2}\,\hat{\mathbf{n}}'(\mathbf{x}) \tag{2.A.49}$$

$$\mathbf{b}_3^{\mathrm{e\,parax}}(t,\mathbf{x},\mathbf{x}') = \frac{\mathbf{j}^{\mathrm{e}}\left(t-\frac{|\mathbf{x}-\mathbf{x}'|}{c},\mathbf{x}'\right)\times\hat{\mathbf{n}}'(\mathbf{x})}{4\pi\varepsilon_0 c|\mathbf{x}-\mathbf{x}'|^2}\times\hat{\mathbf{n}}'(\mathbf{x}) \tag{2.A.50}$$

$$\mathbf{b}_4^{\mathrm{e\,parax}}(t,\mathbf{x},\mathbf{x}') = \frac{\frac{\partial\mathbf{j}^{\mathrm{e}}\left(t-\frac{|\mathbf{x}-\mathbf{x}'|}{c},\mathbf{x}'\right)}{\partial t}\times\hat{\mathbf{n}}'(\mathbf{x})}{4\pi\varepsilon_0 c^2\,|\,\mathbf{x}-\mathbf{x}'\,|}\times\hat{\mathbf{n}}'(\mathbf{x}) \tag{2.A.51}$$

These expressions can be further simplified with the help of standard vector identities, with the result that the field vector density given by Eqs. (2.A.40)–(2.A.42) can be approximated by

$$\mathbf{\Theta}^{\mathrm{e}}\phi^{\mathrm{e}} = \mathbf{\Theta}^{\mathrm{e}}_{\parallel}\phi^{\mathrm{e}} + \mathbf{\Theta}^{\mathrm{e}}_{\perp}\phi^{\mathrm{e}} \tag{2.A.52}$$

where

$$\begin{aligned}\mathbf{\Theta}_{\parallel}^{\mathrm{e\,parax}} &= \frac{c\rho^{\mathrm{e}}\hat{\mathbf{n}}'}{4\pi\varepsilon_0 c|\mathbf{x}-\mathbf{x}'|^2} + \frac{(\mathbf{j}^{\mathrm{e}}\cdot\hat{\mathbf{n}}')\hat{\mathbf{n}}'}{4\pi\varepsilon_0 c|\mathbf{x}-\mathbf{x}'|^2}\\ &\equiv \frac{c\rho^{\mathrm{e}}\hat{\mathbf{n}}'}{4\pi\varepsilon_0 c|\mathbf{x}-\mathbf{x}'|^2} + \frac{\mathbf{j}^{\mathrm{e}}_{\parallel}}{4\pi\varepsilon_0 c|\mathbf{x}-\mathbf{x}'|^2}\end{aligned} \tag{2.A.53}$$

$$\begin{aligned}\mathbf{\Theta}_{\perp}^{\mathrm{e\,parax}} &= \frac{(\mathbf{j}^{\mathrm{e}}\cdot\hat{\mathbf{n}}')\hat{\mathbf{n}}'-\mathbf{j}^{\mathrm{e}}}{4\pi\varepsilon_0 c|\mathbf{x}-\mathbf{x}'|^2} + \frac{\left(\dot{\mathbf{j}}^{\mathrm{e}}\cdot\hat{\mathbf{n}}'\right)\hat{\mathbf{n}}'-\dot{\mathbf{j}}^{\mathrm{e}}}{4\pi\varepsilon_0 c^2\,|\,\mathbf{x}-\mathbf{x}'\,|}\\ &\equiv -\frac{\mathbf{j}^{\mathrm{e}}_{\perp}}{4\pi\varepsilon_0 c|\mathbf{x}-\mathbf{x}'|^2} - \frac{\dot{\mathbf{j}}^{\mathrm{e}}_{\perp}}{4\pi\varepsilon_0 c^2\,|\,\mathbf{x}-\mathbf{x}'\,|}\end{aligned} \tag{2.A.54}$$

For compactness, we introduced the overdot ($\dot{\;}$) notation

$$\dot{\mathbf{j}}^{\mathrm{e}} \equiv \left(\frac{\partial\mathbf{j}^{\mathrm{e}}}{\partial t}\right)_{t=t'} = \frac{\partial\mathbf{j}^{\mathrm{e}}\left(t-\frac{|\mathbf{x}-\mathbf{x}'|}{c},\mathbf{x}'\right)}{\partial t} \tag{2.A.55}$$

in Eq. (2.A.54). We also suppressed obvious arguments.

Using the above expressions (2.A.53) and (2.A.54), we conclude that in the paraxial approximation the formulas (2.A.45) and (2.A.46) can be approximated by

$$\begin{aligned}\mathbf{E}^{\text{parax}}(t,\mathbf{x}) &= \mathbf{E}_{\|}^{\text{parax}}(t,\mathbf{x}) + \mathbf{E}_{\perp}^{\text{parax}}(t,\mathbf{x}) \\ &= \int_{V'} \mathrm{d}^3x' \boldsymbol{\Theta}_{\|}^{\text{eparax}} \phi^{\text{e}}(t,\mathbf{x},\mathbf{x}') + \int_{V'} \mathrm{d}^3x' \boldsymbol{\Theta}_{\perp}^{\text{eparax}}(t,\mathbf{x},\mathbf{x}')\end{aligned} \tag{2.A.56}$$

$$\begin{aligned}\mathbf{B}^{\text{parax}}(t,\mathbf{x}) &= \mathbf{B}_{\perp}^{\text{parax}}(t,\mathbf{x}) = \frac{1}{c}\hat{\mathbf{n}}'(\mathbf{x}) \times \int_{V'} \mathrm{d}^3x' \boldsymbol{\Theta}_{\perp}^{\text{e}} \phi^{\text{e}}(t,\mathbf{x},\mathbf{x}') \\ &= \frac{1}{c}\hat{\mathbf{n}}'(\mathbf{x}) \times \mathbf{E}_{\perp}^{\text{parax}} \phi^{\text{e}}(t,\mathbf{x}) = \frac{1}{c}\hat{\mathbf{n}}'(\mathbf{x}) \times \mathbf{E}^{\text{parax}}(t,\mathbf{x})\end{aligned} \tag{2.A.57}$$

While $\mathbf{E}^{\text{parax}}$ has two mutually perpendicular components, $\mathbf{B}^{\text{parax}}$ has only one component. This component is perpendicular both to $\hat{\mathbf{n}}$ and to the component of $\mathbf{E}^{\text{parax}}$ that, in turn, is perpendicular to $\hat{\mathbf{n}}$. These geometrical properties always hold in the paraxial approximation, even if the far-zone approximation is not applicable. On the other hand, sufficiently far away from a source volume of finite extent, such as an antenna array, the paraxial approximation is always applicable.

A consistently truncated paraxial approximation expression for the electromagnetic linear momentum density is obtained by inserting the paraxially valid approximate expressions (2.A.56) and (2.A.57) for the fields into the exact expression

$$\mathbf{g}^{\text{EM}} = \varepsilon_0 \mathbf{E} \times \mathbf{B} \tag{2.A.58}$$

or, using Eq. (2.A.1),

$$\mathbf{g}^{\text{EM}} = \mathrm{i}\frac{\varepsilon_0}{2c} \mathbf{G} \times \mathbf{G}^* \tag{2.A.59}$$

The result is

$$\mathbf{g}^{\text{EM parax}}(t,\mathbf{x}) = \mathbf{g}_{\perp}^{\text{EM parax}}(t,\mathbf{x}) + \mathbf{g}_{\|}^{\text{EM parax}}(t,\mathbf{x}) \tag{2.A.60}$$

where

$$\mathbf{g}_{\perp}^{\text{EM parax}}(t,\mathbf{x}) = \varepsilon_0 \mathbf{E}_{\|}^{\text{parax}}(t,\mathbf{x}) \times \mathbf{B}_{\perp}^{\text{parax}}(t,\mathbf{x}) \tag{2.A.61}$$

$$\mathbf{g}_{\|}^{\text{EM parax}} \phi^{\text{e}}(t,\mathbf{x}) = \varepsilon_0 E_{\perp}^{\text{parax}}(t,\mathbf{x}) B_{\perp}^{\text{parax}}(t,\mathbf{x}) \hat{\mathbf{n}}'(\mathbf{x}) \tag{2.A.62}$$

The angular momentum density carried by $(\mathbf{E}, \mathbf{B})$ with respect to the fixed moment point $\mathbf{x}_{\text{m}}$ is given by the exact expression [5, section 10.6, 7, 50]

$$\mathbf{h}^{\text{EM}}(t,\mathbf{x};\mathbf{x}_{\text{m}}) = (\mathbf{x}-\mathbf{x}_{\text{m}}) \times \mathbf{g}^{\text{EM}} = \varepsilon_0 (\mathbf{x}-\mathbf{x}_{\text{m}}) \times (\mathbf{E} \times \mathbf{B}) \tag{2.A.63}$$

Using Eq. (2.A.1), we can write this in complex notation as

$$\mathbf{h}^{\text{EM}}(t,\mathbf{x};\mathbf{x}_{\text{m}}) = \mathrm{i}\frac{\varepsilon_0}{2c} (\mathbf{x}-\mathbf{x}_{\text{m}}) \times (\mathbf{G} \times \mathbf{G}^*) \tag{2.A.64}$$

Inserting the paraxially valid approximate expressions (2.A.56) and (2.A.57) for the fields into (2.A.144), we obtain

$$\begin{aligned}\mathbf{h}^{\text{EM parax}}(t,\mathbf{x};\mathbf{x}_{\text{m}}) &= (\mathbf{x}-\mathbf{x}_{\text{m}}) \times \mathbf{g}^{\text{EM parax}}(t,\mathbf{x}) \\ &= \mathbf{h}^{\text{EM parax}}(t,\mathbf{x};\mathbf{0}) - \mathbf{x}_{\text{m}} \times \mathbf{g}^{\text{EM parax}}(t,\mathbf{x})\end{aligned} \tag{2.A.65}$$

where the first term in the second RHS is the angular momentum around the origin $\mathbf{x} = \mathbf{0}$

$$\begin{aligned}\mathbf{h}^{\text{EMparax}}(t,\mathbf{x};\mathbf{0}) &= \varepsilon_0\mathbf{x}\times\left(\mathbf{g}_{\perp}^{\text{EMparax}}(t,\mathbf{x})+\mathbf{g}_{\parallel}^{\text{EMparax}}(t,\mathbf{x})\right)\\ &= \varepsilon_0(\mathbf{B}_{\perp}^{\text{parax}}(t,\mathbf{x})\cdot\mathbf{x})\mathbf{E}_{\parallel}^{\text{parax}}(t,\mathbf{x})\\ &-\varepsilon_0\left(\mathbf{E}_{\parallel}^{\text{parax}}(t,\mathbf{x})\cdot\mathbf{x}\right)\mathbf{B}_{\perp}^{\text{parax}}(t,\mathbf{x})\\ &+\varepsilon_0 E_{\perp}^{\text{parax}}(t,\mathbf{x})B_{\perp}^{\text{parax}}(t,\mathbf{x})\mathbf{x}\times\hat{\mathbf{n}}'(\mathbf{x})\end{aligned} \tag{2.A.66}$$

as follows from Eqs. (2.A.60)–(2.A.62) and the 'bac-cab' vector identity. Recalling that, by definition, the radius vector $\mathbf{x}$ from the source, located at the origin, to the field point is $\mathbf{x} = |\mathbf{x}|\,\hat{\mathbf{n}}'$, we note that $\mathbf{B}_{\perp}\cdot\mathbf{x} = 0$ and $\mathbf{x}\times\hat{\mathbf{n}}' = 0$, which simplifies Eq. (2.A.66) to

$$\mathbf{h}^{\text{EM parax}}(t,\mathbf{x};\mathbf{0}) = -\varepsilon_0\,|\mathbf{x}|\,E_{\parallel}^{\text{parax}}\mathbf{B}_{\perp}^{\text{parax}} \tag{2.A.67}$$

In a spherical polar coordinate system (r,ϑ,φ), where $\mathbf{x} = r\hat{\mathbf{r}}$ and $\hat{\mathbf{r}} = \hat{\mathbf{n}}'$, this becomes

$$\mathbf{h}^{\text{EM parax}}(t,r,\vartheta,\varphi;\mathbf{0}) = -\varepsilon_0 r E_r^{\text{parax}}\mathbf{B}_{\perp}^{\text{parax}} \tag{2.A.68}$$

This shows that in cases when the paraxial approximation can be applied, the angular momentum density with respect to the origin is always proportional to the product of the distance r from the source, the amplitude of the near-zone longitudinal (radial) component of the electric field vector, which goes like r^{-2}, and the amplitude of the transverse component of the magnetic field vector, which goes like r^{-1}, so that the whole expression goes as r^{-2}, as it should. For a single frequency component ω, the angular momentum density (pseudo)vector has one DC component and one component that oscillates along the magnetic field (pseudo)vector at twice the oscillation frequency ω. The cycle average of the latter component vanishes.

Using the paraxial relation (2.A.57) and simple vector algebra, we can express the result in Eq. (2.A.67) entirely in terms of the electric field in the following way:

$$\mathbf{h}^{\text{EM parax}}(t,\mathbf{x};\mathbf{0}) = \frac{\varepsilon_0}{c}\left(r\mathbf{E}_{\parallel}^{\text{parax}}(t,\mathbf{x})\otimes\mathbf{E}_{\perp}^{\text{parax}}(t,\mathbf{x})\right)\times\hat{\mathbf{n}}'(\mathbf{x}) \tag{2.A.69}$$

where $\otimes$ is the dyadic product operator, which for two (three-)vectors $\mathbf{a} = \sum_{i=1}^{3} a_i\hat{\mathbf{x}}_i$ and $\mathbf{b} = \sum_{j=1}^{3} b_j\hat{\mathbf{x}}_j$ is defined as

$$\mathbf{a}\otimes\mathbf{b} = \sum_{i,j=1}^{3}\hat{\mathbf{x}}_i a_i b_j\hat{\mathbf{x}}_j \tag{2.A.70}$$

When the paraxial approximation is not applicable, there exists a certain angular spread in the wave vectors that may require corrections in the geometrical factors but does not change the fact that the asymptotic fall-off is always $1/r^2$.

A.1.2.2 The Far-Zone Approximation

In the far-zone approximation, the integrands (2.A.40)–(2.A.42), (2.A.53) and (2.A.54) can be consistently approximated in order to further simplify the integrals (2.A.45), (2.A.46), (2.A.56) and (2.A.57), respectively. With reference to Figure 2.A.1 and with the choice of the reference point $\mathbf{x}_0'$ as the origin of the spherical polar coordinate used implying that

$$\mathbf{x}-\mathbf{x}_0' = \mathbf{r} = r\hat{\mathbf{r}} \tag{2.A.71}$$

$$\mathbf{x}'-\mathbf{x}_0' = \mathbf{r}' = r'\hat{\mathbf{r}}' \tag{2.A.72}$$

the identity

$$\frac{1}{|\mathbf{x}-\mathbf{x}'|} = \frac{1}{|(\mathbf{x}-\mathbf{x}'_0)-(\mathbf{x}'-\mathbf{x}'_0)|} = \frac{1}{|\mathbf{r}-\mathbf{r}'|} \tag{2.A.73}$$

holds. For configurations such that the shortest distance between V' and V is large, or, more precisely, when

$$\sup r' = \sup|\mathbf{x}'-\mathbf{x}'_0| \ll \inf|\mathbf{x}-\mathbf{x}'_0| = \inf r \tag{2.A.74}$$

the Taylor expansion

$$\begin{aligned}\frac{1}{|\mathbf{x}-\mathbf{x}'|} &= \frac{1}{|\mathbf{r}-\mathbf{r}'|} \\ &= \frac{1}{r} - (\mathbf{r}'\cdot\nabla)\frac{1}{r} - \frac{1}{2}(\mathbf{r}'\cdot\nabla)^2\frac{1}{r} + \ldots \\ &= \frac{1}{r} + \frac{\mathbf{r}'\cdot\hat{\mathbf{r}}}{r^2} + \frac{1}{2}\frac{\hat{\mathbf{r}}\cdot\left(3\mathbf{r}'\otimes\mathbf{r}' - \mathbf{1}_3 r'^2\right)\cdot\hat{\mathbf{r}}}{r^3} + \mathcal{O}\left(r^{-4}\right)\end{aligned} \tag{2.A.75}$$

converges rapidly. This allows us to set $1/|\mathbf{x}-\mathbf{x}'| \approx 1/r$ in Eqs. (2.A.53) and (2.A.54). As a result, the untruncated expressions for the longitudinal and transverse components of the field density vector in the combined paraxial and far-zone approximations can be written as

$$\mathbf{\Theta}_{\parallel}^{\mathrm{e\,parax\,far}} = \frac{c\rho^{\mathrm{e}}\hat{\mathbf{n}}'}{4\pi\varepsilon_0 c r^2} + \frac{\mathbf{j}_{\parallel}^{\mathrm{e}}}{4\pi\varepsilon_0 c r^2} \tag{2.A.76}$$

$$\mathbf{\Theta}_{\perp}^{\mathrm{e\,parax\,far}} = -\frac{\mathbf{j}_{\perp}^{\mathrm{e}}}{4\pi\varepsilon_0 c r^2} - \frac{\dot{\mathbf{j}}_{\perp}^{\mathrm{e}}}{4\pi\varepsilon_0 c^2 r} \tag{2.A.77}$$

Accordingly, the integrals in Eqs. (2.A.56) and (2.A.57) simplify considerably.

Following Thidé et al. [262] and Tamburini et al. [26], we introduce the following shorthand notations:

$$q^{\mathrm{e}}(t') \equiv \int_{V'} \mathrm{d}^3x' \rho^{\mathrm{e}}(t',\mathbf{x}') \tag{2.A.78}$$

$$\mathbf{I}^{\mathrm{e}}(t') \equiv \int_{V'} \mathrm{d}^3x' \mathbf{j}^{\mathrm{e}}(t',\mathbf{x}') \tag{2.A.79}$$

$$\begin{aligned}\dot{\mathbf{I}}^{\mathrm{e}}(t') &\equiv \int_{V'} \mathrm{d}^3x' \dot{\mathbf{j}}^{\mathrm{e}}(t',\mathbf{x}') \\ &\equiv \int_{V'} \mathrm{d}^3x' \frac{\partial \mathbf{j}^{\mathrm{e}}\left(t - \frac{|\mathbf{x}-\mathbf{x}'|}{c}, \mathbf{x}'\right)}{\partial t}\end{aligned} \tag{2.A.80}$$

Then, under the assumption of paraxial as well as far-zone approximation, we can write the perpendicular and parallel components of the fields as follows:

$$\mathbf{E}_{\perp}\phi^{\mathrm{efar}}(t,\mathbf{x}) = -\frac{\mathbf{I}_{\perp}^{\mathrm{e}}(t')}{4\pi\varepsilon_0 c r^2} - \frac{\dot{\mathbf{I}}_{\perp}^{\mathrm{e}}(t')}{4\pi\varepsilon_0 c^2 r} \tag{2.A.81}$$

$$\mathbf{E}_{\parallel}^{\mathrm{parax\,far}}(t,\mathbf{x}) = \frac{q^{\mathrm{e}}(t')\hat{\mathbf{n}}'(\mathbf{x})}{4\pi\varepsilon_0 r^2} + \frac{\mathbf{I}_{\parallel}^{\mathrm{e}}(t')}{4\pi\varepsilon_0 c r^2} \tag{2.A.82}$$

$$\mathbf{B}_{\perp}^{\text{parax far}}(t,\mathbf{x}) = \frac{\mathbf{I}_{\perp}^{\text{e}}(t') \times \hat{\mathbf{n}}'(\mathbf{x})}{4\pi\varepsilon_0 c r^2} + \frac{\dot{\mathbf{I}}_{\perp}^{\text{e}}(t') \times \hat{\mathbf{n}}'(\mathbf{x})}{4\pi\varepsilon_0 c^2 r} \tag{2.A.83}$$

Evaluating the linear momentum density formula (2.A.58) in the paraxial approximation, keeping all terms and approximating each one of them in a consistent way, we see that far away from the source the linear momentum density is

$$\mathbf{g}^{\text{EMparax far}} = \mathbf{g}_{\perp}^{\text{EMparax far}} + \mathbf{g}_{\parallel}^{\text{EM parax far}} \tag{2.A.84}$$

where

$$\mathbf{g}_{\perp}^{\text{EMparax far}} = \frac{1}{16\pi^2\varepsilon_0 c^2 r^3}\left(cq^{\text{e}} + I_{\parallel}^{\text{e}}\right)\left(\frac{\mathbf{I}_{\perp}^{\text{e}}}{r} + \frac{\dot{\mathbf{I}}_{\perp}^{\text{e}}}{c}\right) \tag{2.A.85}$$

$$\mathbf{g}_{\parallel}^{\text{EMparax far}} = \frac{1}{16\pi^2\varepsilon_0 c^2 r^2}\left(\frac{\mathbf{I}_{\perp}^{\text{e}}}{r} + \frac{\dot{\mathbf{I}}_{\perp}^{\text{e}}}{c}\right)\cdot\left(\frac{\mathbf{I}_{\perp}^{\text{e}}}{r} + \frac{\dot{\mathbf{I}}_{\perp}^{\text{e}}}{c}\right)\hat{\mathbf{n}}' \tag{2.A.86}$$

If we express this result in terms of the explicit integrals in Eqs. (2.A.78)–(2.A.80), we find that as the distance r from the source tends to infinity, the linear momentum density vector is given by the consistently truncated expression

$$\begin{aligned}\mathbf{g}^{\text{EM parax far}}(t,\mathbf{x}) &= \frac{1}{16\pi^2\varepsilon_0 c^4 r^3}\int_{V'} \mathrm{d}^3x'\left(c\rho^{\text{e}} + j_{\parallel}^{\text{e}} + \mathcal{O}(r^{-1})\right)\int_{V'} \mathrm{d}^3x'\dot{\mathbf{j}}_{\perp}^{\text{e}} \\ &+ \frac{1}{16\pi^2\varepsilon_0 c^4 r^2}\left(\int_{V'} \mathrm{d}^3x'\dot{\mathbf{j}}_{\perp}^{\text{e}}\cdot\int_{V'} \mathrm{d}^3x'\dot{\mathbf{j}}_{\perp}^{\text{e}} + \mathcal{O}(r^{-1})\right)\hat{\mathbf{n}}'\end{aligned} \tag{2.A.87}$$

To leading order in r, the magnitude of this vector is the scalar

$$\left|\mathbf{g}^{\text{EMparax far}}(t,\mathbf{x})\right| = \frac{1}{16\pi^2\varepsilon_0 c^4 r^2}\left|\int_{V'} \mathrm{d}^3x'\dot{\mathbf{j}}_{\perp}^{\text{e}}\right|^2 + \mathcal{O}(r^{-3}) \tag{2.A.88}$$

that exhibits the well-known r^{-2} asymptotic fall-off.

Starting from the exact expression for the angular momentum density, Eq. (2.A.144), and using the untruncated paraxially approximate expressions (2.A.84)–(2.A.86), noticing that $\mathbf{x} = r\hat{\mathbf{n}}'$, we see that the untruncated paraxially approximate expression for the angular momentum density is

$$\mathbf{h}^{\text{EM parax far}}(t,\mathbf{x};\mathbf{0}) = \frac{cq^{\text{e}}(t') + I_{\parallel}^{\text{e}}(t')}{16\pi^2\varepsilon_0 c^2 r^2}\hat{\mathbf{n}}' \times \left(\frac{\mathbf{I}_{\perp}^{\text{e}}}{r} + \frac{\dot{\mathbf{I}}_{\perp}^{\text{e}}}{c}\right) \tag{2.A.89}$$

Expressing this result explicitly in the integrals in Eqs. (2.A.56) and (2.A.57), and truncating consistently, we obtain

$$\begin{aligned}\mathbf{h}^{\text{EM paraxfar}}(t,\mathbf{x};\mathbf{0}) &= -\frac{1}{16\pi^2\varepsilon_0 c^3 r^2}\int_{V'} \mathrm{d}^3x'\rho^{\text{e}}\int_{V'} \mathrm{d}^3x'\dot{\mathbf{j}}_{\perp}^{\text{e}} \times \hat{\mathbf{n}}' \\ &- \frac{1}{16\pi^2\varepsilon_0 c^4 r^2}\int_{V'} \mathrm{d}^3x' j_{\parallel}^{\text{e}}\int_{V'} \mathrm{d}^3x'\dot{\mathbf{j}}_{\perp}^{\text{e}} \times \hat{\mathbf{n}}' + \mathcal{O}(r^{-3})\end{aligned} \tag{2.A.90}$$

In the first integral in the first term, we may, with the help of the continuity equation, formally express ρ^{e} in $\mathbf{j}^{\text{e}}$ as follows

$$\int_{V'} \mathrm{d}^3x'\rho^{\text{e}}(t',\mathbf{x}') = -\int^{t'} \mathrm{d}t'\int_{V'} \mathrm{d}^3x'\nabla'\cdot\mathbf{j}^{\text{e}}(t',\mathbf{x}') \tag{2.A.91}$$

which, by introducing the shorthand notation

$$\widetilde{\mathbf{j}}^{\mathrm{e}}(\mathbf{x}') = \int^{t'} \mathrm{d}t' \mathbf{j}^{\mathrm{e}}(t', \mathbf{x}') \tag{2.A.92}$$

allows us to write

$$\int_{V'} \mathrm{d}^3x' \rho^{\mathrm{e}}(t', \mathbf{x}') = -\int_{V'} \mathrm{d}^3x' \nabla' \cdot \widetilde{\mathbf{j}}^{\mathrm{e}}(\mathbf{x}') \tag{2.A.93}$$

$$= -\oint_{S'} \mathrm{d}^2x' \hat{\mathbf{n}}' \cdot \widetilde{\mathbf{j}}^{\mathrm{e}}(\mathbf{x}') \tag{2.A.94}$$

where, in the last step, the divergence theorem was applied. We note that we can, in the same way, formally write Eq. (2.A.78) as

$$q^{\mathrm{e}}(t') \equiv -\int_{V'} \mathrm{d}^3x' \nabla' \cdot \int^{t'} \mathrm{d}t' \mathbf{j}^{\mathrm{e}}(t', \mathbf{x}') \tag{2.A.95}$$

$$= -\int^{t'} \mathrm{d}t' \int_{V'} \mathrm{d}^3x' \nabla' \cdot \mathbf{j}^{\mathrm{e}}(t', \mathbf{x}') \tag{2.A.96}$$

By recalling formula (2.A.65) and using expression (2.A.87), we conclude that for an arbitrary source distribution $(\rho^{\mathrm{e}}, \mathbf{j}^{\mathrm{e}})$ located in a volume V' of finite extent, the angular momentum density with respect to an arbitrary fixed moment point $\mathbf{x}_{\mathrm{m}}$ in the paraxial, far-zone approximation is

$$\begin{aligned}\mathbf{h}^{\mathrm{EM\,parax\,far}}(t, \mathbf{x}; \mathbf{x}_{\mathrm{m}}) = &\frac{1}{16\pi^2\varepsilon_0 c^4 r^2}\int_{V'} \mathrm{d}^3x' \left(c\nabla' \cdot \widetilde{\mathbf{j}}^{\mathrm{e}} - j_{\parallel}^{\mathrm{e}}\right) \int_{V'} \mathrm{d}^3x' \dot{\mathbf{j}}_{\perp}^{\mathrm{e}} \times \hat{\mathbf{n}}' \\ &+ \frac{1}{16\pi^2\varepsilon_0 c^4 r^2}\left|\int_{V'} \mathrm{d}^3x' \dot{\mathbf{j}}_{\perp}^{\mathrm{e}}\right|^2 \mathbf{x}_{\mathrm{m}} \times \hat{\mathbf{n}}' + \mathcal{O}(r^{-3})\end{aligned} \tag{2.A.97}$$

or, since $\hat{\mathbf{n}}' \cdot \widetilde{\mathbf{j}}^{\mathrm{e}} = \widetilde{j}_{\parallel}^{\mathrm{e}}$,

$$\begin{aligned}\mathbf{h}^{\mathrm{EM\,paraxfar}}(t, \mathbf{x}; \mathbf{x}_{\mathrm{m}}) = &\frac{1}{16\pi^2\varepsilon_0 c^3 r^2}\oint_{S'} \mathrm{d}^2x' \widetilde{j}_{\parallel}^{\mathrm{e}} \int_{V'} \mathrm{d}^3x' \dot{\mathbf{j}}_{\perp}^{\mathrm{e}} \times \hat{\mathbf{n}}' \\ &- \frac{1}{16\pi^2\varepsilon_0 c^4 r^2}\int_{V'} \mathrm{d}^3x' j_{\parallel}^{\mathrm{e}} \int_{V'} \mathrm{d}^3x' \dot{\mathbf{j}}_{\perp}^{\mathrm{e}} \times \hat{\mathbf{n}}' \\ &+ \frac{1}{16\pi^2\varepsilon_0 c^4 r^2}\left|\int_{V'} \mathrm{d}^3x' \dot{\mathbf{j}}_{\perp}^{\mathrm{e}}\right|^2 \mathbf{x}_{\mathrm{m}} \times \hat{\mathbf{n}}' + \mathcal{O}(r^{-3})\end{aligned} \tag{2.A.98}$$

If the observing volume V is sufficiently far away from the source volume V', the paraxial approximation is always trivially fulfilled. Consequently, Eqs. (2.A.97) and (2.A.98) shows that the magnitude of the angular momentum density *always* has an r^{-2} asymptotic fall-off. See also Thidé et al. [262] and Tamburini et al. [26]. Eqs. (2.A.97) and (2.A.98) generalizes the proof, published already in 1914 by Abraham [42] for the special case when the source is a single electric Hertzian dipole and the origin is chosen as the moment point, to hold for any conceivable combination of charge and current densities, located inside a source volume V' of finite extent, and for an arbitrary moment point $\mathbf{x}_{\mathrm{m}}$.

We note from Eqs. (2.A.45) and (2.A.46) for **E** and **B** that are far away from V', the dominating contribution to the magnitude of the linear momentum density (Poynting vector) comes from the asymptotic approximation of the far-zone $\mathbf{b}_4^{\mathrm{e}}$ term, i.e., the transverse component of the time rate of change of the current density, whereas the dominating contribution to the magnitude of the angular momentum in the same region comes from both the *far-zone* $\mathbf{b}_4^{\mathrm{e}}$ term for **B** and the *near-zone* $\mathbf{b}_1^{\mathrm{e}}$ and $\mathbf{b}_2^{\mathrm{e}}$ terms for **E**, i.e., from both the perpendicular component of the time rate of change of the current density vector and the (time integrated) divergence and the longitudinal component of the current density itself. Hence, the far-zone **E** field has in general no direct, primary influence on the far-zone angular momentum; see, however, formula (2.A.69). This may at first seem like a paradox but is a fact that was established more than a century ago [42]; see also Then and Thidé [78], Thidé et al. [262], and Tamburini et al. [26].

2.A.2 Poincaré Invariants and Conserved Quantities of the EM Field

Symmetries and their associated conservation laws are powerful tools in physics. For instance, if all conserved quantities (constants of motion) of an integrable physical system are known, the system dynamics can be obtained trivially by formulating it in terms of action-angle variables [266]. All in all the electromagnetic field has at least 84 first-order constants of motion [12]. In this appendix we discuss the ten conserved electromagnetic observables that correspond to the ten-dimensional Poincaré symmetry group $P(10)$, i.e. the ten Poincaré invariants, and their associated equations of continuity. They are the scalar quantity energy, the three components of the linear momentum vector, the three components of the angular momentum pseudovector, and the three components of the energy center position vector. Only a small subset of these observables are currently used in radio science, optics, and wireless communications.

Electromagnetic conservation laws can be viewed as constraints on the field dynamics and, in this sense, every conserved quantity is physically significant. The information contained in the conserved quantities corresponds one-to-one to the information contained in the Maxwell–Lorentz postulates [267].

A.2.1 Energy

The energy density $u^{\text{EM}}(t, \mathbf{x})$ carried by the fields generated by the sources in V' is

$$u^{\text{EM}} = \frac{\varepsilon_0}{2}\left(\mathbf{E}\cdot\mathbf{E} + c^2\mathbf{B}\cdot\mathbf{B}\right) \tag{2.A.99}$$

which, according to Eq. (2.A.1) and Table 2.A.1, can be written as

$$u^{\text{EM}} = \frac{\varepsilon_0}{2}\mathbf{G}\cdot\mathbf{G}^* = \frac{\varepsilon_0}{2}|\mathbf{G}|^2 = \frac{\varepsilon_0}{2}\sqrt{(\mathbf{G}\cdot\mathbf{G})(\mathbf{G}\cdot\mathbf{G})^*} \tag{2.A.100}$$

The field energy in an observation volume V away from the source volume V', can be measured with an appropriate energy sensor, e.g. a calorimeter or bolometer, that integrates u^{EM} over V.

With the energy flux vector (Poynting vector) $\mathbf{S}(t, \mathbf{x})$ given by the formula

$$\mathbf{S} = \varepsilon_0 c^2 \mathbf{E} \times \mathbf{B} \tag{2.A.101}$$

or, using Eq. (2.A.1) and Table 2.A.1,

$$\mathbf{S} = \mathrm{i}\frac{\varepsilon_0 c}{2}\mathbf{G} \times \mathbf{G}^* \tag{2.A.102}$$

the energy density balance equation, which follows straightforwardly from the Maxwell–Lorentz Eqs. (2.1)–(2.4), takes the form

$$\frac{\partial u^{\text{EM}}}{\partial t} + \nabla\cdot\mathbf{S} = -\mathbf{j}^{\text{e}}\cdot\mathbf{E} \tag{2.A.103}$$

i.e. a local continuity equation for the field energy density with a sink (loss) term $\mathbf{j}^{\text{e}} \cdot \mathbf{E}$. By using Eqs. (2.A.1), (2.A.100) and (2.A.102), we can write this as

$$\frac{\partial(\mathbf{G}\cdot\mathbf{G}^*)}{\partial t} + \mathrm{i}c\nabla\cdot(\mathbf{G}\times\mathbf{G}^*) = -\frac{1}{\varepsilon_0}\mathbf{j}^{\mathrm{e}}\cdot(\mathbf{G}+\mathbf{G}^*) \tag{2.A.104}$$

The continuity Eq. (2.A.103) shows that $\mathbf{S}$ is the field energy current density and that therefore

$$\mathbf{v}^{\mathrm{EM}}(t,\mathbf{x}) = \frac{\mathbf{S}(t,\mathbf{x})}{u^{\mathrm{EM}}(t,\mathbf{x})} = c^2\frac{\mathbf{g}^{\mathrm{EM}}(t,\mathbf{x})}{u^{\mathrm{EM}}(t,\mathbf{x})} \tag{2.A.105}$$

is the local field energy density velocity [5, 7]. Alternatively, we can express this velocity in terms or the Riemann–Silberstein vector as

$$\mathbf{v}^{\mathrm{EM}} = \mathrm{i}c\frac{\mathbf{G}\times\mathbf{G}^*}{\mathbf{G}\cdot\mathbf{G}^*} \tag{2.A.106}$$

If $\rho^{\mathrm{mech}}(t',\mathbf{x}')$ denotes the local mechanical mass density field and $\mathbf{v}^{\mathrm{mech}}(t',\mathbf{x}')$ the local mechanical velocity field of the charged particles at a point in V', then, in the approximation where statistical mechanical as well as relativistic and quantum physics effects are negligible, the electric current density $\mathbf{j}^{\mathrm{e}}(t',\mathbf{x}')$ carried by the charged particles can be written as $\mathbf{j}^{\mathrm{e}} = \rho^{\mathrm{e}}\mathbf{v}^{\mathrm{mech}}$ and the loss term in Eq. (2.A.103) can therefore be identified as the increase in mechanical energy density of the charges [5, 52]:

$$\frac{\partial u^{\mathrm{mech}}}{\partial t} = \mathbf{j}^{\mathrm{e}}\cdot\mathbf{E} = \mathbf{v}^{\mathrm{mech}}\cdot\rho^{\mathrm{e}}\mathbf{E} \tag{2.A.107}$$

representing dissipation due to Ohmic losses, i.e., transfer of field energy density to mechanical energy density of the current-carrying particles.

This identification makes it possible to introduce the total energy density of the electromechanical system under study (charged particles and their associated fields, e.g., a plasma)

$$u^{\mathrm{sys}}(t,\mathbf{x}) = u^{\mathrm{EM}}(t,\mathbf{x}) + u^{\mathrm{mech}}(t,\mathbf{x}) \tag{2.A.108}$$

and to put Eq. (2.A.103) in the form of an equation of continuity where there are no sources or sinks (losses) [5, 52]:

$$\frac{\partial u^{\mathrm{sys}}}{\partial t} + \nabla\cdot\mathbf{S} = 0 \tag{2.A.109}$$

We let $U(t)$ denote energy (volume integrated energy density). Then, in obvious notation, the system energy is the sum of the energy of the particles and the energy of the field:

$$U^{\mathrm{sys}}(t) = \int_V \mathrm{d}^3x\, u^{\mathrm{sys}}(t,\mathbf{x}) = U^{\mathrm{EM}}(t) + U^{\mathrm{mech}}(t) \tag{2.A.110}$$

Volume integrating the local continuity Eq. (2.A.109) and using the divergence theorem, the corresponding global continuity equation can be written as

$$\frac{\mathrm{d}U^{\mathrm{sys}}}{\mathrm{d}t} + \oint_S \mathrm{d}^2x\,\hat{\mathbf{n}}\cdot\mathbf{S} = 0 \tag{2.A.111}$$

This is the Poynting theorem, also known as the energy theorem in Maxwell's theory. It is utilized, for instance, in radiometry. The field energy itself is utilized in simple, broadband 'on–off' signaling as was used in Marconi's first radio experiments and also in optical fiber communications. The conservation of energy as described by (2.A.111) is a manifestation of the temporal translation invariance symmetry of electrodynamics and describes the kinematics of the system.

A.2.2 Linear Momentum

The electromagnetic linear momentum density $\mathbf{g}^{\mathrm{EM}}(t, \mathbf{x})$ radiated from V' is [5, 6, 50, 52, 268]

$$\mathbf{g}^{\mathrm{EM}} = \varepsilon_0 \mathbf{E} \times \mathbf{B} \tag{2.A.112}$$

or, using Eq. (2.A.1),

$$\mathbf{g}^{\mathrm{EM}} = \mathrm{i}\frac{\varepsilon_0}{2c}\mathbf{G} \times \mathbf{G}^* \tag{2.A.113}$$

In free space the linear momentum density is simply related to the Poynting vector (2.A.101) and (2.A.102) by the Planck relation

$$\mathbf{S} = c^2 \mathbf{g}^{\mathrm{EM}} \tag{2.A.114}$$

We introduce the electromagnetic linear momentum flux tensor $\mathbf{T}(t, \mathbf{x})$ as the negative of Maxwell's stress tensor [5, 6]

$$\mathbf{T} = u^{\mathrm{EM}}\mathbf{1}_3 - \varepsilon_0\left(\mathbf{E}\otimes\mathbf{E} + c^2\mathbf{B}\otimes\mathbf{B}\right) \tag{2.A.115}$$

or, using Eq. (2.A.1),

$$\mathbf{T} = \frac{\varepsilon_0}{2}[\mathbf{G}\cdot\mathbf{G}^*\mathbf{1}_3 - (\mathbf{G}\otimes\mathbf{G}^* + \mathbf{G}^*\otimes\mathbf{G})] \tag{2.A.116}$$

where $\mathbf{1}_3$ is the rank-3 unit tensor and $\otimes$ the tensor (Kronecker, dyadic) product operator, which for two (three-)vectors $\mathbf{a} = \sum_{i=1}^{3} a_i\hat{\mathbf{x}}_i$ and $\mathbf{b} = \sum_{j=1}^{3} b_j\hat{\mathbf{x}}_j$ is defined as

$$\mathbf{a}\otimes\mathbf{b} = \sum_{i,j=1}^{3} \hat{\mathbf{x}}_i a_i b_j \hat{\mathbf{x}}_j \tag{2.A.117}$$

Then the balance equation for linear momentum density that follows from the Maxwell–Lorentz Eqs. (2.1)–(2.4) can be written [5, 6, 52] as

$$\frac{\partial \mathbf{g}^{\mathrm{EM}}}{\partial t} + \nabla\cdot\mathbf{T} = -\mathbf{f} \tag{2.A.118}$$

where

$$\mathbf{f}(t, \mathbf{x}) = \rho^{\mathrm{e}}(t, \mathbf{x})\mathbf{E}(t, \mathbf{x}) + \mathbf{j}^{\mathrm{e}} \times \mathbf{B}(t, \mathbf{x}) \tag{2.A.119}$$

This polar vector $\mathbf{f}$ is the Lorentz force density[3] that behaves like a loss of field linear momentum density. Note that (2.A.119) follows directly from a conservation law, and therefore is a direct consequence of a symmetry of the Maxwell–Lorentz postulates. Hence, the Lorentz force does not have to be postulated as is sometimes done.

By using Eqs. (2.A.1), (2.A.113) and (2.A.116) we can write the local equation of continuity for linear momentum density as

$$\begin{aligned}&\frac{\partial(\mathbf{G}\times\mathbf{G}^*)}{\partial t} - \mathrm{i}c\nabla\cdot\left(\mathbf{G}\cdot\mathbf{G}^*\mathbf{1}_3 - \frac{1}{\varepsilon_0}(\mathbf{G}\otimes\mathbf{G}^* + \mathbf{G}^*\otimes\mathbf{G})\right)\\ &= -\mathrm{i}\frac{c}{\varepsilon_0}[\rho^{\mathrm{e}}(\mathbf{G}+\mathbf{G}^*) - \varepsilon_0\mathbf{j}^{\mathrm{e}} \times (\mathbf{G}-\mathbf{G}^*)]\end{aligned} \tag{2.A.120}$$

3 In certain disciplines, a nonstandard convention is sometimes used in which the term Lorentz force density denotes only the second term in the RHS of Eq. (2.A.119). We follow the standard convention.

In an electromechanical system of charged, massive particles and pertinent fields, e.g. a plasma or a transmitting antenna, the mechanical linear momentum density of the particles is

$$\mathbf{g}^{\mathrm{mech}}(t,\mathbf{x}) = \rho^{\mathrm{mech}}(t,\mathbf{x})\mathbf{v}^{\mathrm{mech}}(t,\mathbf{x}) \tag{2.A.121}$$

The local equation of continuity of linear momentum of an electromechanical system, directly derivable from Maxwell's equations, can therefore be written [5, 6, 52, 53]

$$\frac{\partial \mathbf{g}^{\mathrm{sys}}}{\partial t} + \nabla\cdot\mathbf{T} = \mathbf{0} \tag{2.A.122}$$

where

$$\mathbf{g}^{\mathrm{sys}}(t,\mathbf{x}) = \mathbf{g}^{\mathrm{EM}}(t,\mathbf{x}) + \mathbf{g}^{\mathrm{mech}}(t,\mathbf{x}) \tag{2.A.123}$$

is the system linear momentum density. In the approximation used, the electric current density $\mathbf{j}^{\mathrm{e}}(t,\mathbf{x})$ that is carried by the charged, massive classical particles is directly proportional to their mechanical linear momentum density $\mathbf{g}^{\mathrm{mech}}$:

$$\mathbf{j}^{\mathrm{e}} = \rho^{\mathrm{e}}\mathbf{v}^{\mathrm{mech}} = \frac{\rho^{\mathrm{e}}}{\rho^{\mathrm{mech}}}\mathbf{g}^{\mathrm{mech}} \tag{2.A.124}$$

We therefore conclude that an electromechanical system where the current density varies with time radiates linear momentum density $\mathbf{g}^{\mathrm{EM}}$. Specifically, a thin, linear antenna of a dipole type, say, fed by a current, represented by the term $\nabla \cdot \mathrm{T}$, from a generator (transmitter) via a transmission line will radiate a distribution of linear momentum density $\mathbf{g}^{\mathrm{EM}}$ that propagates away from the antenna.

The classical electromagnetic field linear momentum $\mathbf{p}^{\mathrm{EM}}$ that is localized within a volume V embedded in free space is obtained by volume integrating the linear momentum density $\mathbf{g}^{\mathrm{EM}}$ over V [6, 52, 53]

$$\mathbf{p}^{\mathrm{EM}}(t) = \int_V \mathrm{d}^3x\mathbf{g}^{\mathrm{EM}}(t,\mathbf{x}) = \varepsilon_0\int_V \mathrm{d}^3x[\mathbf{E}(t,\mathbf{x}) \times \mathbf{B}(t,\mathbf{x})] \tag{2.A.125}$$

In practical applications this integration is realised by an appropriate volume integrating linear momentum sensor, e.g., a linear dipole antenna.

The mechanical linear momentum in V is

$$\mathbf{p}^{\mathrm{mech}}(t) = \int_V \mathrm{d}^3x\mathbf{g}^{\mathrm{mech}}(t,\mathbf{x}) = \int_V \mathrm{d}^3x\rho^{\mathrm{mech}}(t,\mathbf{x})\mathbf{v}^{\mathrm{mech}}(t,\mathbf{x}) \tag{2.A.126}$$

which means that the total linear momentum of the system of particles and fields in V is

$$\mathbf{p}^{\mathrm{sys}}(t) = \mathbf{p}^{\mathrm{EM}}(t) + \mathbf{p}^{\mathrm{mech}}(t) \tag{2.A.127}$$

It fulfills the global conservation law [5, 7, 52]

$$\frac{\mathrm{d}\mathbf{p}^{\mathrm{sys}}}{\mathrm{d}t} + \oint_S \mathrm{d}^2x\hat{\mathbf{n}}\cdot\mathbf{T} = \mathbf{0} \tag{2.A.128}$$

This is the linear momentum theorem in Maxwell's theory. It demonstrates that not only the mechanical particles (charges) but also the electromagnetic field itself carries linear momentum (translational momentum) and that the field linear momentum, which is *bilinear* in the fields, couples to the mechanical linear momentum of the charged particles and thereby to the conduction current carried by these particles and *vice versa*.

A.2.2.1 Gauge Invariance

It is common practice to investigate the properties of the electromagnetic observables that we consider here by expressing the fields in terms of their potentials [269]. If the vector potential $\mathbf{A}(t, \mathbf{x})$ is twice continuously differentiable in space, it is well known that it fulfills the vector analytic identity (see, e.g., Thidé [7])

$$\begin{aligned}\mathbf{A}(t,\mathbf{x}) = & -\nabla\int_{V'} \mathrm{d}^3x' \frac{\nabla'\cdot\mathbf{A}(t,\mathbf{x}')}{4\pi \mid \mathbf{x}-\mathbf{x}' \mid} \\ & +\nabla\times\int_{V'} \mathrm{d}^3x' \frac{\nabla'\times\mathbf{A}(t,\mathbf{x}')}{4\pi \mid \mathbf{x}-\mathbf{x}' \mid} \\ & -\oint_{S'} \mathrm{d}^2x'\hat{\mathbf{n}}'\cdot\left(\frac{\mathbf{A}(t,\mathbf{x}')\otimes(\mathbf{x}-\mathbf{x}')}{|\mathbf{x}-\mathbf{x}'|^3}\right)\end{aligned} \tag{2.A.129}$$

The last integral, to be evaluated over a surface S' enclosing the entire volume V' within which $\mathbf{A} \neq \mathbf{0}$ is localized, vanishes if it is evaluated at very large distances and the expression within the big parentheses falls off faster than $1/|\mathbf{x} - \mathbf{x}'|^2 = 1/r^2$, or if it is everywhere perpendicular to the normal unit vector $\hat{\mathbf{n}}'$, or if it has certain (a)symmetry properties[4]. Then, Helmholtz's decomposition is applicable, allowing us to write

$$\mathbf{A}(t,\mathbf{x}) = \mathbf{A}^{\mathrm{irrot}}(t,\mathbf{x}) + \mathbf{A}^{\mathrm{rotat}}(t,\mathbf{x}) \tag{2.A.130}$$

where $\mathbf{A}^{\mathrm{irrot}}$ is the irrotational part of $\mathbf{A}$, fulfilling $\nabla \times \mathbf{A}^{\mathrm{irrot}} = \mathbf{0}$, and $\mathbf{A}^{\mathrm{rotat}}$ is the rotational part, fulfilling $\nabla \cdot \mathbf{A}^{\mathrm{rotat}} = 0$. Since $\mathbf{A}^{\mathrm{rotat}}$ is gauge invariant by definition, it follows that in this case

$$\mathbf{B}(t,\mathbf{x}) = \nabla\times\mathbf{A}(t,\mathbf{x}) = \nabla\times\mathbf{A}^{\mathrm{rotat}}(t,\mathbf{x}) \tag{2.A.131}$$

If we take (2.A.125) as a starting point, and assume that not only the vector potential $\mathbf{A}$ can be Helmholtz decomposed but also $\mathbf{E}$ is so well-behaved that it can be decomposed as

$$\mathbf{E}(t,\mathbf{x}) = \mathbf{E}^{\mathrm{irrot}}(t,\mathbf{x}) + \mathbf{E}^{\mathrm{rotat}}(t,\mathbf{x}) \tag{2.A.132}$$

then the electromagnetic linear (translational) momentum can be represented by the gauge invariant expression

$$\begin{aligned}\mathbf{p}^{\mathrm{EM}}(t) = & \,\varepsilon_0\int_V \mathrm{d}^3x\mathbf{E}^{\mathrm{irrot}}\times\left(\nabla\times\mathbf{A}^{\mathrm{rotat}}\right) \\ & +\varepsilon_0\int_V \mathrm{d}^3x\mathbf{E}^{\mathrm{rotat}}\times\left(\nabla\times\mathbf{A}^{\mathrm{rotat}}\right)\end{aligned} \tag{2.A.133}$$

Using straightforward vector analysis, we obtain the following result [7]

$$\begin{aligned}\mathbf{p}^{\mathrm{EM}}(t) = & \int_V \mathrm{d}^3x\rho^{\mathrm{e}}\mathbf{A}^{\mathrm{rotat}} - \varepsilon_0\int_V \mathrm{d}^3x\mathbf{A}^{\mathrm{rotat}}\times\left(\nabla\times\mathbf{E}^{\mathrm{rotat}}\right) \\ & -\varepsilon_0\oint_S \mathrm{d}^2x\hat{\mathbf{n}}\cdot\mathbf{E}^{\mathrm{rotat}}\otimes\mathbf{A}^{\mathrm{rotat}}\end{aligned} \tag{2.A.134}$$

Whenever Eqs. (2.A.130) and (2.A.132) hold, this gauge invariant expression for $\mathbf{p}^{\mathrm{EM}}$ is exact. If the electric field is such that it cannot be Helmholtz decomposed, every occurrence of $\mathbf{E}^{\mathrm{irrot}}$ and $\mathbf{E}^{\mathrm{rotat}}$ in Eqs. (2.A.133) and (2.A.134) has to be replaced by $\mathbf{E}$. However, the resulting formula would still be exact and gauge independent.

---

4 In fact, the surface integral may also be a vector that is (at most) a function of time t.

If the expression within the big parentheses in the integrand of the last integral in Eq. (2.A.129) is such that this surface integral grows without bounds at very large distances, it is not possible to decompose the vector potential as described by Eq. (2.A.130), and $\mathbf{A}^{\text{rotat}}$ and $\mathbf{E}^{\text{rotat}}$ in Eqs. (2.A.133) and (2.A.134) would have to be replaced by $\mathbf{A}$ and $\mathbf{E}$.

A.2.2.2 First Quantization Formalism

If we restrict ourselves to consider a single temporal Fourier component of the rotational (transverse) component of $\mathbf{E}$ in a region where $\rho^{\text{e}} = 0$, we find that in complex representation

$$\mathbf{E}^{\text{rotat}} = -\frac{\partial \mathbf{A}^{\text{rotat}}}{\partial t} = \mathrm{i}\omega \mathbf{A}^{\text{rotat}} \tag{2.A.135}$$

This allows us to replace $\mathbf{A}^{\text{rotat}}$ by $-\mathrm{i}\mathbf{E}^{\text{rotat}}/\omega$, yielding,

$$\begin{aligned} \langle \mathbf{p}^{\text{EM}} \rangle_t = {} & \operatorname{Re}\left\{ -\mathrm{i}\frac{\varepsilon_0}{2\omega}\int_V \mathrm{d}^3x (\nabla \otimes \mathbf{E}^{\text{rotat}}) \cdot (\mathbf{E}^{\text{rotat}})^* \right\} \\ & + \operatorname{Re}\left\{ \mathrm{i}\frac{\varepsilon_0}{2\omega}\oint_S \mathrm{d}^2x \hat{\mathbf{n}} \cdot \mathbf{E}^{\text{rotat}} \otimes (\mathbf{E}^{\text{rotat}})^* \right\} \end{aligned} \tag{2.A.136}$$

If the tensor (dyadic) $\mathbf{E}^{\text{rotat}} \otimes (\mathbf{E}^{\text{rotat}})^*$ is regular and falls off sufficiently rapidly at large distances or if $\hat{\mathbf{n}} \cdot \mathbf{E}^{\text{rotat}} = 0$ everywhere on S, we can discard the surface integral term and find that the cycle averaged linear momentum carried by the rotational components of the fields, $\mathbf{E}^{\text{rotat}}$ and $\mathbf{B}^{\text{rotat}} \equiv \mathbf{B} = \nabla \times \mathbf{A}^{\text{rotat}}$, is

$$\langle \mathbf{p}^{\text{EM}} \rangle_t = \operatorname{Re}\left\{ -\mathrm{i}\frac{\varepsilon_0}{2\omega}\int_V \mathrm{d}^3x (\nabla \otimes \mathbf{E}^{\text{rotat}}) \cdot (\mathbf{E}^{\text{rotat}})^* \right\} \tag{2.A.137}$$

In complex tensor notation this can be written as

$$\begin{aligned} \langle \mathbf{p}^{\text{EM}} \rangle_t &= -\mathrm{i}\frac{\varepsilon_0}{2\omega}\sum_{i,j=1}^{3}\int_V \mathrm{d}^3x (E_i^{\text{rotat}})^* (\hat{\mathbf{x}}_j \partial_j E_i^{\text{rotat}}) \\ &= -\mathrm{i}\frac{\varepsilon_0}{2\omega}\sum_{i=1}^{3}\int_V \mathrm{d}^3x (E_i^{\text{rotat}})^* \nabla E_i^{\text{rotat}} \end{aligned} \tag{2.A.138}$$

Making use of the field linear momentum operator, Eq. (2.A.6), we can write the expression for the linear momentum of the electromagnetic field in terms of this operator as

$$\langle \mathbf{p}^{\text{EM}} \rangle_t = \frac{\varepsilon_0}{2\hbar\omega}\sum_{i=1}^{3}\int_V \mathrm{d}^3x (E_i^{\text{rotat}})^* \,\hat{\mathrm{p}}^{\text{EM}} E_i^{\text{rotat}} \tag{2.A.139}$$

Expressed in the vector

$$\boldsymbol{\Psi} = \sum_{i=1}^{3} \Psi_i \hat{\mathbf{x}}_i = \sqrt{\frac{\varepsilon_0}{2\hbar\omega}}\mathbf{E}^{\text{rotat}} = \sqrt{\frac{\varepsilon_0}{2\hbar\omega}}\sum_{i=1}^{3} E_i^{\text{rotat}} \hat{\mathbf{x}}_i \tag{2.A.140}$$

this becomes

$$\langle \mathbf{p}^{\text{EM}} \rangle_t = \sum_{i=1}^{3}\int_V \mathrm{d}^3x \Psi_i^* \,\hat{\mathbf{p}}^{\text{EM}} \Psi_i \tag{2.A.141}$$

or, if we assume Einstein's summation convention,

$$\langle \mathbf{p}^{\mathrm{EM}} \rangle_t = \Psi_i | \hat{\mathbf{p}}^{\mathrm{EM}} | \Psi_i \tag{2.A.142}$$

Thus we can represent the cycle (temporal) averaged linear momentum carried by a monochromatic electromagnetic field as a sum of diagonal quantal matrix elements (expectation values) where the rotational (transverse) component of the (scaled) electric field vector behaves as a kind of vector wavefunction.

Methods based on electromagnetic linear momentum are used for radio science and radar applications, and for wireless communications.

A.2.3 Angular Momentum

As is well known from classical mechanics [266], the volumetric density of the mechanical angular momentum of the charged particles around $\mathbf{x}_{\mathrm{m}} = \sum_{i=1}^{3} x_{mi} \hat{\mathbf{x}}_i$ which is a regular but otherwise arbitrary fixed point (the moment point) is

$$\begin{aligned} \mathbf{h}^{\mathrm{mech}}(t, \mathbf{x}; \mathbf{x}_{\mathrm{m}}) &= (\mathbf{x} - \mathbf{x}_{\mathrm{m}}) \times \mathbf{g}^{\mathrm{mech}}(t, \mathbf{x}) \\ &= (\mathbf{x} - \mathbf{x}_{\mathrm{m}}) \times \rho^{\mathrm{mech}}(t, \mathbf{x}) \mathbf{v}^{\mathrm{mech}}(t, \mathbf{x}) \end{aligned} \tag{2.A.143}$$

The volumetric density of the electromagnetic field angular momentum, radiated from V' is defined as [5, section 10.6, 7, 50]

$$\mathbf{h}^{\mathrm{EM}}(t, \mathbf{x}; \mathbf{x}_{\mathrm{m}}) = (\mathbf{x} - \mathbf{x}_{\mathrm{m}}) \times \mathbf{g}^{\mathrm{EM}} = \varepsilon_0 (\mathbf{x} - \mathbf{x}_{\mathrm{m}}) \times (\mathbf{E} \times \mathbf{B}) \tag{2.A.144}$$

Using Eq. (2.A.1), we can write this in complex notation as

$$\mathbf{h}^{\mathrm{EM}}(t, \mathbf{x}; \mathbf{x}_{\mathrm{m}}) = \mathrm{i} \frac{\varepsilon_0}{2c} (\mathbf{x} - \mathbf{x}_{\mathrm{m}}) \times (\mathbf{G} \times \mathbf{G}^*) \tag{2.A.145}$$

Partial differentiation of Eq. (2.A.144) with respect to time yields

$$\begin{aligned} \frac{\partial \mathbf{h}^{\mathrm{EM}}(t, \mathbf{x}; \mathbf{x}_{\mathrm{m}})}{\partial t} &= \frac{\partial}{\partial t} \left((\mathbf{x} - \mathbf{x}_{\mathrm{m}}) \times \mathbf{g}^{\mathrm{EM}} \right) \\ &= \frac{\mathrm{d}\mathbf{x}}{\mathrm{d}t} \times \mathbf{g}^{\mathrm{EM}} + (\mathbf{x} - \mathbf{x}_{\mathrm{m}}) \times \frac{\partial \mathbf{g}^{\mathrm{EM}}}{\partial t} \\ &= (\mathbf{x} - \mathbf{x}_{\mathrm{m}}) \times \frac{\partial \mathbf{g}^{\mathrm{EM}}(t, \mathbf{x})}{\partial t} \end{aligned} \tag{2.A.146}$$

where we used the fact that $\mathrm{d}\mathbf{x}/\mathrm{d}t$ is the energy density velocity $\mathbf{v}^{\mathrm{EM}}$ and, hence, is parallel to $\mathbf{g}^{\mathrm{EM}}$; see (2.A.105). We now use Eqs. (2.A.118) and (2.A.119) to immediately find that

$$\frac{\partial \mathbf{h}^{\mathrm{EM}}(\mathbf{x}_{\mathrm{m}})}{\partial t} + (\mathbf{x} - \mathbf{x}_{\mathrm{m}}) \times \mathbf{f}(t, \mathbf{x}) + (\mathbf{x} - \mathbf{x}_{\mathrm{m}}) \times \nabla \cdot \mathbf{T} = \mathbf{0} \tag{2.A.147}$$

Identifying

$$\frac{\partial \mathbf{h}^{\mathrm{mech}}(t, \mathbf{x}; \mathbf{x}_{\mathrm{m}})}{\partial t} = (\mathbf{x} - \mathbf{x}_{\mathrm{m}}) \times \mathbf{f}(t, \mathbf{x}) = \boldsymbol{\tau}(t, \mathbf{x}; \mathbf{x}_{\mathrm{m}}) \tag{2.A.148}$$

as the Lorentz torque density, and introducing the electromagnetic angular momentum flux tensor

$$\mathrm{M}(t, \mathbf{x}; \mathbf{x}_{\mathrm{m}}) = (\mathbf{x} - \mathbf{x}_{\mathrm{m}}) \times \mathrm{T}(t, \mathbf{x}) \tag{2.A.149}$$

the local (differential) form of the balance equation for angular momentum density can be written [5, 7, 52, 270] as

$$\frac{\partial \mathbf{h}^{\mathrm{mech}}(\mathbf{x}_{\mathrm{m}})}{\partial t} + \frac{\partial \mathbf{h}^{\mathrm{EM}}(\mathbf{x}_{\mathrm{m}})}{\partial t} + \nabla\cdot\mathbf{M}(\mathbf{x}_{\mathrm{m}}) = \mathbf{0} \tag{2.A.150}$$

where the symmetric pseudotensor $\mathbf{M}(\mathbf{x}_{\mathrm{m}})$ represents the electromagnetic angular momentum current density with respect to $\mathbf{x}_{\mathrm{m}}$ and we used the fact that

$$\begin{aligned}(\mathbf{x}-\mathbf{x}_{\mathrm{m}}) \times (\nabla\cdot\mathbf{T}(\mathbf{x}_{\mathrm{m}})) &= \nabla\cdot((\mathbf{x}-\mathbf{x}_{\mathrm{m}}) \times \mathrm{T}(\mathbf{x}_{\mathrm{m}})) \\ &= \nabla\cdot\mathbf{M}(\mathbf{x}_{\mathrm{m}})\end{aligned} \tag{2.A.151}$$

Hence, there is a loss of field angular momentum density to the mechanical angular momentum density, in the form of Lorentz torque density $\boldsymbol{\tau}$, showing that the angular momentum of the electromagnetic field and the rotational dynamics of the charges and currents are coupled to each other.

By using Eqs. (2.A.1), (2.A.116) and (2.A.145) we can write this local equation of continuity for angular momentum density as

$$\begin{aligned}&\frac{\partial[(\mathbf{x}-\mathbf{x}_{\mathrm{m}}) \times (\mathbf{G} \times \mathbf{G}^*)]}{\partial t} - \mathrm{i}c\nabla\cdot\left[(\mathbf{x}-\mathbf{x}_{\mathrm{m}}) \times \left(\mathbf{G}\cdot\mathbf{G}^*\mathbf{1}_3 - \frac{1}{\varepsilon_0}(\mathbf{G}\otimes\mathbf{G}^* + \mathbf{G}^*\otimes\mathbf{G})\right)\right] \\ &= -\mathrm{i}\frac{c}{\varepsilon_0}(\mathbf{x}-\mathbf{x}_{\mathrm{m}}) \times [\rho^{\mathrm{e}}(\mathbf{G}+\mathbf{G}^*) - \varepsilon_0\mathbf{j}^{\mathrm{e}} \times (\mathbf{G}-\mathbf{G}^*)]\end{aligned} \tag{2.A.152}$$

Recalling the definition (2.A.143) of the mechanical angular momentum density, and that the angular momentum density of the system is

$$\mathbf{h}^{\mathrm{sys}}(t,\mathbf{x};\mathbf{x}_{\mathrm{m}}) = \mathbf{h}^{\mathrm{EM}}(t,\mathbf{x};\mathbf{x}_{\mathrm{m}}) + \mathbf{h}^{\mathrm{mech}}(t,\mathbf{x};\mathbf{x}_{\mathrm{m}}) \tag{2.A.153}$$

the continuity Eq. (2.A.147) can be written as

$$\frac{\partial \mathbf{h}^{\mathrm{sys}}(t,\mathbf{x};\mathbf{x}_{\mathrm{m}})}{\partial t} + \nabla\cdot\mathbf{M}(\mathbf{x}_{\mathrm{m}}) = \mathbf{0} \tag{2.A.154}$$

This shows that a volume containing an arbitrary distribution of electric charge density $\rho^{\mathrm{e}}(t',\mathbf{x}')$ and electric current density $\mathbf{j}^{\mathrm{e}}(t',\mathbf{x}')$ does not only radiate linear momentum density (Poynting vector) but also angular momentum density into the surrounding free space; see also Thidé et al. [262]. However, since angular momentum describes two- or three-dimensional rotation, it cannot be sensed by a single linear dipole or similar one-dimensional, purely linear-momentum sensing device.

The total classical electromagnetic angular momentum about an arbitrary moment point $\mathbf{x}_{\mathrm{m}}$ carried by the electromagnetic field in a volume V is (see also Mandel and Wolf [50])

$$\begin{aligned}\mathbf{J}^{\mathrm{EM}}(t;\mathbf{x}_{\mathrm{m}}) &= \int_V \mathrm{d}^3x\,\mathbf{h}^{\mathrm{EM}}(t,\mathbf{x};\mathbf{x}_{\mathrm{m}}) \\ &= \varepsilon_0\int_V \mathrm{d}^3x\,\mathbf{x} \times [\mathbf{E}(t,\mathbf{x}) \times \mathbf{B}(t,\mathbf{x})] \\ &\quad - \mathbf{x}_{\mathrm{m}} \times \varepsilon_0\int_V \mathrm{d}^3x[\mathbf{E}(t,\mathbf{x}) \times \mathbf{B}(t,\mathbf{x})] \\ &= \mathbf{J}^{\mathrm{EM}}(t;\mathbf{0}) - \mathbf{x}_{\mathrm{m}} \times \mathbf{p}^{\mathrm{EM}}(t)\end{aligned} \tag{2.A.155}$$

where, in the last step, Eq. (2.A.125) was used.

The mechanical angular momentum in V is

$$\begin{aligned}\mathbf{J}^{\mathrm{mech}}(t;\mathbf{x}_{\mathrm{m}}) &= \int_V \mathrm{d}^3x\,\mathbf{h}^{\mathrm{mech}}(t,\mathbf{x};\mathbf{x}_{\mathrm{m}}) \\ &= \int_V \mathrm{d}^3x(\mathbf{x}-\mathbf{x}_{\mathrm{m}}) \times \rho^{\mathrm{mech}}(t,\mathbf{x})\mathbf{v}^{\mathrm{mech}}(t,\mathbf{x}) \\ &= \mathbf{J}^{\mathrm{mech}}(t;\mathbf{0}) - \mathbf{x}_{\mathrm{m}} \times \mathbf{p}^{\mathrm{mech}}(t)\end{aligned} \tag{2.A.156}$$

where, in the last step, Eq. (2.A.126) was used.

Hence, the total angular momentum of the system of particles and fields in V is

$$\begin{aligned}\mathbf{J}^{\text{sys}}(t;\mathbf{x}_{\text{m}}) &= \mathbf{J}^{\text{EM}}(t;\mathbf{x}_{\text{m}}) + \mathbf{J}^{\text{mech}}(t;\mathbf{x}_{\text{m}})\\ &= \mathbf{J}^{\text{sys}}(t;\mathbf{0}) - \mathbf{x}_{\text{m}} \times \mathbf{p}^{\text{sys}}(t)\end{aligned} \tag{2.A.157}$$

Clearly, if $\mathbf{p}^{\text{sys}} \neq 0$, then it is always possible to find a moment point $\mathbf{x}_{\text{m}}$ that makes the total system angular momentum vanish. If, on the other hand, $\mathbf{p}^{\text{sys}} = \mathbf{0}$, then the total system angular momentum is independent of moment point $\mathbf{x}_{\text{m}}$.

The total angular momentum of the system of charges, currents and fields fulfills the global conservation law [5, 6, 53]

$$\frac{\mathrm{d}\mathbf{J}^{\text{sys}}(\mathbf{x}_{\text{m}})}{\mathrm{d}t} + \oint_S \mathrm{d}^2x\hat{\mathbf{n}}\cdot\mathbf{M}(\mathbf{x}_{\text{m}}) = \mathbf{0} \tag{2.A.158}$$

i.e., a continuity equation without sources and sinks, showing that for a constant angular momentum flux (including zero) the system's angular momentum is a constant of motion. This is the angular momentum theorem in Maxwell's theory. It shows that not only the mechanical particles (charges) but also the electromagnetic field itself carries angular momentum (rotational momentum) that couples to the mechanical angular momentum of the matter (charged particles) and thereby to the conduction current carried by the charged particles.

Recalling that the conservation law (2.A.158) describes the angular momentum analogs of the linear momentum processes described by the conservation law (2.A.128), we conclude that the use of single monolithic antenna devices based on torque action will allow an alternative way of transferring information. This new way has the added benefit of providing higher information density due to the inherent quantized multistate property of the classical electromagnetic angular momentum.

A.2.3.1 Gauge Invariance

If the vector potential can be decomposed as in SectionA.2.2.1 and if one uses several vector analytic identities and employ partial integration, it is possible to derive an exact, manifestly gauge invariant expression for the total field angular momentum $\mathbf{J}^{\text{EM}}$ that is the sum of two pseudovectors, one independent and one dependent of the moment point $\mathbf{x}_{\text{m}}$. The result is [7]

$$\mathbf{J}^{\text{EM}}(t,\mathbf{x}_{\text{m}}) = \mathbf{\Sigma}^{\text{EM}}(t) + \mathbf{L}^{\text{EM}}(t,\mathbf{x}_{\text{m}}) \tag{2.A.159}$$

where

$$\mathbf{\Sigma}^{\text{EM}}(t) = \varepsilon_0 \int_V \mathrm{d}^3x\mathbf{E}(t,\mathbf{x}) \times \mathbf{A}^{\text{rotat}}(t,\mathbf{x}) \tag{2.A.160}$$

and

$$\begin{aligned}\mathbf{L}^{\text{EM}}(t,\mathbf{x}_{\text{m}}) = &\int_V \mathrm{d}^3x(\mathbf{x}-\mathbf{x}_{\text{m}}) \times \rho^{\text{e}}(t,\mathbf{x})\mathbf{A}^{\text{rotat}}(t,\mathbf{x})\\ &+ \varepsilon_0 \int_V \mathrm{d}^3x(\mathbf{x}-\mathbf{x}_{\text{m}}) \times \left(\left[\nabla\otimes\mathbf{A}^{\text{rotat}}(t,\mathbf{x})\right]\cdot\mathbf{E}(t,\mathbf{x})\right)\\ &- \varepsilon_0 \oint_S \mathrm{d}^2x\hat{\mathbf{n}}\cdot\left[\mathbf{E}(t,\mathbf{x})\otimes(\mathbf{x}-\mathbf{x}_{\text{m}}) \times \mathbf{A}^{\text{rotat}}(t,\mathbf{x})\right]\end{aligned} \tag{2.A.161}$$

The last integral vanishes if the integrand is regular and falls off sufficiently rapidly, becomes perpendicular to $\hat{\mathbf{n}}$, or otherwise causes the surface integral to go to zero at very large distances. However, if the vector potential $\mathbf{A}$ is not regular and/or does not fall off faster than $1/r$ as r tends to infinity, the separation of $\mathbf{J}^{\mathrm{EM}}$ into $\boldsymbol{\Sigma}^{\mathrm{EM}}$ and $\mathbf{L}^{\mathrm{EM}}$ is not gauge invariant and, hence, not unique; see Keller [2, 57, section 7], Białynicki-Birula and Białynicka-Birula [271], Leader [272], Leader and Lorcé [273].

A.2.3.2 First Quantization Formalism

Let us now assume that $\mathbf{A}$ is so well-behaved that the conditions for Helmholtz's decomposition (2.A.130) are fulfilled and, furthermore, that ρ^{e} vanishes in the region of interest. Particularizing to a single temporal Fourier component of the field $\propto \exp(-\mathrm{i}\omega t)$, we can then write $\mathbf{E} = -\partial\mathbf{A}/\partial t = \mathrm{i}\omega\mathbf{A}$ and obtain the following expressions for the cycle averages of $\boldsymbol{\Sigma}^{\mathrm{EM}}$ and $\mathbf{L}^{\mathrm{EM}}$ in complex notation:

$$\langle \boldsymbol{\Sigma}^{\mathrm{EM}} \rangle_t = -\mathrm{i}\frac{\varepsilon_0}{2\omega}\int_V \mathrm{d}^3x(\mathbf{E}^* \times \mathbf{E}) \tag{2.A.162}$$

$$\begin{aligned} \langle \mathbf{L}^{\mathrm{EM}}(\mathbf{x}_{\mathrm{m}}) \rangle_t &= -\mathrm{i}\frac{\varepsilon_0}{2\omega}\sum_{i=1}^{3}\int_V \mathrm{d}^3x E_i^*[(\mathbf{x}-\mathbf{x}_{\mathrm{m}}) \times \nabla]E_i \\ &= \frac{\varepsilon_0}{2\hbar\omega}\sum_{i=1}^{3}\int_V \mathrm{d}^3x E_i^*(-\mathrm{i}\hbar\mathbf{x} \times \nabla)E_i \\ &\quad - \frac{\varepsilon_0}{2\hbar\omega}\mathbf{x}_{\mathrm{m}} \times \sum_{i=1}^{3}\int_V \mathrm{d}^3x E_i^*(-\mathrm{i}\hbar\nabla)E_i \\ &= \langle \mathbf{L}^{\mathrm{EM}}(\mathbf{0}) \rangle_t - \mathbf{x}_{\mathrm{m}} \times \langle \mathbf{p}^{\mathrm{EM}} \rangle_t \end{aligned} \tag{2.A.163}$$

We note that if the cycle averaged electromagnetic field linear momentum $\langle \mathbf{p}^{\mathrm{EM}} \rangle_t \neq \mathbf{0}$, it is always possible to choose the moment point $\mathbf{x}_{\mathrm{m}}$ such that $\langle \mathbf{L}^{\mathrm{EM}}(\mathbf{x}_{\mathrm{m}}) \rangle_t = \mathbf{0}$. But if $\langle \mathbf{p}^{\mathrm{EM}} \rangle_t = \mathbf{0}$, i.e., if the cycle average of the linear momentum of the monochromatic field in question vanishes in a volume V, then the cycle averaged OAM, $\langle \mathbf{L}^{\mathrm{EM}} \rangle_t$, is independent of $\mathbf{x}_{\mathrm{m}}$ in V. In a beam geometry the quantity $\boldsymbol{\Sigma}^{\mathrm{EM}}$ can often serve as a good approximation identified with the SAM carried by the field and $\mathbf{L}^{\mathrm{EM}}$ with the OAM carried by the same field.

The fact that we can express the temporal averages $\langle \mathbf{p}^{\mathrm{EM}} \rangle_t$ and $\langle \mathbf{L}^{\mathrm{EM}} \rangle_t$ in terms of expectation values with the quantal operators for linear and angular momentum, $\hat{\mathbf{p}}^{\mathrm{EM}} = -\mathrm{i}\hbar\nabla$ and $\hat{\mathbf{L}}^{\mathrm{EM}} = -\mathrm{i}\hbar\mathbf{x} \times \nabla$, respectively, hints to the first-quantization character of the Maxwell–Lorentz equations.

As is well known from quantum mechanics, but true also for classical fields, the Cartesian components of $\hat{\mathbf{L}}$ expressed in cylindrical coordinates (ρ, φ, z) are

$$\hat{\mathrm{L}}_x^{\mathrm{EM}} = -\mathrm{i}\hbar\left[\sin\varphi\left(z\frac{\partial}{\partial\rho} - \rho\frac{\partial}{\partial z}\right) + \frac{z}{\rho}\cos\varphi\frac{\partial}{\partial\varphi}\right] \tag{2.A.164}$$

$$\hat{\mathrm{L}}_y^{\mathrm{EM}} = -\mathrm{i}\hbar\left[\cos\varphi\left(z\frac{\partial}{\partial\rho} - \rho\frac{\partial}{\partial z}\right) - \frac{z}{\rho}\sin\varphi\frac{\partial}{\partial\varphi}\right] \tag{2.A.165}$$

$$\hat{\mathrm{L}}_z^{\mathrm{EM}} = -\mathrm{i}\hbar\frac{\partial}{\partial\varphi} \tag{2.A.166}$$

duly recalling that $\hbar$ should be divided out; *cf.* formula (2.A.163). The magnitude squared of the OAM operator is

$$\left|\hat{\mathbf{L}}^{\mathrm{EM}}\right|^2 = \left(\hat{\mathrm{L}}_x^{\mathrm{EM}}\right)^2 + \left(\hat{\mathrm{L}}_y^{\mathrm{EM}}\right)^2 + \left(\hat{\mathrm{L}}_z^{\mathrm{EM}}\right)^2 \tag{2.A.167}$$

with eigenvalues $\hbar^2 l(l + 1)$.

Orbital angular momentum is utilized in photonics and in optical trapping [58, 274–276] but has not yet been exploited fully in wireless communications. The conservation of angular momentum is a consequence of the rotational symmetry of electrodynamics.

References

1 Wheeler, J.A. and Zurek, W.H. (eds.) (2014). *Quantum Theory and Measurement*. Princton Legacy Library. Princeton, NJ, USA: Princeton University Press.
2 Aspelmeyer, M., Gröblacher, S., Hammerer, K., and Kiesel, N. (2010). Quantum optomechanics—throwing a glance. *Journal of the Optical Society of America B* **27** (6): A189.
3 Neuenschwander, D.E. (2011). *Emmy Noether's Wonderful Theorem*. Baltimore, MD: Johns Hopkins University Press.
4 Beckwith, K. and Done, C. (2005). Extreme gravitational lensing near rotating black holes. *Monthly Notices of the Royal Astronomical Society* **359** (4): 1217–1228.
5 Su, F.S. and Mallett, R.L. (1980). The effect of the Kerr metric on the plane of polarization of an electromagnetic wave. *The Astrophysical Journal* **238**: 1111–1125.
6 Białynicki-Birula, I. (1996). Photon wave function. In: *Progress in Optics*, vol. **XXXVI** (ed. E. Wolf), 245–294. Amsterdam, Holland: Elsevier
7 Białynicki-Birula, I. and Białynicka-Birula, Z. (2011). Canonical separation of angular momentum of light into its orbital and spin parts. *Journal of Optics* **13** (6): 064014.
8 Holbourn, A.H.S. (1936). Angular momentum of circularly polarised light. *Nature* **137** (3453): 31–31.
9 Carrara, N. (1949). Torque and angular momentum of centimetre electromagnetic waves. *Nature* **164** (4177): 882–884.
10 Mohanmmadi, S.M., Daldorff, L.K.S., Forozesh, K. et al. (2010). Orbital angular momentum in radio: Measurement methods. *Radio Science* **45** (4): RS4007.
11 Francis E. Low. (1997). *Classical Field Theory. Electromagnetism and Gravitation*. New York, NY, USA: John Wiley & Sons.
12 Ibanescu, M., Fink, Y., Fan, S. et al. (2000). An all-dielectric coaxial waveguide. *Science* **289** (5478): 415–419.
13 Dashti, P.Z., Alhassen, F., and Lee, H.P. (2006). Observation of orbital angular momentum transfer between acoustic and optical vortices in optical fiber. *Physical Review Letters* **96**: 043604.
14 Kerr, R.P. (1963). Gravitational field of a spinning mass as an example of algebraically special metrics. *Physical Review Letters* **11** (5): 237–238.
15 Truesdell, C. (1968). *Essays in the History of Mechanics*. Berlin, Heidelberg, New York: Springer-Verlag.
16 Laporte, O. and Uhlenbeck, G.E. (1931). Application of spinor analysis to the Maxwell and Dirac equations. *Physical Review* **37** (11): 1380–1397.
17 EHT Collaboration et al. (2019). Physical origin of the asymmetric ring. *The Astrophysical Journal Letters* **875** (1): L5.
18 Bai, X., Jin, R., Liang, L. et al. (2015). Generation of OAM radio waves with three polarizations using circular horn antenna array. *International Journal of Antennas and Propagation* **2015**: e132549.

19 Leary, C.C. and Smith, K.H. (2014). Unified dynamics of electrons and photons via *Zitterbewegung* and spin–orbit interaction. *Physical Review A* **89** (2): 023831-1–11.

20 Messiah, A. (1970). *Quantum Mechanics*. Amsterdam, NL: North-Holland.

21 Barnett, S.M. (2014). Optical Dirac equation. *New Journal of Physics* **16** (9): 093008.

22 Tamburini, F., Sponselli, A., Thidé, B., and Mendonça, J.T. (2010). Photon orbital angular momentum and mass in a plasma vortex. *Europhysics Letters* **90** (4): 45001.

23 Białynicki-Birula, I. and Białynicka-Birula, Z. (2009). Why photons cannot be sharply localized. *Physical Review A* **79** (3): 032112.

24 Bateman, H. (1915). *The Mathematical Analysis of Electrical and Optical Wave-Motion on the Basis of Maxwell's Equations*. Cambridge, UK: Cambridge University Press.

25 Kobe, D.H. (1999). A relativistic Schrödinger-like equation for a photon and its second quantization. *Foundations of Physics* **29** (8): 1203–1231.

26 Białynicki-Birula, I. and Białynicka-Birula, Z. (2006). Beams of electromagnetic radiation carrying angular momentum: the Riemann–Silberstein vector and the classical-quantum correspondence. *Optics Communications* **264** (2): 342–351.

27 Weber, H. (1901). *Die partiellen Differential-Gleichungen der mathematischen Physik nach Riemann's Vorlesungen*. Braunschweig: Friedrich Vieweg und Sohn.

28 Anderson, P.W. (1963). Plasmons, gauge invariance, and mass. *Applied Physics Reviews* **130** (1): 439–442.

29 Hammer, C.L. and Good, R.H. (1957). Wave equation for a massless particle with arbitrary spin. *Physical Review* **108** (3): 882–886.

30 Pulecio, J.F., Warnicke, P., Pollard, S.D. et al. (2014). Coherence and modality of driven interlayer-coupled magnetic vortices. *Nature Communications* **5**: 7460.

31 Eyges, L. (1972). *The Classical Electromagnetic Field*. New York, NY, USA: Dover Publications.

32 Jerri, A.J. (1977). The Shannon sampling theorem—its various extensions and applications: a tutorial review. *Proceedings of IEEE* **65** (11): 1565–1596.

33 Lorenz, L.V. (1867). On the identity of the vibrations of light with electrical currents. *The Philosophical Magazine: Series 4* **34** (230): 287–301.

34 Lee, J.H., Foo, G., Johnson, E.G., and Swartzlander, G.A. Jr. (2006). Experimental verification of an optical vortex coronagraph. *Physical Review Letters* **97**: 053901.

35 Molina-Terriza, G., Torres, J.P., and Torner, L. (2007). Twisted photons. *Nature Physics* **3**: 305–310.

36 Philbin, T.G. (2013). Lipkin's conservation law, Noether's theorem, and the relation to optical helicity. *Physical Review Letters* **87** (4): 043843.

37 Red'kov, V.M., Tokarevskaya, N.G., and Spix, G.J. (2012). Majorana–Oppenheimer approach to Maxwell electrodynamics. Part I. Minkowski space. *Advances in Applied Clifford Algebras* **22** (4): 1129–1149.

38 Mignani, R., Recami, E., and Baldo, M. (1974). About a Dirac-like equation for the photon according to Ettore Majorana. *Lettere al Nuovo Cimento* **11** (12): 568–572.

39 Yang, H. and Casals, M. (2014). Wavefront twisting by rotating black holes: Orbital angular momentum generation and phase coherent detection. *Physical Review D* **90** (2): 023014.

40 Majorana, E. (1932). Teoria relativistica di particelle con momento intrinseco arbitrario. *Nuovo Cimento* **9**: 335–344.

41 Einstein, A. and deHass, W.J. (1915). Experimental proof of the existence of Ampère's molecular currents. *Proceedings of the Koninklijke Nederlandse Akademie van Wetenschappen* **18**: 696–708.

42 Roland, H., Good, J., and Nelson, T.J. (1971). *Proceedings of the Koninklijke Nederlandse Akademie van Wetenschappen*. London, UK: Academic Press.

43 Tamburini, F., Mari, E., Thidé, B. et al. (2011). Experimental verification of photon angular momentum and vorticity with radio techniques. *Applied Physics Letters* **99** (20): 204102.

44 Cheng, L., Hong, W., and Hao, Z.-C. (2014). Generation of electromagnetic waves with arbitrary orbital angular momentum modes. *Scientific Reports* **4**: 4814.

45 Heitler, W. (1954). *The Quantum Theory of Radiation. The International Series of Monographs on Physics*, 3e. Oxford, UK: Clarendon Press.

46 Barbuto, M., Toscano, A., and Bilotti, F. (2013). Single patch antenna generating electromagnetic field with orbital angular momentum. *2013 IEEE Antennas Propag. Soc. Int. Symp. (APSURSI)*, 1866–1867.

47 Molina-Terriza, G., Rebane, L., Torres, J.P. et al. (2007). Probing canonical geometrical objects by digital spiral imaging. *Journal of the European Optical Society* **2**: 07014.

48 Kirilyuk, A., Kimel, A.V., and Rasing, T. (2010). Ultrafast optical manipulation of magnetic order. *Review of Modern Physics* **82** (3): 2731–2784.

49 Wang, J., Yang, J.-Y., Fazal, I.M. et al. (2012). Terabit free-space data transmission employing orbital angular momentum multiplexing. *Nature Photonics* **6** (7): 488–496.

50 Siavoush M. Mohammadi, Lars K. S. Daldorff, Jan E. S. Bergman, Roger L. Karlsson, Bo Thidé, Kamyar Forozesh, and Tobia D. Carozzi. (2010). Orbital angular momentum in radio—a system study. *IEEE Transactions on Antennas and Propagation* **58**(2):565–572.

51 Andrews, D.L. (2008). *Structured Light and Its Applications: An Introduction to Phase-Structured Beams and Nanoscale Optical Forces*. Amsterdam, NL: Academic Press.

52 Shi, H. and Bhattacharya, M. (2013). Mechanical memory for photons with orbital angular momentum. *Journal of Physics B: Atomic, Molecular and Optical Physics* **46** (15): 151001.

53 Tamburini, F., Anzolin, G., Umbriaco, G. et al. (2006). Overcoming the Rayleigh criterion limit with optical vortices. *Physical Review Letters* **97** (25): 163903.

54 Coles, M.M., Williams, M.D., Saadi, K. et al. (2013). Chiral nanoemitter array: a launchpad for optical vortices. *Laser & Photonics Reviews* **7** (6): 1088–1092.

55 Anzolin, G., Tamburini, F., Bianchini, A. et al. (2008). Optical vortices with starlight. *Astronomy & Astrophysics* **488**: 1159–1165.

56 Schwinger, J. (1969). A magnetic model of matter. *Science* **165** (3895): 757–761.

57 Bo Thidé. (2011). *Electromagnetic Field Theory*. Mineola, NY, USA: Dover Publications.

58 Konstantin, Y., Bliokh, J.D., and Nori, F. (2014). Conservation of the spin and orbital angular momenta in electromagnetism. *New Journal of Physics* **16** (9): 093037.

59 Donald, Kobe, H. (1999). *Zitterbewegung* of a photon. *Physics Letters A* **253** (1–2): 7–11.

60 Thidé, B., Tamburini, F., Then, H. et al. (2014). Angular momentum radio. In: *Complex Light and Optical Forces VIII*, vol. **8999**, *Proceedings of SPIE*, 89990B–89990B–11. San Francisco, CA, USA: The International Society for Optical Engineering, SPIE.

61 Julius Adams Stratton (1941). *Electromagnetic Theory*. New York, NY, USA: McGraw-Hill Book Co.

62 Tamburini, F., Mari, E., Parisi, G. et al. (2015). Tripling the capacity of a point-to-point radio link by using electromagnetic vortices. *Radio Science* **50** (5): 1–8.

63 R. de Melo e Souza, M. V. Cougo-Pinto, C. Farina, and M. Moriconi. Multipole radiation fields from the Jefimenko equation for the magnetic field and the Panofsky–Phillips equation for the electric field. *American Journal of Physics*, **77**(1):67–72, 2009.

64 Dragoman, D. (2007). Photon states from propagating complex electromagnetic fields. *Journal of the Optical Society of America B* **24** (4): 922–927.

65 Nienhuis, G. (2008). Angular momentum and vortices in optics. In: *Structured Light and its Applications: An Introduction to Phase-Structured Beams and Nanoscale Optical Forces*, 19–61. Amsterdam, NL: Elsevier Inc.

66 Smith, B.J. and Raymer, M.G. (2007). Photon wave functions, wave-packet quantization of light, and coherence theory. *New Journal of Physics* **9**: 414.

67 Mahmouli, F.E. and Walker, S.D. (2013). 4-Gbps uncompressed video transmission over a 60-GHz orbital angular momentum wireless channel. *IEEE Wireless Communications Letters* **2** (2): 223–226.

68 Willner, A.E. (2013). Tb.s^{-1} optical communications using orbital angular momentum. *Microoptics Conference (MOC), 2013 18th*, 1–2.

69 Hui, X., Zheng, S., Hu, Y. et al. (2015). Ultralow reflectivity spiral phase plate for generation of millimeter-wave OAM beam. *IEEE Antennas and Wireless Propagation Letters* **14**: 966–969.

70 Rohrlich, F. (2007). *Classical Charged Particles*, 3e. Singapore: World Scientific.

71 Schwinger, J., DeRaad, L.L. Jr., Milton, K.A., and Tsai, W.-y. (1998). *Classical Electrodynamics*. MA, USA: Perseus Books.

72 Griffiths, D.J. (2017). *Introduction to Electrodynamics*, 4e. Cambridge, UK: Cambridge University Press.

73 Ribarič, M. and Šušteršič, L. (1990). *Conservation Laws and Open Questions of Classical Electrodynamics*. Singapore, New Jersey, London, Hong Kong: World Scientific.

74 Goldstein, H., Poole, C.P. Jr., and Safko, J.L. (2001). *Classical Mechanics*, 3e. Boston, MA, USA: Addison Wesley.

75 Deac, A.M., Fukushima, A., Kubota, H. et al. (2008). Bias-driven high-power microwave emission from MgO-based tunnel magnetoresistance devices. *Nature Physics* **4** (10): 803–809.

76 Gagnon, N., Petosa, A., and McNamara, D.A. (2010). Thin microwave quasi-transparent phase-shifting surface (PSS). *IEEE Transactions on Antennas and Propagation* **58** (4): 1193–1201, *Applied Physics Reviews*.

77 Mendonça, J.T. (2012). Twisted waves in a plasma. *Plasma Physics and Controlled Fusion* **54** (12): 124031.

78 Andrews, D.L. (2014). Photon-based and classical descriptions in nanophotonics: a review. *Journal of Nanophotonics* **8** (1): 081599.

79 Gao, X., Huang, S., Wei, Y. et al. (2014). An orbital angular momentum radio communication system optimized by intensity controlled masks effectively: theoretical design and experimental verification. *Applied Physics Letters* **105** (24): 241109.

80 Bliokh, K.Y. and Nori, F. (2015). Transverse and longitudinal angular momenta of light. *Physical Reports* **592**: 1–38.

81 Swartzlander, G.A. Jr. (2005). Broadband nulling of a vortex phase mask. *Optics Letters* **30** (21): 2876–2878.

82 Elias II., N.M. (2012). Photon orbital angular momentum and torque metrics for single telescopes and interferometers. *Astronomy & Astrophysics* **541**: A101.

83 Emile, O., Brousseau, C., Emile, J. et al. (2014). Electromagnetically induced torque on a large ring in the microwave range. *Physical Review Letters* **112** (5): 053902(4).

84 Bliokh, K.Y., Bekshaev, A.Y., and Nori, F. (2013). Dual electromagnetism: helicity, spin, momentum and angular momentum. *New Journal of Physics* **15** (3): 033026.

85 Barbuto, M., Trotta, F., Bilotti, F., and Toscano, A. (2014). Circular polarized patch antenna generating orbital angular momentum. *Progress in Electromagnetics Research* **148**: 23–30.

86 Aste, A. (2012). Complex representation theory of the electromagnetic field. *Journal of Geometry and Symmetry in Physics* **28**: 47–58.

87 Tennant, A. and Allen, B. (2012). Generation of OAM radio waves using circular time-switched array antenna. *Electronics Letters* **48** (21): 1365–1366.

88 Qiang Bai, A. Tennant, B. Allen, and M.U. Rehman. Generation of orbital angular momentum (OAM) radio beams with phased patch array. *Antennas and Propagation Conference (LAPC), 2013 Loughborough*, 410–413, 2013.

89 Kramers, H.A. (1938). *Quantentheorie des Electrons und der Strahlung*. Leipzig, Germany: Akademische Verlagsgesellschaft.

90 Silberstein, L. (1907). Elektromagnetische Grundgleichungen in bivektorieller Behandlung. *Annalen der Physik* **327** (3): 579–586.

91 Huang, L., Chen, X., Mühlenbernd, H. et al. (2012). Dispersionless phase discontinuities for controlling light propagation. *Nano Letters* **12** (11): 5750–5755.

92 Krenn, M., Fickler, R., Fink, M. et al. (2014). Communication with spatially modulated light through turbulent air across Vienna. *New Journal of Physics* **16** (11): 113028.

93 Keller, O. (2005). On the theory of spatial localization of photons. *Physical Reports* **411** (1–3): 1–232.

94 Mandel, L. and Wolf, E. (1995). *Optical Coherence and Quantum Optics*. New York, NY, USA: Cambridge University Press.

95 Quinteiro, G.F. and Berakdar, J. (2009). Electric currents induced by twisted light in quantum rings. *Optics Express* **17** (22): 20465–20475.

96 Edfors, O. and Johansson, A.J. (2012). Is orbital angular momentum (OAM) based radio communication an unexploited area? *IEEE Transactions on Antennas and Propagation* **60** (2): 1126–1131.

97 Allen, P.J. (1966). A radiation torque experiment. *American Journal of Physics* **34** (12): 1185–1192.

98 Thidé, B., Elias, N.M. II, Tamburini, F. et al. (2011). Applications of electromagnetic OAM in astrophysics and space physics studies. In: *Twisted Photons: Applications of Light With Orbital Angular Momentum*, chapter 9, (eds. J.P. Torres and L. Torner), 155–178. Weinheim, DE: Wiley-Vch Verlag.

99 van Enk, S.J. and Nienhuis, G. (1994). Commutation rules and eigenvalues of spin and orbital angular momentum of radiation fields. *Journal of Modern Optics* **41** (7): 963–977.

100 Dyson, F.J. (2016). Private communication.

101 Kurşunoğlu, B. (1961). Complex orthogonal and antiorthogonal representation of Lorentz group. *Journal of Mathematical Physics* **2** (1): 22–32.

102 Pocklington, H.C. (1897). Electrical oscillations in wires. *Mathematical Proceedings of the Cambridge Philosophical Society* **9**: 324–333.

103 Berestetskii, V.B., Lifshitz, E.M., and Pitaevskii, L.P. (1989). *Quantum Electrodynamics, Course of Theoretical Physics*, 2e, vol. **4**. Oxford, UK: Pergamon Press.

104 Rybicki, G.B. and Lightman, A.P. (2004). *Radiative Processes in Astrophysics. Physics Textbook*. Weinheim: Wiley.

105 Torres, J.P. and Torner, L. (2011). *Twisted Photons: Applications of Light With Orbital Angular Momentum*. Weinheim, DE: Wiley-Vch Verlag, John Wiley and Sons.

106 Leader, E. (2013). The angular momentum controversy: What's it all about and does it matter? *Physics of Particles and Nuclei* **44** (6): 926–929.

107 Leader, E. and Lorcé, C. (2014). The angular momentum controversy: What's it all about and does it matter? *Physical Reports* **541** (3): 163–248.

108 Alexeyev, A.N., Fadeyeva, T., Volyar, A.V., and Soskin, M.S. (1998). Optical vortices and the flow of their angular momentum in a multimode fiber. *Semiconductor Physics Quantum Electronics & Optoelectronics* **1** (1): 82–89.

109 Magueijo, J. and Smolin, L. (2002). Lorentz invariance with an invariant energy scale. *Physical Review Letters* **88** (19): 190403.

110 Allen, L., Barnett, S.M., and Padgett, M.J. (2003). *Optical Angular Momentum*. Bristol, UK: IOP.

111 Lorentz, H.A. (2004). *The Theory of Electrons*, 2e. Mineola, NY, USA: Dover Publications Inc.

112 Gosselin, P., Bérard, A., Mohrbach, H., and Ghosh, S. (2008). Berry phase effects in the dynamics of Dirac electrons in doubly special relativity framework. *Physics Letters B* **660** (3): 267–274.

113 Thidé, B., Lindberg, J., Then, H., and Tamburini, F. (2010). Linear and angular momentum of electromagnetic fields generated by an arbitrary distribution of charge and current densities at rest http://arxiv.org/abs/1001.0954.

114 Bennis, A., Niemiec, R., Brousseau, C., et al. (2013). Flat plate for OAM generation in the millimeter band. *7th Eur. Conf. Antennas Propag. (EuCAP)*, 3203–3207.

115 Bogush, A.A., Krylov, G.G., Ovsiyuk, E.M., and Red'kov, V.M. (2009). Maxwell equations in complex form of Majorana–Oppenheimer, solutions with cylindric symmetry in Riemann S_3 and Lobachevsky H_3 spaces. *Ricerche di Matematica* **59** (1): 59–96.

116 Gibson, G., Courtial, J., Vasnetsov, M. et al. (2004). Increasing the data density of free-space optical communications using orbital angular momentum. In: *Free-Space Laser Communications IV*, vol. **5550** (eds. J.C. Rickling and D.G. Voetz), 367–373. Denver, CO, USA: *Proceedings of SPIE.*

117 Žutić, I., Fabian, J., and Sarma, S.D. (2004). Spintronics: Fundamentals and applications. *Reviews of Modern Physics* **76** (2): 323–410.

118 Papanicolaou, N. and Tomaras, T.N. (1991). Dynamics of magnetic vortices. *Nuclear Physics B* **360** (2-3): 425–462.

119 Bateman, H. (1926). The radiation of energy and angular momentum. *Physical Review* **27** (5): 606–617.

120 Feng, L.-L. and Lee, W. (2001). Gravitomagnetism and the Berry phase of photon in an rotating gravitational field. *International Journal of Modern Physics D* **10** (06): 961–969.

121 Carini, P., Geng, L.-L., Li, M., and Ruffini, R. (1992). Phase evolution of the photon in Kerr spacetime. *Physical Review D* **46** (12): 5407–5410.

122 Tamburini, F. and Vicino, D. (2008). Photon wave function: A covariant formulation and equivalence with QED. *Physical Review A* **78** (5): 052116(5).

123 Fushchich, W.I. and Nikitin, A.G. (1992). The complete sets of conservation laws for the electromagnetic field. *Journal of Physics A: Mathematical and General* **25** (5): L231–L233.

124 Ibragimov, N.H. (2008). Symmetries, Lagrangian and conservation laws for the Maxwell equations. *Acta Applicandae Mathematicae* **105** (2): 157–187.

125 Dicke, R.H. (1946). The measurement of thermal radiation at microwave frequencies. *Review of Scientific Instruments* **17** (7): 268–275.

126 Yamamoto, T., Yamashita, S., and Yajima, S. (2012). Wave function of a photon and the appropriate Lagrangian. *Journal of the Physical Society of Japan* **81** (2): 024402.

127 Opare, K.A. and Kuang, Y. (2014). Performance of an ideal wireless orbital angular momentum communication system using multiple-input multiple-output techniques. *2014 International Conference on Telecommunications and Multimedia (TEMU)*, 144–149.

128 Mohr, P.J. (2010). Solutions of the Maxwell equations and photon wave functions. *Annals of Physics* **325** (3): 607–663.

129 Liu, K., Cheng, Y., Yang, Z. et al. (2015). Orbital-angular-momentum-based electromagnetic vortex imaging. *IEEE Antennas and Wireless Propagation Letters* **14**: 711–714.

130 Tamburini, F., Thidé, B., and Valle, M.D. (2020). Measurement of the spin of the M87 black hole from its observed twisted light. *Monthly Notices of the Royal Astronomical Society: Letters* **492** (1): L22–L27.

131 Good, R.H. (1957). Particle aspect of the electromagnetic field equations. *Physical Review* **105** (6): 1914–1918.

132 Soskin, M.S., Gorshkov, V.N., and Vasnetsov, M.V. (1997). Topological charge and angular momentum of light beams carrying optical vortices. *Physical Review A* **56** (5): 4064–4074.

133 Bozinovic, N., Yang, Y., Ren, Y. et al. (2013). Terabit-scale orbital angular momentum mode division multiplexing in fibers. *Science* **340** (6140): 1545–1548.

134 Beth, R.A. (1936). Mechanical detection and measurement of the angular momentum of light. *Physical Review* **50** (2): 115–125.

135 Zhang, K., Bariani, F., Dong, Y. et al. (2015). Proposal for an optomechanical microwave sensor at the subphoton level. *Physical Review Letters* **114** (11): 113601.

136 Białynicki-Birula, I. and Białynicka-Birula, Z. (2013). The role of the Riemann–Silberstein vector in classical and quantum theories of electromagnetism. *Journal of Physics A: Mathematical and General* **46** (5): 053001.

137 A. Gersten. (1999). Maxwell equations—the one-photon quantum equation. *Foundations of Physics Letters*, **12**(3):291–298.

138 Nyquist, H. (1928). Certain topics in telegraph transmission theory. *Transactions of American Institute of Electrical Engineers* **47** (2): 617–644.

139 Franke-Arnold, S., Barnett, S.M., Yao, E. et al. (2004). Uncertainty principle for angular position and angular momentum. *New Journal of Physics* **6** (103): 1–8.

140 Oppenheimer, J.R. (1931). Note on light quanta and the electromagnetic field. *Physical Review* **38** (4): 725–746.

141 Landau, L.D. and Lifshitz, E.M. (1975). *The Classical Theory of Fields, Course of Theoretical Physics*, vol. **2**. UK: Pergamon, Oxford.

142 Dussaux, A., Georges, B., Grollier, J. et al. (2010). Large microwave generation from current-driven magnetic vortex oscillators in magnetic tunnel junctions. *Nature Communications* **1**: 8.

143 Giannetto, E. (1985). A Majorana-Oppenheimer formulation of quantum electrodynamics. *Lettere al Nuovo Cimento* **44** (3): 140–144.

144 Gao, X., Huang, S., Zhou, J. et al. (2013). Generating, multiplexing/demultiplexing and receiving the orbital angular momentum of radio frequency signals using an optical true time delay unit. *Journal of Optics* **15** (10): 105401.

145 Harwit, M. (2003). Photon orbital angular momentum in astrophysics. *The Astrophysical Journal* **597** (2): 1266–1270.

146 Vogel, A., Martens, M., Weigand, M., and Meier, G. (2011). Signal transfer in a chain of stray-field coupled ferromagnetic squares. *Applied Physics Letters* **99** (4): 042506.

147 Swartzlander, G.A. Jr. (2001). Peering into darkness with a vortex filter. *Optics Letters* **26** (8): 497–499.

148 Murakami, S., Nagaosa, N., and Zhang, S.-C. (2003). Dissipationless quantum spin current at room temperature. *Science* **301** (5638): 1348–1351.

149 Roichman, Y., Sun, B., Roichman, Y. et al. (2008). Optical forces arising from phase gradients. *Physical Review Letters* **100** (1): 013602(4).

150 Lavery, M.P.J., Berkhout, G.C.G., Courtial, J., and Padgett, M.J. (2011). Measurement of the light orbital angular momentum spectrum using an optical geometric transformation. *Journal of Optics* **13** (6): 064006.

151 Dehnen, H. (1973). Gravitational Faraday-effect. *International Journal of Theoretical Physics* **7** (6): 467–474.

152 Yavorsky, M.A. (2013). All-fiber polarization-dependent optical vortex beams generation via flexural acoustic wave. *Optics Letters* **38** (16): 3151–3153.

153 Archibald, W.J. (1955). Field equations from particle equations. *Canad. J. Phys.* **33** (9): 565–572.

154 EHT Collaboration et al. (2019). Array and instrumentation. *The Astrophysical Journal Letters* **875** (1): L2(28).

155 Zhengyi, L., Ohashi, Y., and Kasai, K. (2013). Quasi-LoS MIMO wireless communication with twisted radio wave. *Microwave Conference (EuMC), 2013 European*, 577–580.

156 Carusotto, S., Fornaca, G., and Polacco, E. (1968). Radiation beats and rotating systems. *Nuovo Cimento* **53** (1): 87–97.

157 Juan, P., Treviño, O.L.-C., and Chávez-Cerda, S. (2013). Segmented vortex telescope and its tolerance to diffraction effects and primary aberrations. *Optics Express* **52** (8): 081605–081605.

158 Allen, L., Padgett, M.J., and Babiker, M. (1999). The orbital angular momentum of light. In: *Progress in Optics*, vol. **XXXIX** (ed. E. Wolf), 291–372. Amsterdam, Holland: Elsevier.

159 Smoot, G.F., Bennett, C.L., Kogut, A. et al. (1992). Structure in the COBE differential microwave radiometer first-year maps. *The Astrophysical Journal Letters* **396**: L1–L5.

160 Bohm, D. (1989). *Quantum Theory*. New York: Dover Publications Inc. ISBN 9780486659695.

161 Soper, D.E. (1976). *Classical Field Theory*. New York, NY, USA: John Wiley & Sons, Inc. ISBN 0-471-81368-0.

162 Brunet, C., Vaity, P., Messaddeq, Y. et al. (2014). Design, fabrication and validation of an OAM fiber supporting 36 states. *Optics Express* **22** (21): 26117–26127.

163 Mochizuki, M. (2012). Spin-wave modes and their intense excitation effects in skyrmion crystals. *Physical Review Letters* **108** (1): 017601.

164 Jackson, J.D. and Okun, L.B. (2001). Historical roots of gauge invariance. *Reviews of Modern Physics* **73** (3): 663–680.

165 Wilkinson, D. (1991). The Dicke radiometer and cosmic radiation. *1991 IEEE MTT-S International Microwave Symposium Digest*, Boston, MA, USA, 399–402. IEEE.

166 Thidé, B., Then, H., Sjöholm, J. et al. (2007). Utilization of photon orbital angular momentum in the low-frequency radio domain. *Physical Review Letters* **99** (8): 087701.

167 Cohen-Tannoudji, C., Dupont-Roc, J., and Grynberg, G. (1997). *Photons and Atoms. Introduction to Quantum Electrodynamics*. New York, NY, USA: Wiley & Sons ISBN 0-471-18433-0. Wiley Professional Paperback Edition.

168 Martin, P., Lavery, J., Robertson, D.J. et al. (2013). Efficient measurement of an optical orbital-angular-momentum spectrum comprising more than 50 states. *New Journal of Physics* **15** (1): 013024.

169 Morecroft, J.H., Pinto, A., and Curry, W.A. (2015). *Principles of Radio Communication*. New York, NY, USA: Andesite Press.

170 Wang, J., Yang, J.Y., Fazal, I.M., et al. (2011). Demonstration of 12.8-bit.s^{-1}.hz spectral efficiency using 16-QAM signals over multiple orbital-angular-momentum modes. *Optical Communication (ECOC), 2011 37th European Conference and Exhibition on*, 1–3.

171 Leonov, A.O. and Mostovoy, M. (2015). Multiply periodic states and isolated skyrmions in an anisotropic frustrated magnet. *Nature Communications* **6**: 8275.

172 Gregorius, C., Berkhout, G., Lavery, M.P.J. et al. (2010). Efficient sorting of orbital angular momentum states of light. *Physical Review Letters* **105** (15): 153601(4).

173 Elias, N.M. II (2008). Photon orbital angular momentum in astronomy. *Astronomy & Astrophysics* **492**: 883–922.

174 Litchinitser, N.M. (2012). Structured light meets structured matter. *Science* **337** (6098): 1054–1055.

175 Weber, H. (2013). Angular momentum of low frequency electromagnetic fields. *Optik – International Journal for Light and Electron Optics* **124** (20): 4313–4314.

176 Verma, S.S. (2011). Atomtronics trying to replace electronics. *Electronics For You* **43** (9): 60–64.

177 Maekawa, S., Adachi, H., Uchida, K.-i. et al. (2013). Spin current: Experimental and theoretical aspects. *Journal of the Physical Society of Japan* **82** (10): 102002.

178 Griffiths, D.J. (2012). Resource Letter EM-1: Electromagnetic momentum. *American Journal of Physics* **80** (7): 7–18.

179 Brown, G.H. (1936). The 'Turnstile' antenna. *Electronics* **9**: 14–17.

180 A. I. Sadovskii. Ponderomotorniya djstviya elektromagnitnykh' i svtovykh' voln' na kristally. *Uchen. Zapis. Imp. Yurev. Univ.*, **52**(1):1–126, 1899. (English translation: Ponderomotive action of electromagnetic and light waves on crystals).

181 Dressel, J., Bliokh, K.Y., and Nori, F. (2015). Spacetime algebra as a powerful tool for electromagnetism. *Physical Review* **589**: 1–71.

182 Wolf, S.A., Awschalom, D.D., Buhrman, R.A. et al. (2001). Spintronics: A spin-based electronics vision for the future. *Science* **294** (5546): 1488–1495.

183 Heras, J.A. (1994). Radiation fields of a dipole in arbitrary motion. *American Journal of Physics* **62** (12): 1109–1115.

184 Brataas, A. and Hals, K.M.D. (2014). Spin-orbit torques in action. *Nature Nanotechnology* **9** (2): 86–88.

185 Nieminen, J.V., Vänskä, O., Tittonen, I. et al. (2015). Accessing orbital angular momentum of quantum-ring excitons via directional semiconductor luminescence. *New Journal of Physics* **17** (3): 033046.

186 Padgett, M. and Bowman, R. (2011). Tweezers with a twist. *Nature Physics* **5** (6): 343–348.

187 Williams, M.D., Coles, M.M., Bradshaw, D.S., and Andrews, D.L. (2014). Direct generation of optical vortices. *Physical Review A* **89** (3): 033837.

188 Jefimenko, O.D. (1966). *Electricity and Magnetism. An Introduction to the Theory of Electric and Magnetic Fields*. New York, NY, USA: Appleton-Century-Crofts.

189 Heald, M.A. and Marion, J.B. (1995). *Classical Electromagnetic Radiation*, 3e. Fort Worth, TX, USA: Saunders College Publishing ISBN 0-03-097277-9.

190 Mendonça, J.T., Thidé, B., and Then, H. (2009). Stimulated Raman and Brillouin backscattering of collimated beams carrying orbital angular momentum. *Physical Review Letters* **102** (18): 185005(4).

191 Kuschel, T. and Reiss, G. (2015). Spin orbitronics: Charges ride the spin wave. *Nature Nanotechnology* **10** (1): 22–24.

192 Noether, E. (1918). Invariante Variationsprobleme. *Nachrichten von der Gesellschaft der Wissenschaften zu Göttingen* **1** (3): 235–257. English transl.: *Invariant variation problems, Transport Theory and Statistical Physics*, **1**, 186–207 (1971).

193 Silberstein, L. (1914). *The Theory of Relativity*. London: MacMillan.

194 Ramanathan, A., Wright, K.C., Muniz, S.R. et al. (2011). Superflow in a toroidal Bose–Einstein condensate: An atom circuit with a tunable weak link. *Physical Review Letters* **106** (13): 130401.

195 Aspelmeyer, M., Kippenberg, T.J., and Marquardt, F. (2014). Cavity optomechanics. *Reviews of Modern Physics* **86** (4): 1391–1452.

196 Yu, N. and Capasso, F. (2014). Flat optics with designer metasurfaces. *Nature Materials* **13** (2): 139–150.

197 Then, H., Thidé, B., Mendonça, J.T., et al. (2008). Detecting orbital angular momentum in radio signals. http://arxiv.org/abs/0805.2735.

198 Shannon, C.E. (1949). Communication in the presence of noise. *Proceedings of IRE* **37** (1): 10–21.

199 Balanis, C.A. (2005). *Antenna Theory*, 3e. New York: John Wiley and Sons.

200 Mendonça, J.T., Thidé, B., Bergman, J.E.S., et al. (2008). Photon orbital angular momentum in a plasma vortex. http://arxiv.org/abs/0804.3221.

201 Tamburini, F., Thidé, B., Mari, E. et al. (2012). Reply to comment on 'Encoding many channels on the same frequency through radio vorticity: First experimental test'. *New Journal of Physics* **14** (11): 118002.

202 Panofsky, W.K.H. and Phillips, M. (1962). *Classical Electricity and Magnetism*, 2e. Reading, MA, USA: Addison-Wesley Publishing Company ISBN 0-201-05702-6.

203 Belinfante, F.J. (1940). On the current and the density of the electric charge, the energy, the linear momentum and the angular momentum of arbitrary fields. *Physica* **7** (5): 449–474.

204 Jackson, J.D. (1998). *Classical Electrodynamics*, thirde. New York, NY, USA: Wiley & Sons ISBN 0-471-30932-X.

205 Thidé, B. (2007). Nonlinear physics of the ionosphere and LOIS/LOFAR. *Plasma Physics and Controlled Fusion* **49** (12B): B103–B107.

206 Sipe, J.E. (1995). Photon wave functions. *Physical Review A* **52** (3): 1875–1883.

207 He, H., Friese, M.E.J., Heckenberg, N.R., and Rubinsztein-Dunlop, H. (1995). Direct observation of transfer of angular momentum to absorptive particles from a laser beam with a phase singularity. *Physical Review Letters* **75** (5): 826–829.

208 Then, H. and Thidé, B. (2008). Mechanical properties of the radio frequency field emitted by an antenna array http://arxiv.org/abs/0803.0200.

209 Helmerson, K., Andersen, M.F., Cladé, P. et al. (2009). Vortices and persistent currents: Rotating a Bose-Einstein condensate using photons with orbital angular momentum. *Topologica* **2** (1): 002–1–002–12.

210 Majorana, E. (2009). Research notes on theoretical physics. Unpublished, deposited at the 'Domus Galileana,' Pisa, quaderno 2, p. 101/1; 3, p. 11, 160; 15, p. 16; 17, p. 83, 159.

211 Chang, S. and Lee, S.S. (1985). Optical torque exerted on a homogeneous sphere levitated in the circularly polarized fundamental-mode laser beam. *Journal of the Optical Society of America B* **2** (11): 1853–1860.

212 Vul'fson, K.S. (1987). Angular momentum of electromagnetic waves. *Soviet Physics Uspekhi* **30** (8): 667–674.

213 Kristensen, M., Beijersbergen, M.W., and Woerdman, J.P. (1994). Angular momentum and spin-orbit coupling for microwave photons. *Optics Communications* **104** (4-6): 229–233.

214 Friese, M.E.J., Enger, J., Rubinsztein-Dunlop, H., and Heckenberg, N.R. (1996). Optical angular-momentum transfer to trapped absorbing particles. *Physical Review A* **54** (2): 1593–1596.

215 Higgs, P.W. (1964). Broken symmetries and the masses of gauge bosons. *Physical Review Letters* **13** (16): 508–509.

216 Zagrodzinski, J. (2002). Vortices in different branches of physics. *Physica C* **369** (1–4): 45–54.

217 Schott, G.A. (1912). *Electromagnetic Radiation and the Mechanical Reactions, Arising From It, Being an Adams Prize Essay in the University of Cambridge*. Cambridge, UK: Cambridge University Press.

218 Ferrarese, L. and Merritt, D. (2002). Supermassive black holes. *arXiv:astro-ph/0206222*.

219 Ünal, N. (1997). A simple model of the classical *Zitterbewegung*: Photon wave function. *Foundations of Physics* **27** (5): 731–746.

220 Shi, H. and Bhattacharya, M. (2013). Coupling a small torsional oscillator to large optical angular momentum. *Journal of Modern Optics* **60** (5): 382–386.

221 Salem, M.A. and Caloz, C. (2013). Precision orbital angular momentum (OAM) multiplexing communication using a metasurface. *2013 7th International Congress on Advanced Electromagnetic Materials in Microwaves and Optics (METAMATERIALS)*, 94–96.

222 EHT Collaboration, Akiyama, K., Alberdi, A. et al. (2019). Imaging the central supermassive black hole. *The Astrophysical Journal Letters* **875** (1): L4(52).

223 EHT Collaboration, Akiyama, K., Alberdi, A. et al. (2019). The shadow and mass of the central black hole. *The Astrophysical Journal Letters* **875** (1): L6(44).

224 Sinova, J., Culcer, D., Niu, Q. et al. (2004). Universal intrinsic spin Hall effect. *Physical Review Letters* **92** (12): 126603.

225 Yan Yan, Guodong Xie, Hao Huang, M.J. Lavery, N. Ahemd, Changjing Bao, Yongxiong Ren, A. F. Molisch, M. Tur, Padgett, M., and Willner, A.E. (2014). Demonstration of 8-mode 32-gbit.s^{-1} millimeter-wave free-space communication link using 4 orbital-angular-momentum modes on 2 polarizations. *2014 IEEE International Conference on Communications (ICC)*, 4850–4855.

226 Khan, S.A. (2005). An exact matrix representation of Maxwell's equations. *Physica Scripta* **71**: 440–442.

227 EHT Collaboration et al. (2019). EHT data products. Online image data https://eventhorizontelescope.org/for-astronomers/data.

228 Tamburini, F., Mari, E., Sponselli, A. et al. (2012). Encoding many channels on the same frequency through radio vorticity: First experimental test. *New Journal of Physics* **14** (3): 03301.

229 Yan, Y., Xie, G., Lavery, M.P.J. et al. (2014). High-capacity millimetre-wave communications with orbital angular momentum multiplexing. *Nature Communications* **5**: 4876.

230 Mondal, P.K., Deb, B., and Majumder, S. (2014). Angular momentum transfer in interaction of Laguerre–Gaussian beams with atoms and molecules. *Physical Review A* **89** (6): 063418.

231 Sjöholm, J. and Palmer, K. (2009). Angular momentum of electromagnetic radiation. fundamental physics applied to the radio domain for innovative studies of space and development of new concepts in wireless communications. Master's thesis. Uppsala University, Uppsala, SE.

232 Götte, J.B., O'Holleran, K., Preece, D. et al. (2008). Light beams with fractional orbital angular momentum and their vortex structure. *Optics Express* **16** (2): 993–1006.

233 Tokatly, I.V. (2010). Orbital momentum Hall effect in *p*-doped graphane. *Physical Review B* **82** (16): 161404.

234 Poincaré, M. (1906). Sur la dynamique de l'électron. *Rendiconti del Circolo Matematico di Palermo* **21** (1): 129–175.

235 Andrews, R.W., Peterson, R.W., Purdy, T.P. et al. (2014). Bidirectional and efficient conversion between microwave and optical light. *Nature Physics* **10** (4): 321–326.

236 Williams, M.D., Coles, M.M., Saadi, K. et al. (2013). Optical vortex generation from molecular chromophore arrays. *Physical Review Letters* **111**:(15) 153603.

237 Opare, K.A., Kuang, Y., Kponyo, J.J., et al. (2015). The degrees of freedom in wireless line-of-sight OAM multiplexing systems using a circular array of receiving antennas. *2015 Fifth International Conference on Advanced Computing Communication Technologies (ACCT)*, 608–613.

238 Andrei Bernevig, B., Hughes, T.L., and Zhang, S.-C. (2005). Orbitronics: The intrinsic orbital current in *p*-doped silicon. *Physical Review Letters* **95** (6): 066601.

239 Deng, C., Chen, W., Zhang, Z. et al. (2013). Generation of OAM radio waves using circular Vivaldi antenna array. *International Journal of Antennas and Propagation* **2013**: 847859.

240 Tamburini, F., Thidé, B., Molina-Terriza, G., and Anzolin, G. (2011). Twisting of light around rotating black holes. *Nature Physics* **7** (3): 195–197.

241 Arthur Stanley Eddington (2010). *The Nature of the Physical World*. Whitefish, MT, USA: Kessinger Publishing ISBN 1162789239.

242 EHT Collaboration et al. (2019). The shadow of the supermassive black hole. *The Astrophysical Journal Letters* **875** (1): L1(17).

243 Xie, G., Li, L., Ren, Y., et al. (2014). Performance metrics and design parameters for an FSO communications link based on multiplexing of multiple orbital-angular-momentum beams. *Globecom Workshops (GC Wkshps), 2014*, 481–486.

244 Ren, Y., Li, L., Xie, G., et al. (2014). Experimental demonstration of 16 gbit/s millimeter-wave communications using MIMO processing of 2 OAM modes on each of two transmitter/receiver antenna apertures. *2014 IEEE Global Communications Conference (GLOBECOM)*, 3821–3826.

245 Torner, L., Torres, J.P., and Carrasco, S. (2005). Digital spiral imaging. *Optics Express* **13** (3): 873–881.

246 Maxwell, J.C. (1998). *A Treatise on Electricity and Magnetism*, vol. **2**. Oxford, UK: Oxford University Press (first published 1873).

247 Bagci, T., Simonsen, A., Schmid, S. et al. (2014). Optical detection of radio waves through a nanomechanical transducer. *Nature* **507** (7490): 81–85.

248 Yao, A.M. and Padgett, M.J. (2011). Orbital angular momentum: Origins, behavior and applications. *Advances in Optics and Photonics* **3** (2): 161–204.

249 EHT Collaboration et al. (2019). Data processing and calibration. *The Astrophysical Journal Letters* **875** (1): L3(32).

250 Silberstein, L. (1907). Nachtrag zur Abhandlung über elektromagnetische Grundgleichungen in bivektorieller Behandlung. *Annalen der Physik* **329** (14): 783–784.

251 Bogolyubov, N.N. and Shirkov, D.V. (1959). *Introduction to the Theory of Quantized Fields*. New York: Interscience.

252 Bai, Q., Tennant, A., and Allen, B. (2014). Experimental circular phased array for generating OAM radio beams. *Electronics Letters* **50** (20): 1414–1415.

253 Rosenfeld, L. (1965). *Theory of Electrons*. New York: Dover Publications.

254 Foo, G., Palacios, D.M., Grover, A., and Swartzlander, J. (2005). Optical vortex coronagraph. *Optics Letters* **30** (24): 3308–3310.

255 Moses, H.E. (1959). Solution of Maxwell's equations in terms of a spinor notation: The direct and inverse problem. *Physical Review* **113** (6): 1670–1679.

256 Siegel, D.M. (2003). *Innovation in Maxwell's Electromagnetic Theory: Molecular Vortices, Displacement Current, and Light*. Cambridge, UK: Cambridge University Press ISBN 0521533295.

257 Franke-Arnold, S., Allen, L., and Padgett, M. (2008). Advances in optical angular momentum. *Laser & Photonics Reviews* **2** (4): 299–313.

258 Beth, R.A. (1935). Direct detection of the angular momentum of light. *Physical Review* **48** (5): 471.

259 Dyson, F.J. (1999). Why is Maxwell's Theory so hard to understand? In: *James Clerk Maxwell Commemorative Booklet*. Edinburgh, Scotland, UK: James Clerk Maxwell Foundation.

260 Köksal, K. and Berakdar, J. (2012). Charge-current generation in atomic systems induced by optical vortices. *Physical Review A* **86** (6): 063812.

261 Bochmann, J., Vainsencher, A., Awschalom, D.D., and Cleland, A.N. (2013). Nanomechanical coupling between microwave and optical photons. *Nature Physics* **9** (11): 712–716.

262 Barut, A.O. (1980). *Electrodynamics and Classical Theory of Fields and Particles*. New York, NY, USA: Dover Publications ISBN 0-486-64038-8.

263 Białynicki-Birula, I. (1994). On the wave function of the photon. *Acta Physica Polonica A* **86**: 97–116.

264 Schemmel, P., Pisano, G., and Maffei, B. (2014). Modular spiral phase plate design for orbital angular momentum generation at millimetre wavelengths. *Optics Express* **22** (12): 14712–14726.

265 J. H. Poynting. The wave motion of a revolving shaft, and a suggestion as to the angular momentum in a beam of circularly polarised light. *Proceedings of the Royal Society A* **82**(557):560–567, 1909.

266 Dieter Zeh, H. (2001). *The Physical Basis of The Direction of Time*ISBN 3-540-42081-9, 4e. Springer.

267 Wang, Z.-Y., Xiong, C.-D., and Qiu, Q. (2009). Photon wave function and *Zitterbewegung*. *Physical Review A* **80** (3): 032118.

268 Someda, C.G. (2006). *Electromagnetic Waves*, 2e. Boca Raton, FL, USA: CRC ISBN 0849395895.

269 Esposito, S. (1998). Covariant Majorana formulation of electrodynamics. *Foundations of Physics* **28** (2): 231–244.

270 Abraham, M. (1914). Der Drehimpuls des Lichtes. *Physik. Zeitschr.* **XV**: 914–918.

271 Willner, A.E., Huang, H., Yan, Y. et al. (2015). Optical communications using orbital angular momentum beams. *Advances in Optics and Photonics* **7** (1): 66–106.

272 Vannucchi, F.S., Vasconcellos, Á.R., and Luzzi, R. (2013). Emission of monochromatic microwave radiation from a nonequilibrium condensation of excited magnons. *Applied Physics Letters* **103** (7): 072401.

273 Yan, P., Kamra, A., Cao, Y. et al. (2013). Angular and linear momentum of excited ferromagnets. *Physical Review B* **88** (14): 144413.

274 Dirac, P.A.M. (1931). Quantised singularities in the electromagnetic field. *Proceedings of the Royal Society A* **133** (821): 60–72.

275 F. Tamburini and B. Thidé. (2011). Storming Majorana's Tower with OAM states of light in a plasma. *Europhysics Letters* **96**(6):64005.

276 Cano, E., Allen, B., Bai, Q., and Tennant, A. (2014). Generation and detection of OAM signals for radio communications. *Antennas and Propagation Conference (LAPC), 2014 Loughborough*, 637–640.

Part II

Physical Wave Phenomena of Electromagnetic Vortices

3

Generation of Microwave Vortex Beams Using Metasurfaces

Jia Yuan Yin[1] *and Tie Jun Cui*[2]

[1] *School of Physics and Optoelectronic Engineering, Xidian University, Xi'an, P.R. China*
[2] *State Key Laboratory of Millimeter-Waves, School of Information Science and Engineering, Southeast University, Nanjing, P.R. China*

3.1 Introduction

According to the classical theory of electromagnetic (EM) fields, the EM waves consist of both linear momentum and angular momentum, such as spin angular momentum (SAM) and orbital angular momentum (OAM) [1]. SAM is mainly used to describe the polarization states of the EM waves, which has only two states, while OAM manifests orbital rotation of photons and hence can be in many states. Each photon has an OAM of $l\hbar$, where l denotes the OAM index and can be any integers corresponding to different states. Further, the EM waves with different OAM states are orthogonal to each other; thus, OAM provides a new degree of freedom to EM waves. It can be easily concluded that the OAM-carrying EM waves can enhance the communication capacity without increasing the bandwidth. Therefore, despite the fact that photons carrying OAMs were firstly discovered in the optical frequencies [2], there has been increasing interest to employ OAMs as a new method to address the problems in the wireless communications.

The first radio OAM mode (or twisted wave) simulation was performed in 2007, which provided a theoretical basis for the OAM-based wireless communication [3]. In 2012, the first experiment on wireless radio transmission applying the vortex beam was conducted [4]. In addition, it was demonstrated that the EM waves with different OAM modes can work independently of each other in the wireless communications. That is to say, if the mode of the transmitting OAM wave is designated as l, the spiral phase plates that generate $-l$ mode OAM should be employed to enable the receiving antenna to receive the Gaussian beam [4].

To date, many methods are available to generate the vortex beams carrying OAM in the microwave frequencies. Compared with plane waves, the most noteworthy characteristic of the vortex beams is the helix constant phase front. To realize the phases required by the vortex beams, the most popular method is to use spiral phase plates because of their simple structure [5–7], which are widely used in optics, followed by the microwave bands. Another popular approach is to use antenna arrays [8–12]. When using an antenna array to generate the vortex beams, the design of the phase-shifting network is crucial. Further, single cavities have been of interest [13–15], derived from which half-mode substrate-integrated waveguide antenna was demonstrated for OAM generation [16].

Electromagnetic Vortices: Wave Phenomena and Engineering Applications, First Edition.
Edited by Zhi Hao Jiang and Douglas H. Werner.

Recent progress in metasurface offers new techniques for EM wave manipulations. By properly arranging subwavelength unit cells of metasurface, the phase distribution of EM waves can be redistributed, facilitating the realization of vortex beams. Although various designs of different types for different applications are available, the most important common rule for OAM generation is the introduction of the azimuthal phase term $e^{il\varphi}$ to the EM waves. Considering that the phase design is the key factor, some types of phases involved in metasurface design should be introduced first. The first type of phase is called as resonance phase. Metallic nanoantennas are used at the earliest to realize the phase discontinuity. Light can be coupled to the surface wave propagating through the antenna surface. Because of the resonance of the localized plasmon, these surface waves will be resonant with the free electrons along the metallic surface. This phenomenon is similar to that in an oscillator, such as internal damping and radiation damping. The phase response can reach 2π by combining multiple units. In 2001, Capasso et al. proposed a V-shaped antenna and experimentally verified the 2π phase change [17]. Following this, such a phenomenon is widely used for various purposes such as anomalous refraction, polarization conversion, and optical focusing [18]. The second type is called as dynamic phase, in which the phase control is approached by the optical path difference during the propagation of EM waves [19, 20]. This type of phase is insensitive to the polarization of chiral. The fan changed the distance between nanoslits and the width of slits to control the optical path difference, resulting in beam focusing [21, 22]. The third type is the geometric phase, in which additional phase is introduced by controlling the local polarization changes, which can be realized by rotating the angle of the unit cells. The advantages of this type of phase are the independency of size and resonance of structure and material dispersion. Luo et al. proposed a three-dimensional (3D) color holographic through geometric-phase metasurface [23]. The fourth type is a combination of the dynamic phase and geometric phase. This combination offers a new degree of freedom to control the phase, facilitating continuous manipulations of the polarization and phase [24, 25]. Capasso et al. demonstrated that the polarization can be conversed by tuning the radius and rotation angle of TiO_2 [24]. Moreover, many methods are available to manipulate the phases of EM waves by using different types of metasurfaces.

3.2 Metasurfaces for Vortex-beam Generation

In this section, different types of metasurfaces for vortex-beam generation are reviewed in detail. In addition, some typical examples are discussed for better understanding. The metasurfaces are mainly categorized into five types: the reflective metasurface, the transmission metasurface, the holographic metasurface, the metasurface with an extra feeding network, and the metasurface with the lens. Each type includes different phase designs, as described in the introduction. The same phase designs will be put together to the maximum extent possible in the details below. Finally, a novel method to generate vortex beams using spoof surface plasmon polaritons (SPPs) is introduced because the spoof SPP waveguide can be considered as one-dimensional (1D) metasurface. It is worth noting that this classification method is not rigorous, and we provide only an approximate classification to introduce the characteristics and principles. Each method has its unique characteristics and advantages.

As mentioned above, four types of phases are involved in the design procedure of metasurfaces. However, mainly two methods exist for designing the phase required by OAM in the microwave bands. One method is to vary the structure parameters, which corresponds to the resonance phase. In this case, the desired phase distribution is achieved under the illumination of linearly polarized waves. The other method involves rotating the orientation of the unit cells, corresponding to the geometric phase. This type of metasurface is otherwise called as the Pancharatnam–Berry (PB) metasurface, in which the achromatic phase is realized under the circularly polarized excitation. The geometric-phase metasurfaces exhibit high flexibility in manipulating OAMs because they enable the coupling and interchange between SAM and OAM. If the resonance phase is combined with the geometric phase, the total phase shift originates from the two parts, and hence arbitrary OAM modes can be generated. Each phase has its unique advantages, and the required phase must be chosen based on the actual demands.

Consider a transmission metasurface as an example. Suppose that the propagating direction of the plane wave is along the z-direction, and E_{i} signifies the incident wave, which can be decomposed into two orthogonal components, namely, E_{ix} and E_{iy}. When the wave passes through the metasurface, the transmission coefficients can be expressed as:

$$\begin{bmatrix} E_{tx} \\ E_{ty} \end{bmatrix} = \begin{bmatrix} T_{xx} & T_{xy} \\ T_{yx} & T_{yy} \end{bmatrix} \begin{bmatrix} E_{ix} \\ E_{iy} \end{bmatrix}, \tag{3.1}$$

where T_{xx} and T_{yy} denote the copolarized transmission coefficients under the x- and y-polarized normally incident waves, and T_{xy} and T_{yx} are the cross-polarized transmission coefficients. For a reflection unit cell under the Cartesian coordinate system (see Figure 3.1), which is rotated by an angle of θ,if $T_{yy} = -T_{xx}$ and $T_{yx} = T_{xy} = 0$, the transmission matrix can be derived as [27]:

$$T_l(\theta) = T_{xx} \begin{bmatrix} \cos(2\theta) & \sin(2\theta) \\ \sin(2\theta) & -\cos(2\theta) \end{bmatrix} \tag{3.2}$$

$$T_c(\theta) = T_{xx} \begin{bmatrix} 0 & e^{-2i\theta} \\ e^{2i\theta} & 0 \end{bmatrix}, \tag{3.3}$$

where l and c represent the circumstances under linear and circular polarized waves, respectively.

It is worth noting that, for the circularly polarized waves, an additional phase factor $e^{\pm 2i\theta}$ is introduced, whose form is similar to the azimuthal phase dependence $e^{\pm il\varphi}$. Thus, with the nonzero off-diagonal items in $T_{\mathrm{c}}(\theta)$, the rotated unit cells switch the polarization state from the left-handed (right-handed) circular polarization to the right-handed (left-handed) circular polarization. If we set $\theta = \frac{l\varphi}{2}$, a vortex beam with the OAM mode equal to l appears in the transmitted wave. The value of $\frac{\theta}{2}$ is known as the topology charge q of the metasurface, which determines the index of the OAM mode. However, according to Eq. (3.2), when a linearly polarized wave is used, such a phase factor does not appear. Further, the structure parameters should be varied to obtain the expected phase. The analysis procedure of the reflective metasurface is similar to that of the transmission metasurface, and thus we do not discuss it in detail here.

3.2.1 Reflective Metasurfaces for Vortex-beam Generation

Reflective metasurfaces for the vortex-beam generation have been studied extensively because of their suitability to obtain the expected phases. Reflective OAM generators usually include not only a metasurface but also a feeding antenna. Because the distance between the feed and metasurface must be short for the actual measurement, the incidence cannot be regarded as plane wave. The phase difference caused by different distances between the unit cells and the feed point should be considered during the metasurface design. Consider a reflective metasurface (see Figure 3.1a), which consists of $M \times N$ unit cells and is illuminated by a source located at the position $\overrightarrow{r_f}$. Then, the reflection of EM waves by the metasurface in an arbitrary direction $\hat{u}$ can be expressed as [28]:

$$E(\hat{u}) = \sum_{m=1}^{M} \sum_{n-1}^{N} F(\overrightarrow{r_{mn}} \cdot \overrightarrow{r_f}) A(\overrightarrow{r_{mn}} \cdot \hat{u_0}) A(\hat{u_0} \cdot \hat{u}) \cdot e^{jk_0 \left[\left| \overrightarrow{r_{mn}} - \overrightarrow{r_f} \right| + \overrightarrow{r_{mn}} \cdot \hat{u} \right] + j\varphi_{mn}^{c}}, \tag{3.4}$$

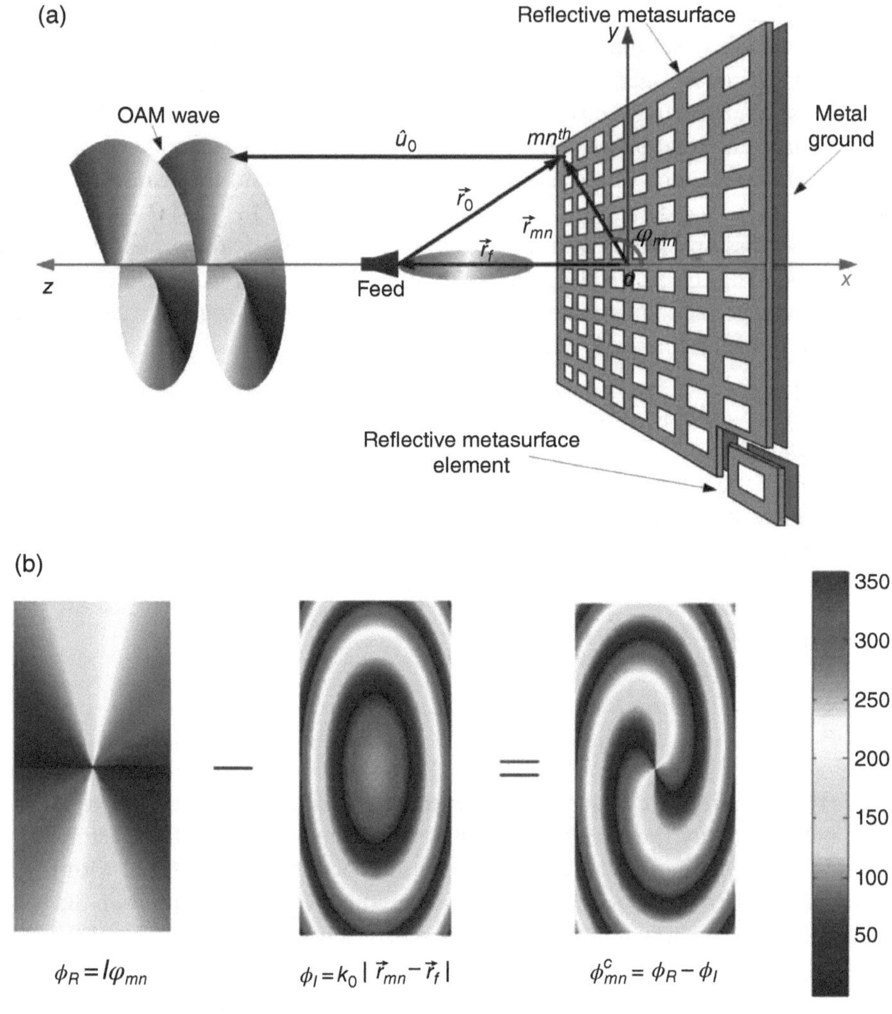

Figure 3.1 (a) Configuration of the reflective metasurface setup for vortex-beam generation. (b) Phase superposition of the design procedure. *Source:* Yu et al. [26]; © AIP Publishing.

where F and A denote the radiation pattern functions of the source and the unit cells, respectively, $\overrightarrow{r_{mn}}$ denotes the position vector of the mnth unit cell, $\hat{u_0}$ represents the expected direction of the reflected EM waves, and k_0 is the propagation constant in vacuum. Then, the phase compensation required by each unit cell can be obtained

$$\varphi_{mn}^c = -k_0\left[\left|\overrightarrow{r_{mn}} - \overrightarrow{r_f}\right| + \overrightarrow{r_{mn}}\cdot\hat{u_0}\right]. \tag{3.5}$$

Once the OAM wave is expected, the phase compensation becomes:

$$\varphi_{mn}^c = l\phi_{mn} - k_0\left[\left|\overrightarrow{r_{mn}} - \overrightarrow{r_f}\right| + \overrightarrow{r_{mn}}\cdot\hat{u_0}\right], l = 0, \ \pm\ 1, \ \pm\ 2\ \ldots, \tag{3.6}$$

where l signifies the mode number of OAM, and ϕ_{mn} is the azimuthal angle of the mnth unit cell. For better understanding, the phase superposition principle is illustrated in Figure 3.1b. The left-most figure shows the required phase distribution of the vortex beam, whereas the rightmost shows

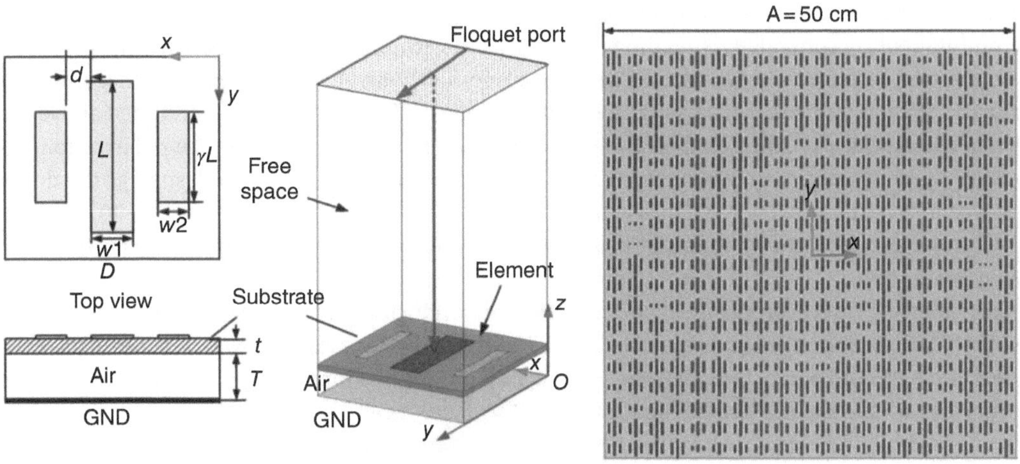

Figure 3.2 Geometry of the triple-dipole unit cell and layout of the designed metasurface. *Source:* Yu et al. [26]; © AIP Publishing.

the phase distribution on the metasurface. The difference between these two is the phase distribution of the wave excited by the feeding antenna, which is given by the figure in the middle.

Using this rule, Ref. [26] shows a metasurface consisting of unit cells with three dipoles printed on the top surface of the substrate (see Figure 3.2). By controlling the length of the dipoles and the ratio between the center dipole and two satellite dipoles, the reflection phase response reaches higher than 360°. This is a typical example for changing the resonance phase by varying the structure parameters of the unit cells. In addition, some other unit cells were designed to realize 360° reflection phase response, such as two concentric annuli [29]. It is demonstrated that the proposed structure has advantages of high efficiency, narrow divergence angle, good stability, and polarization insensitivity. Moreover, the metasurface composed of metallic square loop unit cells exhibits a vortex-beam emitter characteristic [30]. A slightly more complicated structure can manipulate the phase response of the EM waves as well [31]; such a symmetric structure presents the advantage of polarization insensitivity. Another important contribution of this work is the relationship between mode purity and aperture shape. It is proved that a reflective metasurface with a circular aperture and a quarter wavelength unit cell can generate higher OAM modes more efficiently than the conventional half-wavelength unit cell and square aperture reflective metasurface. The proposed method can eliminate the side lobe level, thereby improving the mode purity of the vortex beams for higher modes.

The metallic ground of the above-mentioned three types of metasurfaces is separated from the substrate to obtain the required phase difference. However, their measurements become complicated. When a two-layer reflective structure is used, the metallic ground can be designed to adhere to the substrate [32]. Furthermore, the amplitude response is more stable than that of a single-layer structure. The simulation and measured results of these metasurfaces validate the theoretical analyses discussed above. Researchers are not satisfied with the single polarization manipulation. Then, a cross dipole is selected to manipulate the orthogonally polarized EM waves [33]. By independently controlling the horizontal and vertical arm lengths of the cross dipole, the unit cell exhibits continuous and independent reflection phase behaviors with a phase range of higher than 300° on the x- and y-polarizations, facilitating the generation of vortex beams. Similarly, Joseph cross structure can realize the same functions [34, 35]. Multiple OAM modes in different directions can be simultaneously generated using the phase synthesis with a simple square-shaped structure [36].

In the case of circularly polarized EM waves, another method is required to design the phase response of each unit cell, i.e. varying the rotation angle of the unit cells. As mentioned above, this type of metasurface is called as the PB metasurface, which usually possesses the polarization conversion characteristic. Numerous structures have been proposed and analyzed, such as the double arrow-shaped structure, the N-shaped structure, and the quasi-I-shaped structure [37–40]. All these structures can be rotated by different angles for different phase responses, resulting in the vortex-beam generation.

Particularly, a novel composite of perfectly electric conductor (PEC)- and perfectly magnetic conductor (PMC)-based anisotropic metasurfaces (Figure 3.3) [41] is available, which has two dielectric layers and a ground plane, in which the PEC surface is designed for the x-polarization, whereas the artificial PMC surface is placed beneath the PEC surface. It can be easily said that the x-polarized wave is perfectly reflected back by the PEC surface, whereas the y-polarized wave passing through the strip array is then perfectly reflected back by the PMC surface with a zero-phase shift. The whole metasurface is built by distributing the scatterers with various orientations based on their azimuthal locations. Finally, the vortex beam is generated successfully. Further, a dual-layer meta-atom is proposed [42]. Each layer comprises five metallic dipoles, the length of which is carefully adjusted to tune the resonant frequencies. Because the dipole orientations in each layer are orthogonal, the metasurface exhibits a polarization-insensitive characteristic. It is further demonstrated that, by combining the Aharonov–Anandan and PB phases, the symmetric restriction of the geometric phase can be broken. The expected phase profiles can be easily imposed on the arbitrary polarization states without restricting the orthogonal limitations [43].

Recently, the concept of coding metasurface has been proposed for digital EM wave manipulations, in which two distinct coding particles with opposite phase responses, i.e. 0 and 180°, are regarded as the digital bits '0' and '1' for 1-bit coding. The 2-bit coding metasurface can be extended from this concept, in which the coding particles with 0, 90, 180, and 270° phase responses correspond to the digital bits '00,' '01,' '10,' and '11.' With the help of coding metasurfaces, the design and optimization procedure are significantly simplified. The coding metasurface is additionally a bridge

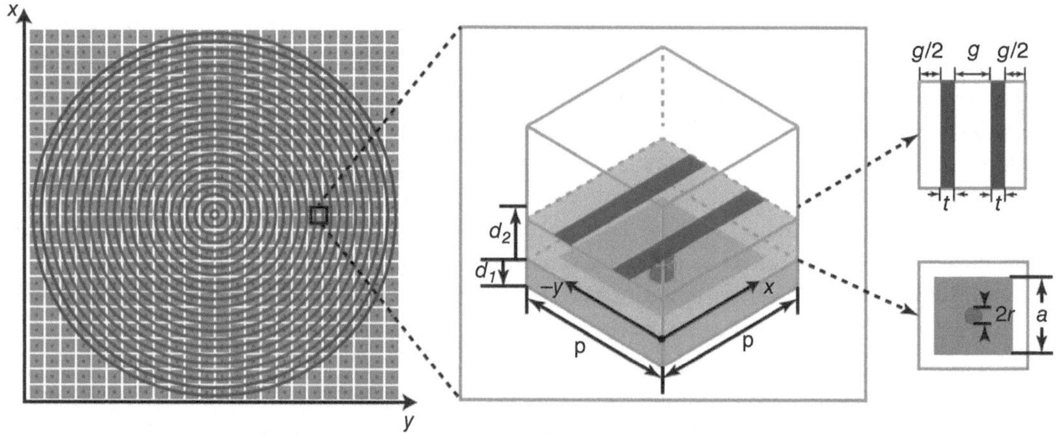

Figure 3.3 PEC–PMC-based unit cell and the metasurface consisting of rotated unit cells. *Source:* Chen et al. [41], with the permission of AIP Publishing.

linking the physical metamaterial particles and digital codes, which enables us to study metamaterials from the perspective of information science. Researchers have been making efforts to design the vortex-beam emitters using the coding metasurfaces. The design procedure is similar to that mentioned above, in which the phase design is always the most important part. For example, a straight rod antenna with different orientation angles can be used to combine the concept of PB phase with the coding metasurface, as shown in Figure 3.4a [44]. When the proposed PB coding metasurface is illuminated normally by a linearly polarized plane wave, it will generate four symmetric beams simultaneously in the upper-half space, which contain two RCP beams in the $-x$ direction and two LCP beams in the $+x$ direction, as shown in Figure 3.4b. Specifically, if illuminated by an RCP plane wave, it creates only two RCP beams in the $-x$ direction, and when illuminated by an LCP plane wave, it creates only two LCP beams in the $+x$ direction. These phenomena show the spin-dependent property of the PB coding metasurface, and such complicated functions are realized only by eight types of unit cells with different orientation angles. Additional

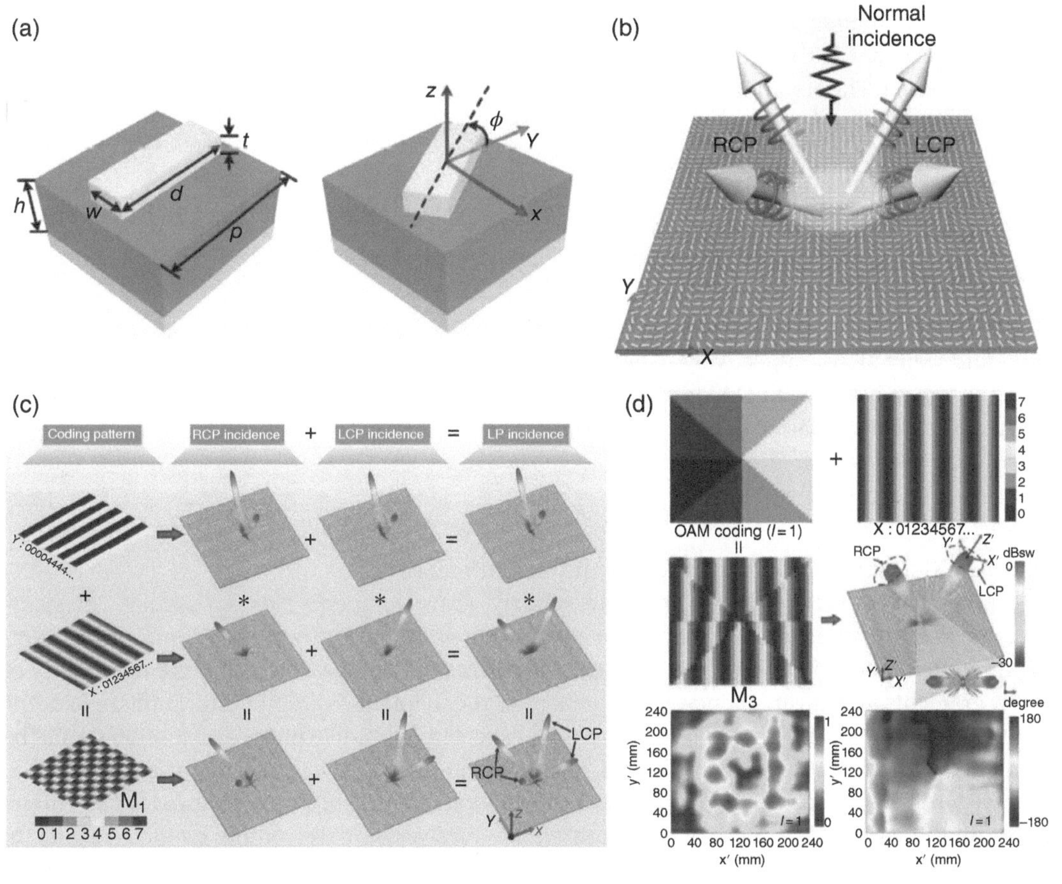

Figure 3.4 Example of the coding metasurface design. (a) Coding unit cell; (b) spin-dependent generation of circularly polarized beams. For RCP incidence, two RCP beams are generated in the $-x$ direction; for LCP incidence, two LCP beams are generated in the $+x$ direction; for LP incidence, two RCP and two LCP beams are generated; (c) phase synthesis of coding patterns and the corresponding radiation patterns; (d) coding patterns of PB coding metasurfaces to control vortex beams with OAM mode $l = 1$ and simulated 3D scatter patterns. *Source:* Zhang et al. [44]; © American Chemical Society.

degrees of freedom to manipulate the EM waves and simplified design procedure are the most significant advantages of this type of metasurface.

When designing, the deviation angle with respect to the z axis can be determined by the generalized Snell's law [17]:

$$\theta = \arcsin(\lambda/\Gamma), \tag{3.7}$$

where λ and Γ represent the wavelength of free space and the period of gradient coding sequence, respectively. Then, the coding metasurfaces with the PB phase can be designed by simply adding several coding patterns. It is worth noting that the convolution of scattering patterns can be easily implemented by calculating the modulus of different coding patterns [45], which allows the PB coding metasurface to manipulate the circularly polarized beams flexibly, which significantly reduces the computational complexity. Figure 3.4c shows some mixed coding patterns, in which $M1$ is obtained by adding the periodic sequence along the y direction with the gradient sequence along the x direction. The far-field patterns of $M1$ under the RCP, LCP, and LP incidences show that the original two symmetric beams in the yoz-plane are deflected toward the $\pm x$ directions. Finally, under the illumination of the LP incident wave, the metasurface generates four symmetrical pencil beams in the upper-half space, including two RCP beams in the $-x$ direction and two LCP beams in the $+x$ direction. The elevation angles and azimuth angles of the four beams can be calculated using the following equations [45]:

$$\begin{cases} \theta = \arcsin\left(\sqrt{\sin^2\theta_1 + \sin^2\theta_2}\right) \\ \varphi = \arctan\left(\dfrac{\sin\theta_2}{\sin\theta_1}\right) \end{cases}, \tag{3.8}$$

where θ_1 and θ_2 are the deviation angles corresponding to the two coding sequences varying along the x and y directions, respectively. Then, if a spiral-like phase shift is introduced by the metasurface, the vortex beams with OAMs can be generated (see Figure 3.4d). The spiral-like phase distribution of the metasurface can be expressed as

$$\phi(x,y) = l \times \varphi = l \times \arctan\left(\frac{y}{x}\right), \tag{3.9}$$

where l is the topological charge and φ is the azimuth angle.

Another type of metasurface is available, namely, space–frequency-domain gradient metasurface, in which new degrees of freedom are achieved by introducing frequency-domain phase gradients to the metasurface [46]. The frequency gradients are obtained by using the derivatives of phase gradients over the frequency, which generates the frequency-variant uniform phase patterns in the working band. Similarly, if a vortex beam is expected, spiral-like phase distribution should be combined with such a coding method [46], as shown in Figure 3.5a[47]. After Fourier transformation [45, 46, 48, 49], specific phase sequences can generate specific beams in the predesigned directions in the far fields (information captured by receivers). This coding method results in information loss at the receivers because it is difficult for the receivers to recognize the whole scattering patterns. For example, the information in side lobes of the pattern is lost in transmission, which is difficult to be recovered [47]. To address the problem, a vector beam modulator carrying the information in orthogonal polarizations and OAM modes was proposed. As a result, the information in the orthogonal polarizations and OAM modes can be transmitted with a low crosstalk. Moreover, such a metasurface is verified to be capable of generating a vortex beam, as shown in

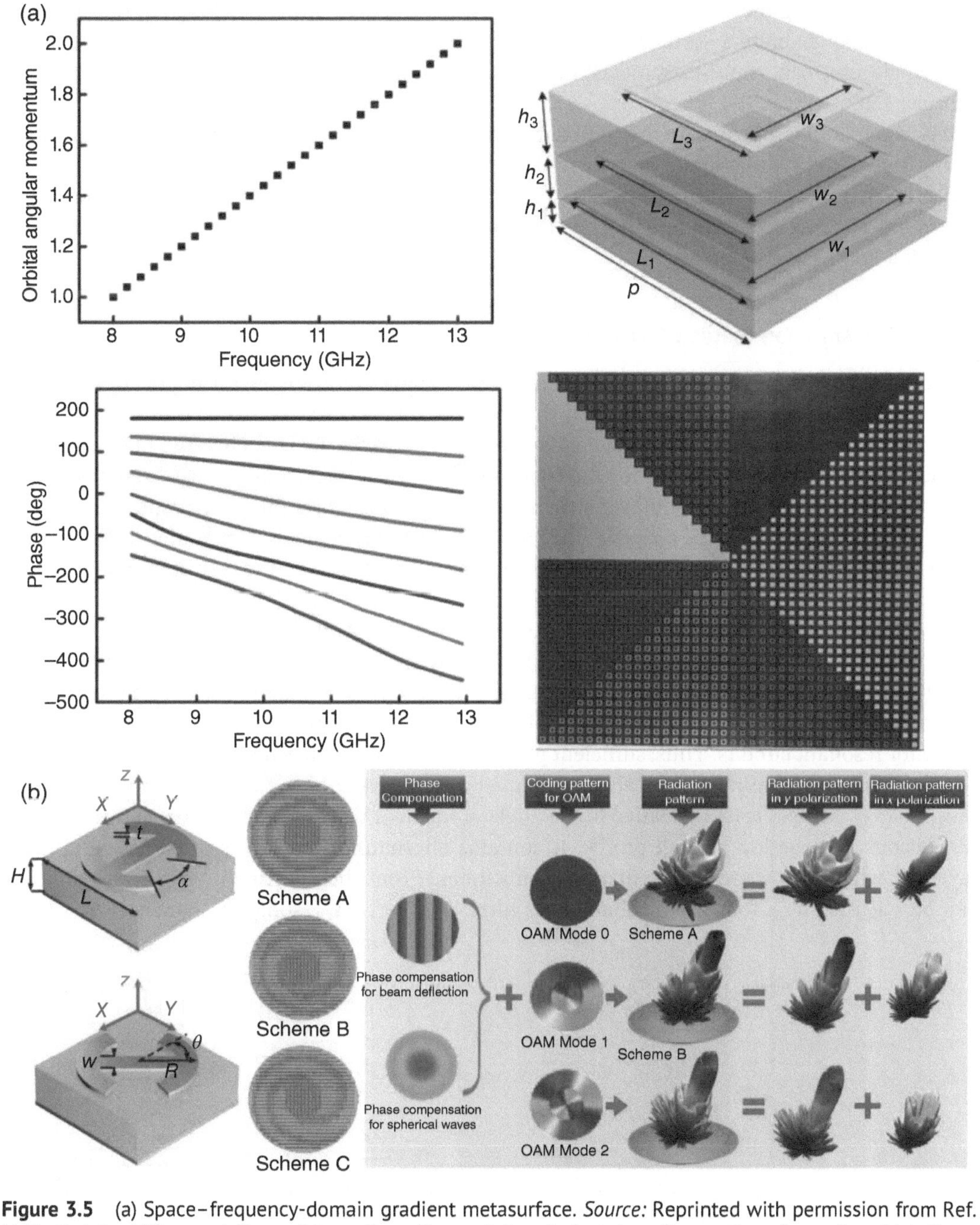

Figure 3.5 (a) Space–frequency-domain gradient metasurface. *Source:* Reprinted with permission from Ref. [46]; © John Wiley and Sons; (b) configurations of the designed coding metasurfaces, in which three schemes are given with the same phase compensation but different coding patterns for OAM modes. Each orthogonal polarization carries an OAM mode to represent the information of the metasurface. The radiation patterns of three schemes are presented, and the components in the x- and y-polarizations are given. *Source:* Ma et al. [47]; © John Wiley and Sons.

Figure 3.5b. Coding metasurface can manipulate the resonance phase by changing the structure parameter and generate a vortex-beam emitter [50, 51]. By integrating different coding unit cells in a single unit cell, multi-task metasurface provides opportunities to improve the information capacity [50].

It is worth noting that most of the coding metasurfaces are composed of metallic resonators, indicating high ohmic dissipation. Dielectric resonators can enhance the efficiency in manipulating EM waves for their intrinsic nonradiative losses. Therefore, using a square dielectric block with a metallic ground, a vortex beam can be generated as well [52], in which the side length of the square block is the key to phase manipulation.

3.2.2 Transmission Metasurfaces for Vortex-beam Generation

Irrespective of whether resonance phase or geometric phase is considered during the design stage, the total phase response range required by the vortex beam is at least 2π, which is not a small number. For the reflective metasurfaces, such phase response is easily achieved because of the existence of the metallic ground. However, for the transmission metasurface, a large phase response range must be realized through a multilayer structure. Various types of multilayer structures are available. The basic phase design of the resonance phase and geometric phase is the same as that for the reflective metasurfaces. The following discussions are mainly based on the type of multilayer structures instead of the phase.

A multilayer structure is shown in Figure 3.6, in which the multilayer mainly means more than one dielectric substrate. The layers are stacked with or without a gap. For example, in Ref. [53], each unit cell consists of four complementary split-ring resonators of two different sizes and orientations. A specific distance exists between the two-layer structures, resulting in two parallel inductor–capacitor resonant circuits. Thus, sufficient phase range can be achieved by varying the orientation of the unit cells. The miniaturized element frequency selective surface (MEFSS) is suitable for this purpose [54]. MEFSS is less sensitive to the angle of incident waves because of the subwavelength dimensions and the low overall profile. It contains alternating four layers of patches and three layers of wire grids separated by dielectric substrates. From the equivalent circuit, the unit cell can be regarded as a fourth-order band-pass filter. Further, the resonance characteristic can be

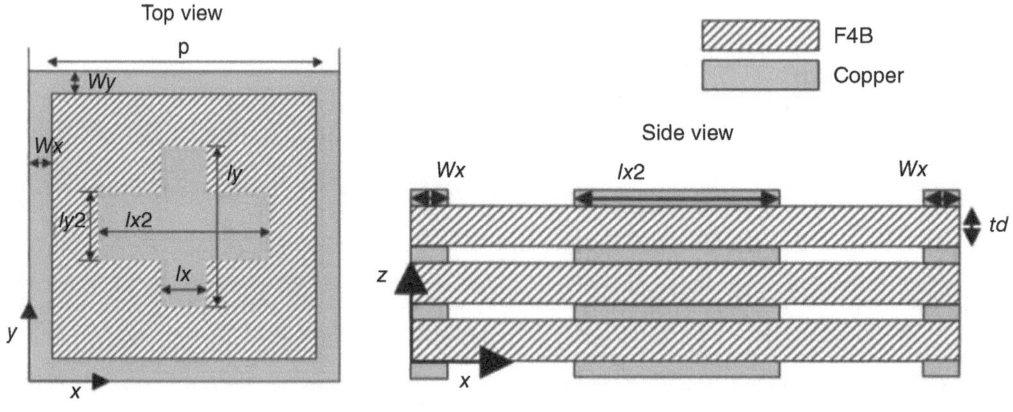

Figure 3.6 Examples of multilayer unit cells combined to form a transmission metasurface for vortex-beam generation. A 360° phase response is realized by changing the dimensions of the unit cells. *Source:* Qi et al. [57]; © Springer Nature.

tuned by changing the dimension of the unit cells. In addition, the designs based on patch antenna are available. The top/bottom layer couples the x- and y-polarized incident waves into the guided mode in the strip-line structure, following which the guided wave is decoupled to the x- and y-polarized waves by the bottom/top layer [55]. Furthermore, polarization conversion of the transmitted wave is realized by selecting the position of the vias in the orthogonal direction. By varying the length of the strip line, the transmission phase can be tuned, allowing for the phase distributions of vortex beams with different OAM modes. Almost the same design is used in Ref. [56]. Dual-mode dual-polarization vortex beams can be generated using this design. It consists of a cascaded metasurface with four metallic layers and three dielectric layers (Figure 3.6) [57]. Cross-type patches are applied to independently manipulate the two orthogonal polarizations. A square loop is introduced to improve the performance of the phase response and transmission amplitude. Similarly, by varying the length of the cross arm, we can obtain the expected phase range. It is verified that the phase variation range of a single-layer structure can be increased from 44° to 122° using the square loop, and the total phase response increases with more numbers of layers. With a cascaded metasurface with four metallic layers and four dielectric layers, a vortex-beam emitter is realized as well [58]. By phase synthesis during the design procedure, multiple functions can be integrated in a single metasurface. The combination of C-shaped split-ring resonators (CSRRs) and C-slot split-ring resonators (CSSRRs) has been proven to exhibit the vortex-beam emitter characteristics [59].

Another type of multilayer structure consists of two identical structures on the top and bottom of the dielectric substrate, which is called as a double-sided structure in this study to differentiate it from the abovementioned structures. The metasurface with unit cell functions as a Huygens surface (with a phase delay of π) under the x-polarized incident wave, where both the electric and magnetic resonances are excited, whereas it acts as a band-pass filter at nonresonant frequencies (without phase delay) under the y-polarized incident wave [60]. The transmission phase of such a structure can be controlled by rotating the angle along the z-axis. The work in Ref. [61] exhibits a similar design. The bilayer meta-atom utilizes a nearly perfect polarization conversion of the two orthogonal transmissions, with the transmission efficiency being as high as 85%. With extra superposition of structures with different dimensions, dual-wavelength responses can be achieved [62]. A bilayer design is used similarly, when each metallic layer consists of a modified complementary split-ring resonator and a circular hole, in the middle of which an electric field-coupled resonator exists. Different resonators correspond to different wavelengths. When the performance of such a double-sided single-layer metasurface is unsatisfactory, the modified principle of the Fabry–Perot cavity was introduced to address the problem of uneven circumferential radiation and posterior lobe radiation [63].

Furthermore, a stereochemical structure is proposed to realize gradient phase, by which the vortex beam is generated [64]. The authors used metallic rods and cylindrical dielectric resonators to construct the metasurface. Despite being a unique idea, the application of such a stereochemical structure requires a large amount of time and effort because no experimental results have been provided.

Let us recall what was demonstrated in Section 2.1: the coding metasurface provides a new degree of EM wave manipulation and simplifies the design procedure, implying that the complicated functions can be easily fulfilled by the coding metasurface. Many reflective metasurfaces that can generate the vortex beams have already been reviewed. A few examples of transmission–reflection-integrated metasurfaces are available [65–67], which are in urgent requirement of rapid developments for modern integration systems. The multilayer structure is still necessary, and active devices are sometimes introduced for better controls of the unit cells and improvement. Such metasurfaces can manipulate not only the reflected waves but also the transmissive waves. Irrespective of the types of waves, multiple functions can be realized. Therefore, more functions are integrated into a single metasurface, according to the expectations of people.

Moreover, a single-layer metasurface with a 2π transmission phase response is not impossible. If varactor diodes are added to the metasurface, the phase response can be satisfied, and the design procedure is simplified as well. The phase response can be fully controlled via the capacitance of the varactor diodes. Thus, only a simple square loop is sufficient for this kind of metasurface [68].

3.2.3 Planar Metasurfaces for Vortex-beam Generation

It is worth noting that all interesting designs mentioned above usually employ standard horns or waveguides as feeding sources. With these feeding sources, the vortex-beam generators become bulky, making it difficult to be integrated. To address this problem, a few approaches have been proposed, among which three efficient methods are mainly available: (i) integrating the source into the same PCB layer of the metasurface, (ii) using a holographic technology, and (iii) adding extra feeding network to realize the expected phase distribution, which will be discussed in detail below.

Ref. [69] presents a novel design that inherently integrates the metasurface with a planar feeding source in the same layer, as shown in Figure 3.7. To control the EM waves from the feed interact with the specially designed metasurface, an extra polarizer is introduced here. The overall process proceeds as follows: a spherical EM wave with linear polarization is generated by the feeding source and then reflected by the polarizer, which is set to fully reflect the incident wave but transmit the orthogonally polarized waves. After being reflected, the EM wave has interactions with the specially designed metasurface. Phase compensation, polarization conversion, and complete reflection of the reflected wave are objectives of the metasurface. Finally, the EM wave is transmitted through the polarizer with the desired phase distribution, resulting in the generation of the OAM mode. The required phase compensation Φ_{ij} of the (i,j)th unit cell can be calculated using the following equations [70]:

$$\begin{cases} \phi_{ij} = k_0\left(\sqrt{R_{ij}^2 + F^2} - F\right) + k_0 R_{ij} \sin\delta + n\varphi_{ij} \\ \delta = \arctan\left(\dfrac{D/2}{Z_{\max}}\right) \end{cases}, \tag{3.10}$$

where k_0 is the wave number in free space, F is the focal length, R_{ij} denotes the distance of the (i,j)th unit cell from the feeding source, and δ is the axicon angle; n and φ_{ij} represent the OAM mode and the azimuthal angle of the (i,j)th unit cell, respectively; and D is the aperture size of the metasurface

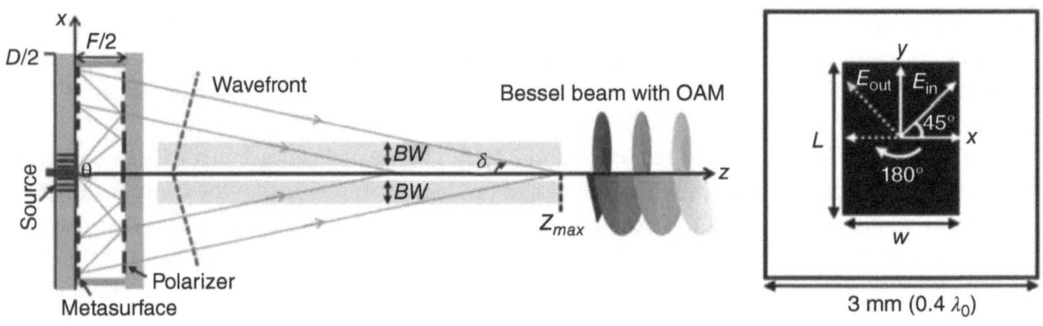

Figure 3.7 Metasurface with the source integrated in the same layer. *Source:* Shen et al. [69]; © AIP Publishing.

and Z_{max} is the maximum nondiffractive propagating distance. The source is designed based on a circular cavity-backed substrate-integrated waveguide. From the simulation and measured results, a nondiffractive first-order Bessel beam carrying OAM is generated.

Another method to generate OAM-carrying beam from the integrated source is to transform the surface wave to the vortex wave through the designed metasurface. A novel ring-surrounded bowtie patch metasurface manipulates the phase of the EM wave, which is excited by the surface wave guided on a single-layer substrate from a monopole [71]. The TM-mode surface wave propagates along the radial direction to the edge of the slab. All unit cells are radially directed as well and are excited by the propagating surface wave under the top of this dielectric substrate. The phase of the scattered electric field can be determined by the propagating phase delay φ_p of the surface wave and the phase control value φ_c of the excited local unit cell. To satisfy the phase requirement of OAM, Φ_c of each unit cell should be carefully designed to compensate the propagating phase delay Φ_p. Further, Φ_c of the (x, y)th position can be calculated as:

$$\phi_c(x,y) = 2n\pi + (l+1)\varphi - \vec{k}_{TM}\cdot\vec{R}(x,y), \tag{3.11}$$

where l is the OAM mode, k_{TM} is the wave vector of the surface wave, and $R(x, y)$ is the vector from the center of the metasurface to the unit cell at the position (x, y). Then, based on the corresponding relation between the phase response and the structure parameters, the metasurface for OAM generation is achieved.

Holography is a powerful tool for vortex-beam generation with low profile. According to the theory of leaky-wave holographic metasurface, the radiated wave (vortex wave in this case) can be reconstructed by exciting the hologram generated by the reference wave and the object wave. Generally, the intensity of the hologram can be expressed as [72]:

$$I = \left|\psi_{ref} + \psi_{obj}\right|^2 = \left|\psi_{ref}\right|^2 + \left|\psi_{obj}\right|^2 + \psi_{ref}\psi_{obj}^* + \psi_{obj}\psi_{ref}^*, \tag{3.12}$$

where ψ_{ref} and ψ_{obj} represent the wave fronts of reference and object waves, respectively. Then, the reference wave (excited source) and the object wave (vortex wave) can be defined as:

$$\begin{cases} \psi_{ref} = e^{-jk_0nx} \\ \psi_{obj} = e^{jl\varphi} \end{cases}, \tag{3.13}$$

where k_0 is the wave number in free space, n is the surface effective refractive index of the whole metasurface, x is the radio distance along the positive x direction from the source located in the metasurface, and l denotes the OAM mode. When the reference wave is applied to excite the hologram, the radiation term $|\psi_{ref}|^2\psi_{obj}$ containing the phase information of the object wave can be generated based on Eq. (3.12). If a conjugate reference wave is used, a conjugate objective wave with $-l$ OAM mode is generated. Both the beams are in the same polarization. Thus, orthogonal two-channel multiplexing OAM radiations can be realized only using one hologram excited by two surface plane waves approaching from opposite directions. Furthermore, a four-channel vortex-beam multiplexer can be designed [72]. Anisotropic holographic metasurfaces are designed to fulfill multiple functions on a single metasurface [73, 74].

An extra feeding network is sometimes useful for the mode reconfiguration and multiplexing. Generally, a feeding network provides the required phase of the vortex beam to the unit cells, as well as the dynamical control. A 4 × 4 metasurface array, combined with a 90° phase-shifting reconfigurable network and a square-driven patch with two switchable feeding points, is demonstrated to generate the OAM mode in a wide-frequency range and dynamically switch different

OAM modes [75]. When a phase shift-feeding network is embedded in the reflective aperture of the metasurface, it can work in the reflect array mode and phased array mode for different frequency ranges [76]; thus, the OAM mode multiplexing is realized. Rotman lens can work as a beamforming network as well [77]. Different patch antennas are fed with different phases. A vortex beam can be generated as long as the total phase distribution satisfies the phase required by the OAM mode. Using a specially designed partially reflecting metasurface, which acts as a Fabry–Perot cavity, the directivity and beam orientation are truly improved.

3.2.4 Metasurfaces for Modified Vortex-beam Generation

It is well known that a vortex beam diverges during transmission, which hinders the reception and identification of the proper objective. Thus, following the proposal of many works for vortex-beam generation, some metasurfaces for converged vortex beams have become a research hotspot. Generally, two methods are available to converge the vortex beams: (i) introducing the lens design into the vortex-beam emitter, the key to which is phase synthesis, and (ii) generating a nondiffracting vortex beam. A high-order Bessel beam, which is regarded as the zeroth-order Bessel beam carrying OAM [78], satisfies the nondiffraction requirement. Therefore, the way of generating the high-order Bessel beams is the key to this method. In Section 2.3, Ref. [69] already provides a generation method for nondiffractive first-order Bessel beam.

When the lens is introduced into the metasurface design, the phase synthesis is used again. The objective is to converge the transmitted beam to a specific point. Therefore, we begin with a plane wave. The phase change introduced by the unit cell at (x, y) to converge a general plane beam can be expressed as:

$$\varphi_1(x,y) = \frac{2\pi\cdot\left(\sqrt{f^2 + (x^2 + y^2)} - |f|\right)}{\lambda}, \tag{3.14}$$

where f is the focal length of the transmitted wave, and λ is the wavelength at the operation frequency. Furthermore, based on the superposition of phase distributions, the rotation angle of each unit cell for the PB metasurface at (x, y) is given by:

$$\theta = \frac{1}{2}l\varphi + \frac{1}{2}\varphi_1 = \frac{1}{2}l\arctan\left(\frac{y}{x}\right) + \frac{1}{2}\left(\frac{2\pi\cdot\left(\sqrt{f^2 + (x^2 + y^2)} - |f|\right)}{\lambda}\right). \tag{3.15}$$

After the rotation angle is determined, the metasurface can be easily designed, and thus the vortex beam is easily generated. This approach was adopted in Refs. [78, 79].

When a high-order Bessel beam is expected, one can integrate the functions of a spiral phase plate that generates the vortex beam with an axicon that generates the Bessel beam. Thus, the phase distribution of the electric field in a plane perpendicular to the propagation direction of the wave can be written as [80]:

$$\phi = \frac{2\pi}{\lambda}r\sin\alpha + l\varphi, \tag{3.16}$$

where r is the distance between the plane and the center of the beam, α is the convergence angle of the zero-order Bessel beam, and l and φ represent the OAM mode and azimuthal angle, respectively. By introducing the phase compensation to the feeding horn antenna, the phase distribution of the designed metasurface is obtained as follows:

$$\phi(x,y) = \frac{2\pi}{\lambda}\sqrt{x^2+y^2}\sin\alpha + l\arctan\left(\frac{y}{x}\right) + \frac{2\pi}{\lambda}\left(\sqrt{x^2+y^2+f^2}-f\right). \quad (3.17)$$

If the PB metasurface is expected, the rotation angle of each unit cell for the PB metasurface at (x, y) is written as:

$$\theta = \frac{1}{2}\left(\frac{2\pi}{\lambda}\sqrt{x^2+y^2}\sin\alpha\right) + \frac{1}{2}l\arctan\left(\frac{y}{x}\right) + \frac{1}{2}\cdot\frac{2\pi}{\lambda}\left(\sqrt{x^2+y^2+f^2}-f\right). \quad (3.18)$$

This method is used in Refs. [80, 81].

3.2.5 One-dimensional Metasurface for Vortex-beam Generation

In this section, a novel method to generate the vortex beam using spoof SPPs is introduced, as shown in Figure 3.8 [82]. Spoof SPP waveguide, which can be regarded as a 1D metasurface, has been intensively investigated because of its subwavelength characteristics, such as field enhancement, low bending loss, and easy integration. Applying a 2D vortex-beam emitter in a 1D design has been making good progress, facilitating its integration in the future communication systems.

Because the vortex beams possess helical phase fronts and azimuthal component of the wave vector, the key to generate a vortex beam is the rotated phase distribution and radiation. To obtain the rotated phase, the conventional spoof SPP waveguide is shaped in circle. Furthermore, some circular patches are placed beside the SPP waveguide to radiate the rotated phases. It is worth noting that the patches here act not only as the radiation elements but also as the phase manipulation

Figure 3.8 Prototype of the vortex-beam emitter using spoof SPP waveguide. (a) Light regions represent the structure at the top of the substrate, whereas dark regions represent the structure at the bottom of the substrate. (b) A magnified image showing the connection between the top and bottom structures. *Source:* Yin et al. [82]; © John Wiley and Sons.

elements. When the circular patches are placed beside the spoof SPP waveguide, a phase lag appears along with energy coupling. By analyzing the relationship between the input electric field E_1 and output electric field E_2 of a single circular patch, it can be found that the phase lag introduced by one circular patch is determined by [82]:

$$\begin{cases} t = \dfrac{E_2}{E_1} = \dfrac{\rho - ae^{i\varphi}}{1 - \rho ae^{i\varphi}}, \\ \phi = \arg(t) \end{cases} \tag{3.19}$$

where ρ and κ are the self-coupling and mutual-coupling coefficients, respectively. Both factors are supposed to be independent of frequency and satisfy $\rho^2 + \kappa^2 = 1$. Suppose that the single-pass amplitude transfer factor is 1. Then, it is noted that, irrespective of the self-coupling coefficient, the effective phase-shift difference between the maximum single-pass phase shift and the minimum single-pass phase shift is always 2π. If the single-pass phase shift after the circular resonator is made to be 2π, an effective phase-shift difference of 2π is achieved between the conditions with and without the circular resonator. Moreover, when the EM waves propagate through the spoof SPP waveguide, the phase change can be calculated as $l' = 2\pi R/\lambda_g = \frac{MR\pi}{p}$, where R is the radius of the loop, p is the period of the unit cell, and M is defined as $M = \frac{kp}{\pi}$, which can be easily read from the dispersion curve of the spoof SPP waveguide. By combining these two parts of the phase, the total phase change after the EM wave propagates through the entire structure can be easily calculated. Because the propagation constant of the spoof SPP is different at different frequencies, the total phase change is different as well, resulting in the appearance of different OAM modes at different frequencies. The proposed method was verified by both numerical calculations and experiments, and the measurement results agree very well with numerical simulations and theoretical predictions. This smaller structure possesses more flexible OAM modes and can be very easily integrated in systems and integrated circuits. This work not only extends the application of spoof SPPs but also simplifies the vortex-beam generation.

3.3 Conclusion

Various works on metasurfaces for microwave vortex-beam generation were reviewed in this chapter. The metasurfaces can achieve wavefront manipulations by using various methods, such as controlling the geometric phase, controlling the resonance phase, integrating the newly designed additional structure, and even using spoof SPP waveguide. Each method exhibits a unique flexibility. Most importantly, with the development of new technologies, the metasurfaces can be multifunctional. Although the vortex-beam generation is the fundamental objective, the functions of multiplexing, demultiplexing, and manipulating the polarizations have become other common objectives. Therefore, metasurface is a very important candidate for the microwave vortex-beam generation.

Acknowledgments

This work was supported by the National Key Research and Development Program of China (2017YFA0700201, 2017YFA0700202, and 2017YFA0700203), the National Natural Science Foundation of China (61631007, 61571117, and 61801354), the Natural Science Foundation of Shaanxi Province (Grant No. 2019JQ-649), and the Fundamental Research Funds for the Central Universities (Grant No. XJS190503).

References

1 Bliokh, K.Y., Bekshaev, A.Y., and Nori, F. (2012). Dual electromagnetism: helicity, spin, momentum, and angular momentum. *New Journal of Physics* **15**: 33026–33053.

2 Allen, L., Beijersbergen, M.W., Spreeuw, R.J.C., and Woerdman, J.P. (1992). Orbital angular momentum of light and transformation of Laguerre Gaussian laser modes. *Physical Review A Atomic Molecular & Optical Physics* **45**: 8185–8189.

3 Thidé, B., Then, H., Sjöholm, J. et al. (2007). Utilization of photon orbital angular momentum in the low-frequency radio domain. *Physical Review Letters* **99**: 087701.

4 Yan, Y., Xie, G., Lavery, M.P.J. et al. (2014). High-capacity millimetre-wave communications with orbital angular momentum multiplexing. *Nature Communications* **5**: 4876.

5 Brasselet, E., Malinauskas, M., Zukauskas, A., and Juodkazis, S. (2010). Photopolymerized microscopic vortex beam generators: precise delivery of optical orbital angular momentum. *Applied Physics Letters* **97**: 211108.

6 Schemmel, P., Maccalli, S., Pisano, G. et al. (2014). Three-dimensional measurements of a millimeter wave orbital angular momentum vortex. *Optics Letters* **39**: 626.

7 Zhang, C. and Ma, L. (2016). Millimetre wave with rotational orbital angular momentum. *Scientific Reports* **6**: 31921.

8 Tennant, A. and Allen, B. (2016). Generation of OAM radio waves using circular time-switched array antenna. *Electronics Letters* **48**: 1365.

9 Bai, Q., Tennant, A., and Allen, B. (2014). Experimental circular phased array for generating OAM radio beams. *Electronics Letters* **50**: 1414–1415.

10 Gao, X., Huang, S., Song, Y. et al. (2014). Generating the orbital angular momentum of radio frequency signals using optical-true-time-delay unit based on optical spectrum processor. *Optics Letters* **39**: 2652.

11 Liu, K., Liu, H., Qin, Y. et al. (2016). Generation of OAM beams using phased array in the microwave band. *IEEE Transactions on Antennas and Propagation* **64**: 3850–3857.

12 Guo, Z. and Yang, G. (2017). Radial uniform circular antenna array for dual-mode OAM communication. *IEEE Antennas and Wireless Propagation Letters* **16**: 404–407.

13 Gambini, F., Velha, P., Oton, C.J., and Faralli, S. (2016). Orbital angular momentum generation with ultra-compact Bragg-assisted silicon microrings. *IEEE Photonics Technology Letters* **28**: 2355–2358.

14 Cai, X.Z., Wang, J., Strain, M.J. et al. (2012). Integrated compact optical vortex beam emitters. *Science* **338**: 363–366.

15 Li, R., Feng, X., Zhang, D. et al. (2014). Radially polarized orbital angular momentum beam emitter based on shallow-ridge silicon microring cavity. *IEEE Photonics Journal* **6**: 1–10.

16 Y. Chen, S. Zheng, C. Hao, et al. (2015). Orbital angular momentum mode multiplexing with half-mode substrate integrated waveguide antenna. Paper presented at the 2015 European Radar Conference (EuRAD).

17 Yu, N., Genevet, P., Kats, M.A., and Aieta, F. (2015). Light propagation with phase discontinuities: generalized laws of reflection and refraction. *Science* **334**: 333–337.

18 Almeida, E., Shalem, G., and Prior, Y. (2016). Subwavelength nonlinear phase control and anomalous phase matching in plasmonic metasurfaces. *Nature Communications* **7**: 10367.

19 Yu, P., Chen, S., Li, J. et al. (2015). Generation of vector beams with arbitrary spatial variation of phase and linear polarization using plasmonic metasurfaces. *Optics Letters* **40**: 3229–3232.

20 Schonbrun, E., Seo, K., and Crozier, K.B. (2011). Reconfigurable imaging systems using elliptical nanowires. *Nano Letters* **11**: 4299–4303.

21 Khorasaninejad, M., Zhu, A.Y., Roques-Carmes, C. et al. (2016). Polarization-insensitive metalenses at visible wavelengths. *Nano Letters* **16**: 7229–7234.

22 Chen, X., Huang, L., Muhlenbernd, H. et al. (2013). Dual-polarity plasmonic metalens for visible light. *Nature Communications* **3**: 1198.

23 Li, X., Chen, L., Li, Y. et al. (2016). Multicolor 3D meta-holography by broadband plasmonic modulation. *Science Advances* **2**: e1601102.

24 Balthasar Mueller, J.P., Rubin, N.A., Devlin, R.C. et al. (2017). Metasurface polarization optics: independent phase control of arbitrary orthogonal states of polarization. *Physical Review Letters* **118** (113901).

25 Devlin, R.C., Ambrosio, A., Rubin, N.A. et al. (2017). Arbitrary spin-to-orbital angular momentum conversion of light. *Science* **358**: 896–901.

26 Yu, S., Li, L., Shi, G. et al. (2016). Design, fabrication, and measurement of reflective metasurface for orbital angular momentum vortex wave in radio frequency domain. *Applied Physics Letters* **108**: 121903.

27 Bouchard, F., Leon, I.D., Schulz, S.A. et al. (2014). Optical spin-to-orbital angular momentum conversion in ultra-thin metasurfaces with arbitrary topological charges. *Applied Physics Letters* **105**: 101905.

28 John, H. and Jose, E. (2008). *Reflectarray Antennas*. Hoboken, NJ, USA: Wiley IEEE.

29 Huang, H. and Li, S. (2019). High-efficiency planar reflectarray with small-size for OAM generation at microwave range. *IEEE Antennas and Wireless Propagation Letters* **18**: 432–436.

30 Yu, S., Long, L., and Na, K. (2017). Generation, reception and separation of mixed-state orbital angular momentum vortex beams using metasurfaces. *Optical Materials Express* **7**: 3312.

31 Akram, Z., Li, X., Qi, Z. et al. (2019). Broadband high-order OAM reflective metasurface with high mode purity using subwavelength element and circular aperture. *IEEE Access* **7**: 71963–71971.

32 Yuzhou, R., Liang, J., Cai, T. et al. (2018). High-performance broadband vortex beam generator based on double-layered reflective metasurface. *AIP Advances* **8**: 095201.

33 Yu, S., Li, L., and Shi, G. (2016). Dual-polarization and dual-mode orbital angular momentum radio vortex beam generated by using reflective metasurface. *Applied Physics Express* **9**: 082202.

34 Yang, J., Zhang, C., Ma, H. et al. (2018). Tailoring polarization states of multiple beams that carry different topological charges of orbital angular momentums. *Optics Express* **26**: 31664–31674.

35 Yang, J., Zhang, C., Ma, H. et al. (2018). Generation of radio vortex beams with designable polarization using anisotropic frequency selective surface. *Applied Physics Letters* **112**: 203501.

36 Yu, S., Li, L., Shi, G. et al. (2016). Generating multiple orbital angular momentum vortex beams using a metasurface in radio frequency domain. *Applied Physics Letters* **108**: 241901.

37 Ran, Y., Liang, J., Tong, C., and Li, H. (2018). High-performance broadband vortex beam generator using reflective Pancharatnam-Berry metasurface. *Optics Communications* **427**: 101–106.

38 Lin, B., Guo, J., Lv, L. et al. (2019). An ultra-wideband reflective phase gradient metasurface using Pancharatnam-Berry phase. *IEEE Access* **7**: 13317–13325.

39 Wang, H., Li, Y., Han, Y. et al. (2019). Vortex beam generated by circular-polarized metasurface reflector antenna. *Journal of Physics D-Applied Physics* **52**: 255306.

40 Guo, W.-L., Wang, G.M., Li, H.P. et al. (2019). Analysis and design of a broadband metasurface-based vortex beam generator. *IEEE Access* **7**: 129529–129536.

41 Chen, M.L.N., Li, J.J., and Sha, W. (2016). Artificial perfect electric conductor-perfect magnetic conductor anisotropic metasurface for generating orbital angular momentum of microwave with nearly perfect conversion efficiency. *Journal of Applied Physics* **119**: 064506.

42 Xu, H.X., Liu, H., Ling, X. et al. (2017). Broadband vortex beam generation using multimode Pancharatnam–Berry metasurface. *IEEE Transactions on Antennas & Propagation* **65**: 7378–7382.

43 Bai, G.D., Ma, Q., Li, R.Q. et al. (2019). Spin-symmetry breaking through metasurface geometric phases. *Physical Review Applied* **12**: 044042.

44 Zhang, L., Liu, S., Li, L.L., and Cui, T.J. (2017). Spin-controlled multiple pencil beams and vortex beams with different polarizations generated by Pancharatnam-Berry coding metasurfaces. *ACS Applied Materials & Interfaces* **9**: 36447–36455.

45 Liu, S., Cui, T.J., and Zhang, L. (2016). Convolution operations on coding metasurface to reach flexible and continuous controls of terahertz. *Advanced Science* **3**: 1600156.

46 Wu, H.T., Wang, D., Fu, X. et al. (2018). Space-frequency-domain gradient metamaterials. *Advanced Optical Materials* **6**: 1801086.

47 Ma, Q., Shi, C.B., Bai, G.D. et al. (2017). Beam-editing coding metasurfaces based on polarization bit and orbital-angular-momentum-mode bit. *Advanced Optical Materials* **5**: 1700548.

48 Cui, T.-J., Liu, S., and Li, L.L. (2016). Information entropy of coding metasurface. *Light-Science & Applications* **5**: e16172.

49 Cui, T.J., Liu, S., and Zhang, L. (2017). Information metamaterials and metasurfaces. *Journal of Materials Chemistry C* **5**: 3644–3668.

50 Bai, G.D., Ma, Q., Shahid, I. et al. (2018). Multitasking shared aperture enabled with multiband digital coding metasurface. *Advanced Optical Materials* **6**: 1800657.

51 Tian, H.W., Jiang, W.X., Li, X. et al. (2019). Generation of high-order orbital angular momentum beams and split beams simultaneously by employing anisotropic coding metasurfaces. *Journal of Optics* **21**: 065103.

52 Shao, L., Zhu, W., Leonov, M.Y., and Rukhlenko, I.D. (2019). Dielectric 2-bit coding metasurface for electromagnetic wave manipulation. *Journal of Applied Physics* **125**: 203101.

53 Chen, M.L.N., Jiang, L.J., and Sha, W. (2017). Ultrathin complementary metasurface for orbital angular momentum generation at microwave frequencies. *IEEE Transactions on Antennas and Propagation* **65**: 396–400.

54 Wang, Y., Zhang, K., Yuan, Y. et al. (2019). Generation of high-efficiency vortex beam carrying OAM mode based on miniaturized element frequency selective surfaces. *IEEE Transactions on Magnetics* **55**: 7501104.

55 Shi, H., Wang, L., Zhao, M. et al. (2018). Transparent metasurface for generating microwave vortex beams with cross-polarization conversion. *Materials* **11**: 2448.

56 Shi, H., Wang, L., Chen, X. et al. (2019). Generation of a microwave beam with both orbital and spin angular momenta using a transparent metasurface. *Journal of Applied Physics* **126**: 063108.

57 Qi, X., Zhang, Z., Zong, X. et al. (2019). Generating dual-mode dual-polarization OAM based on transmissive metasurface. *Scientific Reports* **9**: 97.

58 Liu, K., Wang, G., Li, Z., and Guo, W. (2019). A multi-functional vortex beam generator based on transparent anisotropic metasurface. *Optics Communications* **435**: 311–318.

59 Yan, J., Guo, Y., Pu, M. et al. (2018). High-efficiency multi-wavelength metasurface with complete independent phase control. *Chinese Optics Letters* **16**: 050003.

60 Akram, M.R., Bai, X., Jin, R. et al. (2019). Photon spin hall effect-based ultra-thin transmissive metasurface for efficient generation of OAM waves. *IEEE Transactions on Antennas and Propagation* **67**: 4650–4658.

61 Akram, M.R., Mehmood, M.Q., Bai, X. et al. (2019). High efficiency ultrathin transmissive metasurfaces. *Advanced Optical Materials* **7**: 1801628.

62 Xie, R., Zhai, G., Wang, X. et al. (2019). High-efficiency ultrathin dual-wavelength Pancharatnam-Berry metasurfaces with complete independent phase control. *Advanced Optical Materials* **7**: 1900594.

63 Ma, L., Chen, C., Zhou, L. et al. (2019). Single-layer transmissive metasurface for generating OAM vortex wave with homogeneous radiation based on the principle of Fabry-Perot cavity. *Applied Physics Letters* **114**: 081603.

64 Liu, Y., Liu, C., Song, K. et al. (2018). A broadband high-transmission gradient phase discontinuity metasurface. *Journal of Physics D-Applied Physics* **51**: 095103.

65 Zhang, L., Wu, R.Y., Bai, G.D. et al. (2018). Transmission-reflection-integrated multifunctional coding metasurface for full-space controls of electromagnetic waves. *Advanced Functional Materials* **28**: 1802205.

66 Wu, R.Y., Zhang, L., Bao, L. et al. (2019). Digital metasurface with phase code and reflection-transmission amplitude code for flexible full-space electromagnetic manipulations. *Advanced Optical Materials* **7**: 1801429.

67 Tang, S., Tong, C., Wang, G.-M. et al. (2018). High-efficiency dual-modes vortex beam generator with polarization-dependent transmission and reflection properties. *Scientific Reports* **8**: 6422.

68 Shi, H., Wang, L., Peng, G. et al. (2019). Generation of multiple modes microwave vortex beams using active metasurface. *IEEE Antennas and Wireless Propagation Letters* **18**: 59–63.

69 Shen, Y., Yang, J., Meng, H. et al. (2018). Generating millimeter-wave Bessel beam with orbital angular momentum using reflective-type metasurface inherently integrated with source. *Applied Physics Letters* **112**: 141901.

70 Lemaitre-Auger, P., Abielmona, S., and Caloz, C. (2013). Generation of Bessel beams by two-dimensional antenna arrays using sub-sampled distributions. *IEEE Transactions on Antennas and Propagation* **61**: 1838–1849.

71 Zhang, Y., Yang, L., Hao, G.W. et al. (2018). Transforming surface wave to propagating OAM vortex wave via flat dispersive metasurface in radio frequency. *IEEE Antennas and Wireless Propagation Letters* **17**: 172–175.

72 Li, Y., Yang, L., Hao, G.W. et al. (2019). Four-channel orbital angular momentum beam multiplexer designed with low-profile metasurfaces. *Journal of Physics D-Applied Physics* **52**: 025108.

73 Meng, X., Wu, J., Wu, Z. et al. (2019). Generation of multiple beams carrying different orbital angular momentum modes based on anisotropic holographic metasurfaces in the radio-frequency domain. *Applied Physics Letters* **114**: 093504.

74 Meng, X., Wu, J., Wu, Z. et al. (2019). Design, fabrication, and measurement of an anisotropic holographic metasurface for generating vortex beams carrying orbital angular momentum. *Optics Letters* **44**: 1452–1455.

75 Wu, J., Zhang, Z., Ren, X. et al. (2019). A broadband electronically mode-reconfigurable orbital angular momentum metasurface antenna. *IEEE Antennas and Wireless Propagation Letters* **18**: 1482–1486.

76 Zhang, D., Cao, X., Gao, J. et al. (2019). A shared aperture 1 bit metasurface for orbital angular momentum multiplexing. *IEEE Antennas and Wireless Propagation Letters* **18**: 566–570.

77 Bai, X., Cao, A., Kong, F. et al. (2019). Rotman lens-fed Fabry-Perot resonator antennas for generating converged multi-mode OAM beams. *IEEE Access* **7**: 105768–105775.

78 Li, X., Pu, M., Zhao, Z. et al. (2016). Catenary nanostructures as compact Bessel beam generators. *Scientific Reports* **6**: 20524.

79 Yuan, Y., Ding, X., Zhang, K., and Wu, Q. (2017). Planar efficient metasurface for vortex beam generating and converging in microwave region. *IEEE Transactions on Magnetics* **53**: 2500204.

80 Yang, X., Zhou, Y., and Wang, G. (2019). A wideband transmission metasurface for generating Bessel beam carrying orbital angular momentum. *International Journal of RF and Microwave Computer-Aided Engineering* **29**: e21941.

81 Zhang, K., Yuan, Y., Zhang, D. et al. (2018). Phase-engineered metalenses to generate converging and non-diffractive vortex beam carrying orbital angular momentum in microwave region. *Optics Express* **26**: 1351–1360.

82 Yin, J.Y., Ren, J., Zhang, L. et al. (2018). Microwave vortex-beam emitter based on spoof surface plasmon polaritons. *Laser & Photonics Reviews* **12**: 1600316.

4

Application of Transformation Optics and 3D Printing Technology in Vortex Wave Generation

Jianjia Yi,[1] *Shah Nawaz Burokur*[2]*, and Douglas H. Werner*[3]

[1] *School of Information and Communications Engineering, Xi'an Jiaotong University, Xi'an, P.R. China*
[2] *LEME, UPL, Univ Paris Nanterre, Ville d'Avray France*
[3] *Department of Electrical Engineering, The Pennsylvania State University, University Park, PA, USA*

4.1 Introduction

In this chapter, innovative microwave devices are introduced for vortex wave generation and manipulation, which have been realized using three-dimensional (3D) printing technology. In particular, several designs based on transformation optics are proposed and studied. The first design is a reflection-type vortex-beam generator, which can realize a transformation from plane waves to vortex waves with orbital angular momentum (OAM). An all-dielectric proof-of-concept prototype for a +1 topological charge is physically implemented and the complete characterization of the device is reported. The second design is a medium performing as a collimating lens which was designed by a spatial transformation and fabricated by additive manufacturing technology. The proposed all-dielectric collimating lens enables an increase in the realized gain of vortex waves and may extend the propagation distance, laying the foundation for future vortex wave communications. The third design is an all-dielectric lens that produces omnidirectional OAM beams based on transformation optics. As a demonstration, a lens with omnidirectional radiation in the horizontal plane is designed and simulated with an incident vortex beam carrying the OAM mode $l = +2$. The simulation results prove that the lens can convert a directional to an omnidirectional beam. The proposed methods illustrate the practicality of using transformation optics to achieve a new class of devices for electromagnetic (EM) wave manipulation at microwave frequencies.

4.2 Theoretical Basis of Transformation Optics and 3D Printing

4.2.1 The Concept and Development of Transformation Optics

Transformation optics [1, 2] is a well-known approach to control EM fields in unprecedented ways through the use of judiciously engineered materials with spatially varying parameters. Although the theoretical basis first appeared a few decades ago, the correspondence between the transformed medium and material parameters was not formally established until 2006 [3, 4]. Figure 4.1 shows a schematic diagram of the coordinate transformation principle, in which the grid represents the spatial domain of the EM field, and the solid line represents the propagation path of an EM wave

Electromagnetic Vortices: Wave Phenomena and Engineering Applications, First Edition.
Edited by Zhi Hao Jiang and Douglas H. Werner.

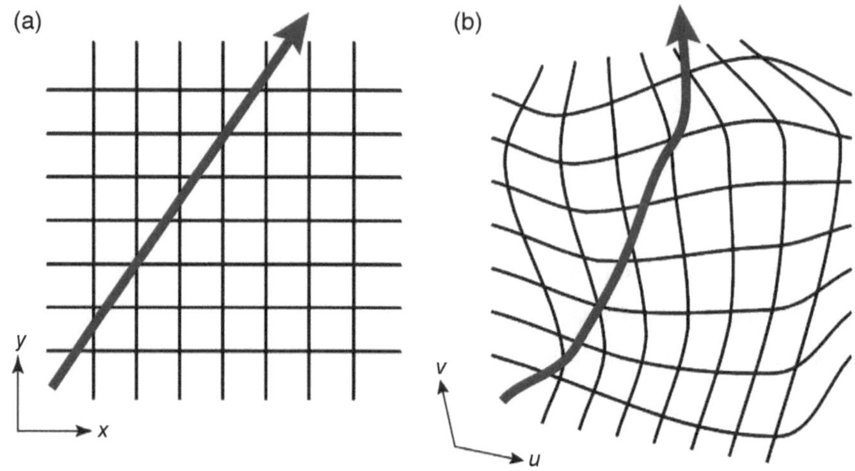

Figure 4.1 Schematic diagram of spatial transformation.

within the spatial domain. As shown in Figure 4.1a, it can be seen that the material in the original space before the transformation is homogeneous and isotropic, so the EM wave propagates along a straight-line path. Figure 4.1b shows the transformed space. It can be seen that the transformed space is deformed compared with the original space. In this case, the space represented by the grid becomes nonuniform and anisotropic.

Assuming that the coordinates of the virtual space and the physical space are (x, y, z) and (x', y', z'), respectively, their relationship can be expressed as:

$$x' = x'(x,y,z), y' = y'(x,y,z), z' = z'(x,y,z), \tag{4.1}$$

The differential form of Maxwell's equations describing EM fields is:

$$\begin{cases} \nabla \times H = \varepsilon \dfrac{\partial E}{\partial t} + J \\ \nabla \times E = -\mu\varepsilon \dfrac{\partial H}{\partial t} \\ \nabla \cdot \varepsilon E = \rho \\ \nabla \cdot \mu H = 0 \end{cases}, \tag{4.2}$$

For isotropic or anisotropic media, the constitutive parameters have the following relations:

$$D = \varepsilon \cdot E, B = \mu \cdot H, \tag{4.3}$$

where ε is the permittivity and μ is permeability. These equations are rewritten in covariant notation in Cartesian coordinates. The invariant form of Maxwell's equations under these conditions is a property of the governing differential equations. The following covariant notation is used for compactness:

$$\begin{cases} \varepsilon^{ijk}\partial_j H_k = \varepsilon^{ij} \dfrac{\partial E_j}{\partial t} + J^i \\ \varepsilon^{ijk}\partial_j E_k = \mu^{ij} \dfrac{\partial H_j}{\partial t} \\ \varepsilon^{ij}\partial_i E_j = \rho \\ \mu^{ij}\partial_i H_j = 0 \end{cases}, \tag{4.4}$$

where, i, j, and k are indices with values ranging from one to three. We note that a particular coordinate is identified by x^i, for example (x, y, z). Moreover, ∂_i is a short-hand notation for $\partial/\partial x^i$. Finally, we note that the completely antisymmetric Levi–Cevita tensor is expressed by ε^{ijk}.

We consider the properties of these equations to lie under a coordinate transformation between x and $x^{i'}$, which is the transformed coordinate (x', y', z'). Therefore, the electric field $\mathbf{E}$ can be transformed as:

$$E_{i'} = A_i^{i'} E_i \tag{4.5}$$

$$E_i = A_i^{i'} E_{i'}, \tag{4.6}$$

where $A_i^{i'}$ is the Jacobian matrix and is given by:

$$A_i^{i'} = \frac{\partial x^{i'}}{\partial x^i}. \tag{4.7}$$

After several steps in the derivation process and subsequent simplifications, the transformed set of equations can be written as:

$$\begin{cases} \varepsilon^{i'j'k'} \partial_{j'} H_{k'} = \dfrac{A_i^{i'} A_j^{j'}}{\left|A_i^{i'}\right|} \varepsilon^{ij} \dfrac{\partial E_{j'}}{\partial t} + \dfrac{A_i^{i'}}{\left|A_i^{i'}\right|} J^i \\ \varepsilon^{i'j'k'} \partial_{j'} E_{k'} = \dfrac{A_i^{i'} A_j^{j'}}{\left|A_i^{i'}\right|} \mu^{ij} \dfrac{\partial H_{j'}}{\partial t} \\ \partial_{i'} \dfrac{A_i^{i'} A_j^{j'}}{\left|A_i^{i'}\right|} \varepsilon^{ij} E_{j'} = \dfrac{\rho}{\left|A_i^{i'}\right|} \\ \partial_{i'} \dfrac{A_i^{i'} A_j^{j'}}{\left|A_i^{i'}\right|} \mu^{ij} H_{j'} = 0 \end{cases}, \tag{4.8}$$

where $\left|A_i^{i'}\right|$ is the determinant of the Jacobian matrix.

When we compare Maxwell's equations before and after the application of a coordinate transformation, we find that there exist transformation rules for the material parameters, current, and charge that will ensure that the transformed equations are identical to the form before the transformation was applied. In other words, if we choose a certain material whose properties are as described below:

$$\begin{cases} \varepsilon^{i'j'} = \dfrac{A_i^{i'} A_j^{j'}}{\left|A_i^{i'}\right|} \varepsilon^{ij} \\ \mu^{i'j'} = \dfrac{A_i^{i'} A_j^{j'}}{\left|A_i^{i'}\right|} \mu^{ij} \\ J^{i'} = \dfrac{A_i^{i'}}{\left|A_i^{i'}\right|} J \\ \rho' = \dfrac{\rho}{\left|A_i^{i'}\right|} \end{cases}, \tag{4.9}$$

then the transformed Maxwell's equations can be rewritten as:

$$\begin{cases} \varepsilon^{i'j'k'}\partial_{j'}H_{k'} = \varepsilon^{i'j'}\dfrac{\partial E_{j'}}{\partial t} + \dfrac{A_i^{i'}}{|A_i^{i'}|}J^i \\ \varepsilon^{i'j'k'}\partial_{j'}E_{k'} = \mu^{i'j'}\dfrac{\partial H_{j'}}{\partial t} \\ \partial_{i'}\varepsilon^{i'j'}E_{j'} = \rho' \\ \partial_{i'}\mu^{i'j'}H_{j'} = 0 \end{cases}, \tag{4.10}$$

Now, we continue with the following transformation between two Cartesian coordinate systems:

$$\varepsilon_C^{i'j'} = \frac{A_i^{i'}A_j^{j'}\varepsilon_c^{ij}}{|A_i^{i'}|}, \tag{4.11}$$

$$\mu_C^{i'j'} = \frac{A_i^{i'}A_j^{j'}\mu_c^{ij}}{|A_i^{i'}|}, \tag{4.12}$$

where ${\varepsilon_c}^{ij}$ and ${\mu_c}^{ij}$ are the components of the permittivity and permeability tensors in the original Cartesian coordinate system. Here, ${\varepsilon_c}^{i'j'}$ and ${\mu_c}^{i'j'}$ are the components of the permittivity and permeability tensors of the transformed media in the new Cartesian coordinate system. The subscript C denotes that the parameters are associated with a Cartesian coordinate system.

From the above formulas, the core idea behind transformation optics is to determine the spatial mapping and spatial coordinate transformation required to control the EM wave propagation path in the desired way, and then calculate the constitutive parameters of the transform media. The final step in the process is to achieve control of the field by realizing the appropriate transformation media. This concept offers a technique that is capable of manipulating incident EM fields by tailoring their complex and desired spatial patterns. It facilitates a diverse range of applications in engineering and applied sciences, such as invisibility cloaking, focusing devices, multibeam lens antennas, and so on.

The traditional transformation optics approach has achieved some of the anticipated results through the design of customized EM materials. However, due to the fact that extreme anisotropy and inhomogeneity of materials may occur when employing the traditional transformation optics approach, difficulties have been encountered in the implementation process, which has brought great challenges to the practical realization of such EM devices. Although the traditional invisibility cloak has been successfully designed, fabricated, and experimentally validated, there are some rather restrictive limitations on its performance. For example, resonant metamaterials have been employed to realize magnetic media, which reduces the effective cloaking bandwidth and introduces a large amount of loss. Therefore, it is desirable to eliminate or minimize the anisotropy of the transforming medium by a judicious selection of coordinate transformations. In other words, it is advantageous to focus efforts on developing transformations that yield EM devices which can be realized by isotropic dielectric materials.

The emergence of quasi-conformal transformation optics (QCTO), proposed by Li et al. in 2008 [5], effectively solved this problem. Through quasi-conformal mappings of the coordinates in the virtual space and the physical space, the anisotropy in the transformed region can be minimized, and the device can be realized by a nonmagnetic, isotropic dielectric material. A simplified solution for calculating the coordinate mapping relationship between quasi-conformal coordinate systems was proposed by Chang et al. [6], which solves the inverse Laplace equation with slip boundaries. Hu et al. [7] proposed an equivalence between coordinate transformation and spatial deformation by using Laplace's equation to determine the deformation of coordinate grids during the transformation process. Neumann and Dirichlet sliding boundary conditions are set at the limits of the transformation zones to establish a quasi-conformal mapping between the physical and the virtual domains.

To ensure that there is no reflection at the boundaries of the mapping region, the wave impedance along the mapping structure must be constant. The formulation requires the calculation of a deformation tensor from the partial derivative of the displacement relative to the original coordinates, so the displacement field must be sufficiently smooth. Thus, it is necessary to have a Dirichlet boundary. The material tensor generated under the quasi-conformal transformation can be expressed as:

$$\frac{AA^T}{|A|} = \mathrm{Diag}\left[M^{-1}, M, (M|A_C|)^{-1}\right]. \tag{4.13}$$

The partial derivative of the mapping in the spatial transform is deduced by solving the Laplacian equations with Dirichlet boundaries:

$$\left(\frac{\partial^2}{\partial {x_1}'^2} + \frac{\partial^2}{\partial {x_2}'^2} + \frac{\partial^2}{\partial {x_3}'^2}\right) U_i = 0, \qquad i = 1, 2, 3, \tag{4.14}$$

where U_i denotes the coordinates in the original space. Through a simple calculation, the required mapping of the material parameters is found to be:

$$\varepsilon = \mu = \begin{bmatrix} 1 & 0 & 0 \\ 0 & 1 & 0 \\ 0 & 0 & |A|^{-1} \end{bmatrix}. \tag{4.15}$$

Consequently, if the polarization direction of the selected electric field is along the z-axis, the distribution of this characteristic parameter can be understood as isotropic over the entire area. The transformed material parameters are determined by the solution of Laplace's equation. The QCTO technique provides great flexibility and convenience for the arbitrary control of EM waves with more conventional materials [8–11].

In general, transformation optics (TO) gives us a powerful design tool to freely regulate the EM wave path, but the main drawback of this method is that it requires anisotropic materials. On the other hand, the QCTO theory provides a pathway for realizing such control over EM waves by properly tailored inhomogeneous isotropic media. The latter approach greatly reduces the difficulty of material realization thereby enabling the creation of a wide range of practical devices.

4.2.2 An Overview of 3D Printing Techniques

Lenses designed by the QCTO method can be realized by using an inhomogeneous all-dielectric material. The size of the metamaterial unit cell is small and continuously changes throughout space, which presents challenges for fabrication. In order to facilitate the processing, a perforated structure is generally used as a structural unit in the design. A mechanical perforation method is typically employed to realize the designed lens structure. With the gradual maturity and widespread popularity of 3D printing technology, there have also been attempts made to utilize it to create such gradually changing unit cell structures.

3D printing, also known as additive manufacturing, is a technology that has been developed for producing 3D products via a layer-by-layer material addition process, which has a broad range of exciting applications. 3D printing integrates digital modeling technology, electromechanical control technology, material science, chemistry, and other cutting-edge technologies. Compared with traditional manufacturing processes, 3D printing does not require that any molds be prepared in advance. Moreover, it is not necessary to use various means of subtractive manufacturing to remove large amounts of material, thereby reducing the assembly process and cost of each component. Therefore, 3D printing has the advantages of rapid processing, material savings, and energy savings. It is very suitable for new product development, personalized rapid speed manufacturing, and small batch as well as complex shape product manufacturing. It has been widely used in creating artistic products, jewelry making, mold processing, construction, clothing, bioengineering, and medicine. Compared with drilling and other conventional machining methods, the 3D printing process is simple and quick, which can reduce the cost and improve the competitiveness of lens-based antenna technology.

The 3D printing materials are the basis of the 3D printing process. The type of materials determines the processing method, degree of difficulty, and whether it can be widely used. At present, more common printing materials include engineering plastics, photosensitive resins, metallic and ceramic materials, etc. These printing materials all have their corresponding 3D printing processing systems. For the dielectric materials commonly used in microwave devices, the most popular printing technologies currently on the market are fused deposition modeling (FDM), stereolithography (SLA), and Polyjet.

FDM technology, depicted in Figure 4.2, is the most widely available in the 3D printing marketplace [12]. A nozzle is used to extrude the molten thermoplastic onto the 3D printing platform layer by layer until the final 3D model is formed. FDM technology does not require the use of lasers, has low equipment operation and maintenance costs, and can utilize a wide range of 3D printing materials, from the more common acrylonitrile butadiene styrene plastic/polylactic acid (ABS/PLA) to composites doped with a variety of enhanced powders, which are generally easy to obtain and inexpensive to engineer. In addition, the process of FDM printing only involves the melting and solidification of thermoplastic materials, which does not depend on such conditions as high temperature and high pressure, and does not produce toxic and harmful substances. Printing materials and supporting materials that are not used in printing or discarded in the process of use can be recycled, processed, and reused. However, the print resolution of FDM technology is usually not high, sometimes laminating on the surface of the printed object, which requires additional polishing to clean the surface of the part. In general, the FDM 3D printing process is also prone to temperature fluctuations, resulting in a slower/faster cooling rate of the thermoplastic filament material and a layered surface. It can be seen that FDM is suitable for making parts with low size requirements, low complexity, and low surface requirements. If there are requirements for higher accuracy and length of 3D printing, then other 3D printing technologies are usually selected.

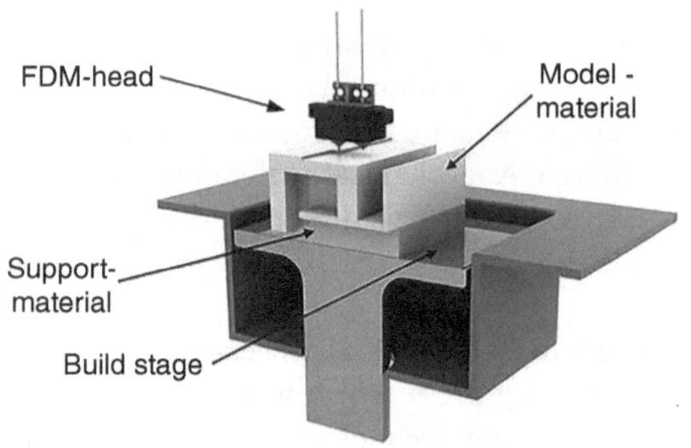

Figure 4.2 Schematic diagram of FDM printing technology.

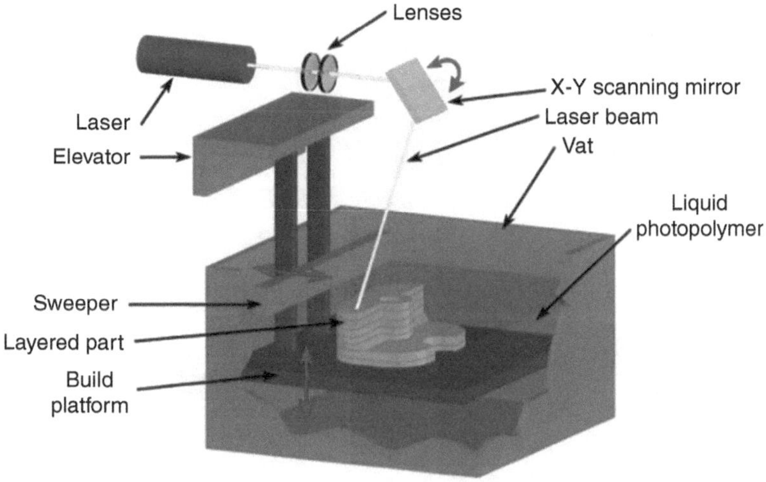

Figure 4.3 Schematic diagram of SLA printing technology.

Although the application of SLA printing technology [13] is not as extensive as FDM, in fact, SLA is the earliest additive manufacturing technology, which is illustrated in Figure 4.3. The main printing material used by SLA is photosensitive resin, which takes advantage of the characteristics of liquid photosensitive resin to quickly cure under the irradiation of an ultraviolet (UV) laser beam. When the process begins, the laser 'pulls' the first layer of printed matter into the photosensitive resin, and then the liquid is solidified in the targeted regions that are illuminated by the laser. A computer-controlled mirror guides the laser to the appropriate coordinates. At this stage, the laser is pointed at the building platform, which starts from a low position and gradually rises. After the first layer is completed, the platform is raised according to the layer thickness (usually about 0.1 mm), and additional resin is allowed to flow under the already printed part. Then, the laser beam cures the next cross-section and repeats the above process until the entire part is printed. The resin that has not been touched by the laser remains in the barrel and can be reused. After the material is polymerized, the platform is raised from the tank to drain the excess resin. At the end of the process, the model is removed from the platform, the excess resin is washed off and then placed in an ultraviolet oven for final curing, after which the object is strengthened and becomes more stable. SLA printing technology has faster processing speeds, shorter production times, and higher printing accuracy. Moreover, it does not require other cutting tools during the printing process. It can process prototypes and molds with complex structural shapes or that are difficult to form using traditional means. However, SLA printing systems are very expensive, the cost of using and maintaining them is very high, and there are very strict requirements on the working environment, which restricts applications in engineering to some extent.

Polyjet [14], a 3D printing technology developed by Objet Geometries in 2000, was the first to allow almost any company of any size to produce complex models quickly and with high quality, as demonstrated in Figure 4.4. Polyjet technology uses an array nozzle to print the mold. The nozzle moves back and forth along the X/Y axis to spray the photopolymer. When the photosensitive polymer material is sprayed onto the table, a lamp will emit UV light along the direction of the nozzle to cure the photosensitive polymer material. After completing the jet printing and curing of one layer, this process will be repeated, layer by layer, until the entire 3D object is formed. Compared with other printing technologies, Polyjet has a very high printing accuracy. The laser spot used in

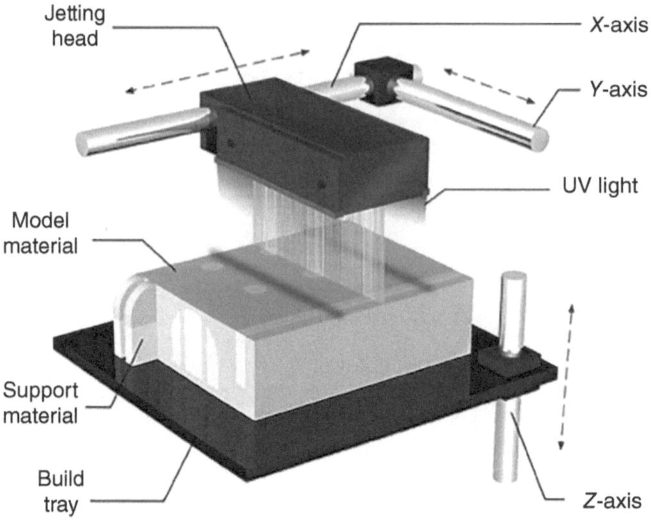

Figure 4.4 Schematic diagram of Polyjet printing technology.

this technology is between 0.06–0.10 mm and has a layer resolution of up to 14 μm. Another feature of Polyjet technology is the rapid molding speed. Each layer of photosensitive polymer material is solidified with UV light immediately after being sprayed, and can be transported and used right away without solidifying afterward. In addition, owing to the multinozzle printing mechanism, Polyjet can simultaneously select a variety of different materials, so as to achieve digital full-color printing. However, it should be noted that Polyjet technology has a high printing cost, and it requires a specially developed photosensitive resin; the workpiece strength and durability of the molded parts are not too high. Hence, it is more suitable for customers that wish to check the product design appearance and structure.

In addition to the aforementioned 3D printing methods , there are some other printing methods available in practice, such as selection laser fusion molding technology, 3D printing rapid prototyping (3DP), continuous liquid interface extraction technology, and so on. Each 3D printing technology has its distinguishing characteristics, and the appropriate method can be selected according to different printing requirements.

4.3 Several Applications of Transformation Optics in Vortex Waves

4.3.1 All-Dielectric Transformed Material for the Generation of OAM Beams

The angular momentum of EM waves can be divided into two classes; namely, spin angular momentum and OAM [15]. Spin angular momentum is related only to the spin of a photon and is characterized by a circular polarization state. OAM, on the other hand, depends on the field spatial distribution and is associated with the beam vorticity and phase singularity, which imparts to the EM wave a helical transverse phase distribution. Therefore, an EM wave carrying OAM modes is referred to as a vortex EM wave that presents a phase quantified by $e^{il\phi}$ in its wavefront, where l is the intrinsic value of OAM ($l = 0, \pm 1, \pm 2$) and ϕ is the transverse azimuthal angle [16]. OAM vortex waves with different topological modes are mutually orthogonal to each other, an important property that has inspired a large number of researchers to explore their intriguing physical properties throughout the entire EM spectrum. OAM vortex waves in the microwave regime can find potential applications in the fields of data transmission, radar imaging, target detection, and sensing [17–19].

Because of the fascinating properties and numerous potential applications of OAM waves, different methods have been proposed to generate OAM beams in the radio frequency (RF) band, including antenna arrays [20], metasurfaces [21], spiral phase plates [22], and helicoidal parabolic antennas [23]. Among these methods, antenna arrays have been extensively studied over the past several years. However, the main drawback is that a complicated feeding network is required to generate a radiated wave with a spiral phase profile. When the desired topological charge of OAM is large, a high number of radiating elements are required, which significantly increases the design and feeding complexities of the system. On the other hand, using a metasurface to generate OAM waves possesses several advantages, such as high gain, a small divergence angle, a low-profile structure, and an easy fabrication process. However, this approach generally suffers from a narrow bandwidth or low efficiency. The spiral phase plate is the most widely used, and is characterized by a straightforward theory and simple structure, while offering high conversion efficiency. Nevertheless, in the RF band, the divergence angle of the generated beam is not convenient for long-distance transmission and the structure responds only to a specific frequency. Finally, the helicoidal parabolic antenna is also an effective device with high efficiency. It is essentially composed of a reflecting rotating phase plate designed as a parabolic dish with a single radial cut introduced on the surface. A spherical wavefront emitted by a feed antenna located at the parabola's focal point is reflected, forming a directive outgoing vortex beam. However, because of the irregular shape and bulkiness of the helicoidal parabolic antenna, its integration in practical communication systems is not very straightforward or convenient. Compared with the optical regime, the generation of vortex beams in the RF band depends mostly either on complex structures that are difficult to integrate or on the feeding and coupling circuits, which lead to narrow bandwidth. To address these shortcomings, it is necessary to pursue a technology that can achieve OAM-wave generation while simultaneously offering a compact form factor, planar device configuration, and broad operational bandwidth.

In this section, according to the concept of spatial transformation, an all-dielectric microwave device that is capable of generating OAM EM waves with mode +1 in the reflection state is presented [24, 25]. By solving Laplace's equation under proper Dirichlet and Neumann boundary conditions, the transformation mapping between the virtual space and physical space zones is established. Thus, for a practical implementation of the device, eight constituent sectors representing different phase variations are extracted from the calculations to cover the total phase range of 2π necessary for vortex-beam generation. Each sector has similar physical dimensions but a different effective-permittivity distribution. The design for the proposed all-dielectric device is presented along with numerical simulations predicting its performance. Furthermore, a reflection-mode prototype capable of generating an EM wave carrying an OAM mode $l = +1$ in the microwave domain is fully implemented by dielectric 3D printing, and exhibits qualitatively good agreement between simulated and measured results. Importantly, because of the use of nonresonant metamaterial structures, the device possesses a substantial broad operational bandwidth.

The spatial-transformation-based generation method is schematically presented in Figure 4.5. A helicoidal parabolic structure composed of several discrete surfaces is considered. One of the surfaces having a central angle of $\varphi = \pi/4$ is shown in Figure 4.5a for illustrative purposes. When an EM beam is incident on the surface, the different regions of the wavefront produce relative phase delays due to the nonplanar spiral structure of the metal parabola, thus achieving the effect of wave

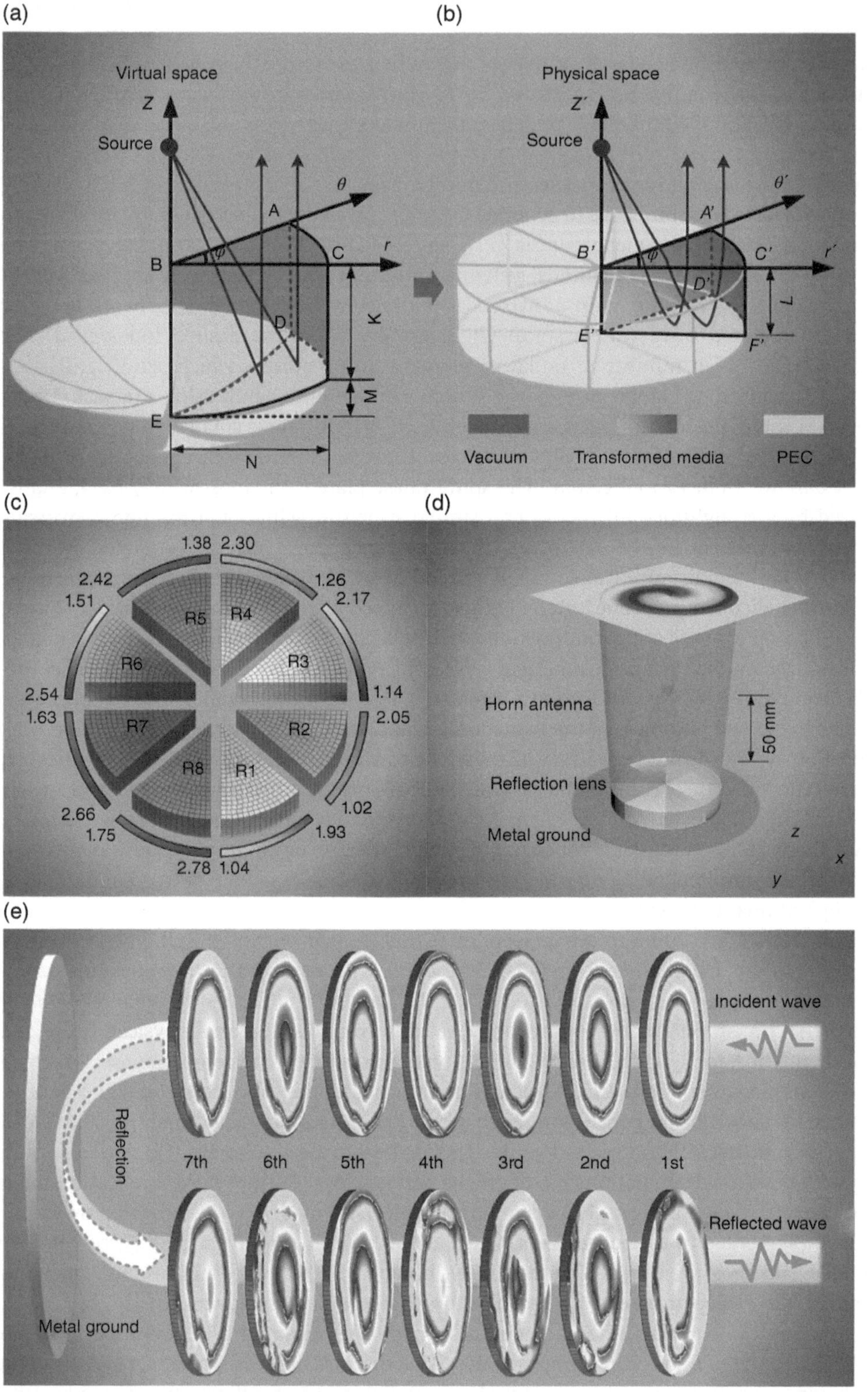

Figure 4.5 Schematic representation of the spatial transformation from virtual space to physical space, where the curved virtual space is transformed into the proposed OAM generation device design: (a) Virtual space and (b) physical space of reflection-type device; (c) calculated permittivity distributions of the OAM generator; (d) operating principle of the device; (e) phase variation of the incident wave after passing through each layer of the vortex-beam generator at 12 GHz.

front twisting. Based on the concept of spatial transformation, a flat device that performs the same function as the helicoidal parabolic reflector is shown in Figure 4.5b.

The curved surface *DEF* and planar surface *DEF* are perfect-electric-conductor boundaries. The thickness of the physical spaces in the reflective device is set as a constant *L*. Neumann and Dirichlet sliding boundary conditions are set at the edges of the device as follows:

$$\begin{cases} r'|_{\mathrm{AB,BC,AC,TD,AD,CF,BE}} = r \\ \hat{n}\cdot\nabla r'|_{\mathrm{EF,DE}} = 0 \end{cases}, \tag{4.16}$$

$$\begin{cases} z'|_{\mathrm{AB,BC,AC}} = 0 \\ z'\Big|_{\mathrm{EF,DE}} = \dfrac{M}{N^2}(r)^2 - (K+M), \\ \hat{n}\cdot\nabla z'|_{\mathrm{BE,AD,CF,DF}} = 0 \end{cases} \tag{4.17}$$

where $\hat{n}$ is the vector normal to the boundaries of the surface.

To obtain the expected mapping between free space and the physical space, the partial differential-equation solver COMSOL Multiphysics is used to solve Laplace's equation subject to the defined boundary conditions. In our design, the geometry parameters denoted in Figure 4.5a are as follows: $N = 105$ mm, $L = 35$ mm, and $M = 19.23$ mm. The physical dimension of each elementary unit is set to $5 \times 5 \times 5$ mm^3. Moreover, each proposed sector is discretized into seven layers along the z axis and each layer consists of 21 circular rings. As a result, the entire device with an oblate cylindrical shape is composed of 1176 different circular rings containing a total of 9576 cubical unit cells. Through simulation and optimization of the variable *K*, eight sectors for the device with different permittivity distributions are used, as presented in Figure 4.5c. According to the continuous permittivity distribution, the entire space is discretized into a plurality of building blocks of similar size. The permittivity of each sector shares a similar distribution but a different range, while the phase variation between adjacent sectors is $\pi/4$. It is worthwhile noting that with ordered combinations of such sectors, the transformed medium can be engineered in an effective manner to generate vortex beams with different modes of OAM. For example, when eight modules are arranged in a clockwise/counterclockwise order, a vortex EM wave with modes of ± 1 can be generated; when modules one, three, five, and seven are arranged in a clockwise/counterclockwise order, a vortex wave with modes of ± 2 can be generated, as shown in Figure 4.6.

Full-wave simulations are performed to verify the functionalities of the OAM-beam-generation device with the finite-element-method-based program HFSS at 12 GHz. A regular X-band horn antenna emitting quasi-spherical waves located along the central axis of the microwave device is used as the feeding source and is placed 50 mm ($2\ \lambda_0$ at 12 GHz, with λ_0 being the free-space wavelength) from the upper surface of the device. The functionality of the reflection-type generator is numerically characterized as illustrated in Figure 4.5d, where the designed transformation medium is backed by a circular reflecting metal plate.

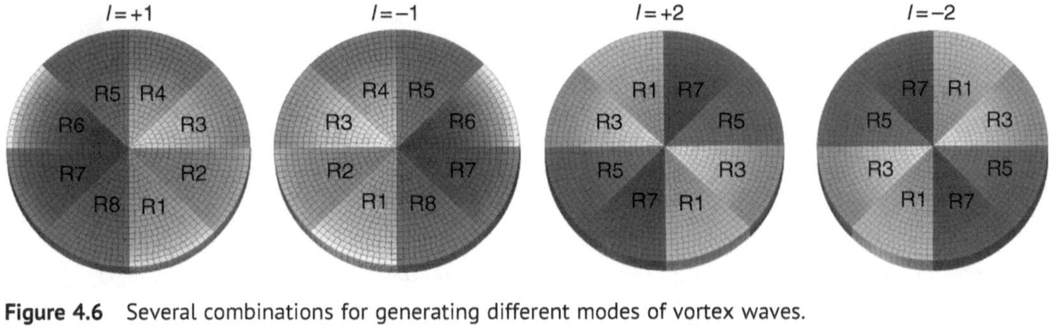

Figure 4.6 Several combinations for generating different modes of vortex waves.

Since the refractive index and wavelength are different in different media, the ray path and propagation period of the incident wave will change as the EM wave propagates through the device. To further observe the phase change of the incident wave inside the transformed medium, the phase profiles are simulated in each layer at 12 GHz and are plotted in Figure 4.5e. It can be clearly observed how the quasi-spherical wave is transformed to the vortex wave inside the device in a step-by-step fashion after being reflected from the metal plate.

Then, the generation lenses with different modes were simulated at the operating frequency of 12 GHz, and vortex waves with modes of ±1 and ±2 were obtained. The phase distribution and 3D radiation patterns are shown in Figure 4.7. The helical phase distribution (see Figure 4.7a), a characteristic of EM waves carrying an OAM mode, is clearly revealed in the simulated data, which indicate that vortex beams with a topological charge $l = \pm 1, \pm 2$ are indeed generated by the permittivity distributions used in the device. Furthermore, simulated 3D far-field radiation patterns of the proposed devices are presented in Figure 4.7b. It can be observed that the center energy of the far-field radiation pattern of the vortex waves of different modes is very small, and a relatively obvious depression is generated, which conforms to the energy distribution characteristics of EM vortex waves.

In order to observe the performance of the lens at different frequencies, the lenses with modes of +1 and +2 were simulated at different operating frequencies. Figure 4.8 presents the phase distributions of the two lenses at different frequencies. Figure 4.8a and b show the phase distribution of the mode +1 and +2 vortex waves at 8–16 GHz. Both lenses exhibit relatively good vortex characteristics. These simulation results verify the feasibility of the proposed reconfigurable multimode vortex wave generation method.

Subsequently, the lens generating the +1 mode vortex wave is further studied. The intensity patterns of the electric field in a plane normal to the direction of propagation as well as characteristics of the far-field patterns in the *E*-plane (*x-o-z* plane containing $\boldsymbol{E}$ and $\boldsymbol{k}$ vectors) and *H*-plane (*y-o-z* plane containing $\boldsymbol{H}$ and $\boldsymbol{k}$ vectors) obtained from numerical simulations are depicted in Figure 4.9. The simulated intensity profiles are illustrated in Figure 4.9a, where an amplitude null in the center of the beam can be clearly seen, which is the main feature of EM waves carrying OAM. The 2D far-field radiation patterns show beams with a hollow main lobe (Figure 4.9b), which is also an intrinsic characteristic of vortex beams, providing further evidence that the generated beam carries an OAM mode. Moreover, the phase and intensity profiles of the near-field distributions for mode $l = +1$ are further compared from 8 to 16 GHz, demonstrating that the OAM generation device is capable of

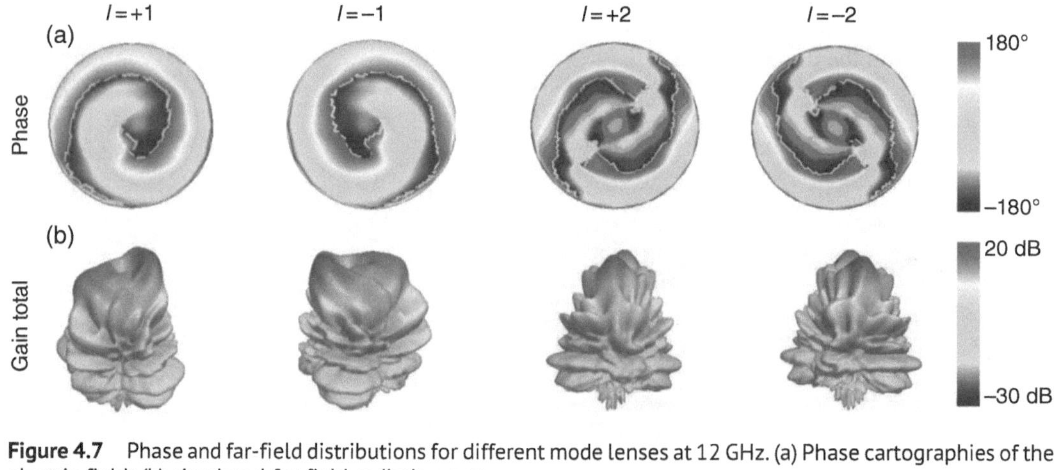

Figure 4.7 Phase and far-field distributions for different mode lenses at 12 GHz. (a) Phase cartographies of the electric field; (b) simulated far-field radiation patterns.

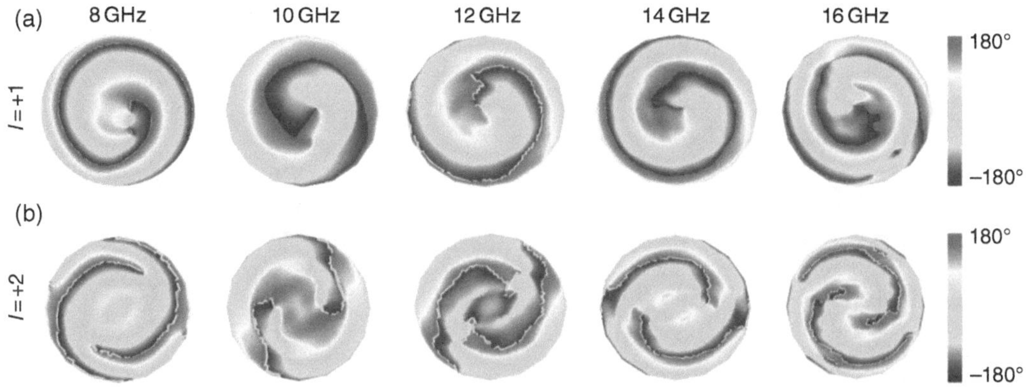

Figure 4.8 Phase distributions of designed devices at 8, 10, 12, 14, and 16 GHz. (a) +1 mode configuration; (b) +2 mode configuration.

retaining performance over a broad frequency range. Theoretically, the achieved bandwidth can be very broad since all-dielectric nonresonant materials are used in the design. However, for realistic discrete structures, the design dimensions and unit-cell periodicity limit the operating bandwidth.

The corresponding power distributions of the proposed device are analyzed based on the purity calculation of OAM modes. Since the azimuthal angle φ is a periodic function, its Fourier conjugate, which is the OAM beam, is a discrete variable. The associating Fourier relationship is given in Ref. [26].

$$A_l = \frac{1}{2\pi}\int_0^{2\pi} \psi(\varphi)d\varphi e^{-jl\varphi}, \tag{4.18}$$

$$\psi(\varphi) = \sum_l A_l e^{jl\varphi}, \tag{4.19}$$

The calculated results at different frequencies are shown in Figure 4.9c. At a frequency of 8 GHz, the total phase deflection after the phase plate does not reach 2π, and therefore in the expansion of the Fourier transform, part of the generated OAM waves are decomposed into modes $l = 0$ and $l = +1$. It can be observed that the expected OAM mode $l = +1$ is relatively dominant at 10, 12, 14, and 16 GHz, which indicates the high quality of the generated OAM beams. To experimentally validate the proposed spatial-transformation approach, a prototype of the reflection-type device generating a vortex beam carrying the OAM mode $l = +1$ was physically realized. To achieve this, first, a 3D model of the all-dielectric device is designed for implementation. To realize a permittivity ranging from 1 to 2.8, two kinds of meta-atom structures are adopted, as illustrated in Figure 4.10c. The elementary meta-atom is composed of three orthogonally crossed cylinders in a cubic region of $5 \times 5 \times 5$ mm^3 (λ_0/5 at 12 GHz). The effective relative permittivity of a unit cell depends on the radius of the cylinders. To ensure a smooth connection between adjacent cells, the meta-atom can be bent and trimmed appropriately. The prototype of the designed device is then fabricated by a Stratasys Objet260 Connex3 polymer jetting 3D printer [27] with the use of a Verowhite dielectric material, which has relative permittivity $\varepsilon_h = 2.8$ in the frequency band of interest. A photograph of the printed flat cylindrical device, which consists of 9576 unit cells, with a height of 35 mm and a base radius of 105 mm, is presented in Figure 4.10b.

Two experimental setups are used to validate the functionality of the OAM-generation device. First, the electric field is scanned with a field-sensing fiber-optic active antenna probe capable of measuring both the intensity and the phase of the electric field through a vector network analyzer

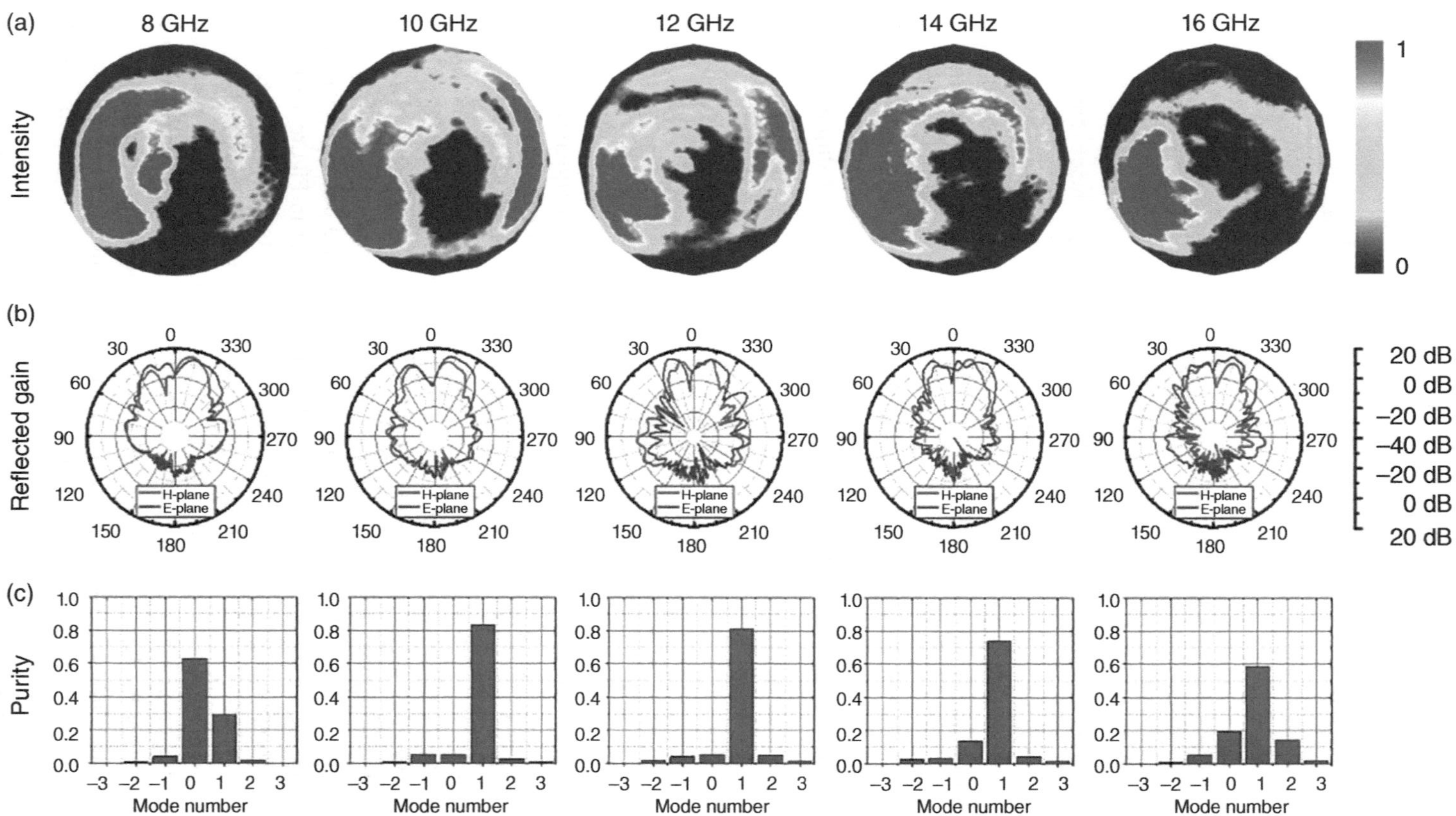

Figure 4.9 Numerical simulation results for the OAM generation device with mode $l = +1$ in a frequency range spanning from 8 to 16 GHz. (a) Intensity distributions of the EM field in the *xoy* plane; (b) 2D far-field radiation patterns in the ***E***- and ***H***-planes of the device, where a characteristic 'bunny ear'-shaped diagram is obtained; (c) calculated mode spectra of the generated $l = +1$ OAM beam.

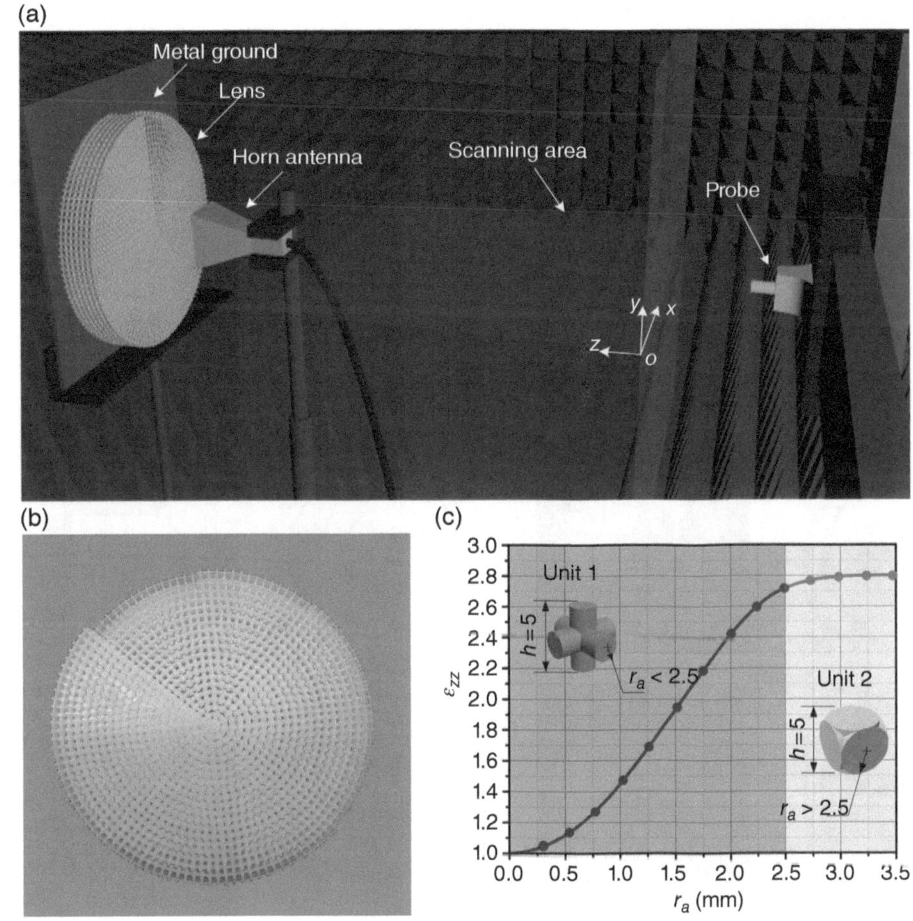

Figure 4.10 (a) Experimental measurement setup of the proposed device in a microwave anechoic chamber; (b) photograph of the fabricated prototype created by Polymer Jetting 3D printing; (c) effective permittivity of two types of unit cells used in the implementation of the device. A parametric analysis is performed to extract the effective permittivity value from the homogenization procedure according to the radius r_a of each unit cell.

(VNA). Hence, at the receiving end, the probe, which is connected to the network analyzer via a coaxial cable, is mounted on two orthogonal computer-controlled linear-motion stages. A quasi-spherical wave is emitted along the z-direction by a regular X-band horn antenna placed 50 mm directly in front of the device. The measurement setup is illustrated in Figure 4.10a. The all-dielectric device, metal plate, and probe are placed in the x-o-y plane. By stepping the field-sensing probe in increments of 2 mm and recording the reflected field intensity and phase at every step, a full 2D spatial-field mapping of the microwave electric field distribution in the x-o-y plane (located at a distance of $z = 500$ mm) over a total scanning surface of 400×400 mm^2 was acquired. Second, far-field antenna radiation patterns were measured in an anechoic chamber, where a horn antenna is used as a receiver to measure the power emitted from the vortex wave generating system. Far-field measurements are performed for elevation angles ranging from −180 to +180°.

The measured phase distributions and intensity patterns at different frequencies are given in Figure 4.11. In Figure 4.11a, when the incident and reflected waves pass through the dielectric device, the phase distribution shows a total 2π phase change in a helical configuration, which is

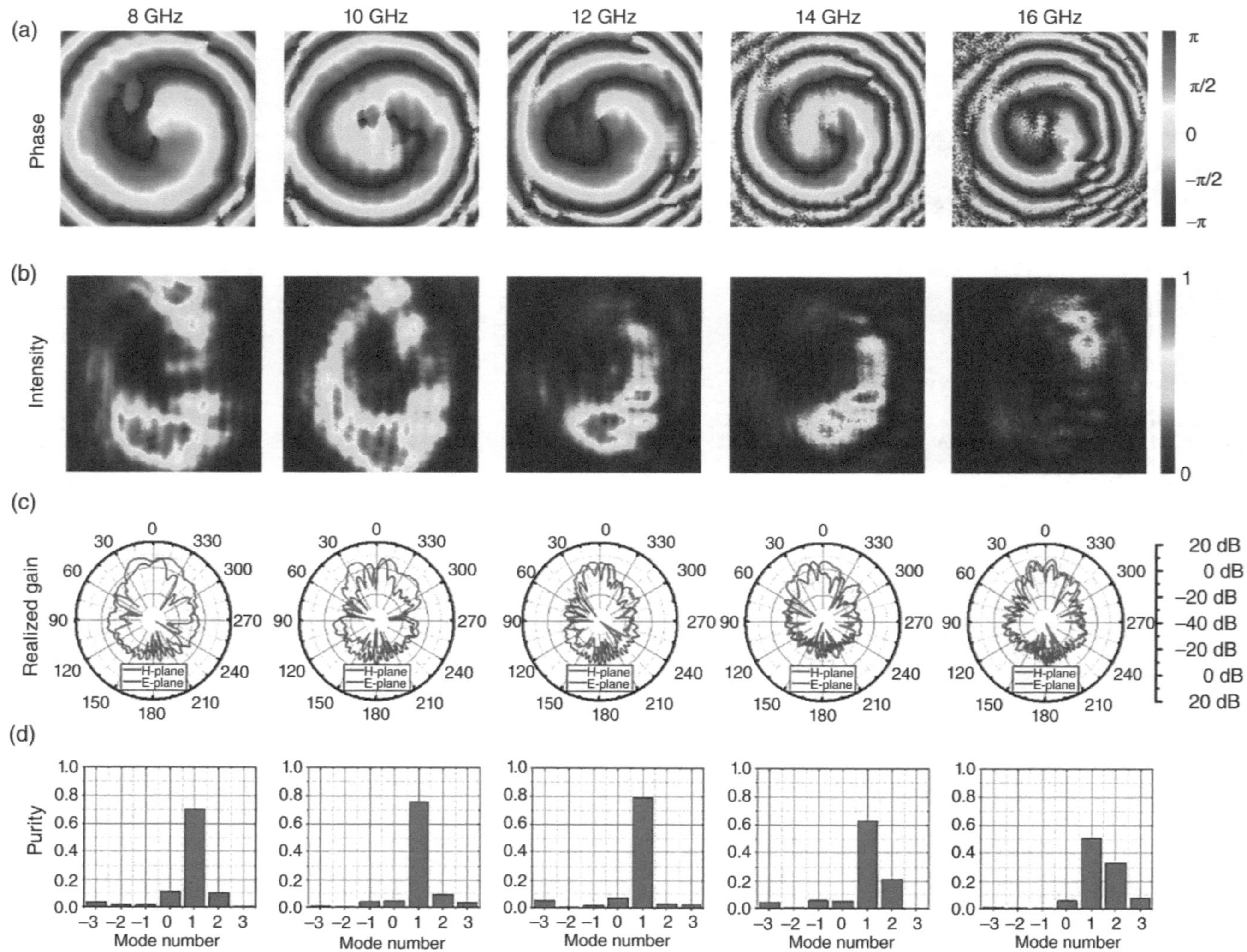

Figure 4.11 Measured performances at 8, 10, 12, 14, and 16 GHz. (a) Phase distributions of the EM field in the *xoy* plane at a distance of 500 mm away from the feed plane. The phase varies from $-\pi$ (light) to π (dark); (b) Intensity distributions of the EM field in the *xoy* plane at a distance of 500 mm away from the feed plane; (c) 2D radiation patterns in the ***H***- and ***E***-planes; (d) experimental mode spectra of the generated $l = +1$ OAM beam.

characteristic of EM waves carrying an $l = +1$ OAM mode. The phase profiles measured from the vortex-beam generator are consistent with those achieved by simulations. As illustrated in Figure 4.11b, the intensity patterns measured with the OAM-generation device have a doughnut shape with a null at the center. Without the transformed material, concentric circular phase patterns and quasi-regular intensity patterns of a plane wave are observed.

Additionally, the measured far-field patterns in the E- and H-planes are presented in Figure 4.11c. It can be clearly seen that the fabricated device produces 'bunny ear'-shaped beams in the frequency band from 8 to 16 GHz. Furthermore, the OAM-mode purity calculated from the experimental data at different frequencies is presented in Figure 4.11d. It can be observed that generated OAM waves with the +1 mode still dominate, validating the efficiency of the designed device. It is important to point out that the intensity distributions of the generated vortex waves are not fully homogeneous and the far-field radiated beams are slightly tilted. These two undesired phenomena can be attributed first to the quasi-isotropic material used to make the device and second to the superposition of other undesired generated modes.

4.3.2 All-dielectric Metamaterial Medium for Collimating OAM Vortex Waves

In wireless communications, EM waves carrying OAM can obviously improve data transmission efficiency due to their particular characteristic of multiple orthogonal modes [28]. However, the divergent peculiarity of vortex waves seriously limits the propagation distance for effective communication.

In this section, a medium performing as a collimating lens for vortex waves is designed by the transformation optics concept and manufactured by additive printing technology [29]. The design concept is presented and the lens is both numerically simulated and experimentally measured at microwave frequencies. The measured near-field amplitude distribution indicates that the divergence angle and the diameter of the doughnut-shaped vortex wave passing through the collimating lens are greatly reduced over a broad frequency range. The far-field antenna gain patterns show that the energy density of the vortex wave is significantly improved, while the phase profile and mode purity show that the topology charge of the vortex wave remains unchanged after passing through the collimating lens. The proposed all-dielectric collimating lens allows to increase the realized gain of vortex waves and may extend the propagation distance, laying the foundation for future vortex wave communications.

The design of the collimating lens is schematically illustrated in Figure 4.12. The color gradient from light to dark represents the direction of propagation of the vortex wave in both spaces. The space enclosed by the conical structure filled with vacuum is initially set as the virtual space (Figure 4.12a). The final desired physical space is the volume enclosed by the cylinder, as shown in Figure 4.12b. A Laguerre–Gaussian (LG) beam possessing a broad main lobe is transformed into a high-order Bessel (HOB) beam with a narrower lobe when it passes through the transformation medium. Using the Fresnel diffraction integral theory, the process of this transformation in Cartesian coordinates can be described as [30]:

$$E_{LG}(x, y, 0) = A\left(\frac{2r^2}{\omega_0{}^2}\right)^{\frac{|l|}{2}} L_p^{|l|}\left(\frac{2r^2}{\omega_0{}^2}\right) e^{\frac{-r^2}{\omega_0{}^2}} e^{il\phi_{(x,y)}}, \tag{4.20}$$

where $r^2 = x^2 + y^2$ and $\phi_{(x,y)} = \arctan\left(\frac{y}{x}\right)$,

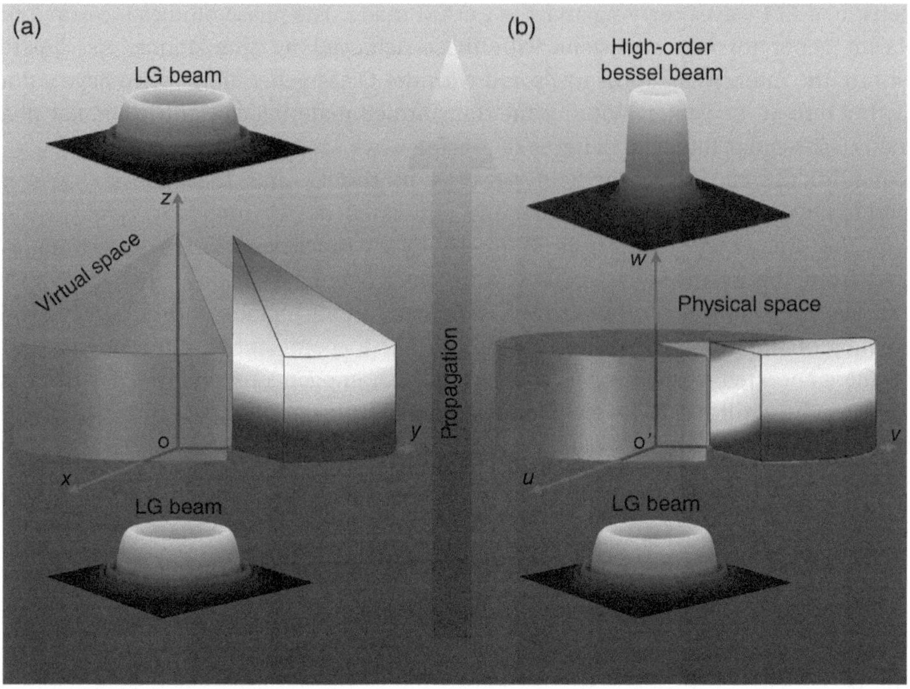

Figure 4.12 Schematic principle of the spatial transformation for designing the collimating lens. (a) Virtual space with uniform air medium; (b) physical space with gradient index medium.

$$
\begin{aligned}
E_{HOB}(x,y,z) &= \frac{e^{ikz}}{i\lambda z}\int_S E_{LG}(x',y',0)e^{-ik_r r'} \times e^{\frac{ik\left((x-x')^2+(y-y')^2\right)}{2z}}dx'dy' \\
&= \frac{ie^{ikz}}{i\lambda z}e^{ik\left(z+\frac{r^2}{2z}\right)}\int_S A\left(\frac{\sqrt{2}r'}{\omega_0}\right)^l e^{il\phi'_{(x,y)}} \times e^{\frac{-r'^2}{\omega_0{}^2}}e^{-ik_r r'}e^{\frac{-ikr'^2}{2z}}e^{\frac{ik(xx'+yy')}{z}}dx'dy',
\end{aligned}
\tag{4.21}
$$

where $E_{LG}(x, y, 0)$ represents the electric field distribution of the LG beam incident on the flat circular base of the axicon. $E_{HOB}(x, y, z)$ represents the electric field distribution of the transformed HOB beam, and $e^{-ik_r r'}$ represents the phase factor of the axicon. Equation (4.21) may be simplified by the principle of stationary phase as:

$$
E_{\mathrm{HOB}}(x,y,z) = \frac{2\pi A k_r}{i\lambda}(-i)^l\sqrt{\frac{1}{kz}}e^{ik\left(z+\frac{r^2+2}{2z}\right)} \times e^{\frac{-(k_r z)^2}{(k\omega_0)^2}}e^{il\phi_{(x,y)}}\left(\frac{\sqrt{2}k_r z}{k\omega_0}\right)^l J_l(rk_r), \tag{4.22}
$$

where $J_l(rkr)$ represents the Bessel polynomial.

In Figure 4.12, the cone at the top of the virtual axicon space is squeezed down into a flat surface to form a cylindrical physical space characterized by a nonlinear permittivity variation. The spatial transformation between the virtual and physical spaces is calculated by solving Laplace's equation with predefined Dirichlet and Neumann boundary conditions [6, 31].

The coordinates of the virtual space are represented by (x, y, z), while the coordinates of the physical space are represented by (u, v, w). The spatial transformation function relationship between the virtual and physical spaces can be expressed as:

$$
u = u(x,y,z), v = v(x,y,z), w = w(x,y,z). \tag{4.23}
$$

In order to further determine the mapping relationship, the boundary conditions are set in the physical space coordinates as:

$$\begin{cases} S_1 = \{S|u^2 + v^2 \leq R^2, w = 0\} \\ S_2 = \{S|u^2 + v^2 = R^2, w \in (0, L)\}, \\ S_3 = \{S|u^2 + v^2 \leq R^2, w = L\} \end{cases} \tag{4.24}$$

$$\begin{cases} (x, y)|_{S_1,S_2} = (u, v) \\ z|_{S_1} = 0 \\ z|_{S_3} = L + L_1 - \sqrt{u^2 + v^2} \cdot L_1/R, \\ \vec{n} \cdot \nabla(x, y)|_{S_3} = 0 \\ \vec{n} \cdot \nabla z|_{S_2} = 0 \end{cases} \tag{4.25}$$

where $\vec{n}$ represents the normal vector to the surface boundaries. L and R represent, respectively, the height and radius of the gradient index medium in the physical space and L_1 represents the height of the cone. The permittivity distribution of the physical space can be calculated by using the Jacobian matrix, which is obtained by solving Laplace's equation according to the invariance of Maxwell's equations:

$$\begin{cases} \left(\frac{\partial^2}{\partial x^2} + \frac{\partial^2}{\partial y^2} + \frac{\partial^2}{\partial z^2}\right) u = 0 \\ \left(\frac{\partial^2}{\partial x^2} + \frac{\partial^2}{\partial y^2} + \frac{\partial^2}{\partial z^2}\right) v = 0 \\ \left(\frac{\partial^2}{\partial x^2} + \frac{\partial^2}{\partial y^2} + \frac{\partial^2}{\partial z^2}\right) w = 0 \end{cases} \tag{4.26}$$

$$\varepsilon = \frac{\varepsilon_r}{\det(J^{-1})}, \mu = 1.$$

In the TO process, the phase factor of the plano-convex axicon lens is mimicked through the use of judiciously designed metamaterials. The gradient refractive index of the flattened transformed lens is therefore equivalent to the phase of the axicon and is able to transform a vortex wave, which is a type of LG beam, into a HOB beam. Similarly, while the direction of propagation of the vortex wave passing through the virtual space does not change, collimation occurs when it passes through the physical space.

Next, the discrete permittivity distribution of the transformed medium is obtained for the collimating lens and is shown in Figure 4.13c. Permittivity discretization is the key to simplify the manufacturing process of the lens. Based on the effective medium model, the permittivity distribution enclosed by the cylindrical medium is divided into 12 flat cylindrical sections. Each flat cylinder is further divided into a plurality of meta-atoms along the radial direction with different heights and radii. A meta-atom is composed of a cylindrical air hole having a radius that can vary from 0 to 1.9 mm made in a cubic dielectric host material block with a height ranging from 2 to 5 mm and with a side length of 5 mm, as depicted in Figure 4.13a. The ratio of the volume of the hole to the volume of the block determines the permittivity. The permittivity, considered to be constant in each building block, is calculated and is taken to be equal to the average permittivity of the block. As such, a cylindrical collimating lens with a diameter of 14.4 cm and a height of 4.8 cm is designed, which consists of 216 rings.

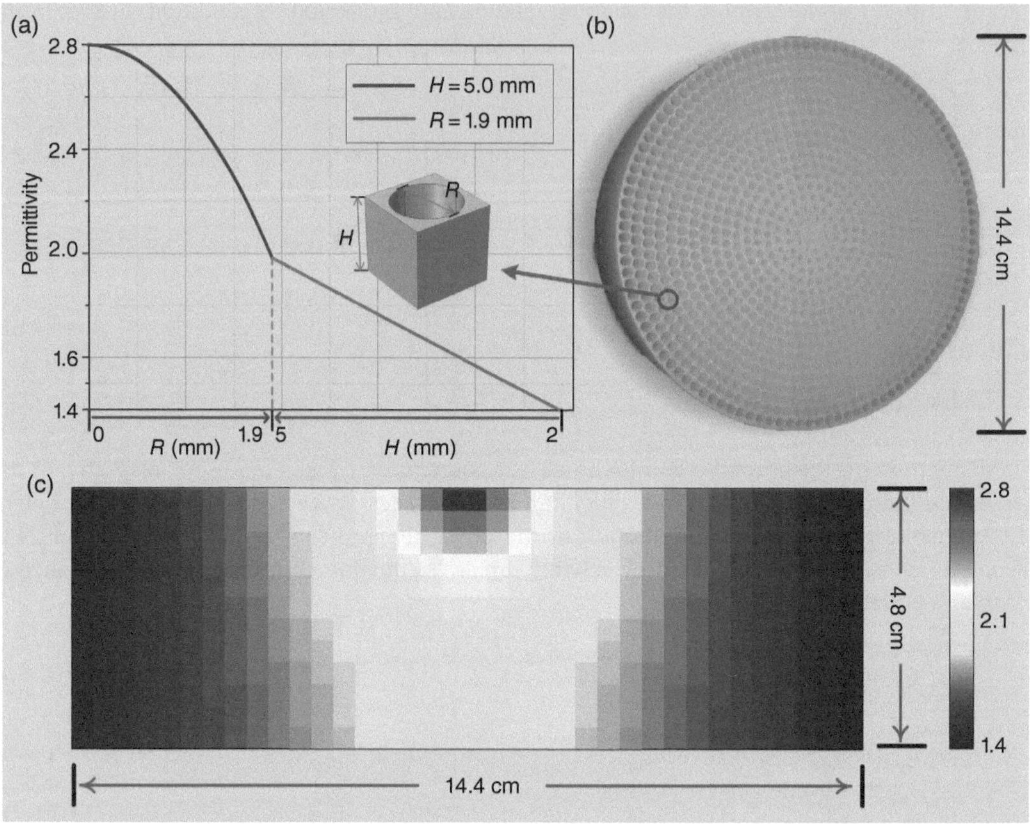

Figure 4.13 (a) Permittivity curve of the engineered unit cells; (b) photograph of the 3D-printed collimating lens prototype; (c) discrete permittivity distribution in a cross-section of the collimating lens.

The unit cells are designed to imitate the permittivity distribution inside the lens. As illustrated by the photograph in Figure 4.13b, the all-dielectric collimating lens was manufactured by dielectric 3D printing technology using a resin material having a permittivity of 2.8. Due to the use of non-resonant all-dielectric building blocks, the losses can be considered low. In each flat cylindrical layer, the meta-atoms are arranged in a concentric configuration and share the same permittivity within an annular region. As shown in Figure 4.13a, when $H = 5.0$ mm, the permittivity gradually decreases from 2.8 to 1.98 as the radius of the cylindrical cavity increases. For a fixed radius $R = 1.9$ mm, the permittivity gradually decreases to 1.4 as the height H of the cubic unit cell decreases. In order to ensure the mechanical stability of the collimating lens structure, the radius of the hole cannot be greater than 1.9 mm. The portion having a permittivity of less than 1.98 is achieved by reducing the height of the cubic host unit. A metamaterial structure respecting all required permittivity values in the collimating lens is then designed.

The proposed collimating lens is illuminated by a set of microstrip patch array antennas designed to generate vortex beams carrying OAM with a topology charge of +1 [32]. Each patch array operates at a different frequency and allows the frequency range spanning from 6 to 14 GHz to be covered. Each array is composed of eight circular metallic patches arranged in a circular configuration with a radius of 0.6 λ on a Rogers 4003(tm) dielectric substrate having a relative permittivity of $\varepsilon_r = 3.55$, a loss tangent of $\tan\delta = 0.0027$, and a thickness of 0.5 mm. The radius of each patch is r_p and

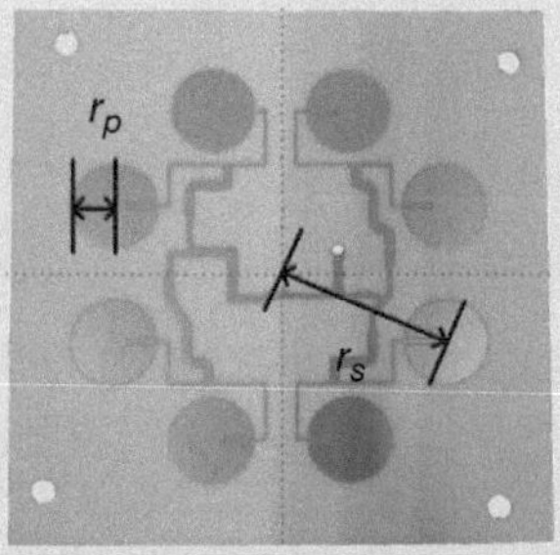

Freq (GHz)	6	8	10	12	14
r_p (mm)	7.8	5.68	4.6	3.89	3.33
r_s (mm)	33.4	25	20	16.7	14.3

Figure 4.14 Photograph of the fabricated prototype of the microstrip patch array used to generate OAM vortex beams. The table shows the different geometrical parameters of the patch arrays used at selected frequencies.

the lateral dimensions of the substrate are 100 × 100 mm. As shown in Figure 4.14, the radii of the circular copper patches in the different feeding sources are 7.8, 5.86, 4.6, 3.89, and 3.33 mm for the corresponding operating frequencies of 6, 8, 10, 12, and 14 GHz. In order to generate the vortex waves with mode +1, feeding lines are designed to create a phase shift of $\pi/4$ between two adjacent patches in the clockwise direction.

The patch arrays were simulated and numerically verified using ANSYS HFSS (High Frequency Structure Simulator), which is a finite-element-based Maxwell's equations commercial solver. The phase distributions obtained from the simulations are presented in Figure 4.15b. It can be observed that the spiral phase distribution covers 2π in the clockwise direction as expected, meaning that OAM mode +1 is generated. Moreover, the normalized 3D antenna radiation patterns are also obtained and presented in Figure 4.15c. The 3D radiation patterns possess a characteristic null in the main beam, which is an intrinsic property of OAM beams.

The microstrip array antennas are fabricated by a conventional printed circuit board technique and measured in a microwave anechoic chamber. Scattering coefficient responses of the different array antennas alone are shown in Figure 4.15a. It can be clearly observed that all five of the designed array antennas operate well at their corresponding frequencies (6, 8, 10, 12, and 14 GHz).

In order to evaluate the influence of the collimating lens on vortex waves with any possible topological mode, simulations of the lens illuminated by different OAM modes are performed. The EM field amplitude distributions (without and with lens) in the xoy plane at $z = 240$ mm for the different considered configurations are presented at 10 GHz in Figure 4.16a and b. The diameter of the energy ring is much smaller after passing through the collimating lens. From the amplitude distributions displayed in Figure 4.16c and d in the xoz plane at $y = 0$ mm, it can be noted that the angle between the two main lobes is 20 and 36° for $l = \pm1$ and $l = \pm2$, respectively. Therefore, the collimating lens is capable of reducing the divergence of vortex waves carrying different modes.

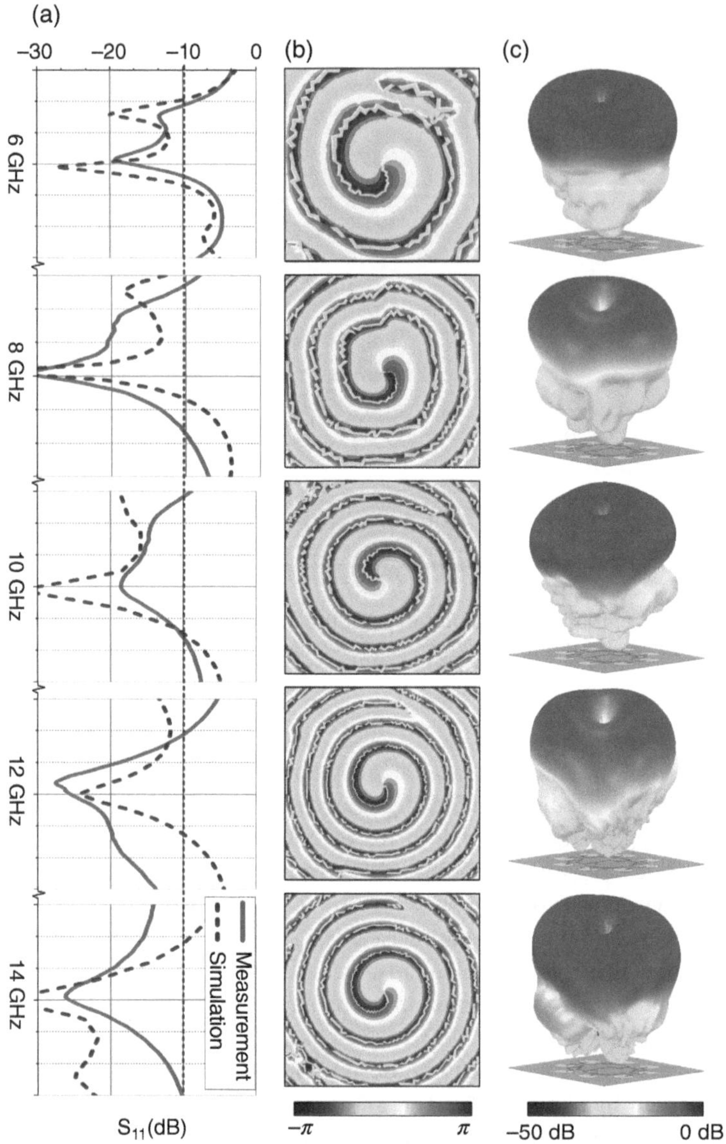

Figure 4.15 Results of OAM array antennas used as a feeding source to illuminate the lens at different frequencies. (a) Simulated and measured reflection responses of the different OAM array antennas used as feeding sources; (b) simulated phase distributions; (c) normalized 3D antenna far-field radiation patterns.

Furthermore, as confirmed in Figure 4.17 from the simulated phase distributions, the helical phase profile is not modified as the OAM vortex wave emitted by the patch array passes through the collimating lens.

The collimating lens is fabricated from a UV-curable liquid photopolymer having a relative permittivity of 2.8 using a Polyjet printing technology based on an Objet Eden260VS 3D printer [27]. The prototype is experimentally measured using the near-field scanning setup illustrated in Figure 4.18. The collimation validation is performed by a measurement system in a microwave anechoic chamber which includes a vortex wave generating module, a receiving fiber-optic-active

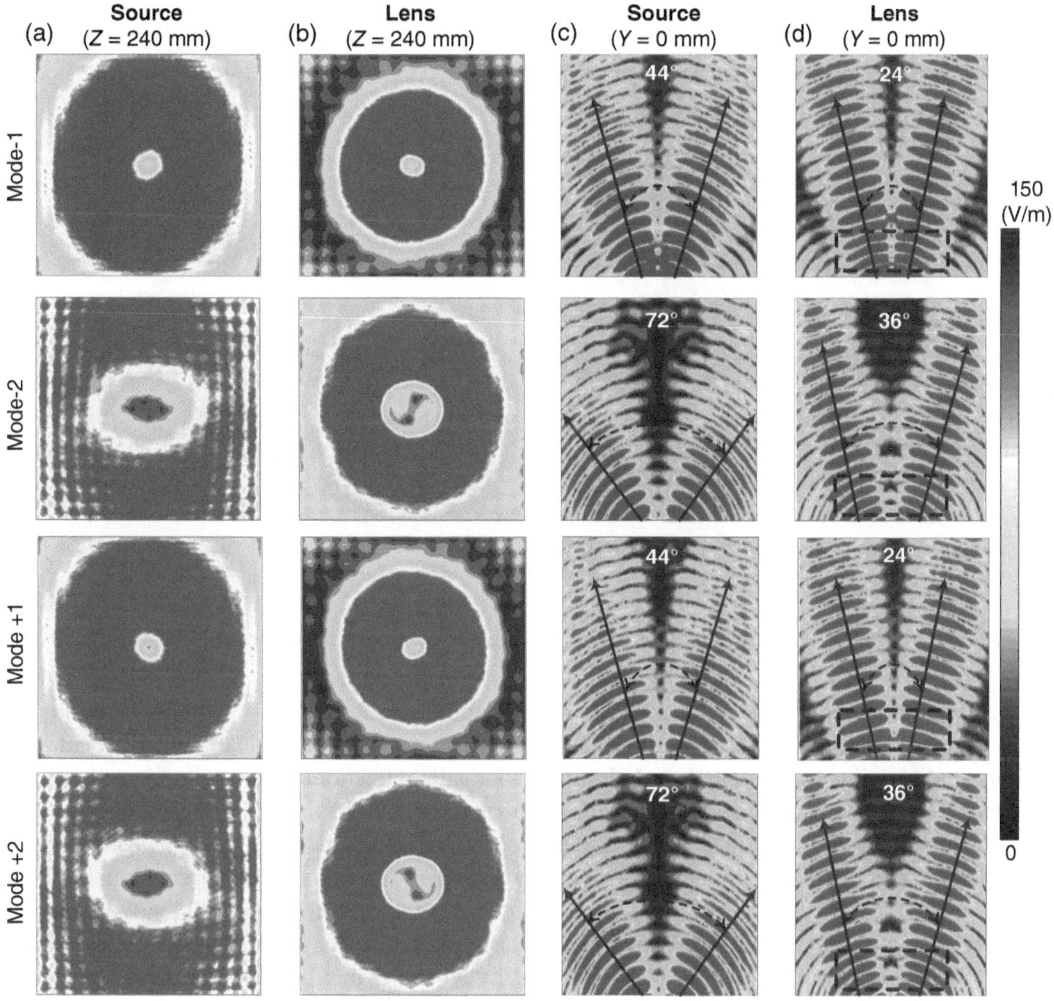

Figure 4.16 Simulation results of OAM array antennas (without lens) and lens-antenna systems at 10 GHz for different topological modes ($l = -1, -2, +1$, and $+2$). (a, b) Amplitude distribution in *xoy* plane at $z = 240$ mm; (c, d) amplitude distribution in *xoz* plane at $y = 0$ mm.

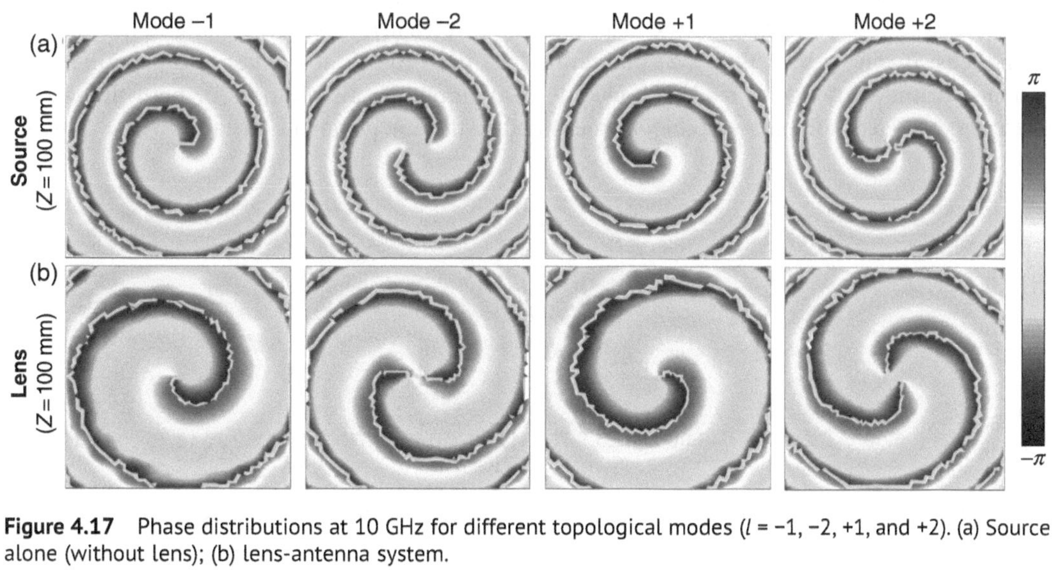

Figure 4.17 Phase distributions at 10 GHz for different topological modes ($l = -1, -2, +1$, and $+2$). (a) Source alone (without lens); (b) lens-antenna system.

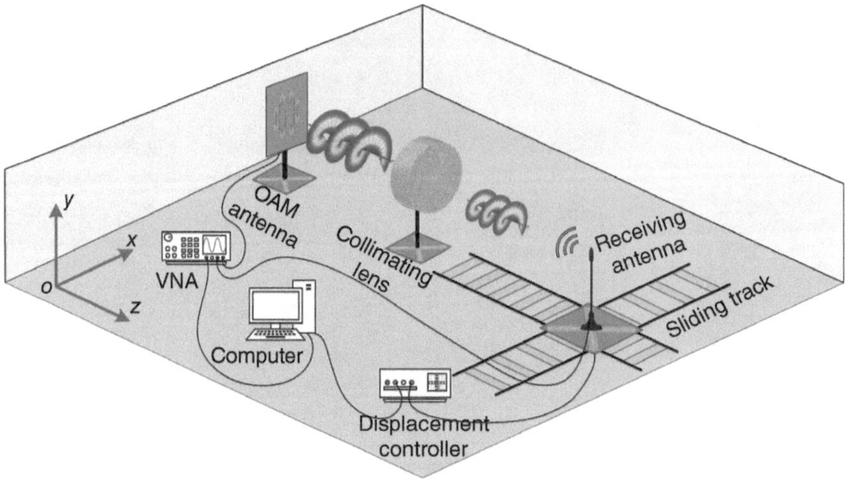

Figure 4.18 Schematic principle illustrating the near-field measurement setup for the lens-antenna system.

antenna module used as a detector, a VNA, and a computer-controlled mobile platform module to allow a scanning electric field over a specific area. The microwave anechoic chamber provides an infinite spatial region as simulated, and eliminates parasitic reflections from the surrounding environment. The vortex wave generating module includes the VNA and vortex wave generating antennas. The antennas designed and simulated in the previous section are fed by the signal generator in the VNA to radiate vortex waves at their corresponding operating frequency. Moreover, the VNA is used to collect the vortex wave signal measured by the detector at the receiving end and also to measure the scattering parameters. The mobile platform module consists of a displacement controller, a sliding carrier, and a slide rail. The displacement controller manipulates the movement of the sliding carrier on the sliding rail to measure the amplitude and phase of the vortex waves at different positions with a step of 2 mm.

The reflection coefficient responses of the different array antennas in the presence of the lens were simulated and measured. From the results displayed in Figure 4.19, it can be observed that the reflection coefficient of the different lens-antenna systems is below −10 dB at the center operating frequency. When comparing the responses of the patch arrays with and without the presence of the lens, it can be seen that the collimating lens has little to no influence on the matching response of the vortex wave feed. The electric field amplitude distribution for both the feeding source and the lens-antenna system was measured in the *xoy* plane at a distance of $z = 240$ mm

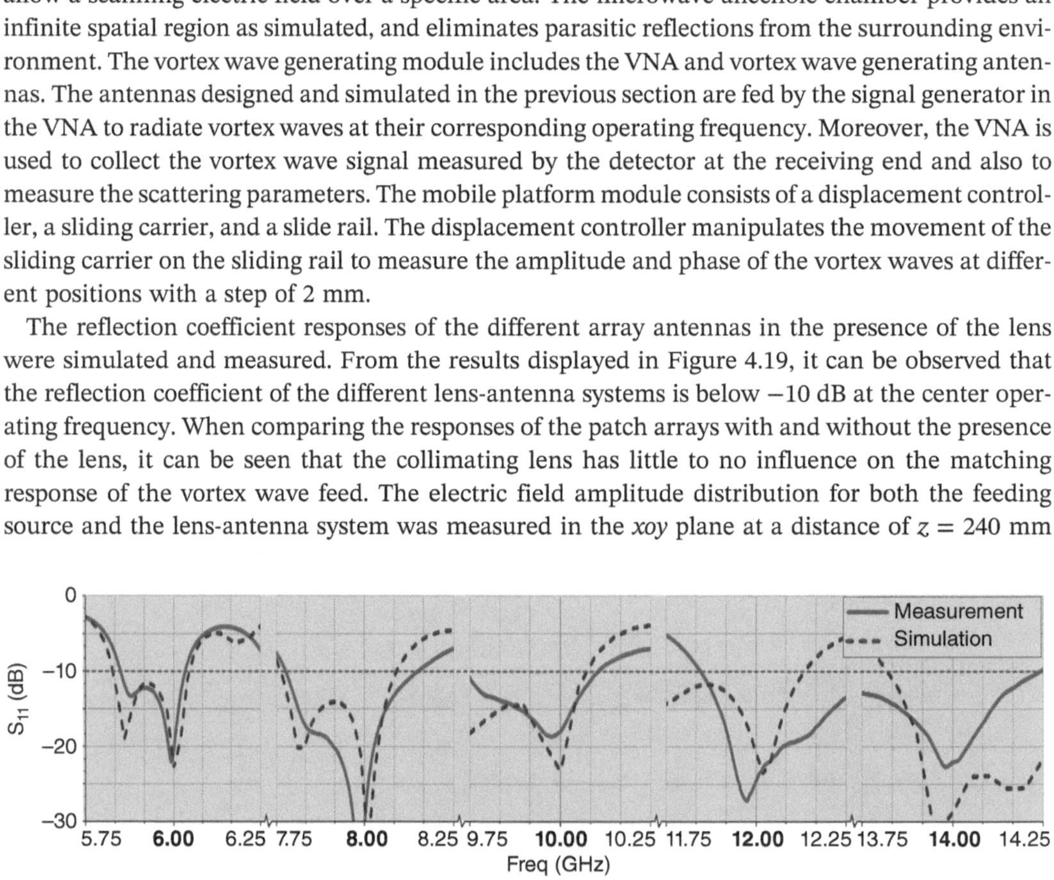

Figure 4.19 Reflection coefficient responses of different feeds in presence of the lens.

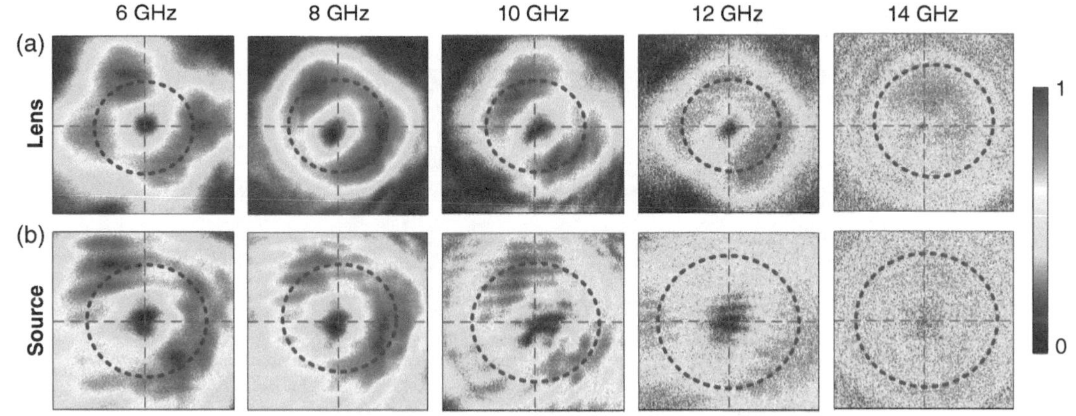

Figure 4.20 Measured electric near-field amplitude distributions in the xoy plane at z = 240 mm from 6 to 14 GHz. (a) Lens-antenna system; (b) Feeding source alone (microstrip array antennas).

and is displayed in Figure 4.20. In the lens-antenna system, the distance between the lens and feeding source was set to 4 cm in order to limit diffraction on the edges of the lens. The diameter of the black dashed circle marked in each electric field distribution map indicates the divergence of the vortex wave. In both cases (with and without the lens), a doughnut-shaped field pattern is measured at all frequencies. However, it is worthwhile to note that the inner diameter of the amplitude ring decreases in the presence of the lens. For example, at 10 GHz, the diameter of the ring decreases from 28 to 21 cm, indicating a collimation of the OAM beam. Furthermore, the collimating lens also focuses on the field and increases the density of the field while reducing the degree of divergence of the vortex wave.

In order to study the influence of the propagation distance on the divergence characteristics of the vortex wave, the electric field distribution is mapped for 10 GHz at different distances along the propagation direction. As shown in Figure 4.21, the diameter of the energy ring increases with propagation distance, for both the source and lens antenna. However, it can be clearly noted that at every measured distance the ring is smaller in the case of the lens-antenna system, confirming the divergence reduction in the OAM wave.

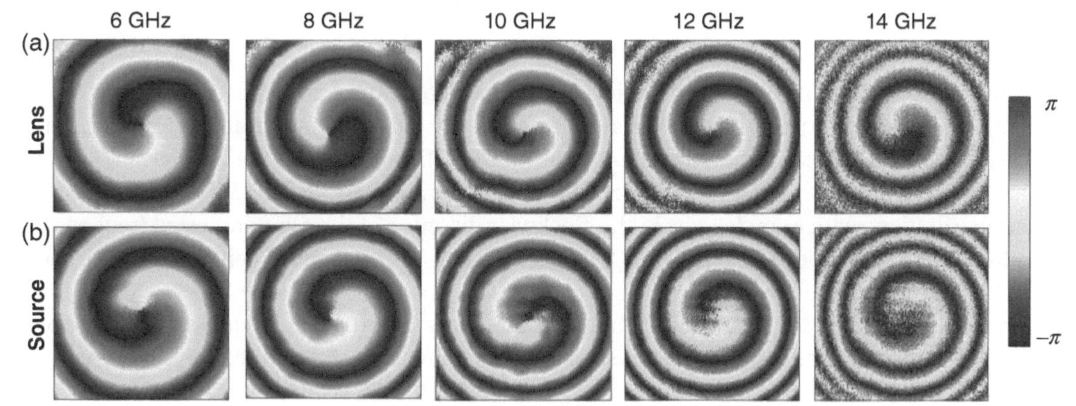

Figure 4.21 Measured near electric field phase distributions from 6 to 14 GHz in the *xoy* plane at z = 240 mm. (a) Lens-antenna system; (b) Feeding source alone.

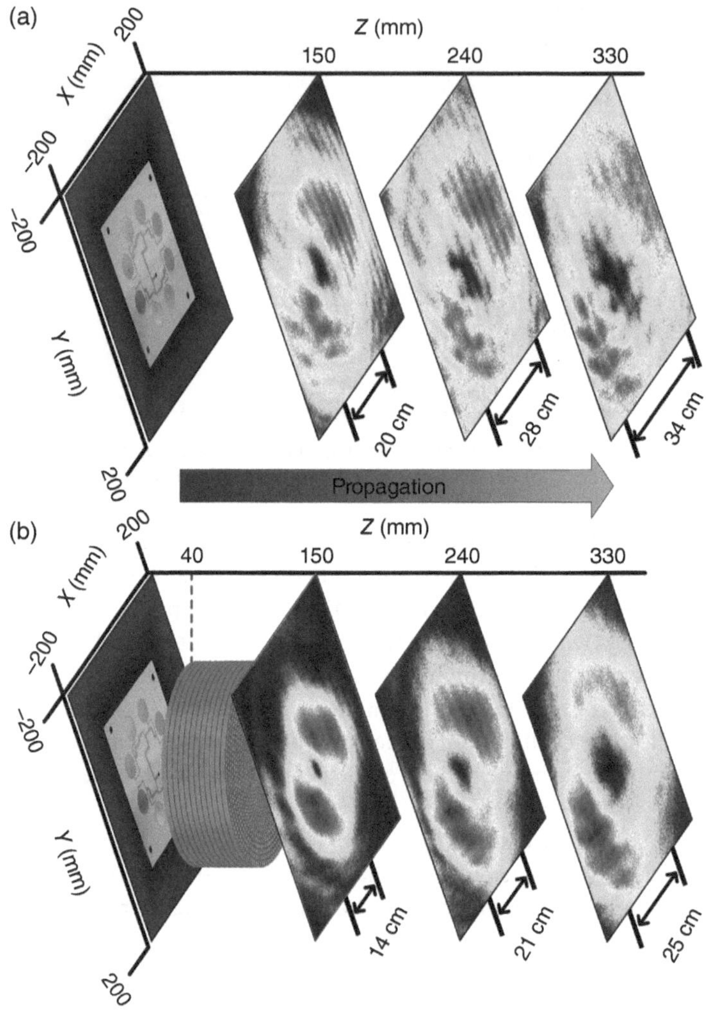

Figure 4.22 Measured electric near-field amplitude distributions at different propagation distances in the *xoy* plane at 10 GHz. (a) Feeding source alone. (b) Lens-antenna system.

Moreover, to ensure that the +1 topological OAM mode emanating from the feeding patch arrays is maintained in the vortex wave after passing through the lens, it is important to investigate the influence of the cylindrical collimating lens on the phase distribution of the vortex wave radiated by the array antennas. The measured phase distributions in the *xoy* plane depicted in Figure 4.22 show phase distributions with a single twist in the clockwise direction at the different tested frequencies, revealing that a vortex beam carrying an OAM mode with $l = +1$ is indeed generated by the array antennas and maintained by the lens.

Measurements were also performed in the *xoz* plane along the propagation direction and are presented in Figure 4.23. Due to the characteristics of the spiral motion of the vortex wave, the energy is radiated as two lobes along the propagation direction with a null at boresight ($\theta = 0°$). The collimating lens significantly reduces the angle between the two main lobes at the different considered frequencies, as shown in Figure 4.23a. A reduction of beam divergence of about 17° was captured at

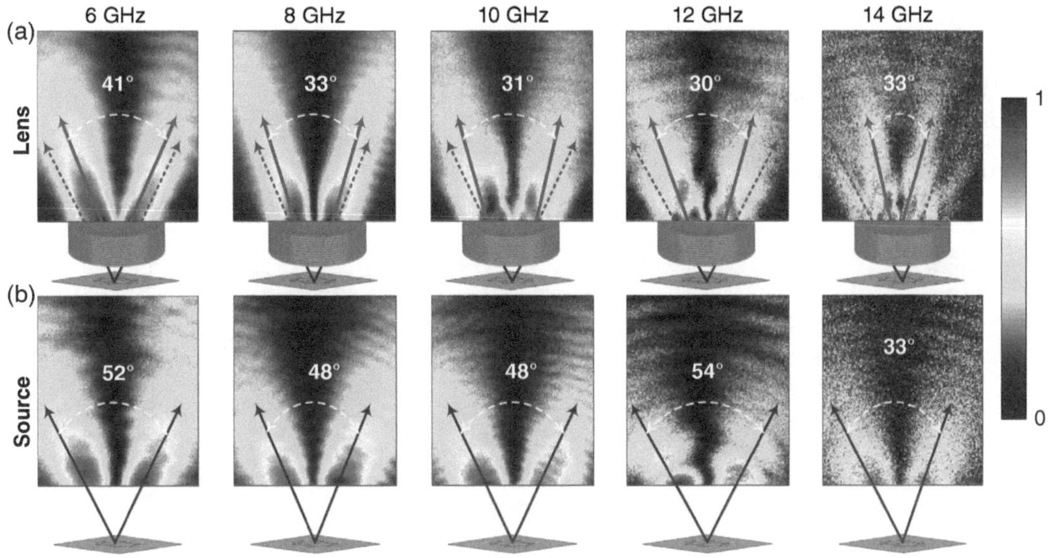

Figure 4.23 Measured near electric field amplitude distributions in the *xoz* plane from 6 to 14 GHz. (a) Lens-antenna system; (b) Feeding source alone.

10 GHz, which is consistent with the 20° observed from the simulated results in Figure 4.16. Far-field antenna measurements were performed in a microwave anechoic chamber. The normalized radiation patterns of the two different configurations (with and without lens) in the two main planes (*E- and H*-planes) are presented in Figure 4.24. Similar to the near-field characteristics, it can be clearly observed that the collimating lens increases the gain of the vortex wave, where this effect is particularly pronounced at 10 GHz. The far-field radiation patterns are consistent with the near-field distributions and confirm the reduction of the divergence angle of the vortex wave by the proposed lens. It is also important to note that the amplitude of the measured far-field patterns is different in the two planes, explaining the slightly irregular near-field doughnut-shaped amplitude distribution as shown in Figure 4.20.

In this study, an antenna radiating vortex OAM waves with topological mode +1 was considered. To show the management of power in different modes, it is necessary to calculate the mode purity of the vortex wave, referring to the proportion of each mode that can be possibly carried by the vortex wave. As shown in Figure 4.25 for a 10 GHz frequency, the ratio of the vortex wave with mode +1 is higher than 75% in both configurations. The data indicate that the collimating lens has little to no influence on the mode purity of the vortex wave radiated by the source.

4.3.3 A Transformation Optics-Based Lens for Horizontal Radiation of OAM Vortex Waves

The schematic principle of transforming a directional vortex wave to an omnidirectional one is illustrated in Figure 4.26. The proposed lens [33], which comprises a transformed medium located above a primary source generating an OAM beam with a topological charge of +2, allows for modifying the elevation angle between the wave vector k and the z-axis from 45 to 90°. The transformation can be regarded as a propagating wave deflecting from directional to omnidirectional and

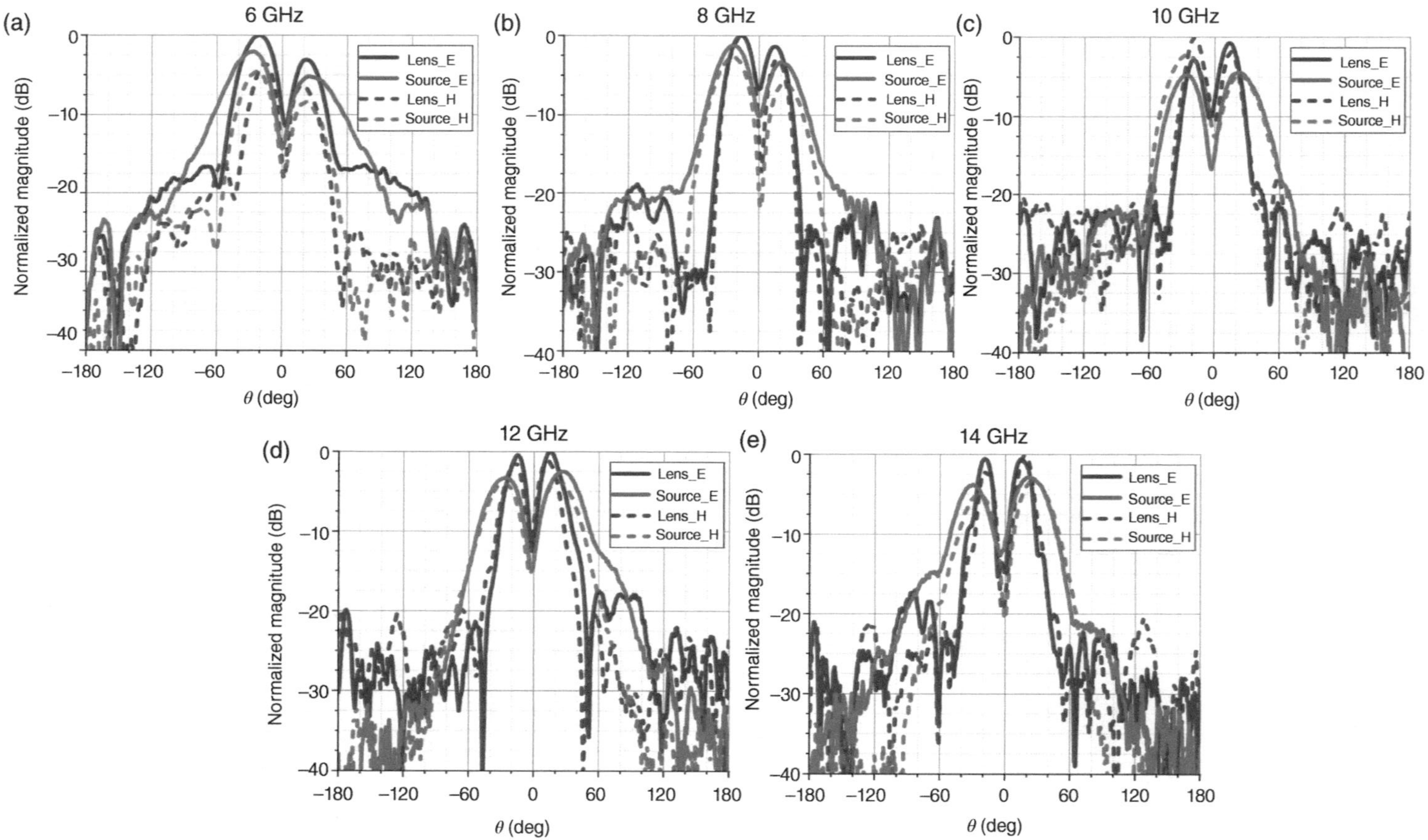

Figure 4.24 (a–e) Far-field normalized radiation patterns without lens and with lens in the frequency band ranging from 6 to 14 GHz.

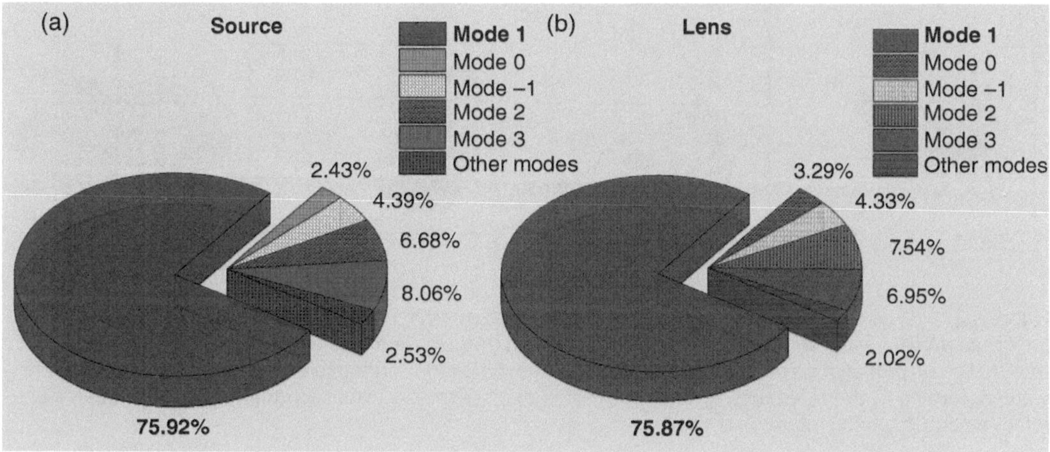

Figure 4.25 Purity mode spectrum at 10 GHz. (a) Feeding source alone; (b) lens-antenna system.

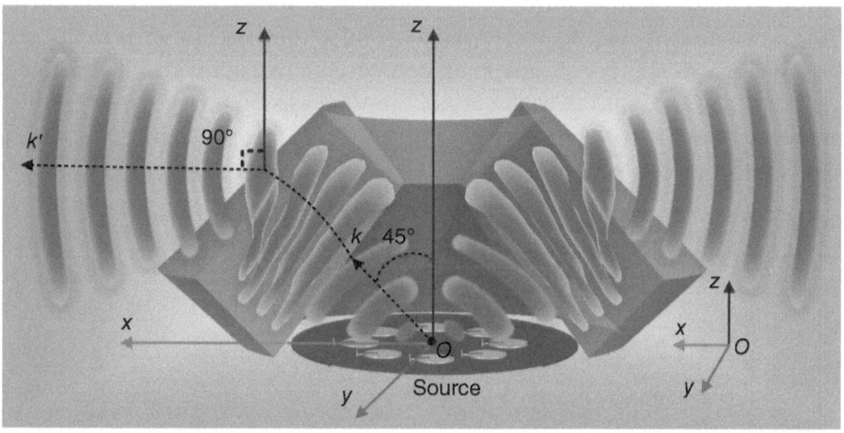

Figure 4.26 Illustration demonstrating the omnidirectional radiation process.

spreading in the horizontal plane. Instead of propagating along the z-axis as with traditional OAM beams, the beam transmitted from the lens propagates mainly in the transverse plane.

In our study, a spatial transformation between the virtual domain and the physical domain is considered. The spatial transformation established between the physical and virtual spaces aims to modify the direction of the EM radiation emitted by a feeding source. The virtual (Figure 4.27a) and the physical (Figure 4.27b) domains are filled with free space (i.e. air) and the transformed medium, respectively. The coordinate system of the virtual domain is denoted by (x, z) and that of the physical domain is (x', z'). The points A', B', C', and D' in the physical domain correspond respectively to the location A, B, C, and D in the virtual domain. The gradient color scheme for different depths of black symbolizes the spatial variation characteristics from the virtual domain to the physical domain. The points C and C' as well as D and D' share the same

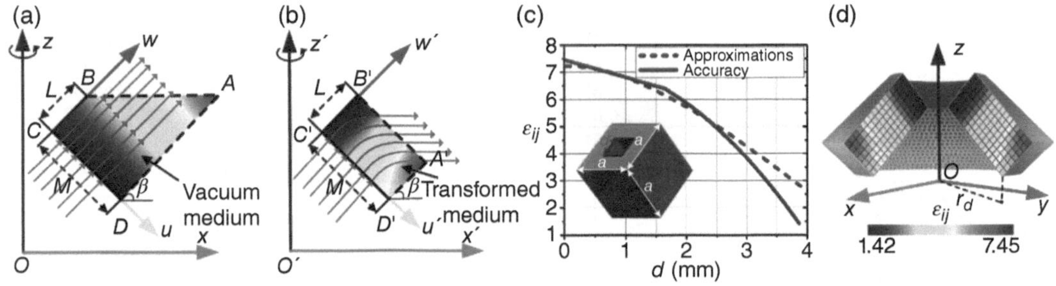

Figure 4.27 Quasi-conformal mapping from virtual domain to physical domain of the omnidirectional lens. (a) Virtual domain (vacuum medium); (b) physical domain (transformed medium). The curved arrows represent the OAM beam propagation direction; (c) a parametric analysis is performed to extract the effective permittivity value according to the side length d of the air hole; (d) a schematic cross-sectional view of the equivalent relative permittivity distribution of the lens.

locations and the lengths of the segments CD and BC are $M = 60$ mm and $L = 30$ mm, respectively. Therefore, the segment AB has a length of $M/\cos(\pi/4)$ and DA has a length that corresponds to $M + L$. The rectangle $A'B'C'D'$ is mapped from the quadrilateral $ABCD$. Finally, the lens model presented in Figure 4.27(d) is obtained by rotating the rectangle $A'B'C'D'$ one revolution around the z-axis.

In order to establish an appropriate mapping between the two domains, the relative coordinate systems at points C and C' are taken as the origin coordinates, and the axes u (u'), v (v'), w (w') are expressed respectively as the coordinate axes. Neumann and Dirichlet sliding boundary conditions are set at the edges of the transformation zones according to:

$$\begin{cases} u|_{B'C',C'D',D'A'} = u' \\ \hat{n}\cdot\nabla u|_{A'B'} = 0 \end{cases} \tag{4.27}$$

$$\begin{cases} w|_{B'C'} = w' \\ w|_{C'D'} = 0 \\ w|_{A'B'} = L + \dfrac{1}{\tan\beta}u', \\ \hat{n}\cdot\nabla w|_{D'A'} = 0 \end{cases} \tag{4.28}$$

where $\hat{n}$ is the normal vector to the boundaries of the outer surface. In addition, mathematical equivalence is expressed by solving Laplace's equation under the applied boundary conditions. Furthermore, the permittivity distribution in the physical domain is obtained by using a partial differential equation solver and calculated from the Jacobian matrix. The continuous permittivity distribution is calculated and the permittivity value varies from 1.37 to 8.25. In order to facilitate the numerical validation and future physical realization of the lens, we discretize the continuous permittivity into 15 rows and 30 columns for a total of 450 values, where the rows corresponding to 1 to 15 are denoted by index i and the columns ranging from 1 to 30 are denoted by index j, as shown in Figure 4.27d.

In order to numerically validate the proposed spatial transformation approach, a 3D model of the proposed lens was adopted for all-dielectric materials with bulk relative permittivity of 8.25, where a cubic structure with a square through-hole is utilized as the unit cell of the lens. The side length of the cube and square hole is respectively $a = 2$ mm and d. Permittivity values ranging from 1.37 to 8.25 are realized by varying d from 0 to 3.87 mm, as shown in Figure 4.27c. The functional relation between the side length of the square hole d_{ij} and the equivalent relative permittivity ε_{ij} of the dielectric unit cells is expressed as:

$$d_{ij} = \sqrt{\frac{a^2 \times (\varepsilon - \varepsilon_{ij})}{\varepsilon - 1}}. \tag{4.29}$$

Approximate permittivity values calculated from Eq. (4.29) are represented by the dashed trace, while accurate values retrieved from the homogenization procedure of a unit cell by using the inversion method are represented by the solid trace (Figure 4.27c). To ensure a smooth connection between adjacent unit cells, the meta-atom can be bent and trimmed appropriately. As shown in Figure 4.27d, the final 3D model of the lens has a conical shape with a height of 70.7 mm and a bottom radius $r_d = 60$ mm. The angle between the sidewalls of the lens and the horizontal plane is 45°.

Full-wave simulations are used to verify the omnidirectional performance of the proposed lens. A series of microstrip array antennas consisting of eight metal patches are used as feeding sources at 8, 11, 14, and 17 GHz. The adjacent metal patches are equipped with a 90° phase shift such that the feeding source is able to generate an OAM beam with a +2 topological charge. The near-zone electric field distributions of the +2 mode OAM beam are shown in Figure 4.28a–d. Obviously, the OAM beams generated from the different sources radiate at 45° from the normal direction. When the sources are placed at 5 mm above the bottom surface of the lens, waves transmitted through the lens present an omnidirectional emission, as shown in Figure 4.28e–h at different frequencies. It can be observed that the waves emitted from the outer surface of the lens are mostly concentrated in the horizontal plane.

The simulated far-field radiation patterns of the different sources and lens-antennas are presented in Figure 4.29. The 3D radiation patterns of the feeding sources alone present directional radiation, as shown in Figure 4.29a–d. Compared to the results without the lens, the main lobes of the vortex waves passing through the lens are directed in the horizontal plane, as shown in Figure 4.29e–h. It can be clearly observed that the angle between the main lobe and the propagation axis is significantly increased from 45 to 90°. Due to the omnidirectional nature of the radiation in the presence of the lens, the maximum gain of the main lobe is slightly lower than that of the source alone. Furthermore, the 2D far-field radiation patterns without/with the lens are presented in Figure 4.29i–l. These radiation patterns numerically validate the characteristics of the omnidirectional radiation according to the modification of the elevation angle.

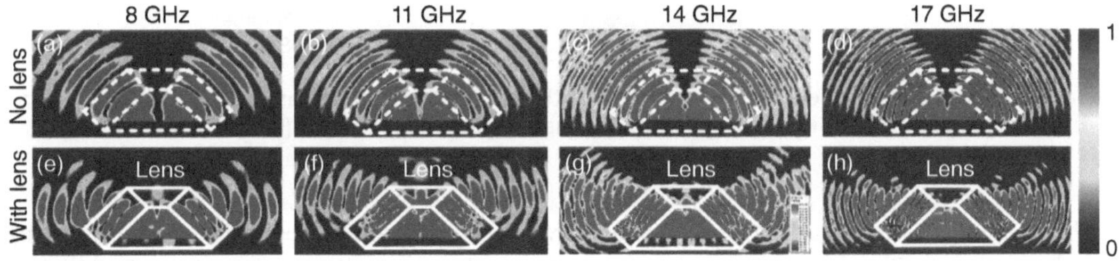

Figure 4.28 E-field amplitude distribution in the x–z plane for an operating frequency spanning from 8 to 17 GHz. (a–d) Source alone; (e–h) source in presence of the lens.

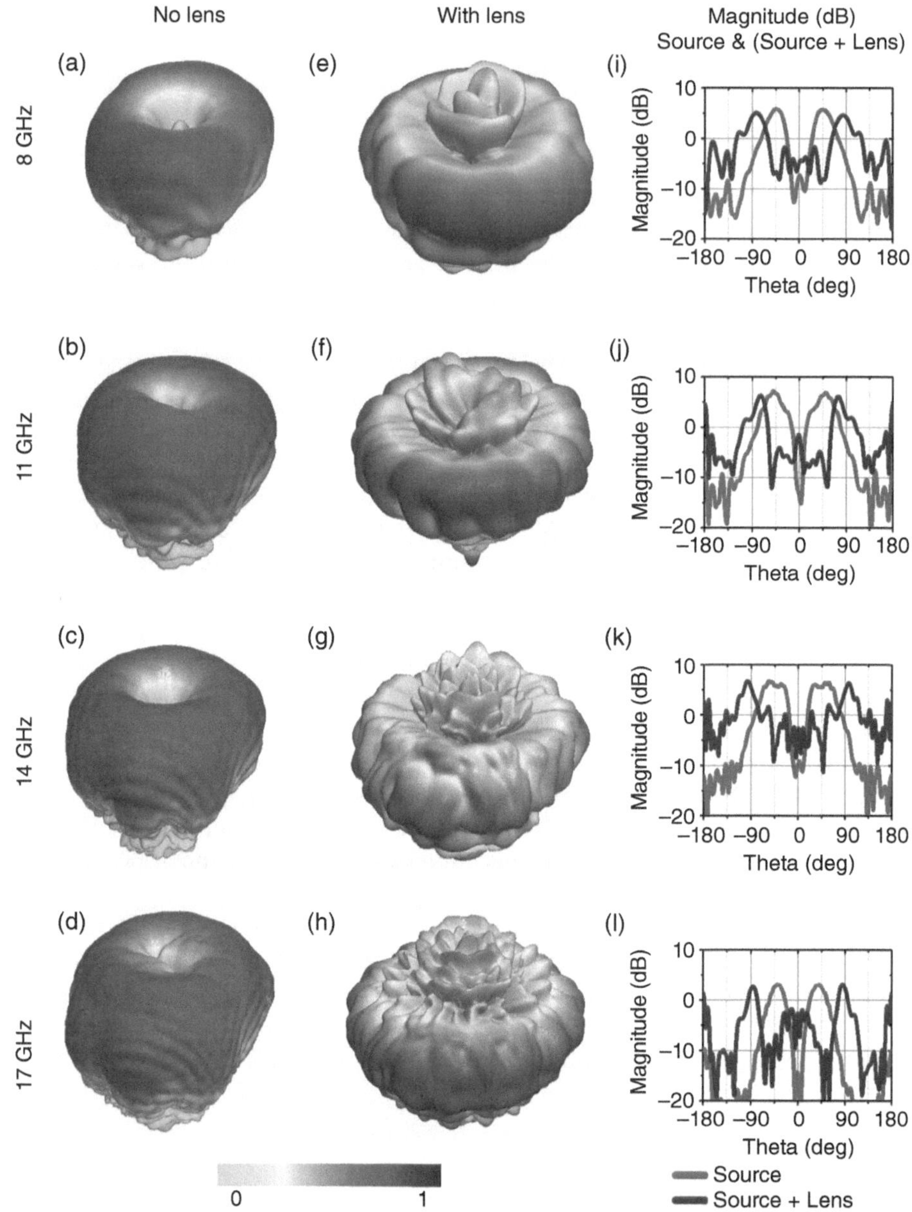

Figure 4.29 Simulated far-field radiation patterns of the different lens-antennas at 8, 11, 14, and 17 GHz. (a–d) 3D radiation patterns without lens; (e–h) 3D radiation patterns with lens; (i–l) 2D radiation pattern cuts without/with lens.

The phase distributions of the feeding sources with and without the omnidirectional radiating lens are shown in Figure 4.30. It can be observed that the phase distribution of the OAM beam radiated from the lens still exhibits the inherent helical state. Two spirals are obtained in both cases, verifying the +2 OAM mode. Though slight distortions appear at 14 GHz with the lens-antenna system, the omnidirectional lens does not change the spiral phase characteristics of the vortex waves over a broad frequency range.

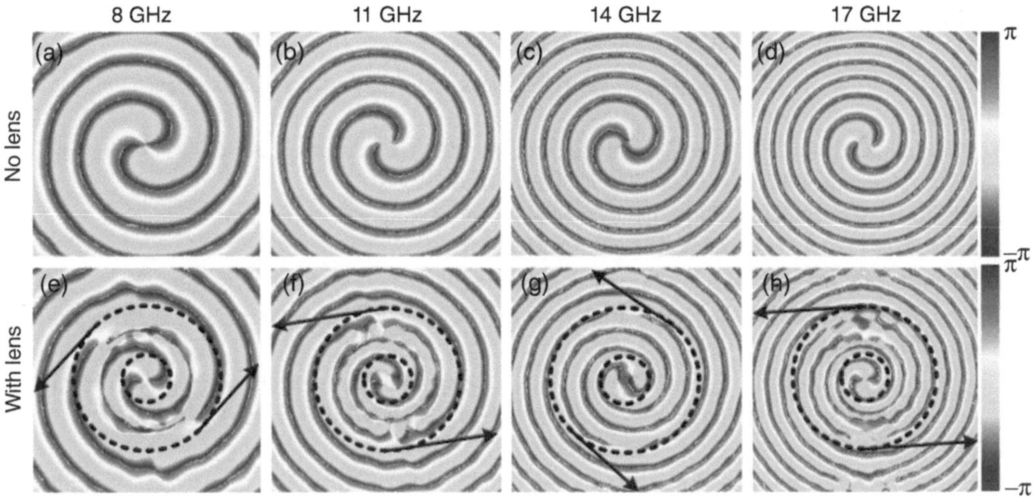

Figure 4.30 Simulated phase distributions of the EM field component in the *x–y* plane at different frequencies. The cross-section is 110 mm above the feed plane, and the phase changes from −π to π. (a–d) Source alone; (e–h) source with lens.

4.4 Conclusions

This chapter is devoted to the application of transformation optics and 3D printing technology to vortex EM wave generation. First, the basic theories of transformation optics and 3D printing are introduced and discussed. Second, several different types of microwave devices were designed by using transformation optics theory in the application field of vortex wave generation and manipulation. The fabrication of microwave devices was carried out by using 3D printing technology. The design methodology based on spatial transformations provides an effective strategy for tailoring EM waves with nearly arbitrary desired properties. In addition, the proposed approach is expected to have a profound impact on future OAM-based systems due to its wide operational bandwidth and ease of manufacturing and integration.

Although transformation optics and 3D printing have opened up a new pathway for further research on vortex waves, there are still some problems that remain to be solved in practical applications. First of all, anisotropic conversion materials are inevitably generated during the process of coordinate transformation. However, for the device implementation, anisotropy is ignored such that only a distribution of ε_{zz} is assumed, which is representative of a quasi-isotropic material. This assumption in the design process will affect the performance of the designed device. In addition, from the perspective of design, it is often assumed that any arbitrary (unconstrained) material properties may be utilized to achieve the desired performance. However, in practice, the actual material properties often need to be constrained to allow fabrication. For example, currently, materials with a high dielectric constant are more challenging to work with, in practice, even though they may have some performance benefits in theory. Finally, the design concept behind transformation optics assumes that EM waves can be arbitrarily controlled, but in practice, there are always some limitations and performance tradeoffs. The examples in this chapter were mainly applied to the generation, convergence, and deflection of vortex EM waves. If further exploration is conducted, however, it is believed that there are many additional scenarios in which transformation optics can be applied.

References

1 Chen, H.Y., Chan, C.T., and Sheng, P. (2010). Transformation optics and metamaterials. *Nature Materials* **9** (5): 387–396.

2 Yi, J.J., Piau, G.P., de Lustrac, A., and Burokur, S.N. (2016). Electromagnetic field tapering using all-dielectric gradient index materials. *Scientific Reports* **6**: 300661.

3 Pendry, J.B., Schurig, D., and Smith, D.R. (2006). Controlling electromagnetic field. *Science* **312** (5781): 1780–1782.

4 Leonhardt, U. (2006). Optical conformal mapping. *Science* **312** (5781): 1777–1780.

5 Li, J. and Pendry, J.B. (2008). Hiding under the carpet: A new strategy for cloaking. *Physical Review Letters* **101** (20): 203901.

6 Chang, Z., Zhou, X.M., Hu, J., and Hu, G.K. (2010). Design method for quasi-isotropic transformation materials based on inverse Laplace's equation with sliding boundaries. *Optics Express* **18** (6): 6089–6096.

7 Hu, J., Zhou, X.M., and Hu, G.K. (2009). Design method for electromagnetic cloak with arbitrary shapes based on Laplace's equation. *Optics Express* **17** (15): 13070–13070.

8 Oliveri, G., Bekele, E.T., Werner, D.H. et al. (2014). Generalized QCTO for metamaterial-lens-coated conformal arrays. *IEEE Transactions on Antennas and Propagation* **62** (8): 4089–4095.

9 Yi, J.J., Burokur, S.N., Piau, G.P., and de Lustrac, A. (2016). Coherent beam control with an all-dielectric transformation optics based lens. *Scientific Reports* **6**: 18819.

10 Chiotellis, N., Zhang, S., Vardaxoglou, Y.C., and Grbic, A. (2020). X Wave radiator implemented with 3-D printed metamaterials. *IEEE Transactions on Antennas and Propagation* **68** (7): 5478–5486.

11 Yi, J.J., Burokur, S.N., and de Lustrac, A. (2015). Conceptual design of a beam steering lens through transformation electromagnetics. *Optics Express* **23** (10): 12942–12951.

12 Most Used 3d Printing Technologies in 2020, (2018). Statistic, Statista, 3 Jun. (2018). www.statista.com/statistics/560304/worldwide-survey-3d-printing-top-technologies/

13 History of 3D Printing (2018). When Was 3d Printing Invented?, All3DP, 10 Dec. 2018. all3dp.com/2/history-of-3d-printing-when-was-3d-printing-invented/#respond

14 Coward, C. (2015). 3D Printing, Penguin.

15 Allen, L., Beijersbergen, M.W., Spreeuw, R.J.C., and Woerdman, J.P. (1992). Orbital angular momentum of light and the transformation of Laguerre-Gaussian laser modes. *Physical Review A* **45** (11): 8185–8189.

16 Tang, J., Ming, Y., Chen, Z.X. et al. (2016). Entanglement of photons with complex spatial structure in Hermite-Laguerre-Gaussian modes. *Physical Review A* **94** (1): 012313.

17 Liu, K., Cheng, Y., Yang, Z. et al. (2014). Orbital-angular-momentum-based electromagnetic vortex imaging. *IEEE Antennas and Wireless Propagation Letters* **14**: 711–714.

18 Yang, T.L., Li, S.H., Xu, O. et al. (2018). Three dimensional SAR imaging based on vortex electromagnetic waves. *Remote Sensing Letters* **9** (4): 343–352.

19 Liu, K., Cheng, Y., Li, X., and Gao, Y. (2019). Microwave-sensing technology using orbital angular momentum: Overview of its advantages. *IEEE Vehicular Technology Magazine* **14** (2): 112–118.

20 Meng, Y.F., Yi, J.J., Burokur, S.N. et al. (2018). Phase-modulation based transmitarray convergence lens for vortex wave carrying orbital angular momentum. *Optics Express* **26** (17): 22019–22029.

21 Zhang, K., Yuan, Y.Y., Ding, X.M. et al. (2019). High-efficiency metalenses with switchable functionalities in microwave region. *ACS Applied Material & Interface* **11** (31): 28423–28430.

22 Chen, Y., Zheng, S., Li, Y. et al. (2015). A flat-lensed spiral phase plate based on phase-shifting surface for generation of millimeter-wave OAM beam. *IEEE Antennas and Wireless Propagation Letters* **15**: 1156–1158.

23 Tamburini, F., Mari, E., Thide, B. et al. (2011). Experimental verification of photon angular momentum and vorticity with radio techniques. *Applied Physics Letters* **99** (20): 204102.

24 Yi, J.J., Cao, X.Q., Feng, R. et al. (2019). All-dielectric transformed material for microwave broadband orbital angular momentum vortex beam. *Physical Review Applied* **12** (2): 024064.

25 Yi, J.J., Liu, C.C., Shi, Z. et al. (2020). Assembled medium: A route to the generation of vortex waves carrying orbital angular momentum with different modes. *Journal of Applied Physics* **128** (4): 044101.

26 Jack, B., Padgett, M.J., and Franke-Arnold, S. (2008). Angular diffraction. *New Journal of Physics* **10**: 103013.

27 Objet connex 3, https://www.stratasys.com/3d-printers/objet260-connex3.

28 Yan, Y., Xie, G.D., Lavery, M.P.J. et al. (2014). High-capacity millimetre-wave communications with orbital angular momentum multiplexing. *Nature Communications* **5**: 4876.

29 Yi, J.J., Guo, M.T., Feng, R. et al. (2019). Design and validation of an all-dielectric metamaterial medium for collimating orbital-angular-momentum vortex waves at microwave frequencies. *Physical Review Applied* **12** (3): 034060.

30 Arlt, J. and Dholakia, K. (2000). Generation of high-order Bessel beams by use of an axicon. *Optics Communications* **177** (1–6): 297–301.

31 Yang, C.F., Yang, J.J., Huang, M. et al. (2012). Electromagnetic concentrators with arbitrary geometries based on Laplace's equation. *Journal of the Optical Society of America A* **27** (9): 1994–1998.

32 Bai, Q., Tennant, A., and Allen, B. (2014). Experimental circular phased array for generating OAM radio beam. *Electronics Letters* **50** (20): 1414–1415.

33 Z. Shi, J. J. Yi, X. Q. Cao, et al. (2019). Omnidirectional radiation lens design of vortex beam carrying orbital angular momentum based on spatial transformation, *2019 13th European Conference on Antennas and Propagation (EuCAP)*, Krakow, Poland, 1–4.

5

Millimeter-Wave Transmit-Arrays for High-Capacity and Wideband Generation of Scalar and Vector Vortex Beams

Zhi Hao Jiang,[1] Lei Kang,[2] Wei Hong[1], and Douglas H. Werner[2]

[1] *State Key Laboratory of Millimeter-Waves, School of Information Science and Engineering, Southeast University, Nanjing, P.R. China*
[2] *Department of Electrical Engineering, The Pennsylvania State University, University Park, PA, USA*

5.1 Introduction

Angular momentum, including spin angular momentum (SAM) and orbital angular momentum (OAM), is one of the fundamental properties associated with electromagnetic waves, in addition to phase, amplitude, and polarization [1, 2]. As demonstrated by Beth in 1936 [3], a circularly polarized (CP) wave intrinsically carries SAM due to its vectorial nature, with a value determined by its spin states ($\sigma_{\pm}$), i.e. the handedness. On the other hand, the OAM is manifested by its annular donut-shaped intensity profile and a helical phase-front distribution of the beam [4]. In detail, a structured beam carrying a well-defined OAM commonly referred to as a 'vortex beam' has a phase dependence of $\exp(-j\ell\varphi)$ in its wavefront, where ℓ is the topological charge and φ is the polar angle. Such a wavefront structure indicates the existence of an optical singularity at the beam center as well as an OAM of $\ell\hbar$ per photon along the beam axis [5], where $\hbar$ is the reduced Plank constant. Since the initial discovery that beams can carry a finite amount of OAM, there has been a tremendous amount of research interest in exploring their intriguing wave behaviors and exploiting their unconventional wavefront for diverse applications throughout the electromagnetic spectrum [6, 7]. OAM related studies have opened up new frontiers and technological advancements in many areas such as astrophysical observation [8, 9], particle trapping and manipulation [10, 11], quantum information processing [12, 13], nonlinear optics [14], microscopy [15], microlasers [16], and communications [17–21], etc..

The vortex beams mentioned above have uniform states-of-polarization on their wavefront, thereby they are referred to as *scalar vortex beams* in a more accurate sense. As a matter of fact, these scalar vortex beams are only special cases of the more generalized *vector vortex beams*, which have a spatially varying polarization distribution across their wavefront [22]. Analogous to the Poincaré sphere (PS), which is widely used for mapping all the states-of-polarizations of a plane wave onto a sphere, higher-order PSs (HOPSs) have been proposed for a straightforward description of all the vector vortex beams that can be represented by the superposition of two CP scalar vortex beams with opposite spin states [23, 24]. By virtue of the inhomogeneous states-of-polarization wavefront, the vector vortex beams can be highly useful for sub-diffraction focusing [25], optical trapping [26], lasers [27], to name but a few.

Electromagnetic Vortices: Wave Phenomena and Engineering Applications, First Edition.
Edited by Zhi Hao Jiang and Douglas H. Werner.

The conventional methods for generating the scalar vortex beams primarily employ transmissive or reflective types of spiral phase plates [28, 29], cylindrical mode converters [30], computer-generated holograms [31], binary gratings [32], gratings based on whispering gallery mode resonators [33], annular arrays of phased antenna elements [34], and so on. For the more complicated vector vortex beams, the traditional approaches for obtaining them require a number of cascaded optical components, including q-plates, waveplates, and polarizers [27, 35]. Thus, these techniques either result in an electrically-large device form factor, necessitate a number of accessory components, or operate only within a narrow bandwidth, thus severely limiting their integration into practical systems.

Recently, various planar generators of scalar vortex beams and vector vortex beams have been proposed and demonstrated based on wavefront engineering enabled by quasi-two-dimensional (2D) metasurfaces [36] and transmit-/reflect-array concepts [37]. Instead of relying on accumulated phase due to optical path difference, the desired aperture field distribution can be constructed by imposing a spatially-variant abrupt phase change on an incident electromagnetic wave via an array of inhomogeneous sub-wavelength building blocks [38]. Several kinds of reflective-mode devices have been reported to provide a single-mode reflected OAM beam from a linearly-polarized incident wave, which utilizes a continuous metallic backing layer behind an array of a nonuniform metallic or dielectric resonator with spatially varying geometries, orientations, and/or dimensions [39–41]. For facilitating architectural simplification of the system and avoiding performance degradation due to feed blockage, transmissive devices are often preferable. For transmissive-type vortex generators, a single-/mixed-mode OAM beam was achieved but with a moderate efficiency or a device profile larger than the wavelength of operation [42–45]. Importantly, a fundamental limitation of these transmit-/reflect-arrays, which rely on dynamic phase delay, is their narrow operational bandwidth, attributed to the strong dispersion of their resonating sub-wavelength building blocks or the Fabry-Pérot resonance arising from the thick dielectric gradient index slabs.

In an attempt to accomplish bandwidth extension, the Pancharatnam–Berry phase, also known as the geometric phase, was exploited that offers dispersionless phase retardation for CP waves [46, 47], where the phase delay is determined purely by the in-plane rotation angle of the unit cells that have anisotropic electromagnetic response. It is physically intriguing as it offers a viable path for generating SAM-dependent OAM beams relying on the spin-to-orbital interaction [48]. By designing geometric-phase devices that operate in the reflective-mode, a high operating efficiency can be achieved for single vortex beam generation [49–52]. By aligning different types of transmissive anisotropic resonators with spatially varying orientation angles in an array [53–56], various transmit-arrays with a sub-wavelength thickness have been implemented for producing a single-mode CP vortex beam. In addition, since the phase delays for CP waves with opposite handedness are locked for the geometric-phase cells, fully independent wavefront manipulation for CP waves with opposite spin states is obscure, which, however, would provide another degree of freedom if possible.

Moreover, as it is noted that the current vortex beam generators can achieve multiplexed beams but only carry a small number of modes, multiplexing methods for obtaining multiple concurrent beams of different wavefronts with a single input is another important topic particularly related to information technology, where the shared-aperture technology was recently applied [57, 58]. The shared-aperture approach has been widely employed in multitasking and multitarget phased arrays for radar applications at microwave frequencies, where the aperture is filled with sub-arrays of radiating elements operating at a specific frequency or polarization. The sub-arrays on the shared aperture, each fulfilling a single functionality corresponding to a specific phase function, can either be separated from each other, i.e. segmented arrays, or have cells sparsely sampled at randomly selected lattice points, i.e. interleaved arrays. Both the segmented and interleaved arrays possess high flexibility for performing multiple independent wavefront construction but suffer from either a poor angular resolution and/or a low efficiency, which leads to a significantly reduced channel capacity as the number of beams increases [59, 60]. Alternatively, harmonic response arrays, which utilize the harmonic diffraction orders, can greatly improve the efficiency and thus enhance the channel capacity [61]. However, this comes at the expense of limited flexibility that is restricted by the harmonic sequence. Therefore, a new array synthesis strategy with simultaneous high channel capacity and large flexibility is highly desirable but has so far remained elusive. It is important to note that such a capability is critical to the successful realization of massive and multitasking scalar and vector vortex beam multiplexing.

In this chapter, we present several high-efficiency and broadband generators of multiple scalar and/or vector vortex beams using transmit-arrays at millimeter-waves, which support either geometric phase delay or even dual-CP phase delay. The chapter starts with a brief description of the vector vortex beams and their representations on the HOPSs. Then, three kinds of low-profile transmit-array unit cells operating at *Ka*-band, *Q*-band, and *K*-band are illustrated, including the design strategy and their electromagnetic response. Based on these designed unit cells, in the third part of the chapter, the designs and experimental validation of various scalar vortex beam and vector vortex beam multiplexers are presented and discussed.

5.2 Vector Vortex Beams and Hybrid-Order PSs

In this section, the basics of vector vortex beams will be presented, including their wavefront properties and synthesis. In addition, the concept of HOPSs, which are extensions of the conventional PS, is introduced and related to the representations of the scalar and vector vortex beams.

The scalar vortex beams denote beams carrying nonvanishing OAM with a uniform state of polarization, either linear, circular, or elliptical, across its annular wavefront. Mathematically, a scalar vortex beam can be represented in general as

$$| \psi_m \rangle = \exp(-j\ell\varphi)(A_x\hat{\mathbf{e}}_x + A_y e^{j\alpha_y}\hat{\mathbf{e}}_y)/\sqrt{2}, \tag{5.1}$$

where ℓ is the topological charge of the OAM mode, A_x, A_y, and α_y determine the polarization of the vortex beam [1]. For particular CP vortex beams, the mode expressions are written as

$$| \boldsymbol{R}_\ell \rangle = \exp(-j\ell\varphi)(\hat{\mathbf{e}}_x - j\hat{\mathbf{e}}_y)/\sqrt{2}, \tag{5.2}$$

$$| \boldsymbol{L}_\ell \rangle = \exp(-j\ell\varphi)(\hat{\mathbf{e}}_x + j\hat{\mathbf{e}}_y)/\sqrt{2}, \tag{5.3}$$

for right-handed circularly-polarized (RCP) and left-handed circularly-polarized (LCP) vortex beams, respectively. In contrast, a vector vortex beam possesses inhomogeneous states-of-polarization on its wavefront and can be represented by a superposition of two scalar CP vortex beams with opposite handedness as

$$\psi_{m,n} = A_m e^{j\alpha_m} | \boldsymbol{R}_m \rangle + A_n e^{j\alpha_n} | \boldsymbol{L}_n \rangle, \tag{5.4}$$

where $|\boldsymbol{R}_m\rangle$ and $|\boldsymbol{L}_n\rangle$ denote RCP and LCP vortex beams with topological charges of m and n, while $A_{m(n)}$ and $\alpha_{m(n)}$ are the amplitude and phase of the two beams, respectively. As a result, the parameters $A_{m(n)}$ and $\alpha_{m(n)}$ determine the state-of-polarization at different positions on the wavefront. For example, when $m = 1$, $n = -1$, $A_m = A_n$, and $\alpha_m = \alpha_n$, the superimposed beam has an electric field distribution illustrated in Figure 5.1a. It has an annular-shaped magnitude profile and linear polarization everywhere on its wavefront with the electric field oriented along the ρ-directions, achieving the radially-polarized vector beam [25]. When letting $\alpha_m = \alpha_n + 180°$ and keeping the other parameters the same, the magnitude profile still has an annular shape. In addition, the state-of-polarization at every point on the wavefront is also linearly-polarized but with the electric field oscillating along the φ-directions, realizing the azimuthally-polarized vector beam, as shown in Figure 5.1b.

It is well-known that for plane waves, all possible states-of-polarization can be represented as a point located on the PS [62]. As shown in Figure 5.2, on such a sphere, the north and south poles represent right-handed and left-handed circular polarizations. All the points on the equator denote

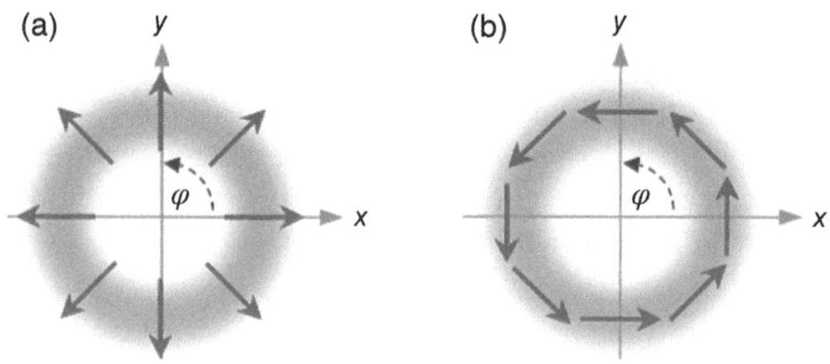

Figure 5.1 Illustration of the vectorial electric field distributions in the transverse plane of (a) radially-polarized and (b) azimuthally-polarized vector vortex beams propagating in the +*z* direction.

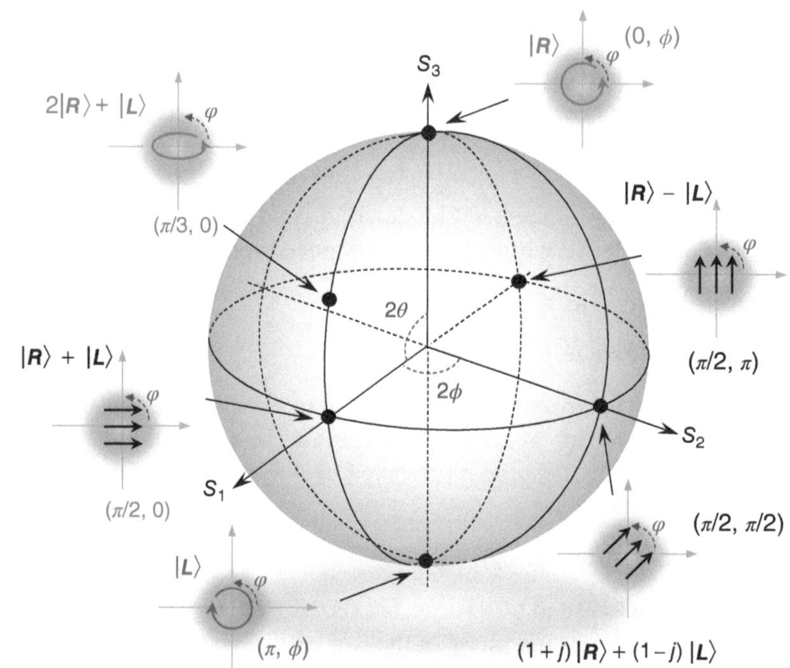

Figure 5.2 PS representation of the states-of-polarization of a plane wave.

linear polarizations with different field orientations. The regions between the poles and the equator are all elliptical polarization states with the elevation angle (θ) and azimuthal angle (ϕ) determining the ellipticity and the orientation angle of the main axis of the polarization ellipse. The modes on the two poles are constitutive components of any other polarization states as represented by

$$\psi(\theta,\phi) = \cos\left(\frac{\theta}{2}\right)\exp\left(j\frac{\phi}{2}\right)|\,\boldsymbol{R}\rangle + \sin\left(\frac{\theta}{2}\right)\exp\left(-j\frac{\phi}{2}\right)|\,\boldsymbol{L}\rangle, \tag{5.5}$$

where $|\,\boldsymbol{R}\rangle = (\hat{\boldsymbol{e}}_x - j\hat{\boldsymbol{e}}_y)/\sqrt{2}$ and $|\,\boldsymbol{L}\rangle = (\hat{\boldsymbol{e}}_x + j\hat{\boldsymbol{e}}_y)/\sqrt{2}$. For the vector vortex beams with inhomogeneous states-of-polarizations across their wavefront, due to the fact that they are synthesized by two CP vortex beams as shown in Eq. (5.4), a class of modified PS can be introduced, which are named as the HOPSs. Thus, all the modes composed by superposition of a particular pair of the mth-order RCP and the nth-order LCP scalar vortex beams with different amplitude ratios and phase difference can be expressed in a way similar to that for states on a PS as

$$\psi_{m,n}(\theta,\phi) = \cos\left(\frac{\theta}{2}\right)\exp\left(j\frac{\phi}{2}\right)|\,\boldsymbol{R}_m\rangle + \sin\left(\frac{\theta}{2}\right)\exp\left(-j\frac{\phi}{2}\right)|\,\boldsymbol{L}_n\rangle. \tag{5.6}$$

When $m = n$, the RCP and LCP vortex beams have the same phase dependence. Consequently, the $\text{HOPS}_{\text{m,n}}$ becomes a sphere containing modes with uniform state-of-polarization on the beam wavefront, thus all the modes on such a sphere are scalar vortex beams. For example, as shown in Figure 5.3a, the scalar vortex modes on the $\text{HOPS}_{1,1}$ are presented. When $m \neq n$, the RCP and LCP vortex beams have different phase variations, and all the modes have inhomogeneous state-of-polarization on the beam wavefront except for the north and south poles of the $\text{HOPS}_{\text{m,n}}$. Relevant examples for modes on the $\text{HOPS}_{-1,1}$, $\text{HOPS}_{1,-1}$, and $\text{HOPS}_{-1,2}$ are shown in Figure 5.3b–d. It should be noted that both the SAM and OAM of the modes mapped on the equators of HOPSs with $m \neq n$ are null since these beams only contain local linear polarizations without a φ-dependent phase variation.

By associating Eqs. (5.4) and (5.6), the spin states and the topological charges of the two beams determine the HOPS that the combined beam ($\psi_{m,n}$) belongs to, while their intensity ratio ($\Delta A = A_m/A_n$) and phase differences ($\Delta\alpha = \alpha_m - \alpha_n$) specify the location of the combined beam on the corresponding HOPS, denoted as $\text{HOPS}_{m,n}(\theta, \phi)$. In the following sections of the chapter, the design, modeling, and experimental validation of transmissive devices for generating multiple concurrent beams on the HOPSs will be discussed in detail.

5.3 Millimeter-Wave Transmit-Array Unit Cell Designs

In this section, the design and numerical results of wideband and low-profile unit cells of millimeter-wave transmit-arrays are presented, including two types of single-CP cells utilizing the geometric phase and a dual-CP cell exploiting both the dynamic and geometric phases.

5.3.1 Ka-Band CP Unit Cell Design

For geometric-phase unit cells, they should possess an anisotropic electromagnetic scattering property similar to that of a transmissive half-wave plate, imparting a spatially-dependent abrupt phase change on the transmitted waves with high transmittance. When an LCP/RCP wave (i.e. $\sigma_- = -1\sigma_- = -1$) illuminates an ideal geometric-phase cell, the power will all pass through with no reflection and absorption loss. At the same time, the polarization of the transmitted wave will be

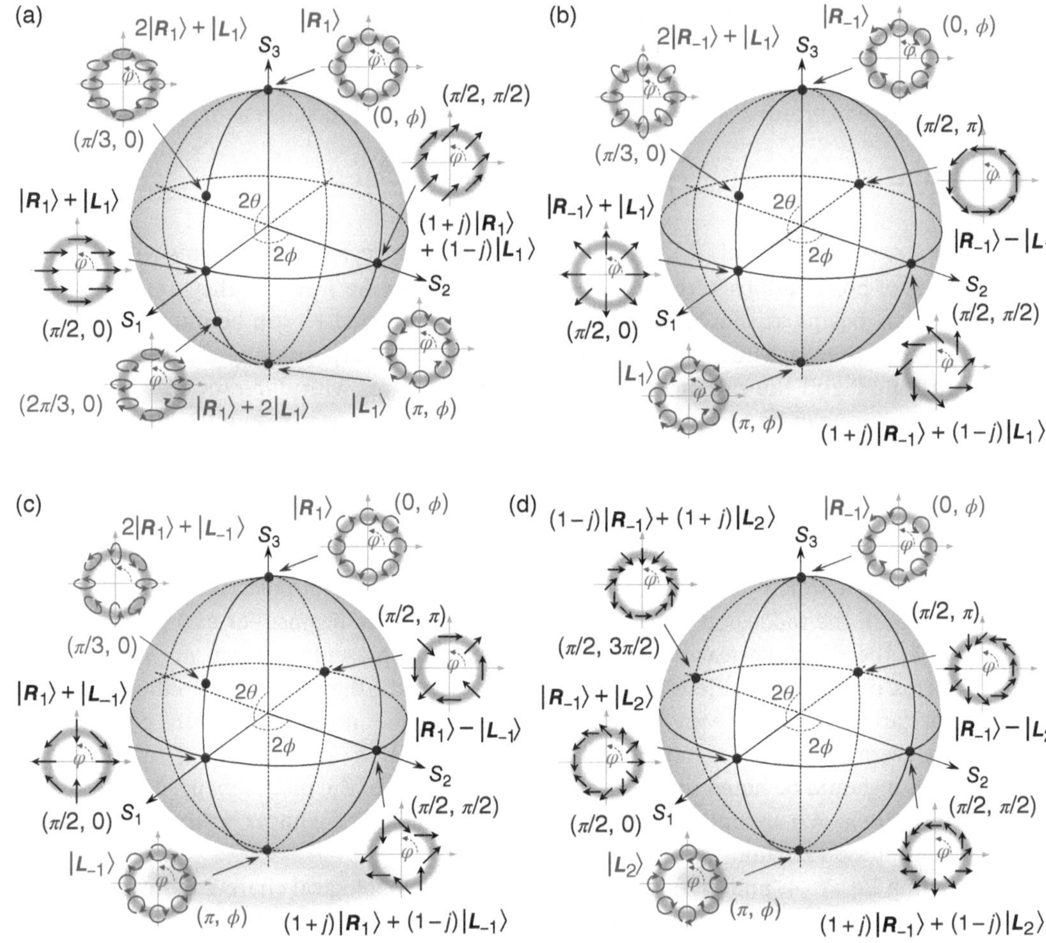

Figure 5.3 Higher-order PS representation of the states-of-polarization of vector vortex beam modes on (a) $HOPS_{1,1}$, (b) $HOPS_{-1,1}$, (c) $HOPS_{1,-1}$, and (d) $HOPS_{-1,2}$.

transformed into an RCP/LCP wave (i.e. $\sigma_+ = +1\sigma_+ = +1$) with an additional phase change of $-2\sigma_\pm\phi_r$, where ϕ_r is the in-plane rotation angle of the cell around its geometrical center [63].

To achieve such a goal, unit cells with a single metallic layer are not sufficient since theoretically, it only provides a maximal total efficiency of 25% [64]. As shown in Figure 5.4a, the unit cell consists of two dielectric spacers and three metallic layers each containing a single elliptical metallic resonator with an I-shaped slot cut out at the center. Miniaturizing the unit cell size down to about $\lambda_0/3$, this slot-cut design is beneficial for improving the resolution of the phase distribution across the transmit-array aperture. The two outer layers have identical patterns, which are different from that of the middle layer, for maintaining a vertically symmetrical structure with a total thickness of around $\lambda_0/3$. As shown in Figure 5.4b, each metallic layer is approximated as a homogenized anisotropic impedance surface with a diagonal surface admittance tensor of $\bar{\bar{Y}}_{so(m)} = \left[Y_{xo(m)}, 0; 0, Y_{yo(m)}\right]$, where $Y_{xo(m)}$ and $Y_{yo(m)}$ can be modeled as a parallel inductor-capacitor (LC) resonator connected in series with a series LC resonator [65]. In detail, the I-shaped slot at the center of the patch can be regarded as a parallel LC resonator with L_p and C_p. The gap between adjacent cells gives the series capacitance C_s, while the metallic parts of the patch are represented by a series inductance L_s. In order to

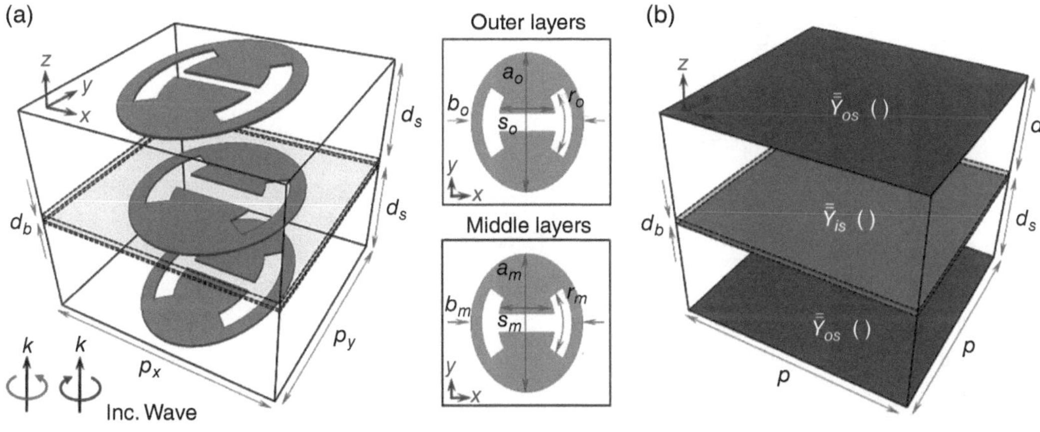

Figure 5.4 (a) Unit cell configuration and (b) the homogenized multilayer anisotropic impedance surface model of the *Ka*-band geometric-phase transmit-array unit cell [56]. *Source*: Jiang et al. [56].

achieve the desirable properties, the responses along the x- and y-directions were made to be capacitive and inductive, respectively. By controlling the circuit parameters (L_p, C_p, L_s, and C_s) via optimizing the geometrical parameters, the phase delay for x- and y-polarized incident waves can be tuned to have a difference of about 180°, while the transmission magnitudes are close to unity. Consequently, the designed cells can efficiently convert LCP incident waves into RCP transmitted waves.

With a targeted frequency of 30 GHz in the *Ka*-band, i.e. $\lambda_0 = 10$ mm, it was found that when $\bar{\bar{Y}}_{so} = \text{diag}[1.05, -1.15]Y_0$ and $\bar{\bar{Y}}_{sm} = \text{diag}[2.15, -1.70]Y_0$, a design can be achieved that provides a remarkably wide operational bandwidth. As shown in Figure 5.5a, the simulated cross-polarized transmittance (T_{LR}=$|t_{LR}|^2$), obtained from a full-wave high-frequency structure simulator solver, is more than 80% over a wide frequency range from 29.3 to 32.1 GHz with a peak value of 94% at 30 GHz. In comparison, the copolarized transmittance (T_{RR}=$|t_{RR}|^2$) is less than 10%, meaning that a majority of the incident RCP wave is transformed into an LCP wave and vice versa due to reciprocity. As shown in Figure 5.5b, cross-polarized CP transmission phase of t_{LR} as a function of rotation angle (ϕ_r) shows that it has a dispersionless linear dependence on the rotation angle at different frequencies, indicating that a stable phase distribution can be formed over a wide bandwidth. The simulated results agree well with the analytical calculations, thus validating the homogenization model applied to the transmit-array unit cell.

To further confirm the unit cell design, the diffraction of an infinite periodic array comprised of one-dimensional phase gradient supercells of eight cells was investigated. As shown in Figure 5.6a, the cells in the supercell have a rotation angle step of $\Delta\phi_r = 22.5°$, which leads to a beam bending in the momentum space with accompanying waves corresponding to other diffraction orders as depicted in Figure 5.6b [66]. When an RCP plane wave is incident normally from the bottom, the transmitted wave is deflected into a direction corresponding to $\theta_t = 20.9°$ and converted into an LCP wave, which is manifested by the snapshots of the electric field components E_x and E_y, displayed in Figure 5.6d. When the incident wave is switched to an LCP state, the transmitted wave becomes RCP with a refraction angle of $\theta_t = -20.9°$, see Figure 5.6e. Importantly, the transmitted wave that undergoes spin-dependent bending has almost the same intensity as that of the incident wave, which can be further confirmed by the diffraction analysis of this gradient array, see Figure 5.6c, indicating that the incident wave is coupled into the first diffraction order mode with an efficiency of higher than 95%.

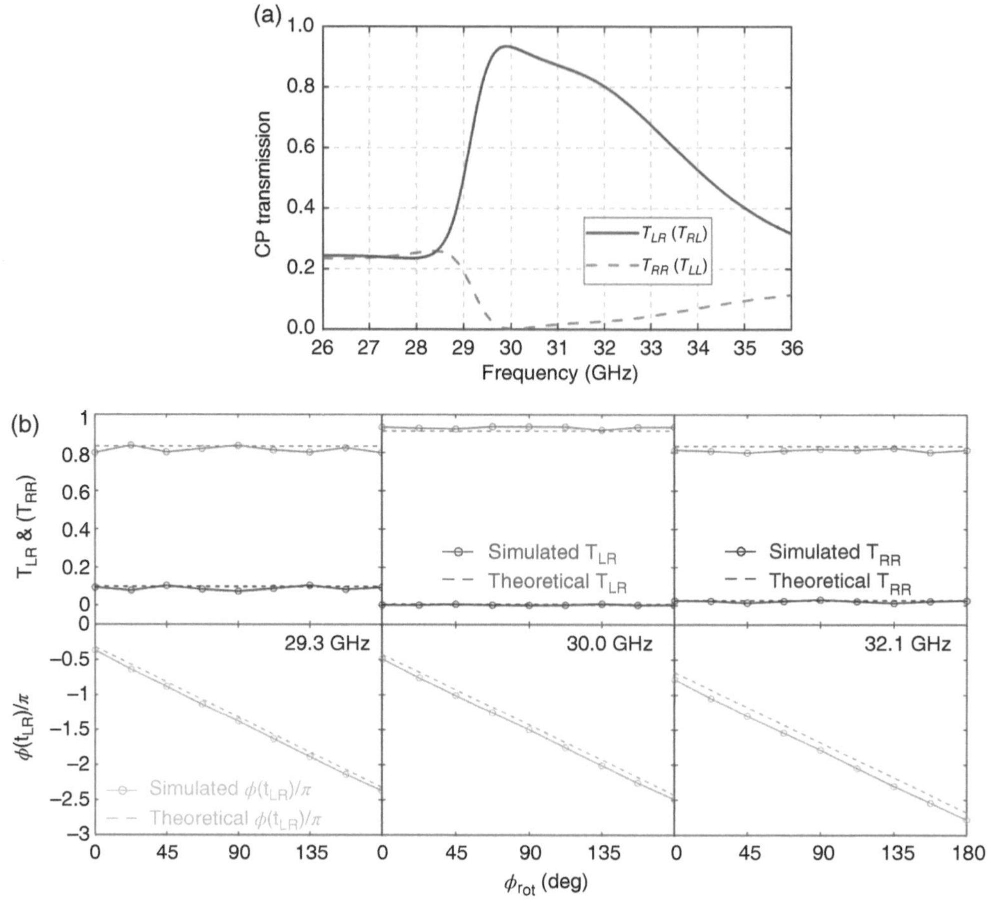

Figure 5.5 (a) Simulated CP transmission as a function of frequency for the *Ka*-band cell with $\phi_r = 0°$. (b) Simulated and theoretically calculated CP transmittance and CP transmission phase of the *Ka*-band cell as a function of rotation angle (ϕ_r) at 29.3, 30, and 32.1 GHz [56]. *Source*: Jiang et al. [56].

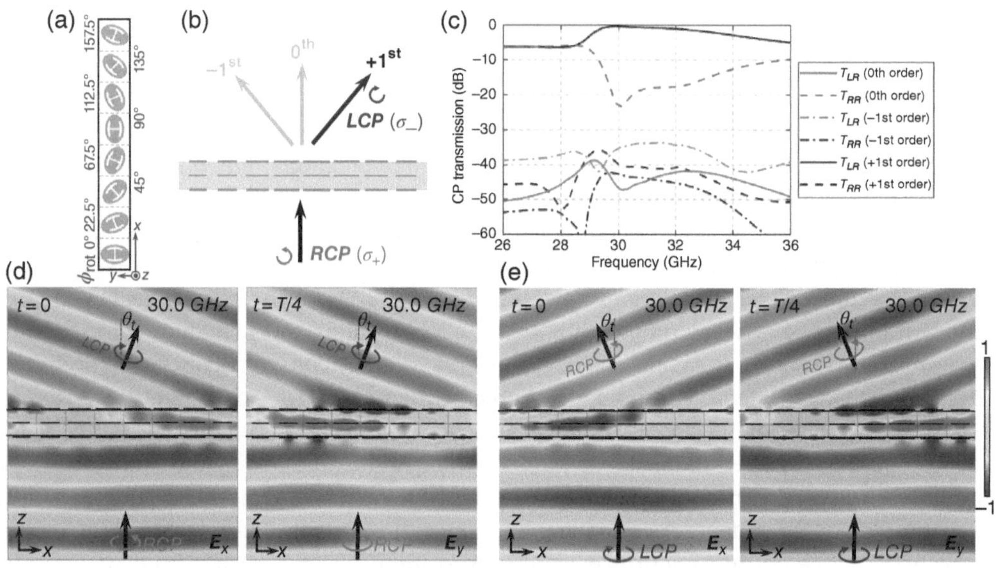

Figure 5.6 (a) Configuration and (b) schematics of diffraction of a supercell of the linear gradient transmitarray with a cell rotation angle step of 22.5°; (c) simulated CP transmission as a function of frequency for different diffraction orders with an RCP incident wave at normal incidence. Snapshots of the simulated electric fields (E_x at $t = 0$ and E_y at $t = T/4$) when (d) an RCP or (e) an LCP wave illuminates an array of unit cells with a gradient rotation angle (ϕ_r).

5.3.2 Q-Band CP Unit Cell Design

The *Ka*-band transmit-array cell utilizes isolated resonators for all the metallic layers which are capacitively coupled to each other. The second design operating at the *Q*-band employs a different strategy – all the metallic layers of the triple-layer unit cell are interconnected with each other. As shown in Figure 5.7a, the outer layers have interconnected thin annular strip meshes with an isolated patch at the center. The middle layer consists of a continuous copper sheet with a circular aperture at the center. A patch with arc-shaped ends was placed inside this circular aperture. The unit cell size p was selected as 2.5 mm, i.e. about $\lambda_0/3$ at the center frequency of 40 GHz, which is smaller than that of conventional reflect-/transmit-arrays and therefore provides a finer phase resolution. Similarly, homogeneous anisotropic impedance surfaces with different dispersive surface admittance tensors of $\overline{\overline{Y}}_{so(m)} = \left[Y_{xo(m)}, 0; 0, Y_{yo(m)}\right]$ were used to model the outer and middle metallic layers, respectively, see Figure 5.7b. In order to account for the frequency dependence, a geometry-induced equivalent circuit was developed to model the dispersive $Y_{x(y)o(m)}$ for the outer and middle anisotropic impedance surface layers, as shown in Figure 5.7c and d. The equivalent circuits can all be simplified to be an inductor ($L_{x(y)o(m)}$) connected in parallel with a series LC resonator containing $C_{x(y)o(m)}$ and $L_{x(y)o(m)p}$, where $C_{x(y)o(m)}$ is expressed as $C_{x(y)o(m)g}C_{x(y)o(m)p}/(C_{x(y)o(m)g} + 2C_{x(y)o(m)p})$ [67]. At the center frequency of 40 GHz, i.e. $\lambda_0 = 7.5$ mm, it was found that the required admittance tensors are $\overline{\overline{Y}}_{so} = \text{diag}[1.05, -1.15]Y_0$ and $\overline{\overline{Y}}_{sm} = \text{diag}[2.15, -1.70]Y_0$, when the dielectric spacers have a thickness of 1.52 mm.

The CP transmission and reflection curves of the optimized realistic structure, obtained from the full-wave simulator, agree well with those calculated using the homogenized model, see Figure 5.8. In particular, the maximum value of T_{LR} reaches 0.94, with a 1-dB bandwidth of about 12.5%. Within this frequency range, the value of T_{RR} is smaller than 0.02, indicating a negligible amount of undesirable copolarized transmitted waves. Meanwhile, $R_{RR}=|r_{RR}|^2$ and $R_{LR}=|t_{LR}|^2$ are all below 0.1 in the majority of the band, ensuring a small reflection loss.

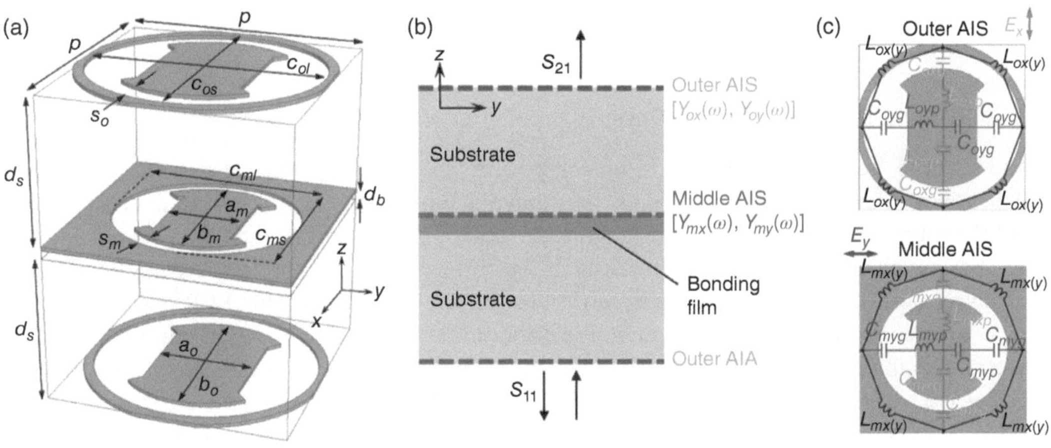

Figure 5.7 (a) Unit cell configuration and (b) the homogenized multilayer anisotropic impedance surface model of the *Q*-band geometric-phase transmit-array unit cell. (c) The equivalent circuits for the outer and middle anisotropic impedance surface layers of the unit cell. The fitted circuit elements are L_{ox} = 0.8 nH, L_{oxp} = 0.26 nH, C_{ox} = 16.5 fF, L_{oy} = 0.7 nH, L_{oyp} = 0.31 nH, C_{oy} = 7.6 fF, L_{mx} = 0.197 nH, L_{mxp} = 6.77 pH, C_{mx} = 95.3 fF, L_{my} = 0.146 nH, L_{myp} = 1.33 pH, C_{my} = 72 fF [67].

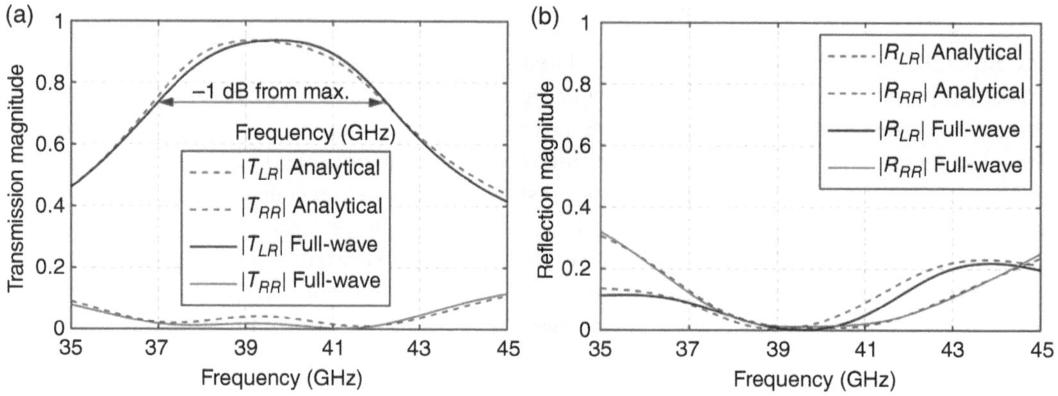

Figure 5.8 CP (a) transmission and (b) reflection magnitudes of the Q-band geometric-phase transmit-array unit cell at normal incidence simulated by a full-wave solver and analytically calculated from the homogenized model [67].

The responses of the proposed Q-band unit cell with a varying rotation angle (ϕ_r) at different angles of incidence are reported in Figure 5.9. It can be seen that it has a high polarization conversion within a wide frequency range around 40 GHz as revealed from the contour plot of T_{LR}. Only at an incident angle of 30°, a slight bandwidth narrowing can be observed. At the same time, T_{RR} remains at a very low level within this field of view. The phase of t_{LR} and t_{RR} as a function of the rotation angle ϕ_r at 37, 40, and 42 GHz for $\theta_{\mathrm{i}} = 0$, 15, and 30° are shown in Figure 5.10. It can be seen that the phase of t_{LR} has a linear dependence with ϕ_r almost regardless of the incident angle. Such a behavior is also not affected by changing the frequency, thus providing a dispersionless phase delay. The analytically calculated transmission phases using the homogenized model are also plotted, showing a strong agreement with the full-wave results, which further substantiates the validity of the model even at oblique incident angles.

5.3.3 K-Band Dual-CP Unit Cell Design

Although the geometric-phase transmit-array unit cells can provide wideband frequency-independent phase delay for CP waves, they have a limitation that the phase delays for RCP and LCP waves always have the same magnitudes and opposite signs. Hence, they do not support true dual-CP phase compensation. Recently, it was proposed at the optical wavelengths that, by combining both the dynamic and geometric phases, the phase delay for RCP and LCP waves can be independently manipulated, enabling an additional degree of freedom for multipolarization wavefront shaping [55]. Planar dual-CP reflect-arrays were then realized at microwave and millimeter-wave frequencies [68, 69].

Here, a transmissive-type dual-CP unit cell is proposed with a low profile and wide operational bandwidth. Figure 5.11a shows the configuration of the unit cell where the antenna-filter-antenna architecture was employed. It consists of five metallic layers separated by four dielectric spacers. The middle layer is a solid ground plane, while the two layers above and below the ground plane are dual-polarized circular stacked patches with four slots cut out on the edges for size reduction.

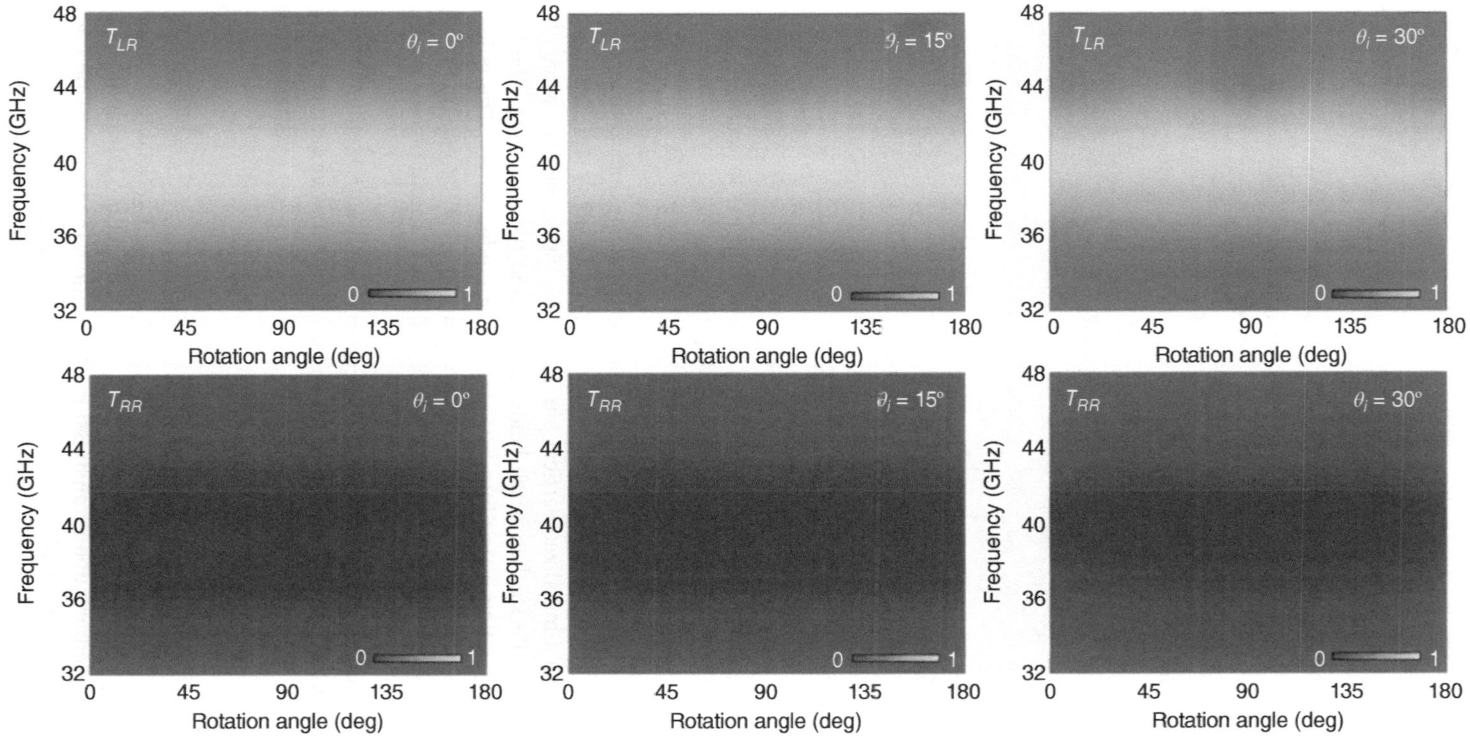

Figure 5.9 Contour plots of simulated T_{LR} and T_{RR} as a function of both frequency and the rotation angle (ϕ_r) of the unit cell for θ_i = 0, 15, and 30° [67].

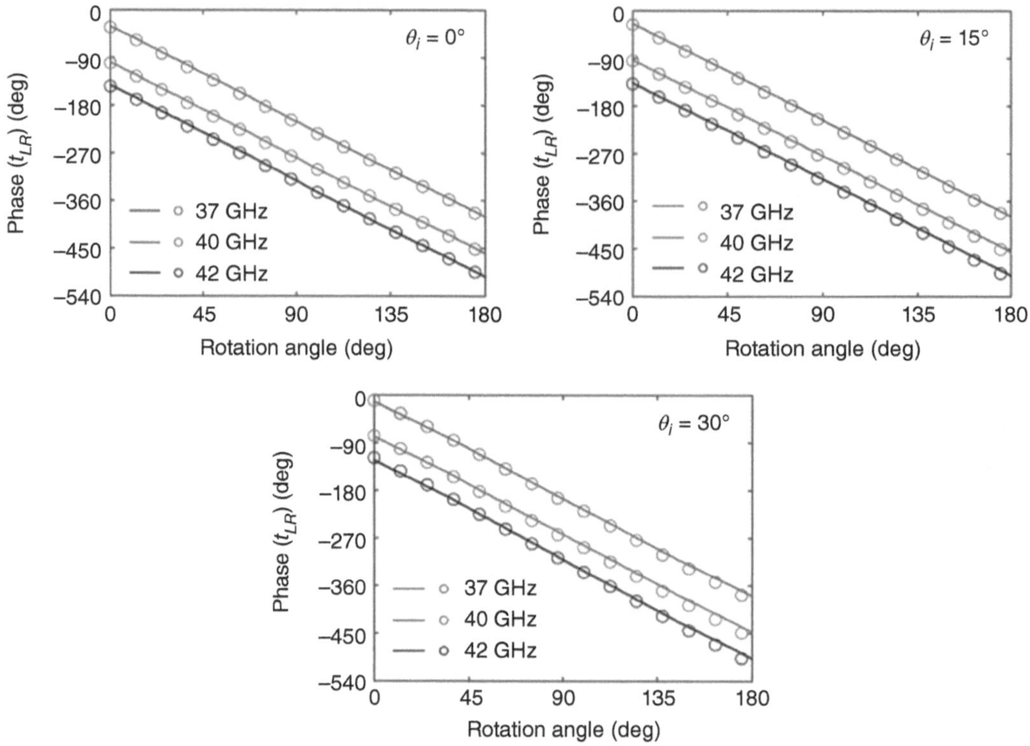

Figure 5.10 Simulated (solid lines) and analytically (circles) calculated phases of t_{LR} at 37, 40, and 42 GHz as a function of the rotation angle (ϕ_r) of the unit cell for θ_i = 0, 15, and 30° [67].

The two sets of stacked patches are both fed by two orthogonal microstrip delay lines at the edge of the inner patches, see Figure 5.11b, while the two microstrip delay lines above and below the ground plane are interconnected via a metalized vertical pin. Basically, for an incident wave illuminating the cell from the bottom, the stacked patch below the ground plane converts the two orthogonally linear-polarized propagating waves into two streams of guided waves. The guided waves then pass through into the stacked patches above the ground plane via the two microstrip delay lines, which are finally reradiated out into the hemisphere above the ground plane, forming the transmitted waves.

In order to obtain the desirable independent dual-CP phase compensation with high efficiency, the transmission matrix of the cell, which relates the incident and transmitted wave vectors, can be written as [55]

$$\bar{\bar{T}} = \begin{bmatrix} t_{xx} & t_{xy} \\ t_{yx} & t_{yy} \end{bmatrix}, \tag{5.7}$$

where the elements are the complex transmission coefficients of the x- or y-polarized transmitted wave due to an x- or y-polarized incident wave propagating in the $+z$ direction. For CP incident waves, it is desirable to flip the handedness upon transmission. In addition, the transmission magnitude is unity, and the desirable transmission phases are ϕ_{RL} and ϕ_{LR} for RCP and LCP transmitted waves, respectively. In order to satisfy such conditions, the following equations can be obtained

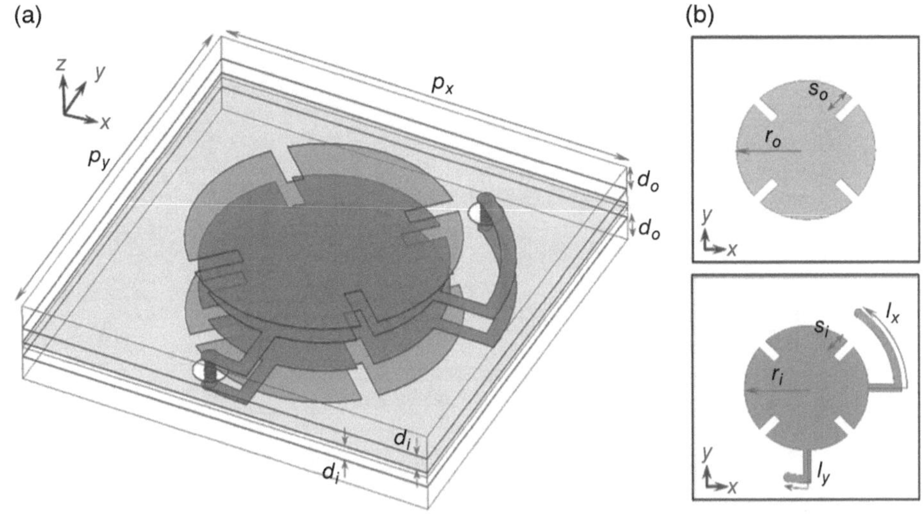

Figure 5.11 (a) Unit cell configuration; (b) the first (fourth) and second (fifth) metallic layers of the *K*-band dual-CP transmit-array unit cell.

$$e^{j\phi_{RL}}\begin{bmatrix}1\\-j\end{bmatrix}=\begin{bmatrix}t_{xx}&t_{xy}\\t_{yx}&t_{yy}\end{bmatrix}\begin{bmatrix}1\\j\end{bmatrix}\tag{5.8}$$

$$e^{j\phi_{LR}}\begin{bmatrix}1\\j\end{bmatrix}=\begin{bmatrix}t_{xx}&t_{xy}\\t_{yx}&t_{yy}\end{bmatrix}\begin{bmatrix}1\\-j\end{bmatrix}\tag{5.9}$$

The incident/transmitted wave vectors $[1,-j]^T$ and $[1,j]^T$ denote RCP and LCP waves, respectively. By solving the linear equation system provided by Eqs. (5.8) and (5.9), the four unknowns, i.e. the four elements of the transmission matrix, are expressed as

$$t_{xx}=\left(e^{j\phi_{LR}}+e^{j\phi_{RL}}\right)/2,\tag{5.10}$$

$$t_{xy}=j\left(e^{j\phi_{LR}}-e^{j\phi_{RL}}\right)/2,\tag{5.11}$$

$$t_{yx}=j\left(e^{j\phi_{LR}}-e^{j\phi_{RL}}\right)/2,\tag{5.12}$$

$$t_{yy}=-\left(e^{j\phi_{LR}}+e^{j\phi_{RL}}\right)/2.\tag{5.13}$$

It can be seen from Eqs. (5.10–5.13) that $t_{xx}=-t_{yy}$ and $t_{yx}=t_{xy}$. By diagonalizing the transmission matrix, it can be deduced that

$$\bar{\bar{T}}=\bar{\bar{M}}_{rot}\bar{\bar{T}}_{diag}\bar{\bar{M}}_{rot}^{-1},\tag{5.14}$$

with

$$\bar{\bar{T}}_{diag}=\begin{bmatrix}e^{\frac{j(\phi_{LR}+\phi_{RL})}{2}}&0\\0&-e^{\frac{j(\phi_{LR}+\phi_{RL})}{2}}\end{bmatrix},\tag{5.15}$$

$$\bar{\bar{M}}_{rot}=\begin{bmatrix}\cos(\Delta\phi/4)&\sin(\Delta\phi/4)\\\sin(\Delta\phi/4)&-\cos(\Delta\phi/4)\end{bmatrix},\tag{5.16}$$

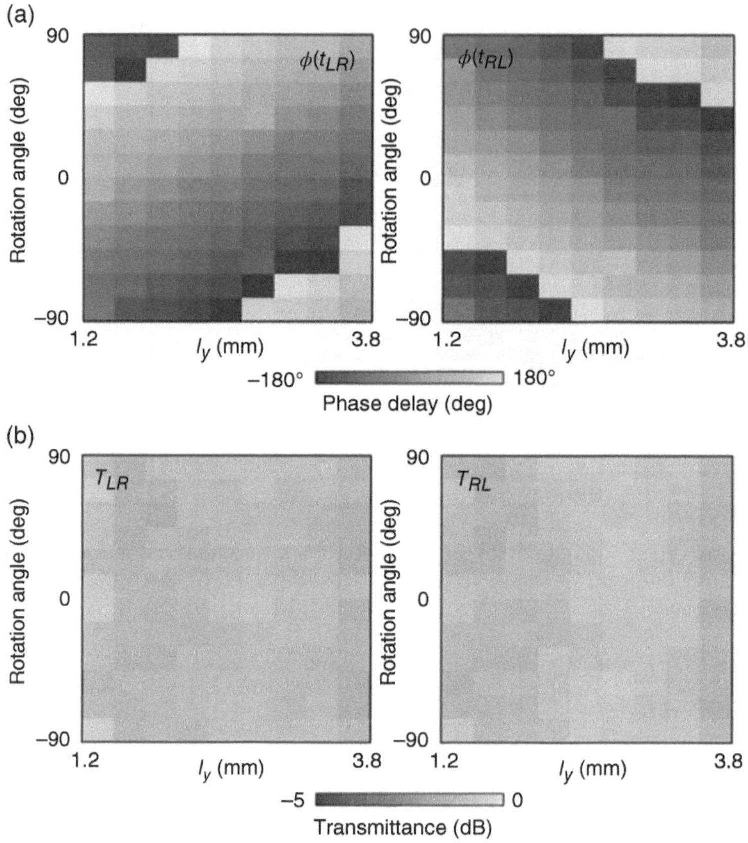

Figure 5.12 Distribution of the simulated (a) cross-polarized CP transmission phases, i.e. $\phi(t_{LR})$ and $\phi(t_{RL})$, and (b) cross-polarized CP transmission amplitudes, i.e. T_{LR} and T_{RL}, as a function of both the delay line length and cell rotation angle at 22 GHz for normal incidence.

where $\Delta\phi = \phi_{LR} - \phi_{RL}$ is the difference of transmission phases for CP waves with opposite spin states. It can be readily observed from Eqs. (5.14–5.16) that, in order to achieve independent dual-CP transmission phase compensation, the proposed unit cell structures should possess scattering properties similar to that of a transmissive half-wave plate but with a gradient dynamic phase determined by the sum of ϕ_{LR} and ϕ_{RL} and a varying rotation angle dependent on the difference between ϕ_{LR} and ϕ_{RL}. To this end, the total lengths of the two microstrip delay lines, which determine the amount of phase delay for the two linear polarizations are denoted as $2l_x$ and $2l_y$, need to have an electrical difference of a half wavelength for each unit cell. In addition, the microstrip delay lines should be confined inside a circle tangential to the periphery of the unit cell since the cells need to be rotated for generating the geometric phases. In order to achieve the required phase difference between the two linear polarizations within a compact area, the microstrip delay lines were bent with their bending sections being curved.

The simulated cross-polarized and copolarized transmission amplitudes and the cross-polarized transmission phases for circular polarization at 22 GHz are shown in Figure 5.12 for both RCP and LCP input waves. It can be seen that when changing the lengths of the microstrip delay lines and the cell rotation angle, the cross-polarized CP transmission amplitude is maintained above −1.5 dB. In contrast, the copolarized CP transmission amplitude is below −15 dB. This observation indicates that the handedness of the CP incident wave is flipped with high efficiency. As the cross-polarized

CP transmission phase shows, $\phi(t_{LR})$ and $\phi(t_{RL})$ exhibit opposite trends versus the in-plane rotation angle of the cells. By utilizing the cells with gradient Berry-phase provided by the different in-plane rotation angles and gradient dynamic-phase offered by the different lengths of the delay lines, independent phase delay for cross-polarized RCP and LCP transmitted waves can be achieved. Specifically, by maintaining the same value of $\phi(t_{LR})$, i.e. the pixels along the diagonal lines in Figure 5.12a, a full 360° can be covered by $\phi(t_{RL})$, and vice versa. This indicates that independent phase delay for the RCP and LCP waves can be achieved, breaking the restrained symmetry of phase delays for CP waves with opposite handedness in cells utilizing only the Berry phase.

5.4 Millimeter-Wave Transmit-Arrays for Vortex Beam Multiplexing

In this section, the generation of massive scalar and vector vortex beams using transmit-arrays synthesized with the three kinds of proposed unit cells is presented. First, a computationally efficient far-field calculation method for the transmit-arrays, which usually contain hundreds or even thousands of unit cells, is illustrated. Then, a scalar vortex beam multiplexer and two kinds of vector vortex beam multiplexers are shown as proof-of-concept examples demonstrating the capabilities of low-profile transmit-arrays for complex wavefront shaping at millimeter-wave frequencies.

5.4.1 Far-Field Pattern Calculation for Transmit-Arrays

The evaluation of electrically-large transmit-arrays composed of sub-wavelength unit cells with 3D features together with a feed antenna can be computationally demanding and even infeasible. Approximated scalar field superposition methods based on array theory and an assumed $\cos^q(\theta)$ model for the radiation pattern of the feeding source have been used extensively [70]. A more accurate yet efficient approach is to utilize, instead of scalar array theory, the vectorial near-field quantities and the scattering parameters of the unit cells. Under the condition that the intercell coupling is weak, a local periodicity condition can be assumed.

The calculation process starts by performing a full-wave simulation of the feed antenna alone, such as a horn or a patch, and extracting out the near-field quantities, i.e. the vectorial electric and magnetic fields ($\boldsymbol{E_f}$, $\boldsymbol{H_f}$), at the aperture of the feed antenna. A plane Ω_f should be selected, which is large enough to cover most of the radiated power, see Figure 5.13a. Next, the electric and magnetic fields ($\boldsymbol{E_i}$, $\boldsymbol{H_i}$) at the plane where the transmit-array is placed, which has a distance

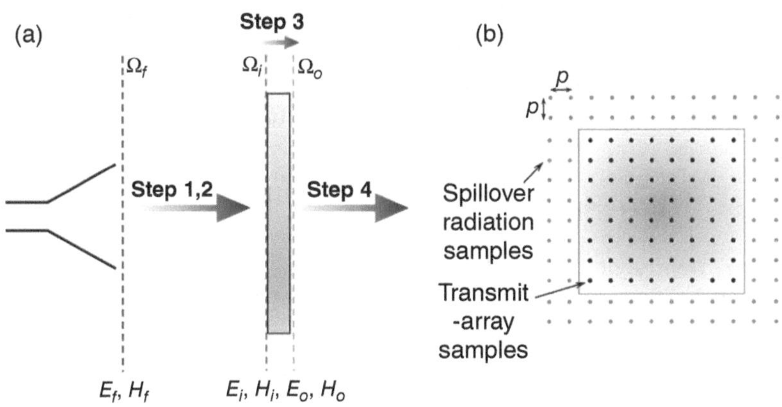

Figure 5.13 (a) Steps for calculating the far-field properties of the transmit-arrays. (b) Scheme for the transmit-arrays model, including spillover radiation [67].

of H away from the feed, are deduced from $\boldsymbol{E_f}$ and $\boldsymbol{H_f}$ by using the free-space dyadic Green's function [71]. A plane Ω_i larger than the size of the transmit-array (Ω_{TA}) should be selected in order to account for the spillover radiation, see Figure 5.13b. The plane Ω_i is discretized into a square grid with a lattice size chosen to be the same as that of the transmit-array unit cell. At each point (x_i, y_i), $\boldsymbol{E_i}$ and $\boldsymbol{H_i}$ are decomposed into transverse electric (TE)- and transverse magnetic (TM)-polarized waves ($\boldsymbol{E_{iTE}}$ and $\boldsymbol{E_{iTM}}$). A lookup table is created for the scattering parameters of the unit cell with various rotation angles ($0° \leq \phi_r \leq 180°$) illuminated by both TE and TM-polarized waves at different angles of incidence ($0° \leq \theta_i \leq 35°$). From this table, the tangential fields ($\boldsymbol{E_{to}}$ and $\boldsymbol{H_{to}}$) at the output face (Ω_o) of the transmit-array can be obtained, where Ω_o has the same area as Ω_i. For the remainder points on Ω_o outside the aperture of the transmit-array, $\boldsymbol{E_{to}}$ and $\boldsymbol{H_{to}}$ can be readily obtained depending on whether the point denotes either free space or a multilayered dielectric substrate. Finally, based on the tangential fields on Ω_o, the far-field quantities $\boldsymbol{E_{far}}(u,v)$ can be calculated using the First Principle of Equivalence [72]. Specifically, the radiated far-fields are expressed as

$$E_\theta^{far} = \frac{jk_0 e^{-jk_0 r}}{4\pi r}\left[P_y \sin\varphi + P_x \cos\varphi + \eta_0 Q_y \cos\theta \cos\varphi - \eta_0 Q_x \cos\varphi \sin\varphi\right], \tag{5.17}$$

$$E_\varphi^{far} = \frac{jk_0 e^{-jk_0 r}}{4\pi r}\left[P_y \cos\theta \cos\varphi - P_x \cos\varphi \sin\varphi - \eta_0 Q_x \sin\varphi - \eta_0 Q_x \cos\varphi\right], \tag{5.18}$$

where k_0 and η_0 are the free-space propagation constant and impedance, respectively. P_x, P_y, Q_x, and Q_y are four spectrum functionals, which are expressed as

$$P_{x(y)} = \iint_S E_{to}^{x(y)}(x,y) e^{jk_0(ux+vy)} dxdy, \tag{5.19}$$

$$Q_{x(y)} = \iint_S H_{to}^{x(y)}(x,y) e^{jk_0(ux+vy)} dxdy, \tag{5.20}$$

with $u = \sin\theta\cos\varphi$ and $v = \sin\theta\sin\varphi$. It should be noted that the calculation of the far-field can be accelerated by applying a 2D fast Fourier transform [73], which will be utilized in the phase-only optimization for synthesizing the transmit-arrays.

5.4.2 Multiplexing of Scalar Vortex Beams

Multiple controllable vortex beam generation with a single transmissive device is of great importance in a wide variety of applications, including high-speed communication systems based on OAM, simultaneous tracking of multiple particles, and so on. By incorporating spiral phase profiles into a harmonic-response spatial light modulator with tailored intensity [74], a spin-dependent dual OAM beam generator was designed. The distribution in the rotation angle $\phi_r(x,y)$ corresponding to the tiling of unit cells across the quasi-circular aperture is represented by

$$\phi_r(x,y) = -arg\left(\sum_{m=1}^{2} A_m e^{j\alpha_m} e^{-j\left[\left(\frac{\pi}{s_y p}\right)y + \ell_m\varphi/2\right]}\right), \tag{5.21}$$

where A_m and α_m represent the amplitude and phase of each desired diffraction order, see Figure 5.14a. For the two transmitted CP vortex beams, which are confined within the y–z plane, the pointing directions correspond to $k_x = 0$, $k_{y1}/k_0 = -\sigma_\pm\lambda_0/s_y p$, and $k_{y2}/k_0 = -\sigma_\pm\lambda_0/s_y p$ with topological charges of ℓ_1 and ℓ_2, respectively. Importantly, the direction and topological charge of each beam, as well as the intensity ratio between the two beams can be independently controlled. Since a beam with a larger topological charge naturally has a lower peak intensity, such a method

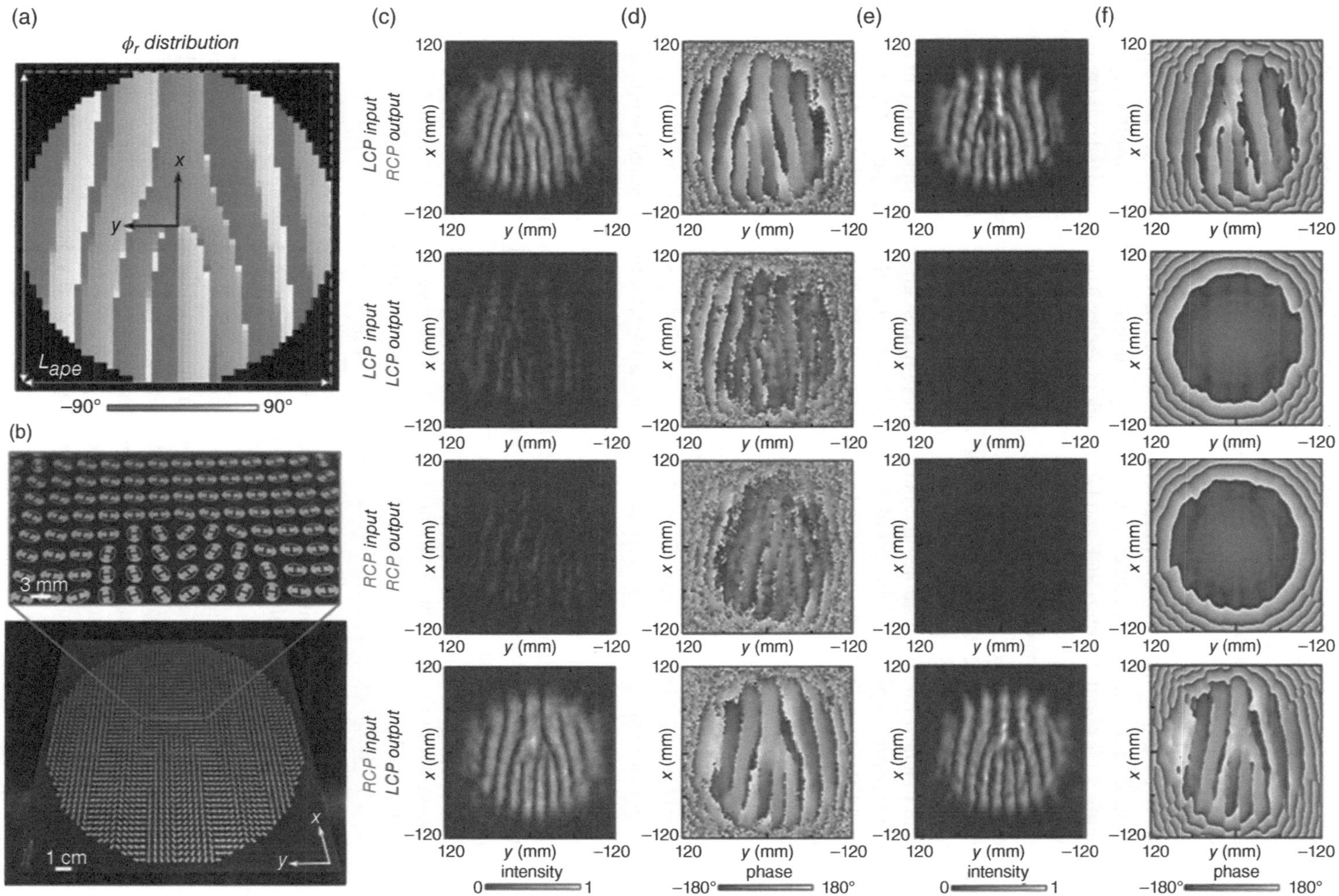

Figure 5.14 (a) Distribution of the rotation angle (ϕ_r) of the *Ka*-band unit cells for generating two CP vortex beams; (b) photograph of the fabricated transmit-array sample. Normalized measured LCP and RCP components of the near-field: (c) intensities and (d) phases at a plane 3 cm away from the output interface of the transmit-array at 29.4 GHz. Normalized simulated LCP and RCP components of the near-field (e) intensities and (f) phases at a plane 3 cm away from the output interface of the transmit-array at 30 GHz [56]. *Source*: Jiang et al. [56].

could be used to balance the signal-to-noise ratio for multiple receivers among the generated vortex beams with different topological charges. The parameter α_m can be optimized to maximize the efficiency of the finite-sized transmit-array and to a certain extent improve the overall beam quality. As a proof-of-concept example, the design parameters were chosen for generating two CP vortex beams pointing at $\theta_{t1} = \sigma_\pm \sin^{-1}[-\lambda_0/8p]$ and $\theta_{t2} = \sigma_\pm \sin^{-1}[\lambda_0/23p]$, while possessing the same peak intensity and topological charges of $\ell_1 = 2\sigma_\pm$ and $\ell_2 = -\sigma_\pm$, respectively.

The fabricated transmit-array for dual vortex beam generation is shown in Figure 5.14b. The normalized measured near-field intensity and phase distributions of the RCP and LCP components at 29.4 GHz, obtained at a plane of $3\lambda_0$ away from the output interface of the transmit-array, are reported in Figure 5.14c and d, for both RCP and LCP incident waves. The copolarized components have a flat distribution with very low peak intensity, confirming that the spin state is almost entirely flipped as the CP incident wave passes through the device. Interference intensity and phase patterns can be observed within the area of the transmit-array, exhibiting a complicated quad-furcating pattern in the region close to the center. From the normalized measured far-field intensity and phase patterns of the LCP and RCP components, see Figure 5.15a and b, two well-separated RCP vortex beams with donut-shaped intensity profiles and helical phase-fronts can be identified with an LCP incident wave, possessing topological charges of $\ell_1 = -2$ and $\ell_2 = +1$ and pointing at $k_{y1}/k_0 = 0.361$ and $k_{y2}/k_0 = -0.127$, respectively. The measured radius ratio between the two modes is about 0.72, which agrees well with the theoretical value of $1/\sqrt{2}$ [75]. When the input wave is RCP, the sign of the topological charges, as well as the directions of the two beams are flipped, due to the inherent spin-dependent property of the geometric phase. Importantly, the intensities of the two vortex beams are about the same, although their modes are different. Nulls can be identified at the center of the two vortex beams, indicating the existence of high-quality optical singularities. The dual vortex beam transmit-array achieves a measured peak efficiency of 71%, see Figure 5.16, with an efficiency > 50% bandwidth of about 13%. As revealed by Figures 5.14–5.16, the measured results agree well with simulation predictions, at both the near-field and far-field.

The geometric-phase transmit-array can be further extended to achieve 2D massive scalar vortex beam multiplexing and demultiplexing. When an incident RCP/LCP beam illuminates the transmit-array at an angle of $(\sigma_\pm k_{xm}/k_0, \sigma_{i\pm} k_{ym}/k_0)$, an LCP/RCP vortex beam will emanate normal to the output interface of the transmit-array with a particular topological charge of $\sigma_\pm \ell_m$, see Figure 5.17a. In such a way, multiple uncorrelated signal beams, each carrying a data stream, can be combined into a single beam with mixed but mutually orthogonal OAM modes, thereby achieving the 2D multiplexing functionality. For the same transmit-array, when it is illuminated by a normally incident LCP/RCP OAM beam carrying a topological charge of $-\sigma_\pm \ell_m$, an RCP/LCP output beam carrying zero OAM can be obtained at an angle of $(-\sigma_\pm k_{xm}/k_0, -\sigma_\pm k_{ym}/k_0)$, see Figure 5.17b. This observation indicates that an incident beam with mixed scalar vortex modes can be separated into directive plane-wave-like beams pointing in angularly resolvable directions, thus enabling the demultiplexing functionality.

As a proof-of-concept example, octa-beam multiplexing/demultiplexing was numerically demonstrated in a transmit-array based on the proposed *Ka*-band unit cell design by exploiting two orthogonal circular polarizations and four vortex modes. Two beams are confined within the x–z and y–z planes, respectively. The topological charge parameters were selected to be $-\ell_1 = \ell_3 = 2$ and $\ell_2 = -\ell_4 = 4$. The normalized simulated far-field intensity and phase patterns indeed show that a scalar vortex beam with a particular topological charge and polarization is obtained at the center

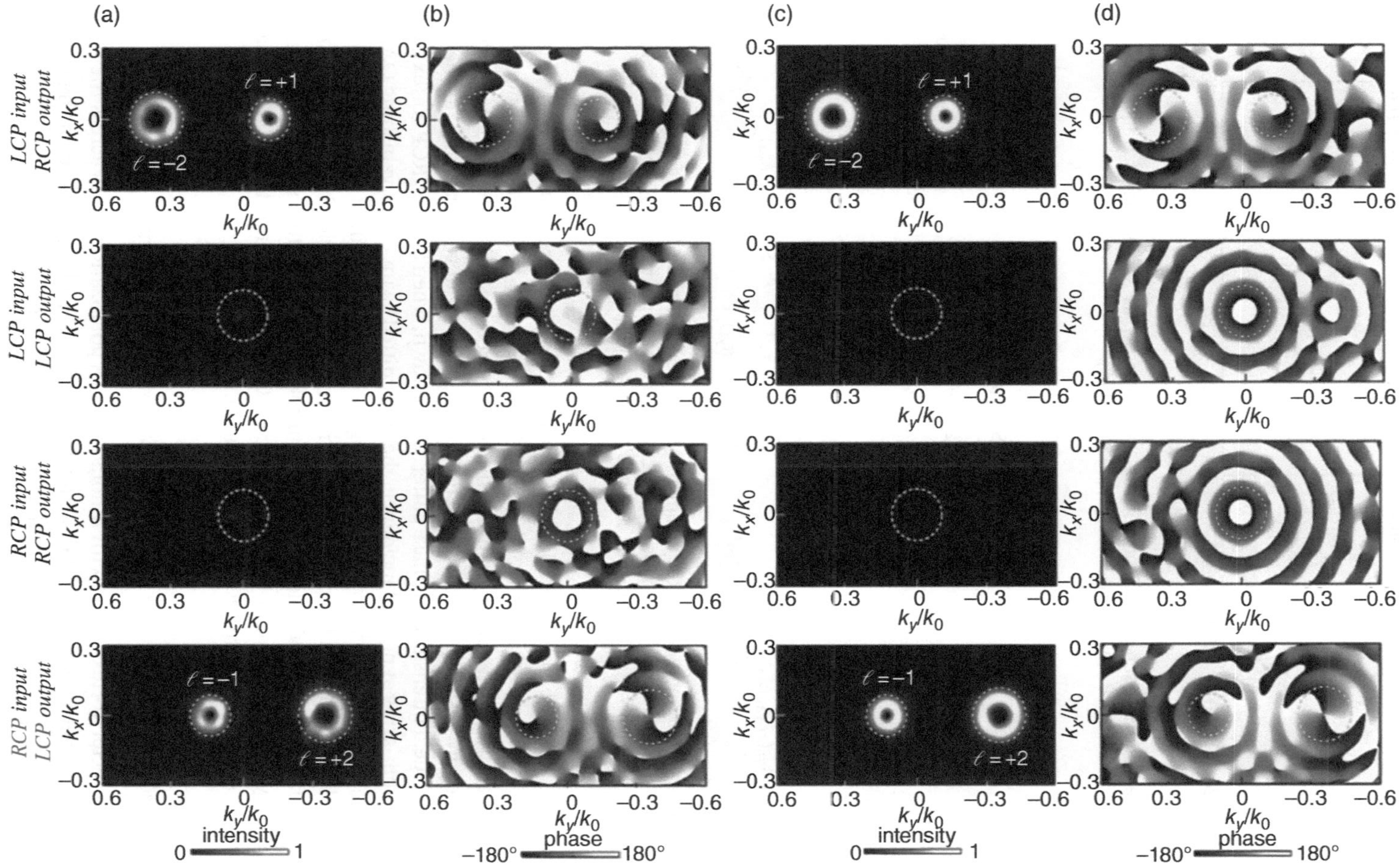

Figure 5.15 Normalized measured LCP and RCP components of the far-field (a) intensities and (b) phases of the transmitted beam. Normalized simulated LCP and RCP components of the far-field (c) intensities and (d) phases of the transmitted beam [56]. *Source*: Jiang et al. [56].

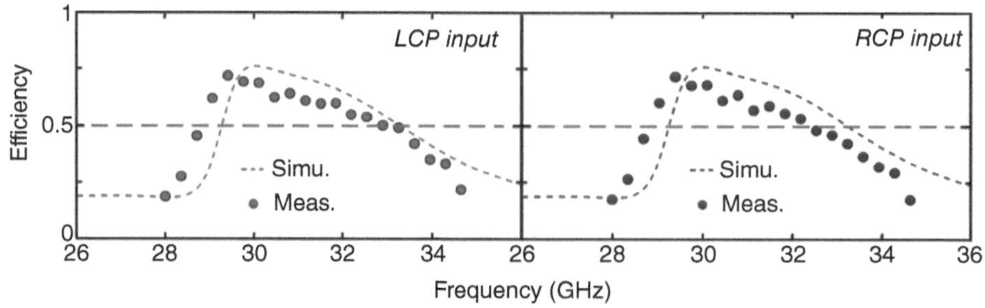

Figure 5.16 Simulated and measured efficiency of the transmit-array as a function of frequency when illuminated by an LCP and an RCP wave [56]. *Source*: Jiang et al. [56].

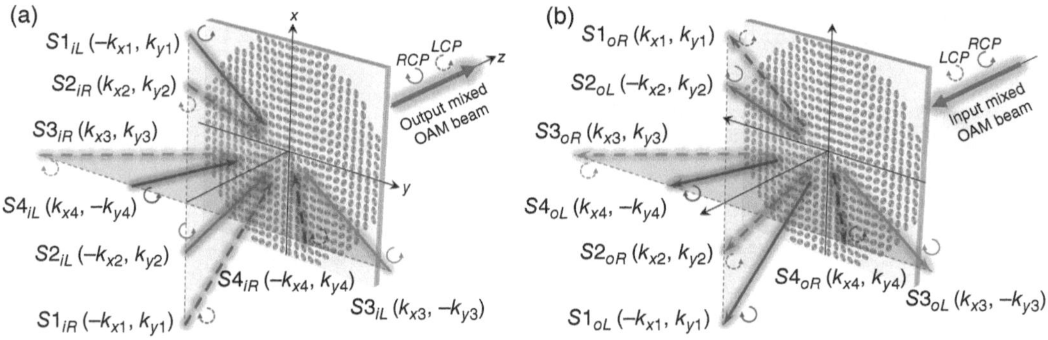

Figure 5.17 Conceptual configurations of the geometric-phase transmit-array for (a) encoding eight CP incident beams into a multiplexing mixed-mode vortex beam and (b) demultiplexing a beam with mixed OAM modes into eight CP transmitted beams [56]. *Source*: Jiang et al. [56].

of the momentum space, see Figure 5.18a. In fact, for each incident beam, four output beams are generated. By enforcing a large angular separation between them, the three remaining undesired beams can be pointed in directions far from the broadside direction that is outside the angular region of interest, thereby ensuring a clean diffraction pattern. For the reverse scenario, as the normalized simulated far-field intensity and phase patterns are shown, when an LCP/RCP vortex beam illuminates the transmit-array at normal incidence, the transmitted beam with a vanishing OAM is refracted into different angles depending on the topological charge and spin of the incident wave, see Figure 5.18b. This confirms that the proposed millimeter-wave geometric-phase transmit-array has the capability of processing multiple signals by multiplexing them into a mixed-mode scalar vortex beam, and by demultiplexing a beam with mixed OAM indices into the original beams and simultaneously separating them in the angular domain.

5.4.3 Multiplexing of Vector Vortex Beams with Symmetry Constraints

In order to generate vector vortex beams with inhomogeneous polarization states across the beam wavefront using the geometric-phase transmit-array, RCP and LCP scalar vortex beams have to be formed pointing at the same directions such that they can interfere. Due to the spin-dependent phase-shifting properties of the geometric-phase cells, a simple linear gradient phase function cannot achieve the superposition of RCP and LCP vortex beams since they are always diffracted into

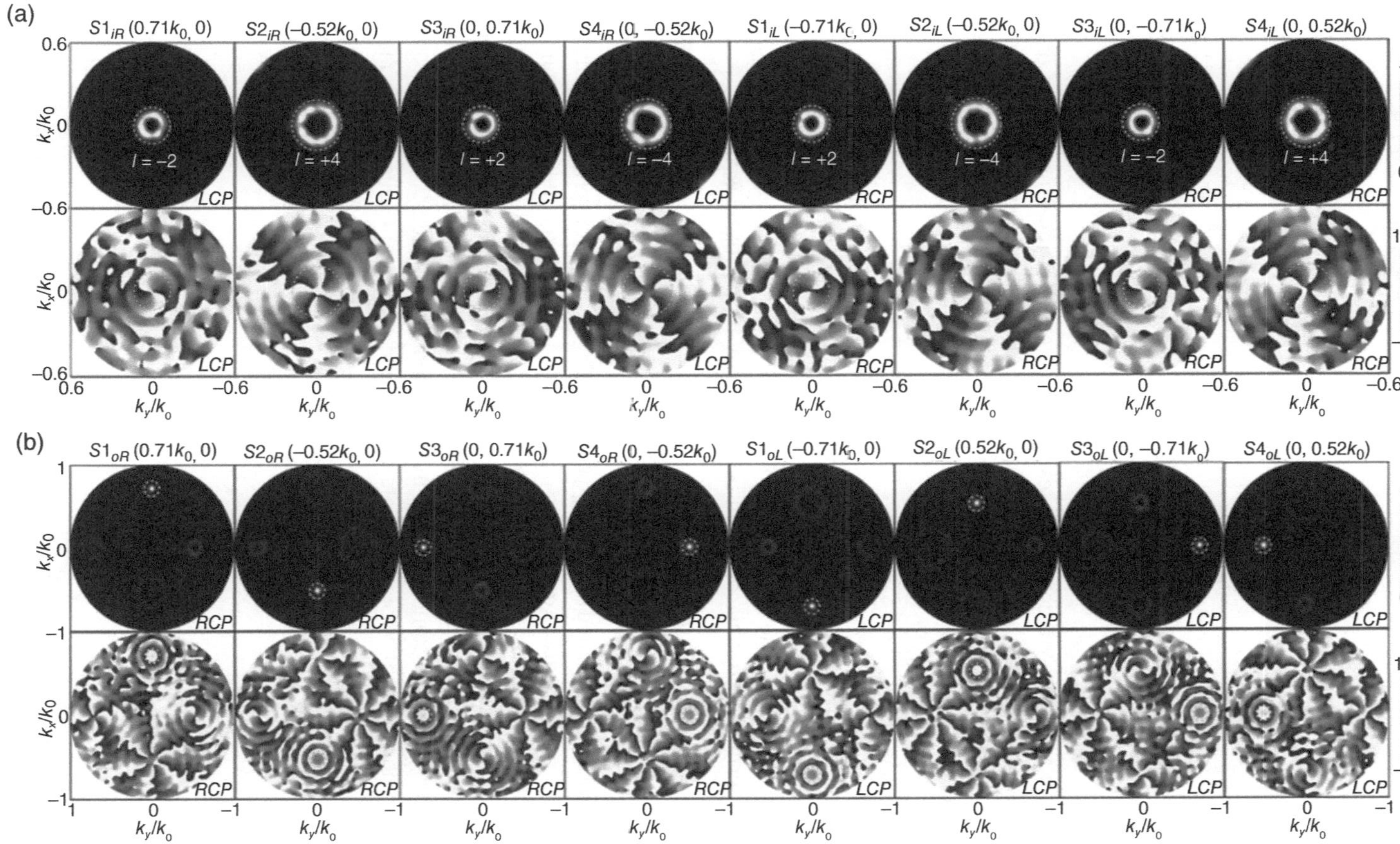

Figure 5.18 (a) Normalized simulated orthogonally-polarized component of the far-field intensities and phases of the transmitted beam at 30 GHz due to incident beams Sm_{iR} and Sm_{iL}, where $m = 1, \ldots, 4$, respectively; (b) normalized simulated orthogonally-polarized component of the far-field intensities and phases of the transmitted beams Sm_{oR} and Sm_{oL} at 30 GHz, where $m = 1, \ldots, 4$, respectively, due to an incident CP vortex beam with a topological charge and polarization corresponding to that shown in (a)[56]. *Source*: Jiang et al. [56].

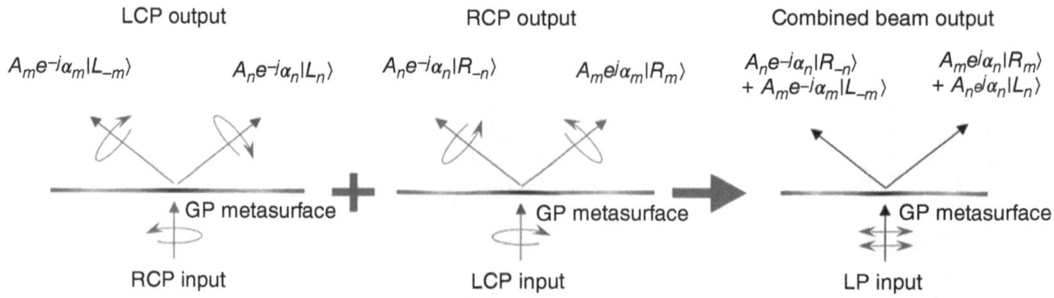

Figure 5.19 Schematics illustrating the use of a single geometric-phase transmit-array multiplexer for producing a pair of vector vortex beams.

opposite directions with a mirror symmetry with respect to the angle of incidence. When only a single geometric-phase metasurface is allowed, a multifurcating phase function needs to be employed. As shown in Figure 5.19, for the simplest case, a bifurcating geometric-phase transmit-array with a cell rotation angle function expressed as

$$\phi_r(x,y) = -arg\left(A_m e^{j\alpha_m} e^{-j\left[\left(\frac{\pi}{s_x p}\right)x + \frac{m\varphi}{2}\right]} + A_n e^{j\alpha_n} e^{-j\left[\left(-\frac{\pi}{s_x p}\right)x + \frac{n\varphi}{2}\right]}\right), \tag{5.22}$$

can produce two symmetrically diffracted beams for both RCP and LCP incident waves with a controllable amplitude ratio and phase difference between them. When it is illuminated by a linearly-polarized wave, in both directions, the diffracted RCP and LCP scalar vortex beams will superimpose with each other, forming two new combined beams exactly in the form given by

$$\psi_{m,n} = A_m e^{j\alpha_m} |\,\boldsymbol{R}_m\rangle + A_n e^{j\alpha_n} |\,\boldsymbol{L}_n\rangle, \tag{5.23}$$

$$\psi^a_{m,n} = A_m e^{-j\alpha_m} |\,\boldsymbol{L}_{-m}\rangle + A_n e^{-j\alpha_n} |\,\boldsymbol{R}_{-n}\rangle, \tag{5.24}$$

yielding two vector vortex beams located on $\text{HOPS}_{\text{m,n}}$ and $\text{HOPS}_{\text{-n,-m}}$, respectively, with their locations determined by $\Delta A = A_m/A_n$ and $\Delta\alpha = \alpha_m - \alpha_n$. Specifically, ΔA governs the latitude and $\Delta\alpha$ controls the longitude. Since the amplitude ratios between the RCP and LCP scalar vortex beams for the two combined beams are ΔA and $1/(\Delta A)$, respectively, whereas the phase difference between the RCP and LCP vortex beams for the two combined beams remains as $\Delta\alpha$, the two combined beams are therefore located on the same meridian but with opposite latitudes.

In such a way, by generalizing the rotation angle distribution of the unit cells to have all the required phase functions for producing multiple pairs of vector vortex beams with different wavefront modes that interfere with each other, massive beam multiplexing can be achieved with independent control of the beam directions, power ratio, and mode orders. The required cell rotation distribution can be expressed as

$$\phi_r(x,y) = -\arg\left(\sum_{q=1}^{\frac{N_c}{2}}\left[A^q_m e^{-j\left(\alpha^q_m + \frac{\pi}{s^q_x p}x + \frac{\pi}{s^q_y p}y + \frac{m_q\varphi}{2}\right)} + A^q_n e^{j\left(\alpha^q_n + \frac{\pi}{s^q_x p}x + \frac{\pi}{s^q_y p}y - \frac{n_q\varphi}{2}\right)}\right]\right), \tag{5.25}$$

where s^q_x and s^q_y control the direction for each pair of beams, while m_q and n_q determine the HOPSs that the produced beams belong to [76]. It should be noted that this approach also encompasses the generation of conventional CP scalar vortex beams when the intensity parameter A_m or A_n is

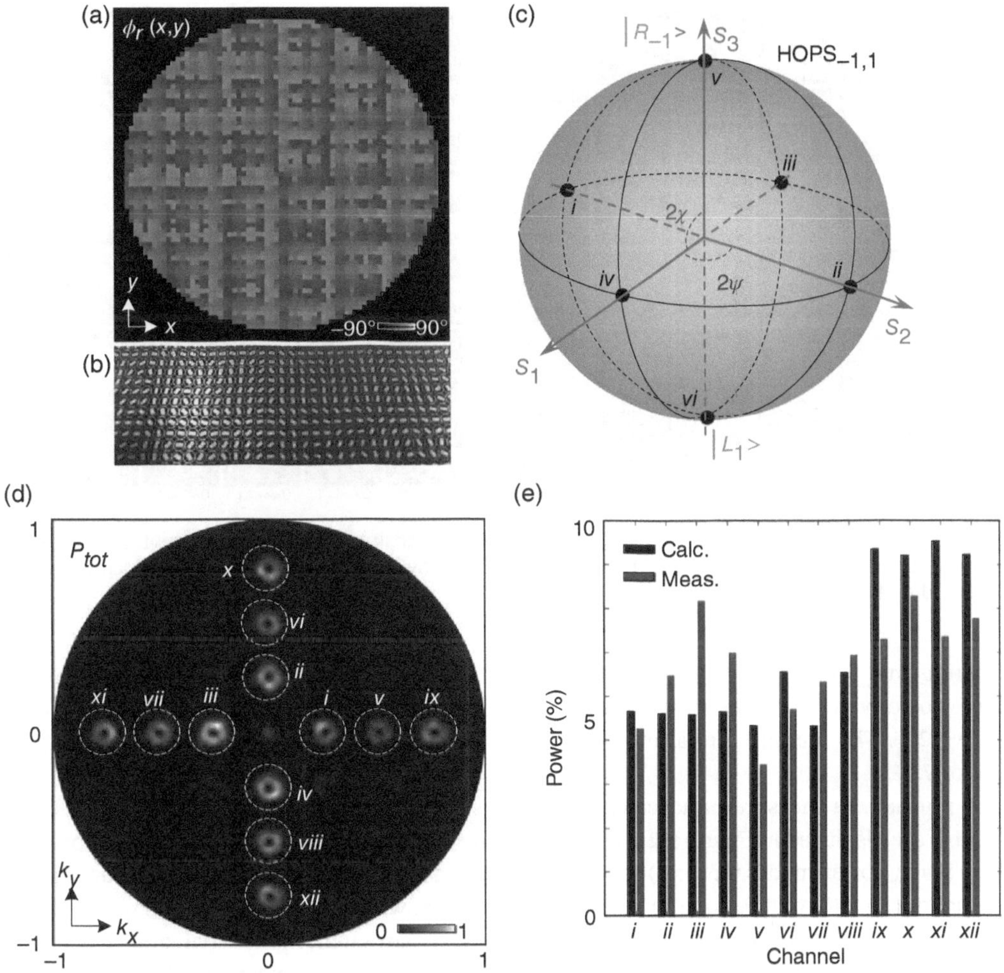

Figure 5.20 (a) Cell rotation angle distribution $\phi_r(x, y)$ and (b) a photograph of a portion of the fabricated transmit-array prototype for generating 12 beams corresponding to (c) six modes on the $\text{HOPS}_{-1,1}$. (d) Measured normalized total power pattern of the beams in momentum space at 40 GHz. (e) The measured and calculated power associated with each beam at 40 GHz.

suppressed to be zero. In addition, the state-of-polarization of the scalar vortex beam is not limited to CP such that an arbitrarily-polarized scalar vortex beam can be obtained by letting the OAM mode parameters m and n be equal to each other.

The first proof-of-concept vector vortex beam multiplexer is a transmit-array, based on the proposed Q-band unit cells, that produces six pairs of beams, which correspond to six modes on the $\text{HOPS}_{-1,1}$, with its cell rotation angle distribution displayed in Figure 5.20a. The six modes are located on the vertices in the $\boldsymbol{x}$-, $\boldsymbol{y}$-, and $\boldsymbol{z}$-directions of the $\text{HOPS}_{-1,1}$, see Figure 5.20c, in which two are CP scalar vortex beams, and four are vector beams with inhomogeneous linear polarization states. A prototype of the transmit-array was fabricated, see Figure 5.20b, and characterized by a near-field scanning system in the anechoic chamber. The measured total power patterns in the momentum space are reported in Figure 5.20d. The patterns have 12 vortex beams with donut-shaped intensity profiles. All the beams have the same radius for their annular amplitude-front,

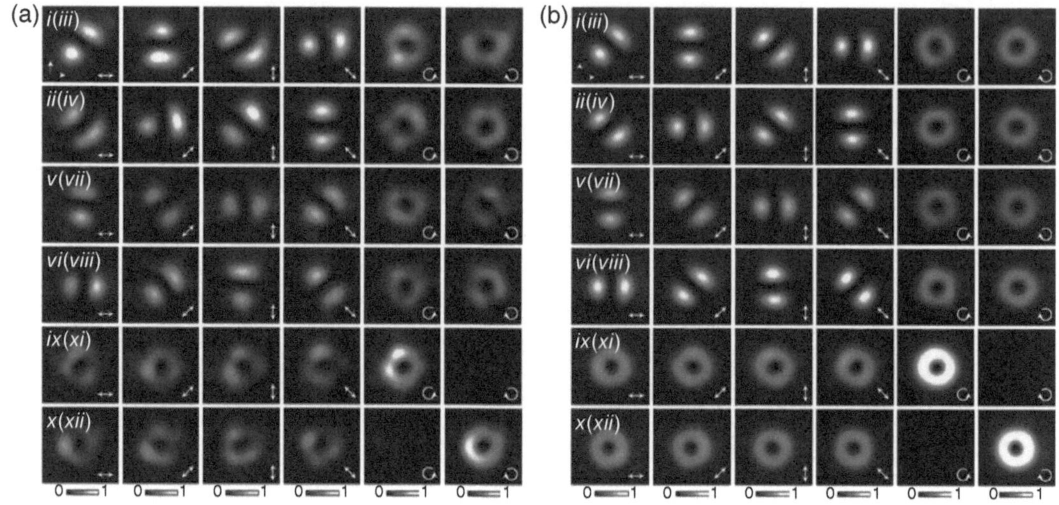

Figure 5.21 Normalized (a) measured and (b) calculated intensity patterns for different polarizations at 40 GHz associated with each vector vortex beam generated by the transmit-array for producing 12 beams corresponding to six modes on the $HOPS_{-1,1}$.

indicating that they have the same magnitude of the OAM mode number. The four outermost beams, corresponding to the north and south poles of the $HOPS_{-1,1}$, have stronger power, see Figure 5.20e. The measured total efficiency is 68% which is slightly smaller than the calculated value of 72%.

The normalized measured and calculated intensity patterns for different polarizations of the 12 beams are displayed in Figure 5.21, which agree well with each other. The four outermost beams are CP scalar vortex modes (*ix* to *xii*) with different spin states. The eight inner beams correspond to four modes (*i* to *viii*) on the equatorial lines of the HOPS, which is a result of a balanced superposition of $|\boldsymbol{R}_{-1}\rangle$ and $|\boldsymbol{L}_1\rangle$ with various phase differences. In particular, modes *v*(*vii*) and *vi*(*viii*) are the well-known azimuthally and radially-polarized cylindrical modes, respectively. It should be noted that these beams have linear polarizations locally across their wavefronts, thus they do not possess any angular momentum, neither the SAM nor the OAM. Due to the fact that $|m - n| = 2$, two nulls can be identified for the linearly polarized component patterns corresponding to these eight beams.

The second proof-of-concept vector vortex beam multiplexing transmit-array, which has a cell rotation angle distribution shown in Figure 5.22a, supports 12 beams corresponding to modes on the vertices of $HOPS_{-2,1}$ and the $HOPS_{-1,2}$ with different OAM mode orders, see Figure 5.22c. The normalized measured total power pattern in the momentum space of the fabricated sample, see Figure 5.22b, is displayed in Figure 5.22d, where 12 vortex beams with donut-shaped intensity profiles can be observed. Similar to the transmit-array supporting 12 beams on the $HOPS_{-1,1}$, the four outermost CP scalar vortex beams, corresponding to the north and south poles of the $HOPS_{-2,1}$ and the $HOPS_{-1,2}$, have stronger power, see Figure 5.22e. The measured total efficiency is 64% which is slightly smaller than the calculated value of 73%.

The normalized measured intensity patterns for different polarizations of the 12 vector vortex beams are displayed in Figure 5.23, which agree well with the calculated results (not shown here). The four outer beams (*ix*, *x*, *xi*, and *xii*) are CP scalar vortex beams with different topological charges and spin states. Since the absolute values of the north and south poles of the two HOPSs are different, the patterns for the beams locating at positions away from the two poles will have

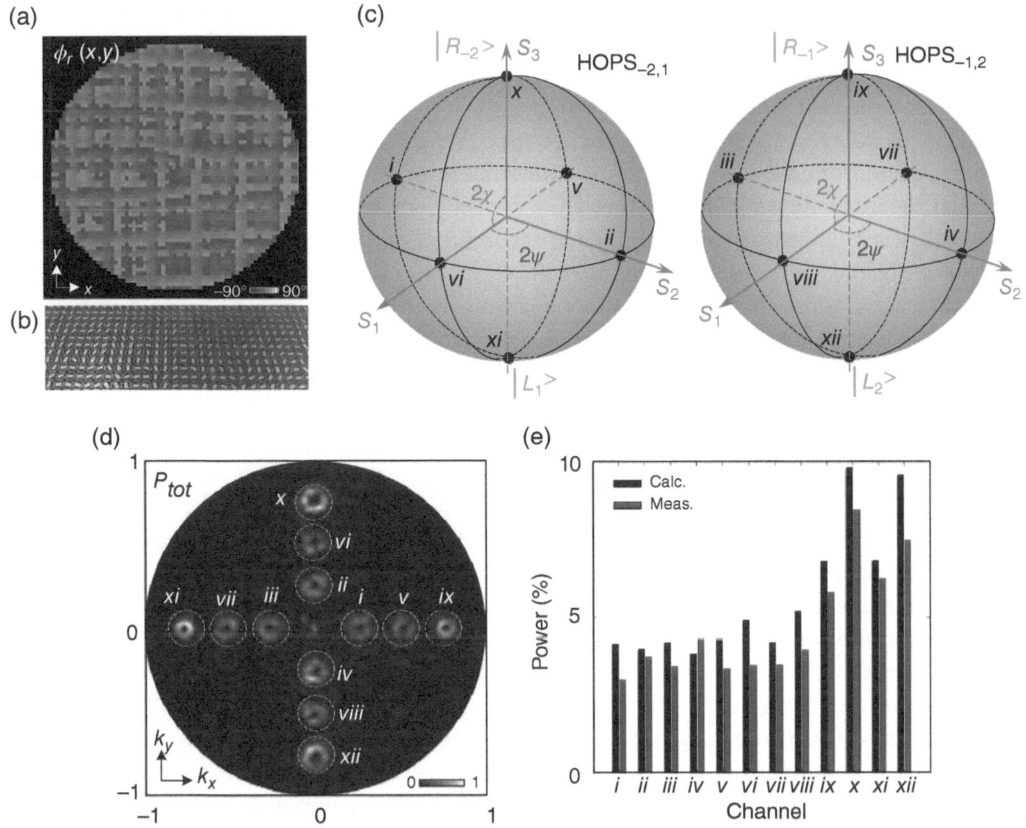

Figure 5.22 (a) Cell rotation angle distribution $\phi_r(x, y)$ and (b) a photograph of a portion of the fabricated transmit-array prototype for generating 12 beams corresponding to (c) 12 modes on the HOPS$_{-2,1}$ and the HOPS$_{-1,2}$. (d) Measured normalized total power pattern of the beams in momentum space at 40 GHz. (e) The measured and calculated power associated with each beam at 40 GHz.

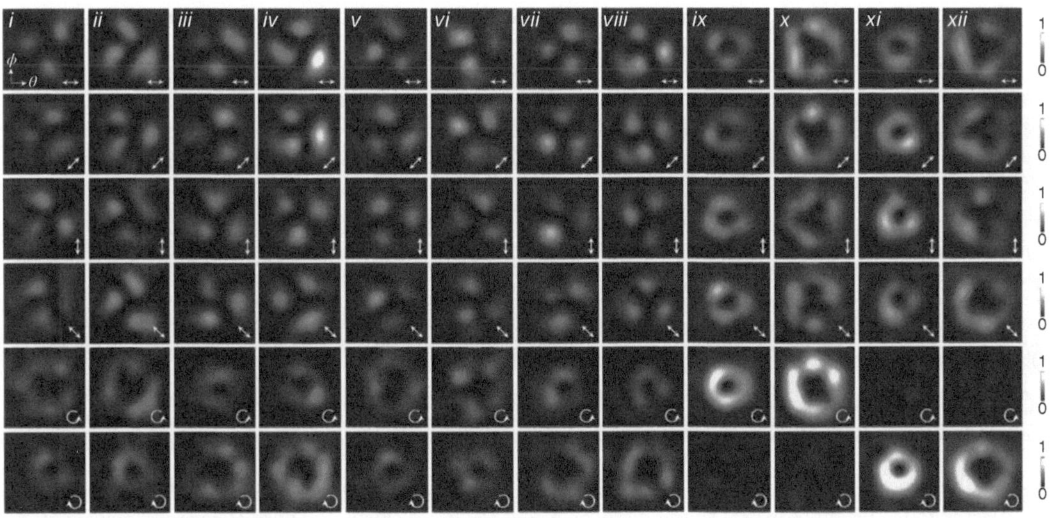

Figure 5.23 Normalized measured intensity patterns for different polarizations at 40 GHz associated with each vector vortex beam generated by the transmit-array for producing 12 beams corresponding to 12 modes on the HOPS$_{-2,1}$ and the HOPS$_{-1,2}$.

inhomogeneous intensity patterns for the linearly-polarized components. The eight inner beams (*i* to *viii*) correspond to modes on the equatorial lines of the two HOPSs, which possess a balanced superposition of $|\boldsymbol{R}_{-2}\rangle$ and $|\boldsymbol{L}_1\rangle$ as well as $|\boldsymbol{R}_{-1}\rangle$ and $|\boldsymbol{L}_2\rangle$ but with various phase differences. Since $|m-n| = 3$, three nulls can be found for the linearly-polarized component patterns of these eight beams. The locations of these nulls are rotated in a counter-clockwise direction as the location of the beam moves from $\phi = 0$ to $270°$ on the equatorial lines of the HOPSs. Small tails can be identified near the intensity peaks for the linearly-polarized components, which are caused by the different radii of the intensity profiles and phase progression rates for the first- and second-order CP scalar vortex beams. It should be noted that the Gouy phase [77], which is a topological charge-dependent propagation phase term that could result in pattern tilting and mode drifting transversely along the latitudes of the HOPSs, was taken into account during the synthesis of the cell rotation distribution of the transmit-array. Consequently, the obtained beams all have polarizations or modes precisely the same as the targeted ones.

5.4.4 Multiplexing of Vector Vortex Beams with Broken Symmetry

The limitation of the transmit-arrays composed of geometric-phase cells for generating vector vortex beams lies in the fact that the two beams with a mirror symmetry about the z-axis in the momentum space must have modes that are also symmetrical about the equator of the HOPS. In other words, if one of the pairs of vector vortex beams is located on the equator, then the other one must have the same mode. This is because the phase delays imposed on the transmitted cross-polarized CP components by the geometric-phase cells always possess opposite signs for RCP and LCP incident waves [63]. By using the proposed K-band dual-CP cells, this symmetry can be broken, thereby offering more flexibility in the multiplexing of vector vortex beams.

As a proof-of-concept example, a quad-beam multiplexer that generates four vector vortex beams on the $\text{HOPS}_{-1,1}$ was synthesized using the dual-CP cell presented in Figure 5.11. The phase delay distribution for both RCP and LCP input waves, i.e. ϕ_{LR} and ϕ_{RL}, are displayed in Figure 5.24a and b, respectively. It can be seen that they are fully independent, which is different from that supported by a transmit-array filled with geometric-phase cells. The normalized calculated total power pattern in momentum space is displayed in Figure 5.24c, where four vortex beams with donut-shaped intensity profiles can be observed with the same radii. The normalized calculated intensity patterns for

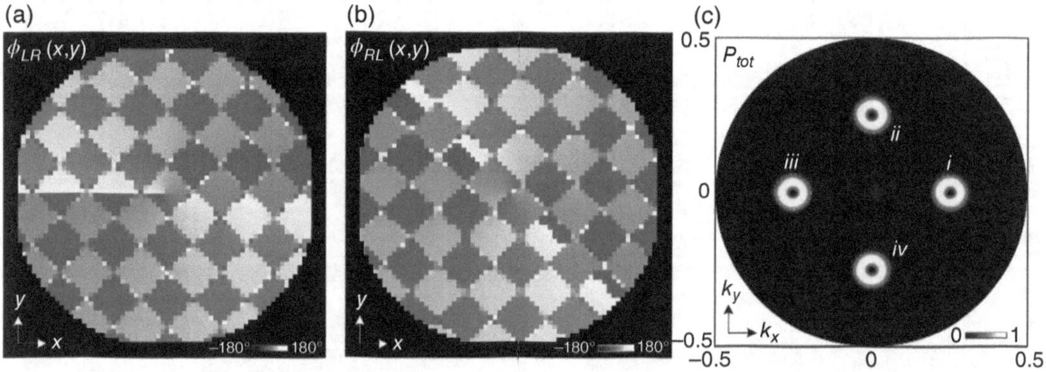

Figure 5.24 Phase delay distributions for both (a) $\phi_{LR}(x,y)$ and (b) $\phi_{RL}(x,y)$ of the transmit-array for generating four vector vortex beams with broken symmetry on the $\text{HOPS}_{-1,1}$; (c) calculated normalized total power pattern of the beams in momentum space at 22 GHz.

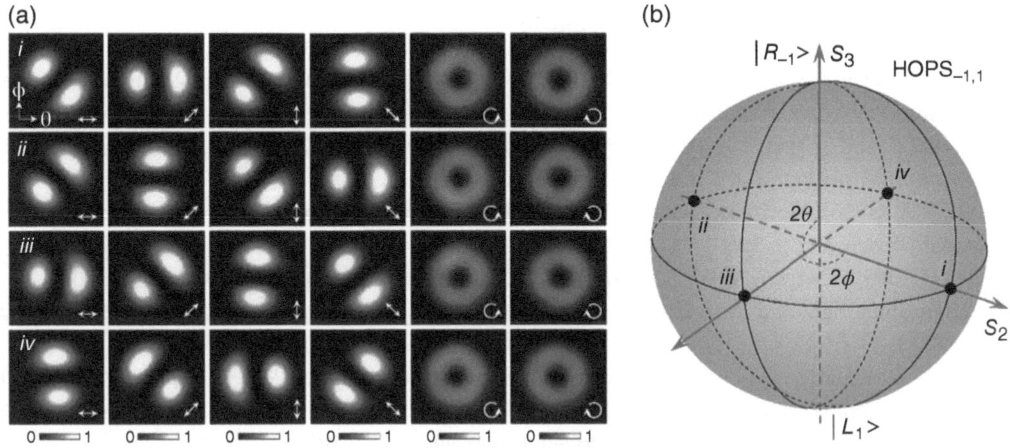

Figure 5.25 (a) Normalized calculated intensity patterns for different polarizations at 22 GHz associated with each vector vortex beam generated by the transmit-array for producing four beams corresponding to (b) four modes on the $HOPS_{-1,1}$.

different polarizations of the four vector vortex beams at 22 GHz are displayed in Figure 5.25a. It can be seen that the state-of-polarization across their wavefront is always linear polarization, implying that they indeed correspond to modes on the equator of a HOPS. Their spatial distribution of the state-of-polarization reveals that they are $HOPS_{-1,1}$ (90°, 90°), $HOPS_{-1,1}$ (90°, 270°), $HOPS_{-1,1}$ (90°, 0°), and $HOPS_{-1,1}$ (90°, 180°), as shown in Figure 5.25b. Unlike the transmit-array presented in Figure 5.20, here the modes that are symmetrical in the momentum space in the $k_x = 0$ or $k_y = 0$ planes are not symmetrical with respect to the equator on the HOPS.

5.5 Conclusion

In this chapter, the fundamentals and mode profiles of the vector vortex beams with spatially varying state-of-polarization are introduced. The designs of several different transmit-array unit cells operating at millimeter-wave frequencies are presented, which support wideband transmissive phase shifting for single- or dual-CP waves. By utilizing the proposed unit cells, a series of transmit-arrays with wide bandwidth and high-efficiency were synthesized and experimentally demonstrated for the generation and multiplexing of scalar vortex beams and vector vortex beams, showing a large degree of freedom in the control of the number, directions, and modes of the beams. These transmit-arrays, which can be scaled readily into terahertz frequencies and even optical wavelengths, offer a promising path towards electromagnetic devices with high-capacity and flexible complex wavefront control.

Acknowledgment

This work was supported partially by the Natural Science Foundation of China (NSFC) under Grants 62122019, 61801109, and 61627801, partially by the Penn State University John L. and Genevieve H. McCain endowed chair professorship, and partially by the High Level Innovation and Entrepreneurial Research Team Program in Jiangsu.

References

1 Allen, L., Barnett, S.M., and Padgett, M.J. (2003). *Optical Angular Momentum*. Abingdon, United Kingdom: Taylor and Francis Group.

2 Torres, J.P. and Torner, L. (2011). *Twisted Photons: Applications of Light with Orbital Angular Momentum*. Hoboken, NJ, United States: Wiley.

3 Beth, R.A. (1936). Mechanical detection and measurement of the angular momentum of light. *Physical Review* **50** (2): 115.

4 Allen, L., Beijersbergen, M.W., Spreeuw, R.J.C., and Woerdman, J.P. (1992). Orbital angular momentum of light and the transformation of Laguerre-Gaussian laser modes. *Physical Review A* **45** (11): 8185.

5 Curtis, J.E. and Grier, D.G. (2003). Structure of optical vortices. *Physical Review Letters* **90** (13): 133901.

6 Thidé, B., Then, H., Sjoholm, J. et al. (2007). Utilization of photon orbital angular momentum in the low-frequency radio domain. *Physical Review Letters* **99** (8): 087701.

7 Geneaux, R., Camper, A., Auguste, T. et al. (2016). Synthesis and characterization of attosecond light vortices in the extreme ultraviolet. *Nature Communications* **7** (1): 12583.

8 Harwit, M. (2003). Photonic orbital angular momentum in astrophysics. *Astrophysics* **597** (2): 1266–1270.

9 Tamburini, F., Thidé, B., Molina-Terriza, G., and Anzolin, G. (2011). Twisting of light around rotating black hole. *Nature Physics* **7** (3): 195–197.

10 He, H., Friese, M., Heckenberg, N.R., and Rubinsztein-Dunlop, H. (1995). Direct observation of transfer of angular momentum to absorptive particles from a laser beam with a phase singularity. *Physical Review Letters* **75** (5): 826–829.

11 Lee, S.-H. and Grier, D.G. (2006). Giant colloidal diffusivity on corrugated optical vortices. *Physical Review Letters* **96** (19): 190601.

12 Mair, A., Vaziri, A., Weihs, G., and Zeilinger, A. (2001). Entanglement of the orbital angular momentum states of photons. *Nature* **412** (6844): 313–316.

13 Fickler, R., Lapkiewicz, R., Plick, W.N. et al. (2012). Quantum entanglement of high angular momenta. *Science* **338** (6107): 640–643.

14 Gariepy, G., Leach, J., Kim, K.T. et al. (2014). Creating high-harmonic beams with controlled orbital angular momentum. *Physical Review Letters* **113** (15): 153901.

15 Yan, L., Gregg, P., Karimi, E. et al. (2015). Q-Plate enabled spectrally diverse orbital-angular-momentum conversion for stimulated emission depletion microscopy. *Optica* **2** (10): 900–903.

16 Miao, P., Zhang, Z., Sun, J. et al. (2016). Orbital angular momentum microlaser. *Science* **353** (6298): 464–467.

17 Gibson, G., Courtial, J., Padgett, M.J. et al. (2004). Free-space information transfer using light beams carrying orbital angular momentum. *Optics Express* **12** (22): 5448–5456.

18 Tamburini, F., Mari, E., Sponselli, A. et al. (2012). Encoding many channels on the same frequency through radio vorticity: First experimental test. *New Journal of Physics* **14** (3): 033001.

19 Yan, Y., Xie, G., Lavery, M.P.J. et al. (2014). High-capacity millimetre-wave communications with orbital angular momentum multiplexing. *Nature Communications* **5** (1): 4876.

20 Wang, J., Yang, J., Fazal, I. et al. (2012). Terabit free-space data transmission employing orbital angular momentum multiplexing. *Nature Photonics* **6** (7): 488–496.

21 Bozinovic, N., Yue, Y., Ren, Y. et al. (2013). Terabit-scale orbital angular momentum mode division multiplexing in fibers. *Science* **340** (6140): 1545–1548.

22 Zhao, Q. (2009). Cylindrical vector beams: From mathematical concepts to applications. *Advances in Optics and Photonics* **1** (1): 1–57.

23 Milione, G., Sztul, H.I., Nolan, D.A., and Alfano, R.R. (2011). Higher-order Poincare sphere, stokes parameters, and the angular momentum of light. *Physical Review Letters* **107** (5): 053601.

24 Yi, X., Liu, Y., Ling, X. et al. (2015). Hybrid-order Poincare sphere. *Physical Review A* **91** (2): 023801.

25 Dorn, R., Quabis, S., and Leuchs, G. (2003). Sharper focus for a radially polarized light beam. *Physical Review Letters* **91** (23): 233901.

26 Grier, D.G. (2003). A revolution in optical manipulation. *Nature* **424** (6950): 810–816.

27 Naidoo, D., Roux, F.S., Dudley, A. et al. (2016). Controlled generation of higher-order Poincare sphere beams from a laser. *Nature Photonics* **10** (5): 327–333.

28 Schemmel, P., Maccalli, S., Pisano, G. et al. (2014). Three-dimensional measurements of a millimeter wave orbital angular momentum vortex. *Optics Letters* **39** (3): 626–629.

29 Tamburini, F., Mari, E., Thidé, B. et al. (2011). Experimental verification of photon angular momemtum and vorticity with radio techniques. *Applied Physics Letters* **99** (20): 204102.

30 Padgett, M.J. and Allen, L. (2002). Orbital angular momentum exchange in cylindrical-lens mode converters. *Journal of Optics B: Quantum and Semiclassical Optics* **4** (2): S17.

31 Heckenberg, N.R., McDuff, R., Smith, C.P., and White, A.G. (1992). Generation of optical phase singularities by computer-generated holograms. *Optics Letters* **17** (3): 221–223.

32 Zhang, N., Yuan, X.C., and Burge, R.E. (2010). Extending the detection range of optical vortices by Dammann vortex gratings. *Optics Letters* **35** (20): 3495–3497.

33 Cai, X., Wang, J., Strain, M.J. et al. (2012). Integrated compact optical vortex beam emitters. *Science* **338** (6105): 363–366.

34 Bai, X.-D., Liang, X.-L., Li, J.-P. et al. (2016). Rotman lens-based circular array for generating five-mode OAM radio beams. *Scientific Reports* **6**: 27815.

35 Chen, S., Zhou, X., Liu, Y. et al. (2014). Generation of arbitrary cylindrical vector beams on the higher order Poincare sphere. *Optics Letters* **39** (18): 5274–5276.

36 Yu, N. and Capasso, F. (2014). Flat optics with designer metasurfaces. *Nature Materials* **13** (2): 139–150.

37 Hum, S.V. and Perruisseau-Carrier, J. (2014). Reconfigurable reflectarrays and array lenses for dynamic antenna beam control: A review. *IEEE Transactions on Antennas and Propagation* **62** (1): 183–198.

38 Yu, N., Genevet, P., Kats, M.A. et al. (2011). Light propagation with phase discontinuities: Generalized laws of reflection and refraction. *Science* **334** (6054): 334–337.

39 Hui, X., Zheng, S., Hu, Y. et al. (2015). Ultralow reflectivity spiral phase plate for generation of millimeter-wave OAM beam. *IEEE Antennas and Wireless Propagation Letters* **14**: 966–969.

40 Yang, Y., Wang, W., Moitra, P. et al. (2014). Dielectric meta-reflectarray for broadband linear polarization conversion and optical vortex generation. *Nano Letters* **14** (3): 1394–1399.

41 Yu, S., Li, L., and Shi, G. (2016). Dual-polarization and dual-mode orbital angular momentum radio vortex beam generated by using reflective metasurface. *Applied Physics Express* **9** (8): 082202.

42 Cheng, L., Hong, W., and Hao, Z.-C. (2014). Generation of electromagnetic waves with arbitrary orbital angular momentum modes. *Scientific Reports* **4**: 4814.

43 Arbabi, A., Horie, Y., Bagheri, M., and Faraon, A. (2015). Dielectric metasurfaces for complete control of phase and polarization with subwavelength spatial resolution and high transmission. *Nature Nanotechnology* **10** (11): 937–944.

44 Sun, J., Wang, X., Xu, T. et al. (2014). Spinning light on the nanoscale. *Nano Letters* **14** (5): 2726–2729.

45 Shalaev, M.I., Sun, J., Tsukernik, A. et al. (2015). High-efficiency all-dielectric metasurfaces for ultracompact beam manipulation in transmission mode. *Nano Letters* **15** (9): 6261–6266.

46 Pancharatnam, S. (1956). Generalized theory of interference, and its applications. Part I. coherent pencils. *Proceedings of Indian Academy of Science A* **44** (6): 398–417.

47 Berry, M.V. (1987). The adiabatic phase and Pancharatnam's phase for polarized light. *Journal of Modern Optics* **34** (11): 1401–1407.

48 Bliokh, K.Y., Rodríguez-Fortuño, F.J., Noril, F., and Zayats, A.V. (2015). Spin–orbit interactions of light. *Nature Photonics* **9** (12): 796–808.

49 Yue, F., Wen, D., Xin, J. et al. (2016). Vector vortex beam generation with a single plasmonic metasurfaces. *ACS Photonics* **3** (9): 1558–1563.

50 Xu, H.-X., Liu, H., Ling, X. et al. (2017). Broadband vortex beam generation using multimode Pancharatnam-Berry metasurface. *IEEE Transactions on Antennas and Propagation* **65** (12): 7378–7382.

51 Yue, F., Wen, D., Zhang, C. et al. (2017). Multichannel polarization-controllable superposition of orbital angular momentum states. *Advanced Materials* **29** (15): 1603838.

52 Yang, L.-J., Sun, S., and Sha, W.E.I. (2020). Ultrawideband reflection-type metasurface for generating integer and fractional orbital angular momentum. *IEEE Transactions on Antennas and Propagation* **68** (3): 2166–2175.

53 Ding, X., Ding, X., Monticone, F. et al. (2014). Ultrathin Pancharatnam-Berry metasurface with maximal cross-polarization efficiency. *Advanced Materials* **27** (7): 1195–1200.

54 Pu, M., Li, X., Ma, X. et al. (2015). Catenary optics for achromatic generation of perfect optical angular momentum. *Science Advances* **1** (9): e1500396.

55 Devlin, R.C., Ambrosio, A., Rubin, N.A. et al. (2017). Arbitrary spin-to-orbital angular momentum conversion of light. *Science* **358** (6365): 896–901.

56 Jiang, Z.H., Kang, L., Hong, W., and Werner, D.H. (2018). Highly efficient broadband multiplexed millimeter-wave vortices from metasurface-enabled transmit-arrays of subwavelength thickness. *Physical Review Applied* **9** (6): 064009.

57 Haupt, R.L. (2005). Interleaved thinned linear arrays. *IEEE Transactions on Antennas and Propagation* **53** (9): 2858–2864.

58 Hong, W., Jiang, Z.H., Yu, C. et al. (2017). Multibeam antenna technologies for 5G wireless communications. *IEEE Transactions on Antennas and Propagation* **66** (12): 6231–6249.

59 Veksler, D., Maguid, E., Shitrit, N. et al. (2015). Multiple wavefront shaping by metasurface based on mixed random antenna groups. *ACS Photonics* **2** (5): 661–667.

60 Maguid, E., Yulevich, I., Yannai, M. et al. (2017). Multifunctional interleaved geometric-phase dielectric metasurfaces. *Light-Science & Applications* **6** (8): e17027.

61 Maguid, E., Yulevich, I., Veksler, D. et al. (2016). Photonic spin-controlled multifunctional shared-aperture antenna array. *Science* **352** (6290): 1202–1206.

62 Balanis, C.A. (1989). *Advanced Engineering Electromagnetics*. Hoboken, NJ, United States: Wiley.

63 Huang, L., Chen, X., Mühlenbernd, H. et al. (2012). Dispersionless phase discontinuities for controlling light propagation. *Nano Letters* **12** (11): 5750–5755.

64 Monticone, F., Estakhri, N.M., and Alù, A. (2013). Full control of nanoscale optical transmission with a composite meta-screen. *Physical Review Letters* **110** (20): 203903.

65 Jiang, Z.H., Sieber, P.E., Kang, L., and Werner, D.H. (2015). Restoring intrinsic properties of electromagnetic radiators using ultra-lightweight integrated metasurface cloaks. *Advanced Functional Materials* **25** (29): 4708–4716.

66 Pfeiffe, C. and Grbic, A. (2013). Metamaterial Huygens' surfaces: Tailoring wave fronts with reflectionless sheets. *Physical Review Letters* **110** (19): 197401.

67 Jiang, Z.H., Kang, L., Yue, T. et al. (2020). Wideband transmit-arrays based on anisotropic impedance surfaces for circularly-polarized single-feed multibeam generation in the Q-band. *IEEE Transactions on Antennas and Propagation* **68** (1): 217–229.

68 Florencio, R., Encinar, J.A., Boix, R.R. et al. (2019). Flat reflectarray that generates adjacent beams by discriminating in dual circular polarization. *IEEE Transactions on Antennas and Propagation* **67** (6): 3733–3742.

69 Jiang, Z.H., Yue, T., and Hong, W. (2020). Low-profile and wideband dual-circularly-polarized reflect-arrays based on rotated metal backed dual-polarized aperture-coupled patch elements. *IEEE Transactions on Antennas and Propagation* **68** (3): 2108–2117.

70 Plaza, E.G., León, G., Loredo, S., and Las-Heras, F. (2015). A simple method for analyzing transmit array lenses. *IEEE Antennas and Propagation Magazine* **57** (2): 131–144.

71 Chew, W.C. (1990). *Waves and Fields in Inhomogeneous Media*. New York, NY, United States: Van Nostrand Reinhold.

72 Zhou, M., Sørensen, S.B., Jørgensen, E. et al. (2011). An accurate technique for calculation of radiation from printed reflectarrays. *IEEE Antennas and Wireless Propagation Letters* **10**: 1081–1084.

73 Prado, D.R., Arrebola, M., Pino, M.R. et al. (2017). Efficient crosspolar optimization of shaped-beam dual-polarized reflectarrays using full-wave analysis for the antenna element characterization. *IEEE Transactions on Antennas and Propagation* **65** (2): 623–635.

74 Albero, J., Moreno, I., Davis, J.A. et al. (2012). Generalized phase diffraction gratings with tailored intensity. *Optics Letters* **37** (20): 4227–4229.

75 Chen, S., Cai, Y., Li, G. et al. (2016). Geometric metasurface fork gratings for vortex-beam generation and manipulation. *Laser & Photonics Review* **10** (2): 322–326.

76 Jiang, Z.H., Kang, L., Yue, T. et al. (2020). A single noninterleaved metasurface for high-capacity and flexible mode multiplexing of higher-order Poincare sphere beams. *Advanced Materials* **32** (6): 1903983.

77 Li, G., Kang, M., Chen, S. et al. (2013). Spin-enabled plasmonic metasurfaces for manipulating orbital angular momentum of light. *Nano Letters* **13** (9): 4148–4151.

6

Twisting Light with Metamaterials

Natalia M. Litchinitser

Department of Electrical and Computer Engineering, Duke University, Durham, NC, USA

6.1 Introduction

The discovery of orbital angular momentum (OAM) in optics has opened a new area of optical physics and led to new understanding of a wide range of phenomena [1–4]. The OAM beams possess an azimuthal phase dependence of $\exp(i\ell\theta)$, where the angle θ is the azimuthal coordinate and the quantized topological charge is denoted by ℓ [5]. The twisting is described by a helical phase distribution $\ell\theta$ that is proportional to the azimuthal angle of a cylindrical coordinate system (r, θ, z), where z is the propagation direction of the wave. Because of this twisting, the light wave along the z-axis cancels out because the value of the angular coordinate is not well defined at $r = 0$, giving rise to a phase singularity, where the amplitude vanishes. On a screen this beam appears as a bright ring with a dark center around which the phase rotates. The continuity of the field at $\theta = 2\pi$ requires that ℓ is an integer number that indicates the number of phase windings around the dark spot. This number is the topological charge of the OAM beam, because it is conserved during propagation. The physical meaning of ℓ corresponds to the velocity of the phase rotation around the singularity, which may be positive for counterclockwise rotation or negative for clockwise rotation.

While doughnut shape intensity profile is one of the typical characteristics of an OAM beam, measuring it in experiment does not prove that the beam carries an OAM. In order to measure the topological charge, one needs to be able to visualize the phase of the beam. Interference experiments offer one of the simplest ways to characterize the phase. Therefore, to determine the value of ℓ, we need to calculate or measure the interference of this beam with a plane or a spherical wave. The results are shown in Figure 6.1, where the patterns corresponding to the interference of a beam with $\ell = \pm 1$ with a tilted plane wave are shown in Figure 6.1d and e, respectively. The interference between a beam with $\ell = \pm 1$ and a spherical wave is shown in Figure 6.1f. In Figure. 6.1g and h, respectively, the patterns are due to the interference between beams with $\ell = \pm 2$ with a plane wave and in Figure 6.1i a beam with $\ell = \pm 2$ with a spherical wave.

Structured light beams find important applications in many areas of modern physics, including light–atom interactions [7], manipulation of microscopic objects [8], imaging [9, 10], optical communication and quantum information systems [11–15], nonlinear waveguiding [16, 17], and blocking the light from a bright star to increase the contrast of astronomical observations [18]. More generally, vortices are investigated in many branches of physics such as fluid mechanics, superconductivity, and Bose–Einstein condensates [19].

Electromagnetic Vortices: Wave Phenomena and Engineering Applications, First Edition.
Edited by Zhi Hao Jiang and Douglas H. Werner.

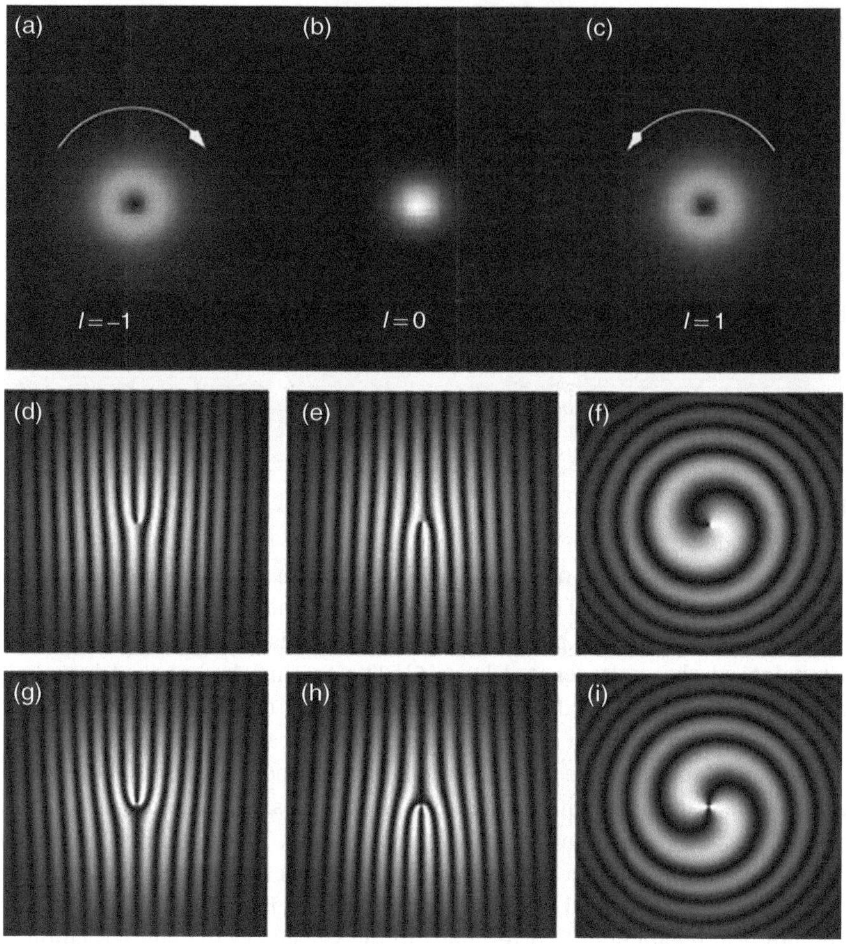

Figure 6.1 (a–c) Intensity of an OAM beam with topological charge $\ell = -1, 0, 1$. (d–f) interference patterns of beams carrying topological charge ℓ and different reference waves ψ. In (d) and (e), $\ell = 1$ and $\ell = -1$, respectively, and ψ is a tilted beam with planar phase. In (f), $\ell = -1$ and ψ has a spherical phase. (g–i) correspond to the same cases as the upper row for $\ell = 2$ and $\ell = -2$, respectively [6]. *Source*: Carpentier and Michinel [6].

Until recently, the majority of approaches for generating OAM beams have been based on macroscopic bulky optical components, such as spiral phase plates [20, 21], cylindrical lens converters [5], q-plates [22], spatial light modulators [23], or specialty fibers [24] (Figure 6.2). With an exception of q-plates, these existing techniques rely on bulk optical components that suit laboratory experiments, but may not be compatible with integrated optics systems that require ultracompact and flat optical components. In the recent years, first steps toward the realization of microscale, planar optical components based on metamaterials (MMs) or metasurfaces (MSs) have been made [29–41]. In this chapter, we discuss passive and active approaches to structured light generation on the nanoscale using engineered MM structures.

MMs are artificial structures that are engineered to have properties that cannot be found in nature. These artificial materials are made of carefully designed building blocks or 'meta-atoms,' which are typically much larger than conventional atoms, but much smaller than the wavelength of incident light, allowing the material to act as an effective medium with the desired optical properties. Meta-atoms are resonant elements, enabling light–matter interactions that are difficult or

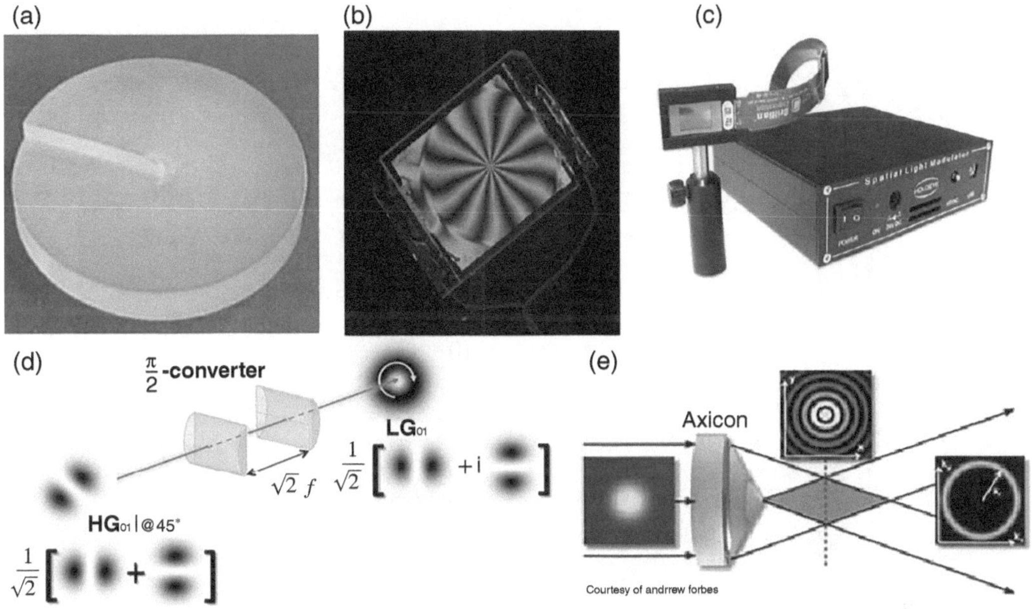

Figure 6.2 Examples of macroscopic vortex generating devices: (a) spiral phase plate [25], *Source*: https://www.st-andrews.ac.uk/~www_pa/Scots_Guide/MMWave/QO/components/spir.html [25]. (b) Q-plate [26], *Source*: Marrucci [26]. (c) spatial light modulators [27], *Source*: Marrucci [27]. (d) pair of cylindrical lenses. [5]. *Source*: Modified from Allen et al. [5], and (e) axicon [28]. *Source*: Dudley et al. [28]. Courtesy of Andrew Forbes.

impossible to realize using conventional atoms and molecules. In particular, the value of the refractive index, or more precisely, the values of the dielectric permittivity and magnetic permeability of a MM, can be tailored to steer the waves in a prescribed trajectory or to impose certain amplitude, phase, or polarization properties onto the transmitted or reflected beams [42–44].

MSs are engineered materials with thicknesses much smaller than incident light wavelength [20, 45–49]. Compared to conventional optical elements, which rely on long propagation distances, MS-based devices facilitate strong light–matter interaction on a subwavelength scale, allowing abrupt changes of beam parameters, as shown in Figure 6.3. In addition, MSs, unlike their three-dimensional analogs, MMs, do not require complicated three-dimensional fabrication techniques and often can be produced in one lithographical step, which makes them very promising for integration on a photonic chip and well suited for mass production.

MMs and MSs representing a new paradigm for the science of light and a variety of applications, such as wavefront modulators, beam steerers, filters, and lenses, permitting subwavelength focusing have been demonstrated. Moreover, they enable novel antennas for automobile industry, mobile phones communication, satellite systems, surface coatings with unique properties, commercial aviation, solar energy, military, medical instruments and many others [50–59]. Figure 6.3 shows several examples of applications of MMs. In automobile industry, microwave and millimeter-wave MMs play an essential role. They may enable adaptive cruise control and safety systems enabling the high field of view and long detection range. In addition, they offer a promising solution for making impedance matching materials to reduce the electromagnetic wave reflection from painted plastic details.

The other field, where MMs can also find applications, is satellite communication systems. MMs allow for affordable, lightweight, and flat satellite antennas that can be mounted on cars, boats, and

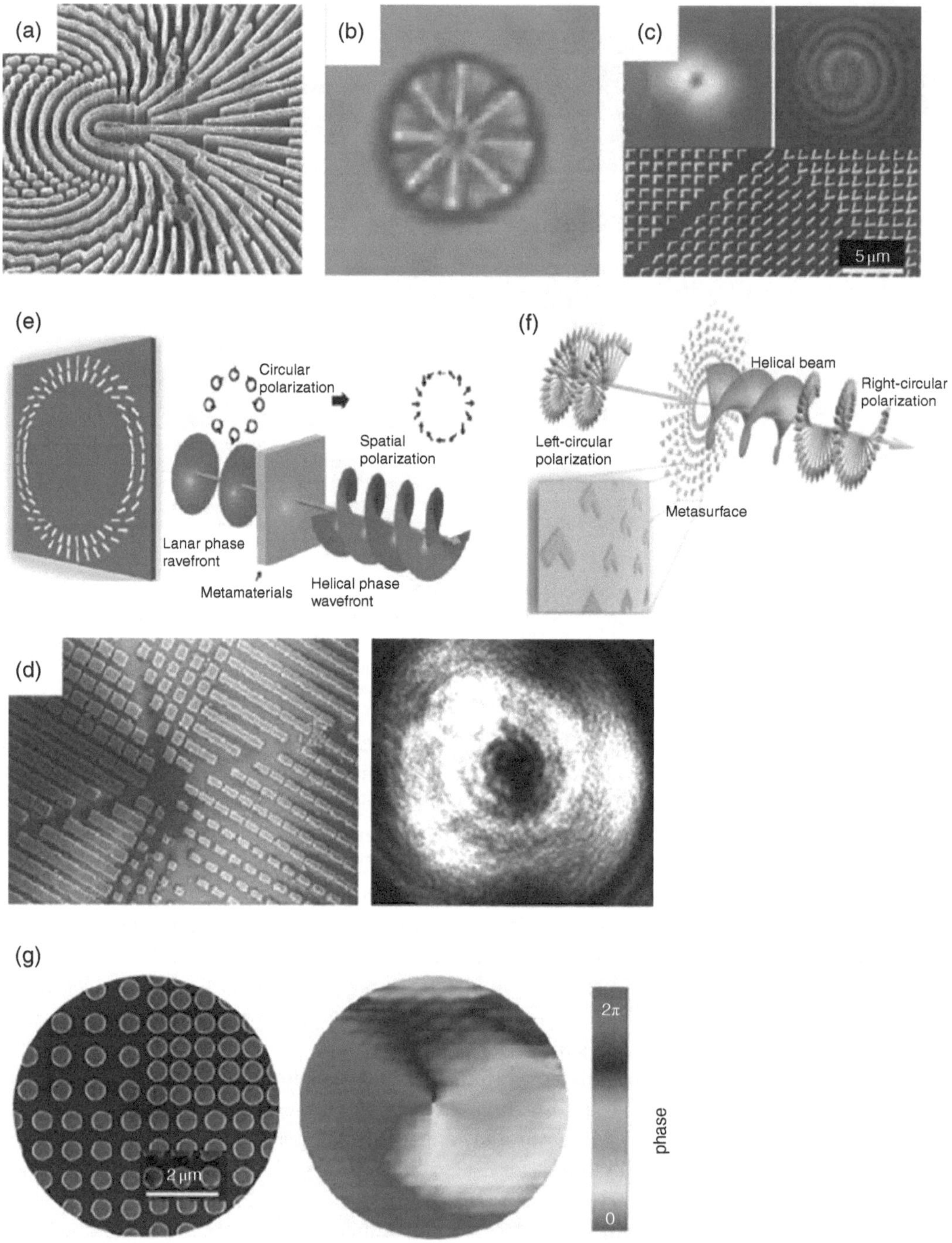

Figure 6.3 Metasurfaces for OAM beams generation [20, 45–49]. *Sources*: Based on Niv et al. [45], Knöner et al. [146], Yu et al. [20], Shalaev et al. [46], Chong et al. [49]. Zhao et al. [47]. Karimi et al. [48].

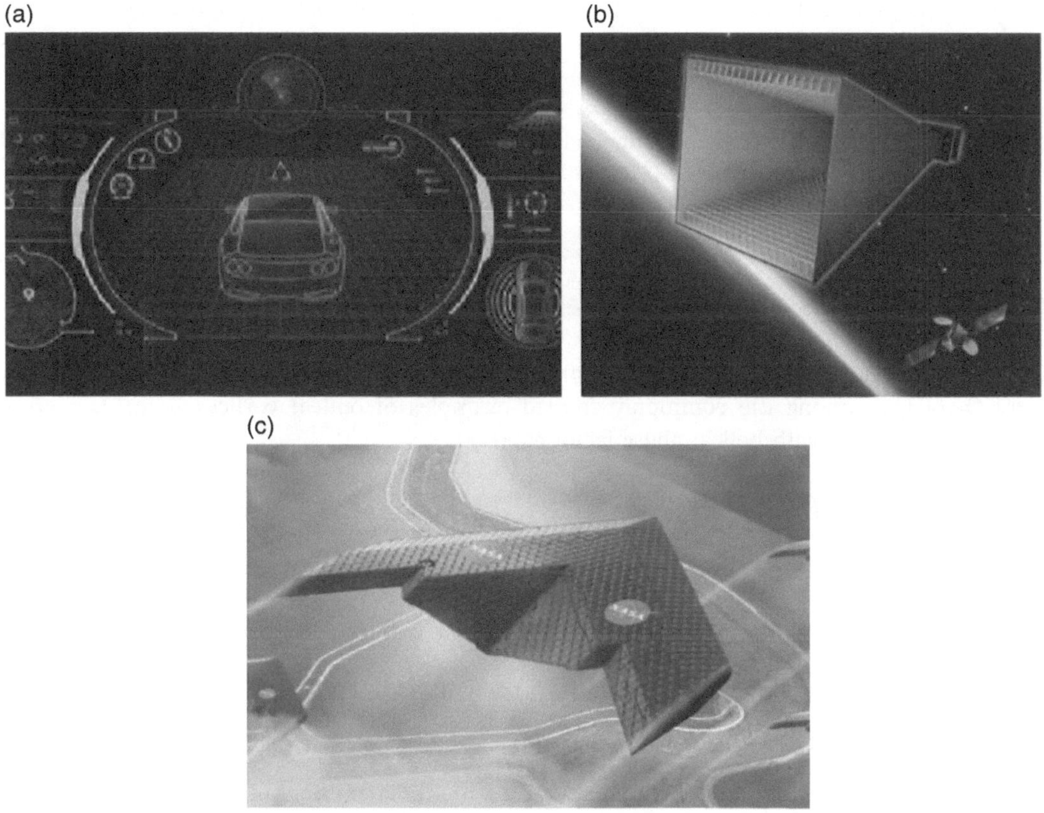

Figure 6.4 (a) Metamaterials applications in automobile industry [57]. *Source*: Jackie Niam/Shutterstock; (b) satellite systems [58]. *Source*: Douglas Werner; (c) and military systems [59]. *Source*: https://www.st-andrews.ac.uk/~www_pa/Scots_Guide/MMWave/QO/components/spir.html [59].

airplanes (Figure 6.4) [57–59]. Moreover, they are likely to facilitate new military application, including radio-frequency MM antennas and shielding materials with reduced thermal, infrared, and radio-frequency signatures. In aviation industry, composite materials can be used on airplane cockpit in order to reject light coming at wide angles and certain frequencies, including the ones of powerful lasers. Finally, MMs find applications in photovoltaic industry, where they boost the efficiency of solar cells by collecting light coming from wide angles and absorbing it from wide spectral range.

On the other hand, the recent progress in the development of MMs with fundamentally new physical properties opens new opportunities for tailoring both linear and nonlinear properties of materials. A majority of photonic MMs that have been demonstrated to date were solid-state materials. However, the concept of meta-atoms can be extended further to realize artificial media with novel electromagnetic properties in liquid [60] or gaseous [61] phases at frequencies ranging from microwave to visible. In particular, at optical frequencies, engineered colloidal suspensions offer a promising platform for engineering polarizabilities and realization of large and tunable nonlinearities. For example, the recent studies have shown that the nonlinearity of colloidal suspensions can be tailored to be either supercritical, when particles have positive polarizability (PP), or saturable, for negative polarizability (NP) particles [62–66]. In this chapter, we also discuss some of the progress made in the field of OAM light–matter interactions in engineered nonlinear colloidal media.

This chapter is organized as follows. In Section 6.2, we discuss passive approaches to twisting light and modifying light–matter interactions with MMs. Section 6.3 focuses on active approaches to the OAM-beam generation using microlasers. In Section 6.4, we discuss novel regimes of nonlinear interactions of the beams with an OAM enabled by engineered nonlinear colloidal suspensions. Section 6.5 summarizes the phenomena and application discussed in this chapter and discuss the outlook for the future studies.

6.2 OAM Beams on the Nanoscale

In 1992, Allen et al. first demonstrated that structured light with helical wavefront (Figure 6.5) can carry OAM [5]. Among the commonly studied examples of optical vortices is the Laguerre–Gaussian (LG) beam with helical phase factor $\exp(i\ell\phi)$

$$\mathrm{LG}_{p\ell} = \sqrt{\frac{2p!}{\pi(p+|\ell|)!}}\frac{1}{w(z)}\left[\frac{\sqrt{2}r}{w(z)}\right]^{|\ell|}\exp\left[-\frac{r^2}{w^2(z)}\right]L_p^{\ell}\left[\frac{2r^2}{w^2(z)}\right]\exp(i\ell\phi)\exp\left[\frac{ik_0r^2z}{2(z^2+z_R^2)}\right]$$
$$\exp\left[-i(2p+|\ell|+1)\tan^{-1}\left(\frac{z}{z_r}\right)\right] \quad (6.1)$$

where $w(0)$ is the LG beam waist radius and $w(z) = w(0)\sqrt{1+\left(\frac{z}{z_R}\right)^2}$ is the radius of $\frac{1}{e}$ amplitude where $z_R = \frac{\pi w_0^2}{\lambda}$ is the Rayleigh range. $(2p+|\ell|+1)\tan^{-1}\left(\frac{z}{z_r}\right)$ is the Gouy phase at z distance and $L_p^{\ell}(x)$ is the associated Laguerre polynomial. ℓ and p are the azimuthal mode number and radial mode number, respectively.

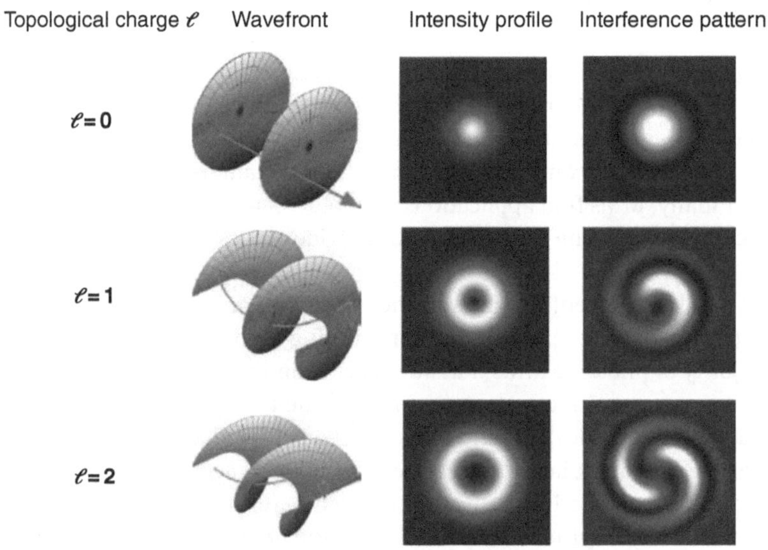

Figure 6.5 Wavefront, intensity profiles, and interference patterns with a Gaussian beam of a structured optical vortex with different topological charge ℓ, the arrows denote the directions of Poynting vector [67]. *Source*: Reproduced from Padgett et al. [67].

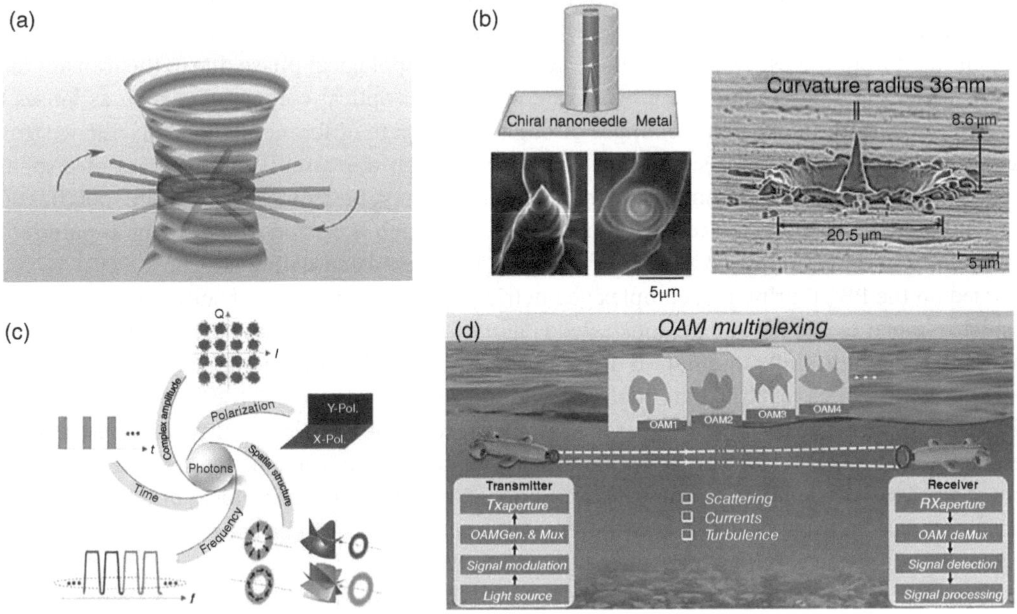

Figure 6.6 Applications of OAM beams. (a) Optical trapping [83]. *Source*: Yan et al. [83]; (b) nanofabrication [84]. *Source*: Toyoda et al. [84]; (c, d) mode-division multiplexing [85, 86]. *Source*: Wang [85], Ren et al. [86]. Licenced under CC BY-4.0.

Unlike the spin angular momentum (SAM) that is related to light polarization with only two possible states of ±1, the OAM, or so-called topological charge, can take any integer number potentially enabling a new degree of freedom for increasing the bandwidth of wireless and fiber communication systems (Figure 6.6) [12–14]. Moreover, the OAM beams are likely to enable a plethora of potential applications such micromanipulation, microscopy, and biomedical science [8, 68, 69] and building higher dimensional quantum encryption systems [70, 71], as shown in Figure 6.5. Many of these new applications call for an ultracompact source of the structured light or beams carrying OAM that can be integrated on a chip or directly on an optical fiber.

Recent progress in nanostructured optical materials has enabled new ways of manipulating intensity, polarization, and phase distribution of light beams. Nanostructures and MMs facilitate a new class of planar optical elements whose optical properties originate from spatial distribution of their refractive index rather than their shape. For example, arrays of nanoslits or nanoholes milled in a metallic film have been demonstrated to produce converging or diverging lenses by realizing different propagation constants in the slits or holes of different sizes [72–81]. On the other hand, a circular metal resonator was shown to produce helical wavefront by emitting scattered guided light from inner grating elements [82].

In addition to unique properties and potential applications of nanostructures for OAM light generation, structured light beams can enable new approaches to nanofabrication (Figure 6.6b). For example, utilizing interactions between 30-ns-pulsed optical vortex laser and metal, it was demonstrated that the total angular momentum of optical vortex can be transferred to tantalum, resulting in the formation of chiral metal nanoneedles, as shown in Figure 6.6b. The chirality of the nanoneedle can be controlled by varying total angular momentum of the optical vortex beam, so that the tip curvature size of the nanoneedle smaller than 40 nm at 1064 nm wavelength is realized.

Let us start with considering an important class of metal–dielectric MSs based on the Pancharatnam–Berry phase optical elements (PBOE) that were first introduced by Bomzon et al. [87]. Light transmitted through these MSs accumulate geometric or topological phase due to the transmission in the 'anisotropic parameter space.' Unlike conventional optical components such as lenses or waveguides that introduce phase changes through optical path differences, PBOEs facilitate strong light–matter interaction on a subwavelength scale through the careful arrangement of resonant optical antennas. Phase accumulation in gradient MSs results from space-variant polarization manipulations, as opposed to requiring propagation through a certain thickness of a material to accumulate sufficient phase. A variety of gradient MSs have been demonstrated.

Based on the PBOE principle, complex geometric-phase MS consisting of subwavelength metal–insulator–metal elements was demonstrated (Figure 6.6). These elements, the wave plates, introduce different phase shifts in different directions. The main difference is that in conventional macroscopic wave plates, such differential phase shifts are introduced by the direction-dependent difference in refractive indices, whereas in MSs, they rely on the two electric dipole scatterers whose resonances are oppositely detuned from a design frequency. The wave plates are oriented so that one of the fast axes depends on the spatial position. By manipulating the local orientations of the nanoantennas, multiple wavefronts with different functionalities can be generated. As a result, the polarization is space varying since it depends on the local

orientation of the wave plates. At different points, the beam travels through a different path on the Poincaré sphere, resulting in a space-variant phase-front modification originating from the geometric phase.

It is noteworthy that such PBOE-based MS addresses one of the most important needs of modern photonics applications, namely, the possibility of realizing ultracompact, multifunctional, integrated optical devices that enable independent control of the phase, amplitude, and polarization. Indeed, based on the concept of shared-aperture antennas [88] that use a single common aperture to perform multiple functions simultaneously (Figure 6.7a, b), several individual functionalities have been combined on the same platform. For example, one such functionality is the conversion of a conventional linearly polarized beam into azimuthally and radially polarized beams that carry OAM. Moreover, space-variant polarization manipulation enables the realization of multiple vectorial optical vortices by the coherent superposition of wavefronts with opposite helicities. Finally, interleaved geometric-phase MSs were used to build a simple and ultracompact spectropolarimeter shown in Figure 6.7c. This device simultaneously measures the polarization state and spectrum of a

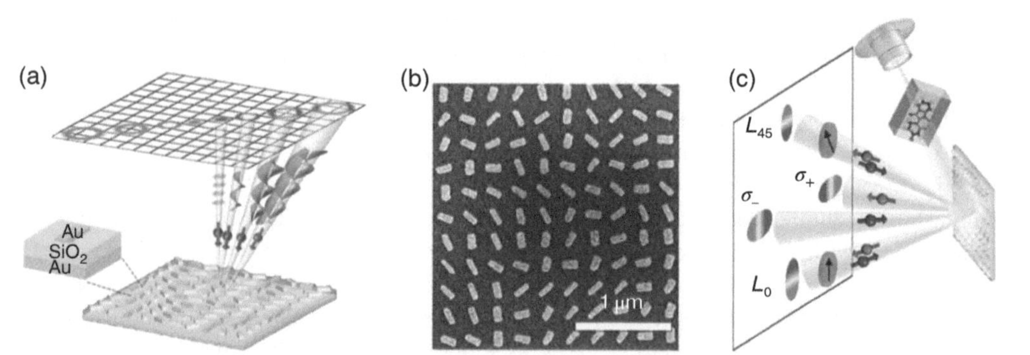

Figure 6.7 (a) Schematic of shared-aperture phased array; (b) scanning electron microscope image of a gap-plasmon MS; (c) schematic setup of the spectropolarimeter MS [88]. *Source:* Maguid et al. [88]

wave transmitted through a semitransparent object. The synergy between the geometric phase and the shared-aperture antenna arrays technology enables multifunctional wavefront shaping using a single, achromatic, and ultrathin nanoscale broadband photonic device.

Despite impressive developments in the field of metal–dielectric MSs, some of the drawbacks include significant losses of metals at optical frequencies and potentially challenging integration with existing semiconductor technologies. An alternative solution focuses on an all-dielectric silicon-based MSs that can control reflected or transmitted beam phase from 0 to 2π [46]. In contrast to plasmonic MSs, fabricated devices made with silicon are compatible with complementary metal–oxide–semiconductor technology. Such silicon-based MSs are likely to enable optical components for NIR photonics, such as flat lenses,

beam deflectors, antireflection coatings, and phase modulators. Potential applications of this broad variety of MSs include drug development, cancer diagnostics, remote sensing, and astrophysics.

An example of an experimental demonstration of an all-dielectric resonant MS with full $0\text{–}2\pi$ phase control at near-infrared wavelength that enables a high-efficiency beam deflector and a light converter for generating an optical vortex beam, carrying an OAM, is shown in Figure 6.8 [46]. In addition to $0\text{–}\pi$ phase control enabled by a single magnetic or electric resonance, overlapping of magnetic and electric resonances in the frequency domain enables an additional phase control with optical impedance matching and, as a result, high efficiency in transmission mode with full 2π phase manipulation. The schematics of light propagation through an infinite array of high refractive-index poly-Si nanoblocks (refractive index $n = 3.67$ with height $h = 270$ nm and dimensions $dx = dy = 650$ nm on top of semi-infinite fused Si substrate (refractive index $n_s = 1.45$), lattice constant $a = 800$ nm is shown in Figure 6.8a miniaturized high efficiency.

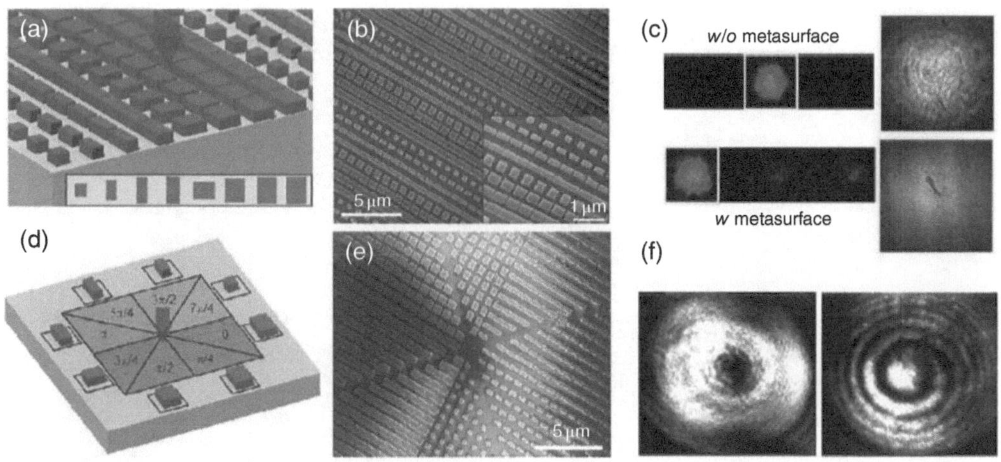

Figure 6.8 (a) Schematics of the proposed beam deflector with the unit cell (shown in the inset) containing eight nanoblocks, each responsible for the phase shift from 0 to 2π with $\frac{\pi}{4}$ increments. *Source*: Shalaev et al. [46]; (b) scanning electron microscopy image of fabricated MSs with $96 \times 96\,\mu$m total size; zoomed-in picture is shown in the inset; (c) beam position without (left top) and with (left bottom) MS; the beam power is refracted to the left side with deflection angle 13.1° and transmission of ~36%. Input and output beam photographs are also shown in the inset (right); (d) schematics. *Source*: Shalaev et al. [46] and (e) scanning electron microscopy image for optical vortex beam converter with eight sectors with different unit cells for each sector of MS. Each sector introduces an additional $\frac{\pi}{4}$ phase shift, thus covering $0\text{–}2\pi$ phase change. (f) Intensity distribution for measured output vortex beam in form of donut shape; (g) OAM and Gaussian beams interference experiment results showing spiral-shaped intensity distribution.

While the majority of all-dielectric MS designed for an OAM-beam generation aimed at producing the OAM beams with particular topological charges fixed by their designs, reconfigurability is one of the desired characteristics allowing switching from one charge to another or from the OAM beam to a beam not carrying an OAM. Recently, electrically tunable q-plates [22] as well as mechanically, electrically, thermally, and optically tunable MSs have been demonstrated (Figure 6.9) [93–103]. Nevertheless, simultaneous realization of ultrafast speed and high efficiency using MSs is still rather challenging.

On the other hand, nonlinear optical interactions offer a promising way for ultrafast, picosecond scale, and efficient optical switching, tuning, and reconfiguration [103–105, 91, 106–109]. Indeed, several material platforms, including silicon and GaAs, have been used to realize nonlinear light–matter interactions in optical MSs. However, silicon and GaAs have both high linear refractive indices and Kerr nonlinearities and low linear losses in the low figure of merit. On the contrary, As_2S_3 chalcogenide (ChG) glass displays a very good nonlinear figure of merit in both the near-infrared and the mid-wave infrared spectral bands [110–112]. Moreover, its excellent nanostructuring properties, which enable patterning with resolution superior to polymer photoresists, enable a resist-free approach that simplifies fabrication telecommunication range, their two-photon absorption coefficients are large, resulting in a of the As_2S_3-based devices to a single-step process. The basic principle of operation of the proposed MS device is illustrated in Figure 6.9. The input Hermite–Gaussian (HG) beam is transmitted through the MS and acquires different phase distribution depending on the input-light intensity (as shown in Figure 6.9a, c). The MS is designed such that, in the low-intensity regime, the phase acquired in the even quadrants is larger than for the odd

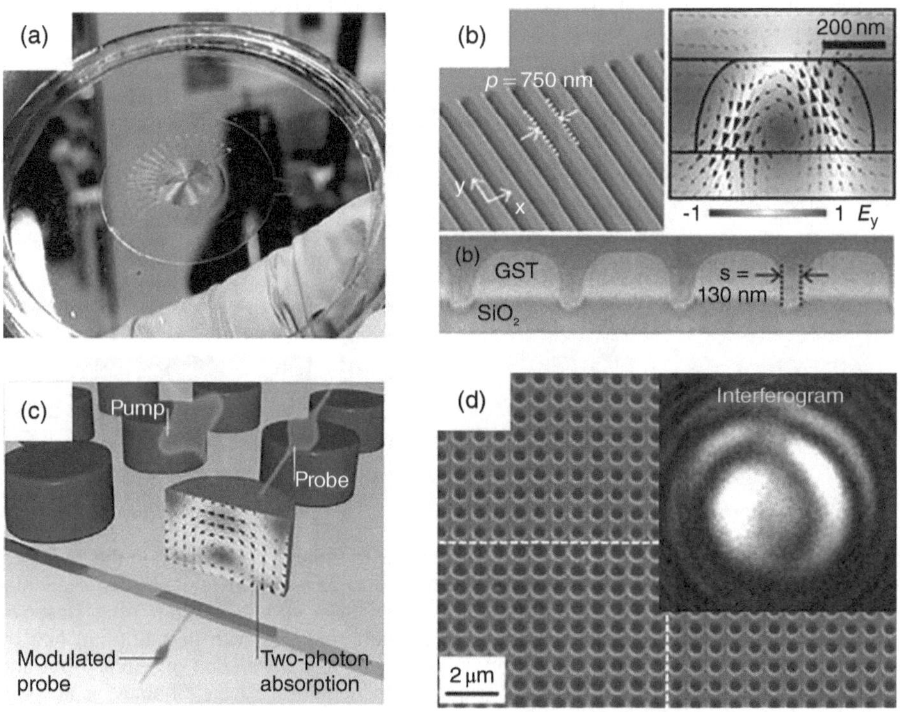

Figure 6.9 Reconfigurable metasurfaces: (a) electrically tunable [89], *Source*: https://www.techbriefs.com/component/content/article/tb/supplements/pit/briefs/33901 [89], (b) phase-change materials based on Ref. [90]. *Source*: Karvounis et al. [90], (c) two-photon absorption based on Ref. [91], *Source*: Shcherbakov et al. [91], and (d) Kerr nonlinearity based on Ref. [92]. *Source*: Xu et al. [92].

quadrants. The phase distribution directly after the MS has a stepwise profile. After propagation, this step-wise phase change smoothens and becomes a continuously varying phase typical for the OAM-carrying beams. Therefore, in the low-intensity regime, the HG beam, upon transmission through our MS, is transformed into a beam that carries OAM. For a high-intensity input beam, the phase introduced by the MS is uniform, and the input HG beam maintains its phase and intensity distribution and does not acquire the OAM. Switching between low- and high-intensity regimes allow for dynamic introduction of the OAM in the beam. The functionality described aforementioned is enabled by a ChG-glass-based MS, illustrated in Figures 6.9d and 6.10b. It consists of two diagonal quarters made of smaller holes and two other diagonal quarters made of larger holes. The desired phase difference between the four quadrants of the MS is obtained by careful optimization of the geometric parameters.

The electric and magnetic field distributions in the holes of the MS for the light interacting with the MS are shown in Figure 6.10b. According to the symmetry of the electric-field distributions to $z = 0$ plane (at the center of the ChG film), there is one odd mode and one even mode at the wavelengths of 1276 and 1289 nm, respectively. These resonances originate from the interaction of the incident wave with the modes of the perforated ChG film. For normal incidence, only the doubly degenerate guided resonances can be excited [113], as only for these modes the symmetry of the in-plane electric-field components allow coupling with the incident plane wave [114].

To illustrate the origin of the guided resonances and characterize them, Figure 6.10d shows the band structure, transmittance, and phase as functions of frequency for the structure with $d = 320$ nm. Figure 6.10b shows the band diagram of the studies structure with the close-up of the spectral range of interest in the vicinity of the center of the first Brillouin zone (Γ point). At the Γ point, both even and odd modes of the MS can be represented as a superposition of the four modes (with the same parity) of a uniform slab. Normally incident light impinging on the MS does not have an in-plane k-vector component and can couple to the modes of the structure located at the Γ point. The coupling efficiency depends on the symmetry of the field distribution of the modes and on its overlap integral with the field of the incident wave. As shown by Fan et al. [113, 114], normally incident light can only couple to the degenerate modes at the Γ point. This degenerates odd and even modes and their spectral position matched with the transmittance dips are shown in Figure 6.9d. The relative spectral position of the even and odd resonances can be controlled by changing the hole diameter.

For $d = 320$ nm, when the resonances are located apart from each other, two dips in transmittance can be seen at wavelengths λ_o and λ_e. In the vicinity of these wavelengths, the phase of the transmitted light changes rapidly by 180°. As the diameter of the hole increases, the central wavelengths of the odd and even guided resonances become closer to each other. When the spectral separation between the resonances is smaller than their width, resonant dips overlap, such that only one dip in transmittance with the corresponding phase change of 360° is observed. For the hole diameter $d = 372$ nm, such resonance is located around the wavelength $\lambda_o = 1255$ nm that was chosen as an operating wavelength for the nonlinear MS. Such combination of the hole diameter d and wavelength λ_o allows for effective modification of the phase of the transmitted light while small intensity-dependent changes of the refractive index of ChG are introduced. The refractive index of the ChG glasses can be efficiently changed as a function of the input-light intensity due to their large nonlinear Kerr response described by $n_{\mathrm{ChG}} = n_0 + \Delta n = n_0 + n_2 I$. This example demonstrates As_2S_3 ChG glass-based MS that enables ultrafast (Kerr nonlinearity based) reshaping of a conventional HG beam with no OAM into an OAM beam, while preserves the original beam's amplitude and phase characteristics at high-intensity levels. Such MS could find applications for a new generation of optical communication systems and optical signal processing.

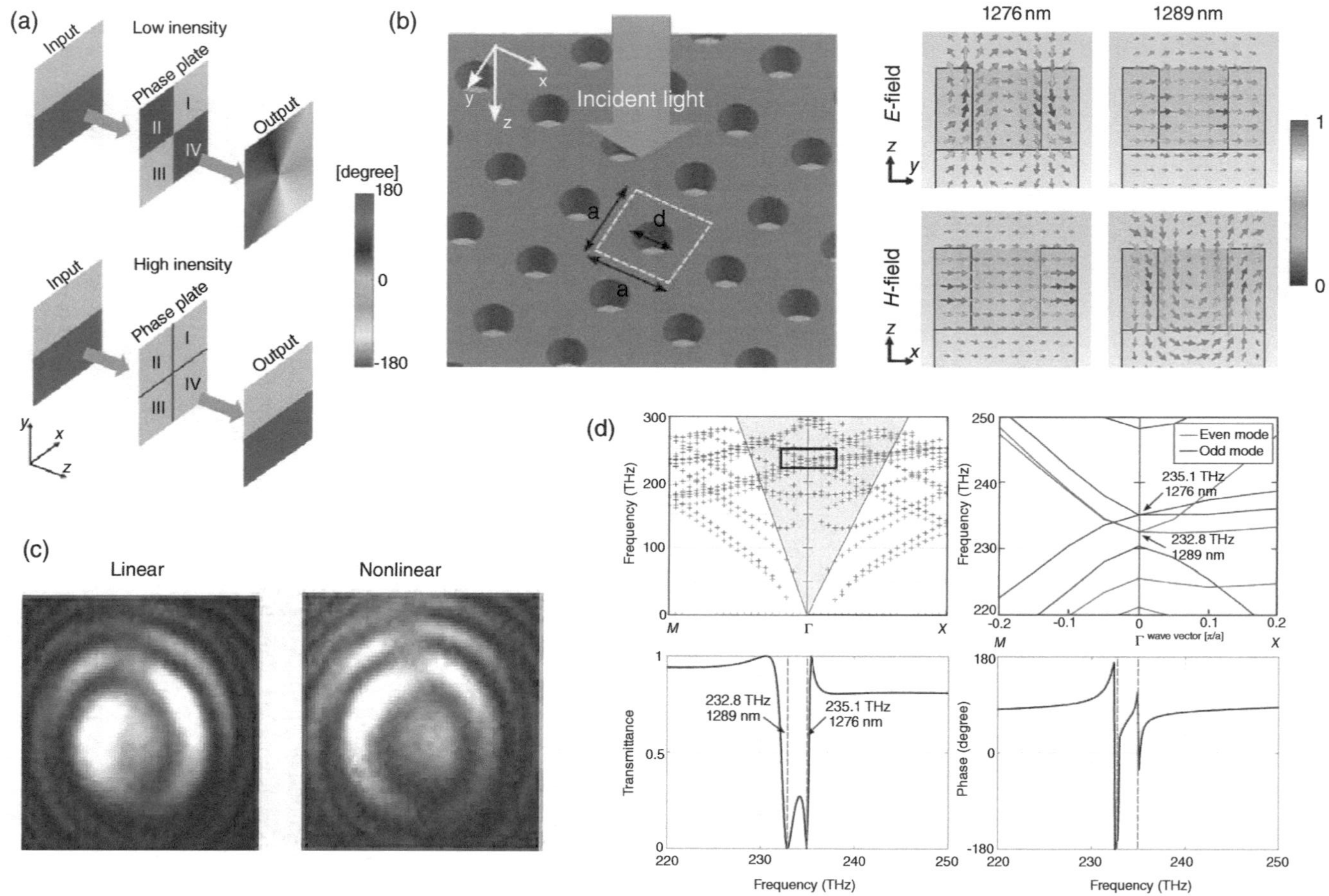

Figure 6.10 (a) Working principle of nonlinear metasurface with reconfigurable output beam [92]. The maps show the phase of the input and output beams and the relative phase introduced by the MS. At low intensity, upon transmission through the MS, the input Hermite–Gaussian beam acquires a nonuniform phase distribution leading to generation of a beam carrying orbital angular momentum. At high intensity, the phase difference introduced by the MS is uniform; the output beam remains Hermite–Gaussian and does not possess OAM; (b) design of the nonlinear reconfigurable MS consisting of a square lattice of circular holes perforated in a film of arsenic sulfide (As_2S_3) ChG glass. The film is deposited on a glass substrate and is covered with a layer of the poly(methyl methacrylate) which also fills the holes; (b) schematic of the MS and distributions of the electric and magnetic fields in the holes made in ChG film frequencies corresponding to the electric and magnetic resonance; (c) experimentally measured interference patterns corresponding to the OAM beam (left) and HG beam (right); (d) top: band structure for the wave vectors along the high symmetry direction MΓ and ΓX in reciprocal space (k_x, k_y) (left). Enlarged version of the band structure inside the black rectangle (right). The degenerate modes are indicated with the arrows and corresponding frequencies. The two degenerate modes at Γ point couple to the electromagnetic wave incident in the direction normal to the surface. Bottom: transmittance (left) and phase (right) for the waves propagating normally to the photonic crystal slab. Two dips in transmittance and two abrupt phase changes in phase at the wavelengths corresponding to the degenerated modes shown in top plots (d). *Source:* Xu et al. [92].

6.3 Active OAM Sources

Up to this point, our discussion focused on passive OAM-beam emitters. However, many photonic applications would benefit from the availability of active OAM laser sources, and therefore, recently several approaches to realization of active devices emerged [115–124]. One approach to creating an OAM laser is based on combining a conventional bulk laser with additional phase-front shaping components [116]. Although straightforward, this approach relies on bulk device technologies and different material platforms, and therefore it is not easily scalable and integratable. Further success in the development of the OAM beam-based optical communication and quantum information technologies necessitates the design of independent micro- and nanoscale laser sources emitting OAM beams with on-demand tunable topological charge and desired polarization properties. In the past few years, significant progress has been achieved in on-chip integrated OAM sources [117–124].

Let us consider a microring cavity-based microlaser, as shown in Figures 6.11c and 6.12c. The microring cavity supports whispering gallery modes (WGMs). These modes circulate inside the cavity and carry large OAM. However, due to the mirror symmetry of a ring cavity, clockwise and counterclockwise eigenmodes can be simultaneously excited and their carried OAMs consequently cancel each other. This is evidenced by the quantized phase, taking discrete values of either 0 or π, azimuthally distributed in the ring, which results from the interference between two counter-propagating WGMs [124]. To observe the OAM of an individual WGM, it is essential to introduce a mechanism of robust selection of either clockwise or counterclockwise mode. In conventional bulk optics, unidirectional ring lasers have been demonstrated by implementing a nonreciprocal isolator in the light path. The optical isolator breaks the reciprocity between counter-propagating waves, facilitating the desired unidirectional flow. This approach, however, is not feasible at the micro- and nanoscale, as the realization of micrometer-sized isolators is challenging.

It is noteworthy that bidirectional emission in ring cavities-based lasers close to the lasing threshold results in spatial hole burning (SHB). In a Fabry–Perot laser cavity, SHB originates from a standing wave pattern formed due to the interference of counter-propagating waves that are reflected at the end cavity mirrors (Figure 6.12a). In a ring cavity, the superposition of the clockwise and counterclockwise traveling waves also result in an interference pattern, leading to the nonuniform gain distribution and the SHB. Such nonuniform gain distribution feeds power to multiple modes, causes mode competition, and leads to irregular lasing emission behavior. Therefore, finding an equivalent to optical isolator on the microscale would not only facilitate the OAM generation but also eliminate the problem of SHB. It was shown that such optical isolation at the microscale can be achieved by tailoring the microlaser resonator at the exceptional point (EP), where the combined index and loss gratings play a similar role of a diode (Figure 6.12b) at the EP operation, the gratings introduce unidirectional mode coupling between counter-propagating modes, which favors laser emission from initial noise in one of the two modes. From basic laser theory [125], it is then known that preferential oscillation in a single traveling wave mode prevents SHB and stochastic mode switching [126, 127]. Another way to understand the unidirectional laser emission based on combined index and loss gratings is as follows. The EP represents the condition when different eigenbranches in a complex spectrum coalesce thus leading to a complete eigenstate degeneracy in both energy and field distributions [128–134]. This implies that at this point, two originally orthogonal clockwise and counterclockwise traveling wave modes collapse into only one traveling wave mode, realizing the desired unidirectional power flow inside a microcavity.

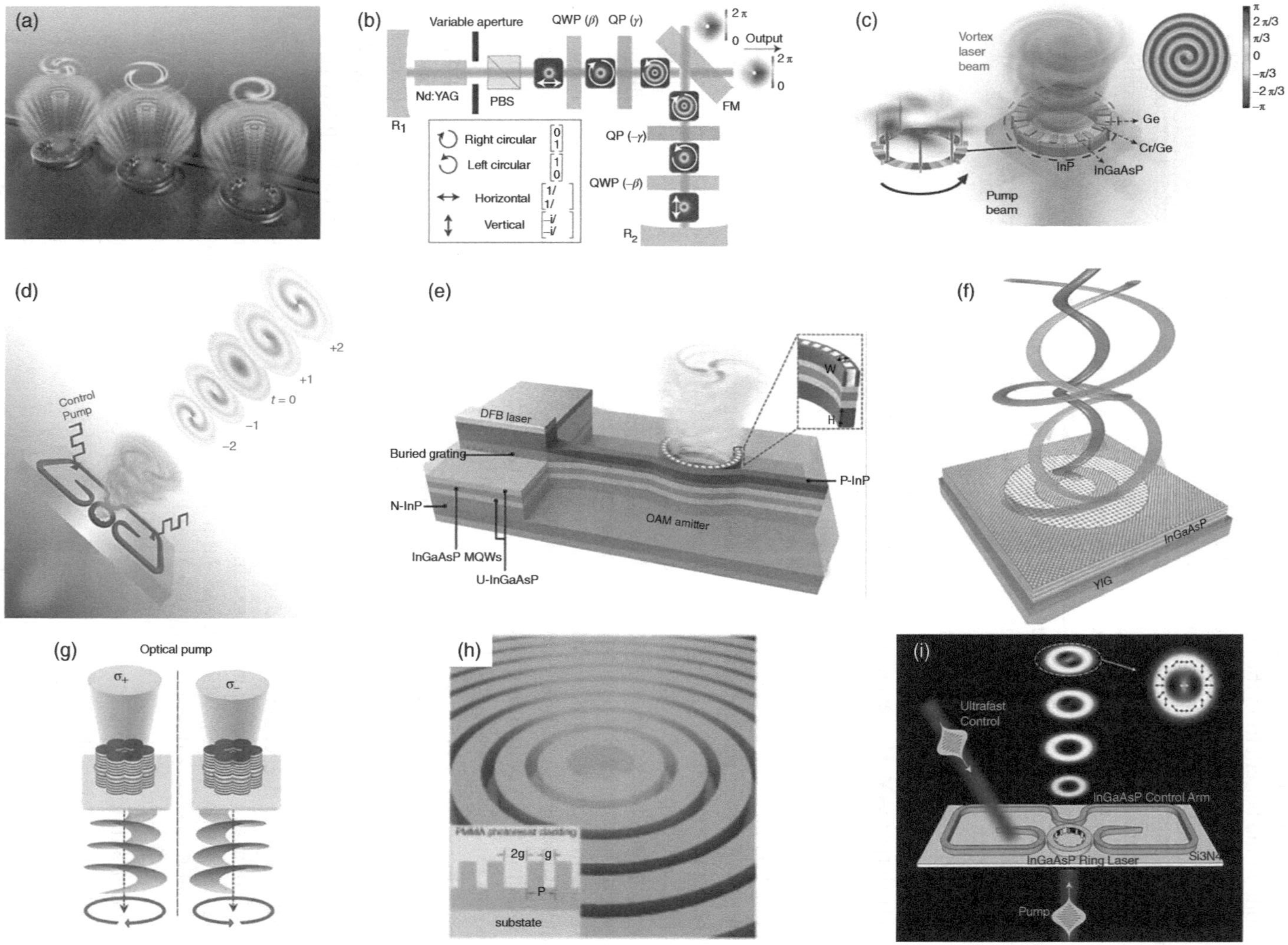

Figure 6.11 From passive to active integrated OAM sources: (a) integrated OAM-beam emitters [115]. *Source*: Modified from Cai et al. [82, 115]; (b) bulk optics based OAM laser [116]. *Source*: Naidoo et al. [116]; (c) OAM microlaser [117]. *Source*: Miao et al. [117]; (d) tunable charge OAM microlaser [118]. *Source*: Zhang et al. [118]; (e) integrated OAM emitter with a distributed feedback laser [119]. *Source*: Zhang et al. [119]. Licenced under CC BY-4.0.; (f) topological OAM laser [120]. *Source*: Bahari et al. [120]; (g) benzene-like OAM lasers [121]. *Source*: Zambon et al. [121]; (h) lead halide perovskite OAM microlaser [122]. *Source*: Sun et al. [122]. Licenced under CC BY-4.0; and (i) fractional OAM microlaser [123]. *Source*: Zhang et al. [123]. Licenced under CC BY-4.0.

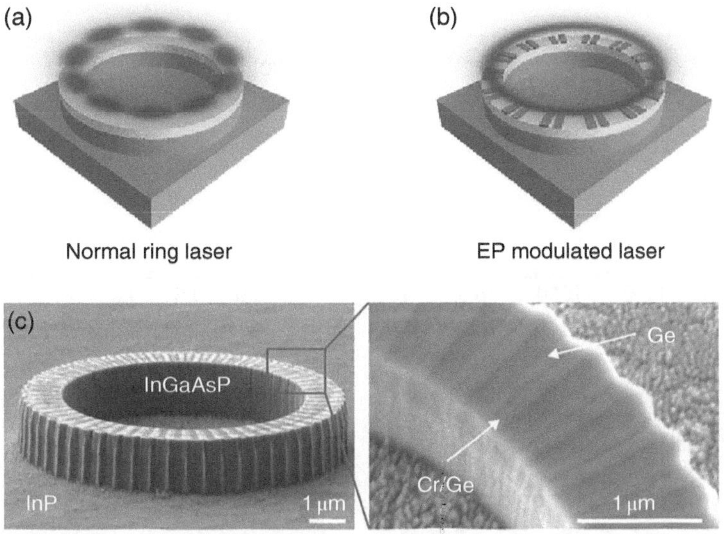

Figure 6.12 Spatial hole burning suppression and OAM lasing in a ring cavity-based laser [117, 124]. (a) Conventional ring laser. *Source*: Miao et al. [117]; (b) exceptional point modulated laser. *Source*: Zhang et al. [124]; and (c) SEM image of the exceptional point modulated OAM laser.

Driven by nonhermiticity (i.e. gain and loss in optics) an EP occurs when multiple eigenstates coalesce into one. The microring laser resonator is designed with 500-nm-thick InGaAsP multiple quantum wells on an InP substrate. The complex refractive-index grating is achieved by placing on top of InGaAsP along the azimuthal direction periodically alternate single-layer Ge and bilayer Cr/Ge structures, corresponding to the refractive index (n') and gain/loss (n'') in the cavity, respectively

$$\Delta n = \begin{cases} in'' \text{ for } 2\pi p/N < \theta < 2\pi\left(p + \frac{1}{4}\right)/N \\ n' \text{ for } 2\pi\left(p + \frac{3}{8}\right)/N < \theta < 2\pi\left(p + \frac{5}{8}\right)/N \end{cases} \tag{6.2}$$

where N denotes the azimuthal number of the targeted WGM and p takes integer values from the set $\{0, N-1\}$. An EP is obtained when the amplitudes of index and gain/loss gratings are set equal (i.e. $n' = n''$). At EP, the Fourier transform of the complex refractive-index modulation is one sided, yielding one-way distributed feedback and robust unidirectional laser emission above threshold. As a result, the counterclockwise WGM unidirectionally circulates in the cavity carrying large OAM. The OAM associated with the unidirectional power flow is extracted upward into free space by introducing sidewall modulations periodically arranged along the microring perimeter [117]. The azimuthal phase dependence of the targeted unidirectional Nth WGM is given by $\varphi = N\theta$.

Finally, in order to force vertical emission, the sidewall modulation is introduced to coherently scatter light, with the phase continuously varying in azimuthal direction, defined by the locations of the scatters and shown in the inset of Figure 6.12c. For M equidistant scatters, the locations of the scatters are given by $\theta_s = 2\pi s/M$, where $s \in \{0; M-1\}$, resulting in the extracted phase $\theta_s = 2\pi sN/M$ that carries OAM. Since the physical phase is measured modulo 2π, we can subtract $2\pi s$ from each of the extracted phases and derive $\varphi_s = 2\pi s\,(N-M)/M$. This equation shows that the extracted phase

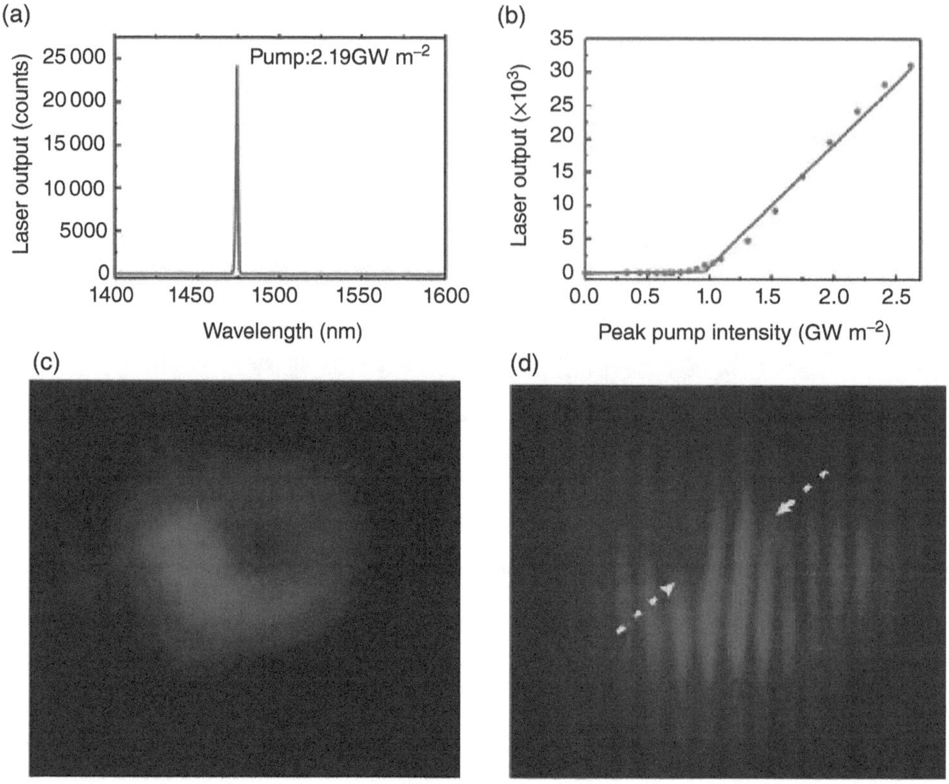

Figure 6.13 (a) Lasing characterization; (b) input–output laser curve, showing a lasing threshold of ~1 GWm^{-2}. *Source*: Miao et al. [117]; (c) far-field intensity distribution of the laser emission exhibiting a doughnut-shaped profile, where the central dark core is due to the phase singularity at the center of the OAM vortex radiation; (d) off-center self-interference of the OAM lasing radiation, showing two inverted forks (marked with arrows) located at two phase singularities, originating from the superposition of central helical and outer quasiplanar phases associated with OAM. Such double-fork pattern confirms the OAM vortex nature of the laser radiation [117]. *Source*: Miao et al. [117].

increases linearly from 0 to $2\pi(N-M)$, therefore creating a vortex beam with topological charge $\ell = N-M$. Figure 6.13 shows the results of optical characterization of the OAM microlaser. The presence of the OAM was demonstrated by the self-interference of two doughnut-shaped beams split from the same lasing emission. In each doughnut beam, helical phase front is more pronounced close to the central singularity area, while in the outer doughnut area, the phase front is relatively uniform, i.e. quasiplanar. If two beams are spatially offset in the observation plane such that the dark center of one beam overlapped with the bright doughnut area of the other and vice versa, the resulting interference patterns between the helical and quasiplanar phase fronts clearly show two inverted forks. The inversion results from the fact that the quasiplanar and helical phases are reversed at the centers of two doughnuts. For both beams, the single fringe split into two at the fork dislocation, thus confirming that the radiation from the microlaser was indeed an OAM beam of topological charge $\ell = -1$.

Finally, we briefly discuss recently reported on-chip integrated laser sources of structured light carrying fractional OAM (FOAM) are highly desirable for the forefront developments of optical communication and quantum information-processing technologies. While a variety of integrated

OAM-beam generators have been previously reported, ultrafast control and sweep of the FOAM light with low-power control, suitable for high-speed optical communication and computing, remains challenging. Recently, fast control of the FOAM from an OAM semiconductor microlaser, based on fast transient mixing of integer laser vorticity induced by a control pulse, was demonstrated. Continuous FOAM sweep between charge 0 and charge +2 is demonstrated in a 100 ps time window with the ultimate speed limit being established by the carrier recombination time in the gain medium.

Integer or fractional OAM beams are traditionally generated by the introduction of a phase variation in the transverse plane of an incident Gaussian beam using carefully aligned optical elements, such as phase plates or MSs. Alternatively, in integrated semiconductor nanophotonic circuits, the generation of vortex beams can rely on a different mechanism: robust selection of chiral resonant modes and their free-space outcoupling with strategic phase matching conditions to convert the in-plane chiral modes into OAM vortex beams. When the chiral modes are generated in active semiconductors (microlasers), thanks to the relatively fast electron-carrier recombination dynamics of the gain medium, a series of high-speed modulation schemes can be applied to maneuver the in-plane chiral modes, which further facilitates ultrafast tunability and reconfigurability in OAM. The FOAM control in the picosecond time scale was achieved using tunable vortex microlaser [123], consisting of a microring resonator and an external coupling loop with two control arms, all made of 200-nm thick InGaAsP multiple quantum wells and embedded in a Si_3N_4 substrate, as shown in Figure 6.11d. The microring resonator supports two chiral modes: counterclockwise (↺) and clockwise (↻) WGMs that are indirectly coupled by the external loop. The in-plane chiral modes are extracted and converted to OAM emissions by an angular grating with M equidistant scatters placed at the inner sidewall of the microring. In Zhang et al. [123], a microring resonator with a diameter of 7 μm and a width of 0.6 μm (shown in Figure 6.11i) was reported. For the wavelength of ~1500 nm, the resonant order of the WGMs is $N = 34$, which, alongside the diffraction order of the angular grating of $M = 35$, yields a total angular momentum of $J = \mp 1$ for extracted emissions from the ↺ and ↻ modes, respectively. The polarization state of extracted fields is in general a superposition of two transverse SAM (i.e. left handed ($s = 1$) and right handed ($s = -1$) circular polarizations) with their amplitudes geometrically dependent on the dimensions of the microring. Therefore, the microlaser emissions contain four components corresponding to the four SAM–OAM combinations, described as

$$I_{\text{out}} = I_{\text{L},-2} + I_{\text{R},0} + I_{\text{L},0} + I_{\text{R},+2} \propto p_{\circlearrowleft}\sigma^2 + p_{\circlearrowleft}\left(1-\sigma^2\right) + p_{\circlearrowright}\left(1-\sigma^2\right) + p_{\circlearrowright}\sigma^2 \quad (6.3)$$

where I_{out} is the total intensity of the microlaser emissions, which is the sum of the intensities of four different spin-OAM components with L/R indicating the left-/right-handed circular polarization and −2, 0, and +2 denoting the associated OAM charge. The four spin-OAM components correspond to the out couplings of four combinations between the ↺/↻ mode and in-plane transverse spins, where $p_{\circlearrowleft}/p_{\circlearrowright}$ denotes the power associated with the ↺/↻ mode inside the cavity, respectively, and σ is the absolute value of the transverse SAM charge to describe the purity of the transverse spin. Equation (6.3) shows that by exciting and reconfiguring the weighting of different OAM components, a precise, continuous FOAM sweeping between charge 0 and charge +2 within 100 ps can be achieved. The associated FOAM charge is calculated by integrating the OAM flux across the whole vector beam with complex polarization state shown in the inset of Figure 6.11i [5, 135].

The WGM microlaser emission is controlled exploiting indirect nonhermitian mode coupling [135–139] via suitable external loops [118], as shown in Figure 6.11i. The effective mode coupling can be turned from completely symmetric (Hermitian) to unidirectional (EP), depending on the gain–loss contrast between the two control arms. This enables actively tune the weighting between the two chiral modes (that is equivalent to the power ratio between p_↺ and p_↻). The ultrafast control is achieved by applying two synchronized femtosecond pulses from the same femtosecond laser (~140 fs) to pump the whole structure. The first is a main pump pulse above the lasing threshold at time T_1 to enable the lasing oscillation in the microring resonator. The second is a control pulse below the lasing threshold at time T_2 to selectively excite gain carriers in one of the two arms. In addition to the control of the peak gain via the pump power, here, the transient carrier dynamics with a carrier lifetime typically of hundreds of picoseconds is the key to the ultrafast control. As the excited carriers relax in time after pulse pumping, the associated gain coefficient also decays together with the gain–loss contrast between the two control arms. Consequently, the nonhermitian-controlled indirect coupling dynamically varies as a function of time, from unidirectional to symmetric. Hence, temporal overlap of the laser action in the microring and the time-varying optical gain in the control arm, controlled by the time delay between the two femtosecond pump pulses, facilitates the ultrafast control of the weighting between the two chiral modes in the microring and thus the weighting between different mode-converted OAM components in Eq. (6.3), yielding dynamical tuning of the FOAM of microlaser emissions with a picosecond resolution.

The ultrafast temporal control of the weighting between the two chiral modes and thus of the ratio between their associated power (p_(↺)/p_(↻)), which is the key dynamical parameter to reconfigure the fraction of each spin-OAM component in the emitted vector beam [see Eq. (6.3)] is achieved by tuning the temporal delay between the main pulse incident on the microring and the control pulse in the left control arm. In particular, the modulation of the carriers, facilitated by the precisely controlled temporal delay of the pulses in both the microring and the left control arm, enables the ultrafast control of the fractional OAM charge, as illustrated in Figure 6.14.

6.4 OAM Light in Engineered Nonlinear Colloidal Systems

The emergence of MMs has a strong potential to enable a plethora of novel nonlinear light–matter interactions and even new nonlinear materials. In particular, nonlinear focusing and defocusing effects are of paramount importance for the manipulation of the minimum focusing spot size of structured light beams necessary for nanoscale trapping, manipulation, and fundamental spectroscopic studies. However, the majority of earlier light–matter interactions in MMs focused on solid-state MMs.

Here, we discuss recent progress in the field of liquid-based MMs, including theoretical studies of plasmonic metafluids made of nanoclusters [140], dumbbell elements [141], and negative-index metafluids [142], as well as experimental demonstration of an isotropic metafluid with a magnetic response [14], colloidal MM made from metallic nanorods [143], and tunable 3D THz meta-liquid crystals (MLC) [144]. Let us briefly discuss an example of tunable liquid MM demonstrated by Liu et al. [144]. In this work, individual meta-atoms are combined into elongated chains and dispersed in a host liquid. Once the external electric field is applied, these chains are rotated by electrophoretic forces, aligning the axis of anisotropy parallel to the field similar to the mesogens of liquid crystals, as shown in Figure 6.15a and b. Due to this analogy, this MM was called a MLC. By changing the direction of the biased electric field, the response of MLCs can be tuned dynamically.

Several realizations of complex meta-atoms or by analogy with LC mesogens of MLCs are shown in Figure 6.15c. These rather complex meta-atoms with subwavelength features enable the possibility of engineering a number of tunable electromagnetic properties not available with

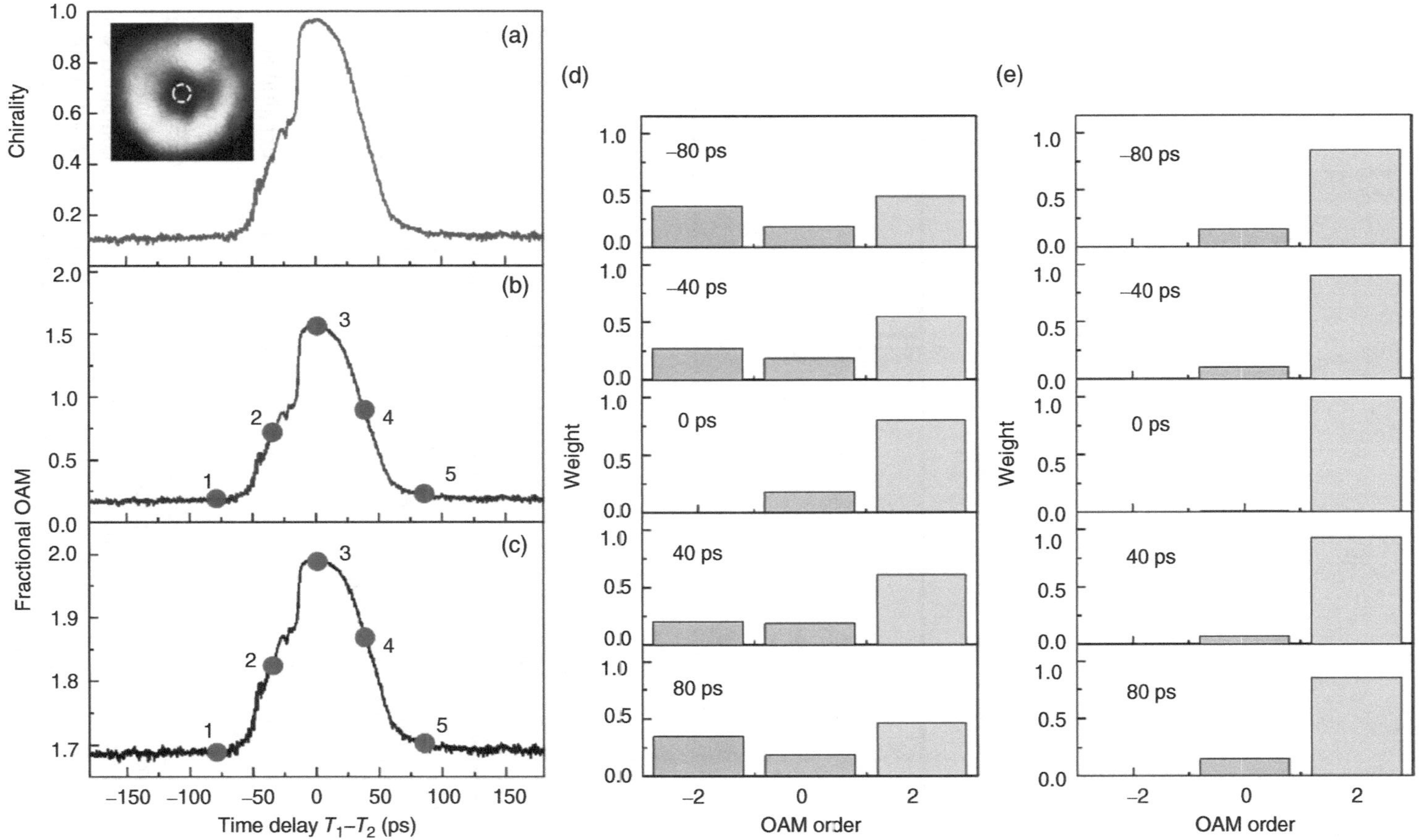

Figure 6.14 (a) Measured chirality of laser emissions as a function of time delay between the pump and control pulses. The inset shows a snapshot of the vortex beam, *Source*: Zhang et al. [123]; (b) measured FOAM charge of laser emissions, which can be continuously tuned between 0.18 and 1.57 in a temporal scale of ~100 ps; (c) measured FOAM charge of laser emissions after filtering out the cross-spin component, which can be continuously tuned between 1.68 and 2 in the same temporal scale; (d) OAM spectra of FOAM laser emissions at different time delays corresponding to the five points marked in (b); (e) OAM spectra of FOAM laser emissions at different time delays corresponding to the five points marked in (c) [123]. *Source*: Zhang et al. [123]. Licensed under CCBY 4.0.

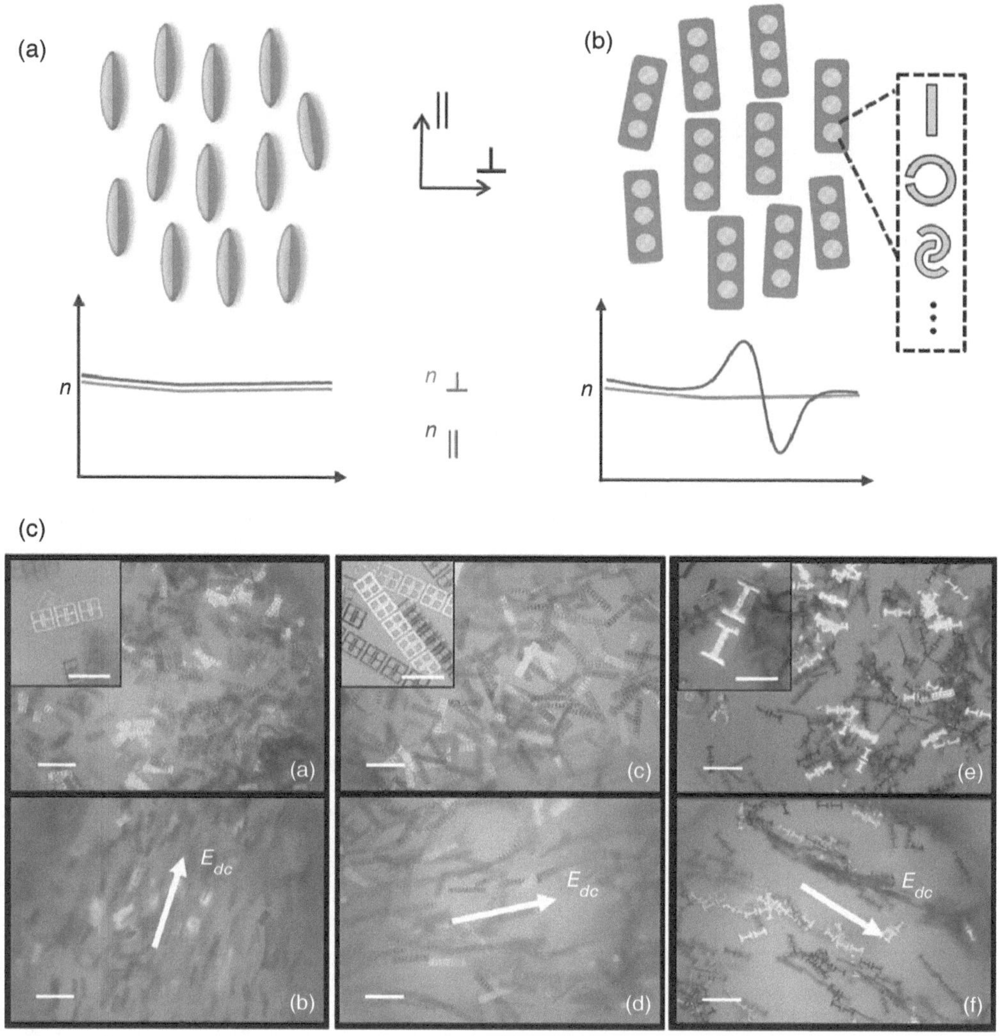

Figure 6.15 Schematic of (a) a conventional liquid crystal in the nematic phase; (b) meta-liquid crystals with meta-mesogens oriented along the external static electric field. *Source*: Liu et al. [144]; and (c) structural changes of the three different meta-mesogens before (top row) and after (bottom row) application of the bias electric field [144]. *Source*: Liu et al. [144].

conventional liquid crystals [145], including large anisotropy, strong magnetic and/or chiral or bianisotropic responses, or spectral features such as electromagnetically induced transparency engineered on demand.

One of the first examples of engineered liquid 'MMs' designed to imprint OAM on a Gaussian laser beam were microscopic diffractive optical elements (MDOEs) that were integrated into a microdevice [146]. Integrated devices consisting of optically driven microrotors combined with MDOEs are essential in a variety of microfluidic and lab-on-a-chip applications. Novel lab-on-a-chip devices can perform a variety of tasks ranging from mixing microscopic amounts of reagents to natural selection on a single enzyme level [147].

The MDOE generates OAM in an incident beam enabling high torques applied to a wide range of objects. Similar to a standard phase hologram, the MDOE induces a phase shift in an incident laser

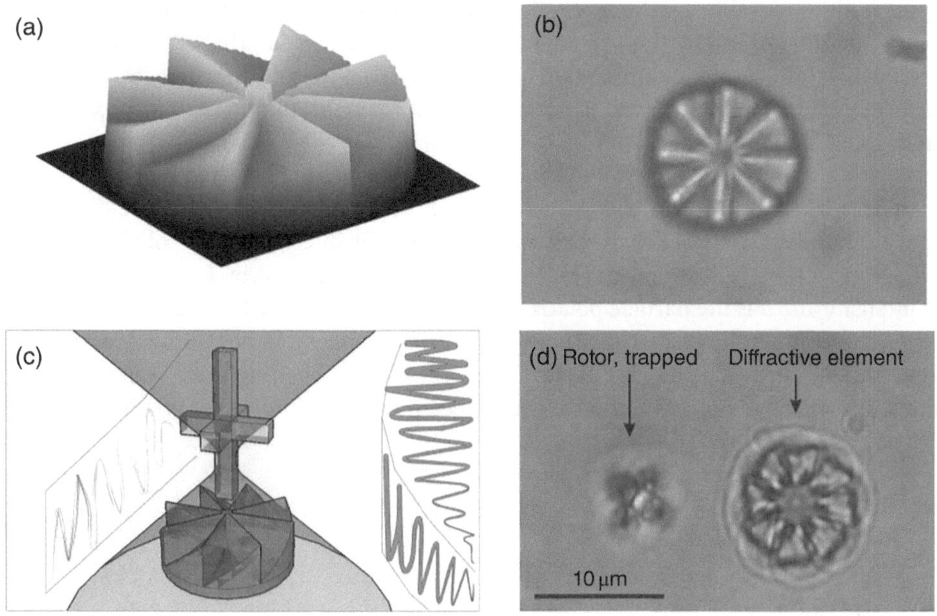

Figure 6.16 (a) A three-dimensional model of an MDOE designed to produce orbital angular momentum carrying LG08 mode components when illuminated with a Gaussian laser beam. *Source*: Knöner et al. [146]; (b) photopolymerized MDOE immersed in water and imaged using bright-field microscopy. *Source*: Knöner et al. [146]; (c) schematic representation of the microrotor trapped in a microchannel and rotating above the MDOE. *Source*: Knöner et al. [146]; (d) bright-field microscopy image of the trapped microrotor and the MDOE. MDOEs can be integrated into microfluidic devices where they could be used to actuate fluids for pumping and mixing or to apply optical torques for biophysical studies [146]. *Source*: Knöner et al. [146].

beam which is proportional to the MDOE's thickness. The MDOE shown in Figure 6.16 consists of eight phase ramps. The phase difference between rays transmitted through the top and the bottom of each ramp is 2π. The splitting of a 16π phase ramp into $8 \times 2\pi$ phase ramps is based on the same principle as employed by the well-known Fresnel lenses.

MDOE offers many advantages over externally generating a beam carrying OAM. For instance, the 3D trapping capability of optical tweezers is generally lost when relatively pure LG modes of higher order are used for trapping because high-order modes have a considerable width in the focal plane with a dark spot in the center caused by the phase singularity. Larger dark spot results into lower field gradient in the axial direction that prevents stable 3D trapping. In the system consisting of engineered, integrated MDOEs trapped rotor is located in the near field of the beam diffracted by the MDOE so that the dark spot in the beam's center has not yet fully developed. Moreover, the MDOE also provides the capability to switch rotation on and off quickly, by moving the MDOE into and out of the beam path without risking release of the trapped particle.

In the reminder of this section, we consider another example of liquid-based engineered optical media—nonlinear colloidal suspensions. Colloidal suspensions are mixtures in which microscopic particles of one substance are dispersed in another. Natural systems including milk and atmospheric fog are examples of colloidal systems. These systems attract significant interest owing to a large number of degrees of freedom for designing their linear and nonlinear properties. Colloidal media with well-defined optical properties have been widely used as model systems in many fundamental and applied studies of light–matter interactions in complex media. Recent progress in the field of engineered nanoscale optical materials with fundamentally new physical properties opens new opportunities for tailoring the properties of colloids. Besides contributing to the fundamental science of light–matter interactions in engineered soft-matter media, this work opens new opportunities for dynamic optical manipulation and transmission of light through scattering media as well as formation of complex optical patterns and light filamentation in naturally existing colloids such as fog and clouds.

Light–matter interactions in colloidal systems can be significantly altered by changing particles' polarizabilities that depend on the particle and the background medium's refractive indices. The optical gradient force defines dynamics of light–matter interactions in colloidal suspensions. Following refs. [148–150], the optical gradient force can be written as by $\vec{F} = \alpha \nabla I/4$, where I is the optical intensity and α is the particle polarizability. Considering a particle near the waist of a Gaussian beam, the formula shows that a particle with PP experiences a force pulling it toward the beam (Figure 6.17a). On the other hand, NP particles experience a repelling force pointing outward, as illustrated in Figure 6.17b. As the intensity of the beam increases, in both cases the beam is guided

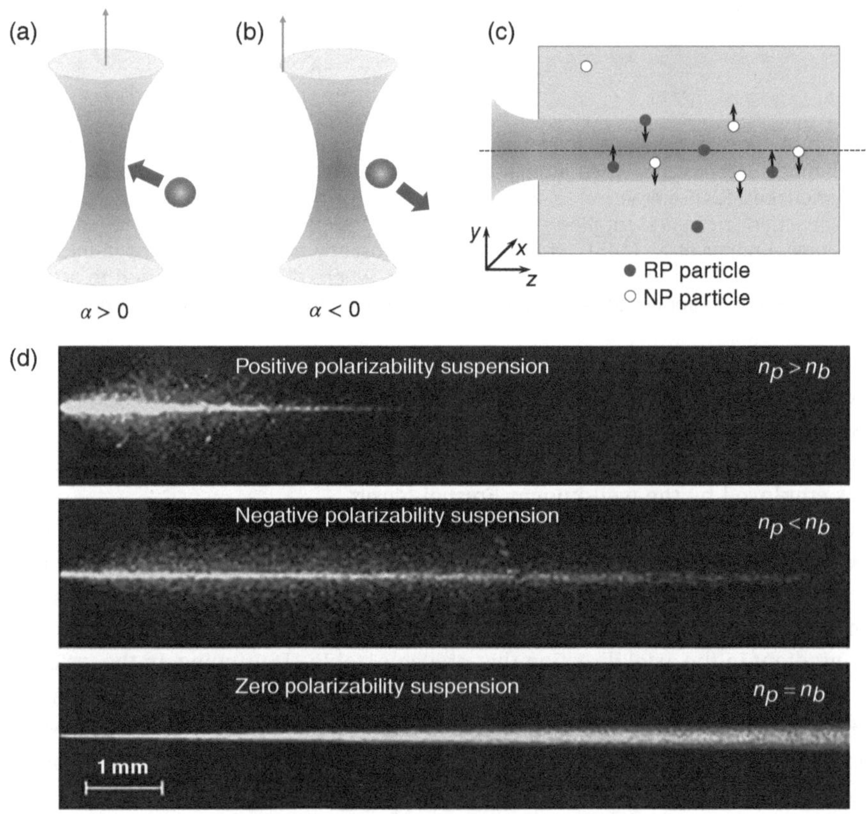

Figure 6.17 Colloidal suspensions offer a unique platform for engineering optical forces, polarizibilities, and nonlinearities: (a) positive polarizability particle with refractive index higher than that of surrounding medium is attracted to the beam; (b) negative polarizability particle with refractive index lower than that of the surrounding medium; and (c) nonlinear colloidal medium with both positive and negative polarizability particles. *Source*: Modified from Man et al. [151]; (d) experimental demonstration of Gaussian beam propagation in different colloidal media: with positive polarizability (top row): catastrophic-self-focusing collapse; negative polarizability (middle row): enhance transmission; and linear propagation dominated by diffraction (bottom row) [151]. *Source*: Man et al. [151]

in a self-induced waveguide that forms due to the balance of nonlinear self-focusing and diffraction induced broadening, as shown in Figure 6.17c.

Nonlinear optics is a fascinating branch of science that focuses on light propagation in media, in which the dielectric polarization responds nonlinearly to the electric and/or magnetic field of light. Although this field has been developing for decades, nonlinear optical materials available to date are still limited by either slow material response time in such phenomena as saturable absorption, photorefractive effect, and thermal nonlinear phenomena or by relatively low and generally band-limited nonlinear susceptibilities responsible for ultrafast nonlinear processes. Over the 50 years of the development of the field of nonlinear optics, significant theoretical progress has been made in understanding the origin and limitations of nonlinear responses. In parallel, a number of practical applications have emerged, such as optical parametric generation, mode locking, and comb generation. After many years of active research, some of the best results have been demonstrated in optical fibers, owing to long propagation length rather than their nonlinear properties. Therefore, the possibility of developing new, engineered nonlinear media for on-demand light propagation and beam shaping is of paramount importance.

Recently, it was shown that the nonlinear response of colloidal suspensions can be tailored at will by judiciously engineering their optical polarizability [151]. Such systems may provide new opportunities for fundamental studies and applications of such nonlinear effects as self-focusing and related beam breakup effects. Figure 6.17d shows the results of experimental studies of nonlinear propagation of Gaussian beam in different colloidal systems.

One of the most significant nonlinear phenomenon that is of particular interest for beam shaping and energy delivery is modulational instability (MI). MI is a phenomenon that reveals itself as the exponential growth of weak perturbations in the presence of an intense pump beam propagating in a nonlinear medium. It plays a key role in such nonlinear optical processes as supercontinuum generation, light filamentation, rogue waves, and ring (or necklace) beam formation. To date, a majority of studies of these phenomena have focused on light–matter interactions in self-focusing Kerr media existing in nature. However, a large and tunable nonlinear response of a colloidal suspension can be tailored at will by judiciously engineering the optical polarizability.

Light self-focusing and filamentation in various nonlinear media have been studied first described by Askaryan [152], followed by many other researchers [153–156]. In particular, transverse spatial periodic breakup of an optical beam due to self-focusing has been shown to be related to supercontinuum generation, filamentation, rogue waves, and necklace beam generation [157–164]. The necklace beams, first studied by Soljačić and Segev [165–167], are ring-shaped beams with intensities that are azimuthally periodically modulated [168–172]. In these studies, it was theoretically predicted and experimentally demonstrated that necklace beams with and without the OAM launched in self-focusing Kerr medium maintain their shape, in spite of the inherent instability in such a medium, and that their diameters increase during propagation at a predictable rate. In another series of studies, self-focusing of optical vortices in Kerr media was investigated and the azimuthal MI for ring-shaped vortices was shown to result in the beam breakup into a ring of filaments [173, 174]. Importantly, a majority of these studies focused on light–matter interactions in self-focusing Kerr medium. Recently, the optical nonlinearity of colloidal suspensions, despite the initial belief that it was of the Kerr type, was found to be exponential [64]. Moreover, this exponential nonlinearity can be saturable or supercritical with intensity, depending on the sign of the particle polarizability in the suspension.

To understand the reason behind these different types of nonlinearity, we note that colloidal suspensions can be created by two kinds of particles: with PPs corresponding to a particle refractive index greater than the background index $n_p > n_b$ or with NP, with the corresponding refractive index of a particle lower than that of the background material $n_p < n_b$ [64]. PP dielectric particles are attracted toward the high-intensity regions of the beam, whereas NP particles are repelled. At the same time, in the PP case, the nonlinear scattering losses increase in the high-intensity region of the beam. In contrast, in the NP case, the nonlinear losses decrease with the increase of intensity since the particle concentration decreases.

Let us consider an optical beam propagating in the nanocolloidal system. The nonlinear Schrödinger equation governing the evolution of the slowly varying electric-field envelope can be written as [62, 63]

$$i\frac{\partial \varphi}{\partial Z} + \frac{1}{2k_0 n_b}\nabla_{\perp}^2 \varphi + k_0(n_p - n_b)V_p \rho_0 e^{\frac{\alpha}{4k_B T}|\varphi|^2}\varphi + \frac{i}{2}\sigma \rho_0 e^{\frac{\alpha}{4k_B T}|\varphi|^2}\varphi = 0 \tag{6.4}$$

where the particle polarizability is denoted by α and $k_B T$ is the thermal energy. V_p is the volume of a particle, ρ_0 is the unperturbed particle concentration, σ is the scattering cross-section, and $k_0 = 2\pi/\lambda_0$ is the wave number. Following the standard linear stability analysis, we assume that the high-intensity optical beam with a topological charge $\ell \in Z$ is accompanied by an azimuthal perturbation

$$E(\theta, z) = \left(|E_0| + a_1 e^{-i(M\theta + \mu z)} + a_2^* e^{i(M\theta + \mu^* z)}\right)e^{i\ell\theta + i\lambda z} \tag{6.5}$$

where $E_0 = E(r = r_\ell, z = 0)$ is the electric-field amplitude of the rotationally invariant steady-state solution with charge m, taken at the average radius r_ℓ, a_1 and a_2 are the amplitudes of the small perturbations, and $M \in Z$ is the deviation from ℓ of the perturbation charge. The topological charge of the perturbation is given by $\ell \pm M$, λ is the propagation constant of the steady-state solution, and μ is the propagation constant correction for the perturbation. The linear stability analysis allows us to calculate the MI gain for the perturbations with the charge $\ell \pm M$ imposed atop the main beam with the charge m and the corresponding averaged radius rm. The gain is given by

$$\text{Im}(\mu) = g_\ell(M) = \frac{M}{2k_0 n_b r_\ell} \times \text{Im}\sqrt{\frac{M^2}{r_\ell^2} - \frac{\alpha}{2k_B T L^2}|E_0|^2 \exp\left(\frac{\alpha}{2k_B T}\right)|E_0|^2} \tag{6.6}$$

where $L^2 = \left(2k_0^2 n_b |n_p - n_b| V_p \rho_0\right)^{-1}$. Here, the particle polarizability is denoted by α, and $k_B T$ is the thermal energy, with the Boltzmann constant k_B and at temperature T. V_p is the volume of a particle, ρ_0 is the unperturbed particle concentration, $k_0 = 2\pi/\lambda_0$ is the wave number, and λ_0 is the free-space wavelength.

Equation (6.6) is used in the following to predict the MI gain for vortices propagating in the nanocolloidal media [175–177]. Experimental setup used to study the propagation of light with the free-space wavelength $\lambda_0 = 532$ nm and power of 6 W, emitted by continuous wave Coherent Verdi 6 laser, followed by a spiral phase plate for OAM conversion, and then focused inside a 10-mm long cuvette filled with the negative-polarizability suspension made of low refractive index 150 nm polytetrafluoroethylene particles ($n_p = 1.35$) dispersed in glycerin water ($n_b = 1.44$) with the volume filling fraction of $\rho_0 = 0.7\%$ is shown in Figure 6.18a.

Figure 6.18b shows intensity profiles, interference patterns, and intensity distributions of the necklace beams formed as a result of MI after the propagation of the OAM beams with topological charges of $\ell = 1$, 2, and 4 in the colloidal medium described earlier. For this set of the parameters, Figure 6.18c shows the gain curves $g_m(M)$ as a function of the perturbation azimuthal index deviation M for different values of the OAM-beam charge m. The analytical predictions are only valid when the perturbation intensity is significantly lower than that of the main beam. Above this limit the dynamics of MI has to be studied using numerical simulations of a three-dimensional NLSE.

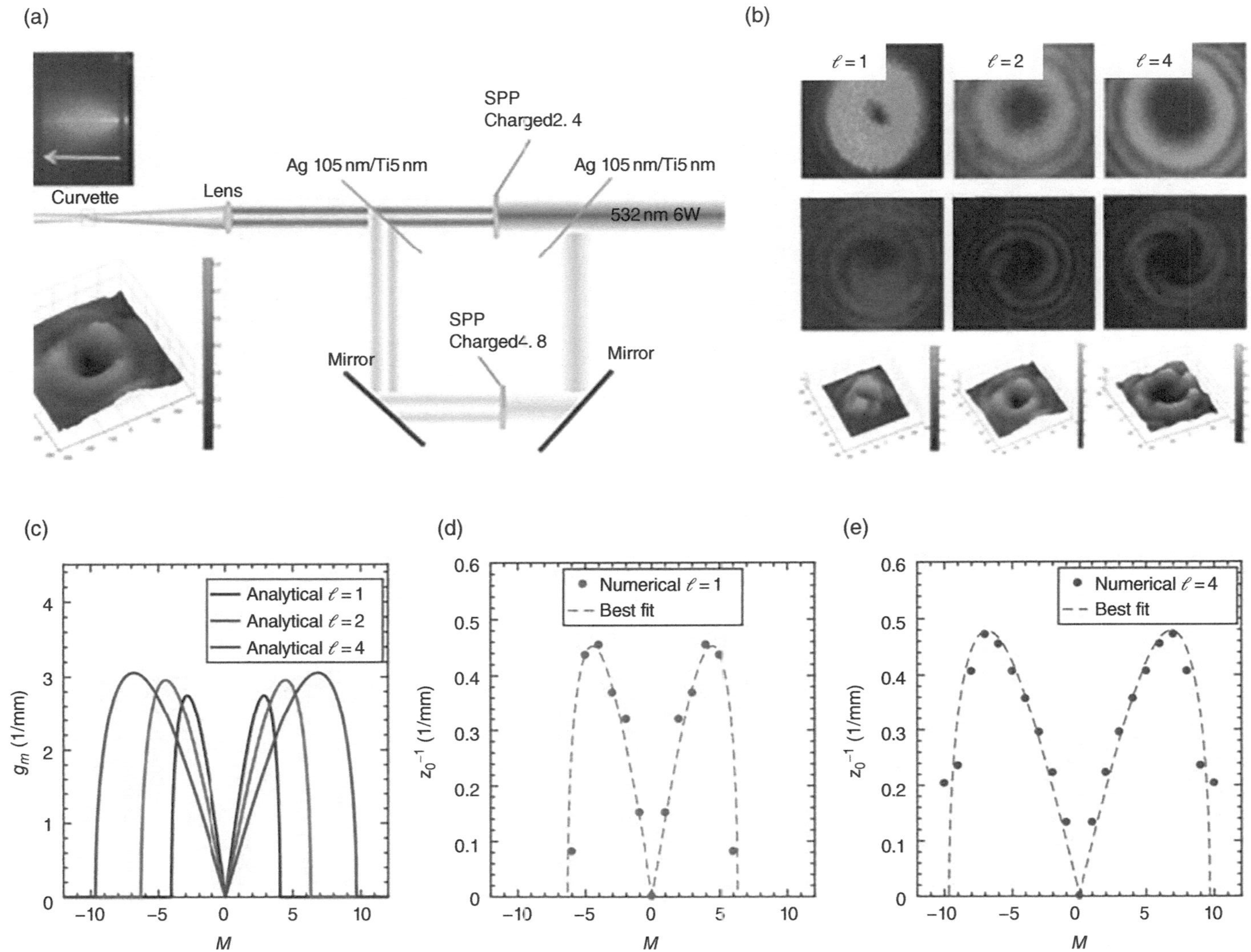

Figure 6.18 Experimental setup (a) and results showing the formation of the necklace beam from an initial vortex beam propagating in a nonlinear colloidal suspension with negative polarizability particles; (b) intensity profiles, interference patterns, and intensity distributions of the resulting necklace beams after the propagation in the colloidal medium; (c) analytically computed instability gain $g_m(M)$ as a function of the perturbation azimuthal index deviation M, for negative polarizability particle-based systems for different topological charges m of the initial steady-state vortex solution; (d), (e) inverse of the beam breakup distance recorded in numerical simulations of seeded MI. Dashed lines show analytical curves with rescaled magnitude [177]. *Source*: Sun et al. [177].

6.5 Conclusion

In this chapter, we discussed fundamental science and potential applications of light–matter interactions at the interface of singular optics and engineered optical media and show that the unique optical properties of optical nanostructures open unlimited prospects to 'engineer' light itself. We discussed recent developments in the field of metal–dielectric and all-dielectric and linear and nonlinear metasurafces that enable new degrees of freedom for manipulation or structured light on the nanoscale using ultracompact, low-loss, fast, and reconfigurable optical components.

Next, we considered active OAM light sources enabled by, for example, nonhermitian photonics design at an EP. In particular, we discussed a microring laser generating a single-mode OAM vortex lasing with the ability to precisely define the topological charge of the OAM mode. Moreover, by harnessing the properties of total momentum conservation, spin–orbit interaction, and optical nonhermitian symmetry breaking, an OAM-tunable vortex microlaser, providing chiral light states of variable integer or fractional topological charges at a single telecommunication wavelength, was demonstrated. These studies may provide a route for the development of the next generation of multidimensional OAM-spin-wavelength division multiplexing technology.

Finally, we discussed recent progress in linear and nonlinear light–matter interactions in engineered colloidal suspensions consisting of NP and saturable nonlinearities. These results are likely to enable a new platform for fundamental studies of nonlinear optical phenomena in engineered media as well as for imaging and light manipulation in scattering media, such as biological and chemical systems.

The synergy of structured light and structured materials has already had a significant impact on the broad variety of studies ranging from fundamentally new regimes of spin–orbit interaction to novel ways of information encoding for the future optical communication systems and has opened unlimited opportunities for the future discoveries.

References

1 Padgett, M.J. (2017). Orbital angular momentum 25 years on. *Optics Express* **25**: 11265–11274.

2 Padgett, M.J. (2014). Light's twist. *Proceedings of The Royal Society A Mathematical Physical and Engineering Sciences* **470**: 20140633.

3 Yao, M. and Padgett, M.J. (2011). Orbital angular momentum: origins, behavior and applications. *Advances in Optics and Photonics* **3**: 161–204.

4 Litchinitser, N.M. (2012). Structured light meets structured matter. *Science* **337**: 1054–1055.

5 Allen, L., Beijersbergen, M.W., Spreeuw, R.J.C., and Woerdman, J.P. (1992). Orbital angular momentum of light and the transformation of Laguerre-Gaussian laser modes. *Physical Review A* **45**: 8185–8189.

6 Carpentier, A.V., Michinel, H., Salgueiro, J.R., and Olivieri, D. (2008). Making optical vortices with computer-generated holograms. *American Journal of Physics* **76**: 916.

7 Babiker, M., Power, W.L., and Allen, L. (1994). Light-induced torque on moving atoms. *Physical Review Letters* **73**: 1239–1242.

8 Friese, M.E.J., Nieminen, T.A., Heckenberg, N.R., and Rubinsztein-Dunlop, H. (1998). Optical alignment and spinning of laser-trapped microscopic particles. *Nature* **394**: 348–350.

9 Furhapter, S., Jesacher, A., Bernet, S., and Ritsch-Marte, M. (2005). Spiral phase contrast imaging in microscopy. *Optics Express* **13**: 689–694.

10 Xu, Y., Sun, J., Walasik, W., and Litchinitser, N.M. (2016). Probing metamaterials with structured light. *Optics Express* **24**: 26249–26254.

11 Tamburini, F., Mari, E., Sponselli, A. et al. (2012). Encoding many channels on the same frequency through radio vorticity: first experimental test. *New Journal of Physics* **14**: 033001.

12 Wang, J., Yang, J.-Y., Fazal, I.M. et al. (2012). Terabit free-space data transmission employing orbital angular momentum multiplexing. *Nature Photonics* **6**: 488–496.

13 Willner, A.E., Huang, H., Yan, Y. et al. (2015). Optical communications using orbital angular momentum beams. *Advances in Optics and Photonics* **7** (66): 66–106.

14 Bozinovic, N., Yue, Y., Ren, Y. et al. (2013). Terabit-scale orbital angular momentum mode division multiplexing in fibers. *Science* **340**: 1545.

15 Mair, A., Vaziri, A., Welhs, G., and Zeilinger, A. (2001). Entanglement of the orbital angular momentum states of photons. *Nature* **412**: 313–316.

16 Carlsson, A.H., Malmberg, J.N., Anderson, D. et al. (2000). Linear and nonlinear waveguides induced by optical vortex solitons. *Optics Letters* **25**: 660–662.

17 Salgueiro, J.R., Carlsson, A.H., Ostrovskaya, E., and Kivshar, Y. (2004). Second harmonic generation in vortex-induced waveguides. *Optics Letters* **29**: 593–595.

18 Foo, G., Palacios, D.M., and Swartzlander, G.A. Jr. (2005). Optical vortex coronagraph. *Optics Letters* **30**: 3308–3310.

19 Matthews, M.R., Anderson, B.P., Haljan, P.C. et al. (1999). Vortices in a Bose-Einstein condensate. *Physical Review Letters* **83**: 2498–2501.

20 Yu, N., Genevet, P., Kats, M.A. et al. (2011). Light propagation with phase discontinuities: generalized laws of reflection and refraction. *Science* **334** (6054): 333–337.

21 Tamburini, F., Mari, E., Thidé, B. et al. (2011). Experimental verification of photon angular momentum and vorticity with radio techniques. *Applied Physics Letters* **99** (20): 204102.

22 Maccalli, S., Pisano, G., Colafrancesco, S. et al. (2013). Q-plate for millimeter-wave orbital angular momentum manipulation. *Applied Optics* **52** (4): 635–639.

23 Jesacher, A., Maurer, C., Schwaighofer, A. et al. (2008). Near-perfect hologram reconstruction with a spatial light modulator. *Optics Express* **16** (4): 2597–2603.

24 Zhang, H., Mao, B., Han, Y. et al. (2019). Generation of orbital angular momentum modes using fiber systems. *Applied Science* **9** (5): 1033.

25 https://www.st-andrews.ac.uk/~www_pa/Scots_Guide/MMWave/QO/components/spir.html.

26 Marrucci, L. (2013). The q-plate and its future. *Journal of Nanophotonics* **7** (1): 078598.

27 http://holoeye.com/spatial-light-modulator.

28 Dudley, A., Lavery, M., Padgett, M., and Forbes, A. (2013). Unraveling Bessel beams. *Optics & Photonics News* **24**: 22–29.

29 Yu, N. and Capasso, F. (2014). Flat optics with designer metasurfaces. *Nature Materials* **13**: 139–150.

30 Yu, N., Genevet, P., Kats, M.A. et al. (2011). Light propagation with phase discontinuities: generalized laws of reflection and refraction. *Science* **334**: 333–337.

31 Kim, S.W., Yee, K.J., Abashin, M. et al. (2015). Composite dielectric metasurfaces for phase control of vector field. *Optics Letters* **40**: 2453–2456.

32 Minovich, A.E., Miroshnichenko, A.E., Bykov, A.Y. et al. (2015). Functional and nonlinear optical metasurfaces. *Laser & Photonics Reviews* **9**: 195–213.

33 Wu, C.H., Arju, N., Kelp, G. et al. (2014). Spectrally selective chiral silicon metasurfaces based on infrared Fano resonances. *Nature Communications* **5**: 1–9.

34 Ni, X.J., Kildishev, A.V., and Shalaev, V.M. (2013). Metasurface holograms for visible light. *Nature Communications* **4**: 1–6.

35 Buchnev, O., Podoliak, N., Kaczmarek, M. et al. (2015). Electrically controlled nanostructured metasurface loaded with liquid crystal: toward multifunctional photonic switch. *Advanced Optical Materials* **3**: 674–679.

36 Sautter, J., Staude, I., Decker, M. et al. (2015). Active tuning of all-dielectric metasurfaces. *ACS Nano* **9**: 4308–4315.

37 Pfeiffer, C., Emani, N.K., Shaltout, A.M. et al. (2014). Efficient light bending with isotropic metamaterial Huygens' surfaces. *Nano Letters* **14**: 2491–2497.

38 Holloway, C.L., Kuester, E.F., Gordon, J.A. et al. (2012). An overview of the theory and applications of metasurfaces: the two-dimensional equivalents of metamaterials. *IEEE Antennas and Propagation Magazine* **54**: 10–35.

39 Lin, D., Fan, P., Hasman, E., and Brongersma, M.L. (2014). Dielectric gradient metasurface optical elements. *Science* **345**: 298–302.

40 Yin, X.B., Ye, Z.L., Rho, J. et al. (2013). Photonic spin Hall effect at metasurfaces. *Science* **339**: 1405–1407.

41 Karimi, E., Schulz, S.A., De Leon, I. et al. (2014). Generating optical orbital angular momentum at visible wavelengths using a plasmonic metasurface. *Light: Science & Applications* **3**: e167.

42 Smith, D.R., Pendry, J.B., and Wiltshire, M.C.K. (2004). Metamaterials and negative refractive index. *Science* **305** (5685): 788–792.

43 Shalaev, V.M. (2007). Optical negative-index metamaterials. *Nature Photonics* **1** (1): 41–48.

44 Litchinitser, N.M., Gabitov, I.R., Maimistov, A.I., and Shalaev, V.M. (2008). Negative refractive index metamaterials in optics. *Progress in Optics* **51**: 51, 1–67.

45 Niv, A., Biener, G., Kleiner, V., and Hasman, E. (2005). Spiral phase elements obtained by use of discrete space-variant subwavelength gratings. *Optics Communications* **251**: 306–314.

46 Shalaev, M.I., Sun, J., Tsukernik, A. et al. (2015). High-efficiency all-dielectric metasurfaces for ultracompact beam manipulation in transmission mode. *Nano Letters* **15**: 96261–96266.

47 Zhao, Z., Wang, J., Li, S., and Willner, A.E. (2013). Metamaterials-based broadband generation of orbital angular momentum carrying vector beams. *Optics Letters* **38**: 932–934.

48 Karimi, E., Schulz, S.A., De Leon, I. et al. (2014). Generating optical orbital angular momentum at visible wavelengths using a plasmonic metasurface. *Light: Science & Applications* **3** (e167).

49 Chong, K.E., Staude, I., James, A. et al. (2015). Polarization-independent silicon metadevices for efficient optical wavefront control. *Nano Letters* **15** (8): 5369–5374.

50 Kadic, M., Milton, G.W., van Hecke, M., and Wegener, M. (2019). 3d metamaterials. *Nature Reviews Physics* **1** (3): 198–210.

51 Buriak, I.A., Zhurba, V.O., Vorobjov, G.S. et al. (2016). Metamaterials: theory, classication and application strategies (review). *Journal of Nano- and Electronic Physics* **8** (4): 04088.

52 Kivshar, Y. (2018). All-dielectric meta-optics and non-linear nanophotonics. *National Science Review* **5** (2): 144–158.

53 Cui, T.J. (2018). Microwave metamaterials. *National Science Review* **5** (2): 134–136.

54 Ziolkowski, R.W., Jin, P., and Lin, C.C. (2011). Metamaterial-inspired engineering of antennas. *Proceedings of IEEE* **99** (10): 1720–1731.

55 McPhedran, R.C., Shadrivov, I.V., Kuhlmey, B.T., and Kivshar, Y.S. (2011). Metamaterials and metaoptics. *NPG Asia Materials* **3**: 100–108.

56 Kadic, M., Buckmann, T., Schittny, R., and Wegener, M. (2013). Metamaterials beyond electromagnetism. *Reports on Progress in Physics* **76** (12): 126501.

57 https://investingnews.com/daily/tech-investing/emerging-tech-investing/metamaterials-automotive-innovation/.

58 https://phys.org/news/2011-01-metamaterials-approach-satellite-antennas.html.

59 https://www.usni.org/magazines/proceedings/2019/october/win-race-invisibility.
60 Liu, M., Fan, K., Padilla, W. et al. (2016). Meta-liquid crystals: tunable meta-liquid crystals. *Advanced Materials* **28**: 1525.
61 Kudyshev, Z.A., Richardson, M.C., and Litchinitser, N.M. (2013). Virtual hyperbolic metamaterials for manipulating radar signals in air. *Nature Communications* **4**: 2557.
62 El-Ganainy, R., Christodoulides, D.N., Musslimani, Z.H. et al. (2007). Optical beam instabilities in nonlinear nanosuspensions. *Optics Letters* **32**: 3185–3187.
63 El-Ganainy, R., Christodoulides, D.N., Rotschild, C., and Segev, M. (2007). Soliton dynamics and self-induced transparency in nonlinear nanosuspensions. *Optics Express* **15**: 10207–10218.
64 Man, W., Fardad, S., Zhang, Z. et al. (2013). Optical nonlinearities and enhanced light transmission in soft-matter systems with tunable polarizabilities. *Physical Review Letters* **111**: 218302.
65 Silahli, S.Z., Walasik, W., and Litchinitser, N.M. (2015). Necklace beam generation in nonlinear colloidal engineered media. *Optics Letters* **40**: 5714–5717.
66 Kelly, T.S., Ren, Y.-X., Samadi, A. et al. (2016). Guiding and nonlinear coupling of light in plasmonic nanosuspensions. *Optics Letters* **41**: 3817–3820.
67 Padgett, M., Courtial, J., and Allen, L. (2004). Light's orbital angular momentum. *Physics Today* **57** (5): 35.
68 Ritsch-Marte, M. (2017). Orbital angular momentum light in microscopy. *Philosophical Transactions of the Royal Society A* **375** (2087): 20150437.
69 Dholakia, K. and Cizmar, T. (2011). Shaping the future of manipulation. *Nature Photonics* **5**: 335–342.
70 Vallone, G., D'Ambrosio, V., Sponselli, A. et al. (2014). Free-space quantum key distribution by rotation-invariant twisted photons. *Physical Review Letters* **113**: 060503.
71 Sit, A., Bouchard, F., Fickler, R. et al. (2017). High-dimensional intracity quantum cryptography with structured photons. *Optica* **4**: 1006–1010.
72 Huang, F.M., Kao, T.S., Fedotov, V.A. et al. (2008). Nanohole array as a lens. *Nano Letters* **8**: 2469–2472.
73 Rogers, E.T.F., Lindberg, J., Roy, T. et al. (2012). A super-oscillatory lens optical microscope for subwavelength imaging. *Nature Materials* **11**: 432–435.
74 Verslegers, L., Catrysse, P.B., Yu, Z., and Fan, S. (2009). Planar metallic nanoscale slit lenses for angle compensation. *Applied Physics Letters* **95**: 071112.
75 Lin, L., Goh, X.M., McGuinness, L.P., and Roberts, A. (2010). Plasmonic lenses formed by two-dimensional nanometric cross-shaped aperture arrays for Fresnel-region focusing. *Nano Letters* **10**: 1936–1940.
76 Gao, H., Hyun, J.K., Lee, M.H. et al. (2010). Broadband plasmonic microlenses based on patches of nanoholes. *Nano Letters* **10**: 4111–4116.
77 Ishii, S., Kildishev, A.V., Shalaev, V.M. et al. (2011). Metal nanoslit lenses with polarization-selective design. *Optics Letters* **36**: 451–453.
78 Ni, X., Ishii, S., Kildishev, A.V., and Shalaev, V.M. (2013). Ultra-thin, planar. *Babinet-inverted plasmonic metalenses. Light: Science & Applications* **2**: e72.
79 Ishii, S., Shalaev, V.M., and Kildishev, A.V. (2013). Holey-metal lenses: sieving single modes with proper phases. *Nano Letters* **13**: 159–163.
80 Dong, J., Liu, J., Hu, B. et al. (2014). Subwavelength light focusing with a single slit lens based on the spatial multiplexing of chirped surface gratings. *Applied Physics Letters* **104**: 011115.
81 Junesch, J. and Sannomiya, T. (2014). Reflection phase and amplitude determination of short-range ordered plasmonic nanohole arrays. *The Journal of Physical Chemistry Letters* **5**: 247–252.
82 Cai, X., Wang, J., Strain, M.J. et al. (2012). Integrated compact optical vortex beam emitters. *Science* **338**: 363–366.

83 Yan, Z.J. and Scherer, N.F. (2013). Optical vortex induced rotation of silver nanowires. *The Journal of Physical Chemistry Letters* **4**: 2937–2942.

84 Toyoda, K., Miyamoto, K., Aoki, N. et al. (2012). Using optical vortex to control the chirality of twisted metal nanostructures. *Nano Letters* **12** (7): 3645–3649.

85 Wang, J. (2016). Advances in communications using optical vortices. *Photonics Research* **4**: B14–B28.

86 Ren, Y., Li, L., Wang, Z. et al. (2016). Orbital angular momentum-based space division multiplexing for high-capacity underwater optical communications. *Scientific Reports* **6** (1): 33306.

87 Bomzon, Z., Biener, G., Kleiner, V., and Hasman, E. (2002). Space-variant Pancharatnam–Berry phase optical elements with computer-generated subwavelength gratings. *Optics Letters* **27**: 1141–1143.

88 Maguid, E., Yulevich, I., Veksler, D. et al. (2016). Photonic spin-controlled multifunctional shared-aperture antenna array. *Science* **352**: 1202.

89 https://www.techbriefs.com/component/content/article/tb/supplements/pit/briefs/33901.

90 Karvounis, A., Gholipour, B., MacDonald, K.F., and Zheludev, N.I. (2016). All-dielectric phase-change reconfigurable metasurface. *Applied Physics Letters* **109**: 051103.

91 Shcherbakov, M.R., Vabishchevich, P.P., Shorokhov, A.S. et al. (2015). Ultrafast all-optical switching with magnetic resonances in nonlinear dielectric nanostructures. *Nano Letters* **15**: 6985–6990.

92 Xu, Y., Sun, J., Frantz, J. et al. (2018). Reconfiguring structured light beams using nonlinear metasurfaces. *Optics Express* **26**: 30930–30943.

93 Chu, C.H., Tseng, M.L., Chen, J. et al. (2016). Active dielectric metasurface based on phase-change medium. *Laser & Photonics Reviews* **10**: 986–994.

94 Li, G., Zhang, S., and Zentgraf, T. (2017). Nonlinear photonic metasurfaces. *Nat. Rev. Mat.* **2**: 17010.

95 Ou, J.Y., Plum, E., Jiang, L., and Zheludev, N.I. (2011). Reconfigurable photonic metamaterials. *Nano Letters* **11**: 2142–2144.

96 Kamali, S.M., Arbabi, E., Arbabi, A. et al. (2016). Highly tunable elastic dielectric metasurface lenses. *Laser & Photonics Reviews* **10**: 1002–1008.

97 Su, Z., Zhao, Q., Song, K. et al. (2017). Electrically tunable metasurface based on Mie-type dielectric resonators. *Scientific Reports* **7**: 1–7.

98 Rahmani, M., Xu, L., Miroshnichenko, A.E. et al. (2017). Reversible thermal tuning of all-dielectric metasurfaces. *Advanced Functional Materials* **27**: 1700580.

99 Wang, Q., Rogers, E.T.F., Gholipour, B. et al. (2016). Optically reconfigurable metasurfaces and photonic devices based on phase change materials. *Nature Photonics* **10**: 60–75.

100 Komar, A., Fang, Z., Bohn, J. et al. (2017). Electrically tunable all-dielectric optical metasurfaces based on liquid crystals. *Applied Physics Letters* **110**: 071109.

101 Boyd, R.W. (2008). *Nonlinear Optics*, 3e. Academic.

102 Krasnok, A., Tymchenko, M., and Alu, A. (2018). Nonlinear metasurfaces: a paradigm shift in nonlinear optics. *Materials Today* **21**: 8–21.

103 Sautter, J., Staude, I., Decker, M. et al. (2015). Active tuning of all-dielectric metasurfaces. *ACS Nano* **9**: 4308–4315.

104 Shadrivov, I.V., Kapitanova, P.V., Maslovski, S.I., and Kivshar, Y.S. (2012). Metamaterials controlled with light. *Physical Review Letters* **109**: 083902.

105 Pandey, A. and Litchinitser, N.M. (2012). Nonlinear light concentrators. *Optics Letters* **37**: 5238–5240.

106 Shcherbakov, M.R., Liu, S., Zubyuk, V.V. et al. (2017). Ultrafast all-optical tuning of direct-gap semiconductor metasurfaces. *Nature Communications* **8**: 17.

107 Liu, S., Sinclair, M.B., Saravi, S. et al. (2016). Resonantly enhanced second-harmonic generation using III–V semiconductor all-dielectric metasurfaces. *Nano Letters* **16**: 5426–5432.

108 Lee, J., Tymchenko, M., Argyropoulos, C. et al. (2014). Giant nonlinear response from plasmonic metasurfaces coupled to intersubband transitions. *Nature* **511**: 65–69.

109 Sanghera, J.S., Shaw, L.B., Florea, C.M. et al. (2010). Nonlinear properties of chalcogenide glass fibers. *International Journal of Applied Glass Science* **1**: 296–308.

110 Eggleton, B.J., Luther-Davies, B., and Richardson, K. (2011). Chalcogenide photonics. *Nature Photonics* **5**: 141–148.

111 Hilton, A.R. (2010). *Chalcogenide Glasses for Infra-red Optics*. McGraw Hill.

112 Vlcek, M. and Jain, H. (2006). Nanostructuring of chalcogenide glasses using electron beam lithography. *J. Optoelectron. Advanced Materials* **8**: 2108–2111.

113 Fan, S. and Joannopoulos, J. (2002). Analysis of guided resonances in photonic crystal slabs. *Physical Review B* **65**: 235112.

114 Paddon, P. and Young, J. (2000). Two-dimensional vector-coupled-mode theory for textured planar waveguides. *Physical Review B* **61**: 2090–2101.

115 Cai, X., Wang, J., Strain, M.J. et al. (2012). Integrated compact optical vortex beam emitters. *Science* **338** (6105): 363–366.

116 Naidoo, D., Roux, F.S., Dudley, A. et al. (2016). Controlled generation of higher-order Poincaré sphere beams from a laser. *Nature Photonics* **10**: 327–332.

117 Miao, P., Zhang, Z., Sun, J. et al. (2016). Orbital angular momentum microlaser. *Science* **353** (6298): 464–467.

118 Zhang, Z., Qiao, X., Midya, B. et al. (2020). Tunable topological charge vortex microlaser. *Science* **368** (6492): 760–763.

119 Zhang, J., Sun, C., Xiong, B. et al. (2018). An InP-based vortex beam emitter with monolithically integrated laser. *Nature Communications* **9**: 2652.

120 Bahari, B., Hsu, L.-Y., Pan, S.H. et al. Topological lasers generating and multiplexing topological light. *arXiv* **1904**: 11873.

121 Zambon, N.C., St-Jean, P., Milićević, M. et al. (2019). Optically controlling the emission chirality of microlasers. *Nature Photonics* **13**: 283–288.

122 Sun, W., Liu, Y., Qu, G. et al. (2020). Lead halide perovskite vortex microlasers. *Nature Communications* **11**: 4862.

123 Zhang, Z., Zhao, H., Gomes Pires, D. et al. (2020). Ultrafast control of fractional orbital angular momentum of microlaser emissions. *Light: Science & Applications* **9**: 179.

124 Zhang, Z., Miao, P., Sun, J. et al. (2018). Elimination of spatial hole burning in microlasers for stability and efficiency enhancement. *ACS Photonics* **5** (8): 3016–3022.

125 Siegman, A.E. (1986). *Lasers*. Melville, NY, USA: University Science Books.

126 Tang, C.L. and Statz, H. (1963). Spectral output and spiking behavior of solid state lasers. *Journal of Applied Physics* **34**: 2289.

127 Sorel, M., Laybourn, P.J.R., Scirè, A. et al. (2002). Alternate oscillations in semiconductor ring lasers. *Optics Letters* **27**: 1992–1994.

128 Berry, M.V. (2004). Physics of nonhermitian degeneracies. *Czechoslovak Journal of Physics* **54**: 1039–1047.

129 Heiss, W.D. (2012). The physics of exceptional points. *Journal of Physics A: Mathematical and Theoretical* **45**: 444016.

130 Wiersig, J., Eberspächer, A., Shim, J.B. et al. (2011). Nonorthogonal pairs of copropagating optical modes in deformed microdisk cavities. *Physical Review A* **84**: 023845.

131 Makris, K.G., El-Ganainy, R., Christodoulides, D.N., and Musslimani, Z.H. (2008). Beam dynamics in PT symmetric optical lattices. *Physical Review Letters* **100**: 103904.

132 Rüter, C.E., Makris, K.G., El-Ganainy, R. et al. (2010). Observation of parity-time symmetry in optics. *Nature Physics* **6**: 192–195.

133 Schomerus, H. (2010). Quantum noise and self-sustained radiation of PT-symmetric systems. *Physical Review Letters* **104**: 233601.

134 Doppler, J., Mailybaev, A.A., Böhm, J. et al. (2016). Dynamically encircling an exceptional point for asymmetric mode switching. *Nature* **537**: 76–79.

135 Jackson, J.D. (1962). *Classical Electrodynamics*. New York: Wiley.

136 Longhi, S. and Feng, L. (2017). Unidirectional lasing in semiconductor microring lasers at an exceptional point. *Photonics Research* **5**: B1–B6.

137 Longhi, S. and Feng, L. (2018). Mitigation of dynamical instabilities in laser arrays via non-Hermitian coupling. *APL Photonics* **3**: 060802.

138 Longhi, S. (2018). Non-hermitian gauged topological laser arrays. *Annals of Physics* **530**: 1800023.

139 Longhi, S., Gatti, D., and Della Valle, G. (2015). Robust light transport in non-hermitian photonic lattices. *Scientific Reports* **5**: 13376.

140 Urzhumov, Y.A., Shvets, G., Fan, J.A. et al. (2007). Plasmonic nanoclusters: a path towards negative-index metafluids. *Optics Express* **15**: 14129.

141 Zharov, A.A., Zharov, A.A., and Zharova, N.A. (2014). Liquid metacrystals. *Journal of the Optical Society of America B* **31**: 559.

142 Fruhnert, M., Mühlig, S., Lederer, F., and Rockstuhl, C. (2014). Towards negative index self-assembled metamaterials. *Physical Review B* **89**: 075408.

143 Sheikholeslami, S.N., Alaeian, H., Koh, A.L., and Dionne, J.A. (2013). A metafluid exhibiting strong optical magnetism. *Nano Letters* **13**: 4137.

144 Liu, M., Fan, K., Padilla, W. et al. (2016). Tunable meta-liquid crystals. *Advanced Materials* **28** (8): 1553–1558.

145 Park, H., Parrott, E.P.J., Fan, F. et al. (2012). Evaluating liquid crystal properties for use in terahertz devices. *Optics Express* **20**: 11899.

146 Knöner, G., Parkin, S., Nieminen, T.A. et al. (2007). Integrated optomechanical microelements. *Optics Express* **15**: 5521–5530.

147 Aharoni, A., Griffiths, A.D., and Tawfik, D.S. (2005). High-throughput screens and selections of enzyme-encoding genes. *Current Opinion in Chemical Biology* **9**: 210–216.

148 Gordon, J.P. (1973). Radiation forces and momenta in dielectric media. *Physical Review A* **8**: 14–21.

149 Ashkin, A. (1970). Acceleration and trapping of particles by radiation pressure. *Physical Review Letters* **24**: 156–159.

150 Zabusky, N.J. and Kruskal, M.D. (1965). Interaction of "solitons" in a collisionless plasma and the recurrence of initial states. *Physical Review Letters* **15**: 240–243.

151 Man, W., Fardad, S., Zhang, Z. et al. (2013). Optical nonlinearities and enhanced light transmission in soft-matter systems with tunable polarizabilities. *Physical Review Letters* **111** (21): 1–5.

152 Askaryan, G.A. (1962). Effects of the gradient of a strong electromagnetic beam on electrons and atoms. *Soviet Physics Uspekhi* **15** (6): 1088–1090.

153 Bespalov, V.I. and Talanov, V.I. (1966). Filamentary structure of light beams in nonlinear liquids. *ZhETF Pisma Redaktsiiu* **3**: 471.

154 Chiao, R.Y., Garmire, E., and Townes, C.H. (1964). Self-trapping of optical beams. *Physical Review Letters* **13**: 479–482.

155 Campillo, A.J., Shapiro, S.L., and Suydam, B.R. (1973). Periodic breakup of optical beams due to selffocusing. *Applied Physics Letters* **23** (11): 628–630.

156 Boyd, R.W., Lukishova, S.G., and Shen, Y.R. (2008). *Self-Focusing: Past and Present: Fundamentals and Prospects*, vol. **114**. Berlin, Germany: Springer Science & Business Media 19, 49.

157 Agrawal, G.P. (2007). *Nonlinear Fiber Optics*. Cambridge, MA, USA: Academic Press.

158 Dudley, J.M., Dias, F., Erkintalo, M., and Genty, G. (2014). Instabilities, breathers and rogue waves in optics. *Nature Photonics* **8**: 755–764.

159 Vinçotte, A. and Bergé, L. (2006). Atmospheric propagation of gradient-shaped and spinning femtosecond light pulses. *Physica D: Nonlinear Phenomena* **223** (2): 163–173.

160 Sukhinin, A. and Aceves, A.B. (2013). Optical UV filament and vortex dynamics. *Journal of Optics* **15** (4): 044010.

161 Haus, J.W., Mozumder, Z., and Zhan, Q. (2006). Azimuthal modulation instability for a cylindrically polarized wave in a nonlinear Kerr medium. *Optics Express* **14** (11): 4757–4764.

162 Desyatnikov, A.S., Kivshar, Y.S., and Torner, L. (2005). Optical vortices and vortex solitons. In: *Progress in Optics*, vol. **47** (ed. E. Wolf). North-Holland: Elsevier.

163 Kartashov, Y.V., Aleshkevich, V.A., Vysloukh, V.A. et al. (2003). Transverse modulational instability of (2 + 1)-dimensional cnoidal waves in media with cubic nonlinearity. *Journal of the Optical Society of America B* **20**: 1273–1284.

164 Litchinitser, N.M., McKinstrie, C.J., Martijn de Sterke, C., and Agrawal, G.P. (2001). Spatiotemporal instabilities in nonlinear bulk media with Bragg gratings. *Journal of the Optical Society of America B* **18** (1): 45–54.

165 Soljačić, M. and Segev, M. (1998). Self-trapping of necklace beams in self-focusing Kerr. *Physical Review E* **62** (2): 2810–2820.

166 Soljačić, M. and Segev, M. (2000). Self-trapping of necklace-ring beams in self-focusing Kerr media. *Physical Review E* **62** (2): 2810–2820.

167 Soljačić, M. and Segev, M. (2001). Integer and fractional angular momentum borne on self-trapped necklace-ring beams. *Physical Review Letters* **86** (3): 420–423.

168 Greenfield, E., Nemirovsky, J., El-Ganainy, R. et al. (2013). Shockwave based nonlinear optical manipulation in densely scattering opaque suspensions. *Optics Express* **21** (20): 23785–23802.

169 Theocharis, G., Frantzeskakis, D.J., Kevrekidis, P.G. et al. (2003). Ring dark solitons and vortex necklaces in Bose-Einstein condensates. *Physical Review Letters* **90** (12): 120403.

170 Yang, J., Makasyuk, I., Kevrekidis, P.G. et al. (2005). Necklacelike solitons in optically induced photonic lattices. *Physical Review Letters* **94** (11): 2–5.

171 He, Y.J., Malomed, B.A., and Wang, H.Z. (2007). Fusion of necklace-ring patterns into vortex and fundamental solitons in dissipative media. *Optics Express* **15** (26): 17502–17508.

172 Kartashov, Y.V., Malomed, B.A., Vysloukh, V.A., and Torner, L. (2009). Stabilization of multibeam necklace solitons in circular arrays with spatially modulated nonlinearity. *Physical Review A* **80** (5): 1–4.

173 Grow, T.D., Ishaaya, A.A., Vuong, L.T. et al. (2006). Collapse dynamics of super-Gaussian beams. *Optics Express* **14** (12): 5468–5475.

174 Gaeta, A.L. (2003). Collapsing light really shines. *Science* **301** (5629): 54–55.

175 Silahli, S.Z., Walasik, W., and Litchinitser, N.M. (2015). Necklace beam generation in nonlinear colloidal engineered media. *Optics Letters* **40**: 5714–5717.

176 Walasik, W., Silahli, S.Z., and Litchinitser, N.M. (2017). Dynamics of necklace beams in nonlinear colloidal suspensions. *Scientific Reports* **7**: 11709.

177 Sun, J., Silahli, S.Z., Walasik, W. et al. (2018). Nanoscale orbital angular momentum beam instabilities in engineered nonlinear colloidal media. *Optics Express* **26** (5): 5118–5125.

7

Generation of Optical Vortex Beams

Yuanjie Yang[1] *and Cheng-Wei Qiu*[2]

[1] *School of Physics, University of Electronic Science and Technology of China, Chengdu, P.R. China*
[2] *Department of Electrical and Computer Engineering, National University of Singapore, Singapore*

7.1 Introduction

Vortex, a structure with a dark core (zero-amplitude centre) and spiral phase, is quite ubiquitous in nature, ranging from the classical macroscopic system to microsystem, such as turbulent liquids and gases, superfluid, and Bose–Einstein condensation [1–3]. Sometimes, vortices are known as nodal points, wave dislocations, phase dislocations, and more commonly phase singularities, because the phase in the centre is undefined. One of the most interesting properties of the vortices is that the change of phase in a circuit enclosing the dark centre always be a multiple of 2π [4, 5]. The study of vortex phenomenon dates back over 190 years, namely, Whewell discovered the phase singularities of the tide waves when he was studying tides in the oceans in 1833 [3, 6]. While, the critical step to understand the full phase singularities was not made until publication of the seminal paper by Nye and Berry [7]. From then, a vibrant field of study on different vortices has grown [2, 5, 8]. In 1989, Coullet introduced the term 'optical vortex' into optics, for the first time, after recognizing the similarity between the mathematical expression of helically phased beams and superfluid vortices [9]. In 1992, Allen et al. pointed out that the optical vortex with $\exp(il\theta)$ phase factor can carry orbital angular momentum (OAM) of $l\hbar$ per photon [10], where the topological charge (the index l) can be any integer value. This means that the value of OAM can be many times greater than that of spin AM which is associated with the polarization [11–13]. The optical vortex beams have found many applications in various fields due to their unique properties, such as optical communication [13–16], optical trapping [17–20], and quantum information [21–24].

In the early days, normally the vortex beams generated by optical elements such as spiral phase plate (SPP), hologram, and mode converter [25–27]. Recently, many advanced apparatuses that effectively and flexibly reshape the phase and amplitude structure are emerging rapidly, such as spatial light modulator (SLM) and digital micromirror device (DMD) [25, 27]. Because of their programmable nature, they are now widely used to produce various types of high-quality vortex beam. Recently, some novel vortex beams are proposed to meet the increasing demand in many applications. Thus, the generation and application of vortex beams promote mutually and develop quickly. In this chapter, some typical and novel vortex beams and their generation approaches are reviewed.

Electromagnetic Vortices: Wave Phenomena and Engineering Applications, First Edition.
Edited by Zhi Hao Jiang and Douglas H. Werner.

7.2 Basic Theory of Optical Vortex

The electric expression of an optical vortex beam along the z axis can be expressed as [28]:

$$E(r, \theta, z) = u_0(r, \theta, z) \exp(il\theta) \exp(-ikz), \tag{7.1}$$

where $u_0(r, \theta)$ is the field amplitude distribution at z plane, z is the propagation distance, k is wavenumber, and l is the topological charge.

The phase distributions of an optical vortex beam for the different topological charges are shown in Figure 7.1. From Figure 7.1a, one can see that the helical phase becomes a plane phase when $l = 0$. Figure 7.1b and c shows the helical phase structure and there exists a phase singularity in the center where the phase value is undefined.

Laguerre–Gaussian (LG) mode, containing the phase factor $\exp(il\theta)$, is a complete and orthogonal set of solutions for the paraxial wave equation, which is inherently a kind of vortex beams. The complex amplitude distribution of the LG beams can be written as:

$$E(r) \propto r^{|l|} \exp(-r^2/w^2) L_p^{|l|}(2r^2/w^2) \exp(il\theta), \tag{7.2}$$

where, w is the half-width of the beam, $L_p^{|l|}(g)$ is an associated Laguerre polynomial, l is the topological charge, and p is the radial mode index.

The intensity and phase distribution of LG beams for different p and l are shown in Figure 7.2. When $p = 0$ and $l = 1$, one can see a donut-shaped intensity pattern in Figure 7.2a and an increase of 2π around the center of the phase pattern in Figure 7.2d. When $p \neq 0$, one can see from Figure 7.2b

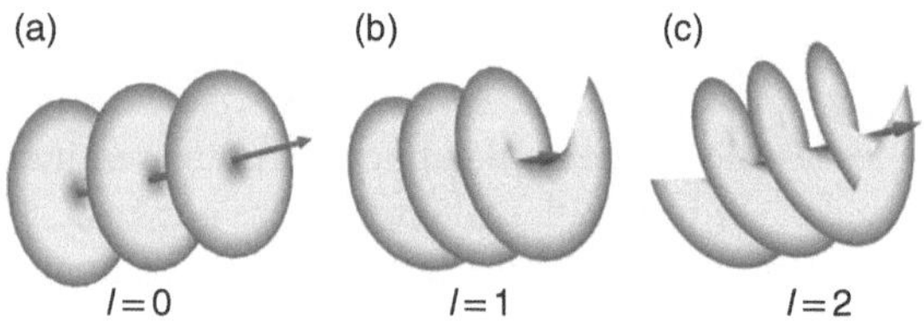

Figure 7.1 The phase distribution of an optical vortex beam for different topological charges.

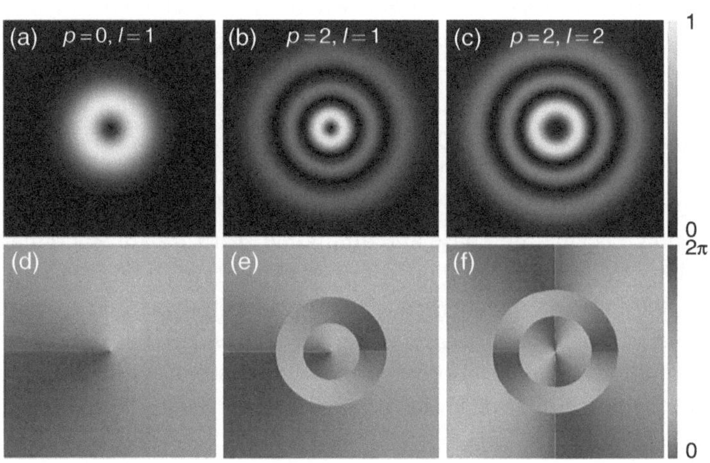

Figure 7.2 The intensity (a–c) and the corresponding phase distribution (d–f) of an LG beam for the different indices.

and c that there exists $p + 1$ concentric rings in the intensity patterns. Figure 7.2e and f shows that the phase changes $2\pi l$ around the singularities.

7.3 Generation of Optical Vortex

7.3.1 Generation of Vortex Beams using Optical Elements

Until now, optical vortex beams have been artificially produced by many approaches. The use of diffractive optical elements including the SPP and computer-generated holograms (CGH) are most widely used.

7.3.1.1 Spiral Phase Plate

The SPP is a transparent plate of refractive index n, whose optical thickness h increases proportionally to the azimuthal angle θ, as shown in Figure 7.3a [29, 30]. The relationship between h and θ can be expressed as [29]:

$$h = h_0 + h_s \frac{\theta}{2\pi} \tag{7.3}$$

where h_0 is the base height of optical element and h_s is the step height. The phase delay can be expressed as:

$$\phi(\theta, \lambda) = \frac{2\pi}{\lambda}\left[\frac{(n - n_0)h_s\theta}{2\pi} + nh_0\right] \tag{7.4}$$

where n_0 is the refraction index of the surrounding medium and λ is the vacuum wavelength. When a plane beam passes through such SPP, a vortex with a topological charge $l = \frac{h_s(n-n_0)}{\lambda}$ is obtained, as shown in Figure 7.3b. Thus, the output beam carries OAM $l\hbar$ per photon.

One can see that the nonhelical beam can directly and effectively transform into a helical beam by SPP. However, the drawbacks are also obvious. For instance, the quality of SPP mainly affects generation of vortex beam. In practice, a smooth helical surface of the SPP is hard to fabricate since it

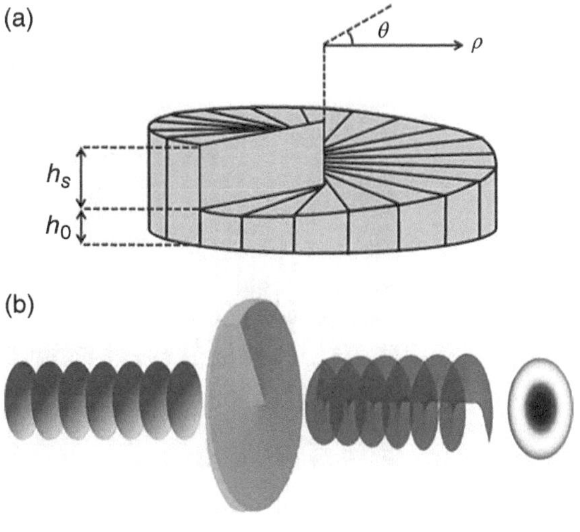

Figure 7.3 (a) The diagram of SPP. *Source:* Beijersbergen et al. [28]; Oemrawsingh et al. [29]; (b) a vortex beam generated by SPP.

needs a high precision of manufacturing technique. The helical surface is approximated by many discrete steps instead resulting into the quality reduction. Besides, one can note that SPP is very sensitive to optical wavelength and only the vortex beam with a fixed topological charge can be produced [31], which is not very effective. In recent years, some approaches have been proposed to overcome them [32]. For example, a flat SPP was proposed whose refractive index increases in the azimuthal direction and the height of the SPP remains unchanged [33].

7.3.1.2 Fork-grating Hologram

Besides the SPP, CGH is another popular method of generating optical vortex beams. The fork-grating (FG) hologram was at first proposed by Bazhenov et al. [34, 35]. The interference pattern between a tilted plane wave and the vortex beam, a FG-like pattern, is recorded as a hologram on a photographic film. Then, if the film is illuminated by a plane wave, a target vortex beam can be reconstructed.

Suppose that the complex amplitude of a target vortex beam and a reference tilted plane wave take the form $E_0 \exp(il\theta)$ and $E_0 \exp(ikx)$, respectively, the intensity distribution of the interference pattern can be given as [36, 37]:

$$I = 2E_0^2[1 + \cos(kx - l\theta)]. \tag{7.5}$$

where E_0 is a real amplitude and $k = \frac{2\pi}{\lambda}$ is the wave number.

Figure 7.4 shows FG for $l = 2$ and $l = 5$ as described in Eq. (7.5), from which we can see the l-pronged fork dislocation corresponds to the topological charge l. When the FG hologram is illuminated by a plane wave, we can get different vortex beams at the different diffracted orders, as illustrated in Figure 7.5.

7.3.1.3 Spiral Zone Plate Holograms

The spiral zone plate (SZP) is another hologram that can be used to generate the vortex beam [38–40]. The generation of a SZP hologram is the same with FG hologram, but the reference tilted plane wave is replaced by a spherical wave. The SZP hologram, interference pattern between a vortex beam and a spherical wave, can be expressed as:

$$I = \left| \exp(il\theta) \exp\left(ik\sqrt{r^2 + z^2}\right) + 1 \right|^2 \tag{7.6}$$

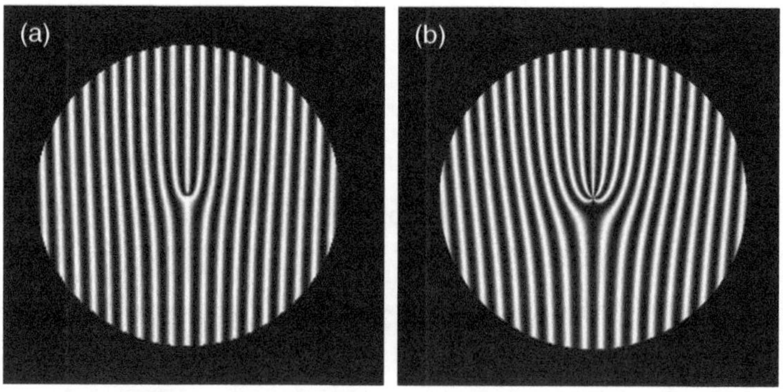

Figure 7.4 A fork-shaped diffraction grating produced by computer simulation. (a) $l = 2$; (b) $l = 5$.

Figure 7.5 The sketch of the fork-shaped hologram that can produce many diffraction orders.

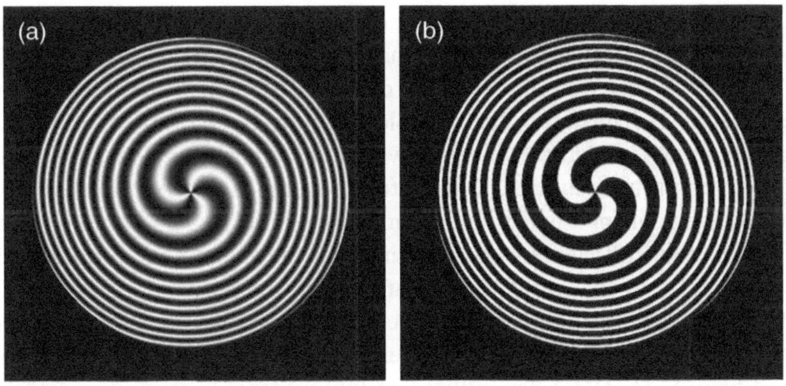

Figure 7.6 (a) Layout structures of the SZP; (b) Binarized SZP.

where l is topological charge, r and z are the transverse and longitudinal coordinates, respectively. The SZP will degrade into a standard Fresnel zone plate in the case of $l = 0$. Figure 7.6 shows an example of computer-generated SZP for $l = 3$. When such a hologram is illuminated by a plane wave, the phase is modulated by the spiral zone and a vortex beam is finally generated in the far field. It is noted that such SZP with l spirals not only produces a single vortex with topological charge l at specific propagation z, but also produces vortices with topological charge of ml at different propagation distances, where m is an integer. This is because of the fact that converging vortex waves can be produced by adjacent zones whose path difference is $m\lambda$ [40].

7.3.2 Generation of Vortex Beams Using Digital Devices

Although the aforementioned optical elements can generate vortex beams, it is noted that the same optical element cannot produce a vortex beam with different topological charges. This means that we have to change to a new SZP if we want to change the topological charge. This is not convenient in the experiment. Fortunately, this problem has been solved recently by the digital devices, such as spatial light modulators (SLM) and DMD [41–43]. SLM is an electro-optical device that can modulate the optical characteristics of light such as phase, amplitude, and polarization. Liquid crystal (LC) SLM is one of the most common devices, which comprises tremendous LC molecules whose transmission properties can be independently modulated by the applied electric field [42]. Based on the working mode, the commercial SLM can further be divided into two types: one is the transmissive type using transparent LC displays and another is the reflective type using liquid crystal on silicon (LCoS) displays. For an LCoS–SLM, the incident light approaches to the LC molecules through the glass substrate and transparent electrode and alignment film. Each LC pixel is controlled by the applied voltage from the corresponding electrode. Therefore, one only needs to encode the grayscale image into computer to manipulate the wavefront of a light beam in real time [44].

Now, we explain how to produce optical vortex beams by SLM. Considering an incident beam as a Gaussian beam, a vortex beam can be produced by imposing a specific phase with topological charge l on the incident beam by SLM. The transfer function of SLM can be written as [44]:

$$\phi_{\text{SLM}} = \text{mod}\left[l\varphi + 2\pi\left(G_x x + G_y y\right), 2\pi\right], \tag{7.7}$$

where mod is the modulus function, the first term is azimuthal phase, the second term is transfer function of the grating, $G_i = \frac{1}{L_i}$ with $i = x, y$ is the grating frequency along the i direction, and L_i is the period of the gratings in i direction. Then, we can generate any vortex beams, including simple vortex beam or complex/composite vortex beams, if we encode the corresponding CGHs on the SLM. Figure 7.7a–e shows the experiment results of measured intensity profiles of different circular optical vortex arrays [41]. The corresponding holograms displayed on the SLM for different topological charges are shown in Figure 7.7a′–e′. As we can see, comparing with static optical elements that only produce specific intensity and phase distribution, the liquid crystal spatial light modulator (LC–SLM) can modulate its phase distribution in the real time. Using SLM to implement the fabrication of aforementioned holograms greatly simplifies the experimental process with high efficiency, while the quality of the vortex beam is affected by the pixels of the SLM.

Recently, DMD was used frequently to generate vortex beams as well. DMD is a microelectromechanical system device derived from the early deformable mirror device. DMD is a binary amplitude (reflective type) SLM. DMD has millions of tiny switchable mirrors and each one can be switched to either +12 or −12 °, which correspond to the on or off states, respectively [43]. The on state micromirrors reflect the beam while the off state ones do not. Millions of DMD pixels ensure that DMD has a powerful ability to tailor the wavefront of light freely. Figure 7.8a illustrated the setup of DMD based microlithography for generating vortex beams [45]. Here, a DMD-based microlithography system was adopted to encode the CGH into LC cells to control the LCs directors,

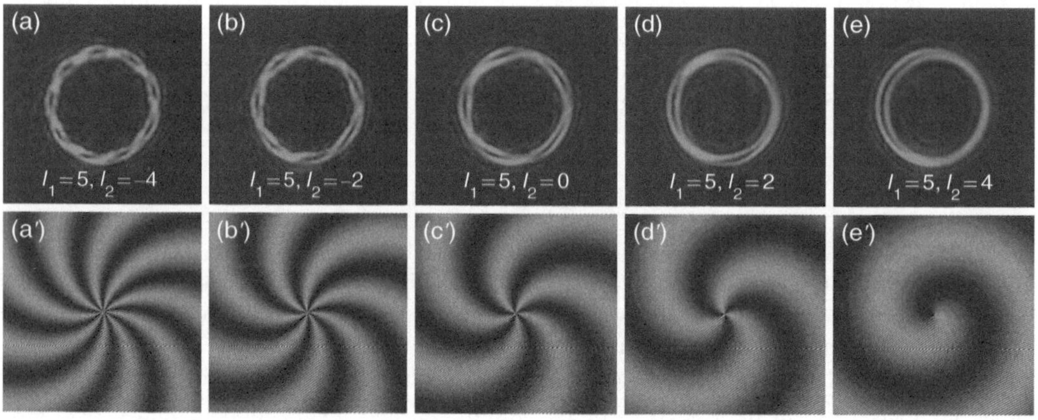

Figure 7.7 Experimentally measured intensity profile of different circular optical vortex arrays (upper row) and the corresponding holograms encoded on the SLM to generate these beams (bottom row) [41]. *Source:* Ma et al. [41].

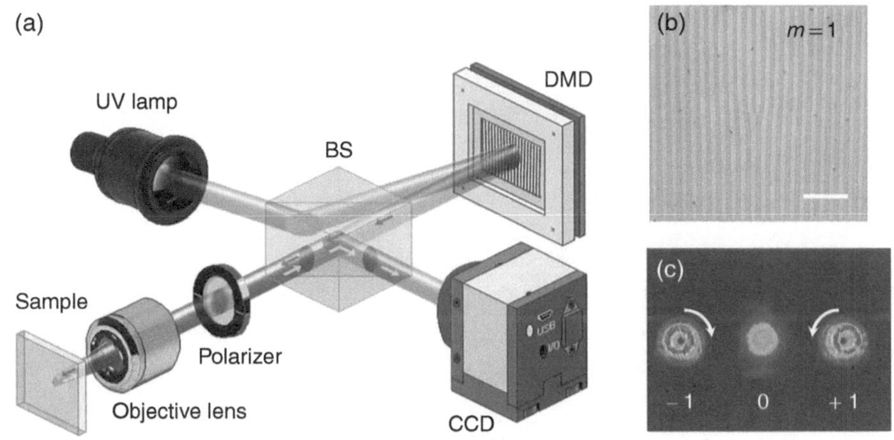

Figure 7.8 (a) Schematic setup of DMD based microlithography. *Source:* Wei et al. [45]; (b) micrograph of orthogonal hybrid aligned nematic fork grating and the corresponding diffraction pattern; (c) BS: beam splitter, CCD: charge coupled device. [45].

and the DMD was used as a dynamic mask. The DMD was illuminated by a uniform beam from an UV lamp, and the beam, carried on the designed fork grating hologram, was polarized and focused by an objective lens. Then the desired microimage was recorded on the substrate. Besides this charged-coupled device (CCD) is used to monitor the image. Figure 7.8b shows a fork grating with topological charge 1, consisting of two complementary hybrid-aligned nematic domains. When such fork grating is illuminated by a laser beam, the diffraction profile is shown in Figure 7.8c. From Figure 7.8c, we can see that there are two vortex rings, corresponding to the ±1st orders of the grating. Moreover, it was shown that the hologram pattern could be erased and then rewritten by the DMD system to generate a new hologram with different topological charges easily [45].

7.3.3 Generation of Vortex Beams Based on Mode Conversion

Through analyzing the relationship between the LG mode and Hermite–Gaussian (HG) mode, we know that the HG modes can transform into LG modes by using specific design of cylindrical lens [45, 46]. This kind of optical elements is called mode converter. When the cylindrical lens is placed at the distance $\sqrt{2}f$, as shown in Figure 7.9a, they can realize a $\frac{\pi}{2}$ mode converter. When the HG mode with indices m and n passes through this mode converter, an LG mode with the radial index $p = \min(m, n)$ and azimuthal index $l = m - n$ is obtained. This phenomenon can be explained from the mathematical relationship between the Hermite and Laguerre polynomial, which is expressed as [45]:

$$u_{mn}^{LG}(x,y,z) = \sum_{k=0}^{N} i^k b(n,m,k) u_{N-k,k}^{HG}(x,y,z) \tag{7.8}$$

where $b\,(n, m, k)$ is a real coefficient. The factor i^k shows that there exists $\frac{\pi}{2}$ phase with a difference between the HG mode and an LG mode realizing the transformation. It is noted that although this method can produce high-quality optical vortex beams, complex optical system and high-order HG beams are still required, which limits its application.

Similarly, as illustrated in Figure 7.9b, the π converter was proposed and the distance between the lenses is $2f$ instead. If an LG mode is incident on the π converter, there will be no OAM exchange between both of the lenses but the sign will be reversed.

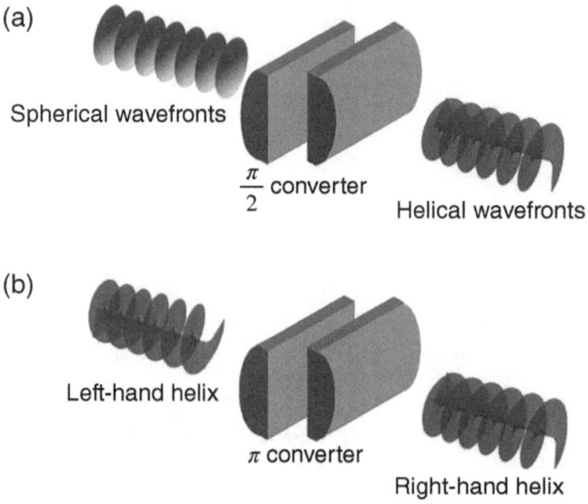

Figure 7.9 Schematic diagram of: (a) $\frac{\pi}{2}$ converter; (b) π converter.

7.3.4 Generation of Vortex Beams Based on the Superposition of Waves

Apart from the conventional methods aforementioned, it is shown that optical vortices can be generated by superposition of several beams, such as plane and Gaussian beams [47, 48]. In 2001, Masajada and Dubik demonstrated that three interferring homogenous plane waves can generate vortices array [47]. Lately, a method was proposed to generate stable optical vortices by using the coherent superposition of multibeams in a radially symmetrical configuration [48]. Suppose that each beamlet is symmetrical with different initial constant phases, and they are located at $(\rho\cos\theta, \rho\sin\theta)$ where, $\theta_j = \frac{\pi(2j-1)}{N}$ is the azimuth angle of the jth beamlet. Each Gaussian beamlet in the configuration at the input plane $z = 0$ is defined as [48]:

$$E_j(x_1, y_1; z = 0) = G_0\left[-\frac{(x_1 - a_j)^2 + (y_1 - b_j)^2}{w_0^2}\, exp\left(i\varphi_j\right)\right], \tag{7.9}$$

where $j = 1, 2, \ldots, N$, denotes the list of the beamlets, $\theta_j = \frac{m\pi(2j-1)}{N}$, w_0 is the beam width, m is the topological charge, and G_0 is the constant.

During propagation, every beamlet gradually interferes with each other in the free space. The linear optical system is denoted by the transfer matrix $\begin{pmatrix} A & B \\ C & D \end{pmatrix}$ and the resulted field distribution of the radial laser array on the output plane could be written as [48]:

$$\begin{aligned} E(x_2, y_2; z) &= \sum_{j=1}^{N} E_j(x_2, y_2; z) \\ &= -G_0\left(\frac{iZ_r}{\xi}\right)\exp\left[ikL_0\right]\exp\left[\frac{ikA\rho^2}{2\xi}\right]\exp\left[\frac{ik\eta\left(x_2^2 + y_2^2\right)}{2}\right]\sum_{j=1}^{N}\exp\left[-\frac{ik\left(a_j x_2 + b_j y_2\right)}{\xi} + i\varphi_j\right] \\ &= -G_0\left(\frac{iZ_r}{\xi}\right)e^{ikL_0}\exp\left[\frac{ikA\rho^2}{2\xi}\right]\exp\left[\frac{ik\eta\left(x_2^2 + y_2^2\right)}{2}\right] \\ &\times\sum_{j=1}^{N}\exp\left[-\frac{ik\rho\{x_2\cos\left[\pi(2j-1)/N\right] + y_2\sin\left[\pi(2j-1)/N\right]\}}{\xi} + i\frac{\pi m}{N}(2j-1)\right], \end{aligned} \tag{7.10}$$

where $Z_r = \frac{kw_0^2}{2}$, $\xi = B - iAZ_r$, and $\eta = \frac{(D - iCZ_r)}{\xi}$.

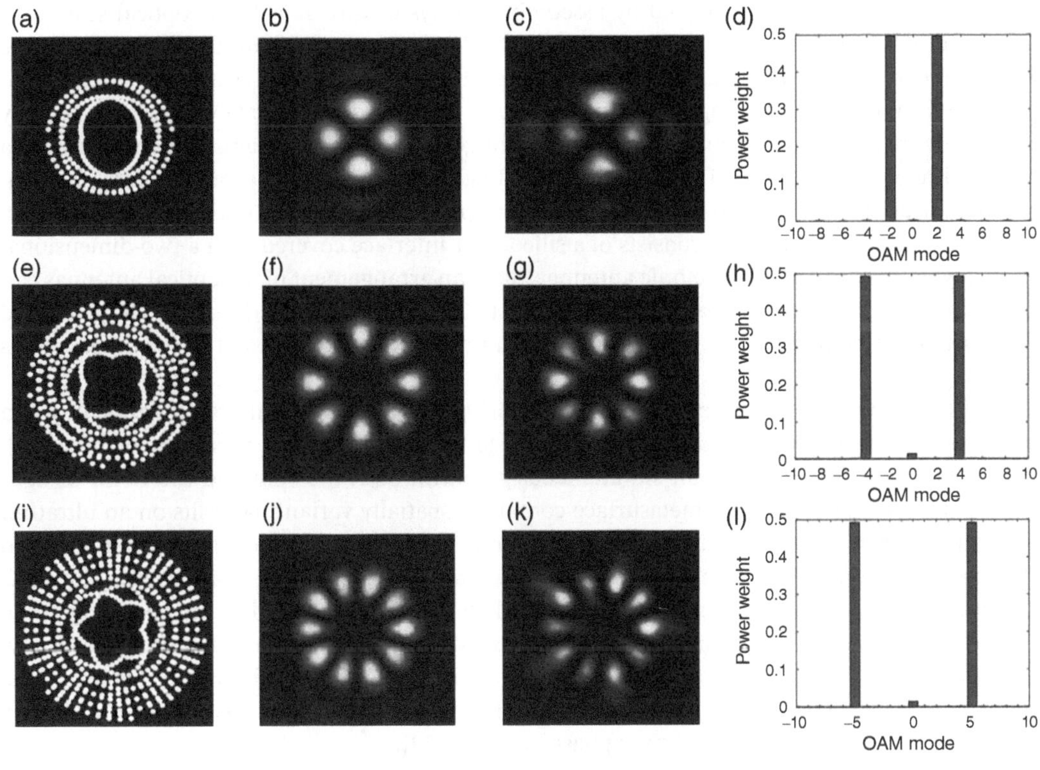

Figure 7.10 Schematics of the generation of compound vortex beams with opposite OAM modes. (a) Simulated mask for $\ell = \pm 2$; (b–d) the corresponding simulated intensity profiles, the experimental results, and OAM spectrum, respectively; (e–h) same with (a–d) but for $\ell = \pm 4$; (i–l) same with (a–d) but for $\ell = \pm 5$.

Based on the Eqs. (7.9) and (7.10), we can know how the optical vortex beam is produced from the coherent superposition of the radial beam arrays.

Recently, a novel technique based on the use of a structured pinhole plate for generating structured beams with a particular OAM spectrum was introduced [49]. The structured pinhole plate has a pair of spiral structures, and each spiral is composed of n pinholes. The azimuthal and radial distributions of the nth pinhole in the mth spiral reads $\alpha_{m,n} = \pm\left(\frac{2\pi n}{N} + \frac{2\pi m}{M}\right)$ and $r_n = \pm\left(\frac{\ell z \lambda \alpha_n}{\pi} + r_0{}^2\right)^{1/2}$, respectively. Figure 7.10a, e, and i shows the binary mask that generates superposition of the modes $\ell = \pm 2$, $\ell = \pm 4$, and $\ell = \pm 5$, respectively. As it is well known that the superposition of two vortex beams of opposite topological charges ($\pm\ell$) generate an intensity pattern consisting of 2ℓ petal-like structures. This is clearly shown in the simulated intensity pattern of the diffracted wave (see Figures 7.10 b, f, and j). Traditionally, petal-like beams are generated, for example, by a coaxial superposition of two vortex beams, here we have shown that a simpler method consisting of a plate with properly arranged pinholes can produce the same effect. The experimental results agree with the simulations very well. The corresponding OAM mode spectra shown in Figures 7.10d, h, and l demonstrate that the generated OAM modes are highly pure.

7.3.5 Generation of Vortex Beams Based on Metasurfaces

Recently, the powerful ability of metasurfaces, a new class of planar optical components to tailor the wavefront of light beam, have been demonstrated in many reported works [50–52].

Metasurfaces are generally produced by assembling arrays of tiny anisotropic optical scatterers, such as nanocavities, nanoparticles clusters, and optical antennas [50, 51]. By using the controllable phase shift between the incident and scattered light of the resonant optical scatterers, we can tailor the phase front freely. Accordingly, metasurfaces have been used to generate vortex beams widely [50–52]. Figure 11a shows the experimental setup for generating vortex beams, which consists of a Mach–Zehnder interferometer [51]. The collimated light beam from a distributed feedback quantum cascade laser was split into two arms by a beam splitter. The beam in one arm is focused on a metasurface phase mask, which consists of a silicon–air interface covered with a two-dimensional (2D) array of V-shaped gold plasmonic antennas. Such an arrangement of the optical antennas can produce a spiral phase that increases linearly from 0 to 2π in the azimuthal direction. The generated vortex beam can be revealed by interference with a reference beam prepared in the other arm as shown in Figure 7.11b.

Recently, a new type of geometric metasurface fork grating combine with the functionality of a metasurface phase mask and that of a linear phase gradient metasurface, was demonstrated [52]. Such metasurface fork grating can simultaneously control both the spin and the OAM of light beam. This fork-type plasmonic metasurface contained spatially variant nanoslits on an ultrathin metal film that can generate vortex beams with different OAM states and split these vortex beams to different diffraction orders according to their spin AM states. Figures 7.12a–d shows the SEM images of the metasurface gratings with topological charge 2 and 3, and the corresponding phase patterns, respectively. Laser beam with the different polarization states are illuminated onto the metasurface fork via a lens, and the generated vortex beams are shown in Figure 7.12e. Figure 7.12e shows that such metasurface can couple spin and OAM and project the vortex beams to different spatial positions via a linear phase gradient [52].

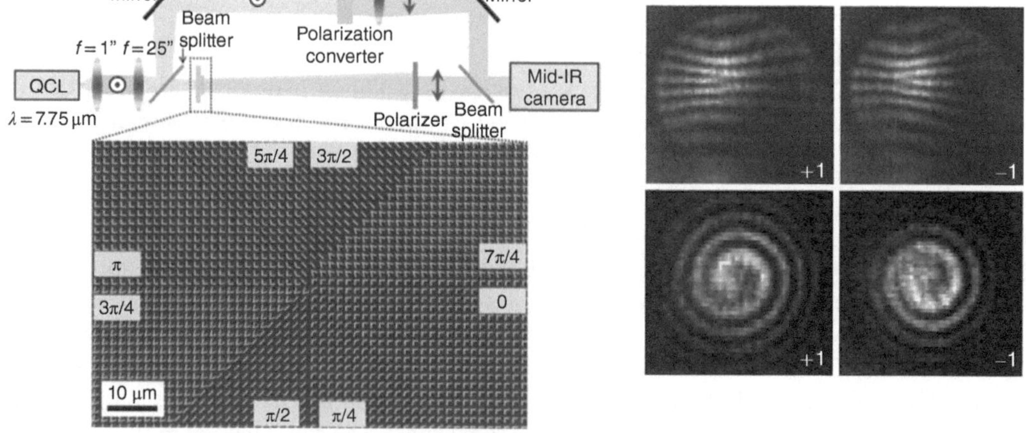

Figure 7.11 (a) Experimental setup based on metasurfaces for generating and characterizing vortex beams. The bottom inset is a scanning electron microscope (SEM) image of a metasurface phase mask; (b) Intensity patterns of the interference between vortex beam, a plane wave reference (top), the interference between vortex beam, and a spherical wave reference (co-propagating Gaussian beam) (top), respectively. [51]. *Source:* Yu et al. [51]. (© [2013] IEEE)

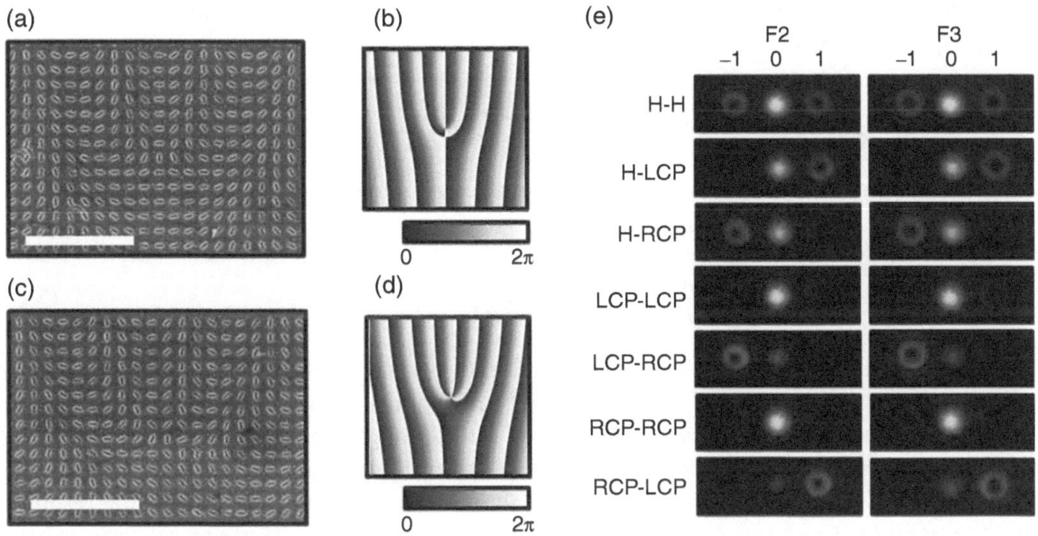

Figure 7.12 (a) SEM image of plasmonic metasurface (F2) with topological charge 2; (b) the corresponding phase pattern of metasurface F2; (c) and (d) same with those in (a) and (b) but for topological charge 3 (F3); (e) produced vortex beams from metasurface fork mask F2 and F3, respectively. H denotes that the incident laser has horizontal polarization; LCP and RCP denote the left and right circularly polarized beams of the incident beams.

7.4 Generation of Novel Vortex Beams

During the past two decades, apart from optical vortex beams aforementioned, the versatile and diverse optical vortex beams are emerging rapidly [53]. The properties of novel vortex beams have found potential applications. Thus, reliable generation methods of such beams have been widely developed. In this section, some novel vortex beams and their generation methods have been summarized.

7.4.1 Perfect Vortex Beam

In some applications, such as micromanipulation, a vortex beam with a small ring, and a large topological charge is required. The conventional vortex beam does not meet requirement since the ring increases with the increasing topological charges. A perfect vortex beams whose radius of ring is independent on the topological charge, was proposed recently [54, 55]. The form of the perfect vortex beam can be expressed as [54]:

$$g_v(\rho,\theta) \propto circ\left(\frac{\rho}{a}\right)\exp(iv\theta)\sum_{n=1}^{\infty}\frac{J_v(\alpha_{v,n}\rho_0/a)}{[J_{v+1}(\alpha_{v,n})]^2}J_v\left(\alpha_{v,n}\frac{\rho}{a}\right) \tag{7.11}$$

where, v is the topological charge, $J_v(\cdot)$ is the vth order Bessel function of the first kind, $\alpha_{v,\,n}$ is the nth zero of function $J_v(\cdot)$, and a is the upper limit of the radial coordinate ρ.

A generation system is sketched in Ref. [54]. A twisted nematic LC–SLM was used to modulate phase distribution. The transmittance of the LC–SLM in experiment can be expressed to an annular topology formed by a set of well-separated transparent and concentric ring slits, of different radii and widths, with periodic azimuthal phase modulation inside. The transmittance of the LC–SLM can be expressed as [54]:

$$t_v(r,\varphi) \propto \sum_{n=1}^{N} \alpha_{v,n}\beta_{v,n} \exp\left(iv(\varphi - \varphi_{v,n})\right)\delta(r - r_{v,n}). \tag{7.12}$$

Through the approximation and complex calculation, a finite-sum approximation of perfect vortex beam can be expressed to replace Eq. (7.11):

$$U_v(\rho,\theta) \propto \exp(iv\theta) \sum_{n=1}^{N} \frac{J_v(\alpha_{v,n}\rho_0/a)}{[J_{v+1}\alpha_{v,n}]^2} J_v\left(\alpha_{v,n}\frac{\rho}{a}\right). \tag{7.13}$$

Figure 7.13a–d shows the simulated results of Eqs. (7.11) and (7.13) with two different values of the topological charge, $l = 1$ and $l = 10$, respectively. It is shown that a finite-sum approximation of such vortex beam [Eq. (7.13)] possesses a brightest central ring and extra weak bright rings. When we observed carefully, it can be seen that only the central ring is just perfect vortex and other undesired extra rings will influence its applications. Figure 7.13e–f shows that experimental results of Eq. (7.13) with $l = 5$ and $l = 10$. We can obviously see that the quality of generated perfect vortex beam is rather poor. The reason is that the actual width of ring slits and not the ideal modulation characteristics of the LC–SLM. Later, the purity of generated perfect vortex beam is improved by use of the width pulse approximation of Bessel function in the LC–SLM [56]. The experimental results of the generated perfect vortex beam of radius = 1.5 mm and different values of topological charge are shown in Figure 7.14.

Besides this, another technique of using the Fourier transform (FT) property of a Bessel beam to generate a perfect vortex beam is proposed [57]. The SLM is designed to form a Bessel Gauss beam. Then, the perfect vortex beam with $l = 4$ can be experimentally formed by optical FT of a fourth order Bessel Gauss beam, as shown in Figure 7.15. The ring radius of the perfect vortex decreases with decreasing a, where a is the axicon parameter.

Recently, an amplitude digital micromirror device (DMD) [58, 59] is introduced to generate a high quality of perfect vortex beam with a large topological charge, as higher as 90. It is demonstrated that the generated perfect vortex beam can maintain rather a high fidelity and does not

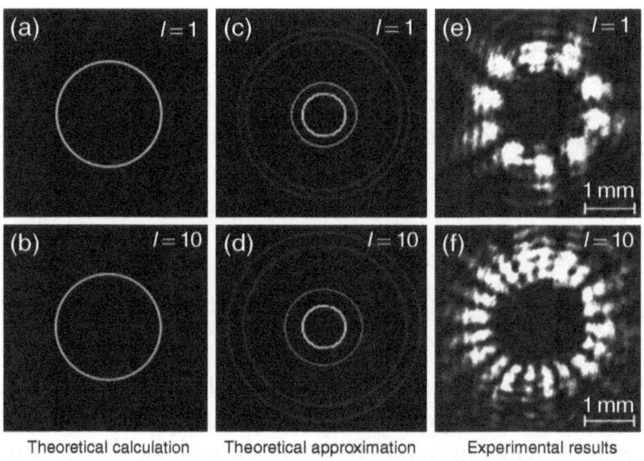

Figure 7.13 The 2-D experimental pattern of the perfect vortex beams with ρ_0 = 1mm, N = 40. (a) Simulated intensity pattern for l = 1; (c) and (e) Corresponding simulated and experimental results for l = 1 using their experimental setup; (b), (d) and (f) The same as (a), (c) and (e) but for l = 10 [54]. *Source:* Ostrovsky et al. [54].

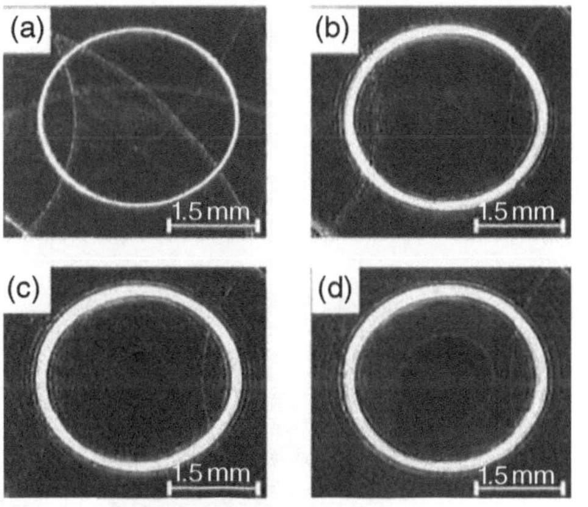

Figure 7.14 The experimental results of the perfect vortex beams with radius $\rho_0 = 1.5$ mm; (a) $l = 1$; (b) $l = 20$; (c) $l = 40$; (d) $l = 60$. [56]. *Source:* García-García et al. [56].

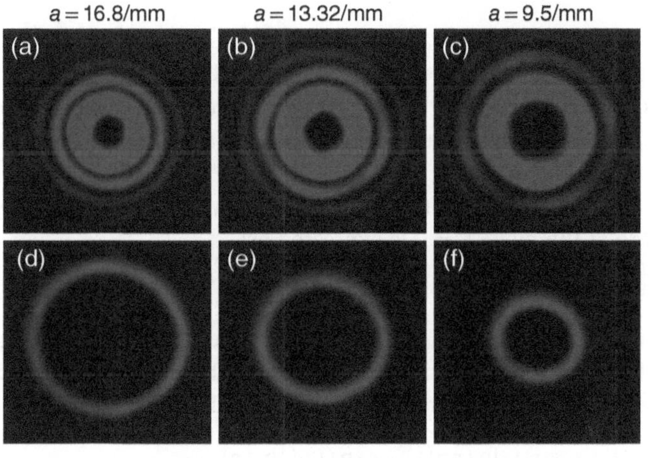

Figure 7.15 Generation of perfect vortex beams through Fourier transformation of Bessel Gauss beams. (a–c) Experimental results of fourth-order Bessel Gauss beams; (d–f) corresponding experimental results of perfect vortex beam. [57]. *Source:* Vaity et al. [57].

produce the undesired extra rings until the topological charge is over 90. The reason is that DMD has higher refreshing frequency and higher fill factor rate of around 90% than SLM, which modulate the amplitude more accurate.

7.4.2 Fractional Vortex Beams

Vortex beams with noninteger topological charges, also called fractional vortex beams, have also attracted interesting attention [60–62]. Numerous results show that fractional vortex beam carries a mixed screw-edge dislocation and has a radial opening in the annular intensity ring. The unique structure and properties, which are different from the integer cases, have been widely studied. The difference between the integer and fractional topological charge deepens the fundamental understanding of the AM. The unique characteristics of such beams are now applied in many practical applications, such as optical trapping [62], quantum information [63], and optical communication [64].

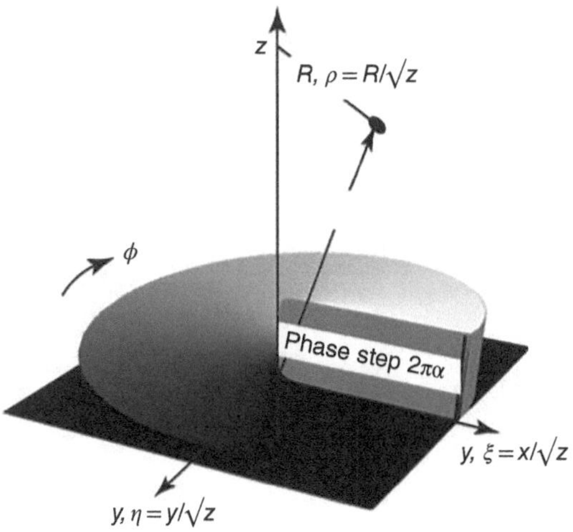

Figure 7.16 Notation and coordinates for a helicodal phase step. [60]. *Source:* Berry [60].

Hitherto, many approaches have been proposed to modulate the fractional topological charge. The conceptually simplest method to yield a fractional-order vortex beam is the SPP with the fractional step height [60]. The notation and coordinates for a helicoidal phase step is illustrated in Figure 7.16.

When a plane wave with unit amplitude passes through the SPP, the wave with a phase singularity of strength $2\pi\alpha$ at the plane $z = 0$ is expressed as:

$$\psi_\alpha(R, 0) = \exp(i\alpha\phi), \tag{7.14}$$

where $R = \{x, y\} = R\{\cos\phi, \sin\phi\}$, $\rho = R/z^{1/2} = \{\xi, \eta\}$. When $\alpha = n$ is an integer, there is a screw dislocation along the z axis $R = 0$. When α is noninteger, there are mixed screw edge dislocations. The corresponding paraxial propagating wave is expressed as [60]:

$$\psi_\alpha(r) = \frac{\exp(i(z + \pi\alpha))\sin(\pi\alpha)}{\pi}\sum_{\infty}^{\infty}\frac{\exp(in\phi)p_n(\rho)}{\alpha - n}, \tag{7.15}$$

where $P_n(\rho)$ is given by Eq. (7.7) in Ref. [60], z is the propagation distance, n is integer topological charge. A density plot of the modulus $|\psi_\alpha = 2.2|$ is shown in Figure 7.17.

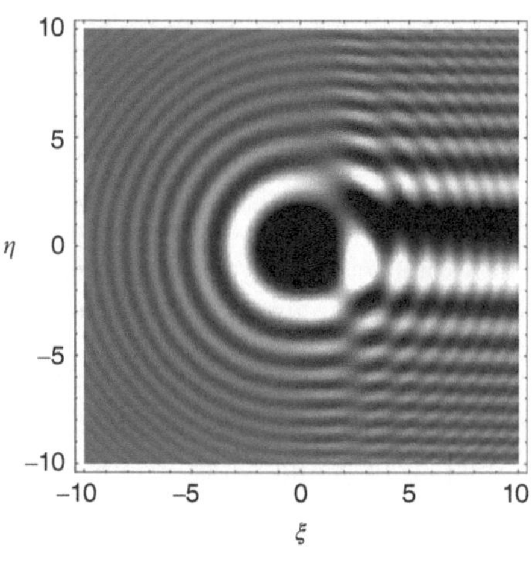

Figure 7.17 A density plot of the paraxial wave $|\psi_\alpha = 2.2|$ in the plane. [60]. *Source:* Berry [60].

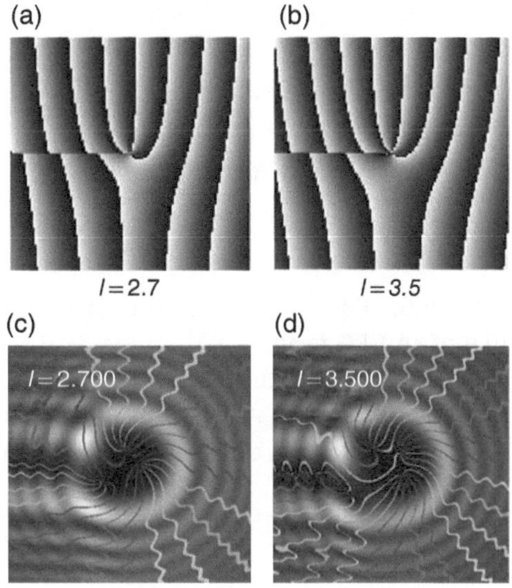

Figure 7.18 Layout structures of CGH for producing fractional vortex beams. (a, b) FG with an additional radial discontinuity for two different topological charge; (c, d) the corresponding intensity distribution. [68]; (a, c) l = 2.7; (b, d) l = 3.5. *Source:* Leach et al. [68].

Other kinds of fractional vortex beams can also be generated by such optical element, such as fractional Bessel beam [65, 66]. In addition, using two such devices, the high-dimensional spatial entanglement of twin photons of fractional OAM is observed [67]. Another widely used method to generate fractional vortex beam is holographic technique. As discussed in Figure 7.4, the CGH displays the l-pronged fork dislocation to produce a vortex beam with topological charge l. Figure 7.18 shows the CGH for producing a vortex beam with fractional topological charge [68], from which we can see that an additional radial discontinuity to the pattern appears along a line $\phi = 0$, due to the phase step of fractional part.

Furthermore, another technique to generate fractional vortex beams is proposed based on the production of subharmonic diffraction orders [69]. Usually, the generated fractional vortex beam is not stable, while a new method by a synthesis of LG modes is used to generate the fractional vortex beam, which shows the structure of such beam is stable on propagation [70]. The aforementioned beams are limited in the visible and near-infrared (NIR) domain. Recently, new generation mechanisms of fractional OAM beams at shorter wavelengths are proposed, such as the EUV and soft X-rays. Hernández García and coworkers investigated the extreme ultraviolet attosecond pulse beams carrying fractional orbital momentum from high-order harmonic generation [71, 72].

7.4.3 Anomalous Vortex Beam

It is well known that the normally vortex beams, such as high-order LG beams and Bessel beams, always maintain the same profile during their propagation. While, an interesting vortex beam named anomalous vortex (AV) beam was proposed in 2013 [73], which can evolve into an elegant LG (ELG) beam in the far field in free space. The electric field of the AV beam at $z = 0$ can be expressed as [73]:

$$E_{n,m}(\rho_0, \theta_0, 0) = E_0 \left(\frac{\rho_0}{\omega_0}\right)^{2n+|m|} \exp\left(-\frac{\rho_0^2}{\omega_0^2}\right) \exp(-im\theta_0), \tag{7.16}$$

where E_0 is a constant, n is the beam order of the AVB, m is the topological charge, and w_0 is the beam waist size of the fundamental Gaussian beam, ρ_0 and θ_0 are the radial and azimuthal coordinates, respectively. If letting $m = 0$ and $n \neq 0$, Eq. (7.16) reduces to the electric field for a hollow Gaussian beams. If letting $n = 0$, it can become an ordinary Gaussian vortex beams.

Using the far-field approximation, the electric field of the AV beam through ABCD optical system can be simplified as [73]:

$$E_{n,m}(\rho,\theta,z) = \frac{i^{m+1}\pi n!}{\lambda z}\exp(-ikz)w_0^{2n+|m|+2}\exp(-im\theta)\left(\frac{\rho}{w}\right)^{|m|}\exp\left(-\frac{\rho_0^2}{w^2}\right)L_n^{|m|}\left(\frac{\rho^2}{w^2}\right). \tag{7.17}$$

From the Eq. (7.17), we can see the far-field complex amplitude of an AV beam is in accord with that of an ELG beam.

Figure 7.19 shows the evolution of an AV beam to ELG beam. In the source plane, we can see that the AVB has a single ring profile and a helical phase. However, in the far field, the amplitude distribution has several rings, which is much different from the LG vortex beams.

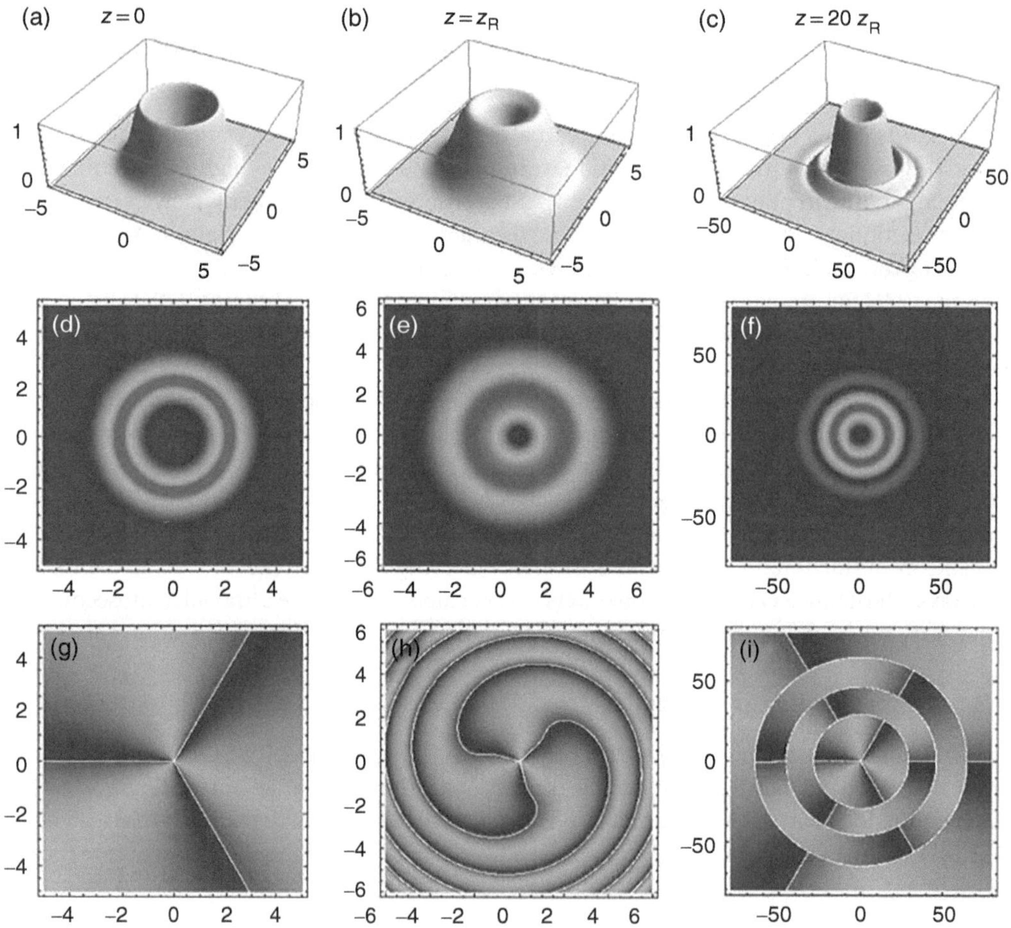

Figure 7.19 Normalized 3D amplitude distribution (top), contour graph for the amplitude (middle) and phase distribution (bottom) of an AVB at different propagation distance: (a) $z = 0$; (b) $z = z_R$; (c) $z = 20z_R$. Here, $m = n = 3$ and the units in the plots is mm. [73]. *Source:* Yang et al. [73].

7.4.4 Vortex Beams with Varying OAM

It is well known that the OAM of the vortex beams generally remain unchanged during the propagation. However, recently, some new features of the novel vortex beams have been found. Recent studies show that OAM can vary not only in the space region [74, 75], but also in time [76]. In 2018, Yang et al. proposed an anomalous Bessel vortex beam with varying topological charge during propagation [74]. The illustration of the generation of such vortex beam using a Fermat's spiral slit is shown in Figure 7.20. The structure of the Fermat's spiral slit can be written as [74]:

$$r_\alpha = \left(r_0^2 + \frac{lz\lambda\alpha}{\pi}\right)^{1/2}, \tag{7.18}$$

where r_0 is the initial radius of the spiral slit, and r_α and α is the radial and the angular coordinates, respectively, z is the propagation distance, l is the topological charge.

Suppose such a spiral slit is placed at $z = 0$ plane, when a plane wave passes through the slit, the transmitted wavelets from the different parts of the slit undergo different optical paths and arrive at the centre of the observation plane. The total phase difference at the point (0, 0, z) can be written as:

$$\int_0^{2\pi} d\theta = \frac{2\pi}{\lambda}\int_0^{2\pi} \frac{\ell\lambda}{2\pi} d\alpha = 2\pi\ell. \tag{7.19}$$

This means that there will be an optical vortex in the center of the observation plane. Moreover, for a given spiral slit, the value of $lz\lambda$ can be regarded as a constant. Therefore, we can see with increasing propagation distance z or λ, topological charge l will decrease. Figure 7.21 shows theoretically and experimentally results of the intensity and phase pattern of the anomalous Bessel vortex beam during propagation from $z = 1\ m$ to $z = 2\ m$. From Figure 7.21, we can see clearly that the OAM of such vortex beam varies continuously during propagation.

It is worth noting that all the aforementioned vortex beams are static, namely, the OAM does not change in time. Recently, Rego et al. introduced a new property of vortex beams, manifested

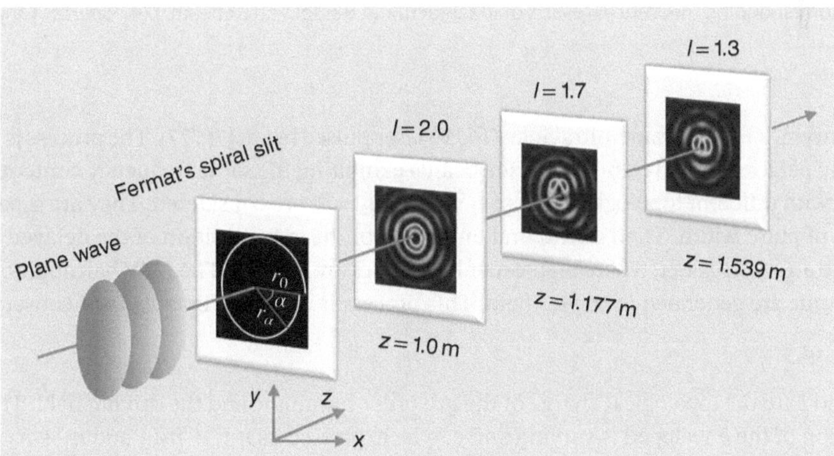

Figure 7.20 The notation of the spiral slit for achieving an anomalous Bessel vortex beam [74]. *Source:* Yang et al. [74].

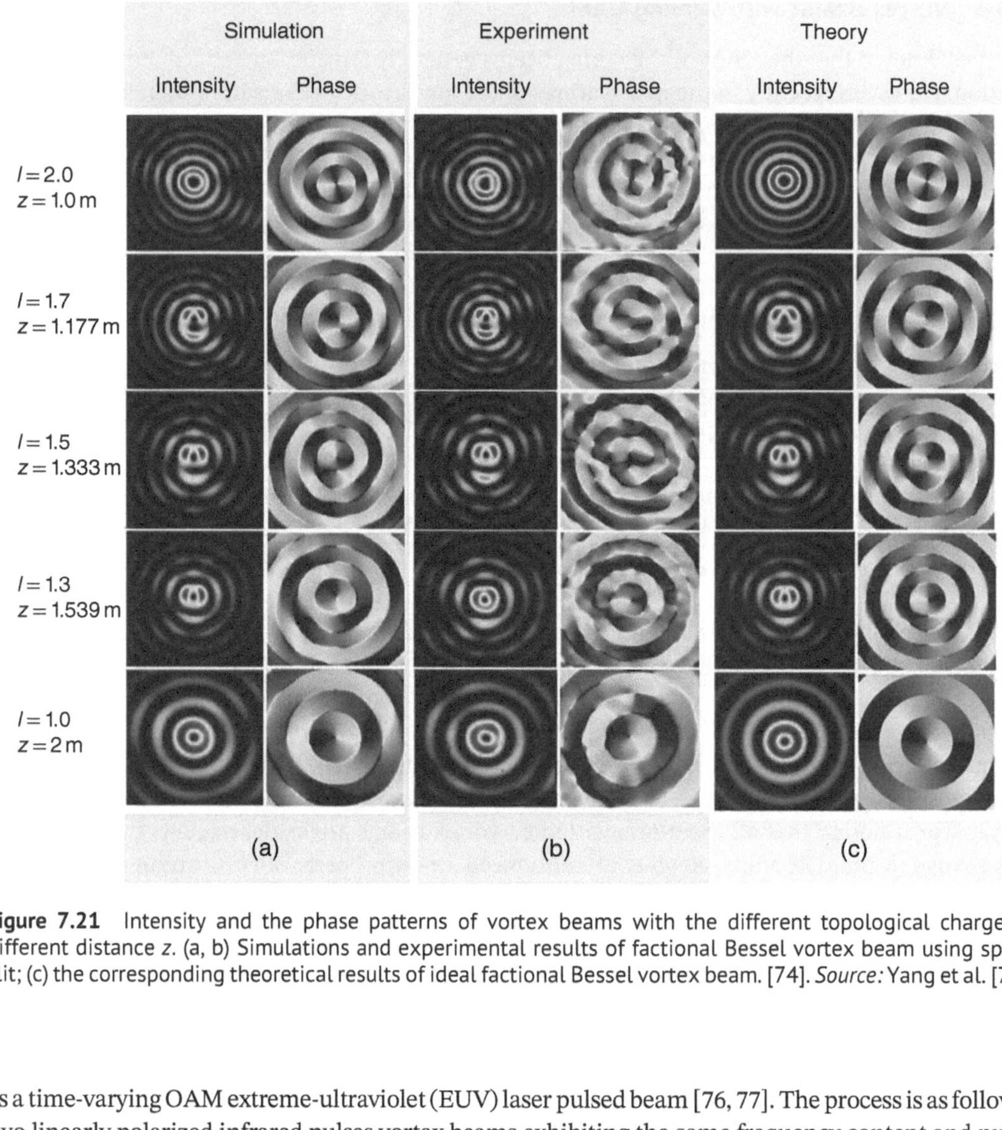

Figure 7.21 Intensity and the phase patterns of vortex beams with the different topological charge at different distance *z*. (a, b) Simulations and experimental results of factional Bessel vortex beam using spiral slit; (c) the corresponding theoretical results of ideal factional Bessel vortex beam. [74]. *Source:* Yang et al. [74].

as a time-varying OAM extreme-ultraviolet (EUV) laser pulsed beam [76, 77]. The process is as follows: two linearly polarized infrared pulses vortex beams exhibiting the same frequency content and pulse width, but with different topological charge $l_1 = 1$ and $l_2 = 2$, are considered. They are separated by a time delay of pulse width. Then a temporal envelopes of the superposition of the delayed beams are focused into an argon gas jet, where high-order harmonics are emitted. The EUV harmonic beams with 17th harmonic are generated in the far field. This process is governed by the OAM conservation law

$$l_q = ql, \tag{7.20}$$

where l_q and l are the topological charge of the qth order harmonic and the driving field. The OAM of superposition of the enveloped is varying and can be expressed as $n_1 l_1 + n_2 l_2$, and $n_1 + n_2 = \text{q}$. Thus, the OAM of EUV harmonic beams with 17th harmonic can vary with time, i.e. from $17l_1 + 0l_2 = 17$ to $0l_2 + 17l_2 = 34$. Such a self-torqued high-harmonic generation beam show a brand new property of vortex beams, and it can be used for laser manipulation on an attosecond time and nanometer spatial scales.

7.5 Conclusion

We have reviewed the advance on optical vortex generation. Some typical generation approaches have been shown, such as diffractive optical element, mode converter, the new techniques based on SLM, DMD, and metasurfaces. In the recent decade, many novel vortex beams are proposed and generated rapidly to meet the special applications. The special properties of novel vortex beams deepen the understanding of the structured beams. Therefore, some typical generation approaches of the novel vortex beams, such as perfect vortex beam, AV beam, vortex beams with time-varying OAM, are summarized. Although great achievements have been made on generating vortex beams thus far, the fast-growing flat optics and digital optics based on metasurfaces will bring more incentives to this field. Besides optical vortex beams, other forms of the vortex beams, such as electron, plasmon, and neutron vortices, have attracted great interests as well [78, 79]. Some methods discussed in this chapter may be available for those vortex beams as well.

References

1 Padgett, M.J. (2017). Orbital angular momentum 25 years on. *Optics Express* **25** (10): 11265.

2 Qiu, C.W. and Yang, Y. (2017). Vortex generation reaches a new plateau. *Science* **357** (6352): 645.

3 Gbur, G.J. (2016). *Singular Optics*, 545. Boca Raton: CRC Press.

4 Dennis, M.R., O'Holleran, K., and Padgett, M.J. (2009). Singular optics: optical vortices and polarization singularities. *Progress in Optics* **53**: 293–363.

5 Uchida, M. and Tonomura, A. (2010). Generation of electron beams carrying orbital angular momentum. *Nature* **464** (7289): 737–739.

6 Whewell, W. (1833). Essay towards a first approximation to a map of cotidal lines. *Philosophical Transactions of the Royal Society of London* **122**: 147–236.

7 Nye, J.F. and Berry, M.V. (1974). Dislocations in Wave Trains. *Proceedings of the Royal Society A: Mathematical, Physical and Engineering Sciences* **336**: 165.

8 Vaughan, J.M. and Willetts, D.V. (1979). Interference properties of a light beam having a helical wave surface. *Optics Communications* **30** (3): 263–267.

9 Coullet, P., Gil, G., and Rocca, F. (1989). Optical vortices. *Optics Communications* **73** (5): 403–408.

10 Allen, L., Beijersbergen, M.W., Spreeuw, R.J., and Woerdman, J.P. (1992). Orbital angular momentum of light and the transformation of Laguerre-Gaussian laser modes. *Physical Review A* **45** (11): 8185.

11 Allen, L., Padgett, M., and Babiker, M. (1999). The orbital angular momentum of light. *Progress in Optics* **39**: 291–372.

12 Barnett, S. and Allen, L. (1994). Orbital angular momentum and nonparaxial light beams. *Optics Communications* **110**: 670–678.

13 Bozinovic, N., Yue, Y., Ren, Y. et al. (2013). Terabit-scale orbital angular momentum mode division multiplexing in fibers. *Science* **340** (6140): 1545–1548.

14 Wang, J. (2016). Advances in communications using optical vortices. *Photonics Research* **4** (5): B14–B28.

15 Wang, J., Yang, J.Y., Fazal, I.M. et al. (2012). Terabit free-space data transmission employing orbital angular momentum multiplexing. *Nature Photonics* **6** (7): 488–496.

16 Fu, S.Y. and Gao, C.Q. (2016). Influences of atmospheric turbulence effects on the orbital angular momentum spectra of vortex beams. *Photonics Research* **4**: B1–B4.

17 He, H., Friese, M.E., Heckenberg, N.R., and Rubinsztein-Dunlop, H. (1995). Direct observation of transfer of angular momentum to absorptive particles from a laser beam with a phase singularity. *Physical Review Letters* **75** (5): 826.

18 Friese, M.E.J., Enger, J., Rubinsztein-Dunlop, H., and Heckenberg, N. (1996). Optical angular-momentum transfer to trapped absorbing particles. *Physical Review A* **54**: 1593–1596.

19 Simpson, N.B., Dholakia, K., Allen, L., and Padgett, M.J. (1997). Mechanical equivalence of spin and orbital angular momentum of light: An optical spanner. *Optics Letters* **22**: 52–54.

20 Padgett, M.J. and Allen, L. (1999). The angular momentum of light: optical spanners and the rotational frequency shift. *Optical and Quantum Electronics* **31** (1): 1–12.

21 Leach, J., Jack, B., Romero, J. et al. (2010). Quantum correlations in optical angle-orbital angular momentum variables. *Science* **329** (5992): 662–665.

22 Fickler, R., Campbell, G., Buchler, B. et al. (2016). Quantum entanglement of angular momentum states with quantum numbers up to 10,010. *Proceedings of the National Academy of Sciences of the United States of America* **113** (48): 13642.

23 Stav, A., Faerman, E., Maguid, D. et al. (2018). Quantum entanglement of the spin and orbital angular momentum of photons using metamaterials. *Science* **361** (6407): 1101–1104.

24 Vaziri, A., Weihs, G., and Zeilinger, A. (2002). Experimental two-photon, three-dimensional entanglement for quantum communications. *Physical Review Letters* **89**: 240401.

25 Yao, A.M. and Padgett, M.J. (2011). Orbital angular momentum: origins, behavior and applications. *Advances in Optics and Photonics* **3** (2): 161–204.

26 Kotlyar, V.V., Kovalev, A.A., and Porfirev, A.P. (2018). *Vortex Laser Beams: Vortex Laser Beams*. CRC Press.

27 Wang, X., Nie, Z., Liang, Y. et al. (2018). Recent advances on optical vortex generation. *Nanophotonics* **7** (9): 1533.

28 Curtis, J.E. and Grier, D.G. (2003). Structure of optical vortices. *Physical Review Letters* **90** (13): 133901.

29 Beijersbergen, M.W., Coerwinkel, R.P.C., Kristensen, M., and Woerdman, J.P. (1994). Helical-wavefront laser beams produced with a spiral phaseplate. *Optics Communications* **112** (5-6): 321–327.

30 Oemrawsingh, S.S.R., Van Houwelingen, J.A.W., Eliel, E.R. et al. (2004). Production and characterization of spiral phase plates for optical wavelengths. *Applied Optics* **43** (3): 688–694.

31 Peele, A.G., McMahon, P.J., Paterson, D. et al. (2002). Observation of an x-ray vortex. *Optics Letters* **27**: 1752–1754.

32 Rotschild, C., Zommer, S., Moed, S. et al. (2004). Adjustable spiral phase plate. *Applied Optics* **43** (12): 2397–2399.

33 Wu, W.B., Sheng, Z.Q., and Wu, H.W. (2019). Design and application of flat spiral phase plate. *Acta Physica Sinica* **68** (5): 054102.

34 Bazhenov, V., Vasnetsov, M., and Soskin, M. (1990). Laser beams with screw dislocations in their wavefronts. *JETP Letter* **52**: 429–431.

35 Bazhenov, V., Soskin, M.S., and Vasnetsov, M.V. (1992). Screw dislocations in light wavefronts. *Journal of Modern Optics* **39**: 985–990.

36 Bekshaev, A.Y. and Karamoch, A.I. (2008). Spatial characteristics of vortex light beam s produced by diffraction gratings with embedded phase singularity. *Optics Communications* **281** (6): 1366–1374.

37 Stoyanov, L., Topuzoski, S., Stefanov, I. et al. (2015). Far field diffraction of an optical vortex beam by a fork-shaped grating. *Optics Communications* **350**: 301–308.

38 Heckenberg, N.R., McDuff, R., Smith, C.P., and White, A.G. (1992). Generation of optical phase singularities by computer-generated holograms. *Optics Letters* **17**: 221–223.

39 Sakdinawat, A. and Liu, Y. (2007). Soft-X-ray microscopy using spiral zone plates. *Optics Letters* **32**: 2635–2637.

40 Saitoh, K., Hasegawa, Y., Tanaka, N., and Uchida, M. (2012). Production of electron vortex beams carrying large orbital angular momentum using spiral zone plates. *Journal of Electron Microscopy* **61**: 171–177.

41 Ma, H., Li, X., Tai, Y. et al. (2017). Generation of circular optical vortex array. *Annalen der Physik* **529**: 1700285.

42 Maurer, C., Jesacher, A., Bernet, S., and Ritsch-Marte, M. (2011). What spatial light modulators can do for optical microscopy. *Laser Photonics Review* **5**: 81–101.

43 Ren, Y., Lu, R., and Gong, L. (2015). Tailoring light with a digital micromirror device. *Annalen der Physik (Berlin)* **527**: 447–470.

44 Rosales-Guzmán, C. and Forbes, A. (2017). *How to Shape Light with Spatial Light Modulators*. Bellingham, Washington: SPIE Press.

45 Wei, B., Hu, W., Ming, Y. et al. (2014). Generating switchable and reconfi gurable optical vortices via photopatterning of liquid crystals. *Advance Materials* **26**: 1590–1595.

46 Beijersbergen, M.W., Allen, L., van der Veen, H., and Woerdman, J.P. (1993). Astigmatic laser mode converters and transfer of orbital angular-momentum. *Optics Communications* **96**: 123–132.

47 Masajada, J. and Dubik, B. (2001). Optical vortex generation by three plane wave interference. *Optics Communications* **198**: 21–27.

48 Wang, G., Wang, L.Q., and Zhu, S.Y. (2009). Formation of optical vortices using coherent laser beam arrays. *Optics Communications* **282**: 1088–1094.

49 Yang, Y., Zhao, Q., Liu, L. et al. (2019). Manipulation of orbital-angular-momentum spectrum using pinhole plates. *Physical Review Applied* **12**: 064007.

50 Yu, N. and Capasso, F. (2014). Flat optics with designer metasurfaces. *Nature Materials* **13**: 139–150.

51 Yu, N., Genevet, P., Aieta, F. et al. (2013). Flat optics: Controlling wavefronts with optical antenna metasurfaces. *IEEE Journal of Selected Topics in Quantum Electroics* **19** (4700423).

52 Chen, S., Cai, Y., Li, G. et al. (2016). Geometric metasurface fork gratings for vortex-beam generation and manipulation. *Laser Photonics Review* **10**: 322–326.

53 Shen, Y., Wang, X., Xie, Z. et al. (2019). Optical vortices 30 years on: OAM manipulation from topological charge to multiple singularities. *Light: Science and Applications* **8**: 90.

54 Ostrovsky, A.S., Rickenstorff-Parrao, C., and Arrizón, V. (2013). Generation of the "perfect" optical vortex using a liquid-crystal spatial light modulator. *Optics Letters* **38** (4): 534–536.

55 Chen, M., Mazilu, M., Arita, Y. et al. (2013). Dynamics of microparticles trapped in a perfect vortex beam. *Optics Letters* **38** (22): 4919–4922.

56 García-García, J., Rickenstorff-Parrao, C., Ramos-García, R. et al. (2014). Simple technique for generating the perfect optical vortex. *Optics Letters* **39** (18): 5305–5308.

57 Vaity, P. and Rusch, L. (2015). Perfect vortex beam: Fourier transformation of a Bessel beam. *Optics Letters* **40** (4): 597–600.

58 Chen, Y., Fang, Z.X., Ren, Y.X. et al. (2015). Generation and characterization of a perfect vortex beam with a large topological charge through a digital micromirror device. *Applied Optics* **54** (27): 8030–8035.

59 Ren, Y.X., Li, M., Huang, K. et al. (2010). Experimental generation of Laguerre-Gaussian beam using digital micromirror device. *Applied Optics* **49** (10): 1838–1844.

60 Berry, M.V. (2004). Optical vortices evolving from helicoidal integer and fractional phase steps. *Journal of Optics A: Pure and Applied Optics* **6** (2): 259.

61 Martinez-Castellanos, I. and Gutiérrez-Vega, J.C. (2015). Shaping optical beams with non-integer orbital-angular momentum: a generalized differential operator approach. *Optics Letters* **40** (8): 1764–1767.

62 Tao, S.H., Yuan, X.C., Lin, J. et al. (2005). Fractional optical vortex beam induced rotation of particles. *Optics Express* **13** (20): 7726–7731.

63 Chen, L.X., Lei, J.J., and Romero, J. (2014). Quantum digital spiral imaging. *Light: Science and Applications* **3** (3): 153.

64 Zhang, W., Wang, L., Wang, W., and Zhao, S. (2019). Propagation property of Laguerre-Gaussian beams carrying fractional orbital angular momentum in an underwater channel. *OSA Continuum* **2** (11): 3281–3287.

65 Wen, J., Wang, L.G., Yang, X. et al. (2019). Vortex strength and beam propagation factor of fractional vortex beams. *Optics Express* **27** (4): 5893–5904.

66 Yang, Z., Zhang, X., Bai, C., and Wang, M. (2018). Nondiffracting light beams carrying fractional orbital angular momentum. *Journal of the Optical Society of America A* **35** (3): 452–461.

67 Dorrah, A.H., Zamboni-Rached, M., and Mojahedi, M. (2016). Controlling the topological charge of twisted light beams with propagation. *Physical Review A* **93** (6): 063864.

68 Leach, J., Yao, E., and Padgett, M.J. (2004). Observation of the vortex structure of a non-integer vortex beam. *New Journal of Physics* **6** (1): 71.

69 Basistiy, I.V., Pas'Ko, V.A., Slyusar, V.V. et al. (2004). Synthesis and analysis of optical vortices with fractional topological charges. *Journal of Optics A: Pure and Applied Optics* **6** (5): S166.

70 Götte, J.B., O'Holleran, K., Preece, D. et al. (2008). Light beams with fractional orbital angular momentum and their vortex structure. *Optics Express* **16** (2): 993–1006.

71 Turpin, A., Rego, L., Picon, A. et al. (2017). Extreme ultraviolet fractional orbital angular momentum beams from high harmonic generation. *Scientific Reports* **7** (1): 43888.

72 Hernández-García, C., Picón, A., San Román, J., and Plaja, L. (2013). Attosecond extreme ultraviolet vortices from high-order harmonic generation. *Physical Review Letters* **111** (8): 083602.

73 Yang, Y., Dong, Y., Zhao, C., and Cai, Y. (2013). Generation and propagation of an anomalous vortex beam. *Optics Letters* **38** (24): 5418–5421.

74 Yang, Y., Zhu, X., Zeng, J. et al. (2018). Anomalous Bessel vortex beam: modulating orbital angular momentum with propagation. *Nanophotonics* **7** (3): 677–682.

75 Wang, H., Liu, L., Zhou, C. et al. (2019). Vortex beam generation with variable topological charge based on a spiral slit. *Nanophotonics* **8**: 317–324.

76 Rego, L., Dorney, K.M., Brooks, N.J. et al. (2019). Generation of extreme-ultraviolet beams with time-varying orbital angular momentum. *Science* **364** (6447): 9486.

77 Hernández, G. C., Rego, L., Román, J. S., Picón, A., Plaja, L. (2017) Non-perturbative twist of attosecond extreme-ultraviolet vortex beams, In: *Society of Photo-Optical Instrumentation Engineers (SPIE) Conference Series*. San Francisco, CA 101021C (23 February 2017).

78 Yang, Y., Thirunavukkarasu, G., Babiker, M., and Yuan, J. (2017). Orbital-angular-momentum mode selection by rotationally symmetric superposition of chiral states with application to electron vortex beams. *Physical Review Letters* **119**: 094802.

79 Jin, Z., Janoschka, D., Deng, J., Ge, L., et. al. (2021). *eLight* **1**: 5.

8

Orbital Angular Momentum Generation, Detection, and Angular Momentum Conservation with Second Harmonic Generation

Menglin L. N. Chen,[1] Xiaoyan Y. Z. Xiong,[1] Wei E. I. Sha[2], and Li Jun Jiang[1]

[1] *Department of Electrical and Electronic Engineering, The University of Hong Kong, Hong Kong, P.R. China*
[2] *Key Laboratory of Micro-nano Electronic Devices and Smart Systems of Zhejiang Province, College of Information Science and Electronic Engineering, Zhejiang University, Hangzhou, P.R. China*

8.1 Orbital Angular Momentum Generation and Detection

It is well known that electromagnetic (EM) waves carry linear momentum. Except for the linear momentum, the angular momentum (AM) which includes both the spin and orbital parts also contributes to the total momentum of EM waves [1]. The spin angular momentum (SAM) is $\pm\hbar$ per photon for circularly polarized waves. The orbital angular momentum (OAM) is $l\hbar$ per photon, where l is known as the OAM index and found to be well defined in Laguerre–Gaussian (LG) modes by Allen et al. since 1992 [2]. While the SAM per photon can take only two possible values, the OAM index can take any integers, fractional, or even rational values [3–5]. This new degree of freedom provided by OAM attracts much attention and has been explored for applications in optical tweezers [6–8], communications [9–11], super-resolution imaging [12], etc.

Although the concept of OAM of light is firstly introduced in optical and quantum regimes, the exploration of its applications is quickly extended to millimeter-wave frequencies [13, 14], and further to radio frequencies [15–17]. As the two most important research areas, the generation and the detection of OAM are being explored extensively. Different principles have been proposed for OAM generation, and the AM conservation law is applicable to all corresponding designs. For example, at quantum regime, the OAM photon states are generated through the parametric down-conversion [18] and spin–orbit coupling [19]. At optical regimes, forked gratings [20, 21], spiral phase plates [22–24], and q plates [25–27] have been proposed. At lower frequencies, by simply scaling the dimensions down, the optical methods still work. However, the devices will inevitably become bulky. Meanwhile, although Maxwell's equations govern behaviors of EM waves from radio to optical frequencies, the devices after scaling could respond distinctively due to the dispersive nature of materials. Therefore, optical designs and radio-wave designs suffer from different challenges. To break these limitations, artificial materials are proposed to implement conventional optical devices without loss of but new functions [28]. Artificial atoms have been designed in various geometries, such as the L-shaped scatterers [28–30], nanorods [31], nanoholes [32], split-ring resonators (SRRs) [33], etc. The artificial atoms respond to EM waves locally and modulate the phase or/and amplitude. As a typical type of modulation approaches, the phase of an EM wave can be controlled by varying the orientation of composed scatterers through the induced

Electromagnetic Vortices: Wave Phenomena and Engineering Applications, First Edition.
Edited by Zhi Hao Jiang and Douglas H. Werner.

geometric phase [34]. In the following, we will introduce several geometric-phase metasurfaces and a photonic crystal (PC) structure for OAM generation at microwave frequencies. As for the OAM detection, a holographic metasurface will be shown. Besides, a modal analysis approach that extracts both the OAM order and amplitude distribution will be explored.

8.1.1 OAM Generation

Light beams with helical wavefronts carry well-defined OAM [2]. Compared with plane waves, vortex beams have an additional phase term, $e^{il\phi}$, where ϕ is the azimuthal angle and l is the OAM index. While different design routes are proposed for different application scenarios, all of them are based on the introduction of the azimuthal phase, $e^{il\phi}$ to EM waves. At radio frequencies, antenna arrays [35, 36], traveling wave and patch antennas [37, 38] that radiate EM waves carrying OAM have been demonstrated. Besides, the phase term can be introduced by engineering scatterers on metasurfaces. Usually, by varying the geometries or orientations of scatterers, a full phase coverage of 2π can be achieved, which is necessary for arbitrary wavefront manipulation. A special type of scatterer covers the phase range from zero to 2π by only varying the orientation. These scatterers construct the so-called geometric-phase metasurfaces. Thus, the desired phase term, $e^{il\phi}$ for OAM generation, is introduced by arranging the scatterers with predesigned orientations.

The phase shift provided by geometric-phase metasurfaces originates from the change in the polarization state along different paths on the Poincaré sphere [39]. First, we model an arbitrary scatterer, Figure 8.1a, by Jones matrix, **J**. **J** connects the incident and scattered (transmitted or reflected) fields, which is expressed as

$$\begin{pmatrix} E_x^s \\ E_y^s \end{pmatrix} = \mathbf{J} \begin{pmatrix} E_x^i \\ E_y^i \end{pmatrix} = \begin{pmatrix} J_{xx} & J_{xy} \\ J_{yx} & J_{yy} \end{pmatrix} \begin{pmatrix} E_x^i \\ E_y^i \end{pmatrix}, \tag{8.1}$$

where the subscripts denote the polarization states and the first/second subscripts of J correspond to the scattered/incident wave.

Jones matrix under circular basis can be rewritten as

$$\begin{aligned} \mathbf{J}^c &= \begin{pmatrix} J_{++} & J_{+-} \\ J_{-+} & J_{--} \end{pmatrix} \\ &= \frac{1}{2} \begin{pmatrix} (J_{xx}+J_{yy}) + i(J_{xy}-J_{yx}) & (J_{xx}-J_{yy}) - i(J_{xy}+J_{yx}) \\ (J_{xx}-J_{yy}) + i(J_{xy}+J_{yx}) & (J_{xx}+J_{yy}) - i(J_{xy}-J_{yx}) \end{pmatrix}, \end{aligned} \tag{8.2}$$

where + and − represent the left circularly polarized (LCP) and right circularly polarized (RCP) components, respectively.

When the scatterer is designed to achieve a perfect conversion between RCP and LCP waves, i.e. $J_{yy} = -J_{xx}$ and $J_{yx} = J_{xy} = 0$, a rotation of the scatterer, Figure 8.1b, will result in the Jones matrix in the following special form:

$$\mathbf{J}(\alpha) = J_{xx} \begin{pmatrix} \cos(2\alpha) & \sin(2\alpha) \\ \sin(2\alpha) & -\cos(2\alpha) \end{pmatrix}, \quad \mathbf{J}^c(\alpha) = J_{xx} \begin{pmatrix} 0 & e^{-2i\alpha} \\ e^{2i\alpha} & 0 \end{pmatrix}. \tag{8.3}$$

We can notice that there are only nonzero off-diagonal items in $\mathbf{J}^c(\alpha)$. Therefore, the rotated scatterer flips the polarization state of the incident circularly polarized wave with an additional phase factor that depends solely on its rotation angle. This phase is called geometric phase. For OAM generation, the geometric phase of each scatterer is set according to its azimuthal position, i.e. equal to $e^{il\phi}$.

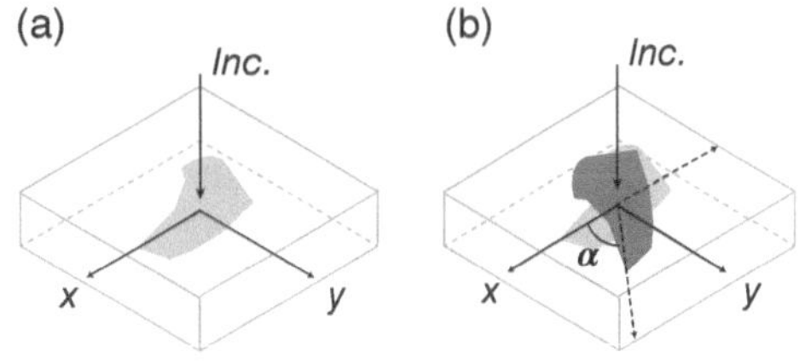

Figure 8.1 Schematic presentation of a scatterer that rotates at an angle of α. (a) Original scatterer; (b) the rotated scatterer.

As described in Eq. (8.3), one can obtain a perfect (100%) conversion if J_{xx} and J_{yy} have the same unit amplitude and π phase difference. When the prerequisite is not strictly satisfied, the co-circularly polarized component needs to be filtered out. Only the cross-circularly polarized component carries OAM.

8.1.1.1 Complementary Metasurfaces

Figure 8.2 shows an ultrathin complementary metasurface that radiates an OAM wave with high efficiency [40]. As shown in Figure 8.2a, a double-layer unit cell consisting of four complementary SRRs in two different sizes and orthogonal orientations is built. The double-layer complementary geometry guarantees the high transmittance. The fundamental resonance of the larger two complementary SRRs is excited by y polarization, while that of the smaller two complementary SRRs is excited by x polarization. By tuning their sizes, the transmitted x and y components exhibit the same amplitude and a π phase difference, i.e. the condition $J_{yy} = -J_{xx}$ is fulfilled. To make $J_{yx} = J_{xy} = 0$, the mirror symmetry with respect to the x–z or y–z plane needs to be maintained [41], which is the case here. Therefore, the unit cell changes the helicity of the incident circularly polarized wave with a conversion efficiency determined by the magnitude of J_{yy} or J_{xx}.

A metasurface is built for OAM generation, as shown on the left panel of Figure 8.2b. The axial rotation angle for each unit cell on the metasurface is equal to 2ϕ. Thus, the phase shift introduced by the unit cell is $e^{\pm 4i\phi}$. The whole metasurface provides an additional phase distribution of $e^{\pm 4i\phi}$. The sign depends on the helicity of the incident circular polarization. The EM responses of the metasurface are obtained by two approaches, namely, the equivalent dipole approximation and the full-wave simulation in CST Microwave Studio (MWS). In the equivalent dipole model, each unit cell is considered as two orthogonal magnetic dipoles, as indicated by the arrows shown on the right panel of Figure 8.2b. Then, the EM response of the whole metasurface is calculated by adding up the responses of all the equivalent magnetic dipoles. The transmitted field through the metasurface under plane wave illumination is shown in Figures 8.2c and d. Since the coupling among unit cells is not considered in the dipole model, the results obtained from the two approaches show some deviations. Overall, the main features of the simulated field distribution from the two approaches keep the same. We can see zero-field modulus (related to phase singularity) at the beam axis, resulting from the ill-defined phase. Meanwhile, there is an 8π phase change in total along a closed path around the beam axis. Consequently, an OAM of order 4 is successfully introduced to the transmitted fields.

8.1.1.2 Quasi-Continuous Metasurfaces

Usually, the scatterers on a geometric-phase metasurface provide discrete phase settings. Inevitably, there will be distortions from the broken periodicity from the scatterers in different orientations and the local-response discontinuity. In this section, we will introduce a quasi-continuous metasurface to generate the OAM with high purity [42, 43].

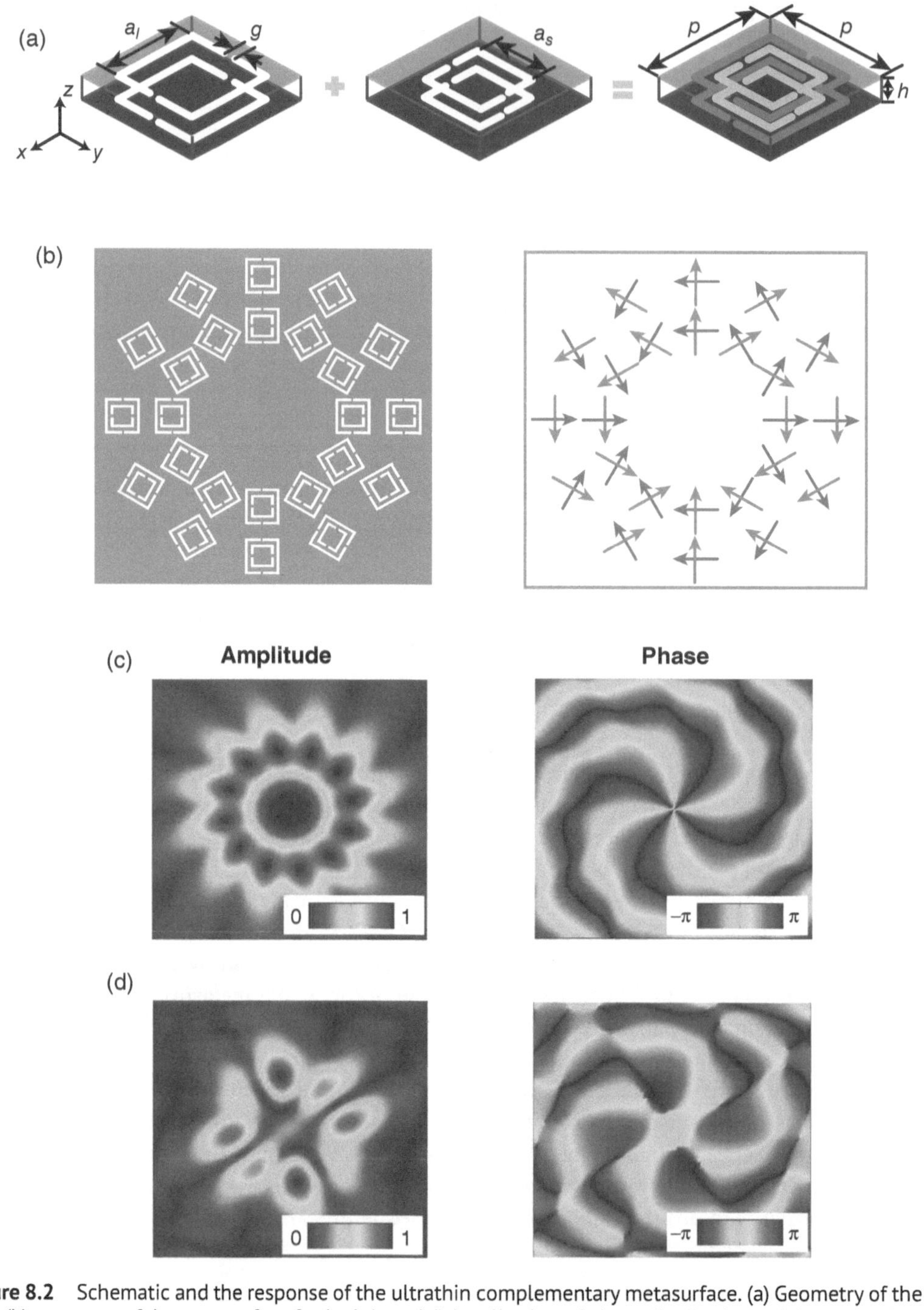

Figure 8.2 Schematic and the response of the ultrathin complementary metasurface. (a) Geometry of the unit cell; (b) geometry of the metasurface for $l = 4$. (c and d) Amplitude and phase distributions of the cross-circularly polarized component of the electric field on a transverse plane. The results in (c) are calculated from the equivalent dipole model with the distributed dipole moments shown in (b). The results in (d) are the full-wave simulation results. *Source:* Chen et al. [40]; © 2017 IEEE.

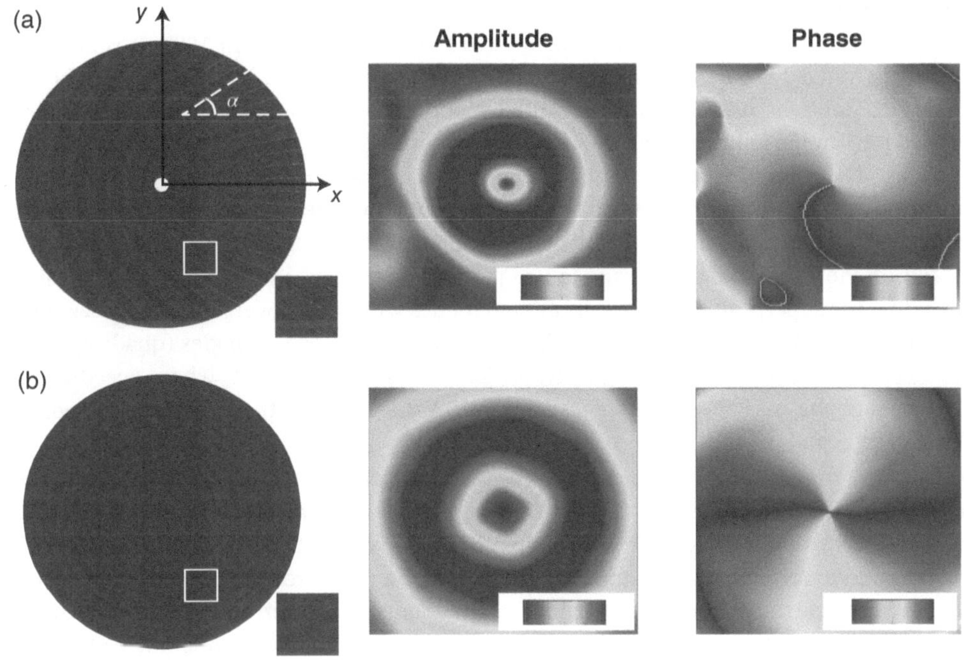

Figure 8.3 The composite PEC–PMC metasurfaces for OAM generation. (a) $l = 1$. (b) $l = 2$. The left panels show the top views of the metasurfaces. α denotes the inclination angle between the metal strip tangent and x axis. The middle and the right panels show the amplitude and phase distributions of the corresponding reflected waves on a transverse plane away from the metasurfaces.

The top views of the two metasurfaces are shown on the left panels of Figure 8.3. Each metasurface is composed of two layers, the top perfect electric conductor (PEC) layer (the dark strips) and the bottom perfect magnetic conductor (PMC) layer (light plane). For practical implementation, the PEC layer can be metallic. A mushroom-like high-impedance surface acts as the PMC layer and is put beneath the metal strips. Any local area on the metasurface can be considered as a rotated composite PEC–PMC scatterer with a rotation angle of α. The PEC layer locally behaves like a parallel-plate waveguide. When the gaps between adjacent metal strips are small enough, the wave component parallel to the strip tangent will be totally reflected with a π phase shift. Then, the penetrated orthogonal component will be reflected by the PMC plane without the phase shift. Therefore, a phase shift of π can be obtained between the two scattered orthogonal components. The local responses of the PEC–PMC scatterers can be considered as the scatterers with $J_{yy} = -1$, $J_{xx} = 1$, along with an additional phase of $e^{\pm 2i\alpha}$. The plus/minus sign is taken when the incident wave is LCP/RCP. In Figure 8.3a, we make $\alpha = \frac{\phi}{2}$ and in Figure 8.3b, $\alpha = \phi$. Therefore, the whole metasurfaces offer phase distributions of $e^{\pm i\phi}$ and $e^{\pm 2i\phi}$, respectively. The patterning of the PEC layer is accomplished with the assistance of grating vectors. It is different from existing design protocols for geometric-phase metasurfaces, where a complicated optimization process of single scatterer is needed.

Then, the metasurfaces are illuminated by an LCP Gaussian wave. The amplitude and phase distributions of the reflected RCP field components at a transverse plane are depicted in Figure 8.3. Clear phase singularities are observed in both cases. We find field distributions of OAM waves with $l = 1$ in Figure 8.3a, and $l = 2$ in Figure 8.3b.

8.1.1.3 Photonic Crystals

The concept of PCs was proposed by Yablonovitch [44] and John [45] in 1987. It is an analog of periodically arranged atoms and therefore, opens up a band gap where no EM state is allowed to propagate. However, a localized state will be supported at the photonic band gap if a defect is introduced [46]. In the following, we will make use of point defect states in dielectric PCs to produce OAM states [47]. Here, the transverse magnetic modes are considered.

By making the radius of one dielectric rod different from the other rods in a PC, point defect states are supported at the frequencies within the photonic band gap. By increasing the radius of the defect rod, monopole, dipole, quadrupole-*xy*, quadrupole-diag, ... modes emerge successively. It is known that the superposition of two vibrational modes with proper weights and phases generates a rotational mode [48]. In Figure 8.4, we show the quadrupole defect modes (quadrupole-*xy* and quadrupole-diag modes) and their superposition. It is noticed that when specific weights ($A = B = 1$) and phases ($\theta = \pm \frac{\pi}{2}$) are chosen, pure OAM states with the spatial phase dependence of $e^{\pm i2\phi}$ can be obtained.

To excite the two quadrupole states with desired weights and phases simultaneously, we conducted an optimization process in CST MWS. The simulation model is shown in Figure 8.5. To be specific, there are dielectric rods sandwiched between two metallic plates. A row of dielectric rods is removed to form a waveguiding channel. There is a defect rod to support two defect states. Right above the defect rod, a hole is opened. The EM energy is transferred from the guided wave in the line defect to localized defect states in the defect rod and then to the unbounded OAM state in free space. Eigen frequencies of the quadrupole states are fixed when the size of the defect rod is fixed. Therefore, at any frequency between the eigenfrequencies, the state can be considered as the superposition of the two quadrupole states. By sweeping the frequency, the weights and phases can be adjusted. The other parameter that needs to be optimized is the height of the rods because it directly determines the modes at the air-dielectric interface. By feeding the structure from the other end of the channel, the helicity of the radiated OAM wave will be changed.

The design is experimentally verified as shown in Figure 8.6. The implementation of the PCs at radio regime adopts the approach in [49]. The PCs were made of alumina rods with $\epsilon_r = 8.5$. They were assembled between two flat aluminum plates, with a cutting hole on the top one (not shown here). The whole structure was surrounded by absorbers to avoid unwanted scattering. A standard linearly polarized horn antenna operating from 6.57 to 9.99 GHz was used to excite the structure. A vertical probe was fixed onto a near-field scanning platform to measure the vertically polarized field component at a transverse plane. Additional absorbers were placed near the horn antenna on the top of the top aluminum plate to prevent direct transmission from the horn antenna to the

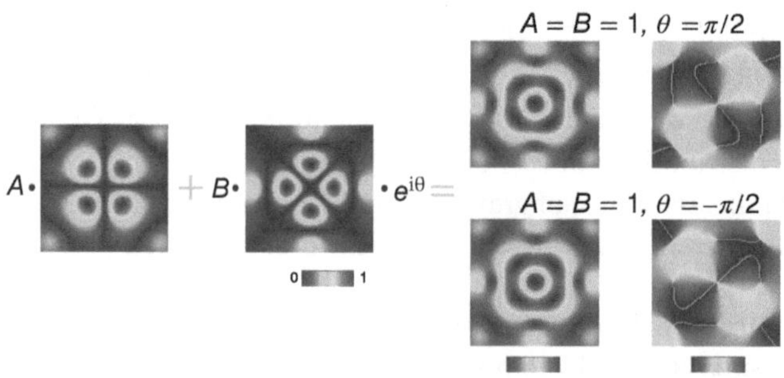

Figure 8.4 The quadrupole-*xy* and quadrupole-diag states (amplitude patterns) and their superposition (amplitude and phase patterns). *A* and *B* are the weights of two states, and *θ* is the relative phase between them.

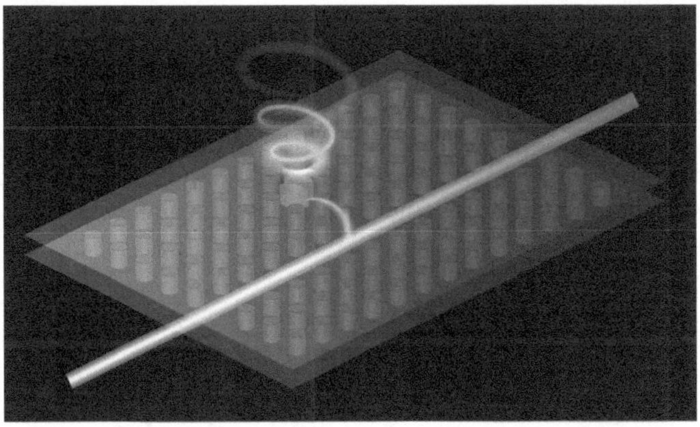

Figure 8.5 A schematic representation of OAM generation in PCs. *Source:* Chen et al. [47]; © 2018 American Physical Society.

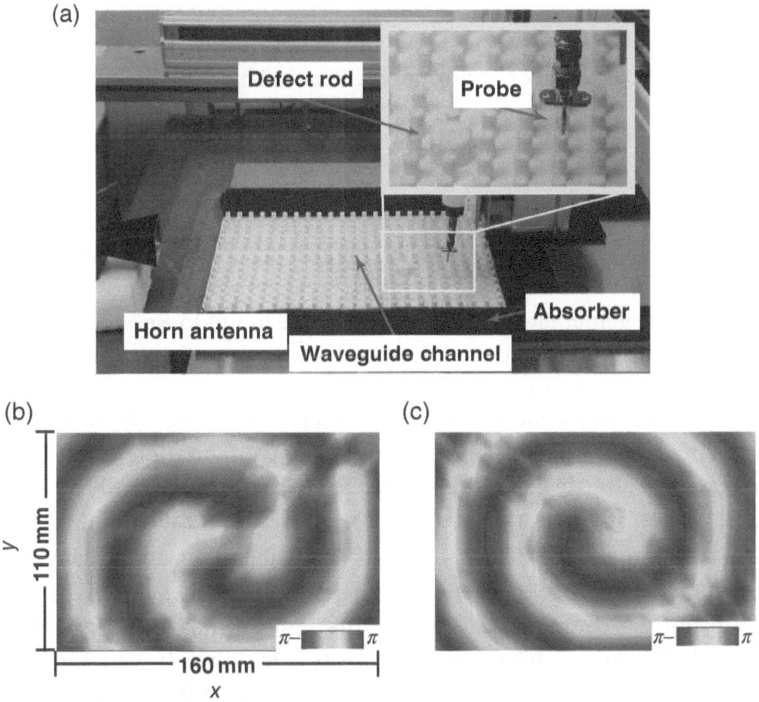

Figure 8.6 The experimental verification of the proposed PC. (a) Experimental setup. The phase distributions of E_z at a transverse plane 30 mm above the structure with (b) quadrupole defect, measurement frequency of $f = 8.75$ GHz and (c) dipole defect, measurement frequency of $f = 9.55$ GHz. *Source:* Chen et al. [47]; © 2018 American Physical Society.

receiving probe. The measured plane was 25 mm away from the structure. The field over a 160 × 110 mm area was measured. The results for the quadrupole and dipole cases are shown in Figure 8.6b, respectively. We can observe the rotating phase of OAM waves of order 2 and 1. The higher-order OAM waves may be produced by continuously increasing the radius or permittivity of the defect rod.

8.1.2 OAM Detection

As a reciprocal process of OAM generation, OAM detection at the receiver side is of equal importance but challenging due to the divergence and spatial-dependence nature of an OAM wave. The approaches for OAM detection can be classified into three categories, mode analysis based on field data [50–52], observation of the rotational Doppler shift [53], and holographic technology [54]. We will discuss the mode analysis approach and holographic technology for the OAM detection at microwave frequencies.

8.1.2.1 Modified Dynamic Mode Decomposition

The dynamic mode decomposition (DMD) was first proposed in hydrodynamics to deal with a spatial-temporal correlated signal and compute its oscillation frequency and decay/growth rate, i.e. the complex ω in $e^{i\omega t}$ [55]. Since then, it has been applied in many fields, such as discovering governing equations in nonlinear system, global power systems, chaos analysis, and the spread of infections [56–59]. Interestingly, there is an analogy between $e^{i\omega t}$ and the phase term of an OAM mode, $e^{il\phi}$ (ω corresponds to l and t corresponds to ϕ). Therefore, as the input of DMD is the time series of data, and the output of DMD is ω, we are able to obtain l by providing the azimuthal series of fields.

We consider only the OAM waves at normal incidence. Thus, the field can be expressed by

$$E(r,\phi) = \sum_{i=1}^{n} A_i(r)e^{il_i\phi} \tag{8.4}$$

where r is the radial position, ϕ is the azimuthal position, l_i is the OAM index, and $A_i(r)$ is the amplitude profile which is solely dependent on r. When the field information of the incident OAM waves is acquired, the complex l_i with the corresponding $A_i(r)$ can be obtained through the modified DMD solver. Details for the implementation of the solver can be found in Ref. [60]. The modes with the nonzero imaginary part of l_i are abandoned. The modes with the only real part of l_i are the effective OAM modes decomposed from the incident wave.

To validate the practical significance of the proposed DMD method, the radiated EM wave from the two uniform circular arrays (UCAs) is analyzed. The configuration of the UCAs is shown in Figure 8.7a. A total of 16 electrical dipole antennas are located equidistantly around the perimeter of each circle. The inner UCA has a phase step of $\frac{\pi}{8}$, as a result, it radiates an OAM wave of order $l_1 = 1$. The phase step of the outer UCA is $\frac{3\pi}{8}$, thus $l_2 = 3$. The full-wave simulated field distribution (amplitude and phase) by CST MWS on a transverse plane 60 mm away from the UCAs is shown in Figure 8.7b. By using the complex field data as the input signal, two OAM modes have been extracted from our DMD solver. The left panel of Figure 8.7c shows the extracted mode amplitudes along the radial coordinate. Clearly, there is a dip at the origin for both modes, which corresponds to the singularity of the OAM waves as in Figure 8.7a. By keeping only the inner or outer UCA, we obtain the amplitude distribution of the $l = 1$ or $l = 3$ mode from CST MWS. The results have good agreements with the extracted results in Figure 8.7c (not shown here). Meanwhile, the angular distributions of the real part of the two extracted mode fields are drawn on the right panel of Figure 8.7c. The OAM index represents the oscillation frequency in angular domain. For the first mode, there is one complete period in the range of 2π along the angular coordinate, implying an OAM index of 1. Similarly, we can see the second mode has an OAM of order 3.

Furthermore, we have tested the proposed DMD solver when a partial angular receiving aperture is used. The field data from 0 to π in Figure 8.7b are collected as the input of our DMD solver.

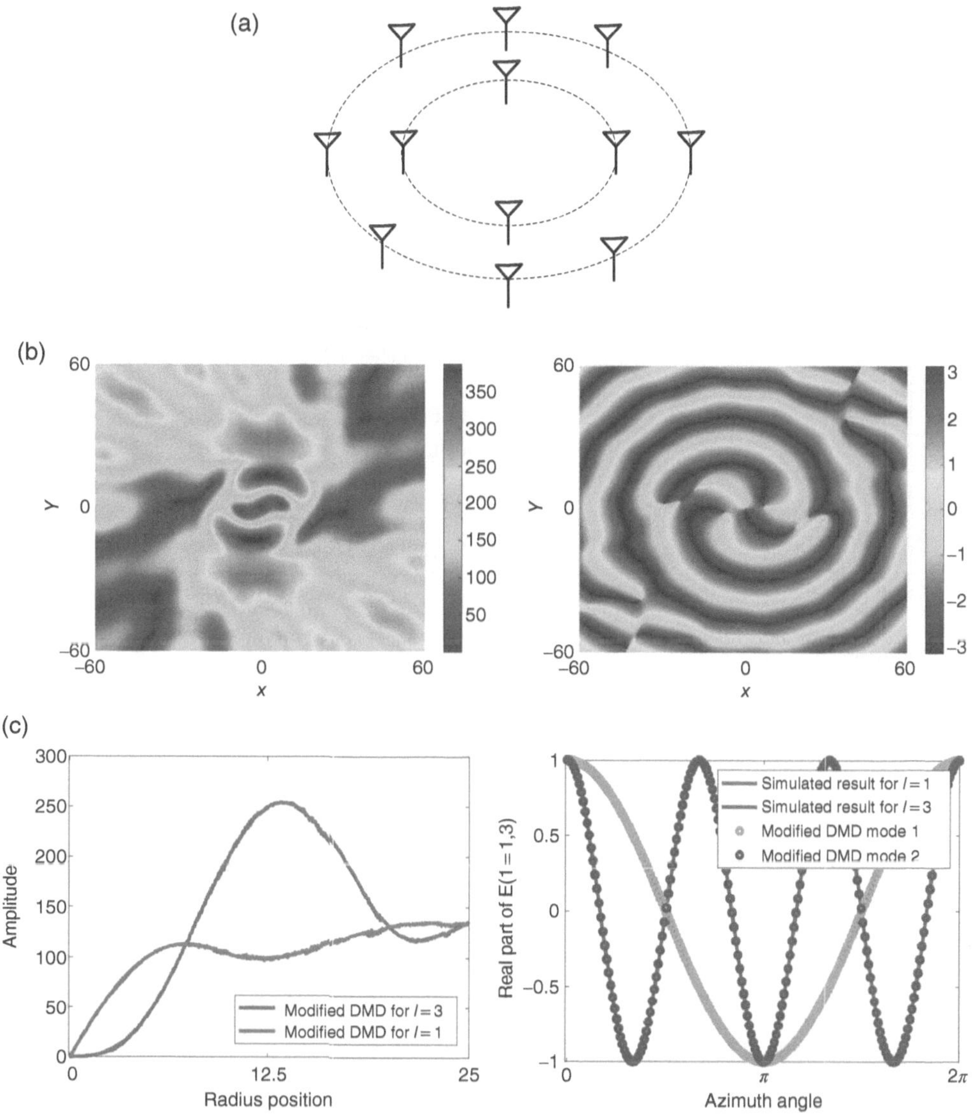

Figure 8.7 Analysis of the radiated EM wave from UCAs. (a) Configuration of the UCAs. The radii of the inner and outer circles are 15 and 60 mm, respectively. (b) The simulated amplitude and phase distributions of the radiated electric field at a transverse plane 60 mm above the UCAs. (c) The extracted mode amplitudes along the radial coordinate and real-space distributions of the electric fields along the angular coordinate. The simulation frequency is 10 GHz. *Source:* Zhang et al. [60]; © 2019 The Optical Society.

Figure 8.8 shows the radial distributions of the decomposed mode amplitudes and the angular distributions of the real part of the mode fields. These results show the consistencies with the results obtained when the data from the whole aperture is used, Figure 8.7c.

Usually, the modal analysis based on the Fourier transform can only find the indices of the OAM modes. By using the proposed DMD solver, both the OAM indices and the corresponding amplitude patterns can be obtained. Furthermore, since it is difficult to measure the whole aperture after long-distance transmission due to the divergence of OAM waves, the proposed algorithm that does not require the data of the whole aperture is more practical.

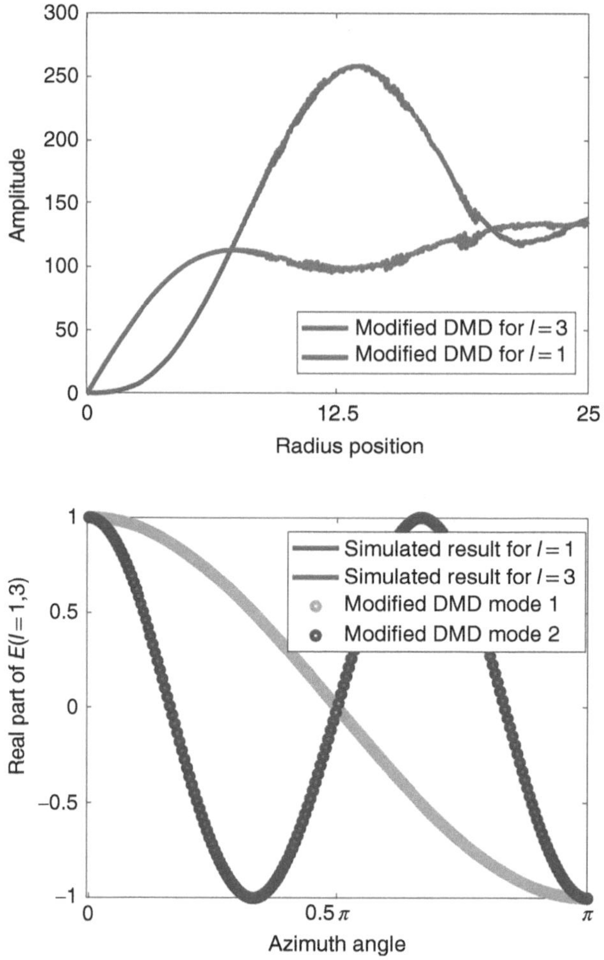

Figure 8.8 The extracted mode amplitudes along with the radial coordinates and real-space distributions of the electric fields along the angular coordinate from the partial angular receiving aperture (0–π) for the case in Figure 8.7. *Source:* Zhang et al. [60]; © 2019 The Optical Society.

8.1.2.2 Holographic Metasurfaces

The proposed DMD method in Section 8.1.2.1 can be employed when both the intensity and phase information are accessible. In an optical regime, it is rather challenging to obtain the phase information. Therefore, the holographic detection approach is proposed. By using the holographic technology, the OAM modes are transferred into a detectable Gaussian mode [18]. We apply this approach to detect OAM modes in microwave frequencies [61]. The implementation of this technology is achieved using the complementary metasurface proposed in Section 8.1.1.1. The detection process is summarized in Figure 8.9. The metasurface converts the incident wave to a Gaussian wave with distinct directivity determined by the incident OAM order. Consequently, by identifying the Gaussian wave, the incident OAM order can be determined.

To start with, we model the metasurface by a transmittance function:

$$t(r,\phi) = \sum_m A_m e^{i\left(l_m\phi + k_{xm}x + k_{ym}y\right)}, \tag{8.5}$$

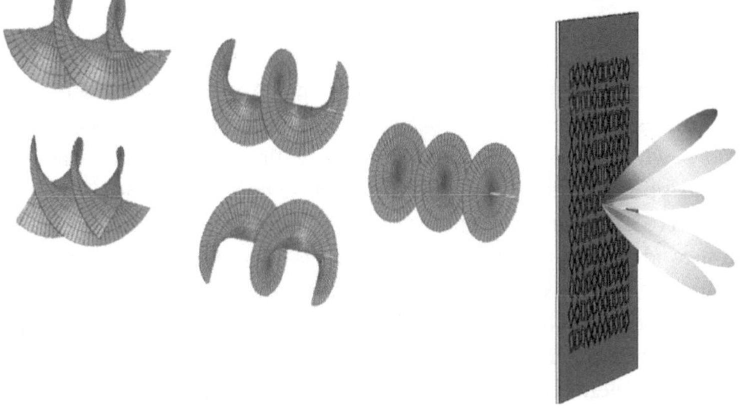

Figure 8.9 Schematic representation of multiple OAM-beam detections by using a single metasurface. *Source:* Chen et al. [61]; © 2018 IEEE.

where r is the radial position, ϕ is the azimuthal position, A_m is the weight of the mth beam, l_m is the corresponding OAM index, and k_{xm}, k_{ym} are the transverse wave numbers of the mth beam.

Then, the far-field response of the metasurface illuminated by a normally incident wave carrying an OAM of order l_0 is calculated by doing the Fourier transform:

$$E = F\{E_{in} \cdot t\} = \sum_{m} A_m F\{E_{\text{OAM}(l_m + l_0)}(k_{xm}, k_{ym})\}. \tag{8.6}$$

Equation (8.6) means that multiple beams are generated at the k-space position (k_{xm}, k_{ym}) with the OAM order of $l_m + l_0$. When $l_M + l_0 = 0$, the beam is Gaussian and its beam axis is at (k_{xM}, k_{yM}). It is known that the OAM wave has a singularity at its beam axis. Therefore, by examining the field intensity at the positions of (k_{xm}, k_{ym}), we can identify the Gaussian beam. Then the incident OAM order l_0 can be determined.

As a proof-of-concept, a five-beam case with $l_m = 2{,}1{,}0, -1, -2$ at the directions of $\theta = 40°$ and $\phi =$ 90, 18, 306, 234, and 162° is demonstrated. The transmittance function $t(r, \phi)$ is calculated first. Then, its phase information is extracted and implemented using the unit cell shown in Figure 8.2a. Full-wave simulated radiation patterns are shown in Figure 8.10. As expected, the maximum radiation directions (axis of the Gaussian beam) when $l_0 = -2, -1{,}0{,}1{,}2$ are at $\theta = 40°$ and $\phi =$ 90, 18, 306, 234, and 162°, respectively.

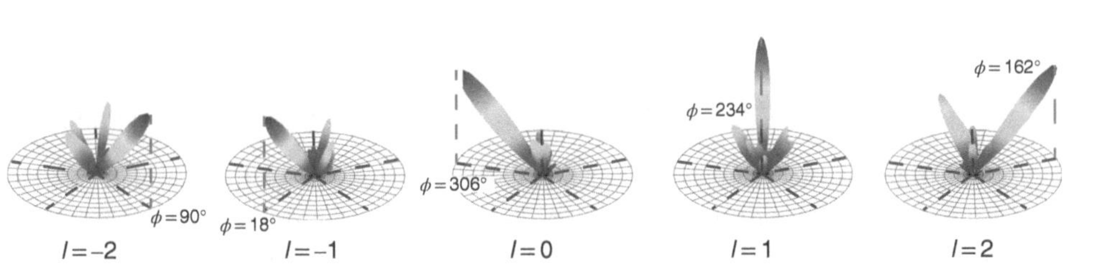

Figure 8.10 Full-wave simulated far-field power patterns when the incident wave carries an OAM of order −2, −1, 0, 1, and 2. *Source:* Chen et al. [61]; © 2018 IEEE.

8.2 AM Conservation: Nonlinear Optics

Nonlinear optical effects arise when electron motion in a strong EM field cannot be considered harmonic. Optical vortex beams carrying orbital and spin AM have also been explored in various nonlinear optical processes recently, e.g. OAM entanglement between photons created by spontaneous parametric down conversion [18], OAM conservation in nonlinear wave mixing [62], and enhanced nonlinear optical activity and nonlinear phase control [63, 64]. The process of second-harmonic generation (SHG), where two photons with fundamental frequency (FF), ω, convert into one photon with second-harmonic (SH) frequency, 2ω, is the nonlinear process most used in studies of OAM, chiefly due to the relative simplicity of its experimental implementation. In the following, we will study the scattering of circularly polarized LG beams from plasmonic and dielectric nanospheres made of centrosymmetric materials, as well as chiral clusters made of such nanospheres, by means of the boundary element method (BEM). In particular, we formulate a general angular momenta conservation law for the nonlinear spin-orbital AM interaction, which includes the quasi-angular momentum of chiral structures with different-order rotational symmetries.

8.2.1 BEM for Nonlinear Optics

The SHG nonlinear process is significantly dependent on the symmetry of both the material being used and the structure being studied [65]. It is forbidden in the bulk of centrosymmetric media under the electric dipole approximation. However, the breaking of inversion symmetry at surfaces results in surface nonlinear scattering [66, 67]. The BEM is proposed for modeling second-order surface nonlinear scattering problems. The proposed method is efficient with a surface discretization, and can employ experimentally tabulated material parameters directly.

The second-order surface nonlinear problem, under undepleted-pump approximation, can be described by the following driven wave equation with the nonlinear polarization of medium as a source term [68]

$$\nabla^2 \mathbf{E}^{(\nu)} + k^{(\nu)2}\mathbf{E}^{(\nu)} = -\frac{\nu^2}{\varepsilon_0 c^2}\mathbf{P}^{(\nu)} \tag{8.7}$$

where $k^{(\nu)}$ is the wave number at frequency ν; $\mathbf{P}^{(\nu)}$ is the surface nonlinear polarization at frequency ν. It has the form $\mathbf{P}^{(\nu)} = \varepsilon_0 \overline{\overline{\chi}}^{(2)} : \mathbf{E}_1^{(\omega_1)}\mathbf{E}_2^{(\omega_2)}$ and $\nu = \omega_1 + \omega_2$ for sum frequency (SF) generation (SFG). The SHG process is a special case of SFG with $\omega_1 = \omega_2$ and $\nu = 2\omega_1$. $\overline{\overline{\chi}}^{(2)}$ is the surface second-order nonlinear susceptibility tensor. $\mathbf{E}_1^{(\omega_1)}$ and $\mathbf{E}_2^{(\omega_2)}$ denote the fundamental fields at frequencies ω_1 and ω_2, respectively. The symbol ':' is a tensor operator functions as $P_i^{(\nu)} = \sum_{j,k}\chi_{ijk}^{(2)}E_{1j}^{(\omega_1)}E_{2k}^{(\omega_2)}$. The surface of nanoparticles made of centrosymmetric materials has an isotropic symmetry with a mirror plane perpendicular to it. The surface susceptibility tensor has only three nonvanishing and independent elements

$$\mathbf{P}^{(\nu)} = \varepsilon_0 \hat{n}\left[\chi_{\perp\perp\perp}^{(2)}\mathbf{E}_{1\hat{n}}\mathbf{E}_{2\hat{n}} + \chi_{\perp\|\|}^{(2)}\left(\mathbf{E}_{1\hat{t}_1}\mathbf{E}_{2\hat{t}_1} + \mathbf{E}_{1\hat{t}_2}\mathbf{E}_{2\hat{t}_2}\right)\right] + \hat{t}_1\chi_{\|\perp\|}^{(2)}\mathbf{E}_{1\hat{n}}\mathbf{E}_{2\hat{t}_1} + \hat{t}_2\chi_{\|\perp\|}^{(2)}\mathbf{E}_{1\hat{n}}\mathbf{E}_{2\hat{t}_2} \tag{8.8}$$

where $\perp$ and $\|$ refer to the orthogonal and tangential components to the nanoparticle surface; $(\hat{t}_1, \hat{t}_2, \hat{n})$ is a system of three orthogonal vectors locally defined on the particle surface. The contributions of tangential and normal components of the surface nonlinear polarization are taken into account by the nonlinear surface electric and magnetic current sources

$$\mathbf{J}_0^{(\nu)} = -i\nu\mathbf{P}_t^{(\nu)} \tag{8.9}$$

$$\mathbf{M}_0^{(\nu)} = \frac{1}{\varepsilon'}\hat{n} \times \nabla_S P_n^{(\nu)} \tag{8.10}$$

where $\mathbf{P}_t^{(\nu)}$ and $P_n^{(\nu)}$ are the tangential and normal components of $\mathbf{P}^{\nu}$. The BEM method is applied to solve the driven wave equation by invoking the Love's equivalence principle.

The domain of the EM field is divided into the interior of the metal domain V_i, the exterior medium V_e, and the interface S. The object is illuminated by the incident wave source $\mathbf{E}^{inc}$. The equivalent currents $\mathbf{J}_e^{(\nu)}, \mathbf{M}_e^{(\nu)}$ positioned on the external page S^+ produce the scattered field in the region V_e and null field in the region V_i while the equivalent currents $\mathbf{J}_i^{(\nu)}, \mathbf{M}_i^{(\nu)}$, defined on the internal side S^-, produce the total field in the region V_i and null field in the region V_e.

$$\left.\begin{array}{l} \mathbf{r} \in V_\ell, \ \mathbf{E}_\ell^{(\nu)}(\mathbf{r}) \\ \mathbf{r} \notin V_\ell, \ 0 \end{array}\right\} = i\nu\mu_\ell \int_S \overline{\mathbf{G}}^{(\nu)}(\mathbf{r}, \mathbf{r}') \cdot \mathbf{J}_\ell^{(\nu)}(\mathbf{r}')d\mathbf{r}' - \int_S \nabla \times \overline{\mathbf{G}}^{(\nu)}(\mathbf{r}, \mathbf{r}') \cdot \mathbf{M}_\ell^{(\nu)}(\mathbf{r}')d\mathbf{r}' \tag{8.11}$$

Where $\overline{\mathbf{G}}^{(\nu)}(\mathbf{r}, \mathbf{r}') = \left(\overline{\mathbf{I}} + k^{(\nu)-2}\nabla\nabla\right) \exp\left(ik^{(\nu)}R\right)/4\pi R$ is the dyadic Green's function at frequency ν. Here, $R = |\mathbf{r} - \mathbf{r}'|$. $\ell = e, i$ denotes the exterior and interior regions of the object, respectively. The magnetic field has a similar representation. The equivalent currents satisfy

$$\mathbf{J}_i^{(\nu)} + \mathbf{J}_e^{(\nu)} = \mathbf{J}_0^{(\nu)} \tag{8.12}$$

$$\mathbf{M}_i^{(\nu)} + \mathbf{M}_e^{(\nu)} = \mathbf{M}_0^{(\nu)} \tag{8.13}$$

The surfaces of the nanostructures are discretized with triangular meshes. The equivalent currents are expanded with Rao–Wilton–Glisson basis functions [69]. A matrix system is then constructed by exploiting the Galerkin testing procedure. A modified Poggio–Miller–Chang Harrington–Wu–Tsai (PMCHWT) formulation [70] is used to ensure accurate solutions even at resonant conditions. The PMCHWT matrix equation can be written as

$$\overline{\mathbf{C}}^{(\nu)} \cdot \mathbf{X}^{(\nu)} = \mathbf{Y}^{(\nu)}, \tag{8.14}$$

the impedance matrix $\overline{\mathbf{C}}^{(\nu)}$ is

$$\overline{\mathbf{C}}^{(\nu)} = \begin{bmatrix} \mathcal{L}_e^{(\nu)} & \mathcal{K}_e'^{(\nu)} & -\mathcal{L}_i^{(\nu)} & -\mathcal{K}_i'^{(\nu)} \\ -\mathcal{K}_e'^{(\nu)} & \eta_e^{-2}\mathcal{L}_e^{(\nu)} & \mathcal{K}_i'^{(\nu)} & -\eta_i^{-2}\mathcal{L}_i^{(\nu)} \\ \mathcal{I} & 0 & \mathcal{I} & 0 \\ 0 & \mathcal{I} & 0 & \mathcal{I} \end{bmatrix}$$

where $\mathcal{L}_\ell^{(\nu)} = i\nu\mu_\ell\overline{\mathbf{G}}^{(\nu)}$ and $\mathcal{K}_\ell'^{(\nu)}$ is the principal value part of the $\mathcal{K}_\ell^{(\nu)}$ operator with $\mathcal{K}_\ell^{(\nu)} = -\nabla \times \overline{\mathbf{G}}^{(\nu)}$. $\eta_\ell = \sqrt{\mu_\ell/\epsilon_\ell}$, with $\ell = e, i$. The vector of unknowns $\mathbf{X}^{(\nu)}$ and the excitation $\mathbf{Y}^{(\nu)}$ are

$$\mathbf{X}^{(\nu)} = \begin{bmatrix} \mathbf{J}_e^{(\nu)} \\ \mathbf{M}_e^{(\nu)} \\ \mathbf{J}_i^{(\nu)} \\ \mathbf{M}_i^{(\nu)} \end{bmatrix}, \quad \mathbf{Y}^{(\nu)} = \begin{bmatrix} \frac{1}{2}\hat{n} \times \mathbf{M}_0^{(\nu)} \\ -\frac{1}{2}\hat{n} \times \mathbf{J}_0^{(\nu)} \\ \mathbf{J}_0^{(\nu)} \\ \mathbf{M}_0^{(\nu)} \end{bmatrix}.$$

Equation (8.14) can be used to solve the fundamental fields as well by setting $\nu = \omega_\alpha$, with $\alpha = 1, 2$ for frequency ω_1 and ω_2, respectively. The driven source now is the incident excitation

$$\mathbf{J}_0^{(\omega_\alpha)} = \hat{n} \times \mathbf{H}^{(\omega_\alpha, inc)} \tag{8.15}$$

$$\mathbf{M}_0^{(\omega_\alpha)} = -\hat{n} \times \mathbf{E}^{(\omega_\alpha, inc)} \tag{8.16}$$

The BEM formulation can model the nonlinear scattering from arbitrarily shaped particles efficiently since it only requires surface discretization. In addition, measured material parameters can be used directly.

8.2.2 Verification of the Algorithm

We first validate the BEM algorithm by comparing SF and SH scattering patterns of spherical nanoparticles with the nonlinear Mie solutions [71, 72]. For SFG, two incident plane waves propagating at different angles with different frequencies are superposed as shown in Figure 8.11. The polarization angles (defined as the angle between $-\hat{\theta}$ and $\mathbf{E}$) are set to zero for both incident waves. In this work, $\chi^{(2)}_{\perp\perp\perp} = 1.59 \times 10^{-18} m^2/V$ and only the normal component $\chi^{(2)}_{\perp\perp\perp}$ of the surface susceptibility tensor is considered, since it is the dominant term of the surface response of metallic nanoparticles. Note that other components are theoretically allowed, but they weakly contribute to the total SH response. Figure 8.12a shows the comparison of SF scattering patterns. Very good agreement is observed. Here, the radius of the sphere is $R = 500$ nm, the wavelengths of the incident waves are $\lambda_1 = 800$ nm, and $\lambda_2 = 3447$ nm, respectively. The opening angle between the source waves is $\beta = 15°$, the index of refraction is set to unity for all wavelengths. The asymmetry in the SF scattering pattern is due to the nonzero opening angle between source beams. For SHG, a single EM field is taken as the source. A gold sphere with radius $R = 50$ nm is excited by a plane wave, propagating along the positive direction of the z-axis and linearly polarized along the x-axis. The exciting wavelength $\lambda = 520$ nm corresponds to the plasmon resonance of the gold spherical particle. The dielectric constant for gold is taken from experimental data [73]. We can see that the SH scattering pattern calculated by the BEM method agrees well with the nonlinear Mie theory result.

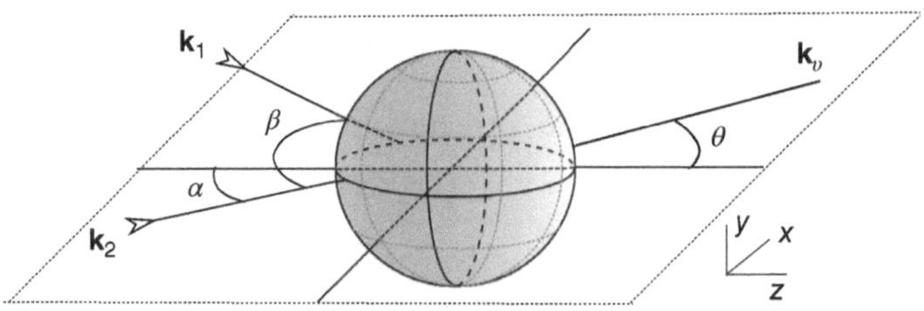

Figure 8.11 Overview of the relevant parameters in the model: the sum-frequency and source waves have wave vectors $\mathbf{k}_\nu$, $\mathbf{k}_1$, and $\mathbf{k}_2$ in order of decreasing frequency. The angle between the propagation direction of the lowest frequency wave and the positive z-axis is α, the opening angle between source waves is β. The sum-frequency scattering pattern is parameterized using the scattering angle θ. *Source:* de Beer and Roke [72]; © 2009 American Physical Society.

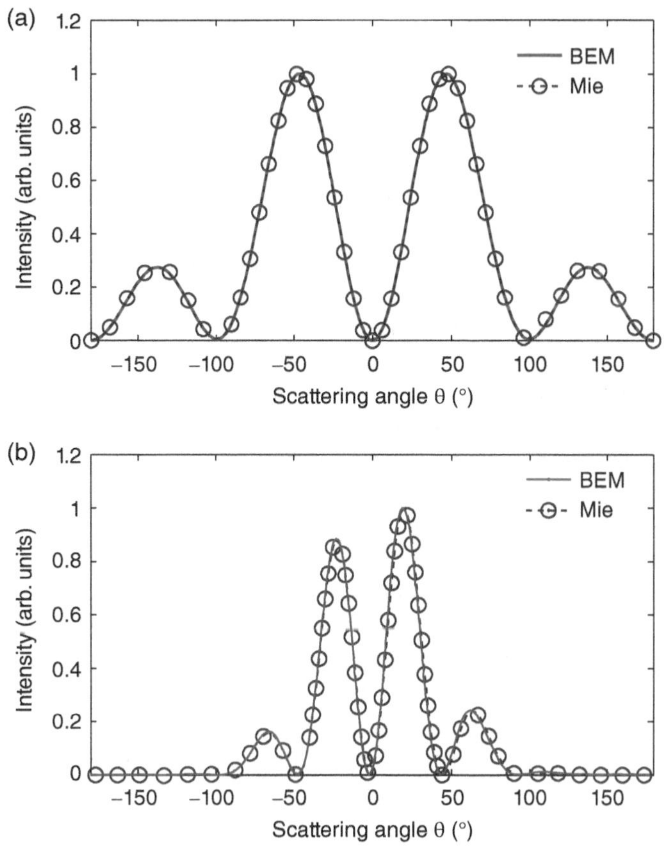

Figure 8.12 Comparisons of SF and SH scattering patterns with the nonlinear Mie solutions. (a) SF scattering pattern. The simulation parameters are R = 500 nm, λ_1 = 800 nm, and λ_2 = 3447 nm; β = 15°. (b) SH scattering pattern. The simulation parameters are R = 50 nm, λ_1 = 520 nm. *Source:* Xiong et al. [74]; © 2017 URSI Radio Science.

8.2.3 Mixing of Spin and OAM

Next, spin-orbital interaction (SOI) in plasmonic and dielectric nanostructures upon the SHG process is studied by using LG beams [75]. The LG beams in the LG_{pl} configuration have a concentric ring-like structure with p radial nodes in the intensity profile and a characteristic azimuthal phase variation of $e^{il\phi}$, where ϕ is the azimuthal coordinate and l is the topological charge of the beam measuring the phase variation, modulo 2π, along a closed curve containing the center of the vortex (l is also called OAM photon number). Generally, optical beams can also contain SAM associated with their polarization states, σ ($\sigma = 0$, for linearly polarized light and $\sigma = 1, -1$ for LCP and RCP light, respectively).

We begin our study with an analysis of the scattering of polarized LG beams from a single gold nanosphere, as the nanosphere possesses well-defined multipole-type optical modes. The gold nanosphere placed at the system origin and embedded in a vacuum is illuminated by an LG beam propagating along the $-z$-axis with its minimum waist in the xy-plane. The LG beam carries a topological charge, l, and it has no node along the radial coordinate (however, it can be zero at $r = 0$), i.e. we consider LG_{0l} beams. In addition, we assume that the beam is circularly polarized, which meaning it also carries SAM. The total AM carried by the circularly polarized LG beam is characterized by the integer $j = l + \sigma$. The proposed BEM method is used to calculate the light scattering at FF and SH frequencies. In our analysis, only the surface contribution to the SHG is considered for centrosymmetric materials [76–78]. The susceptibility components of gold (Au) are $\hat{\chi}^{(2)}_{s,\perp\perp\perp} = 1.59 \times 10^{-18} m^2/V$, $\hat{\chi}^{(2)}_{s,\|\|\perp} = \hat{\chi}^{(2)}_{s,\|\perp\|} = 4.63 \times 10^{-20} m^2/V$, and $\hat{\chi}^{(2)}_{s,\perp\|\|} = 0$ [79].

Due to the SOI, the amplitude and phase of the near-field fundamental and SH longitudinal-field components (E_z) have different features. Figure 8.13a–d and Figure 8.13e–h display the magnitude and phase of longitudinal fundamental and SH fields, respectively, calculated in a section parallel to the xy-plane at the resonant wavelength $\lambda_{FF} = 520$ nm ($\lambda_{SH} = 260$ nm). The radius of the gold sphere is fixed to 100 nm and the observation plane is located 40 nm away from the backside of the sphere (the side furthest from the source).

It can be seen that, for the LCP light ($\sigma = +1$) the amplitude of the scattered fundamental and SH fields vanishes at the axis of the beam due to the phase singularities at this point, (see Figure 8.13a, e). Moreover, the phase profile of the SH field, (Figure 8.13f) has twice as many nodes along a closed path that contains the origin as compared to the profile of the fundamental field, (Figure 8.13b). This means that during the photon conversion process not only has the frequency doubled, but also the photon AM has doubled its value. In other words, both the energy and the total AM are conserved in the nonlinear interaction of light with the nanosphere. In contrast, in the case of the RCP LG_{0l} beam, the amplitudes of the forward scattered fundamental (Figure 8.13c) and SH fields (Figure 8.13g) are no longer zero at the beam axis due to the linear and nonlinear SOI upon resonant light scattering from the Au nanosphere. That is, the phase vorticity related to OAM ($l = 1$) cancels the polarization helicity related to SAM ($\sigma = -1$) in the incident beam, producing at both the FF and SH without phase singularities at the beam axis, (Figure 8.13d, h). This

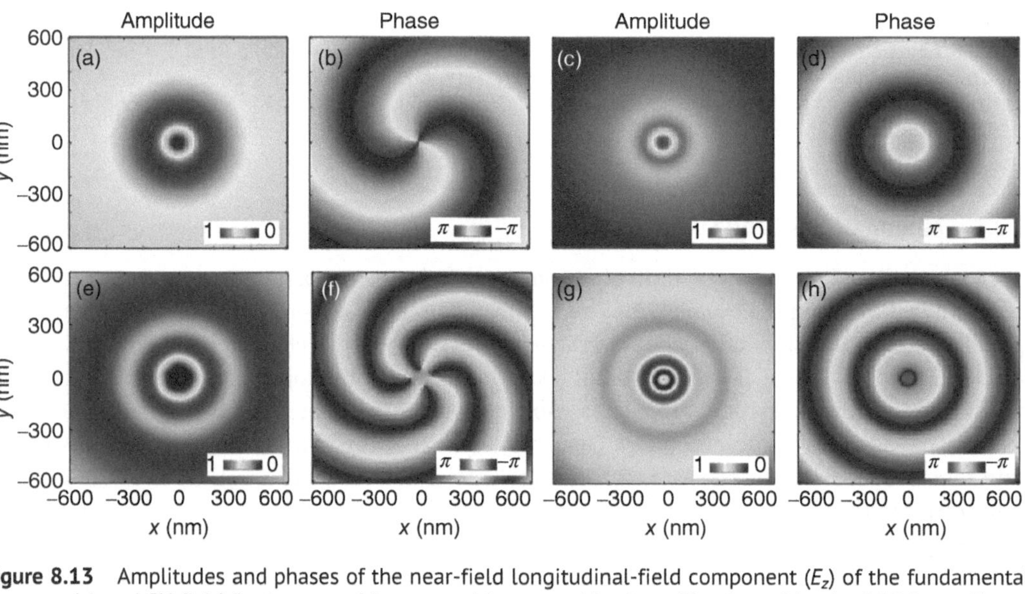

Figure 8.13 Amplitudes and phases of the near-field longitudinal-field component (E_z) of the fundamental (top panels) and SH field (bottom panels) scattered from a gold sphere illuminated by the LCP ($\sigma = +1$) and RCP ($\sigma = -1$) LG beams with the OAM of $l = 1$. The LG beam waist is $w_0 = 2\lambda_{FF}$ and the wavelength of the beam is $\lambda_{FF} = 520$ nm. The radius of the gold sphere is 100 nm. Panels (a), (b), (e), and (f) correspond to LCP beam ($\sigma = +1$), whereas panels (c), (d), (g), and (h) to RCP beam ($\sigma = -1$). *Source:* Xiong et al. [80]; © 2017 American Physical Society.

means that the value of $j = l + \sigma$ is preserved during the light interaction with the nanosphere, i.e. the total AM is conserved. This is not surprising as the sphere has C_∞ symmetry with respect to rotations around the beam axis and consequently, AM cannot be transferred between the beam and nanosphere.

8.2.4 General Angular Momenta Conservation Law

We consider now the scattering of LG beams from planar chiral nanostructures with N-fold rotational symmetry. For the sake of specificity, we set $N = 3$ in our calculations. The chiral nanostructure consists of 10 identical nanospheres arranged in an Archimedes-like spiral configuration, as schematically shown in Figure 8.14. The chiral nanostructures lie in the xy-plane in such a way that their center coincides with the origin of the coordinate system. The LCP and RCP LG beams with the OAM of $l = 4$, LG_{04}, are assumed to propagate along the $-z$-direction with the beam axis passing through the origin as shown in Figure 8.14. The susceptibility components of gold (Au) have been presented above and susceptibility components of silicon (Si) are $\hat{\chi}^{(2)}_{s,\perp\perp\perp} = 6.5 \times 10^{-18} m^2/V$, $\hat{\chi}^{(2)}_{s,\|\|\perp} = \hat{\chi}^{(2)}_{s,\|\perp\|} = 3.5 \times 10^{-19} m^2/V$, and $\hat{\chi}^{(2)}_{s,\perp\|\|} = 1.3 \times 10^{-19} m^2/V$ [81]. The radii of each gold and silicon nanosphere are chosen to be 100 and 500 nm, respectively, so that the clusters as a whole have strong resonances in the studied frequency region.

The SH conversion efficiencies of the plasmonic (gold) and dielectric (silicon) chiral clusters are compared in Figure 8.15. As we can see, the conversion efficiencies of the chiral structures strongly depend on the polarization state of the LG beam. Thus, for the LG_{04} beam with the RCP ($\sigma = -1$), the quantum number of the total AM of the incident beam is $j = l + \sigma = 3$, which matches the rotational symmetry order of the two chiral structures, $N = 3$. This quasi-angular-momentum matching between the incident beam and the chiral scatterers greatly increases the light scattering at the SH.

For the nanostructures with N-fold rotational symmetry, the incident beam with well-defined AM can only scatter into a set of modes with specific values of the total AM. In order to describe how nanostructures with N-fold rotational symmetry affect the transfer of AM from the incident

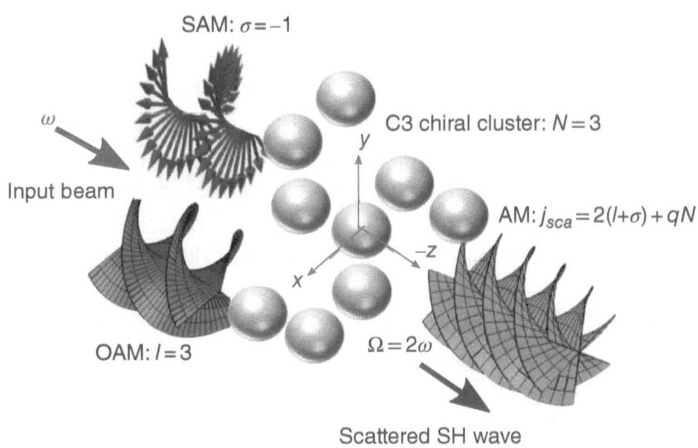

Figure 8.14 Schematic of the nonlinear spin-orbital AM coupling induced by SHG in a chiral cluster with N-fold rotation symmetry. Both the frequency and AM have doubled during the SHG process. In addition, the quasi-angular-momentum, defined as the order of the rotational symmetry of the chiral cluster (N), can be transferred to the scattered SH wave to conserve the total AM. q is an integer and $q = 0, \pm 1, \pm 2, \pm 3, \ldots$. *Source:* Xiong et al. [80]; © 2017 American Physical Society.

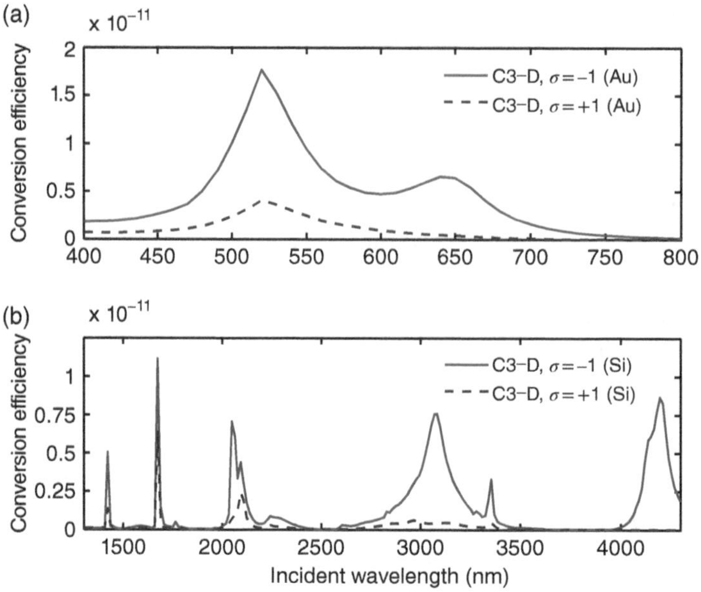

Figure 8.15 SH conversion efficiencies of chiral clusters illuminated by LCP and RCP LG_{04} beams. (a) Plasmonic (gold) chiral cluster. (b) Dielectric (silicon) chiral cluster. *Source:* Xiong et al. [80]; © 2017 American Physical Society.

field to the scattered field during both linear and nonlinear interaction processes, we formulate a general AM conservation law. It can be written as:

$$j_{sca} = s(l + \sigma) + qN, \tag{8.17}$$

where l and σ are the orbital and SAM numbers of the incident beam, respectively, s denotes the order of harmonic generation ($s = 1$ for linear processes, $s = 2$ for SH generation, etc.), N is the quasi-angular-momentum number possessed by the nanostructure with N-fold rotational symmetry, q is an integer $q = 0, \pm 1, \pm 2, \pm 3, \ldots$, and j_{sca} is the total AM number of the scattered field. Equation (8.17) implies that the total angular momenta of the scattered fundamental or SH fields depend not only on the angular momenta of the incident beam but also on the rotational symmetry properties of the scatterer. The quasi-angular-momentum of the scatterer can be transferred to the scattered wave in order to conserve the total AM. Note that this is the rotational analog of the conservation of momentum upon the scattering of EM waves diffraction gratings, when the quasi-momentum of the grating must be taken into account when describing the light scattering from such optical structures.

To validate this AM conservation law, the total AM of the scattered fundamental and SH fields from the chiral cluster are analyzed by the multipole expansion method [82]. Figure 8.16 shows the relative contributions of the electric and magnetic multipoles to the total AM of the scattered fields. For the incident LG beam with $l = 4$ and $\sigma = -1$, the quantum numbers characterizing the total optical angular momentum are $j_{sca} = 0, \pm 3, \pm 6, \pm 9, \ldots$, for both the scattered fundamental and SH fields. For the incident LG beam with $l = 4$ and $\sigma = +1$, the scattered fundamental field has the quantum numbers $j_{sca} = \ldots, -7, -4, -1, 2, 5, 8, \ldots$, whereas the scattered SH field has the quantum numbers $j_{sca} = \ldots, -11, -8, -5, -2, 1, 4, 7, 10, \ldots$. These predictions, based on Eq. (8.17), completely agree with the multipole decomposition weights of the total AM of the scattered fundamental and SH fields presented in Figure 8.16.

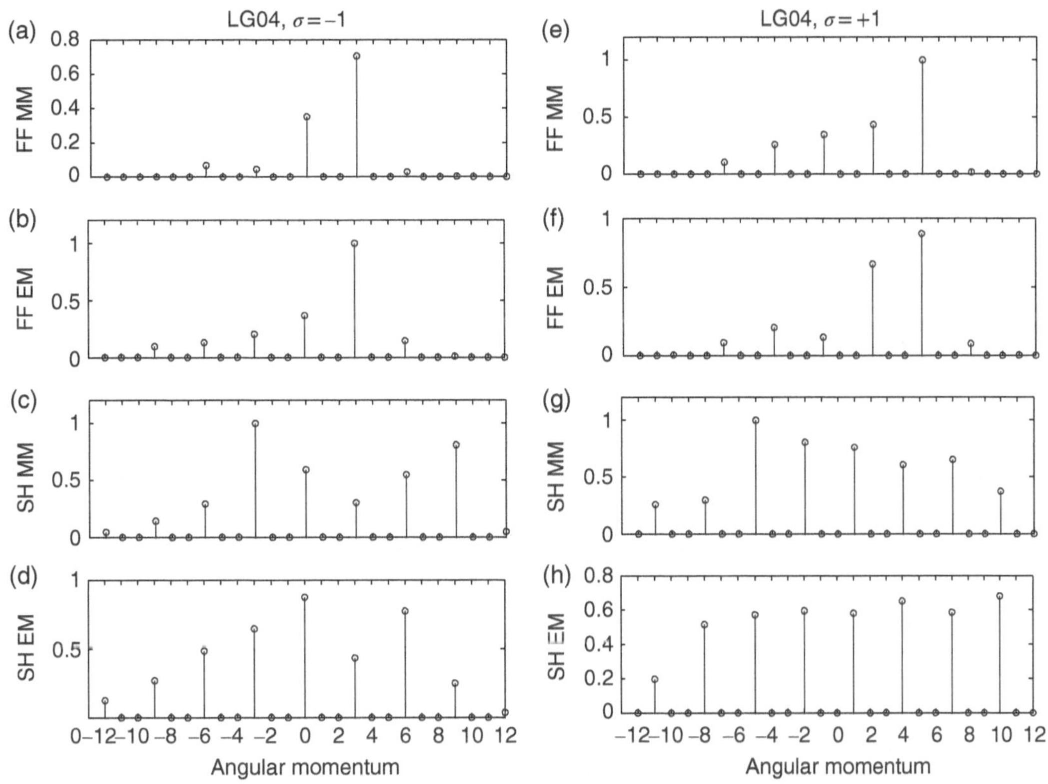

Figure 8.16 Multipolar decompositions of the fundamental and SH scattered waves from the silicon chiral cluster in terms of AM channels. MM and EM denote magnetic multipole and electric multipole, respectively. The silicon chiral cluster is illuminated by the RCP LG_{04} beam (left panels) and LCP LG_{04} beam (right panels). (a), (b), (e), and (f) correspond to scattered waves at FF, whereas (c), (d), (g), and (h) correspond to wave scattering at SH. *Source:* Xiong et al. [80]; © 2017 American Physical Society.

In this section, a numerical solution based on the BEM is proposed for the second-order surface nonlinear scattering problems. We investigate the characteristics of the SOI mediated by plasmonic and dielectric nanospheres, chiral clusters made of such nanospheres and with specific rotational symmetry properties, and its relation to the properties of the linear and SHG scattering processes. The doubling of the AM and frequency during the SH nonlinear spin–orbit interaction process has been demonstrated. A general angular momenta conservation law, including the quasi-angular-momentum of chiral nanostructures with *N*-fold rotational symmetry, has been formulated to describe how the rotational symmetry properties affect the transfer of optical angular momentum from incident beams to both the scattered fundamental and SH fields.

8.3 Conclusion

In summary, we have shown several engineering routes to generate and detect OAM modes at microwave regime. Furthermore, we have established a BEM method and concluded a general angular momenta conservation law to understand linear and nonlinear SOI processes. In the future, there are still several open questions to be answered for OAM research: (i) How to enhance the bandwidth and purity of the generated OAM modes simultaneously? (ii) What is a fundamental channel limit for OAM detection? (iii) How to realize OAM multiplexing in a complex non-line-of-sight environment for future 5G/6G communication? (iv) How to model the high-frequency scattering of OAM waves from electrically large objects? (v) How to understand the topological nature of OAM and connect it to photonic topological insulators?

References

1 Bliokh, K.Y., Bekshaev, A.Y., and Nori, F. (2013). Dual electromagnetism: helicity, spin, momentum and angular momentum. *New Journal of Physics* **15** (3): 033126.
2 Allen, L., Beijersbergen, M.W., Spreeuw, R.J.C., and Woerdman, J.P. (1992). Orbital angular-momentum of light and the transformation of Laguerre-Gaussian laser modes. *Physical Review A* **45** (11): 8185–8189.
3 Oemrawsingh, S.S.R., Ma, X., Voigt, D. et al. (2005). Experimental demonstration of fractional orbital angular momentum entanglement of two photons. *Physical Review Letters* **95** (24): 240501.
4 O'Dwyer, D.P., Phelan, C.F., Rakovich, Y.P. et al. (2010). Generation of continuously tunable fractional optical orbital angular momentum using internal conical diffraction. *Optics Letters* **18** (16): 16480–16485.
5 Huang, K., Liu, H., Restuccia, S. et al. (2018). Spiniform phase-encoded metagratings entangling arbitrary rational-order orbital angular momentum. *Light-Science & Applications* **7** (3): 17156–17156.
6 He, H., Friese, M.E.J., Heckenberg, N.R., and Rubinsztein-Dunlop, H. (1995). Direct observation of transfer of angular momentum to absorptive particles from a laser beam with a phase singularity. *Physical Review Letters* **75** (5): 826.
7 Simpson, N.B., Dholakia, K., Allen, L., and Padgett, M.J. (1997). Mechanical equivalence of spin and orbital angular momentum of light: an optical spanner. *Optics Letters* **22** (1): 52–54.
8 Grier, D.G. (2003). A revolution in optical manipulation. *Nature* **424** (6950): 810–816.
9 Gibson, G., Courtial, J., Padgett, M.J. et al. (2004). Free-space information transfer using light beams carrying orbital angular momentum. *Optics Express* **12** (22): 5448–5456.
10 Willner, A.E., Huang, H., Yan, Y. et al. (2015). Optical communications using orbital angular momentum beams. *Advances in Optics and Photonics* **7** (1): 66–106.
11 Ren, Y.X., Wang, Z., Liao, P.C. et al. (2016). Experimental characterization of a 400 Gbit/s orbital angular momentum multiplexed free-space optical link over 120 m. *Optics Letters* **41** (3): 622–625.
12 Tamburini, F., Anzolin, G., Umbriaco, G. et al. (2006). Overcoming the Rayleigh criterion limit with optical vortices. *Physical Review Letters* **97** (16): 163903.
13 Yan, Y., Xie, G.D., Lavery, M.P.J. et al. (2014). High-capacity millimetre-wave communications with orbital angular momentum multiplexing. *Nature Communications* **5**: 4876.
14 Hui, X.N., Zheng, S.L., Chen, Y.L. et al. (2015). Multiplexed millimeter wave communication with dual orbital angular momentum (OAM) mode antennas. *Scientific Reports* **5**: 10148.
15 Thide, B., Then, H., Sjoholm, J. et al. (2007). Utilization of photon orbital angular momentum in the low-frequency radio domain. *Physical Review Letters* **99** (8): 087701.
16 Mahmouli, F.E. and Walker, S.D. (2013). 4-Gbps uncompressed video transmission over a 60-GHz orbital angular momentum wireless channel. *IEEE Wireless Communications Letters* **2** (2): 223–226.
17 Liu, K., Cheng, Y.Q., Yang, Z.C. et al. (2015). Orbital-angular-momentum-based electromagnetic vortex imaging. *IEEE Antennas and Wireless Propagation Letters* **14**: 711–714.
18 Mair, A., Vaziri, A., Weihs, G., and Zeilinger, A. (2001). Entanglement of the orbital angular momentum states of photons. *Nature* **412** (6844): 313–316.
19 Nagali, E., Sciarrino, F., De Martini, F. et al. (2009). Quantum information transfer from spin to orbital angular momentum of photons. *Physical Review Letters* **103** (1): 013601.

20 Bazhenov, V.Y., Vasnetsov, M.V., and Soskin, M.S. (1990). Laser-beams with screw dislocations in their wave-fronts. *Journal of Experimental and Theoretical Physics Letters* **52** (8): 429–431.

21 Chen, S.M., Cai, Y., Li, G.X. et al. (2016). Geometric metasurface fork gratings for vortex-beam generation and manipulation. *Laser & Photonics Review* **10** (2): 322–326.

22 Beijersbergen, M.W., Coerwinkel, R.P.C., Kristensen, M., and Woerdman, J.P. (1994). Helical-wavefront laser beams produced with a spiral phaseplate. *Optical Communications* **112** (5–6): 321–327.

23 Kotlyar, V.V., Almazov, A.A., Khonina, S.N. et al. (2005). Generation of phase singularity through diffracting a plane or gaussian beam by a spiral phase plate. *Journal of the Optical Society of America A* **22** (5): 849–861.

24 Niv, A., Biener, G., Kleiner, V., and Hasman, E. (2005). Spiral phase elements obtained by use of discrete space-variant subwavelength gratings. *Optical Communications* **251** (4–6): 306–314.

25 Marrucci, L., Manzo, C., and Paparo, D. (2006). Optical spin-to-orbital angular momentum conversion in inhomogeneous anisotropic media. *Physical Review Letters* **96** (16): 163905.

26 Karimi, E., Piccirillo, B., Nagali, E. et al. (2009). Efficient generation and sorting of orbital angular momentum eigenmodes of light by thermally tuned q-plates. *Applied Physics Letters* **94** (23): 231124.

27 Cardano, F., Karimi, E., Slussarenko, S. et al. (2012). Polarization pattern of vector vortex beams generated by q-plates with different topological charges. *Applied Optics* **51** (10): C1–C6.

28 Yu, N.F., Genevet, P., Kats, M.A. et al. (2011). Light propagation with phase discontinuities: generalized laws of reflection and refraction. *Science* **334** (6054): 333–337.

29 Genevet, P., Yu, N.F., Aieta, F. et al. (2012). Ultra-thin plasmonic optical vortex plate based on phase discontinuities. *Applied Physics Letters* **100** (1): 013101.

30 Karimi, E., Schulz, S.A., De Leon, I. et al. (2014). Generating optical orbital angular momentum at visible wavelengths using a plasmonic metasurface. *Light-Science & Applications* **3** (5): e167.

31 Georgi, P., Schlickriede, C., Li, G.X. et al. (2017). Rotational Doppler shift induced by spin-orbit coupling of light at spinning metasurfaces. *Optica* **4** (8): 1000–1005.

32 Jin, J.J., Luo, J., Zhang, X.H. et al. (2016). Generation and detection of orbital angular momentum via metasurface. *Scientific Reports* **6**: 24286.

33 Wang, W., Li, Y., Guo, Z.Y. et al. (2015). Ultra-thin optical vortex phase plate based on the metasurface and the angular momentum transformation. *Journal of Optics* **17** (4): 045102.

34 Chen, M.L.N., Jiang, L.J., and Sha, W.E.I. (2018). Orbital angular momentum generation and detection by geometric-phase based metasurfaces. *Applied Science* **8** (3): 362.

35 Mohammadi, S.M., Daldorff, L.K.S., Bergman, J.E.S. et al. (2010). Orbital angular momentum in radio-a system study. *IEEE Transactions on Antennas and Propagation* **58** (2): 565–572.

36 Comite, D., Valerio, G., Albani, M. et al. (2017). Exciting vorticity through higher order bessel beams with a radial-line slot-array antenna. *IEEE Transactions on Antennas and Propagation* **65** (4): 2123–2128.

37 Zheng, S.L., Hui, X.N., Jin, X.F. et al. (2015). Transmission characteristics of a twisted radio wave based on circular traveling-wave antenna. *IEEE Transactions on Antennas and Propagation* **63** (4): 1530–1536.

38 Barbuto, M., Trotta, F., Bilotti, F., and Toscano, A. (2014). Circular polarized patch antenna generating orbital angular momentum. *Progress in Electromagnetic Research* **148**: 23–30.

39 Berry, M.V. (1987). The adiabatic phase and Pancharatnam's phase for polarized light. *Journal of Modern Optics* **34** (11): 1401–1407.

40 Chen, M.L.N., Jiang, L.J., and Sha, W.E.I. (2017). Ultrathin complementary metasurface for orbital angular momentum generation at microwave frequencies. *IEEE Transactions on Antennas and Propagation* **65** (1): 396–400.

41 Menzel, C., Rockstuhl, C., and Lederer, F. (2010). Advanced Jones calculus for the classification of periodic metamaterials. *Physical Review A* **82** (5): 053811.

42 Chen, M.L.N., Jiang, L.J., and Sha, W.E.I. (2016). Artificial perfect electric conductor-perfect magnetic conductor anisotropic metasurface for generating orbital angular momentum of microwave with nearly perfect conversion efficiency. *Journal of Applied Physics* **119** (6): 064506.

43 Chen, M.L.N., Jiang, L.J., and Sha, W.E.I. (2019). Quasi-continuous metasurfaces for orbital angular momentum generation. *IEEE Antennas and Wireless Propagation Letters* **18** (3): 477–481.

44 Yablonovitch, E. (1987). Inhibited spontaneous emission in solid-state physics and electronics. *Physical Review Letters* **58** (20): 2059–2062.

45 John, S. (1987). Strong localization of photons in certain disordered dielectric superlattices. *Physical Review Letters* **58** (23): 2486–2489.

46 Yablonovitch, E., Gmitter, T.J., Meade, R.D. et al. (1991). Donor and acceptor modes in photonic band-structure. *Physical Review Letters* **67** (24): 3380–3383.

47 Chen, M.L.N., Jiang, L.J., and Sha, W.E.I. (2018). Generation of orbital angular momentum by a point defect in photonic crystals. *Physical Review Applied* **10** (1): 014031.

48 Landau, L.D. and Lifshitz, E.M. (1977). *Quantum Mechanics: Non-Relativistic Theory*. Oxford: Pergamon Press.

49 Yang, Y.T., Huang, X.Q., and Hang, Z.H. (2016). Experimental characterization of the deterministic interface states in two-dimensional photonic crystals. *Physical Review Applied* **5** (3): 034009.

50 Mohammadi, S.M., Daldorff, L.K.S., Forozesh, K. et al. (2010). Orbital angular momentum in radio: measurement methods. *Radio Science* **45** (4): 1–14.

51 Schulze, C., Dudley, A., Flamm, D. et al. (2013). Measurement of the orbital angular momentum density of light by modal decomposition. *New Journal of Physics* **15** (7): 073025.

52 Hui, X.N., Zheng, S.L., Zhang, W.T. et al. (2016). Local topological charge analysis of electromagnetic vortex beam based on empirical mode decomposition. *Optics Express* **24** (5): 5423–5430.

53 Zhang, C. and Ma, L. (2017). Detecting the orbital angular momentum of electro-magnetic waves using virtual rotational antenna. *Scientific Reports* **7** (1): 4585.

54 Genevet, P., Lin, J., Kats, M.A., and Capasso, F. (2012). Holographic detection of the orbital angular momentum of light with plasmonic photodiodes. *Nature Communications* **3**: 1278.

55 Schmid, P.J. (2010). Dynamic mode decomposition of numerical and experimental data. *Journal of Fluid Mechanics* **656**: 5–28.

56 Barocio, E., Pal, B.C., Thornhill, N.F., and Messina, A.R. (2015). A dynamic mode decomposition framework for global power system oscillation analysis. *IEEE Transactions on Power Systems* **30** (6): 2902–2912.

57 Brunton, S.L., Proctor, J.L., and Kutz, J.N. (2016). Discovering governing equations from data by sparse identification of nonlinear dynamical systems. *Proceedings of the National Academy of Sciences* **113** (15): 3932–3937.

58 Brunton, S.L., Brunton, B.W., Proctor, J.L. et al. (2017). Chaos as an intermittently forced linear system. *Nature Communications* **8**: 19.

59 Proctor, J.L. and Eckhoff, P.A. (2015). Discovering dynamic patterns from infectious disease data using dynamic mode decomposition. *International Health* **7** (2): 139–145.

60 Zhang, Y., Chen, M.L.N., and Jiang, L.J. (2019). Analysis of electromagnetic vortex beams using modified dynamic mode decomposition in spatial angular domain. *Optics Express* **27** (20): 27702–27711.

61 Chen, M.L.N., Jiang, L.J., and Sha, W.E.I. (2018). Detection of orbital angular momentum with metasurface at microwave band. *IEEE Antennas and Wireless Propagation Letters* **17**: 110–113.

62 Tabosa, J.W.R. and Petrov, D.V. (1999). Optical pumping of orbital angular momentum of light in cold cesium atoms. *Physical Review Letters* **83** (24): 4967.

63 Chen, S., Zeuner, F., Weismann, M. et al. (2016). Giant nonlinear optical activity of achiral origin in planar metasurfaces with quadratic and cubic nonlinearities. *Advanced Materials* **28** (15): 2992–2999.

64 Li, G., Chen, S., Pholchai, N. et al. (2015). Continuous control of the nonlinearity phase for harmonic generations. *Nature Materials* **14** (6): 607–612.

65 Shen, Y.R. (1984). *The Principle of Nonlinear Optics*. Inc.: John Wiley & Sons.

66 Sipe, J.E., So, V.C.Y., Fukui, M., and Stegeman, G.I. (1980). Analysis of second-harmonic generation at metal surfaces. *Physical Review B* **21** (10): 4389.

67 Ciraci, C., Poutrina, E., Scalora, M., and Smith, D.R. (2012). Origin of second-harmonic generation enhancement in optical split-ring resonators. *Physical Review B* **85** (20): 201403.

68 Boyd, R.W. (1992). *Nonlinear Optics*. Academic Press: New York.

69 Rao, S.M., Wilton, D.R., and Glisson, A.W. (1982). Electromagnetic scattering by surfaces of arbitrary shape. *IEEE Transactions on Antennas and Propagation* **30** (3): 409–418.

70 Medgyesi-Mitschang, L.N., Putnam, J.M., and Gedera, M.B. (1994). Generalized method of moments for three-dimensional penetrable scatterers. *Journal of the Optical Society of America A* **11** (4): 1383–1398.

71 Dadap, J.I., Shan, J., and Heinz, T.F. (2004). Theory of optical second-harmonic generation from a sphere of centrosymmetric material: small-particle limit. *Journal of the Optical Society of America B* **21** (7): 1328–1347.

72 de Beer, A.G.F. and Roke, S. (2009). Nonlinear Mie theory for second-harmonic and sum-frequency scattering. *Physical Review B* **79** (15): 155420.

73 Johnson, P.B. and Christy, R.W. (1972). Optical constants of the noble metals. *Physical Review B* **6** (12): 4370.

74 Xiong, X.Y.Z., Jiang, L.J., Sha, W.E.I. et al. (2017). Sum-frequency and second-harmonic generation from plasmonic nonlinear nanoantennas. *URSI Radio Science Bulletin* **360**: 43–49.

75 Al-Jarro, A., Biris, C.G., and Panoiu, N.C. (2016). Resonant mixing of optical orbital and spin angular momentum by using chiral silicon nanosphere clusters. *Optics Express* **24** (7): 6945–6958.

76 Zhang, Y., Grady, N.K., Ayala-Orozco, C., and Halas, N.J. (2011). Three-dimensional nanostructures as highly efficient generators of second harmonic light. *Nano Letters* **11** (12): 5519–5523.

77 Butet, J., Brevet, P.-F., and Martin, O.J.F. (2015). Optical second harmonic generation in plasmonic nanostructures: from fundamental principles to advanced applications. *ACS Nano* **9** (11): 10545–10562.

78 Dadap, J.I., Shan, J., Eisenthal, K.B., and Heinz, T.F. (1999). Second-harmonic Rayleigh scattering from a sphere of centrosymmetric material. *Physical Review Letters* **83** (20): 4045.

79 Krause, D., Teplin, C.W., and Rogers, C.T. (2004). Optical surface second harmonic measurements of isotropic thin-film metals: gold, silver, copper, aluminum, and tantalum. *Journal of Applied Physics* **96** (7): 3626–3634.

80 Xiong, X.Y.Z., Al-Jarro, A., Jiang, L.J. et al. (2017). Mixing of spin and orbital angular momenta via second-harmonic generation in plasmonic and dielectric chiral nanostructures. *Physical Review B* **95** (16): 165432.

81 Falasconi, M., Andreani, L.C., Malvezzi, A.M. et al. (2001). Bulk and surface contributions to second-order susceptibility in crystalline and porous silicon by second-harmonic generation. *Surface Science* **481** (1–3): 105–112.

82 Jackson, J.D. (2007). *Classical Electrodynamics*. John Wiley & Sons.

Part III

Engineering Applications of Electromagnetic Vortices

9

Orbital Angular Momentum Based Structured Radio Beams and its Applications[1]

Xianmin Zhang,[1,2] *Shilie Zheng,*[1,2] *Wei E. I. Sha,*[1] *Li Jun Jiang,*[3] *Xiaowen Xiong,*[1] *Zelin Zhu,*[1] *Zhixia Wang,*[1] *Yuqi Chen,*[1] *Jiayu Zheng,*[1] *Xinyue Wang*[1], *and Menglin L. N. Chen*[3]

[1] *Key Laboratory of Micro-nano Electronic Devices and Smart Systems of Zhejiang Province, College of Information Science and Electronic Engineering, Zhejiang University, Hangzhou, P. R. China*
[2] *Ningbo Research Institute, Zhejiang University, Ningbo, P. R. China*
[3] *Department of Electrical and Electronic Engineering, The University of Hong Kong, Hong Kong, P. R. China*

9.1 Introduction

Angular momentum is the basic physical parameter of the electromagnetic (EM) wave manifesting the rotational symmetry of the EM field. It consists of two parts, one is spinning angular momentum (SAM), which is an intrinsic property connecting to the polarization state, and the other is orbital angular momentum (OAM), which is an extrinsic property describing the phase singularity and the beam vorticity. OAM was first found in the optical domain [1], and the applications based on OAM have ranged from optical manipulation to high-capacity communications [2]. Over the last decade, it was found that the photon OAM is not restricted in optical domain, and can be used in radio domain [3, 4]. Besides the upsurging interest in communications [5–7], OAM has also evoked great interests in EM-sensing applications, such as super-resolution imaging [8], single-frequency computational imaging [9], and rotational Doppler effect [10], etc.

The most attracting property of the OAM is its unlimited eigenstates that are orthogonal to one another. Theoretically, in an ideal case where the monolithic multidimensional antennas at the transmitting and receiving ends can make direct use of the rotation degree of freedom, OAM is an unused natural degree of freedom in radio domain which is the same as that in optical domain. However, the explorations for OAM-based applications in the radio domain have encountered obstructs because of the on-axis phase singularity and serious divergence of OAM-carrying radio wave. The traditional OAM-carrying wave features a helical wavefront and a doughnut-like intensity profile with a central dark zone. Owing to the beam divergence, the radius of the dark zone increases with the increasing transmitting distance. It is not yet available in a practical situation since an unbearably large aperture of receiving antenna or arrays in far field is required to recover the transmitting signals for demultiplexing the carried information utilizing the orthogonality. Furthermore, the divergence of OAM beam scales approximately with the square root of mode number $|l|$ [11], indicating that the radio waves carrying different OAM modes might have different divergence angles. These features make it difficult to overlap the mainlobes of multiple OAM-carrying

1 This work was supported by the National Basic Research Program of China (973 program) under Grant 2014CB340005, and the Natural Science Foundation of China under Grant 61371030 and 61571391.

Electromagnetic Vortices: Wave Phenomena and Engineering Applications, First Edition.
Edited by Zhi Hao Jiang and Douglas H. Werner.

waves, thereby resulting in the unbalanced power across the different OAM-mode–based communication links at the receiving end and leading the multiplexed links to be infeasible [12]. These unique properties result in many challenges for OAM-based applications in radio frequency region.

In the recent times, structured light has become synonymous with OAM. The richness of the topic as well as its myriad of applications has attracted much attention. The superposition of composite optical vortex has been proposed to constitute the structured beams (SBs) that are indeed emerging as a new frontier in the modern photonics [13]. Structured radio beams with diverse radiation patterns exhibit huge potentials in many applications. We have proposed a novel form of OAM radio wave, termed as plane spiral OAM (PS–OAM) EM wave, to tackle the difficulties stemmed from the singularity-caused dark zone and beam divergence [14, 15]. The PS–OAMs are the special OAM modes propagating in a two-dimensional (2D) transverse plane. There is no singularity along the propagating path and all PS–OAM modes have the same radiation pattern. It will greatly benefit the mode grouping and multibeam synthesis. Recently, we proposed a PS–OAM-based structured radio beam construction scheme, in which several PS–OAM waves are superposed together to acquire the anticipated radiation pattern and phase distribution in the azimuthal domain [16]. Since the azimuthal phase distributions of $\exp(-jl\varphi)$ for different OAM modes can be consisted into one complete set of the eigenmode for the EM waves, various structured radio waves can be reconstructed by manipulating the intensity and phase of the OAM-mode waves.

The remaining of the chapter is organized as follows. At first, the PS–OAM is reviewed and the construction of structured radio beams based on PS–OAM is analyzed. The basic characteristics are discussed. Then, the antennas for structured radio beam are proposed and investigated. At final, the potential applications of the structured radio beams for radar detection, multiple input multiple output (MIMO) system, and spatial field digital modulation (SFDM) are explored.

9.2 PS–OAM Based Structured Beams

9.2.1 Plane Spiral OAM

Figure 9.1a shows the schematic of the PS–OAM wave, for comparison, the traditional OAM wave is also shown in Figure 9.1b. Instead of propagating along the main axis with the helical phase fronts in the three-dimensional space ($\rho \times \theta \times \varphi$ in spherical coordinates) as the traditional OAM wave, the plane spiral OAM wave propagates mainly along the transverse direction with a 2D structure ($\rho \times \varphi$, $\theta = 90°$), and a spiral phase distribution in the radial plane.

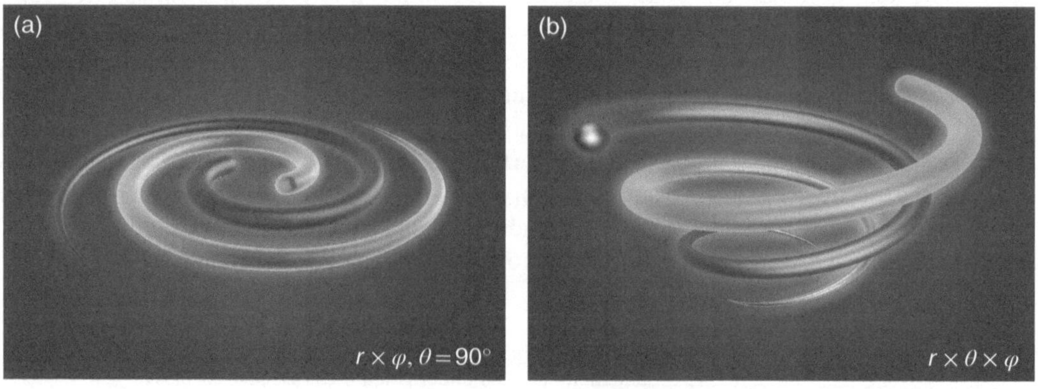

Figure 9.1 The schematic of OAM waves: (a) Plane spiral OAM (PS–OAM) wave; (b) general OAM wave.

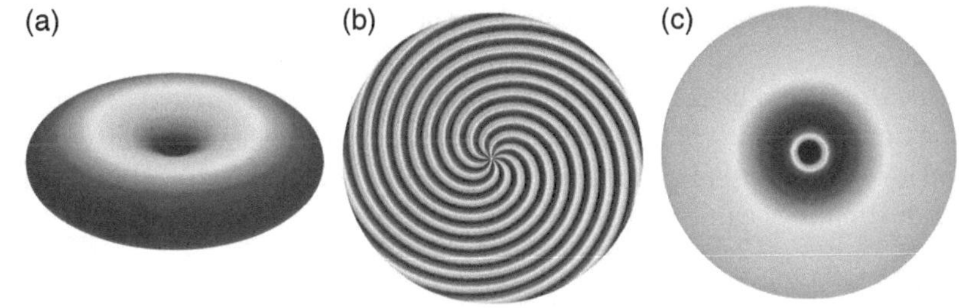

Figure 9.2 Distributions of PS-OAM wave (l = 9): (a) radiation pattern; (b) phase distribution; (c) amplitude distribution.

A traveling wave circular loop antenna (CLA) with radius a and current distribution of $I = I_0 \exp(-jl\varphi)$ can be used to generate the OAM-carrying waves [17], where I_0 is current amplitude, φ is the azimuth angle, and l is the OAM mode number. The far-field radiation of this antenna is determined by the OAM mode number l and the radius a. Through optimization of these two parameters, a hollow disk-like radiation pattern can be obtained as shown in Figure 9.2a and the radiated power concentrates near the equatorial plane. The phase and amplitude distributions are shown in Figure 9.2b and c, respectively.

To some extent, it can be regarded as a quasi-2D structure. This kind of wave possesses the typical characteristics of the OAM-carrying waves, i.e., the spiral phase distribution. Hence, the advantages based on the OAM property can be inherently kept, while the problem of the wide dark zone for the long-distance transmission can be automatically avoided. Furthermore, as the phase singularity of the field is in the site of the antenna itself, the divergence for the different OAM mode waves are the same, the different vortex radius is no longer a problem in the far-field region. Compared with the complicated polarization evolution for the OAM wave generated by the traveling wave circular loop antenna (CLA) [17], the polarization state of the PS-OAM wave is a bit simple. It is linearly polarized in the far-field region. Thus, the overlapping of different OAM modes can be easily achieved. The vorticity and orthogonality of PS-OAM waves can also be guaranteed.

The generation of PS-OAM waves has been reported using the traveling wave circular slot antenna [18, 19], the substrate-integrated waveguide (SIW) [20], the gradient-index metasurface [21], and the circular cylindrical conformal microstrip antenna array [22], etc. The MIMO system based on the vorticity of PS-OAM waves [23, 24], the single-frequency computational-imaging based on orthogonality of PS-OAM modes [9], the vector velocity measurement in 2D plane based on the Doppler effect of PS-OAM waves [10], and the 360° beam steering based on the superposition of PS-OAM waves [25, 26] have been explored. Recently, the PS-OAM wave has also been investigated in the optical domain [26].

9.2.2 Structured Radio Beam

Periodic complex exponentials play a central role in the treatment of signals and systems, because they can serve as extremely useful building blocks for many other signals. In spatial domain, PS-OAM waves possess azimuthal phase distribution of $\exp(-jl\varphi)$. The azimuthal angle φ is periodic, which is linked to the OAM mode number l by Fourier transform. It can be described as:

$$A_l = \frac{1}{2\pi}\int_{-\pi}^{\pi} \phi(\varphi) e^{jl\varphi} d\varphi \tag{9.1}$$

$$\phi(\varphi) = \sum_{l=-\infty}^{+\infty} A_l e^{-jl\varphi} \tag{9.2}$$

where, $A_l = |A_l| e^{j\theta_l}$ is a complex representing amplitude $|A_l|$ and initial phase θ_l for PS–OAM mode l.

Thus, the structured radio beam in azimuthal domain can be constructed in terms of the spiral harmonics. The transverse of far-field radiation pattern of structured radio beam can be mathematically described in terms of the spiral harmonics $\exp(-jl\varphi)$, i.e. a beam with arbitrary angular intensity and phase distributions can be obtained by superposed PS–OAM modes.

$$S(\varphi) = \sum_{n} A_n e^{-jl_n\varphi} \tag{9.3}$$

Figure 9.3 shows the schematic of the structured radio beam constructed by four PS–OAM modes of $l = 1,2,3,4$. The azimuthal amplitude distributions of the four PS–OAM modes and the constructed radio beam are shown in the left, while the corresponding phase distributions are shown in the right. When the PS–OAM waves are superposed together, beams with various intensity and phase distributions can be achieved. It is shown that the constructed radio beam has a directional radiation pattern and a specific phase distribution.

As discussed in 9.2.1, the PS–OAM waves owe the unique characteristics. The transverse propagation avoids the different divergence for different OAM-carrying waves, while the simple polarization and easy amplitude control are also helpful to superpose multiple PS–OAM waves. Figure 9.4 shows the normalized beam patterns at the transverse plane by grouping 2, 8, and 16 consecutive PS–OAM waves, respectively. In the calculation, the weight of each PS–OAM is supposed to be equal for simplicity. In fact, the weight of each PS–OAM can be complex, including the amplitude and phase, and the OAM modes can be of arbitrary order. Clearly, the beam pattern changes when the number of the superposed PS–OAM waves varies. The main lobe becomes narrower and the number of the sidelobes increases with the increasing components

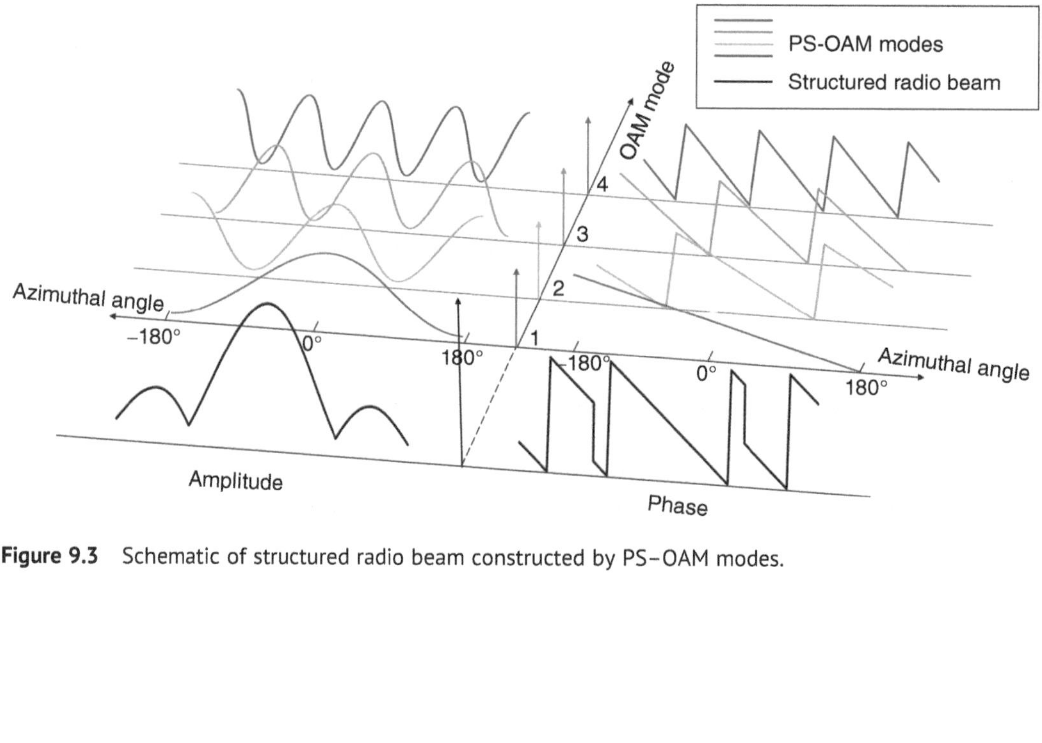

Figure 9.3 Schematic of structured radio beam constructed by PS–OAM modes.

Figure 9.4 Normalized beam pattern for 4 PS-OAM modes of $l = \{1, 2\}$ (solid line), 8 PS-OAM modes of $l = \{1, 2, 3, ..., 8\}$ (dash line), and 16 PS-OAM modes of $l = \{1, 2, 3, ..., 16\}$ (dotted line).

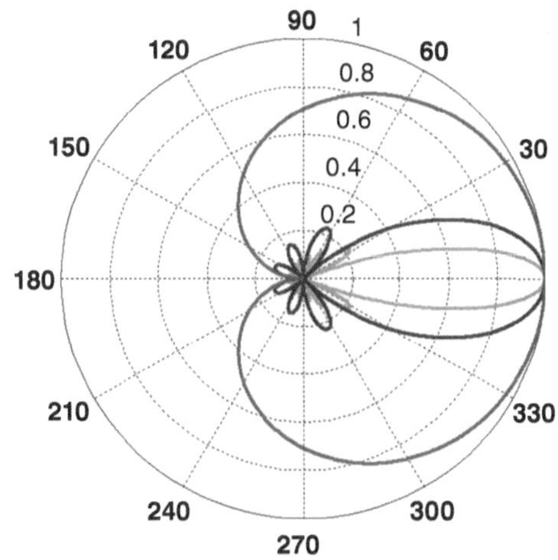

of PS-OAM waves. When much more PS-OAM modes are combined together, a highly directional beam can be synthesized.

The angular intensity variations of the radio beams can be easily achieved by the PS-OAM modes grouping. Suppose, there are N PS-OAM modes with a constant mode interval Δl, each of them has the same amplitude. Then, the constructed radio beam can be described as [27]:

$$S(\varphi) = \sum_{n=0}^{N-1} A_n e^{-jl_n\varphi} = \frac{\sin\left(\frac{N\cdot\Delta l}{2}\varphi\right)}{\sin\left(\frac{\Delta l}{2}\varphi\right)} \cdot e^{-j\left[l_0 + \frac{(N-1)\cdot\Delta l}{2}\right]\varphi} \tag{9.4}$$

where, l_1 is the first OAM mode among the superposed PS-OAM waves. Although, the intensity patterns for the superposition of multiple PS-OAM modes with same amount and same mode interval are identical, as shown in Figure 9.5a, their phase distributions are different according to Eq. (9.4). $\sin(N \cdot \Delta l/2 \cdot \varphi)/\sin(\Delta l/2 \cdot \varphi)$ is the amplitude factor, which is dependent on N and Δl, but is independent of l_0. It describes directional gain of the constructed radio beam that provides an efficient solution to generate high-gain beams. $\exp\{-j[l_0 + (N-1) \cdot \Delta l/2]\varphi\}$ is the phase factor depending on N, Δl, and l_0. Interestingly, the phase of the superposed beam changes linearly along the azimuthal angle, the slope is the average of all the superposed OAM modes. By selecting the appropriate OAM mode l_0, the phase characteristics of the constructed radio beams can also be manipulated. Figure 9.5b shows the identical intensity pattern with the different phase distributions of two groups of OAM mode. It is clearly shown that both the two OAM groups owe different periods of phase change. For clarity, only the phase variations within the mainlobe region are presented in Figure 9.5b.

In addition, it is really significant to find that orthogonality will exist between these hybrid waves. Suppose that there are two OAM mode groups A and B, which are composed of m PS-OAMs with equal weight, mode group A is a superposition of modes of $\{l_{a1}, l_{a2}, ..., l_{am}\}$, and mode group B is a superposition of the PS-OAM modes of $\{l_{b1}, l_{b2}, ..., l_{bm}\}$. It can be easily proved that if $\{l_{a1}, l_{a2}, ..., l_{am}\} \cap \{l_{b1}, l_{b2}, ..., l_{bm}\} = 0$ then mode group A and mode group B are ideally orthogonal to each other. As we know that more PS-OAM modes superposition results in narrow mainlobe, it can say that the hybrid wave of group A is approximately orthogonal to that of group B within the mainlobe region.

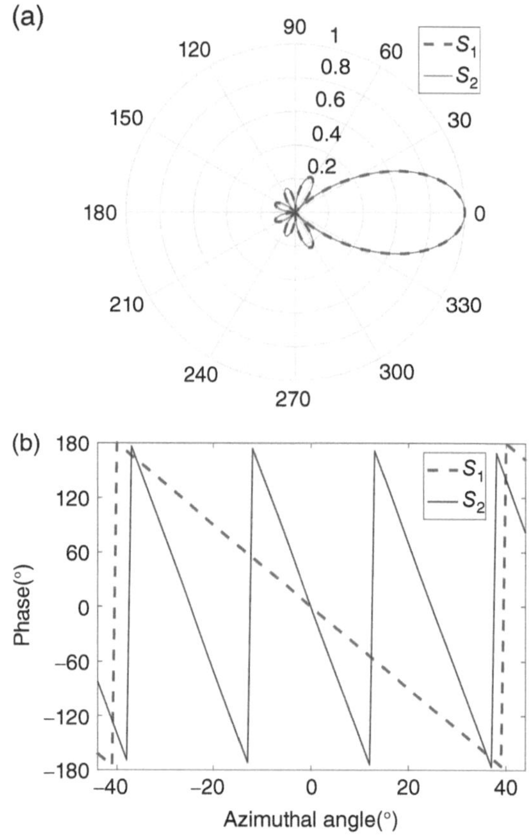

Figure 9.5 Structured radio beams with 8 PS-OAM modes. (a) Normalized beam pattern; (b) the phase distribution within the main lobe. S_1 = {1,2,3,4, 5,6,7,8} (dash line), S_2 = {11,12,13,14,15,16,17,18} (solid line).

9.3 Antennas for Structured Beams

9.3.1 Antennas for PS-OAM Waves

A traveling wave circular slot antenna based on the ring resonant cavity can generate a single mode OAM wave when a 90°hybrid coupler is used to excite the antenna and the phase changes along a circle of $2\pi l$ in the cavity [17]. The amplitude distribution of OAM wave with mode number l in the zenith direction satisfies:

$$A(\theta) \propto J_l[ka \sin(\theta)], \tag{9.5}$$

where, k is the wavenumber and a is the radius of the traveling wave circular slot. The radiation direction can be calculated by Eq. (9.5). A main feature of PS-OAM is that the radiation direction is along the transverse plane regardless of the mode number l. There are three ways to help the radiation concentrate on the radial plane.

The first method is to reduce the radius a of the ring resonant cavity. When $ka \sin(\theta)$ is less than the first maximum point of $J_l(\cdot)$, then the radiation direction of the different modes are all toward to $\theta = 90°$. The ring DRA is proposed to generate PS-OAM in this way as shown in Figure 9.6 [19].

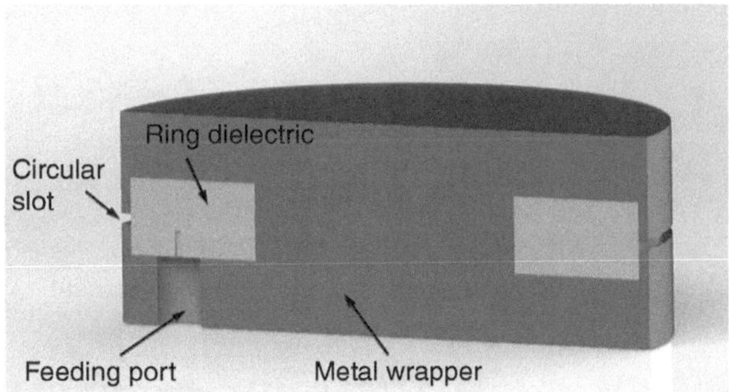

Figure 9.6 The section view of the ring dielectric resonator antenna (DRA).

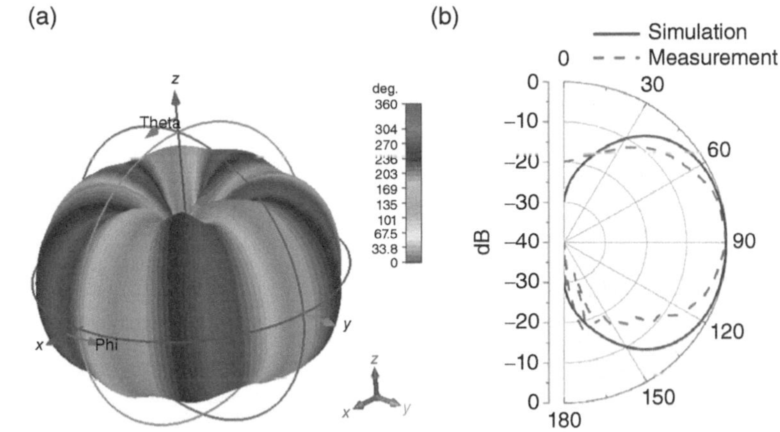

Figure 9.7 Electric far-field radiation patterns of the ring DRA. (a) The simulated three-dimensional (3D) pattern; (b) the simulated (solid line) and measured results (dash line) for the *E*-plane (*xz*-plane) patterns.

Figure 9.7a shows the simulated 3D radiation pattern. Clearly, the generated wave propagates mainly along the transverse plane. Figure 9.7b shows the simulated and measured 2D far-field radiation patterns in the transverse plane, respectively.

The second method is to load the ring horn [15, 18]. A prototype with OAM states $l = \pm 3$ for 10 GHz operation is shown in Figure 9.8a, while Figure 9.8b presents the internal structure of the fabricated antenna. The simulated and measured far-field patterns are shown in Figure 9.9.

The third method is to load a rotating parabolic reflector (RPR) [28]. Figure 9.10 shows the cross-section of the antenna. It consists of a coaxial resonator group (CRG) and an RPR. The CRG has four resonators, and each resonator can generate a pair of levorotatory and dextrorotatory modes so it can generate eight OAM waves with mode number $l = \pm 1, \pm 2, \pm 3, \pm 4$. Thanks to the RPR, all OAM waves are reflected near the transverse plane as shown in Figure 9.11b.

Figure 9.12 shows the simulated results of the amplitude and phase distributions of each mode at the E-plane. It is seen that the PS–OAM beams have good-phase linearity and amplitude uniformity in the azimuthal domain.

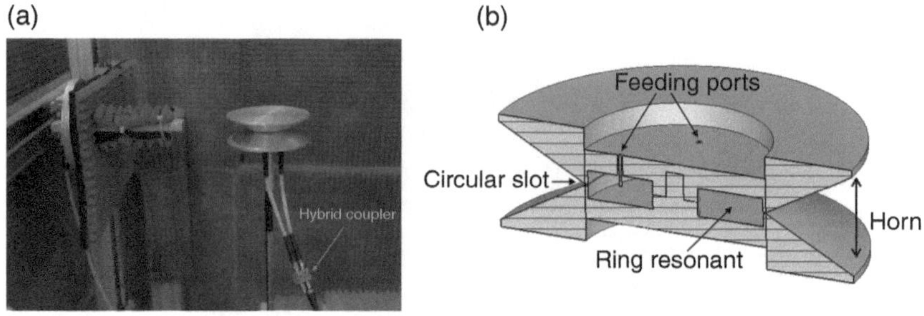

Figure 9.8 (a) An antenna prototype for generation of PS–OAM waves of l = ±3 with two feeding ports; (b) schematic of ring horn loaded with traveling-wave circular slot antenna.

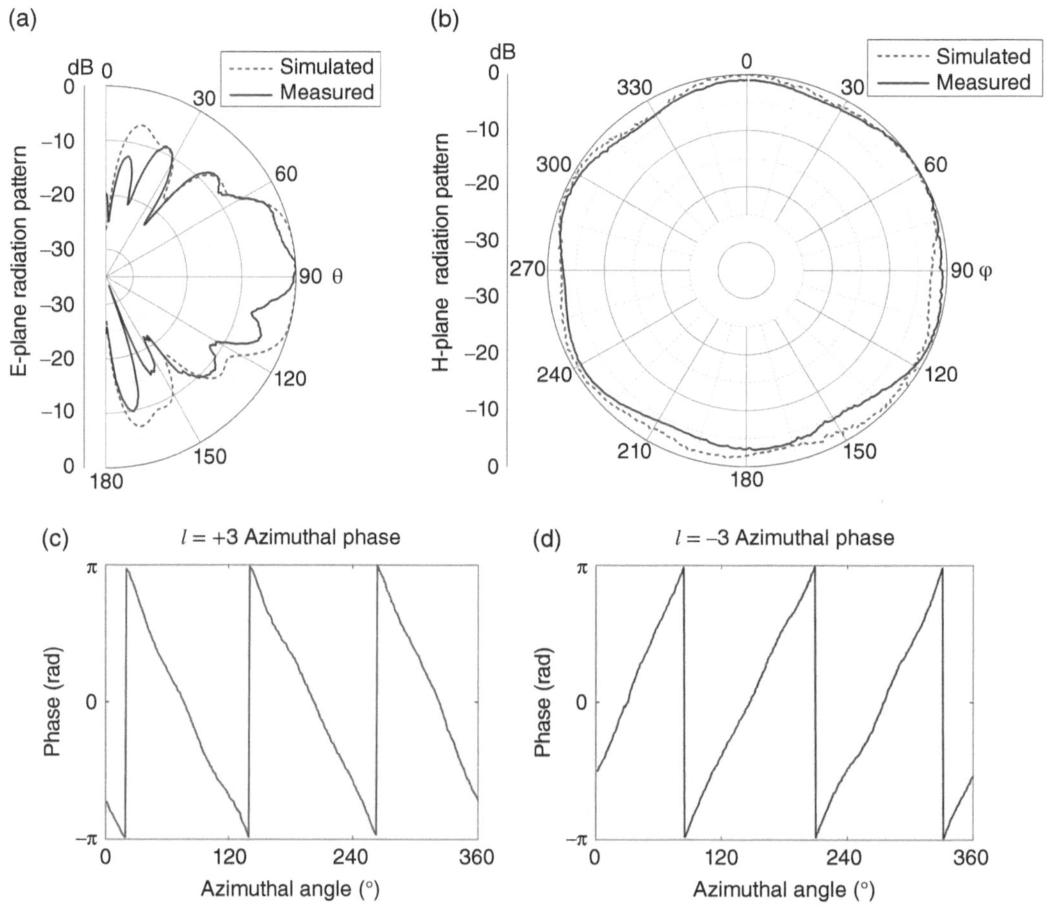

Figure 9.9 Far-field radiation of ring horn loaded traveling wave circular slot antenna for l = ±3. (a) *E*-plane radiation pattern; (b) *H*-plane radiation pattern; (c and d) phase distributions with azimuthal angle.

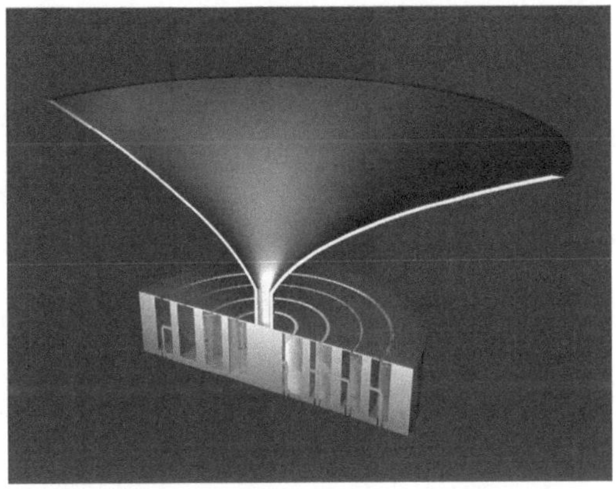

Figure 9.10 The cross-section of the antenna.

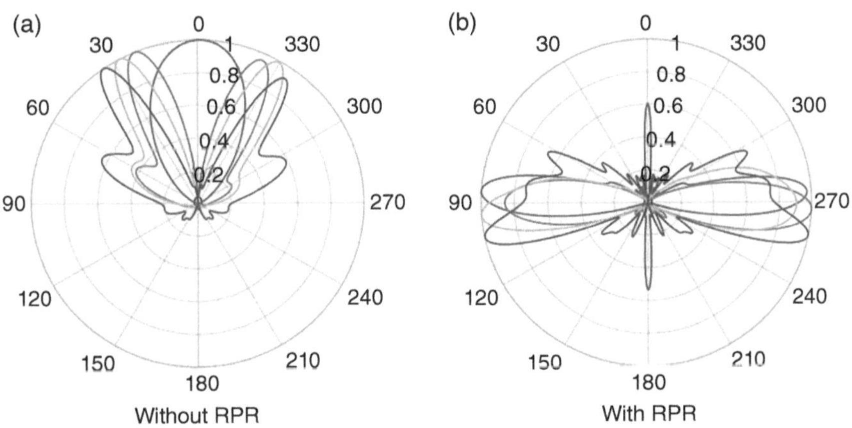

Figure 9.11 Simulated patterns at E-plane with or without RPR.

Using the designed antenna, eight PS–OAM modes can be generated independently. Hence, this antenna can be used to realize the structured radio beams. Figure 9.13 shows the pencil beam constructed by all the eight modes with the different phases, where the difference between adjacent modes is $\Delta\varphi$. It is seen that the main lobe direction is dependent on $\Delta\varphi$. The PS–OAM-based beam scanning has the ability of omni-directional 360° scanning without distortion in azimuthal domain as expected [25].

9.3.2 SIW-based Compact Antenna

SIW is a planar-guided wave structure formed by the upper and lower metal layers and periodic metallic via holes through substrate. Figure 9.14 is the structure of the proposed antenna. Figure 9.14b shows the geometry of a single mode SIW antenna, which comprises a circular patch, a dielectric substrate, and a metal ground. An inner circle of metallic via holes is used here as perfect

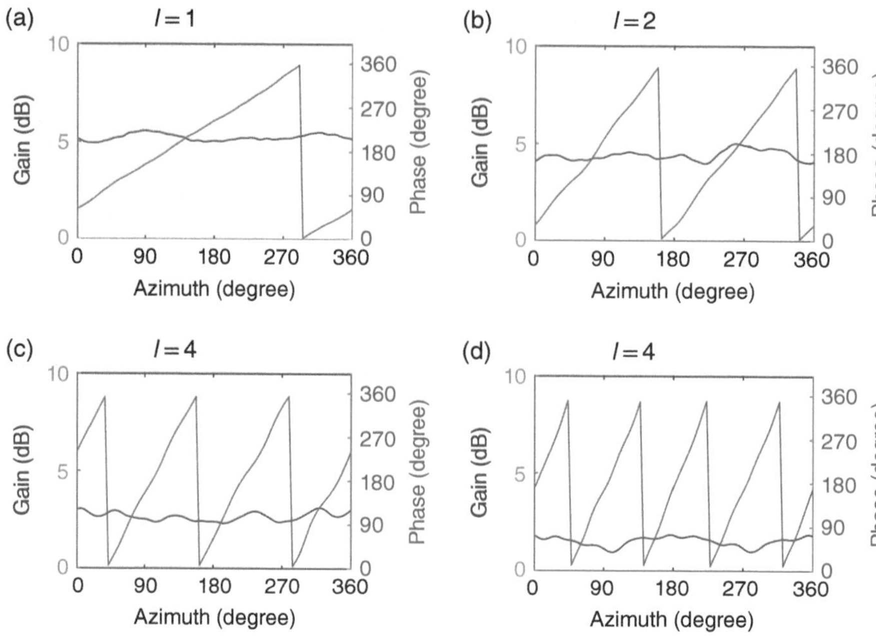

Figure 9.12 Gain and phase distributions of the PS–OAM modes at the *H*-plane ($\theta = 90°$).

electric conducting, while the outer side wall of the antenna is treated as perfect magnetic conducting. Rogers 3202 with the relative permittivity of 3.02 and thickness of 0.75 mm is chosen as the dielectric material for its low-loss tangent. The radius of each circular patch is proportional to the OAM mode l and the guided wavelength λ_g. The radii of the circular patches are 12.8, 15.42, 18.38, and 21.48 mm for $l = \pm 3, \pm 4, \pm 5, \pm 6$, respectively.

In order to obtain a more uniform amplitude distribution, four coaxial feeds with the same amplitude are used here. The phase difference between two adjacent feed points is ±90°. The angle α between two adjacent feed points is related to the OAM mode l, which can be expressed as follows [29]:

$$\alpha = \frac{(2m+1)\pi}{2l}, \quad m = 0, 1, 2, \ldots, l-1 \tag{9.6}$$

where, l is the OAM mode and m is a nonnegative integer. Four coaxial feeds should be placed on one circle as evenly as possible.

To generate a multi-PSOAM beam, eight ring resonators are placed symmetrically on the two sides of the center plane, and each resonator corresponds to one OAM state, as shown in Figure 9.14a. Since the thickness of the SIW structure is quite thin, the superposition of different PS–OAM beams can be achieved by simply stacking the resonators. By exciting eight resonators simultaneously, a structured radio wave with $l = \{\pm 3, \pm 4, \pm 5, \pm 6\}$ can be achieved.

Based on the structure aforementioned, the proposed PS–OAM antenna is modeled and simulated. In order to present the quality of OAM beams quantitatively, the OAM spectrum is investigated by the Fourier transformation [30], which is shown in Figure 9.15. With good performances of intensity and phase distributions, the purities of all the modes are greater than 90%.

Figure 9.16 shows the calculated and simulated beam scanning schematic by the phase manipulation of eight PS–OAM waves, which are presented by dotted and solid lines, respectively. It is obvious that the beams can rotate 360° with the same radiation pattern. The simulated results are well matched with the calculated ones.

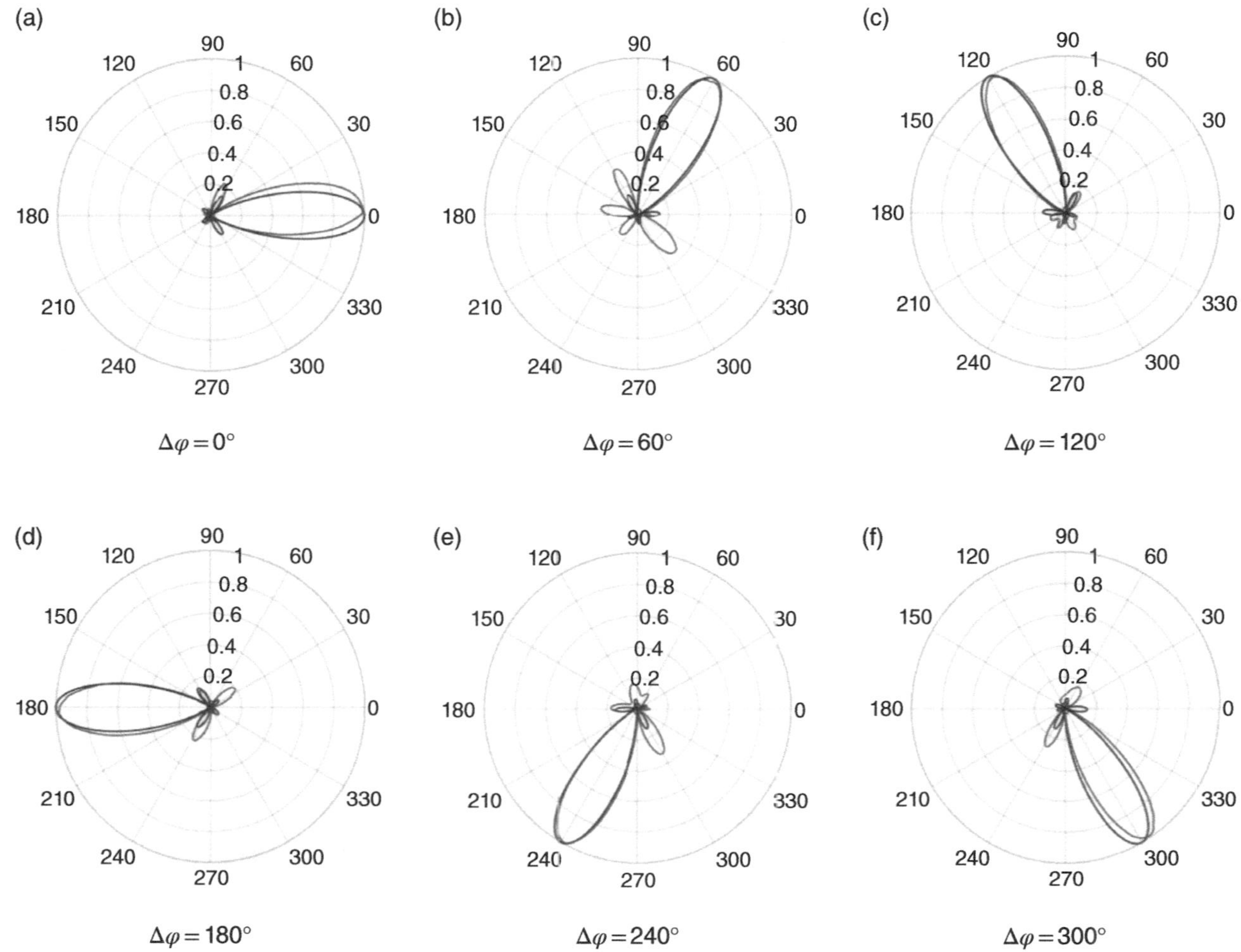

Figure 9.13 Normalized theoretical (solid line) and measured (dash line) patterns of the pencil beams constructed by eight modes.

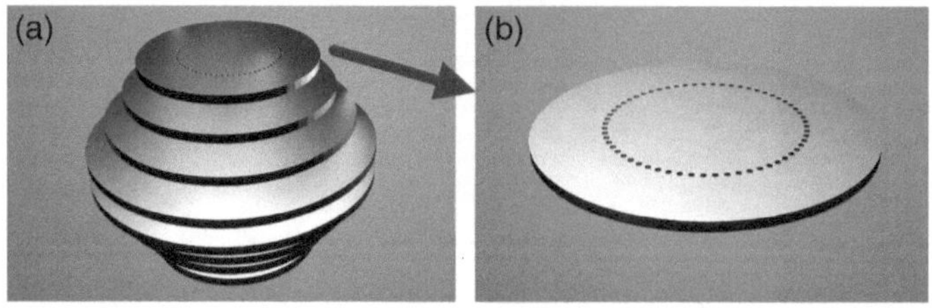

Figure 9.14 Structure of the proposed antenna: (a) multi-mode SIW antenna; (b) single mode SIW antenna.

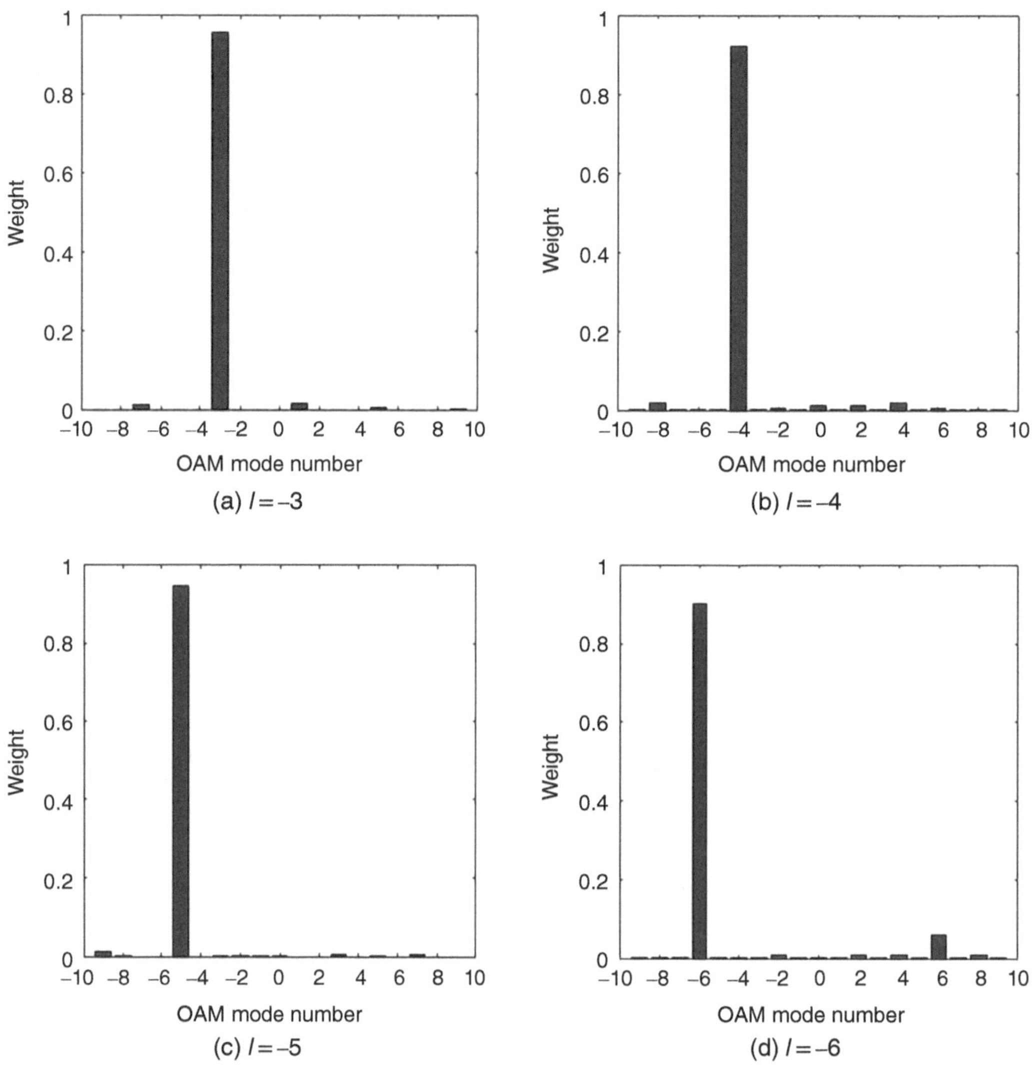

Figure 9.15 OAM spectra of the different modes.

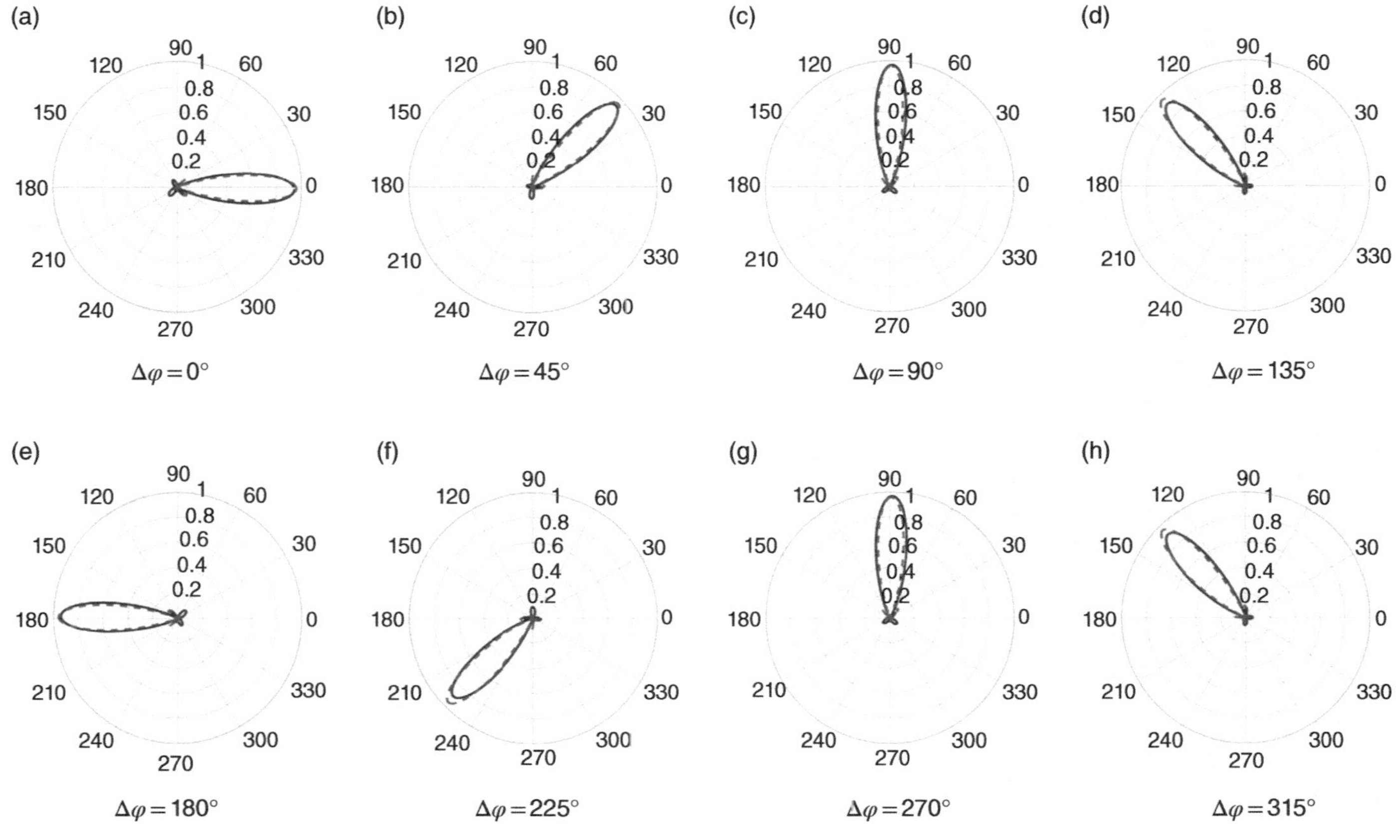

Figure 9.16 Beam scanning schematic provided by the manipulation of eight PS–OAM waves of OAM modes $l = \{\pm 3, \pm 4, \pm 5, \pm 6\}$.

9.3.3 Partial Arc Transmitting Scheme

No matter in the applications of OAM-multiplexing communication or constructing structured radio beam with the OAM mode grouping, the generation of higher order OAM modes has always been a bottleneck. The implemented methods described earlier for constructing structured radio beam are to generate several OAM modes and then superpose them together. However, it is only suitable for constructing the lower order OAM modes. Being aim at realizing a structuted beam with the higher equivalent OAM mode order, it will be hard to generate several single higher order OAM modes, not to mention how to superpose them in practice. In an optical domain, a restriction of the angular range within an OAM light beam profile will modify the OAM spectrum and generate OAM sidebands around the dominant mode [31]. Because of the paraxial approximate property in an optical domain, the OAM spectrum of the transmitted light through the angular mask presents the typical sin c envelope, i.e., a symmetrical distribution. The transmitted light beam shows a linear phase distribution in the mainlobe. Drawing on this idea, in radio domain, the partial slotted curved waveguide leaky-wave antenna is proposed [32], which is designed based on the circular traveling wave antenna [17], and can be regarded as part of it. The quasi-symmetrical OAM components form an OAM mode group that shows a high-gain beam with a linear phase distribution in the mainlobe.

As shown in Figure 9.17, an antenna model that can generate structured radio beams with the higher equivalent OAM mode order $l_e = \pm 40$ at a frequency of 60 GHz is simulated. Figure 9.18 shows the directivity and phase distribution around azimuthal angle φ within the mainlobe of the structured beam at $\theta = 24°$ plane. The beams inherit the spiral phase distribution, i.e., the vorticity. The change of phase period is 9° in azimuth, which means that the equivalent OAM mode order is

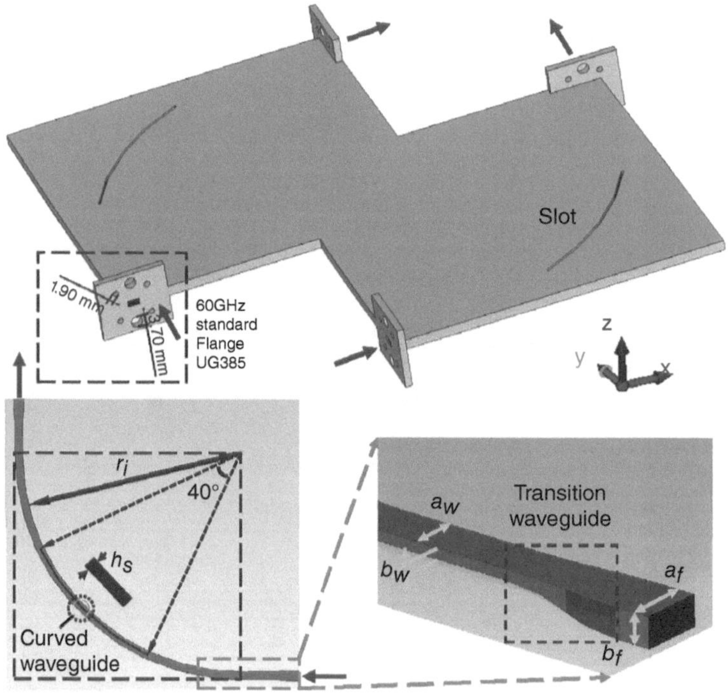

Figure 9.17 Configuration of the partial slotted curved waveguide leaky-wave antenna.

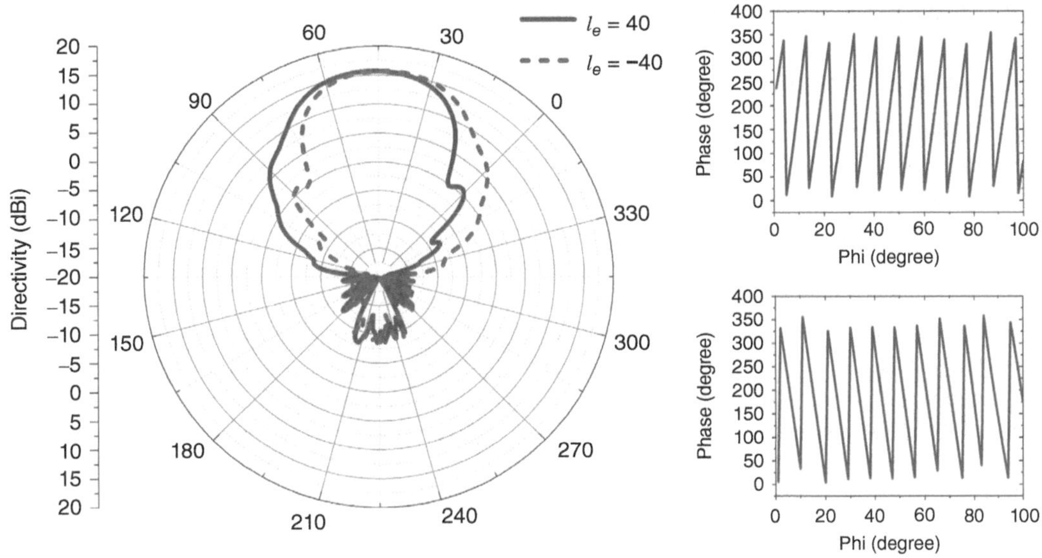

Figure 9.18 The directivities and phase distributions within the mainlobe for $l_e = \pm 40$ at $\theta = 24°$ plane.

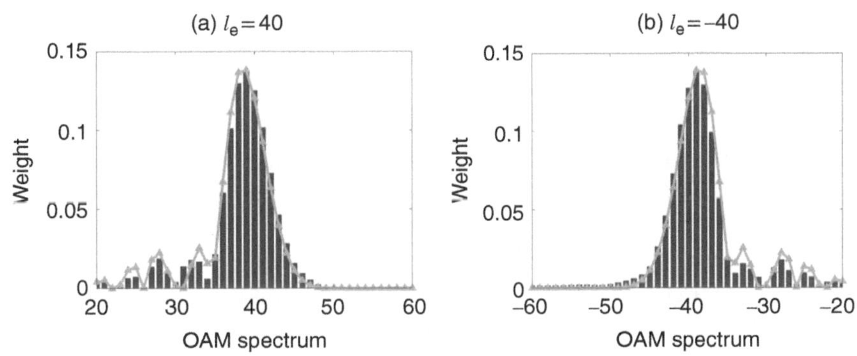

Figure 9.19 The OAM spectrum analysis of the simulated results (bar graph) and the theoretical results (envelope line), $\theta_0 = 24°$.

about $l_e = 40$ or $l_e = -40$. The OAM spectrum of the SBs generated by the partial arc transmitting scheme can be given by [33]:

$$W(l) = \text{sinc}\left[\frac{\varphi_s(l - l_{\text{center}})}{2}\right] J_l[ka \sin(\theta_0)] \tag{9.7}$$

where, φ_s represents the angle corresponding to the arc length of slot, a refers to the radius of the slot, and θ_0 is the divergence angle of the structured beam. The OAM spectrum analysis of the simulated and theoretical results are illustrated in Figure 9.19. The OAM spectrum components of the two beams with $l_e = \pm 40$ nearly present a quasi-symmetrical distribution around the central mode $l_{\text{center}} = \pm 40$, which results in the good phase linearity of the structured beam within its mainlobe.

Such a partial arc transmitting scheme can generate the structured radio beams with higher order OAM modes. The proposed method is much simpler for practical system implementation.

9.4 Potential Applications

9.4.1 Radar Detection

Beam scanning plays an important role in radar target detection. The structured radio waves can be pencil beams with the different phase distribution and orthogonal to each other. The beam steering of the constructed radio beams can be implemented by manipulating the phase of the PS–OAM modes. There are two advantages compared with the phased array-based beam scanning scheme. One is that the radiation pattern of the structured radio beam constructed by PS–OAM modes remains constant when rotating in the azimuthal direction. The other is that it employs phase shift instead of time delay that offers a simple solution for wideband applications.

For OAM-based radar detection and imaging, the structured radio beam steering can be applied. As shown in Figure 9.20, supposing the transmitting antenna transmits the structured beam composed of PS–OAM modes $\{l_1, l_2, ..., l_N\}$, the phases of different PS–OAM modes are controlled so that the direction of the synthesized beam is rotated. When the beam hits the detection object, the reflected echo is received by a single omnidirectional antenna located at the center of the transmit antenna. The signals received during the scanning can be written as [34]:

$$\begin{aligned} S_1 &= A_1 e^{-jl_1\varphi_0} + A_2 e^{-jl_2\varphi_0} + \dots + A_N e^{-jl_N\varphi_0} \\ S_2 &= A_1 e^{-jl_1(\varphi_0 + \Delta\varphi)} + A_2 e^{-jl_2(\varphi_0 + \Delta\varphi)} + \dots + A_N e^{-jl_N(\varphi_0 + \Delta\varphi)} \\ &\dots \\ S_M &= A_1 e^{-jl_1[\varphi_0 + (M-1)\Delta\varphi)]} + A_2 e^{-jl_2[\varphi_0 + (M-1)\Delta\varphi]} + \dots \\ &\quad + A_N e^{-jl_N[\varphi_0 + (M-1)\Delta\varphi]} \end{aligned} \tag{9.8}$$

Utilizing orthogonal mode decomposition, the reflected echo of each PS–OAM mode, $A_1, A_2, \dots,$ A_N, can be obtained. We can employ the compressive sensing techniques [9] or super-resolution imaging techniques [35], to achieve the object detection and imaging. Compared to the detection techniques with each OAM mode, the constructed radio beam technique causes higher signal-to-noise ratio (SNR) due to the radiation gain of the pencil beam.

The results of Figure 9.21 show that when the structured beam contains only eight modes PS–OAM beams, then the detection system cannot distinguish the two detection targets well with 7° interval of azimuthal angle, while when the mode number of PS–OAM beams increases to 12, then there are two indications that appear in the spatial spectrum calculated by the scattering echo. The azimuthal angles of the two targets are estimated to be 44.47 and 51.53°, respectively, and the

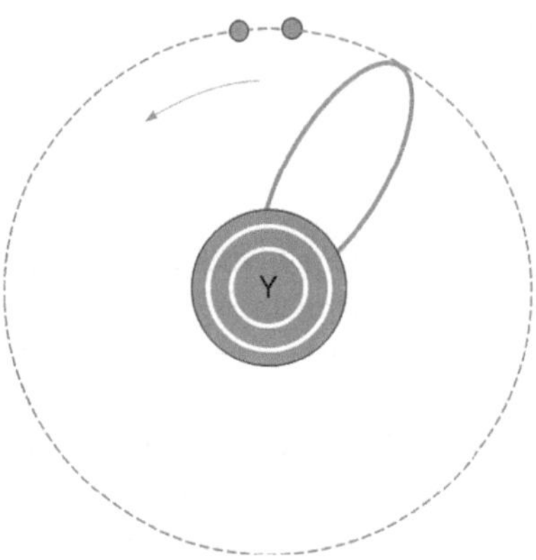

Figure 9.20 Beam steering of structured radio beam for radar detection.

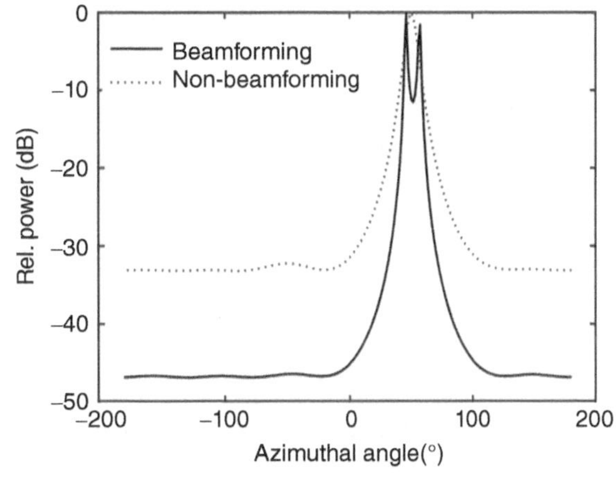

Figure 9.21 Spatial spectrum of the imaging system under different topological charge numbers.

estimation error is only 0.06°. So, the performance of the imaging system based on the plane-structured beam can be improved by increasing the number of topological charges contained in the mode group.

9.4.2 MIMO System

The structured radio beam can be applied to the transmitting end in an LoS–MIMO system [36]. The common high-gain beams used in the MIMO systems just improve the SNR, and the differences among channels are caused by different propagation paths. Whereas the SBs cannot only improve the SNR, but also decrease the spatial correlations between subchannels which benefits from the abundant phase information.

The proposed MIMO system is shown in Figure 9.22. For the sake of illustration, the model is illustrated as a 2×2 LoS–MIMO system. At the transmitting end, the antennas for SBs are placed coaxially, and then radiate into the free space from the same point. At the receiving end, two general horn antennas are placed at the symmetry positions along the propagation axis. Focusing on the link performance, several considerations are assumed:

- The number of the SBs is equal to the number of horn antennas.
- As a contrast, the transmitting aperture of the existing MIMO system is $l_{max}\lambda/\pi$, where l_{max} is the highest order OAM mode among the SBs.

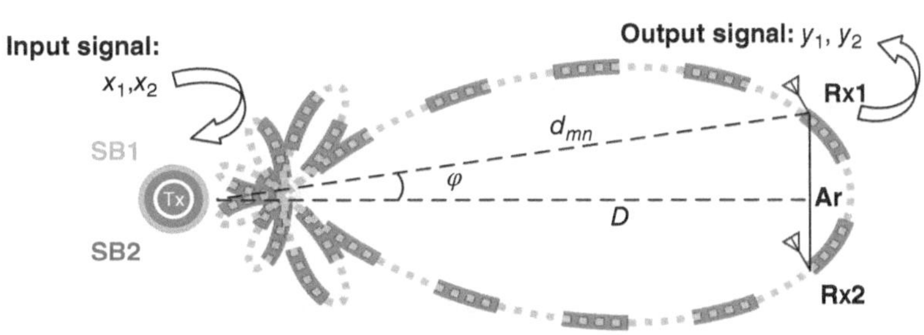

Figure 9.22 System configuration of the MIMO system. d_{mn}: the propagation distance between the mth SB and the nth Rx.

The channel model of the proposed system could be expressed as a narrow band MIMO system, which is as follows [37]:

$$\mathbf{y} = \mathbf{H}_{\mathrm{SB}}\mathbf{x} + \mathbf{n} \tag{9.9}$$

where, $\mathbf{H}_{\mathrm{SB}} \in \mathrm{C}^{2\times 2}$ is the transfer matrix consisting of the several transfer functions. Given a propagation distance d_{mn} between a pair of SB and Rx, the transfer function in $\mathbf{H}_{\mathrm{SB}}$ can be denoted by [36]:

$$h_{\mathrm{SB}}^{mn} = \beta \frac{\lambda}{4\pi d_{mn}} e^{-jkd_{mn}} \left[\frac{1}{\sqrt{P^m}} \sum_{p^m=0}^{P^m-1} e^{-j\left(l_0^m + p^m \Delta l\right)\varphi} \right] \tag{9.10}$$

where, β refers to the antenna performance parameter, k is the free-space wavenumber, and $\lambda/(4\pi d)$ depicts the propagation path loss. The term in the square brackets indicates that the structured beam antennas take the place of general antennas at the transmitting end. l_0^m and P^m are the first mode and the total number of modes in the mth structured beam, respectively. The performance of channel capacity can be analyzed utilizing singular value decomposition (SVD) algorithm [37].

Figure 9.23 presents the capacity gain of existing MIMO system, PS–OAM based MIMO system (PSOAM–MIMO), and structured beam based MIMO system (SB–MIMO) in a LoS scenario. The multiplexed SBs are SB^a {1, 2, ..., 9, 10}, SB^b {11, 12, ...,19,20}, SB^c {− 10, −9, ..., −2, −1}, and SB^d {− 20, − 19, ..., −12, − 11}. Each SB has the same certain intensity pattern. The capacity gains are simulated numerically at a receiving SNR of 30 dB, no matter how long the relative distance D is.

Benefitting from the SNR improvement due to the directionality of the structured beam, the proposed scheme has a better peak gain of about 2.6. Besides this, to achieve a same capacity gain such as 2, the proposed scheme can propagate much longer, e.g. multiplexed SBs are SB^a and SB^b, the propagating distance is raised from 84λ of MIMO system to 328λ. Interestingly, for the same receiving aperture and same mainlobe gain, if the difference value of equivalent OAM mode order Δl_e between multiplexed SBs is larger, and the correlation of subchannels is going to be lower. For

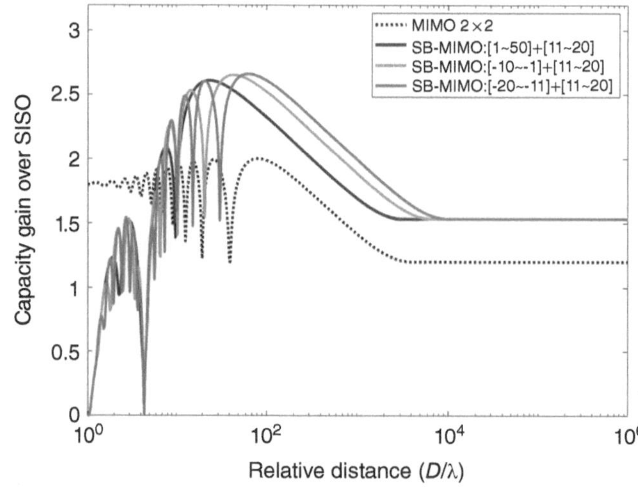

Figure 9.23 Capacity gain over SISO system for existing MIMO system, PS–OAM–MIMO system, and SB–MIMO system. The receiving SNR is 30 dB and the Rayleigh distance of $2A_r^2/\lambda$ is approximately 82λ.

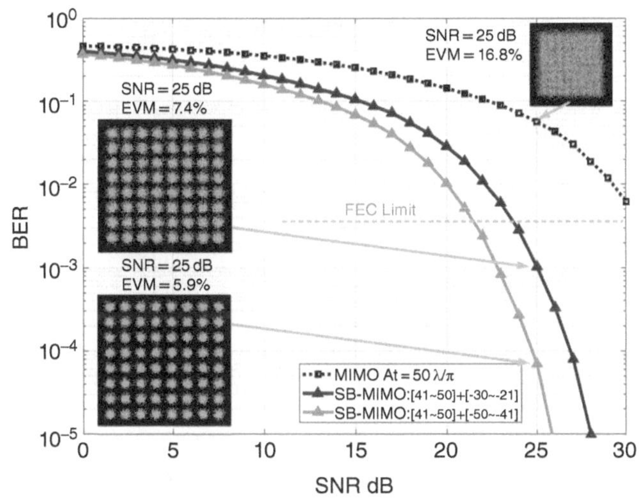

Figure 9.24 The BER curves and the receiving constellations with EVM.

example, the $\Delta l_e = 21$ between SB^b and SB^c, the $\Delta l_e = 31$ between SB^b and SB^d are all larger than $\Delta l_e = 10$ between SB^a and SB^b. Thus the propagating distance is raised to 655 and 993λ further for a capacity gain of 2. In the virtue of the link level simulations, the bit error ratio (BER) performance is shown in Figure 9.24, where the modulation format is 64 QAM with a spectral efficiency of 12 bit/s/Hz. In addition, the received constellations with EVM are plotted. One pair is SB^e {41, 42, ...,49,50} and $SB^f\{-30, -29, ..., -22, -21\}$, Δl_e^{ef} is 71. The other pair is SB^e and $SB^g\{-50, -49, ..., -42, -41\}$, Δl_e^{eg} is 91. The A_r of all curves are $50\lambda/\pi$, and the propagation distance D is 100 m at a frequency of 10 GHz. The BER of the proposed scheme shows better performance than existing MIMO system. Because of larger Δl_e, the latter pair could reach FEC limit at a lower SNR. In this case, for the same receiving SNR of 25 dB, the EVM of SB^e and SB^g, SB^e and SB^f are 5.9, 7.4%, respectively, which are all the better than MIMO 16.8%.

9.4.3 Spatial Field Digital Modulation

Recently, a new kind of wireless communication system has been proposed [38–40]. Different from the conventional communication methods that modulate the information on the time dimension of the EM field, this new method modulates the information on its spatial dimension. Hereinafter, it is called SFDM.

SFDM, as pointed out by the previous studies, is very suitable for the low probability of intercept communication [41]. Moreover, few active microwave devices are needed in the receiving end [38], which is able to reduce the noise and the receiver complexity. Benefitted by the cooperative receiving of signals with multiple receivers, SFDM system has strong antinoise capability and is capable of operating at a low-SNR condition.

Figure 9.25 shows the architecture of the SFDM system based on the PS-OAM based structured radio beam [42]. The binary data, namely, the information stream to be sent, is input to OAM mode decoder, which is actually an antenna controller. And then, OAM mode decoder controls multi-mode OAM antenna to generate the corresponding pattern. At the receiving end, a receiver array with a certain spatial position distribution is adopted. Each receiver measures the intensity or/and

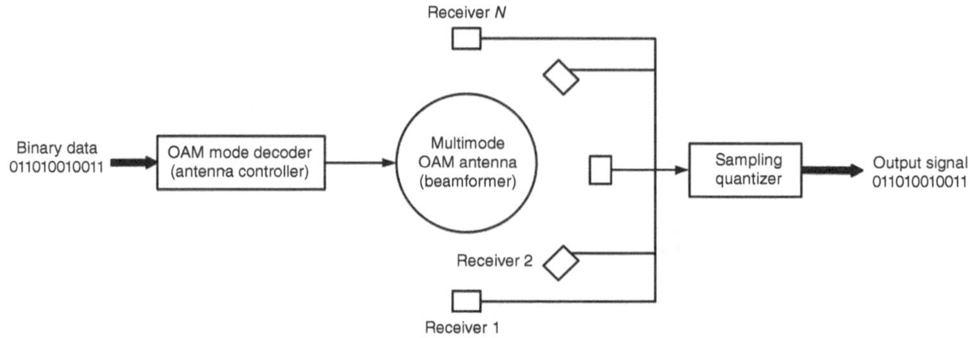

Figure 9.25 The architecture of the SFDM system based on OAM mode group.

the phase of the field at its position and inputs them to the sampling quantizer to make a decision. Finally, the sampling quantizer outputs the demodulated signal.

By superposing multiple OAM modes, i.e., using the OAM mode group, variations of the radiation patterns can be obtained. We call each pattern a 'state.' The pattern in azimuth direction of PS–OAM or the conventional OAM beams with the same divergence angle can be described as follows:

$$f(\varphi) = \sum_{l} A_l e^{\,j(l\varphi + \varphi_{0,l})} \tag{9.11}$$

where φ is azimuth angle, A_l is the intensity of mode l, and $\varphi_{0,l}$ is the phase of mode l at $\varphi = 0$, i.e., the spatial initial phase.

In the OAM-based SFDM system, the binary data are truncated into a certain length and then mapped to the corresponding state. If we select 16 states for transmission information, the binary data will be truncated into 4 bits binary code, which is hereinafter called as the information code.

In the receiving end, at the time of making a decision, a feasible criterion of Euclidean distance is adopted. It can be expressed as:

$$d = \arg\min_{s} \left\{ \sum_{k=1}^{N} [f_s(\varphi_k) - g(\varphi_k)]^2 \right\} \tag{9.12}$$

where, d is the result of the decision, s is the number of the state, $f_s(\varphi_k)$ is the expected response of the receiver, k is located at the azimuth angle φ_k, it is also the value of the pattern function at φ_k, $g(\varphi_k)$ is the measurement of receiver k at φ_k, and N is the total number of receivers. The pattern function shall be normalized according to the maximum value and the receiver array measurement sequence shall be normalized according to the maximum value of the expected response as well. Using Eq. (9.12), the state which has the minimum Euclidean distance to the receiver measurement sequence can be obtained, and it is the output result. This process corresponds to the sampling quantizer in Figure 9.25.

It is worth noting that there may be some states that have a slight difference between the corresponding patterns. Hence, we need to select the states that are easier to be distinguished. To quantify this difference, an N-dimensional Euclidean space in which each state occupies a point is used, the point can be the vector $(f_s(\varphi_{k1}), f_s(\varphi_{k2}), \ldots, f_s(\varphi_{kN}))$. In this Euclidean space, the closer the two states are, the more likely they are to be confused. A greedy algorithm can be used to recursively execute this state selection process in case of determining the first selected state.

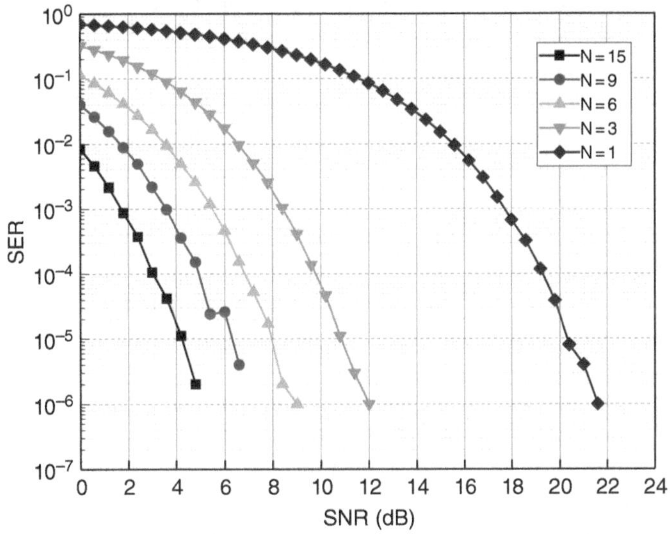

Figure 9.26 SER *vs* SNR with different number of receivers *N*, given the number of states is 16.

The simulation results show that the antinoise capability of SFDM can be improved greatly with the phase information. Actually, one of the advantages of OAM-based SFDM is the ability to control the spatial phase distribution.

The variation of SER with SNR with different number of receivers is shown in Figure 9.26. As the number of receivers decreases, it becomes increasingly difficult to distinguish states, and the SER increases. When the number of receivers is reduced to 1, the benefit of cooperative receiving disappears and the antinoise performance deteriorates seriously.

9.5 Conclusion

We extend the concept of structured light to radio-frequency EM wave. The complex spatial phase and amplitude distribution in a specific beam can be constructed by grouping OAM mode waves in space. This novel kind of beam inherits the inherent vorticity and orthogonality of OAM wave, and has three advantages. Its amplitude directional gain can improve the SNR of the communication link. The phase vortex characteristic can enrich the difference between subchannels in MIMO system. The quasiorthogonal property between beams can simplify the signal processing at the receiving end. Therefore, a spatial structured radio beam is expected to overcome the problems of traditional OAM beam and expand the applications of OAM characteristics in the radar and communication field.

References

1 Allen, L., Beijersbergen, M.W., Spreeuw, R.J.C., and Woerdman, J.P. (1992). Orbital angular momentum of light and the transformation of laguerre-gaussian laser modes. *Physical Review A* **45**: 8185–8189.

2 Padgett, M. (2017). Orbital angular momentum 25 years on. *Optics Express* **25**: 11265.

3 Thidé, B., Then, H., Sjöholm, J. et al. (2007). Utilization of photon orbital angular momentum in the low-frequency radio domain. *Physical Review Letters* **99**: 087701.

4 Mohammadi, S.M., Daldorff, L.K.S., Bergman, J.E.S. et al. (2010). Orbital angular momentum in radio—a system study. *IEEE Transactions on Antennas Propagation* **58** (2): 565–572.

5 Tamburini, F., Mari, E., Sponselli, A. et al. (2012). Encoding many channels on the same frequency through radio vorticity: first experimental test. *New Journal of Physics* **14**: 033001.

6 Hui, X., Zheng, S., Chen, Y. et al. (2015). Multiplexed millimeter wave communication with dual orbital angular momentum (OAM) mode antennas. *Scientific Reports* **5**: 10148.

7 Zhang, W., Zheng, S., Hui, X. et al. (2017). Mode division multiplexing communication using microwave orbital angular momentum: An experimental study. *IEEE Transactions on Wireless Communications* **16**: 1308–1318.

8 Liu, K., Cheng, Y., Gao, Y. et al. (2017). Super-resolution radar imaging based on experimental OAM beams. *Applied Physics Letters* **110**: 164102.

9 Chen, Y., Zheng, S., Jin, X. et al. (2017). Single-frequency computational imaging using OAM-carrying electromagnetic wave. *Journal of Applied Physics* **121**: 184506.

10 Zheng, J., Zheng, S., Shao, Z., and Zhang, X. (2018). Analysis of rotational doppler effect based on radio waves carrying orbital angular momentum. *Journal of Applied Physics* **124**: 164907.

11 Padgett, M.J., Miatto, F.M., Lavery, M.P.J. et al. (2015). Divergence of an orbital-angular-momentum-carrying beam upon propagation. *New Journal of Physics* **17**: 023011.

12 Andersson, M., Berglind, E., and Björk, G. (2015). Orbital angular momentum modes do not increase the channel capacity in communication links. *New Journal of Physics* **17**: 043040.

13 Rubinsztein-Dunlop, H., Forbes, A., Berry, M.V. et al. (2017). Roadmap on structured light. *Journal of Optics* **19**: 013001.

14 S. Zheng, Z. Zhang, Y. Pan, et al., (2015). Plane spiral orbital angular momentum electromagnetic wave, *Asia-Pacific Microwave Conference (APMC)*, Nanjing, China (6–9 December 2015).

15 Z. Zhang, S. Zheng, J. Zheng, et al., (2016). Plane spiral orbital angular momentum wave and its applications, *IEEE MTT-S International Microwave Symposium (IMS)*, San Francisco, CA, USA (22–27 May 2016).

16 X. Zhang and S. Zheng (2017). Grouping plane spiral electromagnetic waves for structured RF beams, *Sixth Asia-Pacific Conference on Antennas and Propagation (APCAP)*, Xi'an, China (16–19 October 2017).

17 Zheng, S., Hui, X., Jin, X. et al. (2015). Transmission characteristics of a twisted radio wave based on circular traveling-wave antenna. *IEEE Transactions on Antennas and Propagation* **63**: 1530–1536.

18 Zhang, Z., Zheng, S., Jin, X. et al. (2017). Generation of plane spiral OAM waves using traveling-wave circular slot antenna. *IEEE Antennas and Wireless Propagation Letters* **16**: 8–11.

19 R. Dong, Y. Chen, S. Zheng, et al., (2017). Generation of plane spiral orbital angular momentum microwave with ring dielectric resonator antenna, *Sixth Asia-Pacific Conference on Antennas and Propagation (APCAP)*, Xi'an, China (16–19 October 2017).

20 Wang, X., Zheng, S., Yu, X. et al. (2019). A compact PSOAM antenna based on substrate integrated waveguide. *Journal of Communications and Information Networks* **4**: 18–24.

21 L. Hua, S. Zheng, X. Yu, et al., (2018). Transformation of OAM waves to plane spiral OAM waves based on gradient-index meta-surface, *Asia-Pacific Microwave Conference (APMC)*, Kyoto, Japan (9 November 2018).

22 Q. Ma, S. Zheng, J. Zheng, et al., (2018). Realization of structured electromagnetic waves based on plane spiral orbital angular momentum waves using circular cylindrical conformal microstrip antenna array, *Asia-Pacific Microwave Conference (APMC)*, Kyoto, Japan (9 November 2018).

23 Zhang, Z., Zheng, S., Zhang, W. et al. (2017). Experimental demonstration of the capacity gain of plane spiral OAM-based MIMO system. *IEEE Microwave and Wireless Components Letters* **27**: 757–759.

24 Zheng, S., Dong, R., Zhang, Z. et al. (2017). Non-line-of-sight channel performance of plane spiral orbital angular momentum MIMO systems. *IEEE Access* **5**: 25377–25384.

25 Zheng, S., Chen, Y., Zhang, Z. et al. (2018). Realization of beam steering based on plane spiral orbital angular momentum wave. *IEEE Transactions on Antennas and Propagation* **66**: 1352–1358.

26 S. Zeng, Y. Zhang, J. Zhu, et al., (2019). 360° on chip optical beam steering based on superposition of planar spiral orbital angular momentum waves, *Conference on Lasers and Electro-Optics Europe European Quantum Electronics Conference (CLEO/Europe-EQEC)*, Munich, Germany (23–27 June 2019).

27 Wang, Z., Zheng, S., Xiong, X. et al. (2020). Structure radio beam construction in azimuthal domain. *IEEE Access* **8**: 9395–9402.

28 Zhu, Z., Zheng, S., Xiong, X. et al. (2021). A compact pattern reconfiguration antenna based on multimode plane spiral OAM. *IEEE Transactions on Antennas and Propagation* **69**: 1168–1172.

29 Chen, Y., Zheng, S., Chi, H. et al. (2016). Half-mode substrate integrated waveguide antenna for generating multiple orbital angular momentum modes. *Electronics Letters* **52**: 684–686.

30 Molina-Terriza, G., Torres, J.P., and Torner, L. (2001). Management of the angular momentum of light: preparation of photons in multidimensional vector states of angular momentum. *Physical Review Letters* **88**: 013601.

31 Jack, B., Padgett, M., and Franke-Arnold, S. (2008). Angular diffraction. *New Journal of Physics* **10**: 103013.

32 Xiong, X., Zheng, S., Zhu, Z. et al. (2020). Direct generation of OAM mode-group and its application in LoS-MIMO system. *IEEE Communication Letters* **24**: 2628–2631.

33 Z. Zhu, S. Zheng, X. Xiong, et al., (2020). Theoretical analysis of the non-uniform OAM source based on an equivalent source transformation, arXiv:2006.00908.

34 X. Zhang, J. Zheng, and S. Zheng (2018). Structured radio beam for radar detection, *IEEE Asia-Pacific Conference on Antennas and Propagation (APCAP)*, Auckland, New Zealand (5–8 August 2018).

35 Lin, M.T., Liu, P.G., Gao, Y., and Liu, J.B. (2016). Super-resolution orbital angular momentum based radar targets detection. *Electronics Letters* **52**: 1168–1170.

36 Xiong, X., Zheng, S., Zhu, Z. et al. (2020). Performance analysis of plane spiral OAM mode-group based MIMO system. *IEEE Communication Letters* **24**: 1414–1418.

37 Edfors, O. and Johansson, A.J. (2011). Is orbital angular momentum (OAM) based radio communication an unexploited area? *IEEE Transactions on Antennas and Propagation* **60**: 1126–1131.

38 Cui, T.J., Liu, S., Bai, G.D., and Ma, Q. (2019). Direct transmission of digital message via programmable coding metasurface. *Research* **2019**: 2584509.

39 Chepala, A., Ding, Y., and Fusco, V.F. (2019). Multimode circular antenna array for spatially encoded data transmission. *IEEE Transactions on Antennas and Propagation* **67**: 3863–3868.

40 Basar, E. (2018). Orbital angular momentum with index modulation. *IEEE Transactions on Wireless Communications* **17**: 2029–2037.

41 Zhou, J., Zheng, S., Yu, X. et al. (2019). Low probability of intercept communication based on structured radio beams using machine learning. *IEEE Access* **7**: 169946–169952.

42 Y. Chen, X. Xiong, Z. Zhu, et al., (2020). Orbital angular momentum mode-group based spatial field digital modulation: Coding scheme and performance analysis, IEEE International *Conference on Communications Workshops (ICC Workshops)*, Dublin, Ireland (7–11 June 2020).

10

OAM Multiplexing Using Uniform Circular Array and Microwave Circuit for Short-range Communication

Kentaro Murata and Naoki Honma

Faculty of Science and Engineering, Iwate University, 4-3-5 Ueda, Morioka, Iwate, Japan

10.1 Introduction

Orbital angular momentum (OAM) multiplexing in a radio frequency (RF) band has been actively studied in various academic and industrial fields for applications to long-distance, high-capacity, wireless backhaul solutions for the fifth generation wireless or newer systems. Various methodologies have been proposed to generate the OAM transmission states, such as the use of spiral phase plates, spiral reflectors, and metamaterials, but these approaches cannot generate multiple OAM modes simultaneously [1]. To improve the spectral efficiency, the use of coaxial uniform circular arrays (UCAs), and discrete Fourier transform (DFT)/inverse DFT (IDFT) beamformers has been proven effective, as they can allow the conduction of multiple signals simultaneously at the same frequency with orthogonal OAM modes. From an electromagnetic viewpoint, it is an interesting phenomenon that the UCAs generate, and decode the multiple orthogonal OAM beams with vortex phase fronts at the different rotational velocities. Conversely, the use of spatial degrees-of-freedom (DOF) can be interpreted simply as a concept that is included in line-of-sight multiple-input multiple-output (LOS MIMO) technologies.

Numerous research studies have been published on LOS MIMO from the late 1990s to the late 2000s. In general, the MIMO scheme yields better performances in the multipath-rich environments, whereby a Rayleigh channel matrix with low-channel correlation is observed. A complete LOS condition likely leads to highly correlated channels that result in a poorer MIMO channel capacity. However, it has been shown that a maximum channel capacity could be achieved when transmitting (Tx) and receiving (Rx) MIMO antennas are optimally arranged so that the normalized channel matrix is unitary, i.e. all the eigenvalues of the channel correlation matrix are equal [2]. Furthermore, the optimal antenna spacing (or antenna separation) was analytically derived based on the geometric channel model for uniform linear arrays (ULAs) [3, 4] and uniform rectangular arrays (URAs) [5] that consisted of an arbitrary number of antennas. These studies proved the effectiveness of the LOS MIMO from an electromagnetic perspective.

However, there are three drawbacks related to the UCA configuration intended for OAM multiplexing.

- **Divergent Beams:** Given that higher-order OAM modes form the divergent beams with a null point on coaxis, the larger aperture Tx/Rx arrays are required to converge and capture the beams for long-distance transmission [6–9]. Although the converged beam can be formed by utilizing

Electromagnetic Vortices: Wave Phenomena and Engineering Applications, First Edition.
Edited by Zhi Hao Jiang and Douglas H. Werner.

the condenser lens or parabolic reflectors, the high-complexity of the antenna system is inevitable.

- **Large Array Occupation Area:** Compared with the conventional LOS MIMO and one-dimensional ULAs, the OAM multiplexing occupies a two-dimensional area because of the UCA configuration. Therefore, with the increase in the transmission distance, the size of the UCA in the OAM-multiplexing system becomes larger at a faster rate compared to ULA LOS MIMO to keep the optimal antenna spacing to satisfy the orthogonal channel matrix condition. As stated in [10], there is less merit associated with the OAM multiplexing in terms of channel capacity versus array size.
- **Limit of Spatial DOF:** The coaxial UCA configuration can, in principle, generate an infinite number of the OAM states. However, to the best of the authors' knowledge, there is no prior published study that introduced an analytical solution of the optimal antenna spacing for UCAs with more than four elements. Indeed, the nonexistence of the optimal antenna spacing was numerically confirmed in [11]. This means that the OAM multiplexing cannot fully capitalize on the potential of the spatial DOF in the higher-order cases.

As indicated earlier, OAM multiplexing is in fact intrinsically unsuitable for long-distance and higher-order transmission. Conversely, as suggested in [12], this type of multiplexing may be used in the short-range applications intended for large-capacity data transfer dealing with high-definition images/movies and high-resolution music. This is a reasonable solution considering the unique features of OAM multiplexing and specifications required for short-range wireless communication, as described later.

- **Frequency-independent Mode Orthogonality:** Assuming short-range communications with the use of a small mobile terminal, it is not preferable to increase the capacity by spatial multiplexing with large-scale MIMO implementation. Even if this is possible, the multiplexing order would be limited. Thus, the combination with the wideband transmission technology is essential to boost the capacity. In this regard, orthogonal OAM multiplexing states can, in principle, be maintained independently of frequency so the DOFs of space and frequency can be effectively utilized. However, note that another measure is needed for the deterioration of the OAM beam gain and mode orthogonality owing to the frequency characteristics of the antenna and beamforming circuit.
- **Low-complexity and Low-power Consumption:** Unlike conventional MIMO signal processing with channel state information (CSI) feedback, OAM multiplexing ideally enables parallel stream transmission only with fixed DFT/IDFT beamformers owing to the nature of the circulant channel matrix observed in the coaxial UCA configuration. In particular, the DFT/IDFT beamformers can be implemented with a passive microwave circuit called Butler matrix. Thus, this allows multiple signal separation with (i) low-power consumption owing to the lossless property of the Butler matrix and (ii) simpler RF frontends with the narrower dynamic range and smaller quantization resolution, or ultimately, no successive digital signal processing. Both are great benefits for the short-range communication terminals with power and design constraints.
- **Robustness to Displacement and Rotation:** The OAM mode orthogonality can be maintained regardless of the transmission distance and coaxial rotation as long as the cyclicity in the channel matrix, i.e. rotational symmetry of the UCAs holds. This can partially cover the positional gaps caused in touch-and-go applications involving station and mobile terminals. However, the inter-mode interference will likely occur in the practical scenarios owing to the positional gap that ruins the rotational symmetry of the UCA configuration. Fortunately, the residual interference may be efficiently cancelled with low-computational overhead by supportive analog or digital beamforming at the station side based on the linear channel variations observed in LOS scenarios [13, 14].

This chapter focuses on the short-range applications of UCA-based OAM multiplexing. At first, a basic OAM multiplexing system is presented, and its multiplexing mechanism is explained from the linear algebraic viewpoint. Second, the optimal OAM array antenna configuration is discussed to yield a maximum achievable rate for given short-range situations. Furthermore, the design theory of DFT/IDFT beamformers is systematically explained. At last, the key challenges are summarized as the conclusions of this chapter.

Following are the notations used throughout this chapter. Specifically, j and e are an imaginary unit and a Napier's constant, $|\cdot|$ denotes an absolute of a complex scalar, the superscripts ^{T}, ^{H} and $^{-1}$ denote transpose, conjugate transpose and inverse, $\det(\cdot)$ and $||\cdot||_F$ denote the determinant and the Frobenius norm of a matrix, $[\cdot]_{nm}$ denotes the nth row, mth column components of a matrix, and $\boldsymbol{O}_N$ and $\boldsymbol{I}_N$ are the Nth zero and identity matrices.

10.2 OAM Multiplexing System and its Mechanism

In this section, the basic configuration of the OAM multiplexing system is presented, and its multiplexing mechanism is elaborated from the linear algebraic viewpoint. As shown in Figure 10.1, the OAM multiplexing system is divided into three parts: coaxial UCAs, circulant channel matrix, and DFT/IDFT beamformers. The key points of each part are explained as follows:

10.2.1 Coaxial UCA Configuration

The array configuration for OAM multiplexing is illustrated in Figure 10.2. For simplicity purposes, the same number of antennas, N, and the same array radius, r, are assumed for both Tx/Rx circular arrays, but the Tx/Rx radii can indeed be different. Both Tx/Rx arrays are associated with uniform angular intervals equal to $\phi = 2\pi/N$, as shown in Figure 10.2a and b, respectively. The UCAs are coaxially arranged at the transmission distance of D, as shown in Figure 10.2c.

Regarding the array element, any type of antenna is acceptable as long as the coaxial rotational symmetry of the entire array is retained. As shown in Figure 10.2a, each element of the Tx UCA is labeled with the antenna numbers Tx 1 to Tx N in an anticlockwise manner. Likewise, the corresponding element in the Rx UCA is labeled with the antenna number Rx 1 to Rx N, but the Rx UCA may be rotated uniaxially by an arbitrary angle $\Delta\phi$, as shown in Figure 10.2b.

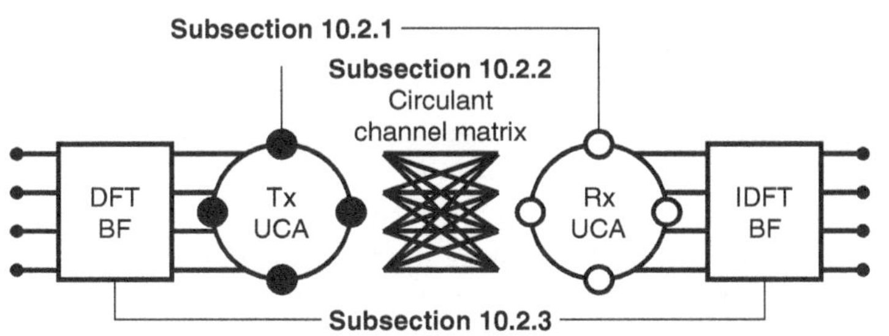

Figure 10.1 OAM multiplexing system using coaxial UCAs and DFT/IDFT beamformers.

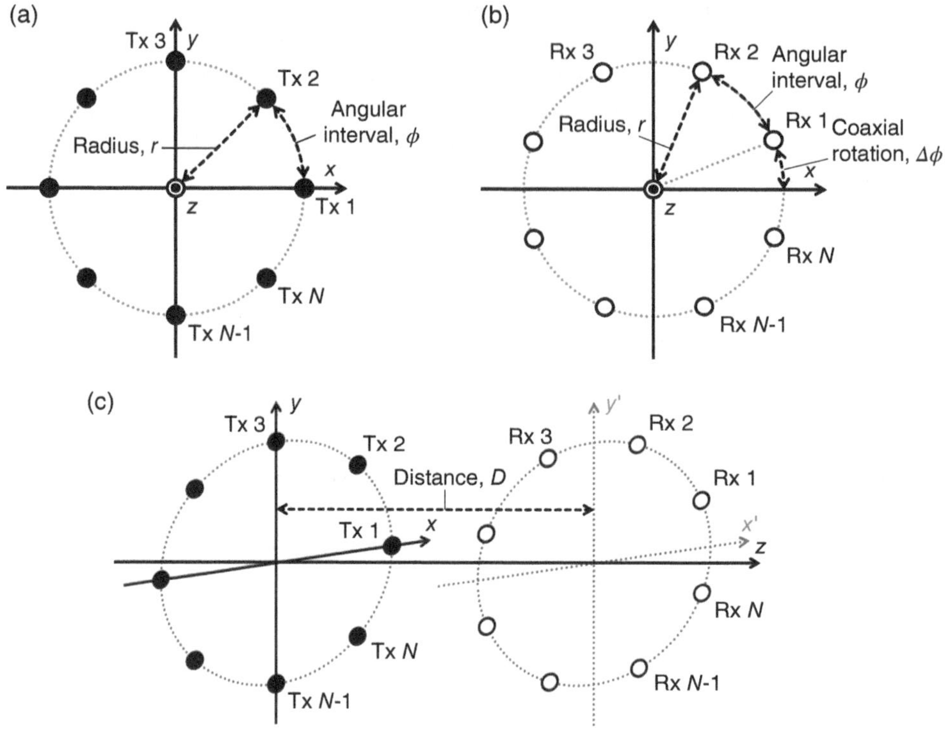

Figure 10.2 Coaxial UCA configuration; (a) Tx UCA; (b) Rx UCA; (c) Tx/Rx UCAs.

10.2.2 Circulant Channel Matrix

A special form of the channel matrix is observed in the aforementioned coaxial UCA configuration which is explained further. The channel matrix $\boldsymbol{H} \in \mathbb{C}^{N\times N}$ of the Tx/Rx UCAs is defined as:

$$\boldsymbol{H} = \begin{bmatrix} h_{11} & h_{12} & \cdots & h_{1N} \\ h_{21} & h_{22} & & h_{2N} \\ \vdots & & \ddots & \vdots \\ h_{N1} & h_{N2} & \cdots & h_{NN} \end{bmatrix}. \tag{10.1}$$

In Eq. (10.1), each element represents a channel response between a certain antenna pair of the Tx/Rx UCAs, and the first and second numbers of the subscript denote the Tx/Rx antenna numbers, respectively.

If a point-wave source is assumed as an array element, and the entire array lies in a perfect LOS, the channel response between the Tx/Rx antenna pair can be expressed based on the information of the relative geometric positions of the antennas. Accordingly, a specific regularity appears in the channel matrix. For example, let the channels from Tx 1, 2, $\cdots$, $(N-1)$, and N to Rx 1 be defined as the complex constants c_1, c_2, $\cdots$, c_{N-1}, c_N, respectively. The relative positions of Rx 2 to Tx 2, 3, $\cdots$, N, 1 are identical to that for Rx 1 owing to the cyclicity of the UCA configuration. That is, the channels for Rx 2 can also be given using the aforementioned complex constants. This is also true for the other Rx antennas. Based on the above regularity, each element in the channel matrix $\boldsymbol{H}$ given in Eq. (10.1) can be replaced as:

$$\boldsymbol{H} = \begin{bmatrix} c_1 & c_2 & \cdots & c_N \\ c_N & c_1 & & c_{N-1} \\ \vdots & & \ddots & \vdots \\ c_2 & c_3 & \cdots & c_1 \end{bmatrix}. \tag{10.2}$$

From Eq. (10.2), one may notice that the channel matrix is given as a circulant matrix where each row vector is shifted by one element to the right relative to the preceding row vector. Further, it should be noted that the cyclicity in the channel matrix is retained independent of the array radius, transmission distance, and coaxial rotation in the coaxial UCA configuration.

10.2.3 DFT/IDFT Beamformers

Because of the nature of the circulant matrix, the channel matrix given in Eq. (10.2) can be decomposed as follows [15]:

$$\boldsymbol{H} = \boldsymbol{W}_N \Lambda \boldsymbol{W}_N^H. \tag{10.3}$$

In Eq. (10.3), $\boldsymbol{W}_N \in \mathbb{C}^{N\times N}$ is a DFT matrix, while its conjugate transpose $\boldsymbol{W}_N^H$ is an IDFT matrix. $\boldsymbol{W}_N$ consists of N column vectors as indicated below:

$$\boldsymbol{W}_N = [\boldsymbol{w}_1 \ \boldsymbol{w}_2 \ \cdots \ \boldsymbol{w}_N] \tag{10.4}$$

where, the nth column vector $\boldsymbol{w}_n$ is given as:

$$\boldsymbol{w}_n = \frac{1}{\sqrt{N}}\left[1 \quad \omega_n \quad \omega_n^2 \quad \cdots \quad \omega_n^{N-1}\right]^T. \tag{10.5}$$

w_n in Eq. (10.5) is the root of unity, and is expressed as:

$$w_n = e^{-j\frac{2\pi(n-1)}{N}}, \tag{10.6}$$

which represents a twiddle factor moving in $2\pi/N$ angular intervals on a unit circle on a complex plane. $\Lambda \in \mathbb{C}^{N\times N}$ is a diagonal matrix with its nth eigenvalue, λ_n being its nth diagonal component. The nth eigenvalue, λ_n, is expressed with the use of the circulant channel components and roots of unity as:

$$\lambda_n = \sum_{m=1}^{N} c_m \omega_n^{m-1}. \tag{10.7}$$

The channel matrix when the DFT/IDFT beamformers are applied to Tx and Rx UCAs, $\boldsymbol{H}_{\text{OAM}} \in \mathbb{C}^{N\times N}$, is expressed by multiplying DFT and IDFT matrices in Eq. (10.3) as:

$$\boldsymbol{H}_{\text{OAM}} = \boldsymbol{W}_N^H \boldsymbol{H} \boldsymbol{W}_N = \Lambda \tag{10.8}$$

where in the unitarity of the DFT matrix, $\boldsymbol{W}_N \boldsymbol{W}_N^H = \boldsymbol{W}_N^H \boldsymbol{W}_N = \boldsymbol{I}_N$, is used. In this way, the channel matrix can be diagonalized, that is, orthogonal OAM multiplexed streams are constructed.

From the aforementioned analysis, the decisive differences of OAM multiplexing using coaxial UCA configuration from conventional LOS MIMO using ULAs and URAs are summarized as follows:

- The channel matrix always becomes a circulant matrix independent of array radius, transmission distance, and coaxial rotation

- The circulant channel matrix is always diagonalized by the fixed DFT and IDFT beamformers that are uniquely determined only by the antenna number

Based on the aforementioned understandings; in the next section, we will discuss the potential applications of the OAM multiplexing technology.

10.3 OAM Multiplexing for Short-range Communications

As mentioned in the introduction, the application of OAM multiplexing in short-range communications is justified based on considerations of its beneficial features and the required specifications for short-range communication terminals. In this section, the optimal OAM system design is discussed to offer high-quality wireless communication.

10.3.1 Achievable Rate

At first, to evaluate wireless communication performance by OAM multiplexing, the achievable rate is introduced as an evaluation index in this section. The rate achieved by OAM multiplexing, R_{OAM}, is calculated by considering the possible intermode interference as:

$$R_{\mathrm{OAM}} = \sum_{n=1}^{N} \log_2 \left(1 + \frac{\left|[\boldsymbol{H}_{\mathrm{OAM}}]_{nn}\right|^2 \frac{P_{\mathrm{S}}}{N}}{\sum\limits_{\substack{m=1 \\ m \neq n}}^{N} \left|[\boldsymbol{H}_{\mathrm{OAM}}]_{nm}\right|^2 \frac{P_{\mathrm{S}}}{N} + P_{\mathrm{N}}} \right). \tag{10.9}$$

In Eq. (10.9), P_{S} and P_{N} denote the total Tx signal power and Rx noise power, respectively, and an equal power is allocated to each mode. The fractional term in Eq. (10.9) represents the signal-to-interference-plus-noise ratio (SINR), and its numerator represents the signal power of the nth mode stream corresponding to the nth diagonal element of $\boldsymbol{H}_{\mathrm{OAM}}$. The first and second terms of the denominator represent the total power of the intermode interference, i.e. the nondiagonal elements of $\boldsymbol{H}_{\mathrm{OAM}}$, to the nth mode stream, and the noise power, respectively. Again, only when the channel matrix maintains the cyclicity expressed by Eq. (10.2) the channel matrix is perfectly diagonalized as indicated in Eq. (10.8). Correspondingly, Eq. (10.9) can be rewritten as:

$$R_{\mathrm{OAM}} = \sum_{n=1}^{N} \log_2 \left(1 + \frac{P_{\mathrm{S}}}{NP_{\mathrm{N}}} |\lambda_n|^2 \right). \tag{10.10}$$

For comparison with OAM multiplexing, the channel capacity of the MIMO eigenmode transmission is also evaluated. This can be interpreted as the maximum achievable rate in a specific MIMO system. The eigenmode transmission proceeds as follows: subject to the assumption that the Tx side knows CSI, the singular value decomposition (SVD) of the channel matrix gives:

$$\boldsymbol{H} = \boldsymbol{U} \Sigma \boldsymbol{V}^H \tag{10.11}$$

where, $\boldsymbol{U} \in \mathbb{C}^{N \times N}$ and $\boldsymbol{V} \in \mathbb{C}^{N \times N}$ are unitary matrices with singular vectors, and $\Sigma \in \mathbb{R}^{N \times N}$ is a diagonal matrix with the nth singular value, σ_n, being its nth diagonal element. By applying $\boldsymbol{V}$

and $\boldsymbol{U}^H$ as Tx/Rx beamformers to the channel matrix $\boldsymbol{H}$, the diagonalized MIMO channel matrix $\boldsymbol{H}_{\text{MIMO}}$ can be obtained as:

$$\boldsymbol{H}_{\text{MIMO}} = \boldsymbol{U}^H \boldsymbol{H} \boldsymbol{V} = \Sigma, \tag{10.12}$$

where, the unitary property of $\boldsymbol{U}^H\boldsymbol{U} = \boldsymbol{I}_N$ and $\boldsymbol{V}^H\boldsymbol{V} = \boldsymbol{I}_N$ is used. Given that the SVD in Eq. (10.11) is always valid for any matrix, the MIMO eigenmode transmission can diagonalize the channel matrix regardless of its entries, i.e. even if there is no cyclicity in the channel matrix, unlike OAM multiplexing. Thus, similar to Eq. (10.10), the channel capacity of the MIMO eigenmode transmission under the equal power allocation, R_{MIMO}, is calculated as:

$$R_{\text{MIMO}} = \sum_{n=1}^{N} \log_2\left(1 + \frac{P_{\text{S}}}{NP_{\text{N}}}\sigma_n^2\right). \tag{10.13}$$

Reviewing Eqs. (10.10) and (10.13), the set of the absolute eigenvalues, $|\lambda_n|$, and the set of the singular values, σ_n, have exactly the same values if the channel matrix is circulant. Therefore, there is indeed no difference between the achievable rate by OAM multiplexing and the channel capacity by MIMO eigenmode transmission in the coaxial UCA configuration in a perfect LOS case.

10.3.2 Array Topology

In this section, the array topology that is suitable for OAM multiplexing is discussed. For example, Figure 10.3 displays possible four element array topologies intended for OAM multiplexing.

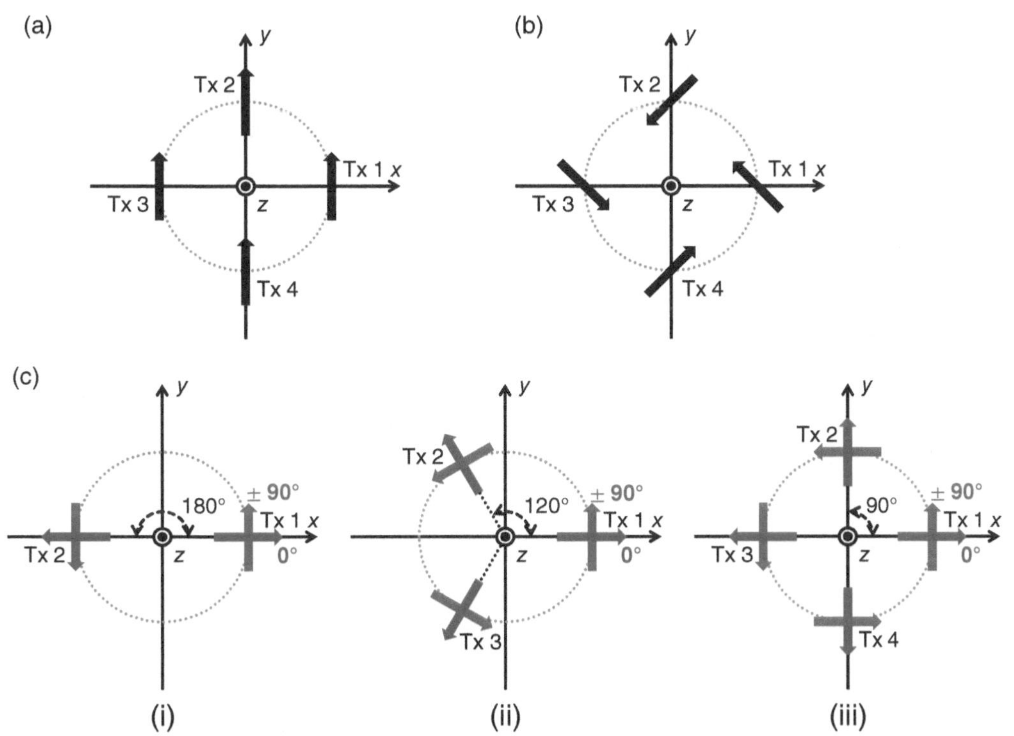

Figure 10.3 Various types of coaxial UCA configurations: (a) single-polarized array; (b) dual-polarized array; (c) circular-polarized arrays for (i) $N = 2$, (ii) $N = 3$, and (iii) $N = 4$.

First, Figure 10.3a shows a single-polarized array. This topology achieves a polarization match among all pairs of the Tx/Rx antennas offering a higher total channel gain. However, considering the physical antenna shape, the coaxial rotational symmetry of the entire array cannot be rigorously maintained unlike point-wave source assumptions. This spoils the cyclicity in the channel matrix and results in the deterioration of the orthogonality of the OAM mode.

The second candidate is a dual-polarized array, as shown in Figure 10.3b. Even though there are several array patterns that involve dual polarization, as introduced in [12], a sequential array configuration is used as an example. Similar to a windmill, this configuration can maintain rotational symmetry holding the cyclicity in the channel matrix. While the sequential array inherently offers ideally isolated streams between cross-polarized pairs owing to the perpendicular bisected arrangements with respect to each other, this also indicates lower total channel gains compared with single-polarized arrays.

To resolve the above trade-off between the total channel gain and the cyclicity in the channel matrix, the employment of a circular-polarized array is the best way. Figure 10.3c shows the sequentially arranged circular-polarized arrays that consist of cross-polarized elements with $\pm 90°$ phase shifts for two, three, and four element cases. Note that positive and negative signs correspond to left- and right-handed circular polarizations (LHCP and RHCP), respectively. The circular-polarized array can realize a polarization match among all Tx/Rx antenna pairs by maintaining rotational symmetry that offers better OAM multiplexing performance than the other topologies. For reference, single-fed microstrip antennas (MSAs) with notches can also be used as circular-polarized elements to simplify the antenna and feeding system configuration, as shown in Figure 10.4.

In the following section, the achievable rates by OAM multiplexing in the aforementioned three topology schemes with four elements each are compared to clarify the superiority of the circular-polarized topology. For the electromagnetic simulations, the method of moments (MoM) was used. In the simulation, a thin, half-wavelength dipole antenna was used as the array element. Regarding the circular-polarized array topology, cross dipoles with small offsets were used and fed with $+90°$ phase shifts, i.e. LHCP. Unless specified, the transmission distance, coaxial rotation, array radius, and Tx-signal-to-Rx-noise power ratio were set to $1.0\,\lambda$, $0°$, $0.5\,\lambda$, and, 30 dB, respectively, where λ is the wavelength.

Figure 10.5a–c shows the achievable rates of the aforementioned three topologies when the transmission distance, coaxial rotation, and array radius are varied independently. In the figures, the achievable rates of OAM multiplexing are plotted for different types of markers. In addition, the channel capacity of the MIMO eigenmode transmission is plotted with different types of lines for comparison.

At first, by focusing on Figure 10.5a, almost the same rate was achieved for all the topologies in the small D region. This was because only the pairs of antennas that faced each other were strongly coupled, and each pair could form independent streams regardless of the topology. By increasing D,

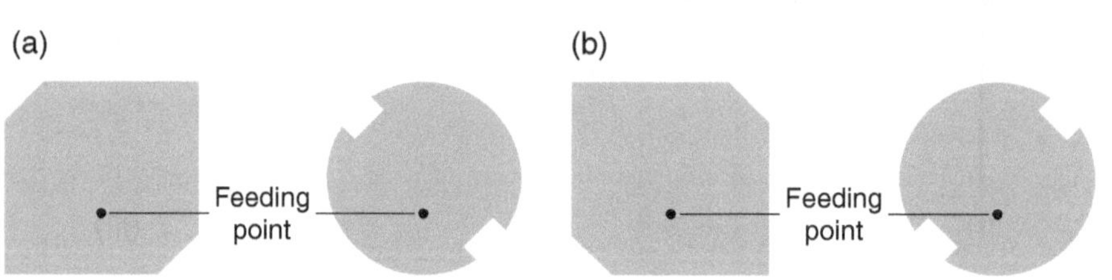

Figure 10.4 Examples of circular-polarized MSAs: (a) LHCP; (b) RHCP.

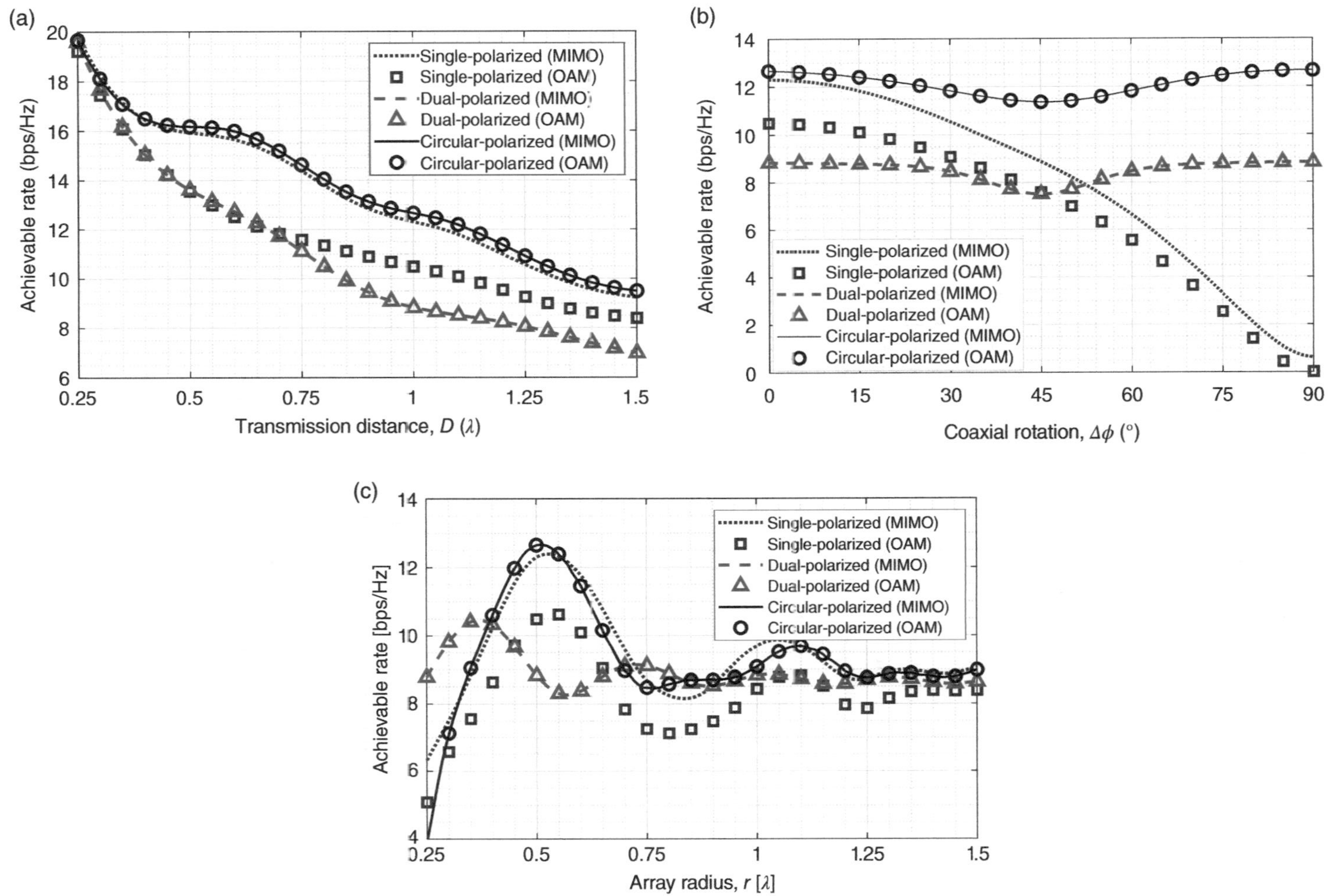

Figure 10.5 Achievable rates of different configurations versus (a) transmission distance; (b) coaxial rotation; (c) array radius.

all topologies experienced rate reductions owing to the distance attenuation. There are two points that should be emphasized. The first point is that while OAM yielded exactly the same rate as that of MIMO in circular- and dual-polarized arrays, a deviation was confirmed in the single-polarized case. This deviation was attributed to the intermode interference caused by the noncirculant channel matrix in the single-polarized topology case. The second point pertains to the fact that while comparable MIMO rates were obtained in single- and circular-polarized topologies, a noticeable difference was confirmed in the dual-polarized topology. This difference was attributed to the polarization loss between the orthogonally polarized antennas that resulted in lower total channel gain.

The above two points are also applied to Figure 10.5b and c. Specifically, Figure 10.5b shows that as $\Delta\phi$ approaches 90°, the rate of the single-polarized topology drastically decreases owing to the polarization mismatch. Incidentally, a slight rate degradation can be observed at $\Delta\phi = 45°$ in the cases of the dual- and circular-polarized topologies. This degradation was induced by the specific mode that excited adjacent antenna pairs in antiphase, and it produced blind spots at the middle angle between the adjacent pair where the electromagnetic field was cancelled out.

From the aforementioned results, the superiority of circular-polarized topology was obvious. Therefore, in the rest of the analysis, the circular-polarized topology is used.

Finally, we focus on the increasing and decreasing tendency following increases in r in Figure 10.5c. The characteristic tendency can be understood as follows:

- In large r regions, the facing antenna pair generates independent streams but channels between noncorresponding antennas vanish. As a result, the total rate converges to N times to that of a single stream.
- In small r regions, strong mutual coupling occurs among closely spaced antennas, and the channel matrix becomes highly correlated. These two negative effects degrade the rate.
- There exists an optimal array radius, r_{opt}, for a certain transmission distance that maximizes the rate.

Thus, the next subsection discusses the optimal channel condition for OAM multiplexing, and yields an optimal array radius derived based on geometric analysis.

10.3.3 Optimal Array Radius

This subsection first explains the optimal channel matrix condition that offers a maximum achievable rate by OAM multiplexing. As proved in the previous subsection, the OAM multiplexing yields exactly the same rate as that of MIMO eigenmode transmission in the circular-polarized topology given that under ideal conditions there is no intermode interference. Therefore, if it is assumed that the (i) signal-to-noise power ratio (SNR) is high and (ii) the channel matrix has a full rank, i.e. all the eigenvalues are nonzero; the achievable rate by OAM multiplexing can be approximated by ignoring the unit term as:

$$\begin{aligned} R_{\text{OAM}} &\approx \sum_{n=1}^{N} \log_2\left(\frac{P_{\text{S}}}{NP_{\text{N}}}|\lambda_n|^2\right) \\ &= N\log_2\frac{P_{\text{S}}}{NP_{\text{N}}} + \log_2\prod_{n=1}^{N}|\lambda_n|^2. \end{aligned} \tag{10.14}$$

Assuming that the total channel gain is constant, that is,

$$||\boldsymbol{H}||_F^2 = \sum_{n=1}^{N} |\lambda_n|^2 = \text{const.}, \tag{10.15}$$

and considering that the arithmetic–geometric mean inequality is given as:

$$\prod_{n=1}^{N} |\lambda_n|^2 \leq \left(\frac{1}{N}\sum_{n=1}^{N} |\lambda_n|^2\right)^N = \text{const.}, \tag{10.16}$$

the upper bound of the achievable rate by OAM multiplexing in Eq. (10.13) is now clear. Further, the upper bound can be achieved subject to the condition that all absolute eigenvalues are equal, i.e. $|\lambda_1| = |\lambda_2| = \cdots = |\lambda_N|$. This condition is equivalent to the fact that the normalized channel matrix is unitary.

As it can be observed from the previous results of Figure 10.5c, the maximum achievable rate is obtained at a specific array radius for a certain transmission distance that almost satisfies the aforementioned upper bound condition. Therefore, the optimal array radius is analyzed next based on the geometric channel model.

The geometric model assumes a point-wave source and plane-wave propagation. It determines the channel response as a complex exponential function of the relative positions of the Tx/Rx antennas that represent the phase rotation owing to the path length, but ignores the distance attenuation. Correspondingly, even though the geometric model assumes a far-field configuration, it is actually valid to some extent, even in the short-range, as it will be numerically verified in the end of this section. Additionally, in the circular-polarized topology, an additional phase rotation occurs that is caused by the antenna rotation owing to the sequential arrangement and should be considered separately from the geometric factor.

Specifically, the channel response between the mth Tx and nth Rx antennas, h_{nm}, in the coaxial UCA configuration (as shown in Figure 10.2c) is defined as the multiplication of three factors as:

$$h_{nm} \sim p_{\mathrm{R},n} g_{nm} p_{\mathrm{T},m}. \tag{10.17}$$

At first, $p_{\mathrm{T},m}$ and $p_{\mathrm{R},n}$ in Eq. (10.17) represent the respective phase rotation factors attributed to the Tx/Rx antenna rotations. For simplicity, the following analysis deals with the case in which the coaxial rotation of the Rx array, $\Delta\phi$, is zero. Assuming the existence of spatial uniformity of polarization, and ignoring the antenna offset from the coaxis, $p_{\mathrm{T},m}$ and $p_{\mathrm{R},n}$ can be simply expressed as complex exponential functions of the mth Tx and nth Rx antenna rotations with respect to the first ones (see Figure 10.6) based on the Jones calculus [16] as:

$$p_{\mathrm{T},m} = \begin{cases} e^{-j(m-1)\phi} & \text{(LHCP)} \\ e^{j(m-1)\phi} & \text{(RHCP)} \end{cases} \tag{10.18}$$

and

$$p_{\mathrm{R},n} = \begin{cases} e^{j(n-1)\phi} & \text{(LHCP)} \\ e^{-j(n-1)\phi} & \text{(RHCP)} \end{cases}, \tag{10.19}$$

respectively.

The symbol g_{nm} in Eq. (10.17) represents the phase rotation attributed to the path length between the mth Tx and nth Rx antennas. The path length between the mth Tx and nth Rx antennas, d_{nm}, is expressed by the function of transmission distance, D, array radius, r, and angular interval, ϕ, that can be approximated with the use of the far-field assumption given that $D \gg r$ as:

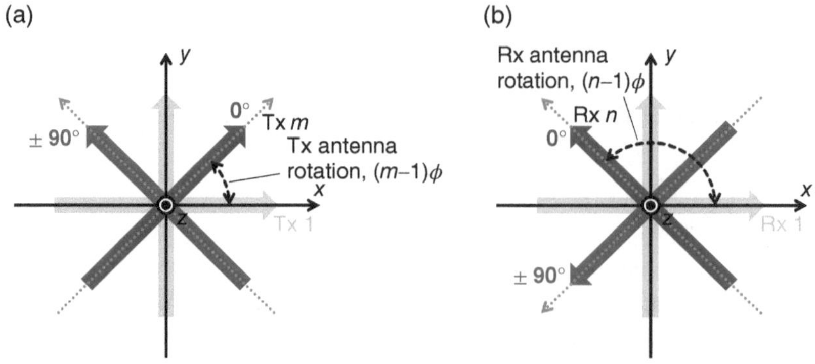

Figure 10.6 Antenna rotation: (a) *m*th Tx antenna; (b) *n*th Rx antenna.

$$
\begin{aligned}
d_{nm} &= \sqrt{D^2 + 2r^2(1 - \cos\phi_{nm})} \\
&\approx D + \frac{r^2}{D}(1 - \cos\phi_{nm})
\end{aligned}
\tag{10.20}
$$

where, $\phi_{nm} = (n - m)\phi$ and the first-order Taylor series approximation is used in the second line. Accordingly, g_{nm} is expressed using Eq. (10.20) as:

$$
\begin{aligned}
g_{nm} &= e^{-jkd_{nm}} \\
&\approx e^{-jk\left(D + \frac{r^2}{D}\right)} e^{jk\frac{r^2}{D}\cos\phi_{nm}}
\end{aligned}
\tag{10.21}
$$

where, k is the wavenumber given as $k = 2\pi/\lambda$. In Eq. (10.21), only the second complex exponential function of the angular interval ϕ_{nm} makes the relative phase difference among the channel responses.

Based on the above, the channel matrix, $\boldsymbol{H}$, can be expressed by the multiplication of the three matrices as:

$$
\boldsymbol{H} \sim \boldsymbol{P}_{\mathrm{R}}\boldsymbol{G}\boldsymbol{P}_{\mathrm{T}} \tag{10.22}
$$

where, $\boldsymbol{P}_{\mathrm{T}} \in \mathbb{C}^{N\times N}$ and $\boldsymbol{P}_{\mathrm{R}} \in \mathbb{C}^{N\times N}$ are diagonal matrices whose mth and nth diagonal elements are given by Eqs. (10.18) and (10.19), respectively. Given that both $\boldsymbol{P}_{\mathrm{T}}$ and $\boldsymbol{P}_{\mathrm{R}}$ are unitary, they have no impact on the product of the absolute square of all eigenvalues, $|\lambda_n|^2$, i.e., the determinant of $\boldsymbol{H}\boldsymbol{H}^H$ or $\boldsymbol{H}^H\boldsymbol{H}$. This can be easily proven by using the following two basic properties of the determinant: (i) the determinant of the product is equal to the product of the determinants and (ii) the product of the determinants is commutative. Thus, the optimal channel condition is independent of the above two antenna rotation factors, and is only dependent on the geometric factor, $\boldsymbol{G} \in \mathbb{C}^{N\times N}$. The geometric factor $\boldsymbol{G}$ is defined as:

$$
\boldsymbol{G} = \begin{bmatrix} g_{11} & g_{12} & \cdots & g_{1N} \\ g_{21} & g_{22} & & g_{2N} \\ \vdots & & \ddots & \vdots \\ g_{N1} & g_{N2} & \cdots & g_{NN} \end{bmatrix}. \tag{10.23}
$$

The substitution of Eq. (10.21) in Eq. (10.23) and the consideration of the channel cyclicity given in Eq. (10.2) yield:

$$\mathbf{G} = e^{-jk\left(D+\frac{r^2}{D}\right)} \begin{bmatrix} e^{jk\frac{r^2}{D}} & e^{jk\frac{r^2}{D}\cos\phi} & \cdots & e^{jk\frac{r^2}{D}\cos(N-1)\phi} \\ e^{jk\frac{r^2}{D}\cos(N-1)\phi} & e^{jk\frac{r^2}{D}} & & e^{jk\frac{r^2}{D}\cos(N-2)\phi} \\ \vdots & & \ddots & \vdots \\ e^{jk\frac{r^2}{D}\cos\phi} & e^{jk\frac{r^2}{D}\cos 2\phi} & \cdots & e^{jk\frac{r^2}{D}} \end{bmatrix} \tag{10.24}$$

where, only the matrix in the square brackets will be considered for the analysis of the optimal conditions because the complex exponential coefficient outside the square brackets does not affect the eigenvalue distribution.

Again, the optimal condition used to maximize the achievable rate is based on the fact that all the eigenvalues of the channel correlation matrix are equal, i.e. the normalized channel matrix is a unitary matrix. In consideration of the properties of the circulant matrix, if a row or column vector in the circulant matrix is a constant amplitude zero autocorrelation (CAZAC) sequence, i.e. the cyclic autocorrelation of the vector is zero, then the circulant channel matrix becomes unitary. Using the first row vector of the matrix in Eq. (10.24) as an example, the relevant condition can be written as:

$$\sum_{p=0}^{N-1} e^{jk\frac{r^2}{D}\cos 2\pi\frac{p}{N}} e^{-jk\frac{r^2}{D}\cos 2\pi\frac{p+q}{N}} = \begin{cases} N & (q=0) \\ 0 & (q\neq 0) \end{cases} \tag{10.25}$$

where, $p, q = 0, 1, \cdots, N-1$ and $\phi = 2\pi/N$. It is obvious that Eq. (10.25) is always equal to N when $q = 0$ for any r owing to the complex conjugate relation. That is, consideration of the system with at most $(N-1)$ equations of the variable r when $q \neq 0$, allows the estimation of the optimal radius r_{opt} that satisfies the unitary matrix condition. The derivations of the optimal radii for $N = 2$, 3 and 4 are as follows:

For $N = 2$, the unitary condition is given as in Eq. (10.25) that can be simplified in the case of $q = 1$ after the basic manipulation of the complex exponential function as:

$$\cos 2k\frac{r^2}{D} = 0. \tag{10.26}$$

Although there is an infinite number of phase conditions that satisfies Eq. (10.26) owing to the periodicity of the cosine function; the solution which yields the smallest optimal radius is given as:

$$2k\frac{r^2}{D} = \frac{\pi}{2}. \tag{10.27}$$

The optimal radius for $N = 2$ can be obtained as:

$$r_{\mathrm{opt}} = \frac{1}{2}\sqrt{\frac{\lambda D}{2}} \tag{10.28}$$

where, $k = 2\pi/\lambda$.

In the case at which $N = 3$, identical conditions are obtained for Eq. (10.25) for both the cases of $q = 1$ and 2 as:

$$\cos\frac{3}{2}k\frac{r^2}{D} = -\frac{1}{2}. \tag{10.29}$$

The phase condition that satisfies Eq. (10.29) is expressed as:

$$\frac{3}{2}k\frac{r^2}{D} = \frac{2}{3}\pi, \tag{10.30}$$

which yields the smallest optimal radius as:

$$r_{\text{opt}} = \frac{\sqrt{2\lambda D}}{3}. \tag{10.31}$$

In the same manner, for $N = 4$, the unitary conditions are obtained from Eq. (10.25) as:

$$\begin{aligned} &\cos k\frac{r^2}{D} = 0 \qquad (q = 1, 3) \\ &\cos 2k\frac{r^2}{D} = -1 \quad (q = 2) \end{aligned}, \tag{10.32}$$

and the phase conditions for Eq. (10.32) are given as:

$$\begin{aligned} &k\frac{r^2}{D} = \frac{\pi}{2} \quad (q = 1, 3) \\ &2k\frac{r^2}{D} = \pi \quad (q = 2) \end{aligned}, \tag{10.33}$$

respectively. The smallest optimal radius that simultaneously satisfies both conditions in Eq. (10.33) can be obtained as:

$$r_{\text{opt}} = \frac{\sqrt{\lambda D}}{2}. \tag{10.34}$$

For reference, the optimal radii derived earlier are listed in Table 10.1. Interestingly, the results are the same as those derived from the geometric model without the consideration of the antenna rotation in the literature. However, note that, as mentioned in the introduction, an optimal radius that satisfies the aforementioned condition does not exist, as numerically verified for $N > 4$ by [11].

To verify the validity of the theoretical derivation presented above, the theoretical optimums in Table 10.1 are compared with those obtained based on MoM simulations. In the simulation, the

Table 10.1 Optimal array radius.

Antenna number, N	Optimal array radius, r_{opt}
2	$\frac{1}{2}\sqrt{\frac{\lambda D}{2}}$ [a]
3	$\frac{\sqrt{2\lambda D}}{3}$ [b]
4	$\frac{\sqrt{\lambda D}}{2}$ [c]

The same results can be derived even without any consideration of the antenna rotation, which is shown in the following references:
a [3, 4];
b [11];
c [5, 17].

parameter of the array radius is swept with the transmission distance fixed to a certain value, and the optimal radius used to maximize the approximate achievable rate in Eq. (10.14) is determined based on the following criteria:

$$\begin{aligned} r_{\text{opt}} &= \arg\max_{r} R_{\text{OAM}} \\ &\approx \arg\max_{r} \prod_{n=1}^{N} |\lambda_n|^2. \end{aligned} \tag{10.35}$$

Figure 10.7a–c shows the variations of the optimal radii as a function of the transmission distance for $N = 2$, 3, and 4, respectively. In the figures, the theoretical optimums shown in Table 10.1 are plotted with the use of solid lines, and the simulated electromagnetic results are plotted with the use of circles. As it can be observed from the figures, in the far range $D \geq 10\,\lambda$, the theoretical and simulated values almost match. The theoretical values also estimate the optimal radius accurately, even in the cases of short-/mid-ranges from $1\,\lambda \leq D \leq 10\,\lambda$ in all the studied cases. Possible reasons for the deviations and underestimations of the theoretical values in very short range $D \leq 1\,\lambda$ include: (i) the spherical wave effects owing to the antenna spacings that are comparable to the transmission distance and (ii) the mutual coupling owing to the closely spaced antennas, both of which were not considered in the geometrical analysis. However, the optimal radii are predicted with a satisfactory accuracy based on theory, even in the short range. Hence, the above theoretical optimal outcomes can be used as the initial design guidelines of the UCA for OAM multiplexing.

10.3.4 Butler Matrix

This subsection explains the basic characteristics and implementation of the Butler matrix as the DFT/IDFT beamformers for OAM multiplexing. The following subsection mainly describes the second-, third-, and fourth-order DFT/IDFT beamformers, and introduces a systematic approach to extend them to higher orders.

At first, a 180° hybrid coupler is described for second-order DFT/IDFT beamformers. The 180° hybrid coupler is a passive microwave circuit with two inputs and two outputs, as shown in Figure 10.8a. The ideal hybrid is lossless, and its inputs In 1 and In 2 are perfectly matched and isolated. This is also true for its outputs Out 1 and Out 2. In the circuit, symbols 'Σ' and 'Δ' denote 'sum (or sigma)' and 'difference (or delta)' ports, respectively. If a signal is input to In 1 of the sum port, it is equally split into Out 1 and Out 2 in phase. By contrast, if a signal is input to In 2 of the difference port, it is also equally split but is output is out of phase. If the two signals are simultaneously input to In 1 and In 2, then their sum and difference are respectively output to Out 1 (sum port) and to Out 2 (difference port). These explain the nomenclature used for these ports. Owing to the structural symmetry of the hybrid, same input–output characteristics are obtained even if inputs and outputs are swapped. The hybrid can be implemented based on the transmission line theory as a rat-race coupler, as shown in Figure 10.8b, whereby, Z_0 represents a reference impedance defined to be equal to 50 Ω throughout the chapter.

The phase characteristics of the hybrid are summarized in Figure 10.8c. Considering its in-phase and out-of-phase signal flow, one may notice that its operation is exactly the same as the butterfly computation in the fast Fourier transform (FFT) algorithm. That is, the hybrid can be used as a fundamental component of the Butler matrix to realize the 2^Pth-order DFT/IDFT beamformers where P is a natural number.

In next, the third-order Butler matrix is introduced as third-order DFT/IDFT beamformers. Compared with the 2^Pth-order Butler matrix, there is only a limited number of studies that deal

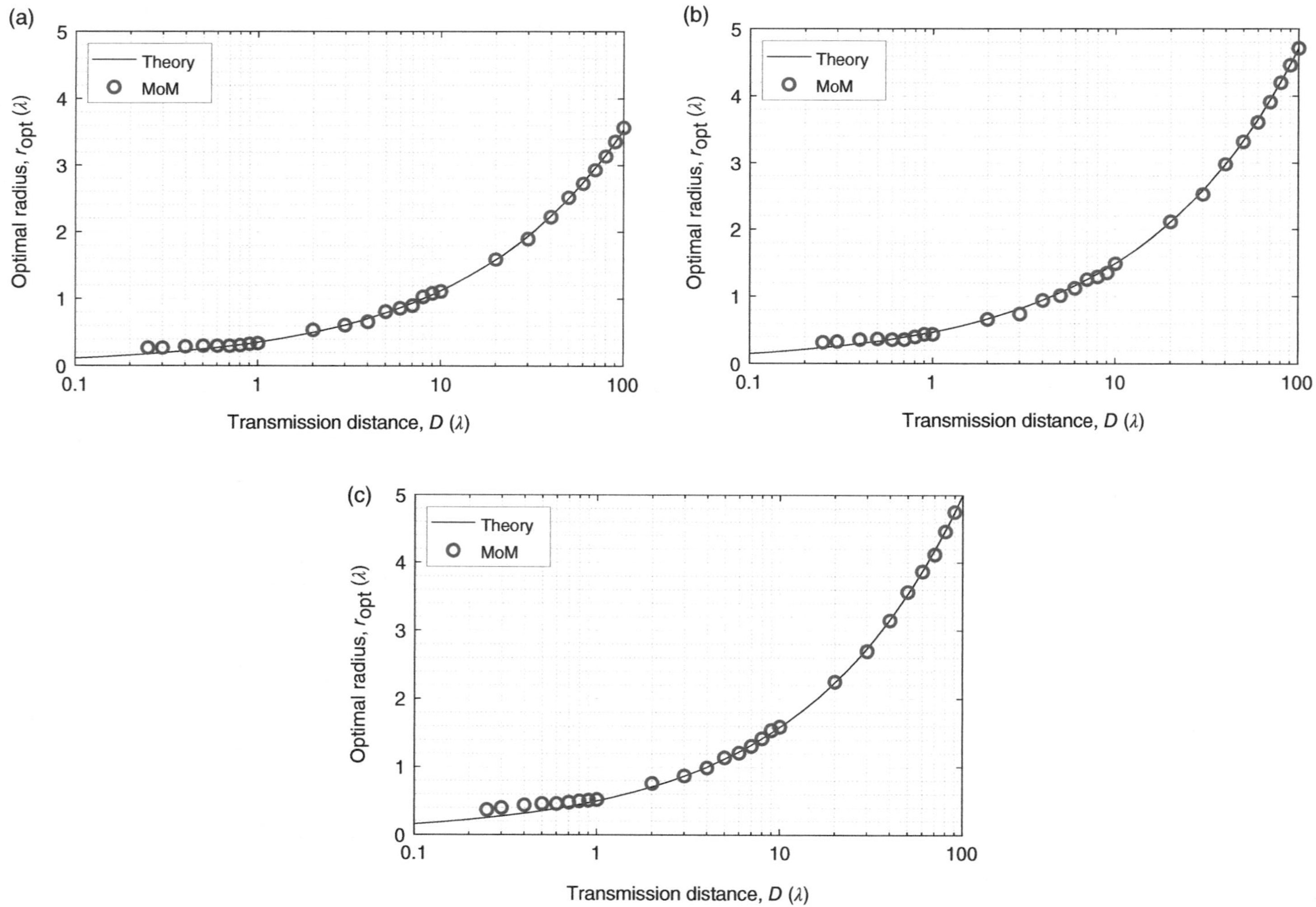

Figure 10.7 Optimal radius versus transmission distance for (a) $N = 2$, (b) $N = 3$, and (c) $N = 4$.

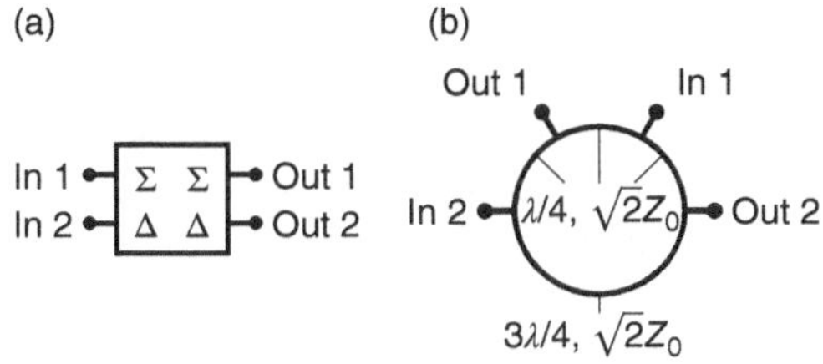

(c)

Phase [°]	In 1	In 2
Out 1	0	0
Out 2	0	180

Figure 10.8 Schematic of a 180° hybrid coupler: (a) symbol; (b) circuit configuration; (c) phase characteristics.

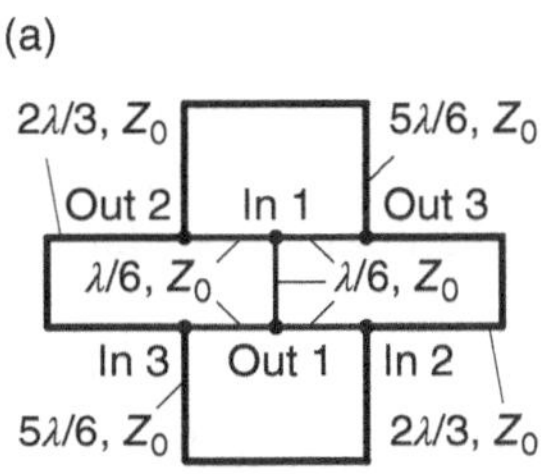

(b)

Phase [°]	In 1	In 2	In 3
Out 1	0	0	0
Out 2	0	−120	120
Out 3	0	120	−120

Figure 10.9 Third-order Butler matrix: (a) circuit configuration; (b) phase characteristics.

with the third-order Butler matrix. For example, the use of the third-order Butler matrices has been demonstrated in the fabrication of a three-port coupler, through vias, and transmission lines [18], and in symmetric and asymmetric couplers [19] based on the use of multilayer technologies. Conversely, coplanar fabrication techniques can also be employed, as shown in Figure 10.9a. This configuration ideally achieves both full-port matching and input/output isolation, and works as a DFT beamformer. An input signal to each port is lossless and evenly divided to three outputs with a specific phase difference, as shown in Figure 10.9b. When used as an IDFT beamformer, the port pairs of In 2/In 3 or Out 2/Out 3 have to be swapped to perform conjugate transpose operations.

The fourth-order Butler matrix can be configured with four hybrids and a 90° phase shifter as shown in Figure 10.10a. The phase shifter corresponds to the premultiplied root of unity that causes specific phase delays in the butterfly diagram. Figure 10.10b shows its phase characteristics as a DFT beamformer. An IDFT beamformer can be implemented by swapping the port pairs of In 2/In 4 or Out 2/Out 4.

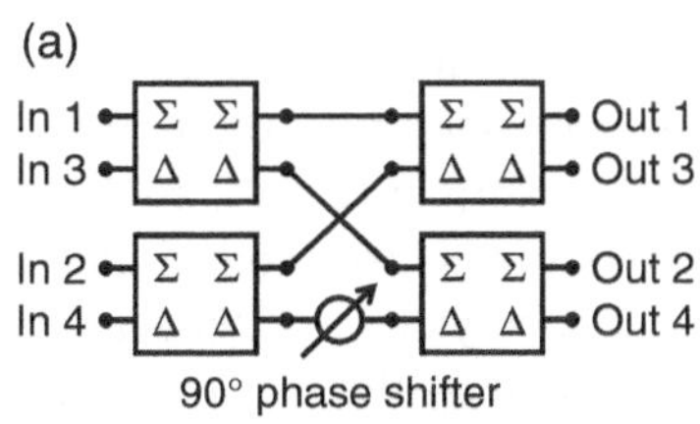

(b)

Phase [°]	In 1	In 2	In 3	In 4
Out 1	0	0	0	0
Out 2	0	−90	180	90
Out 3	0	180	0	180
Out 4	0	90	180	−90

Figure 10.10 Fourth-order Butler matrix: (a) circuit configuration; (b) phase characteristics.

To evaluate the achievable rate versus the frequency characteristics in the next subsection, the frequency characteristics of the Fourth-order Butler matrix are analyzed. In the simulation, rat-race couplers and a $\lambda/4$ delay line are respectively used as the hybrids and 90° phase shifter, whereby, it is assumed that interconnections are frequency independent. Its frequency characteristics are calculated by considering the transmission line length and line impedance of the circuit for a $\pm 10\%$ frequency deviation.

Figure 10.11 shows reflections and couplings, and Figures 10.12 and 10.13 show the amplitudes and phases of the transmissions for each input port in the S-parameter matrix (the output characteristics are omitted because of the structural symmetry and reciprocity of the circuit). For example, the subscript letters of the S-parameters 'I' and 'O' denote the input and output ports, respectively, while the subscript numbers denote the corresponding port numbers. Note that, the phase of each S-parameter is normalized by the phase between a relevant input to Out 1. The figures confirm that both reflections and couplings are less than −20 dB, and the deviations of the transmission amplitude (ideally 6 dB) and the transmission phase are approximately ± 0.5 dB and $\pm 20°$, respectively. These results validate the fact that the Butler matrix maintains a satisfactory operation in a broad frequency band.

The lower-order Butler matrices listed above can be extended to higher-order matrices. The 2^Pth-order Butler matrix can be decomposed into two 2^{P-1}th-order matrices followed by 2^{P-1} hybrids and $(2^{P-1} - 1)$ phase shifters, as illustrated in [12]. Likewise, the 2^P3^Q-order Butler matrix (Q is a natural number) can be implemented by the cascade interconnection of the minimum required numbers of third-order Butler matrices, hybrids, and phase shifters that cause appropriate phase changes, as illustrated in [20].

10.3.5 Performance Evaluation

At the end of this section, the achievable rate is evaluated in the presence of three deterioration factors: (i) Rx array misalignment along with a subaxis x' as in Figure 10.2, Δx; (ii) Rx array inclination with respect to the same subaxis, $\Delta\theta$ (right-handed rotation is positive); (iii) frequency deviation, Δf. In the following subsection, the fourth-order OAM multiplexing and MIMO eigenmode transmission performance are evaluated. The transmission distance was set to 1.0 λ, the array radius was set to the optimum shown in Table 10.1 for the given distance, and the Tx-signal-to-Rx-noise power ratio was varied from 0 to 60 dB.

Figure 10.14a–c shows the achievable rates versus misalignment, Δx, inclination, $\Delta\theta$, and frequency deviation, Δf, respectively. In the figure, the MIMO and OAM characteristics are plotted with the use of lines and a curved surface.

Regarding the misalignment characteristics in Figure 10.14a, the deterioration of the MIMO performance is gradual even though the Rx array is shifted by the same distance as the transmission distance to the orthogonal subaxis direction. This is attributed to the fact that although the distance attenuation increases owing to the array misalignment, the Tx and Rx arrays are still in the range of the half-power beam width of the dipole used as the array element. Additionally, regarding the inclination characteristics in Figure 10.14b, the MIMO performance starts to drop from the inclination $\Delta\theta = 45°$ as it approaches 90°. This is because of the polarization mismatch owing to the array inclination. Conversely, the OAM performance suddenly deteriorates even with small misalignment and inclination. This rapid deterioration is caused by the occurrence of the intermode interference owing to the destruction of the rotational symmetry of the array configuration. The impact of the intermode interference becomes dominant compared with the Rx noise power as

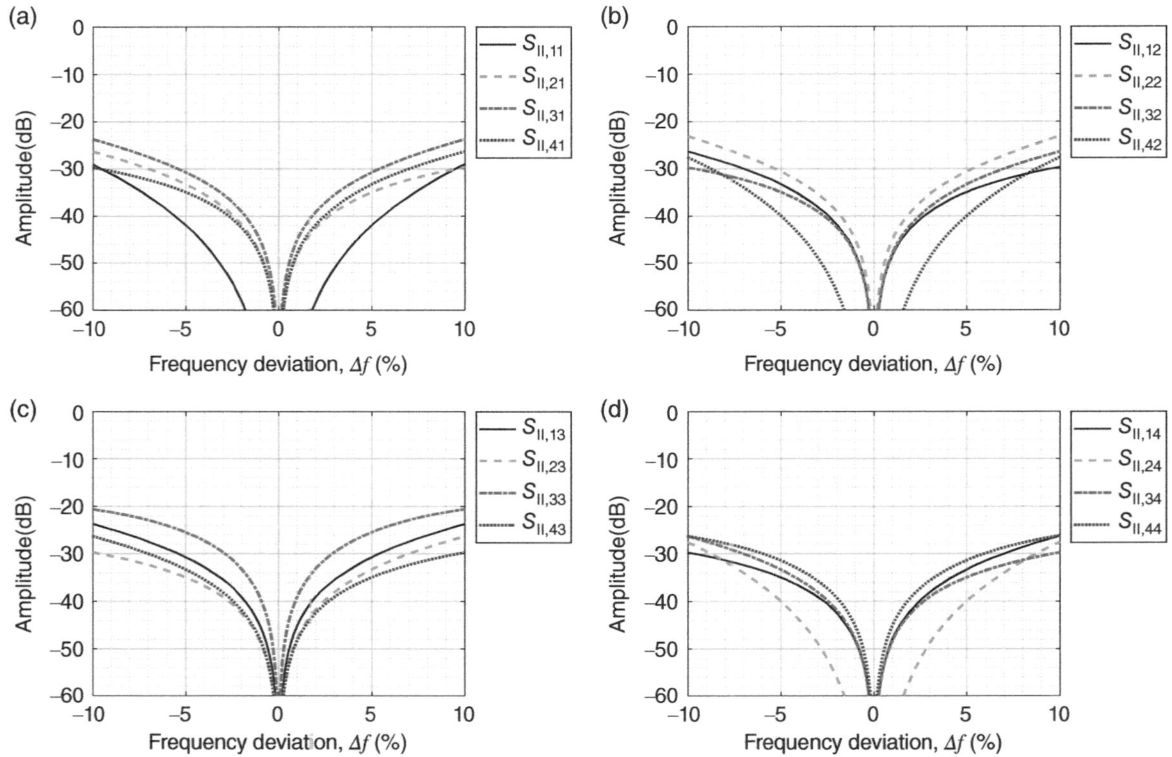

Figure 10.11 Reflection and coupling of the fourth-order Butler matrix: (a) In 1; (b) In 2; (c) In 3; (d) In 4.

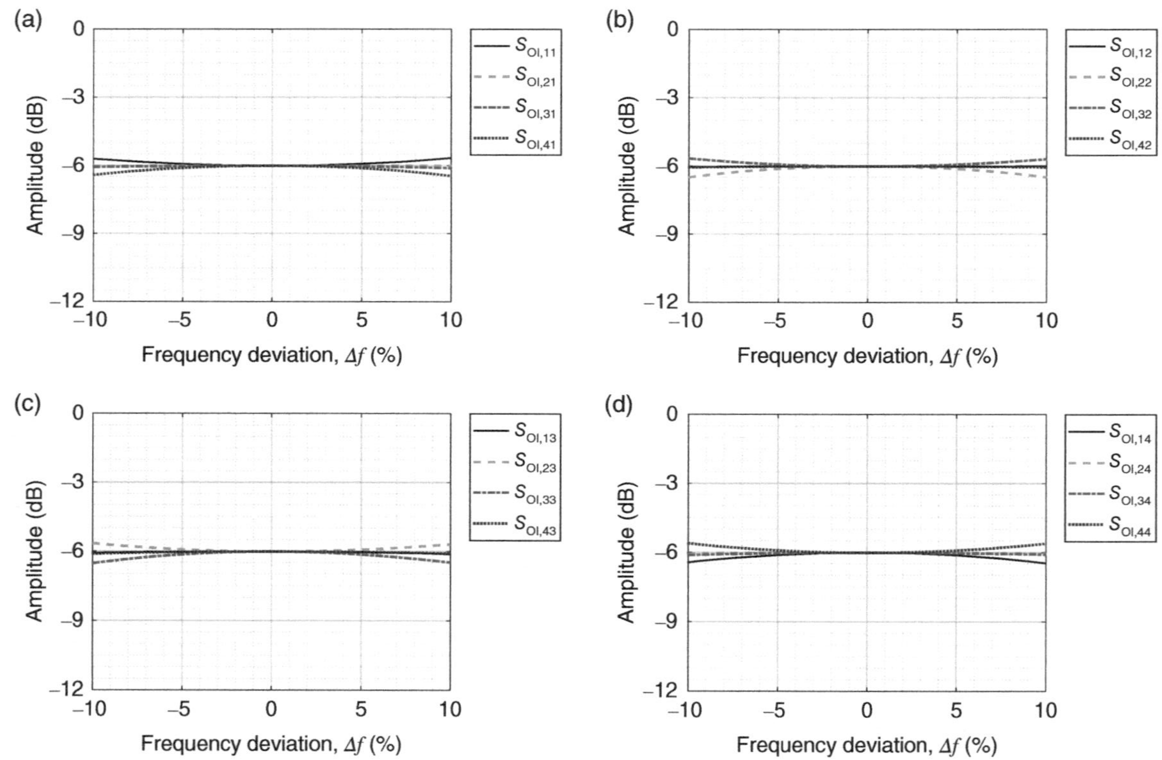

Figure 10.12 Transmission amplitude of the fourth-order Butler matrix: (a) In 1; (b) In 2; (c) In 3; (d) In 4.

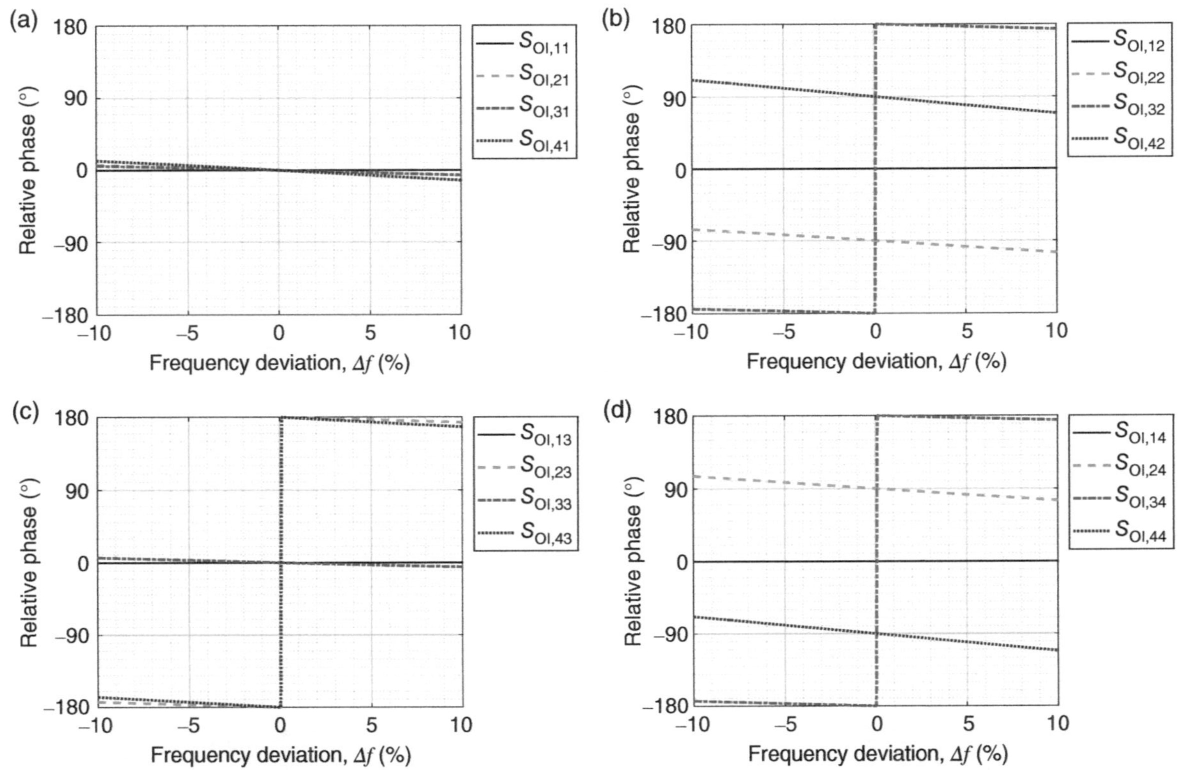

Figure 10.13 Transmission phase of the fourth-order Butler matrix: (a) In 1; (b) In 2; (c) In 3; (d) In 4.

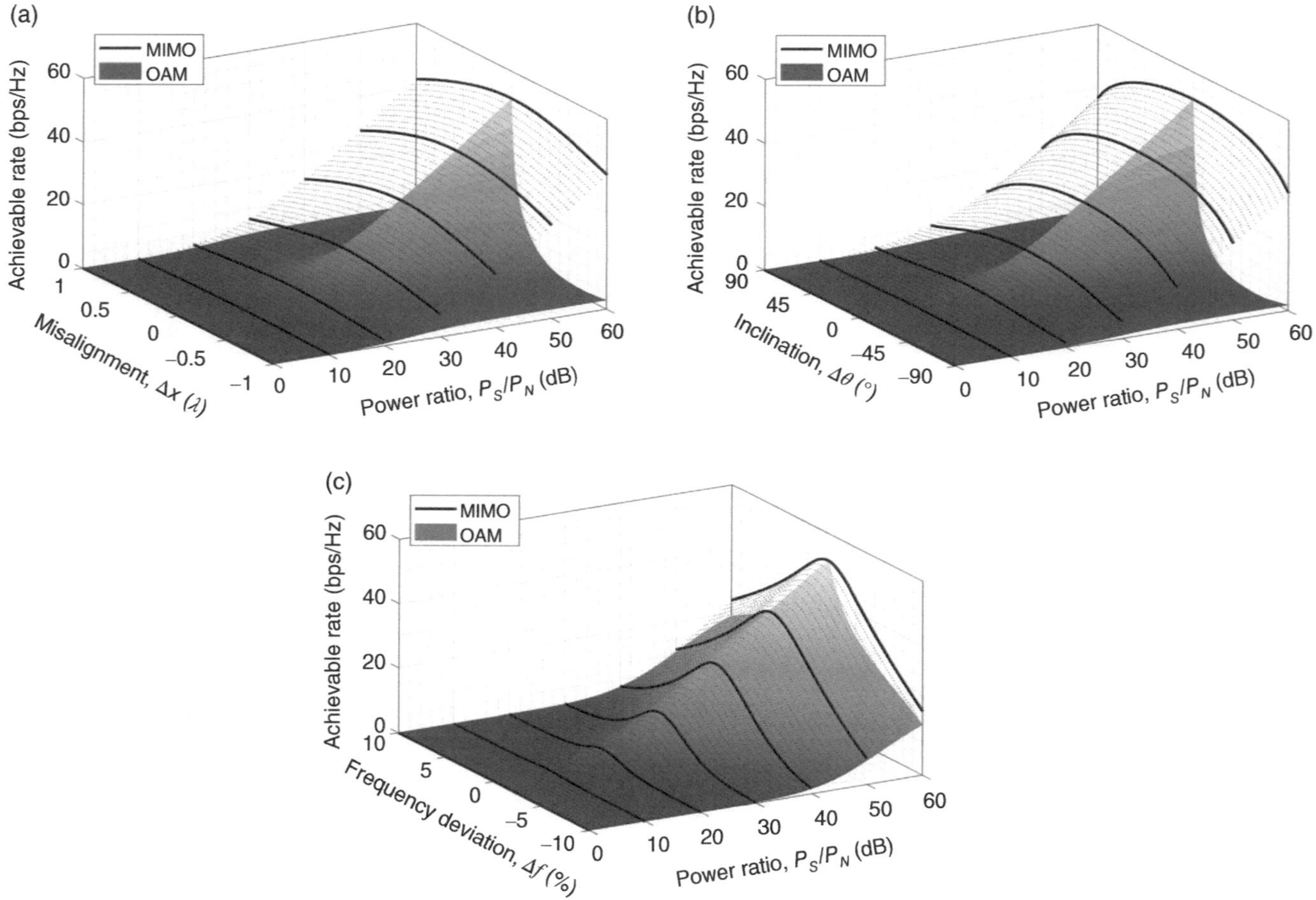

Figure 10.14 Achievable rate characteristics versus (a) misalignment, Δx; (b) inclination, $\Delta\theta$; (c) frequency deviation, Δf.

the Tx signal power increases, as indicated in Eq. (10.9). Therefore, the higher the Tx signal power, the greater the difference between MIMO and OAM performances.

The frequency characteristics of the cross dipoles, quadrature hybrid couplers used for the CP generation, and the Tx/Rx Butler matrices used as the DFT/IDFT beamformers, are considered in the simulation, as shown in Figure 10.14c. The figure confirms that even the MIMO performance significantly degrades as the frequency deviation increases. This is mainly caused by the deterioration of the dipole gain owing to its bandwidth limitation. However, the performance difference between MIMO and OAM is not so severe unlike the misalignment and inclination characteristics. This is attributed to the wideband operation of the Butler matrix as shown in the previous subsection that yields and maintains a fair performance for the DFT/IDFT beamformers in the analysis range.

The simulation results listed above show that the array positional gap has a considerable impact on the OAM performance rather than on the circuit frequency characteristics. However, realistically, the manufacturing error of the circuit is negligible. Thus, it is desirable to perform supportive beamforming at either the Tx or Rx sides for intermode interference cancellation.

10.4 Conclusion and Key Challenges

The OAM multiplexing technology, which will ideally enable the transmission of parallel streams with a passive configuration, can bring unique benefits to short-range wireless communication. Notably, low-power consumption and system simplification, achieved by eliminating complicated signal processing, are some of the most significant advantages of small short-range communication terminals with severe power constraints. Furthermore, the frequency independence of the orthogonality of the OAM modes enables the maximum utilization of both spatial and frequency resources, to realize high-speed large-capacity communication.

On the other hand, since the Butler matrix, used for OAM mode generation, is generally bulky, its miniaturization is highly sought after, for example, by employing monolithic microwave-integrated circuit technology. However, due care should be taken, during circuit design, to ensure broadband characteristics. The gain of the OAM beam depends on the frequency characteristics of the antenna element itself. Thus, small antenna design is also essential to achieve both high gain and fair axial ratio over a wide band. In addition, countermeasures against possible positional deviation during actual operation are also indispensable. In particular, the development of the simple beamforming methods, based only on the information of the relative displacement, is anticipated so that the simplicity of the OAM multiplexing system is maintained.

Recently, new OAM multiplexing applications have been suggested in two emerging wireless communication fields: (i) full-duplex communication (or simultaneous transmission and reception) [21–23] and (ii) simultaneous wireless information and power transmission (SWIPT) [24, 25].

Full-duplex requires not only stream isolation but also Tx–Rx port isolation to decode the desired signals experiencing strong self-interference correctly. As suggested in the above literature, mode selection is one of the most promising candidates to effectively suppress the self-interference by means of employing a few OAM modes that inherently have high-port isolation. To secure more spatial DOF, the combination of antenna cancellation, based on careful antenna design or decoupling measurements, and subsequent analog-and-digital cancellation is required.

The concept of SWIPT includes both far-field and near-field applications. For the near-field applications, as discussed in this chapter, the matching conditions for the Tx and Rx ports likely change due to the strong coupling between very closely spaced Tx and Rx antennas. To solve this problem, a novel network theory was proposed to realize simultaneous decoupling and matching between Tx and Rx ports considering the near-field interaction [26, 27]. However, the high sensitivity of the near-field coupling makes practical circuit implementation challenging and is, therefore, still open to further research.

In the future, various forms of short-range wireless communication will coexist during the development of an advanced IoT society, such as vehicular-to-vehicular (including drones), sensor-to-sensor, and chip-to-chip communications. The authors hope that the OAM multiplexing technology can contribute to the establishment of the future of seamlessly connected communication networks.

References

1 Chen, R., Zhou, H., Moretti, M. et al. (2020). Orbital angular momentum waves: Generation, detection and emerging applications. *IEEE Communications Surveys Tutorials* **22** (2): 840–868.

2 Driessen, P.F. and Foschini, G.J. (1999). On the capacity formula for multiple input-multiple output wireless channels: a geometric interpretation. *IEEE Transactions on Communications* **47** (2): 173–176.

3 Sarris, I. and Nix, A.R. (2007). Design and performance assessment of high-capacity MIMO architectures in the presence of a line-of-sight component. *IEEE Transactions on Vehicular Technology* **56** (4): 2194–2202.

4 Bøhagen, F., Orten, P., and Øien, G.E. (2007). Design of optimal high-rank line-of-sight MIMO channels. *IEEE Transactions on Wireless Communications* **6** (4): 1420–1425.

5 Bøhagen, F., Orten, P., and Øien, G.E. (2007). Optimal design of uniform rectangular antenna arrays for strong line-of-sight MIMO channels. *EURASIP Journal on Wireless Communications and Networking* **2007**: 45084.

6 Wang, P., Li, Y., and Vucetic, B. (2014). Millimeter wave communications with symmetric uniform circular antenna arrays. *IEEE Communications Letters* **18** (8): 1307–1310.

7 Nguyen, D.K., Pascal, O., Sokoloff, J. et al. (2015). Antenna gain and link budget for waves carrying orbital angular momentum. *Radio Science* **50** (11): 1165–1175.

8 Craeye, C. (2016). On the transmittance between OAM antennas. *IEEE Transactions on Antennas and Propagation* **64** (1): 336–339.

9 Xu, J. (2017). Degrees of freedom of OAM-based line-of-sight radio systems. *IEEE Transactions on Antennas and Propagation* **65** (4): 1996–2008.

10 Oldoni, M., Spinello, F., Mari, E. et al. (2015). Space-division demultiplexing in orbital-angular-momentum-based MIMO radio systems. *IEEE Transactions on Antennas and Propagation* **63** (10): 4582–4587.

11 Zhou, L., Ohashi, Y., and Yoshida, M. (2015). Low complexity millimeter-wave LOS-MIMO systems with uniform circular arrays for small cells wireless backhaul. *IEICE Transactions on Communications* **E98.B** (11): 2348–2358.

12 Murata, K., Honma, N., Nishimori, K. et al. (2017). Analog eigenmode transmission for short-range MIMO based on orbital angular momentum. *IEEE Transactions on Antennas and Propagation* **65** (12): 6687–6702.

13 Sakamoto, K., Arai, M., Hiraga, K. et al. (2015). A novel ZF-based fast beamforming method for short-range MIMO transmission. *IEEE Wireless Communications Letters* **4** (5): 557–560.

14 Song, X., Cvetkovski, D., Rave, W. et al. (2019). Sequential channel equalization in strong line-of-sight MIMO communication. *IEEE Transactions on Wireless Communications* **18** (1): 340–356.

15 Gray, R.M. (2006). Toeplitz and circulant matrices: A review. *Foundations and Trends® in Communications and Information Theory* **2** (3): 155–239.

16 Jaeckel, S., Börner, K., Thiele, L., and Jungnickel, V. (2012). A geometric polarization rotation model for the 3-D spatial channel model. *IEEE Transactions on Antennas and Propagation* **60** (12): 5966–5977.

17 Liu, L., Hong, W., Wang, H. et al. (2007). Characterization of line-of-sight MIMO channel for fixed wireless communications. *IEEE Antennas and Wireless Propagation Letters* **6**: 36–39.

18 Ding, K., Fang, X., Wang, Y., and Chen, A. (2014). Printed dual-layer three-way directional coupler utilized as 3× 3 beamforming network for orthogonal three-beam antenna array. *IEEE Antennas and Wireless Propagation Letters* **13**: 911–914.

19 Zhu, H., Sun, H., Jones, B. et al. (2019). Wideband dual-polarized multiple beam-forming antenna arrays. *IEEE Transactions on Antennas and Propagation* **67** (3): 1590–1604.

20 Shelton, J. and Kelleher, K. (1961). Multiple beams from linear arrays. *IRE Transactions on Antennas and Propagation* **9** (2): 154–161.

21 Zhang, Z., Yuan, Y., Cang, J. et al. (2017). An orbital angular momentum-based in-band full-duplex communication system and its mode selection. *IEEE Communications Letters* **21** (5): 1183–1186.

22 Zhang, Y. and Li, J. (2018). Analyses and full-duplex applications of circularly polarized OAM arrays using sequentially rotated configuration. *IEEE Transactions on Antennas and Propagation* **66** (12): 7010–7020.

23 Zhang, Y. and Li, J. (2019). An orbital angular momentum-based array for in-band full-duplex communications. *IEEE Antennas and Wireless Propagation Letters* **18** (3): 417–421.

24 Mishra, D. and Alexandropoulos, G.C. (2018). Jointly optimal spatial channel assignment and power allocation for MIMO SWIPT systems. *IEEE Wireless Communications Letters* **7** (2): 214–217.

25 Al-Baidhani, A., Vehkapera, M., and Benaissa, M. (2019). Simultaneous wireless information and power transfer based on generalized triangular decomposition. *IEEE Transactions on Green Communications and Networking* **3** (3): 751–764.

26 Murata, K., Honma, N., Nishimori, K. et al. (2016). Simultaneous decoupling and matching technique for short-range MIMO. *IEICE Transactions on Communications* **E99-B** (8): 1846–1858.

27 Phang, S., Ivrlač, M.T., Gradoni, G. et al. (2018). Near-field MIMO communication links. *IEEE Transactions on Circuits and Systems I: Regular Papers* **65** (9): 3027–3036.

11

OAM Communications in Multipath Environments

Xiaoming Chen and Wei Xue

School of Electronic and Information Engineering, Xi'an Jiaotong University, Xi'an, P. R. China

11.1 Introduction

Since wireless communication was demonstrated by Guglielmo Marconi in the early 1900s, it has undergone decades of evolution and development. Compared with the traditional wired communication, wireless communication has several advantages, e.g. cost effectiveness, high mobility, reliability, and convenience. Nowadays, wireless communications have become the fastest growing and most vibrant technologies. Some of the commonly used wireless communication systems are broadcast systems (i.e. AM radio, FM radio, TV, and paging), wireless local area networks (LANs), bluetooth communication, cellular communication, infrared communication, satellite communication, to name a few. Overall, wireless communication systems provide us almost all the essential services in our daily lives.

Despite the remarkable evolution of wireless communications, there are still many technical challenges. One of the biggest challenges we are facing today is the ever-increasing demand for data rates and frequency spectrum congestion. As a result, multiple-input multiple-output (MIMO) spatial multiplexing (i.e. transmission of multiple data streams over the spatial domain using the same time-frequency resource) was proposed to alleviate the problem. And recently, orbital angular momentum (OAM) multiplexing (i.e. transmission of multiple data streams via different OAM modes using the same time–frequency resource) was proposed for the same purpose. The latter has aroused some controversy. To date, there are still concerns about the feasibility of OAM communication. A major concern is that the (conventional) OAM communication is vulnerable to multipath. In this chapter, we are going to revisit the feasibility of OAM multiplexing in multipath environments. To that end, we first introduce some basics about wireless communications in multipath environments.

11.1.1 Fading in Wireless Propagation

In wireless communications, the propagation distance can be any value from a few meters (e.g. bluetooth) to thousands of kilometers (e.g. satellite communication). There are various factors that affect the transmission of electromagnetic waves, such as propagation (path) loss, large-scale (macroscopic) fading, and small-scale (microscopic) fading [1]. We will discuss them separately in the sequel.

Electromagnetic Vortices: Wave Phenomena and Engineering Applications, First Edition.
Edited by Zhi Hao Jiang and Douglas H. Werner.

11.1.1.1 Pass Loss

For a specific propagation path, the free-space path loss can be interpreted as the ratio of the power of the received signal to the power of the transmitted signal. The pass loss is a function of propagation distance. It can be expressed using the famous Friis Transmission equation [2]:

$$\frac{P_r}{P_t} = \frac{\lambda^2 G_t(\theta_t, \phi_t) G_r(\theta_r, \phi_r)}{(4\pi d)^2}, \tag{11.1}$$

where P_t and P_r are the powers of the transmitted and received signals, respectively, G_t and G_r are the gains of the transmitting and receiving antennas, respectively, λ represents the wavelength, and d is the propagation distance. Obviously, the received power is inversely proportional to the square of the propagation distance.

Fading refers to the fluctuations of the power of the received signal in the presence of multipath, it reflects the characteristics of the multipath channel. According to the rate of the fluctuations over propagation distance, fading can be classified into large- and small-scale fadings.

11.1.1.2 Large-Scale Fading

During the propagation, there may be many obstacles (e.g. buildings, hills) between the transmitter and the receiver. Obviously, the signal strength decreases due to the blockage of the obstacles. Due to the mobility of the transceivers or in general multipath environments (taking into account of different realizations of similar scenarios), the obstacle blockages are usually modeled as a random process, resulting in signal strength variation over long distances (i.e. typically over dozens of wavelengths [3]). This phenomenon is called large-scale fading. Since the large-scale fading is caused by the shadowing effects of the obstacles, it is also known as shadowing.

11.1.1.3 Small-Scale Fading

Analogous to the large-scale fading, the small-scale fading refers to signal variation over a relatively short range (i.e. within tens of wavelength [3]). The bandwidth of a practical communication system is always finite, which means that the communication system has limited resolution in the time-delay domain [4]. Small-scale fading occurs when multipath components with similar delays yet different phases (or in the jargon of channel modeling, subpaths within the same cluster) at the receiver add up to each other constructively or destructively. When there are large enough of such subpaths, the random signal tends to follow Gaussian distribution due to the central limit theorem [5].

The propagation channel can be divided into two parts: the line of sight (LoS) component and the multipath components. The LoS is a propagation mechanism when the transmitter and receiver are in view of each other (without any obstacles in between them). The LoS path is also called the direct path. Besides the LoS path, the signal can propagate along different paths with different mechanisms (i.e. reflection, scattering, and diffraction). These paths are collectively referred to as multipath. Specifically, reflection occurs when the signal encounters a large solid (e.g. a concrete wall), whose size is much larger than the wavelength of the signal. Diffraction occurs when the signal encounters an edge or a corner (e.g. an edge of a wall), whose size is larger than the wavelength of the signal. Scattering occurs when the signal impinges on a rough surface or tiny objects of size smaller than the wavelength of the signal. For convenience, the three mechanisms are sometimes referred to as scattering in the propagation literature. Figure 11.1 illustrates the ratio of the received power to the transmitted power versus the propagation distance for the combined effects of path loss, shadowing, and multipath.

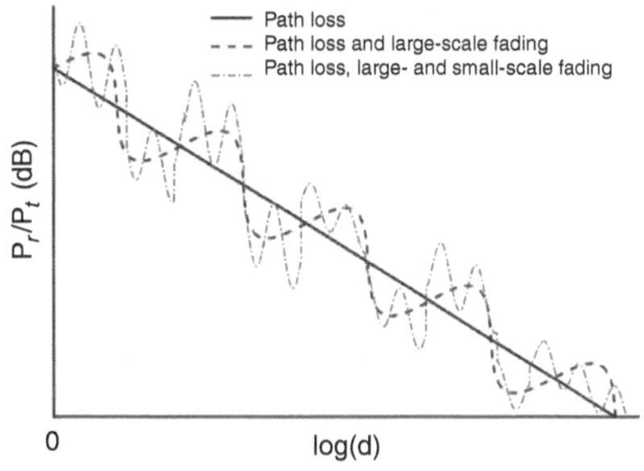

Figure 11.1 Illustration of path loss, shadowing, and multipath versus propagation distance.

11.1.2 Diversity and Multiplexing

Spatial diversity and multiplexing are the two most important techniques in a MIMO system. In general, the diversity scheme tends to improve the reliability of the communication system by introducing redundancy (e.g. using multiple antennas to receive the same signal), whereas the spatial multiplexing aims at data rate enhancement by transmitting multiple data streams simultaneously over the same bandwidth. Although diversity and multiplexing are two distinct schemes, they are often used jointly in the MIMO system. For instance, for a 4×2 MIMO system (i.e. an MIMO system with four receiving antennas and two transmitting antennas) operating in full spatial multiplexing mode (i.e. the transmitter sends two independent data streams from the two transmitting antennas), by employing a conventional linear spatial equalizer at the receiver, the MIMO system can enjoy not only data rate enhancement but also higher reliability in that there is extra degree-of-freedom at the receiver to decode the signals (i.e. four antennas are used to decode the two signals, resulting in a multiplexing order of two and a diversity order of two [1]).

Fading has strong effects on the wireless communications, spatial diversity is a powerful technique to combat the multipath fading effects and improve the reliability of communication systems. The basic idea of diversity is sending multiple copies of the same signal across different independent fading paths. Since the independent propagation paths fade differently, the signal copies propagating along independent fading paths have a low probability of suffering deep fade simultaneously. By combining the received signals from independent fading paths using certain diversity algorithms, the fading effect can be effectively mitigated.

Apart from spatial diversity, diversity schemes can be implemented in other domains as listed below:

- **Polarization diversity:** Utilizing two or three antennas with orthogonal polarizations.
- **Pattern diversity:** Multiple directional antennas whose beamwidths are toward different directions.
- **Frequency diversity:** Utilizing different carrier frequencies that are separated by a certain bandwidth. Commonly, the separation bandwidth ought to be no smaller than the coherence bandwidth (i.e. the statistical bandwidth over which the channel frequency response can be regarded as flat).
- **Time diversity:** The same signal is transmitted repeatedly with a certain time intervals. The time interval should be larger than the coherence time (i.e. the statistical time interval over which the channel can be regarded as static).

Note that spatial diversity, polarization diversity, and pattern diversity sometimes are collectively referred to as antenna diversity. Polarization and pattern diversities can be regarded as special usages of the spatial channel. Therefore, in this chapter, we use spatial diversity and antenna diversity interchangeably unless otherwise specified.

As mentioned before, the major challenge of wireless communication systems is the ever-increasing data rate demand and the spectrum congestion. Thus, how to increase the channel capacity or the spectral efficiency is a more important issue. Spatial multiplexing is the key to this problem. The basic idea of spatial multiplexing is sending independent data streams through the spatial channel simultaneously and over the same bandwidth. In this way, the spectral efficiency can be greatly improved.

Like diversity, multiplexing can be implemented in different domains. Some of the commonly used multiplexing techniques are briefly introduced below:

- **Space division multiplexing (SDM):** The space channel is divided into different subchannels by using multiple antennas (located at different positions). Depending on if precoding is used at the transmitter, the multiple antennas can transmit multiple signals either jointly or separately. This scheme is also known as spatial multiplexing as we discussed before.
- **Frequency division multiplexing (FDM):** The frequency bandwidth is divided into several subchannels. Each subchannel is allocated to a specific signal. These subchannels can be separated by a certain frequency guard band so that they do not overlap with each other or they can be arranged in a smarter way so that there is no interference between different subchannels (subcarriers) even though different subchannels overlap with each other. The latter is called orthogonal frequency division multiplexing (OFDM), which obviously has better spectral efficiency than the conventional FDM. Note that in optical communications, multiple optical carriers can be multiplexed onto a single optical fiber by utilizing different wavelengths of light. This scheme is referred to as wavelength division multiplexing (WDM). Yet it is essentially the same as FDM.
- **Time division multiplexing (TDM):** The time-domain channel is divided into several time slots, and each time slot is allocated to a specific signal (or user in the jargon of cellular communications). Each user can only transmit data within its own time slot.

Obviously, FDM and TDM cannot improve the spectral efficiency in that they realize multiplexing by using the frequency or time resource. Hence, we focus on the spatial multiplexing in this chapter.

11.1.3 MIMO Systems

Based on the numbers of transmitting antennas and receiving antennas, communication systems can be classified into four categories: single-input single-output (SISO) systems ($M_T = M_R = 1$), single-input multiple output (SIMO) systems ($M_T = 1$, $M_R > 1$), multiple-input single-output (MISO) systems ($M_T > 1$, $M_R = 1$), and MIMO systems ($M_T > 1$, $M_R > 1$). SISO, SIMO, and MISO can be interpreted as special cases of MIMO. MIMO is an important technology in wireless communication that utilizes multiple antennas at both transmitter and receiver to transmit multiple data streams via the spatial channels.

Figure 11.2 shows the basic architecture of the MIMO system with M_T transmitting antennas and M_R receiving antennas. The multiple spatial channels make it possible for the MIMO system to improve the reliability or the data rate of the wireless link. Specifically, if the same signal is sent across different spatial channels, the MIMO system can be used to achieve spatial diversity. In

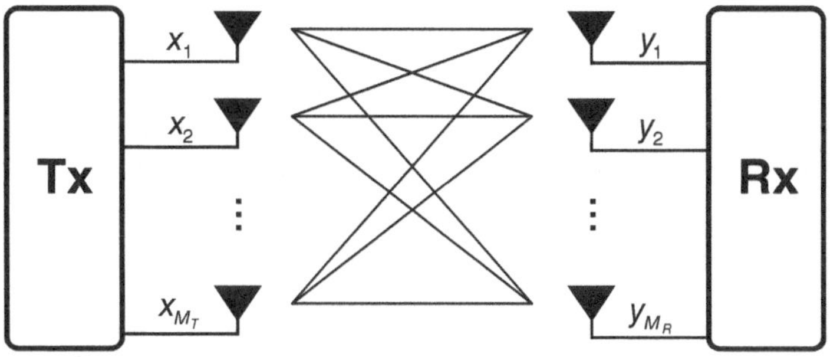

Figure 11.2 MIMO system.

theory, one can get a diversity order of $M_T \times M_R$ when all the $M_T \times M_R$ channels fade independently. If different signals are sent across the spatial channels, the MIMO system can be used to achieve spatial multiplexing. Provided that all the $M_T \times M_R$ channels fade independently, the multiplexing gain is the minimum value of M_T and M_R [4]. MIMO systems can enhance diversity gain or multiplexing gain or both. Thus, MIMO systems have attracted considerable interest and become one of the most important wireless techniques.

11.2 OAM Communication in Line-of-sight Environment

Besides the conventional MIMO techniques, OAM multiplexing was proposed to enhance the channel capacity due to the theoretically unlimited number of orthogonal OAM modes, e.g. [6–9]. Different OAM generation methods exist in the literature. For instances, OAM modes generation has been demonstrated using spiral phase plates (SPPs) [10, 11], uniform circular arrays (UCAs) [12, 13], and metasurfaces [14, 15]. In spite of various forms of OAM generations, in essence, all the methods generate OAM waves via discrete Fourier transform (DFT). The conventional OAM detection can be realized similarly by performing inverse DFT (IDFT) of the received signals. Hence, OAM multiplexing is achieved by transmitting data in the spatial–frequency domain. In this aspect, OAM does not really offers a new degree-of-freedom as compared with the conventional MIMO system [16, 17]. If we consider DFT and IDFT as special forms of MIMO precoding and decoding, OAM multiplexing is basically a special form of the MIMO technique. For better understanding of the mechanism of OAM multiplexing, in this section, we first discuss OAM multiplexing in LoS environment, as presented in the majority of the OAM literature. Discussions of OAM multiplexing in a multipath environment are deferred to the next section.

11.2.1 Conventional OAM Multiplexing

It is known that the angular momentum and angular position are related by the Fourier series [18], which results in discrete values of the angular momentum. The Fourier relationship between the angular distribution $\Psi(\varphi)$ and the angular momentum distribution (OAM spectrum) ψ_l is given as [18]:

$$\psi_l = \frac{1}{2\pi}\int_0^{2\pi} \Psi(\varphi)e^{-jl\varphi}d\varphi \tag{11.2}$$

$$\Psi(\varphi) = \sum_{l=-\infty}^{+\infty} \psi_l e^{jl\varphi} \tag{11.3}$$

where φ is the azimuthal angle. The purity of the generated OAM mode can be calculated based on the Fourier relationship. Moreover, the detection and demultiplexing of the OAM waves can also be realized based on the Fourier relationship. Assuming the transmitted OAM wave is in a pure OAM state l_1, the field in the cylindrical coordinate can be expressed as [19, 20]:

$$E_{l_1}(\rho, \varphi, z) = A_{l_1}(\rho, z) e^{-jl_1\varphi} \tag{11.4}$$

where $A_{l_1}(\rho, z)$ is a function of the radial position ρ and the propagation distance z. Then, the demultiplexing device (e.g. UCA and SPP) at the receiving side should have a phase distribution of $e^{jl_2\varphi}$ (i.e. OAM mode $-l_2$). For the conventional whole angular aperture receiving (WAAR) scheme, the received field can be written as:

$$U = \int_0^{\rho} A_{l_1}(\rho, z)\rho d\rho \int_0^{2\pi} e^{-j(l_1 - l_2)\varphi} d\varphi \tag{11.5}$$

The integration in Eq. (11.5) will be zero when $l_1 \neq l_2$ and a nonzero value when $l_1 = l_2$. That is the transmitted OAM wave with mode l_1 can only be detected when $l_1 = l_2$. Obviously, all the OAM modes are orthogonal to each other and can be simply demultiplexed in the WAAR scheme.

Besides the helical phase distribution (i.e. $e^{-jl\varphi}$), it is well known that the OAM waves have ring-shaped intensity with a null area in the center (usually referred as the dark zone). The higher the order of OAM mode is, the larger the radius of the dark zone becomes. Moreover, the radius of the dark zone grows larger as the propagation distance increases due to the divergence of OAM waves. Since the size of the receiving aperture should be comparable to the dark zone in order to detect the OAM mode, it can be very large for long distance transmission. Obviously, this makes the detection of OAM waves rather difficult and inconvenient. To alleviate this problem, a partial angular aperture receiving (PAAR) scheme was proposed for OAM multiplexing systems [19]. Instead of using the whole angular aperture (i.e. 2π), only part of the angular aperture is used. In this way, the size of the receiving aperture can be reduced efficiently. Assuming an angular aperture of $2\pi/m$, Eq. (11.5) can be rewritten as:

$$U = \int_{\rho_1}^{\rho_2} A_{l_1}(\rho, z)\rho d\rho \int_0^{2\pi/m} e^{-j(l_1 - l_2)\varphi} d\varphi = \frac{1}{m}\int_{\rho_1}^{\rho_2} A_{l_1}(\rho, z)\rho d\rho \int_0^{2\pi} e^{-j(l_1 - l_2)\varphi'/m} d\varphi' \tag{11.6}$$

where $\varphi' = m\varphi$. Different from the WAAR scheme, the integration in Eq. (11.6) will not always be zero when $l_1 \neq l_2$. In other words, not all the OAM modes are mutually orthogonal in the PAAR scheme. However, if $l_1 - l_2 = km$ (where k is an integer) is satisfied, Eq. (11.6) will be zero when $k \neq 0$. For an angular aperture of $2\pi/m$ in the PAAR scheme, the OAM modes satisfying $l_k - l_1 = km$ are mutually orthogonal and can be used for multiplexing.

Compared with the WAAR scheme, the PAAR scheme is much more efficient and cost-effective. However, it is difficult to obtain large OAM modes even using the state-of-the-art OAM generation technique, which makes the aperture of $2\pi/m$ still bulky to some extent. Based on the PAAR scheme and sampling receiving, an improved scheme named partial aperture sampling receiving (PASR) scheme is proposed [20] (as shown in Figure 11.3).

Multiple OAM waves ($l = l_1, l_2, ..., l_{N_A}$) are multiplexed at the transmitting side. The N_A receiving antennas are uniformly distributed along a circular arc of $2\pi/m$ with a radius of R_0, each receiving antenna can be regarded as a sampling point. Further demultiplexing is realized through a digital phase-shift (DPS) component. Supposing two OAM modes (e.g. $l = l_1, l_2$) are multiplexed, the correlation between the two sampled OAM modes can be calculated using

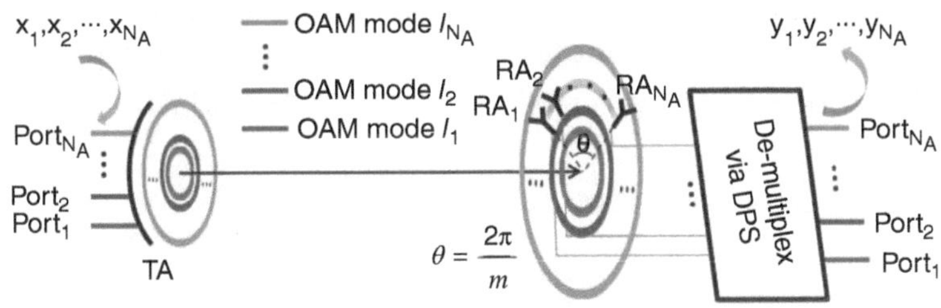

Figure 11.3 Architecture of the PASR-based OAM multiplexing system. RA: receiving antenna, DPS: digital phase-shift [20]. *Source:* Zhang et al. [20]. (© [2016] IEEE)

$$\sum_{n=0}^{N_A-1} E_{l_1}\left(\frac{2\pi n}{mN_A}\right) E^*_{l_2}\left(\frac{2\pi n}{mN_A}\right) = A_{l_1}A^*_{l_2}\sum_{n=0}^{N_A-1} e^{-j\frac{2\pi n}{N_A}\frac{l_1-l_2}{m}} \tag{11.7}$$

Same as the PAAR scheme, when $l_1 - l_2 = km$ is satisfied, the two sampled OAM waves are orthogonal to each other, i.e. Eq. (11.7) is zero. It is known that the resolvable OAM mode l is restricted by the angle between two adjacent receiving antennas with respect to the center of the circle β, i.e. $|l| < \frac{\pi}{\beta}$. Similarly, the OAM mode l that can be resolved should satisfy $|l| < \frac{mN_A}{2}$ in the PASR scheme.

Taking the PASR scheme as an example to demonstrate the conventional OAM multiplexing system. For a clear demonstration of the PASR scheme, Eq. (11.4) can be written in detail as

$$E_l(\rho, \varphi, z) = \frac{\lambda\alpha_0}{4\pi\sqrt{\rho^2+z^2}} J_l\left(\frac{k_0 D\rho}{2\sqrt{\rho^2+z^2}}\right) e^{-jl\varphi} \tag{11.8}$$

where α_0 is a constant determined by the configuration of the OAM multiplexing system, λ is the wavelength, k_0 is the wavenumber (i.e. $k_0 = \frac{2\pi}{\lambda}$), D is the diameter of the transmitting aperture, and $J_l(x)$ is the Bessel function of the first kind of order l. For the configurations shown in Figure 11.3 (i.e. N_A OAM modes satisfying $l_k - l_0 = km$ ($k = 1, 2, 3, \ldots, N_A$) are multiplexed at the transmitting side, and N_A receiving antennas are uniformly distributed along a circular arc of $\frac{2\pi}{m}$), the received signal $\mathbf{y} \in \mathbb{C}^{N_A \times 1}$ can be modeled as:

$$\mathbf{y} = \mathbf{T}^H\mathbf{H}\mathbf{x} + \mathbf{n} \tag{11.9}$$

where $\mathbf{T}$ and $\boldsymbol{T}^H \in \mathbb{C}^{N_A \times N_A}$ (where ^{H} denotes the conjugate transpose operator) are the normalized multiplexing and demultiplexing matrices, respectively, $\boldsymbol{H} \in \mathbb{C}^{N_A \times N_A}$ is the transfer function matrix whose (n, k)th element $H(n,k)$ corresponds to the subchannel from the kth transmitting OAM source to the nth receiving antenna, $\mathbf{x} \in \mathbb{C}^{N_A \times 1}$ is the transmitted signal, and $\boldsymbol{n} \in \mathbb{C}^{N_A \times 1}$ is the additive white Gaussian noise (AWGN).

According to the PASR scheme and the field of OAM waves, the (n, k)th element of matrix $\mathbf{T}$ and $\mathbf{H}$ can be written as

$$t(n,k) = \frac{1}{\sqrt{N_A}} e^{-j2\pi\frac{(n-1)l_k}{mN_A}} \tag{11.10}$$

$$H(n,k) = E_{l_k}\left(\rho_0, \frac{2\pi(n-1)}{mN_A}, z_0\right) = \frac{\lambda\alpha_0}{4\pi\sqrt{\rho_0^2+z_0^2}} J_{l_k}\left(\frac{k_0 D\rho_0}{2\sqrt{\rho_0^2+z_0^2}}\right) e^{-j\frac{2\pi(n-1)}{N_A}\frac{l_k}{m}} \tag{11.11}$$

where l_k represents the kth transmitted OAM mode, ρ_0 is the radius of receiving arc, and z_0 is the distance between the transmitting and the receiving antennas. Combining Eqs. (11.10) and (11.11), $\mathbf{H}$ can be rewritten as $\mathbf{H} = \mathbf{T}\boldsymbol{\Lambda}$(where $\boldsymbol{\Lambda}$ is a diagonal matrix of $diag\{\beta_1, \beta_2, \cdots, \beta_{N_A}\} \in \mathbb{C}^{N_A \times N_A}$). Hence, we have:

$$\mathbf{y} = \mathbf{T}^H\mathbf{T}\boldsymbol{\Lambda}\mathbf{x} + \mathbf{n} = \boldsymbol{\Lambda}\mathbf{x} + \mathbf{n} \tag{11.12}$$

In the LoS near-field environment, the diagonal matrix $\boldsymbol{\Lambda}$ is full rank. The capacity of the PASR-based OAM multiplexing system is given by [3]:

$$C = \sum_{k=1}^{N_A} \log_2\left(1 + \beta_k^2 \frac{P_k}{\sigma_0^2}\right) \tag{11.13}$$

where P_k is the power allocated to the kth OAM channel according to the water-filling principle and σ_0^2 is the fixed noise variance at each receiving antenna.

Following the architecture of the PASR-based OAM multiplexing system (cf. Figure 11.3), relevant measurements are conducted. Four independent quadrature phase shift keying (QPSK) signals at an intermediate frequency (IF) carrier of 800 MHz are generated by a software-defined radio (SDR) platform, and converted to the carrier at 10 GHz. The QPSK signals are then modulated into 40 Mbit s^{-1} OFDM signals with 20 MHz bandwidth. Four OAM modes (i.e. $l = -6, -2, +2$, and $+6$) are generated. The four OAM waves carrying OFDM–QPSK signals are multiplexed and transmitted by a specially designed 4-OAM-mode antenna. The distance between the transmitting and the receiving apertures is 10 m. At the receiving side, four identical horn antennas are uniformly distributed along a circular arc of $\frac{\pi}{2}$ with a radius of 45 cm. These four sampled OAM signals are transformed into amplified IF signals and converted to the digital baseband, finally demultiplexed by the DPS component.

The inter-channel interference is calculated as the ratio of the power received from other channels to that of the transmitted channel, corresponding results are shown in Figure 11.4a. The averaged inter-channel interference for OAM modes $l = -6, -2, +2$, and $+6$ is about −16.9 dB, −16.5 dB, −18.1 dB and −13.6 dB, respectively. The interference is mainly caused by imperfect antenna alignment and OAM wave generation. Figure 11.4b shows the measured constellations and error vector magnitudes (EVMs) [21] when the signal-to-noise ratio (SNR) is 16 dB. Figure 11.4c shows the raw bit error rate (BER) performances of different OAM modes as functions of SNR. It can be seen that all the four OAM channels are able to achieve a raw BER below 3.8×10^{-3} (at sufficient SNR), which is the level that enables extremely low block error rates through the application of efficient forward error correction (FEC) codes (i.e. FEC limit) [9, 22]. The BER performance of the OAM channel $l = +6$ is worse than other channels, this is in agreement with the measured inter-channel interference and EVM.

The PASR-based OAM multiplexing system is established by combining OAM antennas with phase-shift components. Compared with conventional UCA-based systems, the receiving end of the PASR-based system is much more compact and simplified. Moreover, the conventional IDFT method is utilized to demultiplex the OAM waves, thus the channel capacity can be enhanced without additional complex computation. However, some consecutive OAM modes are discarded in order to obtain the specific sets of orthogonal OAM modes. As the size of the aperture decreases, OAM modes with higher-order are required. Overall, the number of OAM channels available for multiplexing is still very limited.

To partially address the dilemma of long-distance transmission and finite aperture, another OAM multiplexing system using compound angular lens (CALs) [23] will be introduced next.

The architecture of the CAL-based OAM multiplexing system is shown in Figure 11.5. The CAL module is composed of SPPs with different OAM modes and dielectric lenses with different focal

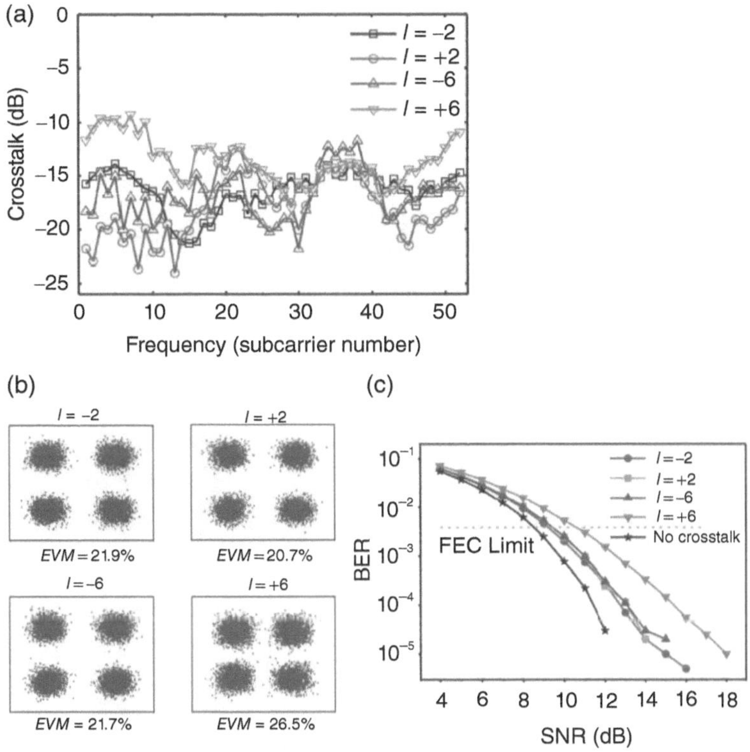

Figure 11.4 The measured (a) inter-channel interference; (b) constellations; and (c) BERs when SNR = 16 dB for the four multiplexed OAM modes (i.e. $l = \pm 2$ and ± 6) [20]. *Source:* Zhang et al. [20]. (© [2016] IEEE)

lengths. The plane wave passing through the CAL will be transformed into OAM waves with different modes and converged at different areas. Obviously, the multiplexed OAM modes can be separated by the receiving antennas locating at different energy convergence areas. Note that when multiple OAM modes are multiplexed, the energies of different OAM modes distribute over concentric circles with different radii due to different beam divergence angles. As a result, the SPPs and dielectric lens in the CAL module are all ring-shaped.

Since the OAM channels with different modes are spatially separated with isolated receptions, the inter-channel interference is basically eliminated. Hence, the BER performance of the CAL-based system is very close to the AWGN channel and outperforms the UCA-based systems. Compared with the PASR-based system, all consecutive OAM modes can be utilized in the CAL-based system. Moreover, the computational complexity is reduced in that the IDFT operation at the receiving side is no longer needed. More details about the designs of the CAL-based OAM multiplexing system and measurement results can be found in [23].

11.2.2 OAM Multiplexing with Spatial Equalization

Conventional OAM multiplexing systems are realized by utilizing the orthogonality of OAM modes. Here, we introduce an OAM multiplexing system that adopts independent OAM antennas as the elements of the transmitting uniform linear array (ULA) [7]. Especially, the used OAM antennas can generate a new form of OAM waves (i.e. plane spiral OAM, PSOAM), which

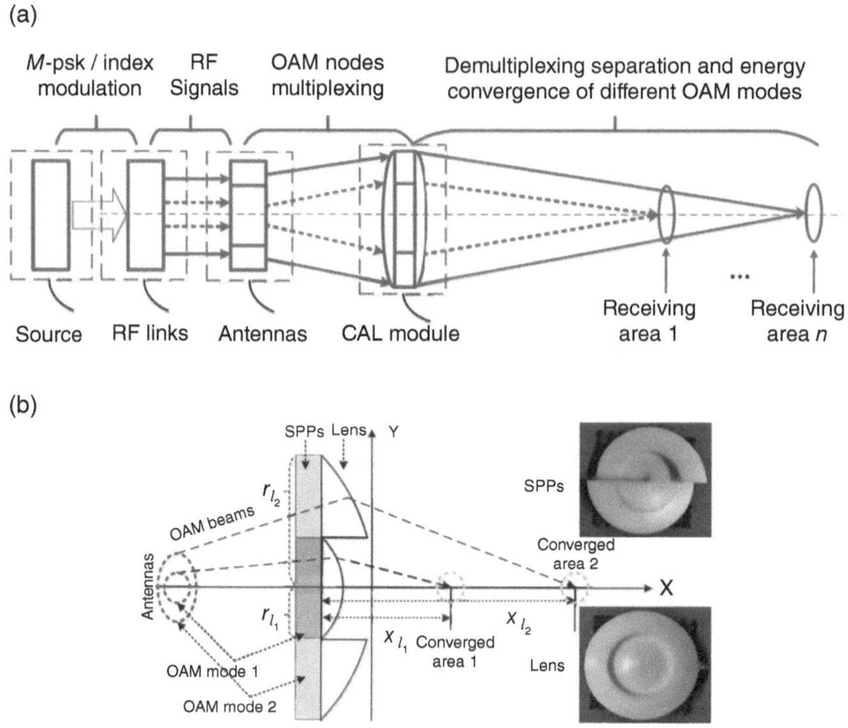

Figure 11.5 Architecture of (a) the CAL-based OAM multiplexing system and (b) the CAL module [23]. *Source:* Zhao and Zhang [23]. (© [2019] IEEE)

propagate along the transverse direction. As mentioned before, the OAM wave has a ring-shaped intensity with a dark zone in the beam center, and the dark zone becomes larger for higher OAM modes. Moreover, the dark zone will increase with the propagation distance. Hence, the receiving aperture should be large enough to capture the effective OAM beams, which limits not only the coverage of the communication system but also the maximum OAM modes that can be used for multiplexing. Besides the PAAR and PASR techniques, the PSOAM-based MIMO system offers a new solution to this problem. However, it should be noted that all these methods have certain requirements on the transmitted OAM modes.

Figure 11.6 shows the 4×4 PSOAM-based MIMO system [7]. The transmitter is a ULA consisting of PSOAM antennas, while the receiver is a ULA consisting of horn antennas. The carrier frequency is 10 GHz, the inter-element separation is 5.83λ (i.e. 0.175 m), and the propagation distance is 233λ (i.e. 7 m). The OAM modes of the PSOAM antennas are set with an OAM mode interval of ΔL. The positive (negative) sign of ΔL represents the increasing (decreasing) of the transmitted OAM modes. Here, $\Delta L = \pm 6$ are adopted to investigate the influence of the sign on the PSOAM-based MIMO system. Correspondingly, the transmitted OAM modes are −9, −3, +3, +9 ($\Delta L = +6$) and +9, +3, −3, −9 ($\Delta L = -6$). At the receiving side, the zero-forcing algorithm is applied to migrate the inter-stream interference.

Denoting the channel transfer function matrix of the PSOAM-based MIMO system as $\mathbf{H}$, the capacity of the system is related to the eigenvalues of $\mathbf{H}$. Figure 11.7 shows the eigenvalues of $\mathbf{H}$ of the PSOAM-based MIMO system and the conventional one as functions of propagation distance (the typos in the original figure [7] was corrected). Different eigenvalues represent different

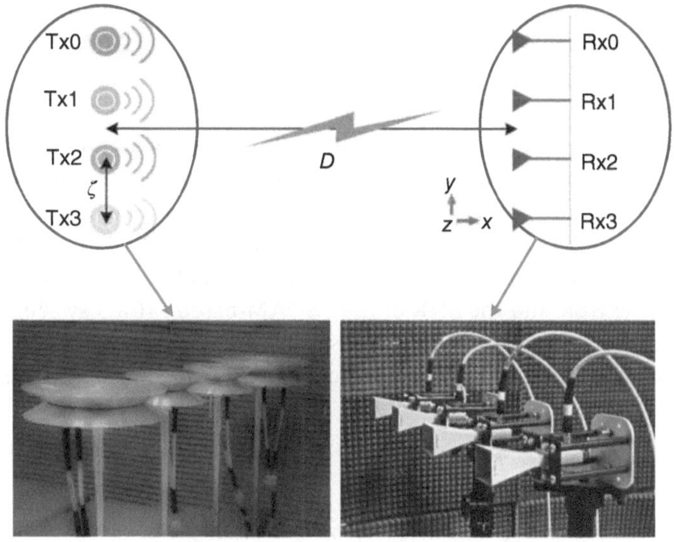

Figure 11.6 A 4 × 4 PSOAM-based MIMO system [7]. *Source:* Zhang et al. [7]. (© [2017] IEEE)

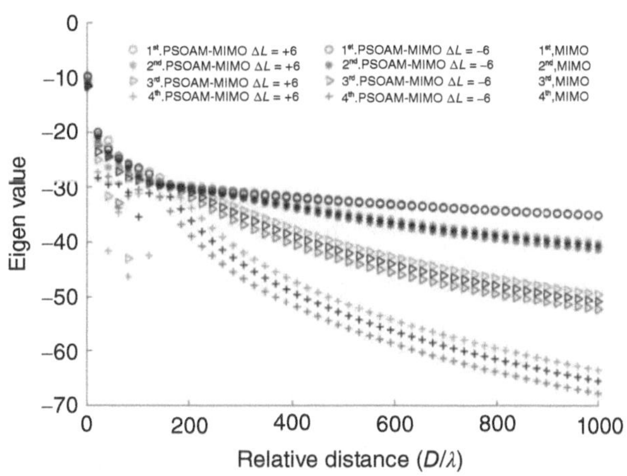

Figure 11.7 Eigenvalues of **H** of the PSOAM-based and conventional MIMO systems as functions of propagation distance [7]. *Source:* Zhang et al. [7]. (© [2017] IEEE)

subchannel gains. It can be seen that, for the same subchannel order (except for the 1st subchannel, in which case all the eigenvalues are about the same), the eigenvalue of the PSOAM-based MIMO system with $\Delta L = +6$ is the largest, while that of the PSOAM-based MIMO system with $\Delta L = -6$ is the smallest. Therefore, the PSOAM-based MIMO system with $\Delta L = +6$ has the highest capacity.

To facilitate the comparison of the performance of the OAM-based and conventional MIMO systems, an equivalent model is established from the view of wave vector [8]. Utilizing this model, an OAM-based MIMO system can be equivalent to a conventional MIMO system with an increased or decreased element spacing. Specifically, the sign of an OAM mode denotes the direction of the azimuthal wave vector $\overrightarrow{k_\varphi}$, ΔL determines the element spacing of the equivalent MIMO system. It is verified that a larger $|\Delta L|$ leads to a lower spatial correlation and more independent subchannels, which indicates a higher capacity. It should be noted that the sign of ΔL has a significant influence on the capacity of the OAM-based MIMO system when $|\Delta L|$ is small. Figure 11.8 shows the equivalent MIMO systems of the PSOAM-based MIMO systems with $\Delta L = \pm 6$. It can be seen that the equivalent element spacing of the PSOAM-based MIMO system with $\Delta L = +6$ is larger than the conventional MIMO system. Naturally, a higher capacity is achieved for the PSOAM-based MIMO system. However, the PSOAM-based MIMO system with $\Delta L = -6$ performs worse due to the decreased equivalent element spacing. The results are in agreement with the simulated eigenvalues.

The BER and SNR are two commonly used metrics to evaluate the performance of a communication system, thus the BER and the SNR of the PSOAM-based MIMO system with $\Delta L = \pm 6$ are investigated through both simulations and measurements. Corresponding results are shown in Figure 11.9. From the simulated BERs, it can be concluded that the PSOAM-based MIMO system

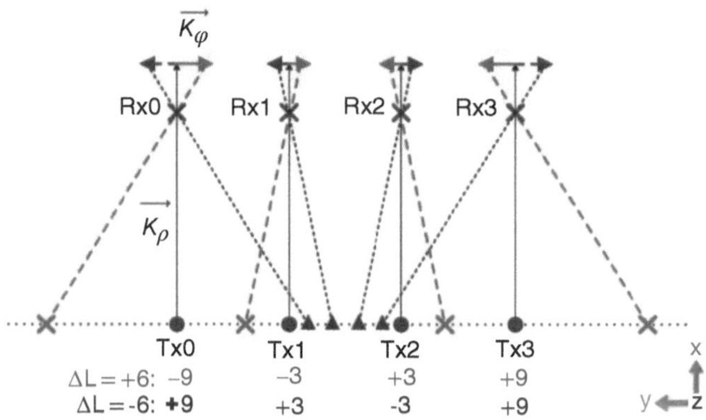

Figure 11.8 Equivalent MIMO systems of the PSOAM-based MIMO systems with $\Delta L = \pm 6$. The PSOAM transmitting antennas and receiving horn antennas are denoted by black dots and black crosses, respectively. The equivalent MIMO antennas of the PSOAM-based MIMO system with $\Delta L = +6$ and $\Delta L = -6$ are denoted by the long and short dashed lines, respectively [7]. *Source:* Zhang et al. [7]. (© [2017] IEEE)

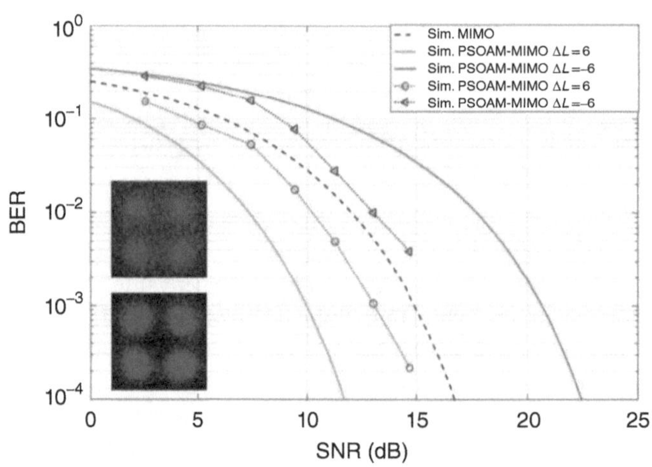

Figure 11.9 Simulated and measured BER of the PSOAM-based and conventional MIMO systems as functions of propagation distance [7]. *Source:* Zhang et al. [7]. (© [2017] IEEE)

with $\Delta L = +6$ performs better than the conventional MIMO system, while the PSOAM-based MIMO system with $\Delta L = -6$ performs worse (i.e. a larger BER). For the same simulated BER of 10^{-4}, the PSOAM-based MIMO system with $\Delta L = +6$ has a performance gain of about 5 dB over the conventional MIMO system and 10.8 dB over the PSOAM-based MIMO system with $\Delta L = -6$. The measured BERs of the two PSOAM-based MIMO systems show the same trend with the corresponding simulated BERs. For the same measured SNR of 9.42 dB, the BER of the PSOAM-based MIMO system with $\Delta L = +6$ is 2.1 times lower than that of the conventional MIMO system and 4.4 times lower than that of the PSOAM-based MIMO system with $\Delta L = -6$. The constellations of the received QPSK signals of the two PSOAM-based MIMO systems at 9.42 dB SNR are also shown in Figure 11.9. It can be seen that the EVMs of the system with $\Delta L = +6$ is much smaller than that of the system with $\Delta L = -6$.

It should be noted that there exist different degrees of discrepancies between the simulated and measured BERs at different SNRs. This is mainly caused by two reasons. The purity of the generated OAM wave is not 100% (i.e. the desired OAM spectrum expands to other OAM modes), which inevitably leads to interference. The other reason is the environment is not an ideal LoS case and the multipath tends to increase the crosstalk.

The spatial multiplexing of the 4 × 4 PSOAM-based MIMO system is investigated and an equivalent model of the OAM-based MIMO system is briefly introduced. The PSOAM-based MIMO system with a positive mode interval performs better than the conventional MIMO system, while the PSOAM-based MIMO system with a negative mode interval performs worse. If the mode interval is large enough, the PSOAM-based MIMO system provides a better performance regardless of the sign of the mode interval. Overall, the PSOAM-based spatial multiplexing with a certain mode interval performs better than the conventional spatial multiplexing, however, obvious performance degradation can be observed in the presence of multipath.

Next, we introduce a system combining OAM multiplexing with conventional spatial multiplexing, which can increase the system capacity with certain space constraints [9].

The architecture of the system combining the OAM multiplexing with the conventional spatial multiplexing is shown in Figure 11.10. There are *N* pairs of transmitting–receiving apertures, each

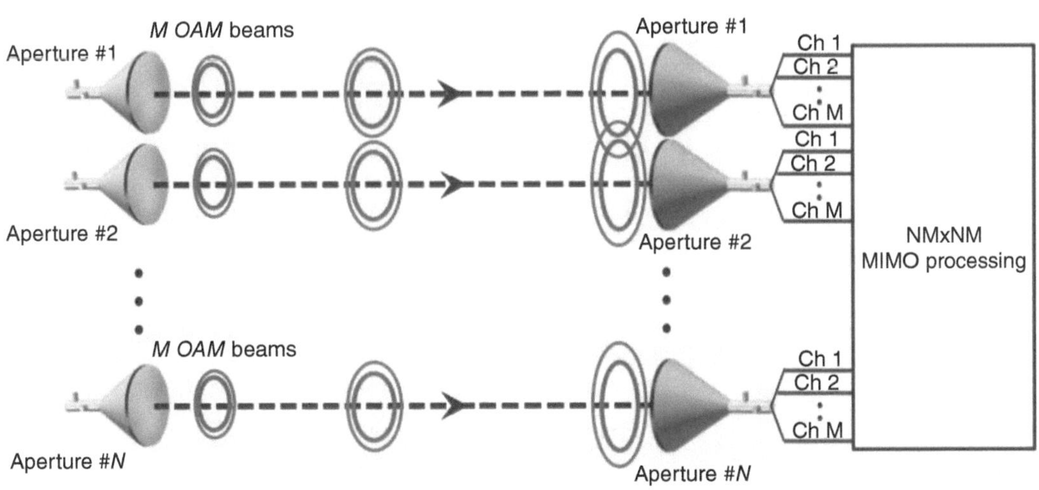

Figure 11.10 Architecture of the system combining the OAM multiplexing with the conventional spatial multiplexing [9]. *Source:* Ren et al. [9]. (© [2017] IEEE)

transmitting aperture T_i ($i = 1, 2, ..., N$) transmits M multiplexed OAM waves. Thus, there exist total NM OAM channels. The received signal of each receiving aperture R_i ($i = 1, 2, ..., N$) is a superposition of the signals from different OAM channels, and it can be expressed as:

$$\mathbf{y}_i = \sum_{j=1}^{N} \mathbf{H}_{i,j}\mathbf{x}_j + \mathbf{n}_i \tag{11.14}$$

where $\mathbf{y}_i = (y_{i,1}, y_{i,2}, ... y_{i,M})^T$ is the received signal at the ith receiving aperture (in which $y_{i,m}$ is the received signal of the mth OAM channel, and ^{T} denotes the transpose operator), $\mathbf{x}_j = (x_{j,1}, x_{j,2}, ... x_{j,M})^T$ is the transmitted signal at the jth transmitting aperture (in which $x_{j,m}$ is the transmitted signal of the mth OAM channel), $\mathbf{n}_i$ is the noise of the ith receiving aperture, and $\mathbf{H}_{i,j}$ is the channel transfer function from transmitting aperture j to receiving aperture i with a dimension of $M \times M$. The total channel transfer matrix can be written as:

$$\mathbf{H} = \begin{bmatrix} \mathbf{H}_{1,1} & \mathbf{H}_{1,2} & \cdots & \mathbf{H}_{1,N} \\ \mathbf{H}_{2,1} & \mathbf{H}_{2,2} & \cdots & \mathbf{H}_{2,N} \\ \vdots & \vdots & \cdots & \vdots \\ \mathbf{H}_{N,1} & \mathbf{H}_{N,2} & \cdots & \mathbf{H}_{N,N} \end{bmatrix}_{NM \times NM} \tag{11.15}$$

It should be stressed that the architecture presented here is mainly applicable to the scenarios of short-range LoS communications. If the purity of the generated OAM wave is 100% and there are perfect alignments between the transmitting apertures and receiving apertures, $\mathbf{H}_{i,i}$ is actually a diagonal matrix due to the orthogonality between the M OAM waves. In this case, the crosstalk is only determined by the overlaps between the OAM waves from different transmitting apertures, and related to the aperture spacing and propagation distance.

Figure 11.11 shows the experimental setup of a 28-GHz multiplexing system with $N = M = 2$, i.e. each transmitted signal contains two multiplexed OAM waves ($l = +1$ and $l = +3$). At the

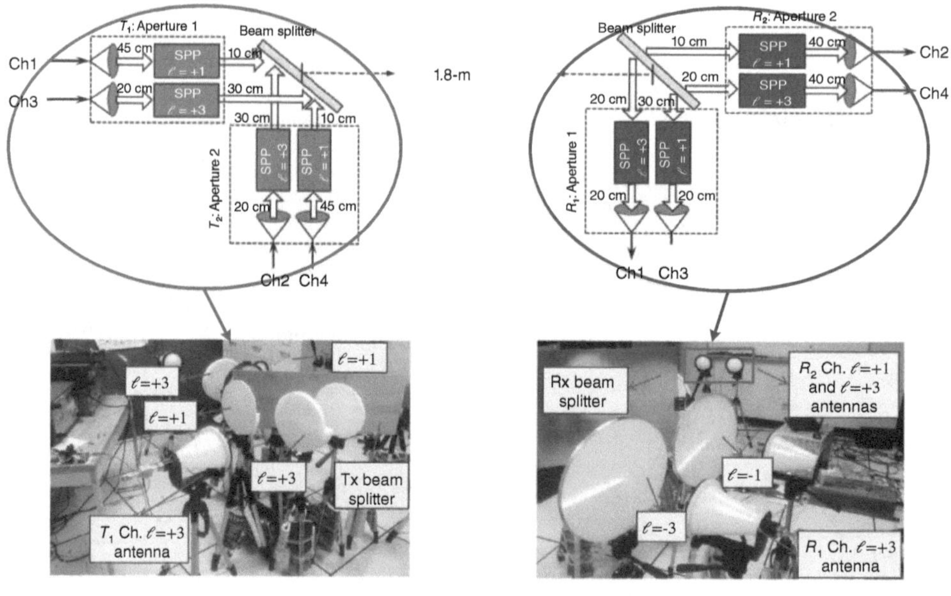

Figure 11.11 Measurement setup of the multiplexing system where each transmitted signal contains two multiplexed OAM waves (i.e. $l = +1$ and $l = +3$) [9]. *Source:* Ren et al. [9]

transmitting side, two independent 1 Gbaud 16-QAM signals are generated. Each of the two signals is amplified and split into two branches. One branch has an 8.3 ns (i.e. 8.3 symbol periods for a 1 Gbaud 16-QAM signal) delay compared to the other. The two 16-QAM signals are fed to two lensed horn antennas with diameters of 15 cm. In this way, the four 16-QAM signals fed to four channels can be regarded as independent. The SPP technique [10, 11] is used to convert the Gaussian beams to OAM beams (i.e. $l = +1$ and $l = +3$). The SPPs have circular apertures with diameters of 30 cm. Obviously, the size of the SPP is larger than that of the lensed horn antennas, which alleviates the effects caused by the truncated aperture. The two OAM waves at each transmitting aperture are combined into a single wave using a beam splitter. The beam splitter is located after the SPPs with an angular difference of 45°. The distance between the centers of the two-beam splitters is 1.8 m. The distances between the antennas, SPPs, and the beam splitters are marked in Figure 11.11. It can be seen that the layout is not totally symmetric. For the convenience of expression, the antennas, SPPs, and the beam splitters at the transmitting and the receiving sides are grouped and denoted as $\frac{T_1}{T_2}$ and $\frac{R_1}{R_2}$.

The superposition of the OAM beams with $l = +1$ and $l = +3$ at T_1 or T_2 is verified by both simulation and measurement, as well as the superposition of four OAM beams at 0.5 m or 1 m away from the transmitter. The measurements show good agreements with the simulations [9]. Before evaluating the system performance, crosstalk between the OAM channels should be characterized first. Instead of the actual 1 Gbaud 16-QAM signal, a continuous-wave (CW) signal at 28 GHz is transmitted by each channel to have a quick measurement of the crosstalk. Specifically, the CW signal is fed to channel 1 (i.e. OAM $l = +1$ at T_1) while there are no inputs for the other channels. Once the signal is transmitted, the received power at each receiving antenna is recorded utilizing a radio frequency (RF) spectrum analyzer. The measurement procedure is repeated for all the channels (i.e. channels 1–4). The transferred power of each OAM channel is finally obtained and shown in Table 11.1. Compared with the case that the signal is transmitted through the channels with $l = +1$, the received power of the noncoaxial channels is larger when the signal is transmitted through the channels with $l = +3$. This is because the OAM wave with higher mode diverges faster, which in turn results in a larger overlapping area at the receiving side. In addition, the power leakage can also be observed for the coaxial channels with $l = +1$ and $l = +3$, owing to the imperfection of the generated OAM waves, multiplexing, and setup misalignment. The total crosstalk of a specified channel is defined as the ratio of the total received power from all channels to the received power of this channel. The corresponding results are also listed in Table 11.1. Obviously, the total crosstalk of the channels with $l = +3$ is higher than that of the channels with $l = +1$, due to the increased overlapping areas between the OAM waves as explained before.

Table 11.1 Transferred power and crosstalk of each OAM channel [9].

	Aperture T_1 $l = +1$	Aperture T_1 $l = +3$	Aperture T_2 $l = +1$	Aperture T_2 $l = +3$	Total Crosstalk
Aperture R_1 $l = +1$	−31.2 dBm	−49.2 dBm	−52.8 dBm	−55.3 dBm	−15.7 dB
Aperture R_1 $l = +3$	−51.1 dBm	−31.8 dBm	−42.3 dBm	−47.6 dBm	−8.8 dB
Aperture R_2 $l = +1$	−54.3 dBm	−45.2 dBm	−28.8 dBm	−50.3 dBm	−14.8 dB
Aperture R_2 $l = +3$	−43.6 dBm	−44.3 dBm	−49.2 dBm	−29.2 dBm	−11.1 dB

Source: Ren et al. [9]. (© [2017] IEEE)

To mitigate the channel interference and demultiplex the transmitted signals, the multi-modulus algorithm (MMA) based adaptive equalization is utilized [24, 25]. The measured BERs of different cases are shown in Figure 11.12, i.e. the BERs for channels $l = +1$ and $l = +3$ when only one pair of the transmitter/receiver apertures is activated (case 1), the BERs for all the channels without MMA-based adaptive equalizer (case 2), and the BERs for all the channels with MMA-based adaptive equalizer (case 3). For all the cases, the theoretical BER of a 16-QAM signal is plotted as a reference. For case 1, the constellations of the received 1 Gbaud 16-QAM signals for channel $l = +1$ in R_1 (SNR = 28.7 dB) and R_2 (SNR = 31.3 dB) are also plotted. Since there is no interference from other apertures, a BER of 3.8×10^{-3} (i.e. the FEC limit) can be achieved for all the channels. For cases 2 and 3, the channels suffer from the interferences caused by both the coaxial and noncoaxial channels. Without the MMA-based adaptive equalizer, the measured BERs of all the channels are way larger than the FEC limit due to the strong interference, as shown in Figure 11.12b. However, the measured BERs of all the channels decrease and fall below the FEC limit when the MMA-based adaptive equalizer is adopted. The discrepancy between the measured and theoretical BERs may be caused by imperfect heterodyne detection and signal generation.

For illustration of the improvement by the equalizer, the constellations of the received 1 Gbaud 16-QAM signals for channel $l = +1$ and $l = +3$ in both apertures are shown in Figure 11.13. Corresponding SNRs and EVMs are also given in the same figure. It can be seen that the equalizer can significantly improve the EVM performance and the channel $l = +1$ in R_2 has the best performance after equalization. The results are in agreement with the BERs. It should be noted that the performance will degrade as the propagation distance increases due to the divergence of the OAM wave.

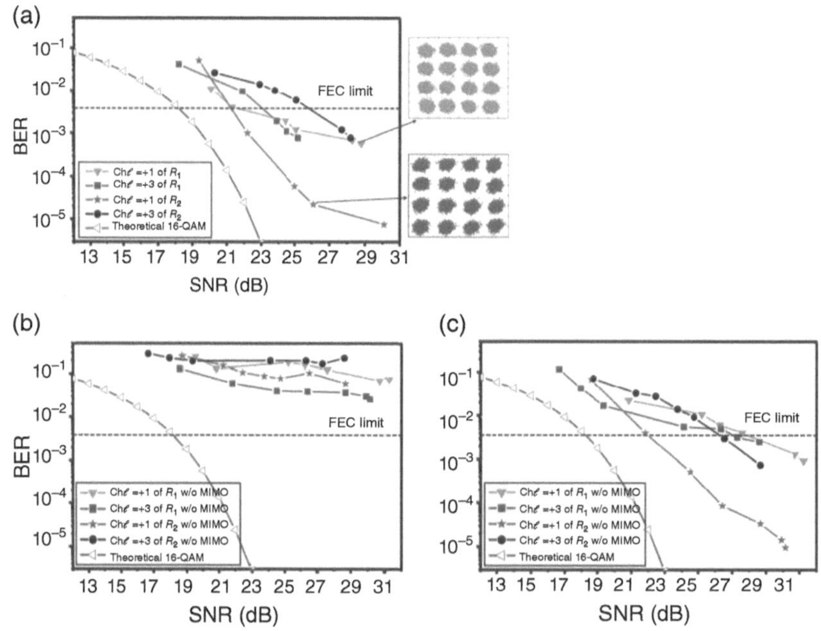

Figure 11.12 Measured BERs as functions of SNR of different cases: (a) BERs for channels $l = +1$ and $l = +3$ for aperture R_1 (when T_2 is not fed) and R_2 (when T_1 is not fed); (b) BERs for all the channels without MMA-based adaptive equalizer; and (c) BERs for all the channels with MMA-based adaptive equalizer [9]. *Source:* Ren et al. [9]. (© [2017] IEEE)

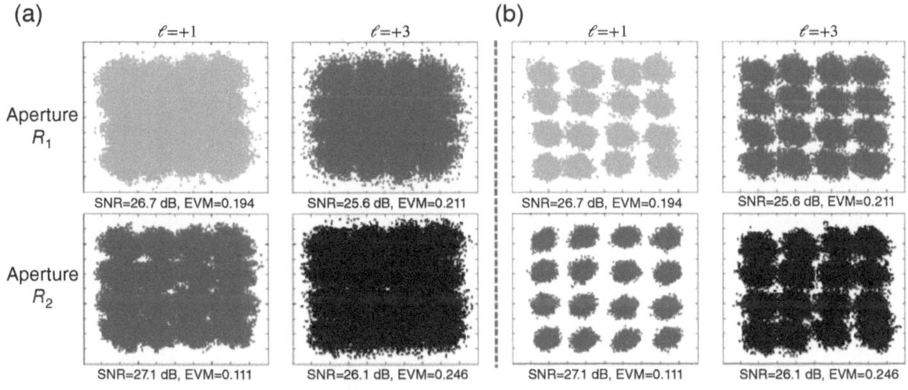

Figure 11.13 Constellations of received 1 Gbaud 16-QAM signals for channel $l = +1$ and $l = +3$ in both apertures: (a) without MMA-based adaptive equalizer and (b) with MMA-based adaptive equalizer [9]. *Source:* Ren et al. [9]. (© [2017] IEEE)

11.3 OAM Multiplexing in Multipath Environment

OAM multiplexing has been demonstrated for capacity enhancement in pure LoS (free space) environments. However, compared with free space, multipath is a more realistic scenario for wireless communications. In this section, we focus on the performance of OAM multiplexing in sparse and rich multipath environments.

11.3.1 Specular Reflection

Multipath propagation brings not only fading (introduced in Section 11.1.1) but also interference to the multiplexing system. Interference has mainly two forms: inter- and intra-channel interferences. The multipath components of the same channel result in intra-channel interference, while the multipath components of different channels result in the inter-channel interference. Moreover, the orthogonality of the OAM waves will be destroyed in multipath environments.

The specular reflection caused by a reflector parallel to the link (shown in Figure 11.14) is adopted to analyze the effects of the inter- and intra-channel interferences on OAM communications. The SPP technique is used to generate and detect the OAM waves at 28 GHz. The signal is transmitted over the OAM channel of mode l_1, while the receiver can only receive the signal over the OAM channel of mode l_2. Supposing the SPPs can generate and recover OAM waves perfectly, the signal can only be received when $l_1 = l_2$ for the LoS path. Taking reflection into consideration, the reflected OAM wave can be treated as an OAM wave transmit by an imaging source of mode $-l_1$. Combining the LoS path and reflection path, the received signal is a mix of OAM waves of modes $\pm l_1$.

To investigate the reflection effect on OAM communication, simulations with OAM mode +3 are firstly conducted. The intensity, phase, and OAM spectrum of the LoS OAM wave, the reflected OAM wave, and the total received wave are shown in Figure 11.15. It can be seen that the LoS wave is the ideal OAM wave of mode +3, as all the power is concentrated in mode +3. However, the reflected wave is not a single OAM wave of mode −3 due to the 'misalignment' of the reflected wave and the receiver. As a result, the spectrum of the reflected wave expands to other OAM modes, and the power of each mode is nearly the same. Since the LoS wave is mixed with the reflected wave at

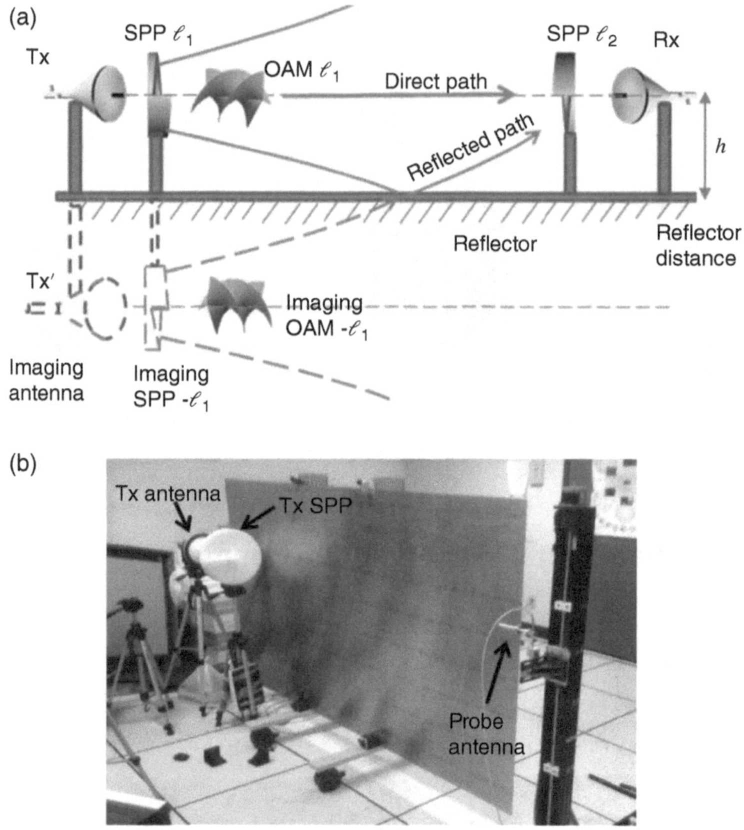

Figure 11.14 (a) Draft and (b) practical measurement setups to the analysis of the effects of multipath on anOAM channel (l_1) using specular reflection from a parallel ideal reflector [22]. *Source:* Yan et al. [22]. (© [2015] IEEE)

the receiver, the OAM mode +3 carried in the spectrum of the reflected wave will cause the intra-channel interference, while the other OAM modes carried in the spectrum of the reflected wave will cause the inter-channel interference. As can be seen from Figure 11.15, the intensity of the combined wave shows a fringing pattern and the phase is also distorted due to the inter- and intra-channel interferences caused by reflection. Measurements are also performed to verify the simulation results. Good agreements are observed [22].

11.3.1.1 Intra-channel Interference

To verify the intra-channel interference, we configure $l_1 = l_2 = l$ in the setup shown in Figure 11.14. Measurement results of different OAM modes are shown in Figure 11.16. It can be seen that the received power decreases as the OAM mode increases. This is in accordance with the divergence property of OAM waves. Moreover, the received power fluctuates much more severely for higher OAM mode or smaller reflector distance. In other words, the interference of the reflected wave is severer for higher OAM modes or smaller reflector distances.

Figure 11.17 shows the simulation results of the intra-channel interference for OAM waves of modes 0, +1, and +3 as functions of the reflector distance. The square line curve represents the received power of OAM mode l that comes from the LoS path. It can be seen that the power of

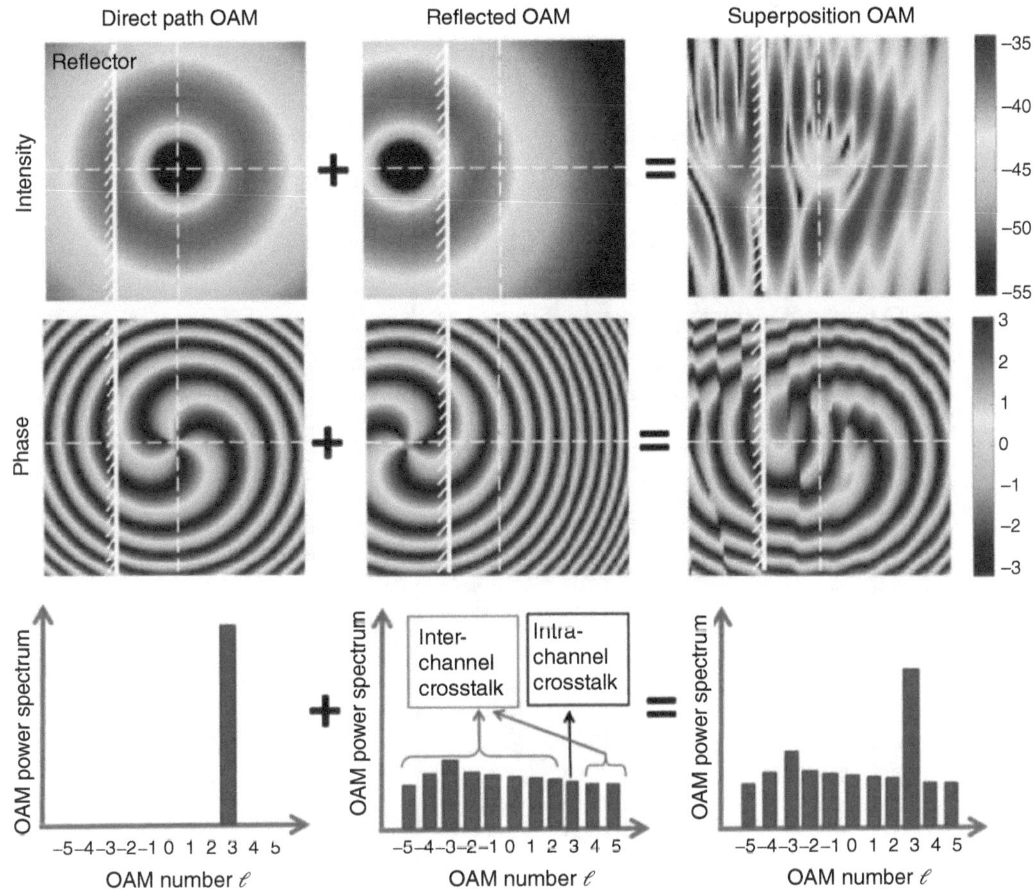

Figure 11.15 Intensity, phase, and OAM spectrum of the LoS OAM wave, the reflected OAM, and the total received wave [22]. *Source:* Yan et al. [22]. (© [2015] IEEE)

Figure 11.16 Normalized received power of the signal for different l_1 (i.e. l_1 = 0, 1, 2, 3) as functions of the reflector distance [22]. *Source:* Yan et al. [22]. (© [2015] IEEE)

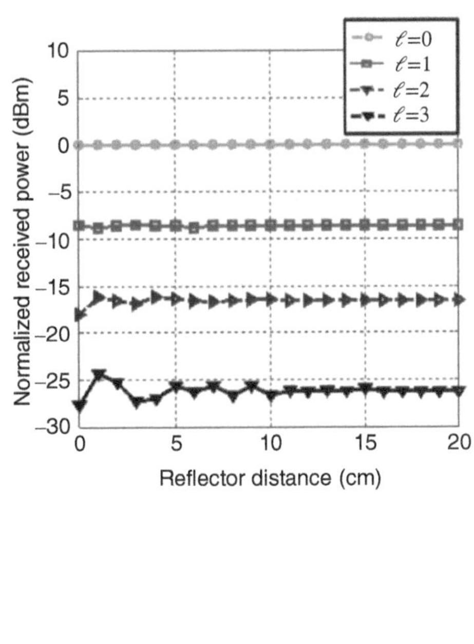

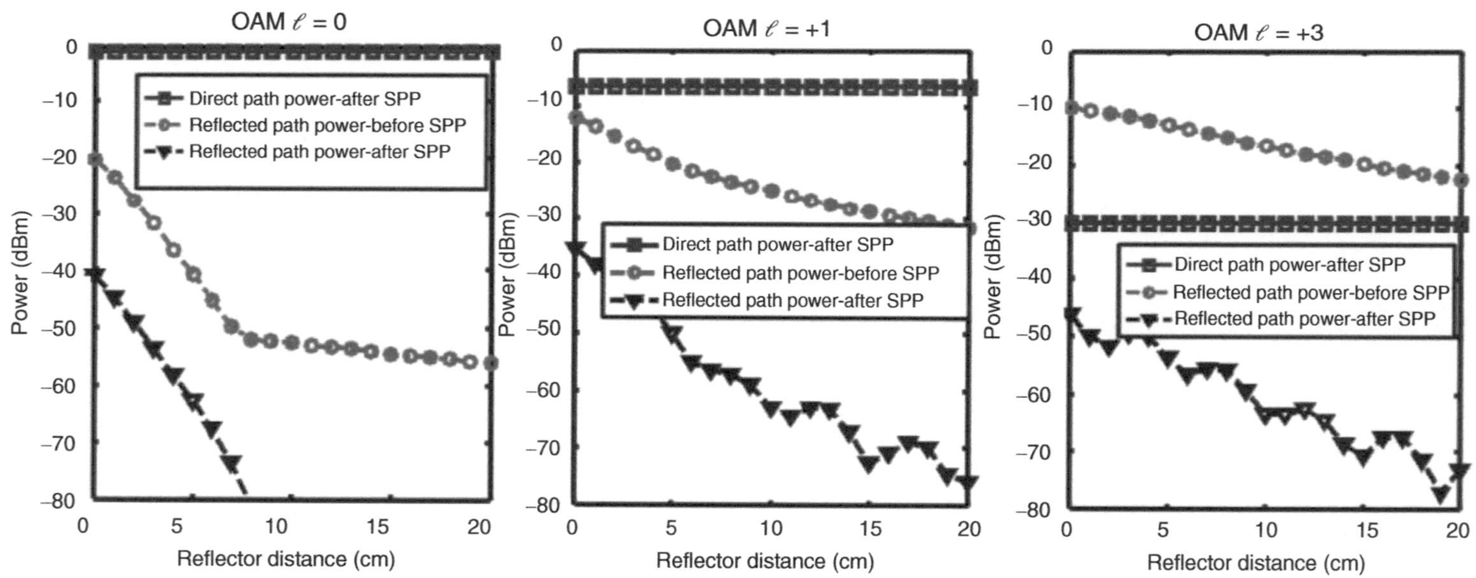

Figure 11.17 Simulation results of the intra-channel interference for OAM waves of mode 0, +1, and +3 as functions of the reflector distance [22]. *Source:* Yan et al. [22]. (© [2015] IEEE)

the LoS path decreases as the OAM mode increases due to the divergence property of OAM waves. The circle line curve represents the received power of all OAM modes that come from the reflected path. The power of the reflected path is higher for a higher OAM mode. As the OAM mode increases, the beam spreads and becomes larger, thus more power will be reflected. The triangle line curve represents the received power of OAM mode l that comes from the reflected path. Obviously, the triangle line curves are much closer to the square line curves as the OAM mode increases. In other words, the reflected path contributes more to the total received power of the mode l as the OAM mode increases, that is the intra-channel interference becomes more serious as the OAM mode increases. Measurements are also performed to verify the simulation results. Good agreements are observed [22].

BER and SNR are used as performance metrics to investigate the multipath effects on OAM communications. A 16-QAM signal with a data rate of 1 Gbaud is used in the measurements. As shown in Figure 11.16 that the intra-channel interference is severe when the distance between the aperture and the reflector is within 10 cm. Since the multipath delay is much smaller than the symbol duration (i.e. the coherence bandwidth of the channel is much larger than the bandwidth of the signal), there is little inter-symbol interference (ISI). In other words, the OAM channel can be treated as a flat-fading channel. Figure 11.18 shows the measured BER and SNR as functions of the reflector distance for OAM modes +1 and +3. The BER and SNR fluctuate mildly as the reflector distance varies for OAM mode +1. However, the BER and SNR fluctuate severely as the reflector distance varies for OAM mode +3. Thus, we can conclude that OAM communication is more affected by the intra-channel interference for high OAM modes.

11.3.1.2 Inter-channel Interference

To verify the inter-channel interference, l_1 is fixed while l_2 is set to different values. Figure 11.19 shows the received power of different OAM modes (i.e. $l_2 = \pm1, \pm3$) as functions of reflector distance when the transmitted OAM modes are +1 and +3 (i.e. $l_1 = +1, +3$). It can be seen that the received powers of the other modes (i.e. $l \neq l_1$) are large at a small reflector distance, and the differences between the received power of the transmitted OAM mode and the other modes become larger as the distance increases. This means the inter-channel interference becomes smaller as the reflector distance increases. Moreover, the received power of the other modes is much closer to

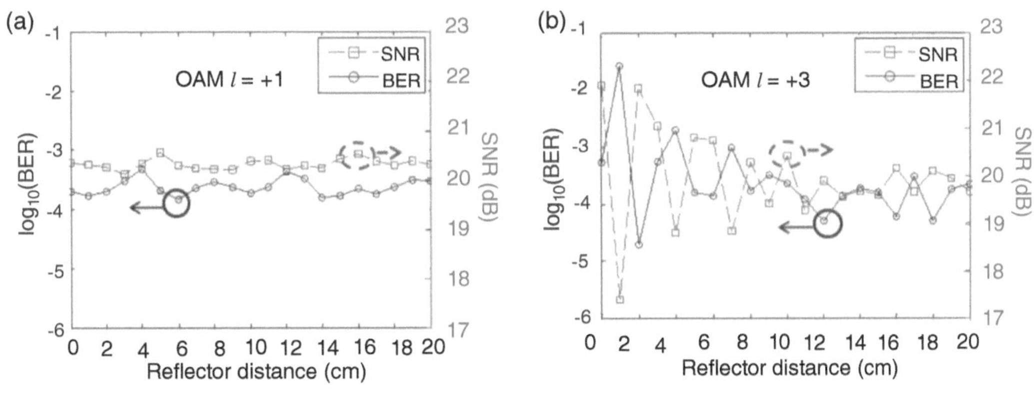

Figure 11.18 Measured BER and SNR as functions of the reflector distance for (a) l = +1 and (b) l = +3 [22]. *Source:* Yan et al. [22]. (© [2015] IEEE)

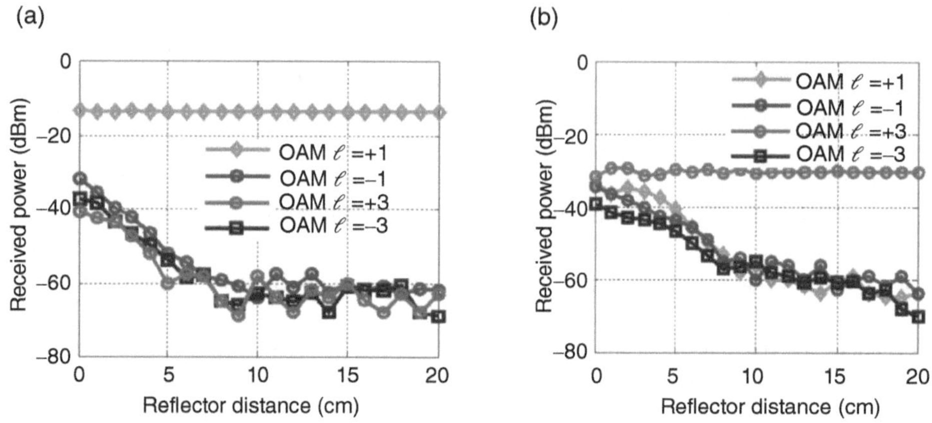

Figure 11.19 Received power from different OAM channels (i.e. $l_2 = \pm1, \pm3$) as functions of the reflector distance when the transmitted OAM mode is (a) $l_1 = +1$ and (b) $l_1 = +3$ [22]. *Source:* Yan et al. [22]. (© [2015] IEEE)

the transmitted mode for higher transmitted OAM mode, which means the inter-channel interference has more effects on OAM communication for high OAM modes.

Finally, two OAM modes ($l_1 = +1$ and $l_1 = +3$) are multiplexed, i.e. each of them carries an independent 16-QAM signal with a data rate of 1 Gbaud. In this way, the inter- and intra-channel interferences are introduced simultaneously to the OAM channel. Figure 11.20 shows the measured BER as a function of the reflector distance for $l_2 = +1$ and $l_2 = +3$. Combining Figures 11.18a and 11.20a, it can be seen that the BER performance of OAM channel +1 deteriorates significantly. This indicates the inter-channel interference from OAM channel +3 to +1 has strong effects on the BER performance, especially when the reflector distance is small. However, the variation of the BER performance of OAM channel +3 is small (cf. Figures 11.18b and 11.20b), which means the inter-channel interference from OAM channel +1 to +3 is relatively low. In other words,

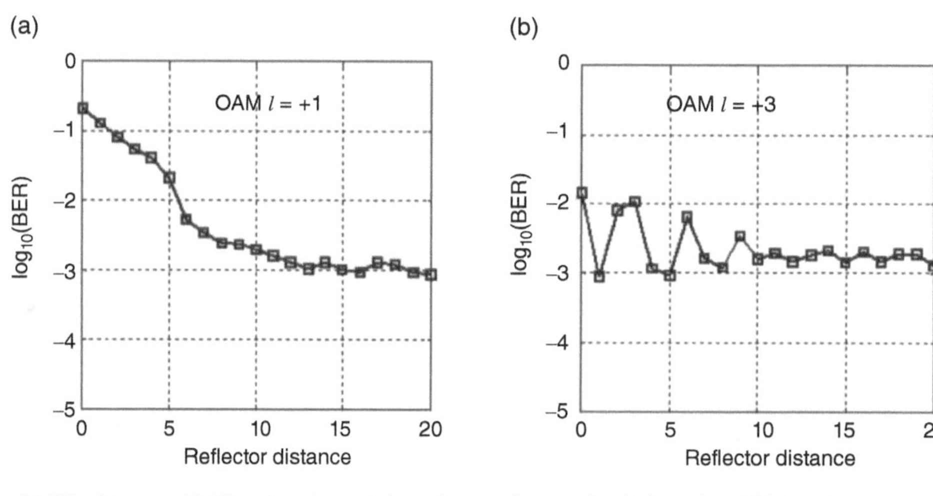

Figure 11.20 Measured BER as functions of the reflector distance for (a) $l_2 = +1$ and (b) $l_2 = +3$ when $l_1 = +1$ and $l_1 = +3$ are multiplexed (extracted from [22]). *Source:* Yan et al. [22]. (© [2015] IEEE)

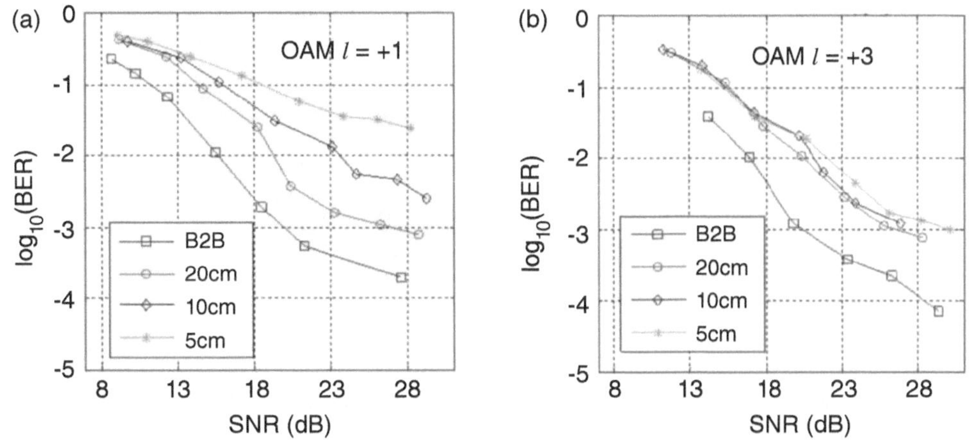

Figure 11.21 Measured BERs as functions of SNR at different reflector distances when l_1 = +1 and l_1 = +3 are multiplexed for (a) l_2 = +1 and (b) l_2 = +3 [22] (where B2B refers to the case without reflector). *Source:* Yan et al. [22]. (© [2015] IEEE)

for OAM channel +3, the intra-channel interference affects the BER performance more significantly than the inter-channel interference does.

The measured BERs as functions of SNR at different reflector distances are shown in Figure 11.21. (Note that the typos in the original figure in [22] are corrected). Since the inter-channel interference of OAM channel +3 has significant effects on the OAM channel +1, the BER of the OAM channel +1 decreases as the reflector distance decreases. For OAM channel +3, the BER is insensitive to reflector distance. This is in agreement with the findings that the intra-channel interference affects more than the inter-channel interference.

Since the divergence angle increases with increasing OAM mode order, more power will be reflected for higher OAM mode. As a result, the OAM channel with a higher OAM mode suffers more intra-channel interference and causes more inter-channel interference to other OAM modes.

The multipath effects caused by an infinite reflector (i.e. the ground plane) are also investigated through extensive simulations [27], where the LoS link is not always parallel to the ground plane. As expected, the infinite ground plane introduced interferences to the OAM channel of the LoS path, and the interferences become much more significant as the OAM transmitter gets closer to the ground and the OAM mode grows larger. It is found that the effects on the OAM communication system strongly depend on the orientation of the OAM beam. Moreover, the strongest and weakest communication links can be found based on the amplitudes of the compound electric field. More details about the simulations and results can be found in [27].

11.3.2 Indoor Environment

The intra- and inter-channel interferences caused by multipath have been analyzed in the presence of specular reflection. To further investigate the multipath effects on OAM multiplexing, more realistic cases of the indoor environment are introduced and analyzed.

11.3.2.1 Inter-Symbol Interference (ISI)

In a multipath environment, the received OAM signal is composed of multiple signals with different propagation delays. When the delays are much smaller than the symbol duration, the time

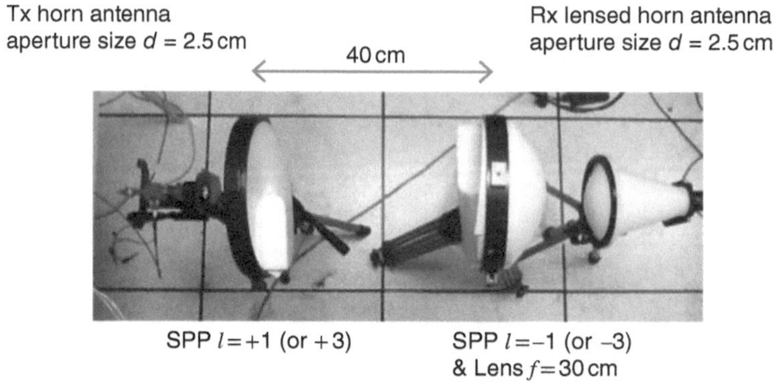

Figure 11.22 Measurement setup to investigate the frequency-selective OAM channel [28]. *Source:* Yan et al. [28].

dispersion of the channel can be neglected and the channel can be treated as a flat-fading channel. Otherwise (i.e. when the delays are comparable or larger than the symbol duration), there exists ISI and the channel becomes frequency-selective. The flat-fading channel has been discussed in Section 11.3.1, while the frequency-selective fading channel is the focus of this section.

Figure 11.22 shows the measurement setup for investigating the frequency-selective OAM channel. SPP technique is used to generate and detect the OAM waves at 28 GHz. A horn antenna with an aperture size of 2.5 cm is adopted as the transmitting antenna, and a horn lensed antenna with an aperture diameter of 15 cm is adopted as the receiving antenna. The distance between the two SPPs is 40cm. Due to the limitation of the measurement equipment, the transmitter and the receiver can generate and detect up to 2 Gbaud QPSK signal, or OFDM signal with 4 subchannels (i.e. 0.5 Gbaud QPSK signal each subchannel).

To confirm the ISI caused by the multipath, impulse response of the OAM channels is measured for OAM modes $l = 0, +1$, and $+3$. The impulses are observed using a real-time oscilloscope and the corresponding results are shown in Figure 11.23. As can be seen, only one pulse can be observed when $l = 0$; several pulses with lower amplitude following the main pulse can be observed when $l = +1$; for $l = +3$, there is a second pulse with relatively high amplitude following the main pulse, and the delay difference is about 15 cm (0.5 ns). We can conclude that the ISI is more serious for a higher OAM mode.

Since the multipath delay is about 0.5 ns, ISI should be observed when the baud rate is equal to or larger than 1 Gbaud. Figure 11.24 shows the EVM of the QPSK signal as a function of baud rate when $l = 0, +1$, and $+3$. The EVM varies mildly and the BER stays zero for all baud rates when $l = 0$. This is because the multipath effect is not significant for $l = 0$. The EVM and BER increase obviously as the baud rate increases when $l = +1$ and $+3$. This is because the ISI becomes severe when the symbol duration becomes shorter as the baud rate increases. In addition, the frequency selective fading can be observed for OAM channels +1 and +3 from Figure 11.25.

OFDM technique is commonly used to cope with frequency-selective fading, by converting a single frequency-selective fading channel into several flat-fading channels. Moreover, ISI can be mitigated by combining it with the cyclic prefix (CP) [3, 28, 29]. In order to show the performance of OFDM over OAM channels, three scenarios (i.e. a single QPSK channel, 4 OFDM subchannels

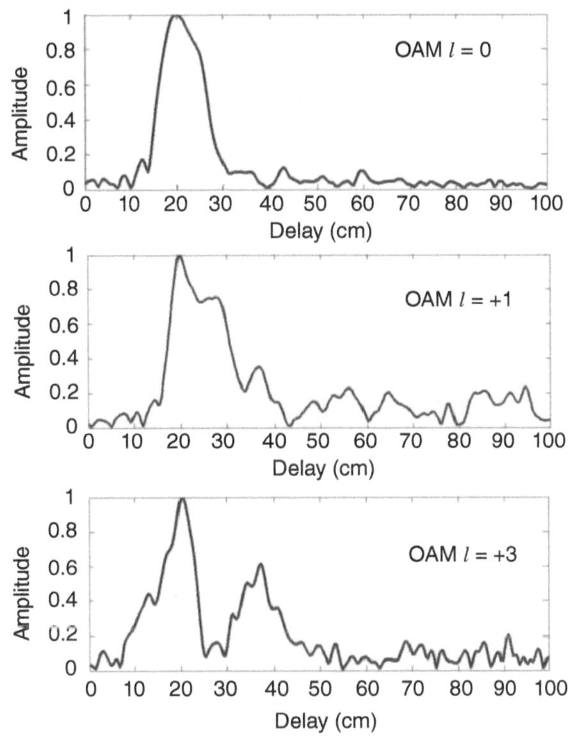

Figure 11.23 Impulse response of OAM channels when l = 0, +1, and +3 [28]. *Source:* Yan et al. [28]. (© [2016] IEEE)

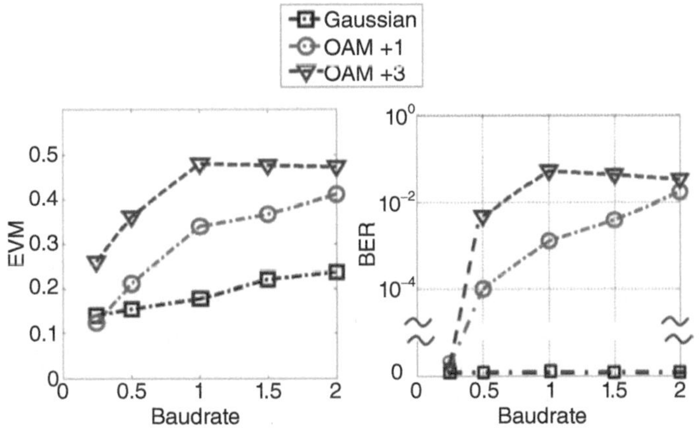

Figure 11.24 EVM and BER of the QPSK signal as a function of baud rate when l = 0, +1, and +3 [28]. *Source:* Yan et al. [28]. (© [2016] IEEE)

without CP, and 4 OFDM subchannels with 25% CP) are measured for OAM channels $l = 0, +1$, and $+3$.

It has been concluded that OAM channel $l = 0$ suffers little multipath effect from the previous discussion, thus OFDM has little improvements without doubt [28]. The measured EVM and BER of three scenarios for OAM channels $l = +1$ and $+3$ are shown in Figures 11.26 and 11.27. Compared with the single QPSK channel, the measured EVM and BER are improved when OFDM is

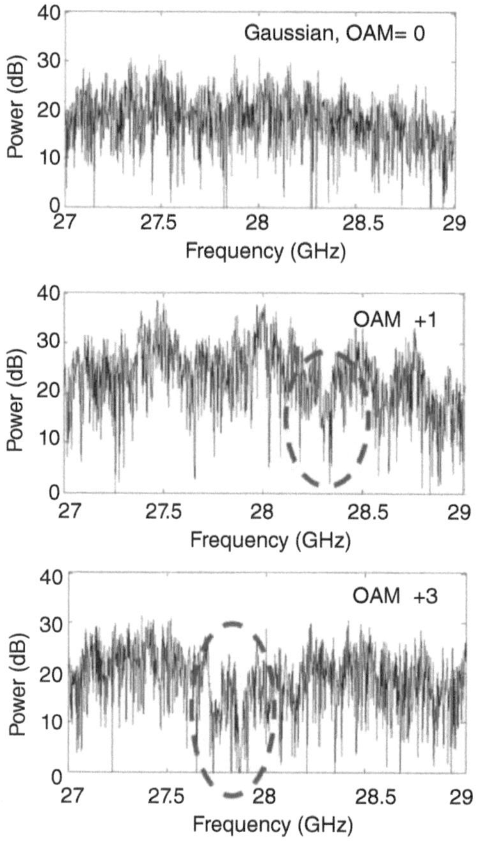

Figure 11.25 Spectra of the OAM channels $l = 0$, +1, and +3 when 2 Gbaud QPSK is used [28]. *Source:* Yan et al. [28]. (© [2016] IEEE)

used. Moreover, adding CP provides further improvements. Comparing Figures 11.26 and 11.27, it can be seen that OFDM and CP provide more improvement for OAM channel +3 than OAM channel +1. It should be noted that the improvements are different for different subchannels and baud rates, this is because different subchannels have different degrees of frequency-selectivity at different baud rates. The spectra of OFDM channels at different baud rates of $l = +3$ can be found in [28].

The OAM channel with higher OAM mode suffers more ISI and frequency-selectivity due to the divergence of OAM waves. Combining OFDM with CP can mitigate the effects of ISI and frequency-selectivity. However, it should be noted that if the subchannels of OFDM are too small, the improvements cannot be guaranteed.

11.3.2.2 Antenna misalignment

To investigate the effect of the antenna misalignment and the beamwidth in the indoor environment, further measurements in lecture room and corridor are conducted [30].

Figure 11.28 shows the measurement configuration in the lecture room. The lecture room has a size of 15.87 × 16.33 × 3.86 m. The transmitting and the receiving antennas are oriented toward each other and located at a height of 1.44 m. The uniform circular array antennas are adopted to generate and detect the OAM waves at 5.8 GHz. The transmitter generates an OAM wave of mode +1. The receiver detects OAM wave of modes ±1 along three parallel paths (i.e. paths 0, 1, and 2). Obviously, path 1 is the aligned path, while the other two paths are misaligned paths.

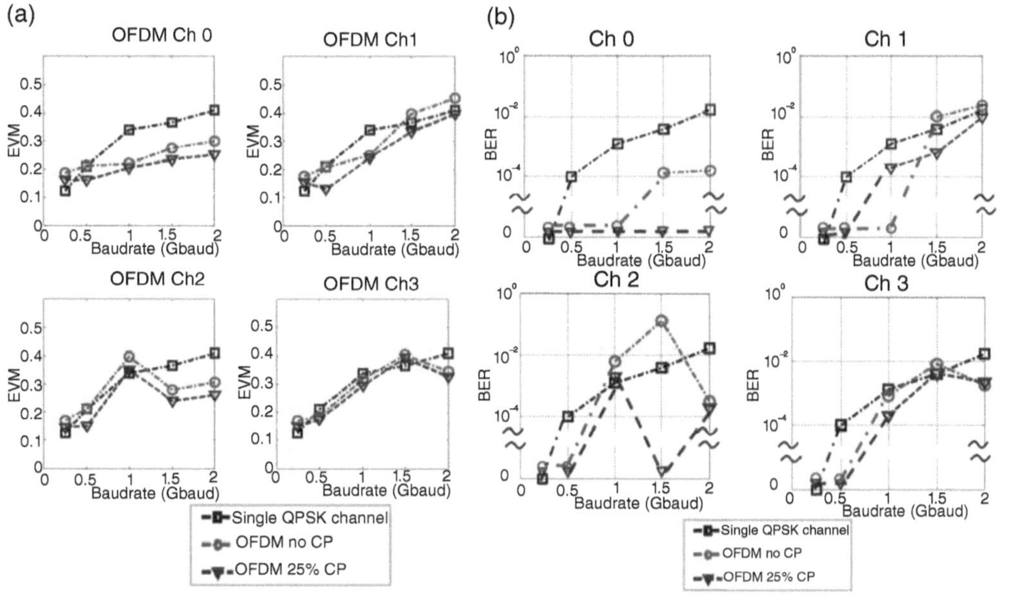

Figure 11.26 Measured (a) EVM and (b) BER of three scenarios (i.e. a single QPSK channel, 4 OFDM subchannels without CP, and 4 OFDM subchannels with 25% CP) for OAM channels $l = +1$ [28]. *Source:* Yan et al. [28]. (© [2016] IEEE)

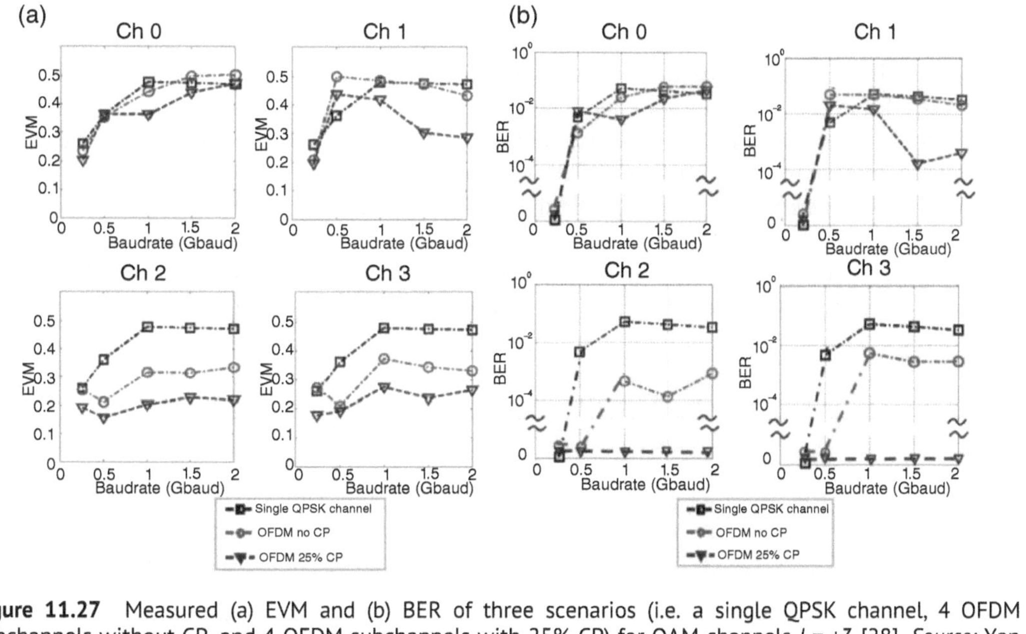

Figure 11.27 Measured (a) EVM and (b) BER of three scenarios (i.e. a single QPSK channel, 4 OFDM subchannels without CP, and 4 OFDM subchannels with 25% CP) for OAM channels $l = +3$ [28]. *Source:* Yan et al. [28]. (© [2016] IEEE)

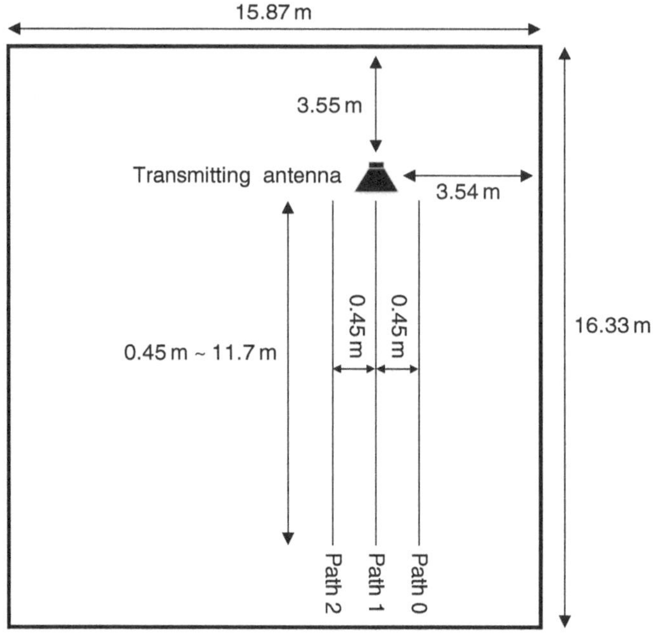

Figure 11.28 Measurement configurations in a lecture room [30]. *Source:* Boseok et al. [30]. Licenced under CC BY-4.0.

Measurement results are shown in [30]. For path 1, the received power of mode +1 is larger than that of mode −1 when the distance between the transmitting and receiving antennas is smaller than 1.8 m. However, the received powers of modes ±1 are indistinguishable for paths 0 and 2 regardless of the distance. Obviously, the antenna misalignment has strong effects on the OAM channels in a multipath environment. Even for path 1, the received power of mode +1 decreases as the distance increases and becomes indistinguishable with the received power of mode −1 when the distance is larger than 1.8 m. The results are in agreement with that of the LoS case, that is, the OAM communication is not suitable for long-distance communication. Due to the interference caused by scattering, the situation is worse for a multipath environment.

Figure 11.29 shows measurement configuration in a corridor. The corridor has a size of 60 × 2.51 × 2.57 m. The transmitting and the receiving antennas are oriented toward each other and located at a height of 1.29 m. Analogous to the lecture room scenario, path 1 is aligned, while the other two paths are misaligned. The signal of mode +1 is transmitted and mode ±1 is detected along

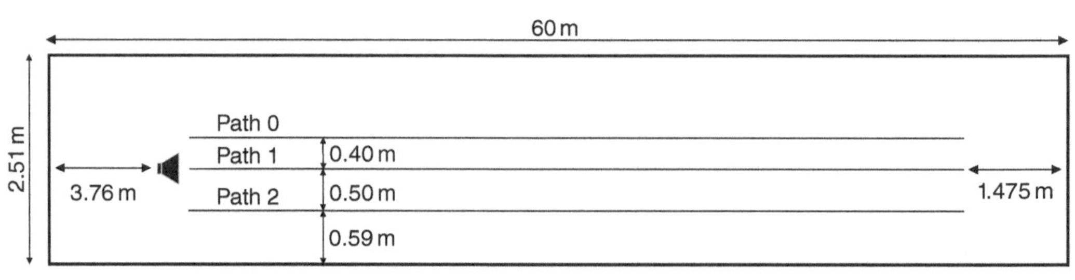

Figure 11.29 Measurement configurations in a corridor [30]. *Source:* Boseok et al. [30].

the three paths. Measurement results are shown in [30]. The received powers of modes ±1 are also indistinguishable for paths 0 and 2. Similarly, the antenna misalignment has strong effects on the OAM channels. Moreover, the received power of mode +1 becomes indistinguishable from the received power of mode −1 when the distance is larger than 1.2 m for path 1. Compared with the results of the lecture room, the distance for a valid OAM communication is smaller in the corridor. For the same propagation distance, more power will be reflected in a narrower space due to the divergence of the OAM waves. Hence, the multipath effects are more significant in a narrower space.

The distance for viable OAM communication in a multipath environment is restricted by reflections, thus reducing the reflected power is necessary for increasing the coverage of (conventional) OAM communication. As antenna beamwidth can be specified by field of view (FOV) angles of the radiation pattern, simulations of different FOV angles (i.e. 3, 5, and 10°) are conducted. The results show that the coverage of the OAM communication increases as the FOV angles decreases. For the same propagation distance, less power will be reflected for a smaller FOV angle. In other words, a smaller FOV angle results in a larger coverage range.

In the multipath indoor environment, the OAM communication links will be totally destroyed by antenna misalignment. The communication link can only be established along the aligned (LoS) path. However, the coverage of the OAM communication is restricted by the space of the room due to reflections. For LoS communication in the multipath environment, decreasing the beamwidth is an effective method to increase the coverage range.

11.3.3 Highly Reverberant Environments

We have investigated the OAM communication system in the specular reflection and indoor environments, both situations indicate the multipath effects destroy the orthogonality of the OAM waves and induce serious interference. Moreover, it is shown that the communication link can only be established along the aligned (LoS) path over a certain distance in the multipath environment. Different with the previous sections, we study the OAM communication system in a rich multipath (highly reverberant) environment using the example of a reverberation chamber [34] in this section.

The reverberation chamber (RC) is an electrically large conducting cavity with arbitrarily shaped metallic mode stirrers (as depicted in Figure 11.30), which can change the boundary conditions of the electromagnetic field (stir the electromagnetic modes) in the RC through mechanical rotations [31]. From the view of statistics, the stirred electromagnetic field in the RC is random,

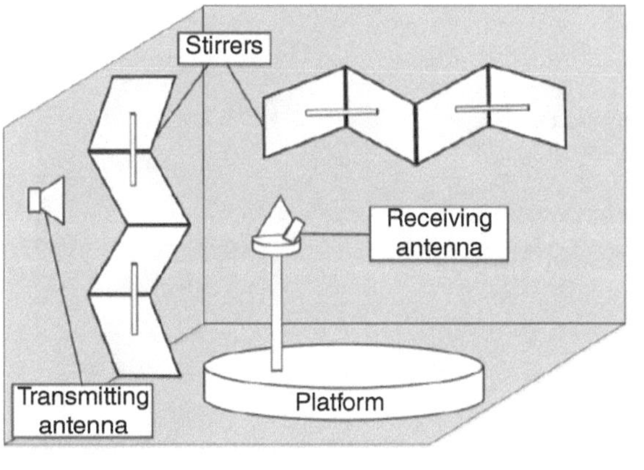

Figure 11.30 An illustration of the reverberation chamber.

homogeneous, and isotropic. To stir the electromagnetic field more efficiently, source stirring (e.g., antenna stirring, polarization stirring, and platform stirring) and frequency stirring techniques are commonly used combining with mechanical stirring.

RC is initially used for electromagnetic compatibility tests, such as measurements of radiated immunity of components or systems, radiated emissions, shielding effectiveness (SE) of materials, or shielded enclosures. Due to RC's particular properties (e.g. low cost, high efficiency, and good repeatability), it has been extended to various over-the-air (OTA) tests, such as measurements of antenna efficiency, free space reflection coefficient, diversity gain, capacity of MIMO antennas, total radiated power (TRP), total isotropic sensitivity (TIS), and throughput of wireless devices [31].

Since the RC measurements are usually studied from the view of statistics, it is important to obtain enough independent sample numbers. A larger independent sample number means smaller measurement uncertainty. Essentially, all the stirring techniques are aiming at increasing the number of independent samples. Very recently, OAM has been applied as a new source stirring technique to measurement uncertainty reduction [32]. There exist several methods to evaluate the measurement uncertainty of the RC, including the standard deviation (STD) method, the independent sample number method, and the Rician K-factor-based method [33].

Different from EMC/EMI tests, the OTA tests usually have specific requirements on the channel characteristics (e.g. delay spread, Doppler spread, and K-factor) in the RC. The delay spread (coherence bandwidth) can be decreased (increased) by loading the RC with absorbers. The Doppler spread (coherence time) can be controlled by varying the rotation speed of the stirrers or platform. The K-factor can be controlled by loading the RC with absorbers or adjusting the LoS component using one-directional antenna or both antennas.

A 2 × 2 OAM-based MIMO system is established in an RC (as shown in Figure 11.31) and the measurements are conducted from 5.0 GHz to 5.2 GHz. The RC has a size of 1.5 × 1.44 × 0.92 m^3, and contains two mechanical stirrers (one vertical and one horizontal) and one turn-table platform. The slot antennas designed for mobile phone applications are used as receiving antennas [35]. As can be seen from Figure 11.31, there are eight slot antennas actually. However, we selected two farthest separated slot antennas to a minimum the mutual coupling and the crosstalk between the receiving antennas. Two horn antennas are used as transmitting antennas and located on the form supporters. The metasurface technique is adopted to generate OAM waves [36]. There are two OAM

Figure 11.31 Measurement setup of the OAM-based 2 × 2 MIMO system in the RC [34]. *Source:* Chen et al. [34].

metasurfaces and each metasurface can generate two OAM modes (i.e. $l = \pm 1$ or $l = \pm 2$). The two metasurfaces are placed on the top of the two horn antennas.

It is worth noting that the OAM metasurfaces are not utilized in the receiving antennas. This is quite different from the conventional OAM communication system, which adopts OAM-based antennas at both transmitting and receiving sides. The orthogonality of OAM waves will be totally destroyed in the highly reverberant environment, thus there is no need to use OAM-based antennas at the receiving side.

Due to the limitation of the measurement equipment, we can only determine one subchannel of the MIMO system from a single measurement. We need to conduct the measurement four times to determine all the subchannels. During the measurement of all the subchannels, the two stirrers move step-wise and simultaneously to 20 angles. At each stirrer position, the platform moves step-wise to 20 angles. The stirring positions (i.e. the 20 angles) of both stirrers and platform are evenly distributed over one complete rotation. At each stirring position, all the S-parameters are recorded by a vector network analyzer (VNA). Hence, we have 400 samples for each subchannel. To keep the scattering environment is the same for each subchannel, the initial positions of the stirrer and the platform should be the same.

To compare the OAM-based MIMO system (OAM multiplexing) with the nonOAM MIMO system (conventional spatial multiplexing), the whole measurement procedure is repeated when the OAM metasurfaces are not utilized to the transmitting antennas. Note that the OAM metasurfaces are still placed in the RC to keep the loading condition of the RC the same for both measurements of OAM-based and nonOAM MIMO systems. In the latter case, we have a conventional 2×2 MIMO system. For the conciseness, the measurement setup of this case in not shown here.

The ergodic capacity of an $M_T \times M_R$ MIMO system can be calculated using [37]:

$$C(f) = E\left\{ \log_2 \left[\det\left(\mathbf{I} + \frac{\gamma}{M_T} \mathbf{H}(f)\mathbf{H}^H(f) \right) \right] \right\}, \tag{11.16}$$

where $\mathbf{H}(f)$ denotes the channel transfer function of the $M_T \times M_R$ MIMO system at frequency f, the superscript ^{H} denotes conjugate transpose, $\mathbf{I}$ is the identity matrix, γ denotes the reference SNR, det is the determinant operator, log is the logarithm operator, and E denotes the expectation over the random channel $\mathbf{H}$.

For the cases here, $M_T = M_R = 2$ and $C(f)$ is calculated using 400 samples (cf. Section 11.3.3.2). In a well-stirred RC, $\mathbf{H}(f)$ follows the Gaussian distribution with a zero mean. In other words, the channel is a Rayleigh fading channel in the well-stirred RC. The measured channel transfer function matrix is normalized by the square root of the average power transfer function of the reference measurement (i.e. $\mathbf{H} = \frac{\mathbf{H}_{\text{measured}}}{\sqrt{P_{\text{ref}}}}$). Hence, $\mathbf{H}(f)$ in Eq. (11.16) is the normalized channel transfer function actually.

To investigate the effects of the rich multipath only, all the other factors are removed. Especially, the insertion loss of the OAM metasurfaces is measured separately and calibrated out. Figure 11.32 shows the measured ergodic capacities of the 2×2 OAM-based and nonOAM MIMO systems when $\gamma = 15$ dB. The OAM orthogonality is totally destroyed in the highly reverberant environment. However, the capacities of the OAM-based and nonOAM MIMO systems are almost the same over the whole testing bandwidth. This means the information carried by the OAM waves is not lost in the highly reverberant environment. This is a rather surprising result and quite different from the previous findings. From the view of ray physics, the original transmitted wave can be treated as a single ray, however, the received waves are a superposition of a large number of rays due to the scattering of the stirrers and the walls during the propagation. Due to the random scattering in the RC, the

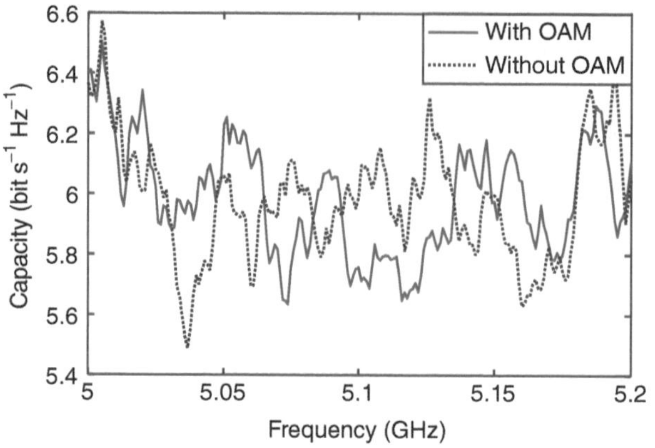

Figure 11.32 Ergodic capacities of the 2 × 2 OAM-based and nonOAM MIMO systems at 15 –dB reference SNR when γ = 15 dB [34]. *Source:* Chen et al. [34]. (© [2020] IEEE)

channel transfer function can be expressed as a random matrix **H**. The amplitude and phase of the element of **H** will be changed when the OAM metasurfaces are used. If the imperfect designs are neglected, the changes can be expressed as a block-diagonal and unitary matrix **M** [38]. Thus, the random matrix **H** is redistributed and transformed into a new random matrix **HM**. It has been proved that $E[\mathbf{HH}^H] = E[\mathbf{HM}(\mathbf{HM})^H]$ in [38], which explains the measurement results. From another view point, as mentioned in the previous section, OAM waves of different modes can be used to increase the independent sample number. This indicates the OAM waves can offer independent spatial samples, although the orthogonality is destroyed. In other words, the independence of the OAM waves exists regardless of the wavefronts in the RC. This also helps to explain the comparable capacities. The small discrepancy between the two capacities is mainly because the performance of the nonideal OAM metasurface is different at different frequencies. Even though the capacities are comparable, the interferences are inevitable due to the received signal is a mix of two transmitted signals. The equalization technique can be used to detect signals by diagonalizing the channel matrix **H**.

As discussed in Section 11.3.2, OFDM and CP techniques can mitigate the effect of frequency-selective fading on OAM channels effectively when each OFDM subchannel is flat-fading. Hence, further simulations are conducted for OAM-based and nonOAM MIMO–OFDM systems. Supposing the bandwidth of the channel is 200 MHz (i.e. 5.0–5.2 GHz). The coherence bandwidth of the channel in the RC is about 2 MHz. Obviously, the number of the OFDM subchannels should be larger than 100 to ensure each OFDM subchannel is flat-fading. To be safe, 512 OFDM subchannels are adopted in the simulation. The length of the CP is set to 128 (i.e. 640 ns), which is sufficiently larger than the delay spread of the channel. A 64-QAM signal is transmitted through the OFDM subchannels. Two independent data streams are sent simultaneously and convolved with the channel impulse response. In this way, the received signals are obtained. At the receiving side, the zero-forcing equalizer is adapted to detect the transmitted data streams [1]. The simulated uncoded BER performances of the OAM-based and nonOAM MIMO–OFDM systems are shown in Figure 11.33. (The BER here is calculated as the ratio of the number of falsely detected bits to that of the totally transmitted bits.) It can be seen that the BER of the OAM-based and nonOAM MIMO–OFDM systems are indistinguishable, which indicates the OAM multiplexing has the same performance as the conventional spatial multiplexing. We use a 2 × 2 MIMO system to transmit two data streams,

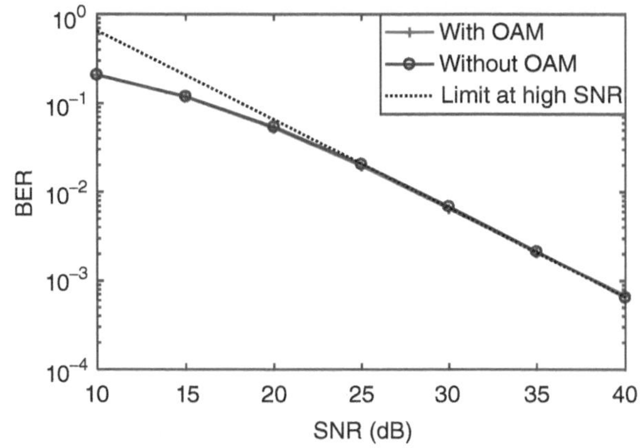

Figure 11.33 Uncoded BER performances of the 2 × 2 OAM-based and nonOAM MIMO–OFDM systems [34]. *Source:* Chen et al. [34]. (© [2020] IEEE)

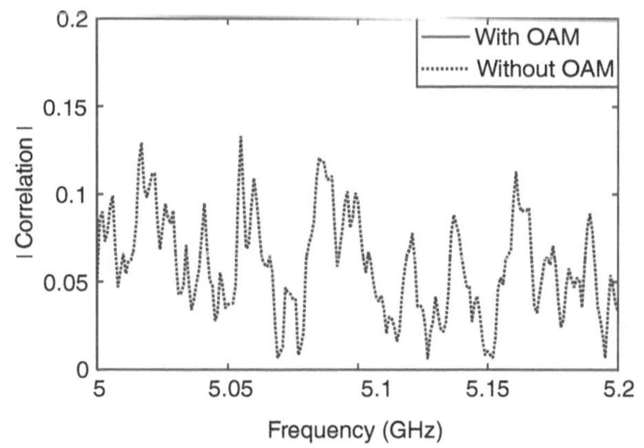

Figure 11.34 Correlation magnitudes between the transmitted signals of the 2 × 2 OAM-based and nonOAM MIMO systems [34]. *Source:* Chen et al. [34]. (© [2020] IEEE)

thus there is no spatial diversity (i.e. the diversity order is one). In this case, the limitation of the BER is about $K\gamma^{-1}$ [1] for high SNR, where K is a constant and γ represents the SNR. As can be seen, the simulated BERs are below the limitation as expected, and become indistinguishable with the limitation when $\gamma > 20$ dB.

Figure 11.34 shows the correlation magnitudes between the two transmitted signals of the OAM-based and the nonOAM MIMO systems. It can be seen that the correlations are independent with frequency. Combining Figures 11.32 and 11.34, a relatively higher correlation corresponds to a lower capacity in the frequency domain. Overall, the correlations fluctuate in a rather small range, which is in accordance with the identical BER performance. Note that the spatial channel in the RC follows the Kronecker structure, that is the correlations at the receiving side are independent with that at the transmitting side [39]. The correlations at the receiving side are basically smaller than 0.1 for both the OAM-based and the nonOAM MIMO systems. As discussed in previous sections, the conventional OAM multiplexing depends on the orthogonality of the OAM waves. Moreover, the SNR of the channel is limited due to the interference, which makes the modulation cannot be higher than 16-QAM. However, the measurements and simulations in a highly reverberant environment prove that 64-QAM can be supported for OAM multiplexing when combined with zero-forcing equalization. This is because the zero-forcing equalizer can mitigate the interference caused by both multipath and intrinsic crosstalk between different OAM modes. Nevertheless, compared with conventional OAM-based detection, the spatial equalizer results in higher computation complexity, especially for a large number of data streams.

The OAM multiplexing in a highly reverberant environment is investigated in this section. It is shown that the OAM multiplexing combined with zero-forcing equalization and OFDM could not only cope with the interference caused by rich multipath but also support higher-order modulation (e.g., 64-QAM). Moreover, the OAM multiplexing system exhibits comparable performance to the conventional MIMO system. All in all, the OAM multiplexing can also be used in a multipath environment.

11.4 Conclusion

In this chapter, the OAM multiplexing technique in both LoS and multipath environments has been discussed. Corresponding techniques to mitigate the interference caused by multiplexing are also introduced. The OAM multiplexing technique performs well in short-range LoS situation, especially when proper MIMO-based signal processing methods (e.g. equalization) are used to mitigate the interference caused by multiple channels. Moreover, the OAM multiplexing technique and conventional spatial multiplexing technique can be used simultaneously to provide design flexibility (i.e. increasing the capacity with reasonable space consumption). However, the divergence property of the OAM waves leads to obvious performance degradation of the OAM multiplexing system over long-distance LoS communication. In the multipath environment, the scattering of the transmitted OAM wave will cause strong inter- and intra-channel interferences, and the influences are more significant for higher OAM modes. The OAM multiplexing in a rich multipath environment is also investigated at the end of this chapter. Different from previous researches on OAM multiplexing in multipath environment, it is proved that the OAM multiplexing technique combined with proper equalization (e.g. zero-forcing) and OFDM could not only cope with the interference caused by rich multipath but also support higher-order modulation (e.g. 64-QAM). Moreover, the OAM multiplexing system exhibits comparable performance to the conventional MIMO system in a rich multipath environment. Overall, the OAM multiplexing can also be used in a multipath environment and, therefore, should not be confined to LoS scenarios only.

References

1 Paulraj, A., Nabar, R., and Gore, D. (2003). *Introduction to Space-Time Wireless Communications*. Cambridge: Cambridge University Press.

2 Balanis, C.A. (2005). *Antenna Theory: Analysis and Design*. New York: Wiley-Interscience.

3 Goldsmith, A. (2005). *Wireless Communications*. Cambridge: Cambridge University Press.

4 Tse, D. and Viswanath, P. (2005). *Fundamentals of Wireless Communication*. Cambridge: Cambridge University Press.

5 Grimmett, G. and Stirzaker, D. (2001). *Probability and Random Processes.* New York: Oxford University Press.

6 Yan, Y., Xie, G., Lavery, M.P.J. et al. (2014). High-capacity millimetre-wave communications with orbital angular momentum multiplexing. *Nature Communications* **5**: 4876.

7 Zhang, Z., Zheng, S., Zhang, W. et al. (2017). Experimental demonstration of the capacity gain of plane spiral OAM-based MIMO system. *IEEE Microwave and Wireless Components Letters* **27** (8): 757–759.

8 Zhang, Z., Zheng, S., Chen, Y. et al. (2016). The capacity gain of orbital angular momentum based multiple-input-multiple-output system. *Scientific Report* **6**: 25418.

9 Ren, Y., Li, L., Xie, G. et al. (2017). Line-of-sight millimeter-wave communications using orbital angular momentum multiplexing combined with conventional spatial multiplexing. *IEEE Transaction on Wireless Communications* **16** (5): 3151–3161.

10 Zhao, Y. and Zhang, C. (2019). Compound angular lens for radio orbital angular momentum coaxial separation and convergence. *IEEE Antennas and Wireless Propagation Letters* **18** (10): 2160–2164.

11 Tamburini, F., Mari, E., Thide, B. et al. (2011). Experimental verification of photon angular momentum and vorticity with radio techniques. *Applied Physics Letters* **99**: 204102.

12 Guo, Z. and Yang, G. (2017). Radial uniform circular antenna array for dual-Mode OAM communication. *IEEE Antennas and Wireless Propagation Letters* **16**: 404–407.

13 Mohammadi, S.M., Daldorff, L.K.S., Bergman, J.E.S. et al. (2010). Orbital angular momentum in radio—a system study. *IEEE Transactions on Antennas and Propagation* **58** (2): 565–572.

14 Shi, H., Wang, L., Peng, G. et al. (2019). Generation of multiple modes microwave vortex beams using active metasurface. *IEEE Antennas and Wireless Propagation Letters* **18** (1): 59–63.

15 Chen, M., Jiang, L., and Sha, W.E.I. (2017). Ultrathin complementary metasurface for orbital angular momentum generation at microwave frequencies. *IEEE Transactions on Antennas and Propagation* **65** (1): 396–400.

16 Andersson, M., Berglind, E., and Björk, G. (2015). Orbital angular momentum modes do not increase the channel capacity in communication links. *New Journal of Physics* **17**: 1–7.

17 Edfors, O. and Johansson, A. (2012). Is orbital angular momentum (OAM) based radio communication an unexploited area? *IEEE Transactions on Antennas and Propagation* **60** (2): 1126–1131.

18 Yao, E., Franke-Arnold, S., Courtial, J. et al. (2006). Fourier relationship between angular position and optical orbital angular momentum. *Optics Express* **14** (20): 9071–9076.

19 Zheng, S., Hui, X., Zhu, J. et al. (2015). Orbital angular momentum mode demultiplexing scheme with partial angular receiving aperture. *Optics Express* **23** (9): 12251–12257.

20 Zhang, W., Zheng, S., Chen, Y. et al. (2016). Orbital angular momentum-based communications with partial arc sampling receiving. *IEEE Communications Letters* **20** (7): 1381–1384.

21 Richter, T., Palushani, E., Schmidt-Langhorst, C. et al. (2012). Transmission of single-channel 16-QAM data signals at terabaud symbol rates. *IEEE Journal of Lightwave Technology* **30** (4): 504–511.

22 Yan, Y., Li, L., Xie, G. et al. (2015). Experimental measurements of multipath-induced intra- and inter-channel crosstalk effects in a millimeter-wave communications link using orbital-angular-momentum multiplexing. *IEEE International Conference on Communication (ICC)*, London. 1370–1375.

23 Zhao, Y. and Zhang, C. (2019). Compound angular lends for radio orbital angular momentum coaxial separation and convergence. *IEEE Antennas and Wireless Propagation Letters* **18** (10): 2160–2164.

24 Yang, J., Werner, J.-J., and Dumont, G.A. (2002). The multimodulus blind equalization and its generalized algorithms. *IEEE Journal on Selected Areas in Communications* **20** (5): 997–1015.

25 Yuan, J.-T. and Tsai, K.-D. (2005). Analysis of the multimodulus blind equalization algorithm in QAM communication systems. *IEEE Transaction on Communications* **53** (9): 1427–1431.
26 Georgiadis, A. (2004). Gain, phase imbalance, and phase noise effects on error vector magnitude. *IEEE Transactions on Vehicular Technology* **53** (2): 443–449.
27 Wang, L., Park, W., Yang, C. et al. (2020). Wireless communication of radio waves carrying orbital angular momentum (OAM) above an infinite ground plane. *IEEE Transaction on Electromagnetic Compatibility* **62** (5): 2257–2264.
28 Yan, Y., Li, L., Xie, G. et al. (2016). OFDM over mm-wave OAM channels in a multipath environment with intersymbol interference. *IEEE Global Communications Conference (GLOBECOM)*, Washington, DC. 1–6.
29 Chen, R., Yang, W., Xu, H., and Li, J. (2018). A 2-D FFT-based transceiver architecture for OAM-OFDM systems with UCA antennas. *IEEE Transactions on Vehicular Technology* **67** (6): 5481–5485.
30 Boseok, J., Hayeon, K., and Haengseon, L. (2018). Indoor propagation of electromagnetic waves with orbital angular momentum at 5.8GHz. *International Journal of Antennas and Propagation* **2018**: 1–9.
31 Chen, X., Tang, J., Li, T. et al. (2018). Reverberation chambers for over-the-air tests: an overview of two decades of research. *IEEE Access* **6**: 49129–49143.
32 Chen, X., Xue, W., Shi, H. et al. (2019). Improving field uniformity using source stirring with orbital angular momentum modes in a reverberation chamber. *IEEE Microwave and Wireless Components Letters* **29** (8): 560–562.
33 Chen, X. (2019). Measurement uncertainty of RC and its reduction techniques for OTA tests: a review. *IET Microwaves, Antennas & Propagation* **13** (15): 2598–2604.
34 Chen, X., Xue, W., Shi, H. et al. (2020). Orbital angular momentum multiplexing in highly reverberant environments. *IEEE Microwave and Wireless Components Letters* **30** (1): 112–115.
35 Li, J., Zhang, X., Wang, Z. et al. (2019). Dual-band eight-antenna array design for MIMO applications in 5G mobile terminals. *IEEE Access* **7**: 71636–71644.
36 Shi, H., Wang, L., and Zhao, M. (2018). Transparent metasurface for generating microwave vortex beams with cross-polarization conversion. *Materials* **11** (12): 1–10.
37 Foschini, C.J. (1996). Layered space-time architecture for wireless communication in a fading environment when using multi-element antennas. *Bell Labs Technical Journal* **1**: 41–59.
38 Chen, M., Jiang, L., and Sha, W.E.I. (2017). Ultrathin complementary metasurface for orbital angular momentum generation at microwave frequencies. *IEEE Transactions on Antennas and Propagation* **65** (1): 396–400.
39 Chen, X. (2014). Throughput modeling and measurement in an isotropic-scattering reverberation chamber. *IEEE Transactions on Antennas Propagation* **62** (4): 2130–2139.

12

High-capacity Free-space Optical Communications Using Multiplexing of Multiple OAM Beams

Alan E. Willner,[1] Runzhou Zhang,[1] Kai Pang,[1] Haoqian Song,[1] Cong Liu,[1] Hao Song,[1] Xinzhou Su,[1] Huibin Zhou,[1] Nanzhe Hu,[1] Zhe Zhao,[1] Guodong Xie,[1] Yongxiong Ren,[1] Hao Huang[1], and Moshe Tur[2]

[1] *Department of Electrical and Computer Engineering, University of Southern California, Los Angeles, CA, USA*
[2] *School of Electrical Engineering, Tel Aviv University, Ramat Aviv, Israel*

12.1 Introduction

There has been interest in utilizing free-space optical (FSO) link to increase the capacity for line-of-sight (LOS) wireless data transmission [1]. Due to the higher carrier frequency of the lightwave, FSO communications generally hold the promise of having both higher capacity and lower probability of intercept (LPI) than radio-frequency (RF) and millimeter-wave techniques [2]. Moreover, the FSO system can be potentially utilized in various scenarios, including atmospheric air [3] and underwater environment [4]; for different platforms, including fixed point-to-point links [5], and manned and unmanned aerial platforms [6].

A typical method of increasing the transmission capacity in FSO optical communication systems is the multiplexing of multiple independent data channels. For example, multiple independent data channels can be located on different wavelengths, polarizations, or spatial channels, corresponding to wavelength-division multiplexing (WDM) [5, 7], polarization-division multiplexing (PDM) [8], and space-division multiplexing (SDM) [9, 10], respectively.

As shown in Figure 12.1, one special case of SDM is the utilization of orthogonal spatially overlapping and copropagating spatial modes, known as mode-division-multiplexing (MDM) [9]. In such a case, each mode can carry an independent data channel, and the orthogonality enables efficient (de)multiplexing among multiple modes with low inherent inter-modal crosstalk [9]. There are several different types of orthogonal modal basis sets that are potential candidates for such MDM systems [9, 10]. One such set is the orbital angular momentum (OAM) mode [12, 13].

As shown in Figure 12.2, at the transmitter (Tx), each OAM beam carries an independent data stream, and multiple OAM beams are then multiplexed, spatially overlapped, and co-propagated in the free space. At the receiver (Rx), different OAM beams can be demultiplexed with low crosstalk and the carried data channel can be decoded using coherent detection. The system capacity and spectral efficiency can be increased N-fold by multiplexing N OAM modes. Furthermore, the utilization of OAM multiplexing is compatible with the WDM and PDM [11], giving rise to the potential of an FSO communications system with a high capacity > Tbit/s [15, 16].

Due to the unique spatial phase structure of the OAM beams, there are certain degrading effects in and on different environments and platforms which produces challenges to the deployment of

Electromagnetic Vortices: Wave Phenomena and Engineering Applications, First Edition.
Edited by Zhi Hao Jiang and Douglas H. Werner.

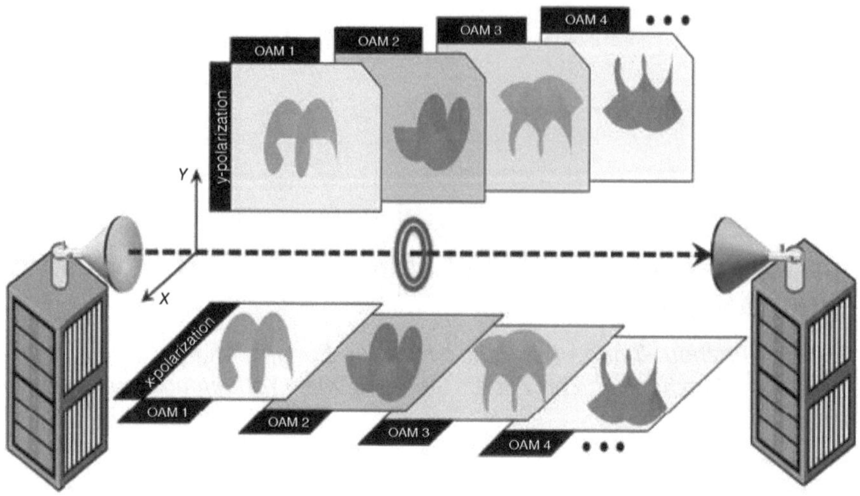

Figure 12.1 Concept of a building-to-building communications employing OAM multiplexing. *Source:* Yan et al. [11], © 2014 Macmillan Publishers.

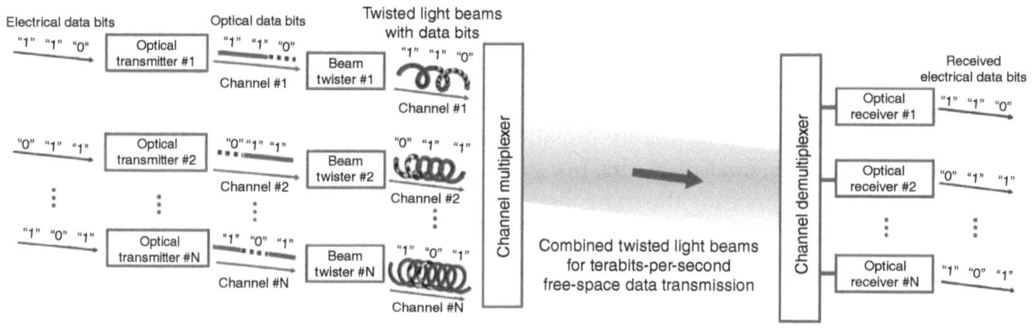

Figure 12.2 Conceptual schematic of a communication system employing OAM multiplexing. *Source:* Willner et al. [14], © 2015 Optical Society of America.

OAM systems. With respect to the OAM-multiplexed FSO data transmission, in either atmospheric or underwater environments, the inhomogeneous distribution of refractive index (i.e. turbulence effects) causes phase distortion to the wavefront of the OAM beams and induces inter-channel crosstalk in the multiplexed data channels. For special application scenarios, such as aerial platforms, the precise alignment between the Tx and the Rx is required to ensure low crosstalk between the transmitted OAM beams.

In this chapter, various degrading effects for the OAM systems will be discussed in Section 12.2. We will introduce the early proof-of-concept demonstrations of OAM multiplexing systems without severe environmental degrading effects in Section 12.3. The demonstrated crosstalk mitigation approaches for OAM systems will be discussed in Section 12.4. Moreover, we will include the experimental demonstrations of OAM-based data transmission for the aerial platform and for the underwater environment in Sections 12.5 and 12.6, respectively.

12.2 Challenges for an OAM Multiplexing Free-space Optical Communication System

OAM multiplexing can help increase the capacity and spectral efficiency of a communication system. However, as shown in Figure 12.3, there are various potential challenges in designing such an OAM-based multiplexing system, especially in terms of longer distance, and several important issues need to be carefully considered, including beam divergence and misalignment which might induce power loss and crosstalk in long-distance OAM multiplexed systems [17]. Moreover, in a practical FSO communication system, atmospheric turbulence is another critical challenge, which can distort the phase front of the light beam [18, 19]. This could be more important for OAM communications, since the efficient demultiplexing of OAM beams relies on their helical phase front. In addition, the beam path can be partially blocked by obstructions which can also distort the helical phase front [20].

A simulation model was employed to investigate the effects of the beam divergence and misalignment on the performance of the OAM multiplexing system as shown in Figure 12.4a [17]. In this model, independent data streams are carried by different Gaussian beams at the same wavelength and a spiral phase plate is then utilized to convert one of the Gaussian beams to a data-carrying OAM beam. Different OAM beams are multiplexed and coaxially transmitted through free space and the propagation of these OAM beams is numerically calculated using the Kirchhoff–Fresnel

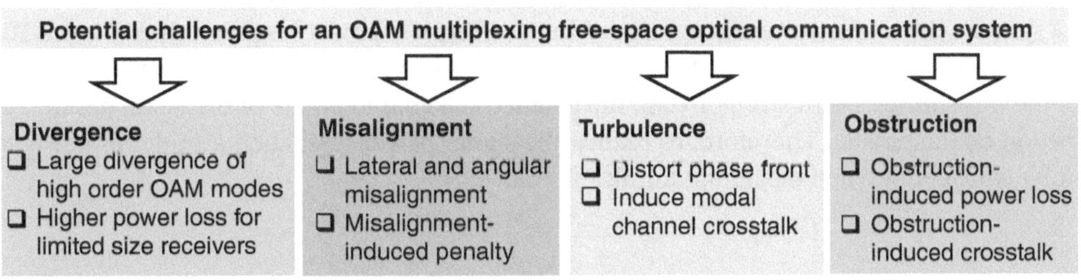

Figure 12.3 Potential challenges for an OAM multiplexing FSO communication link.

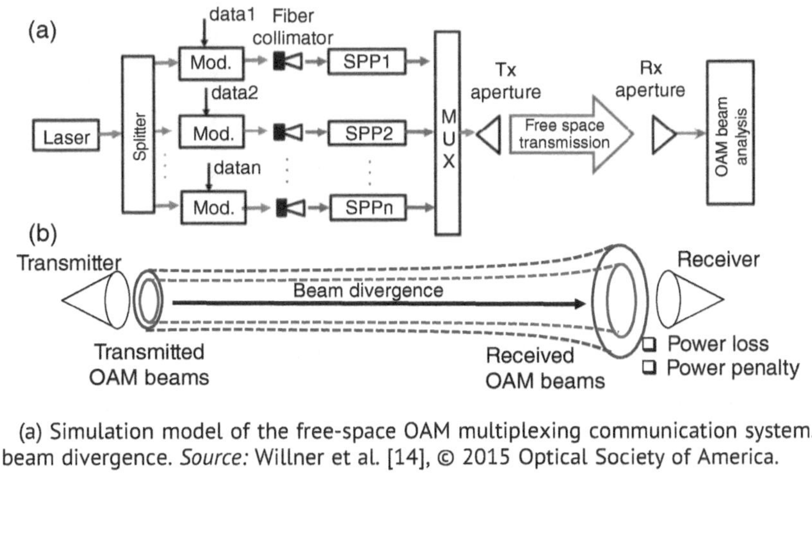

Figure 12.4 (a) Simulation model of the free-space OAM multiplexing communication system. (b) Concept of the OAM beam divergence. *Source:* Willner et al. [14], © 2015 Optical Society of America.

diffraction integral [21]. At the receiver side, the power loss and crosstalk are analyzed by studying the power distribution among the different OAM modes.

12.2.1 Beam divergence

For a communication link, it is generally preferred to collect as much signal power as possible at the receiver to ensure a sufficient signal-to-noise ratio (SNR). Based on the diffraction theory, a collimated OAM beam diverges while propagating in free space. Since available optical elements usually have limited-size apertures, the system, therefore, suffers power loss when the diverged OAM beams become too large at the receiver, as shown in Figure 12.4b.

Figure 12.5a presents the beam divergence of different OAM beams with the same beam size of 3 cm. As expected, the spot size at the receiver increases with the transmission distance. In addition, an OAM beam with a higher-order has a larger received beam size over a given distance. Besides the OAM mode order and transmission distance, the received beam size is also related to the transmitted beam size. As shown in Figure 12.5b, when the link distance is 100 m, there is a minimum received beam size for each OAM beam when varying the transmitted beam size. On the other hand, the aperture size of a practical system might be limited due to the component cost. As a result, the diverged beam might be too large to be fully collected, thereby inducing signal power loss. Figure 12.5c shows the power loss of the OAM 3 with various receiver aperture sizes. Given a fixed transmitted beam size, the power loss of a 1 km link is higher than that of a 100-m link for the same transmitted beam size due to the larger beam divergence. Moreover, it is also noted that in a 100-m link, a system with a transmitted beam size of 3 cm suffers less power loss than that of 1 and 10 cm. This is because when the transmitted beam size is 1 cm, it diverges faster than 3-cm beam. However, when the transmitted beam size is 10 cm, the geometrical characteristic of the beam dominates its diffraction characteristic. Therefore, to reduce the signal power loss, such a trade-off needs to be carefully considered when designing a link.

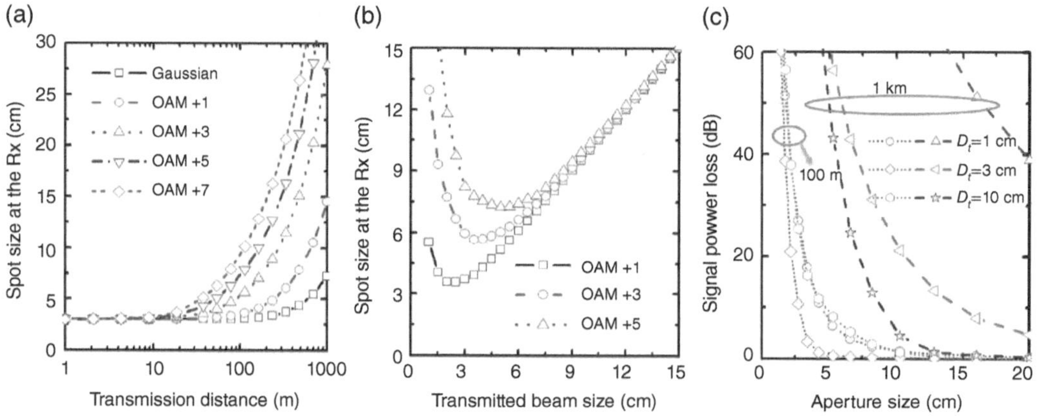

Figure 12.5 Simulation results. (a) Simulated spot size (diameter) of different orders of OAM beams as a function of transmission distance when the spot size of the transmitted beam is 3 cm. (b) Simulated spot size of different orders of OAM beams as a function of the transmitted beam size for a 100-m link. (c) Simulated power loss as a function of aperture size when only $\ell = 3$ beam is transmitted; Dt: transmitted beam size. *Source:* Willner et al. [14], © 2015 Optical Society of America.

12.2.2 Misalignment

The efficient multiplexing and de-multiplexing of OAM beams are enabled by the orthogonality between beams with different OAM mode orders. This orthogonality is ensured when all OAM beams are coaxially propagated and the receiver is perfectly aligned on-axis with the beams, as shown in Figure 12.6a1. However, due to the jitter and vibration of the transmitter/receiver platform, there might be some misalignments between the transmitter and the receiver, including lateral displacement and angular error, as shown in Figure 12.6b1, b2, respectively. As a result, these misalignments would degrade the orthogonality due to a phase mismatch and cause power to be coupled into other modes, i.e., severe intermodal crosstalk, as shown in Figure 12.6a2.

Figure 12.7 presents the effects of the lateral displacement and the receiver angular error when only OAM +3 is transmitted with a beam size of 3 cm. Given a fixed link distance of 100 m, as the lateral displacement or the angular error increases, the power coupling into other modes increases, whereas the power on OAM +3 decreases. This is because a larger displacement causes a larger mismatch between the received OAM beams and the receiver. The power coupled into OAM +2 and OAM +4 is greater than that of OAM +1 and OAM +5 due to their smaller mode spacing with respect to OAM +3. This indicates that a system with larger mode spacing is more tolerant to the lateral displacement and the angular error. One of the important concerns is the ratio of the power coupled to the nearest mode (here, OAM +4 is used) to the power on the desired mode (i.e., OAM +3), which is named relative crosstalk XT-1. Figure 12.7c and d shows the relative crosstalk XT-1 as a function of the lateral displacement or the angular error with various transmission distances and transmitted beams sizes. The results show that a larger beam size at the receiver will result in two opposing effects: (i) a smaller lateral displacement-induced crosstalk because the differential phase change per unit area is smaller and (ii) a larger tilt phase error-induced crosstalk because the phase error scales with a larger optical path delay.

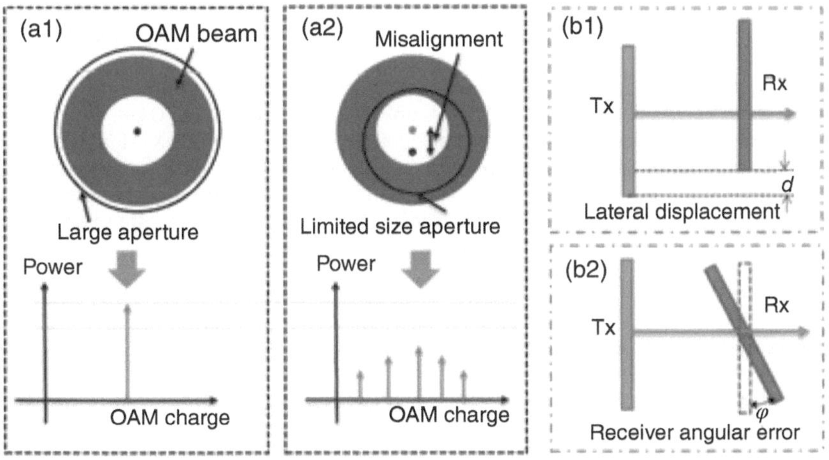

Figure 12.6 (a) Two different cases of alignment between the transmitter and receiver: (a1) a perfectly aligned system, (a2) a system with misalignment and a limited-size aperture; (b) two types of misalignments: (b1) lateral displacement and (b2) receiver angular error. Tx, transmitter; Rx, receiver. *Source:* Willner et al. [22], © 2017 the Royal Society.

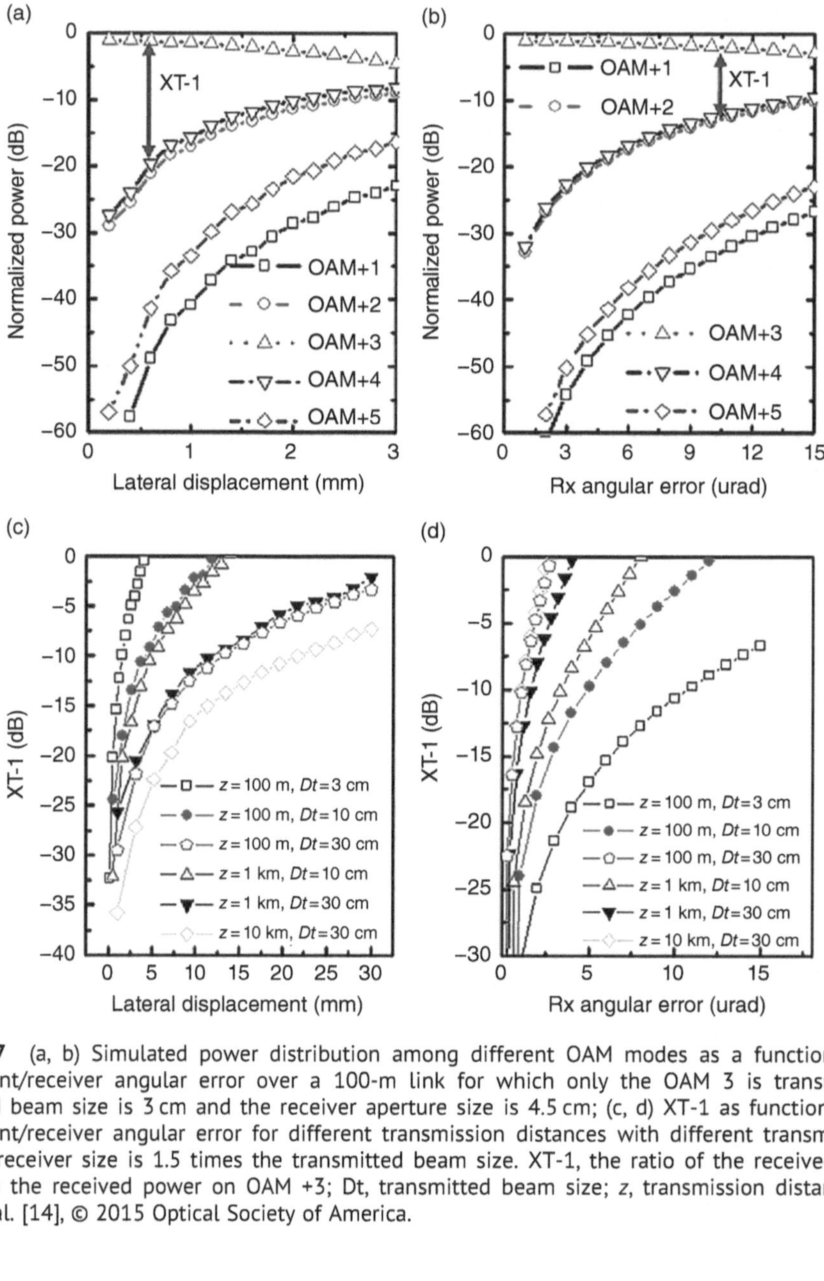

Figure 12.7 (a, b) Simulated power distribution among different OAM modes as a function of lateral displacement/receiver angular error over a 100-m link for which only the OAM 3 is transmitted. The transmitted beam size is 3 cm and the receiver aperture size is 4.5 cm; (c, d) XT-1 as functions of lateral displacement/receiver angular error for different transmission distances with different transmitted beam sizes. The receiver size is 1.5 times the transmitted beam size. XT-1, the ratio of the received power on OAM +4 to the received power on OAM +3; Dt, transmitted beam size; *z*, transmission distance. *Source:* Willner et al. [14], © 2015 Optical Society of America.

12.2.3 Atmospheric Turbulence Effects

In addition to the beam divergence and the misalignment, atmospheric turbulence is another important issue that needs to be considered for an OAM multiplexed communication system. It is known that inhomogeneities in the temperature and pressure of the atmosphere can lead to variations of the refractive index along the transmission path [23, 24]. In an OAM multiplexed link, the orthogonality among multiple copropagating OAM beams depends on the helical phase front. The turbulence-induced refractive index inhomogeneities can easily distort the phase front of OAM

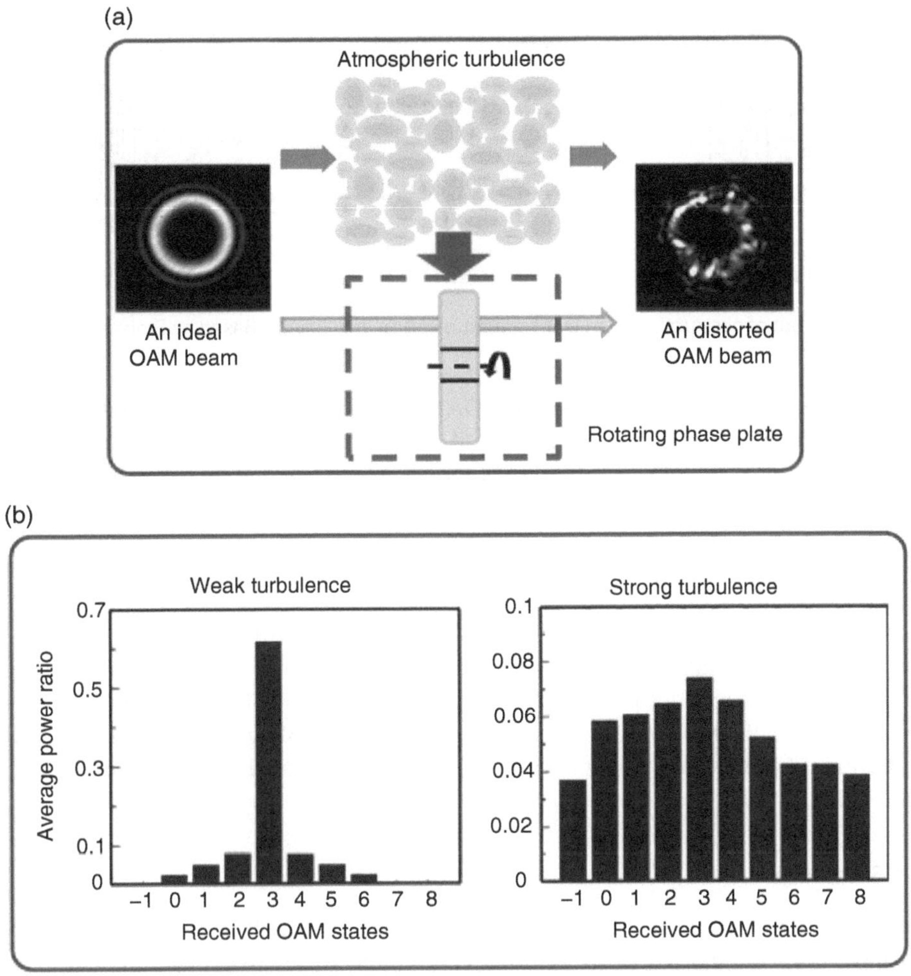

Figure 12.8 (a) A rotating phase plate is used as an atmospheric turbulence emulator. An ideal OAM beam emerges as a distorted OAM after passing through the phase plate; (b) measured power distribution of an OAM beam after passing through the turbulence emulator with different strengths. *Source:* Willner et al. [14], © 2013 Optical Society of America.

beams, thereby causing both degradation of a single OAM channel and intermodal crosstalk between different data channels with different OAM orders, as shown in Figure 12.8a [25–27].

The effects of atmospheric turbulence on the performance of OAM multiplexed systems have been experimentally evaluated in the laboratory by several ways [19, 24]. Figure 12.8a presents the concept of using a rotating phase plate to emulate the turbulence effects. The phase screen plate is mounted on a rotation stage and placed in the middle of the optical path. On the rotating plate, the pseudorandom phase distribution obeys Kolmogorov spectrum statistics. The strength of the emulated turbulence is generally dependent on Fried coherence length r_0, and the beam size that is incident on the plate. In order to evaluate the turbulence effects, the modal crosstalk is characterized by measuring the power of the distorted beam in each OAM mode. Figure 12.8b presents the normalized power distribution among the neighboring OAM modes under weak or strong turbulences when an OAM +3 beam is transmitted. It is shown that under the weak turbulence, the

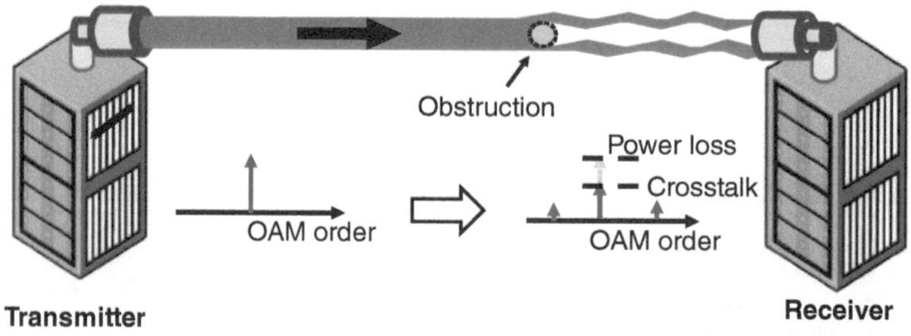

Figure 12.9 Concept of obstruction in an OAM communication link. *Source:* Based on Franke-Arnold et al. [20].

majority of the power is still in the transmitted OAM mode (i.e., OAM +3), and only a small part of power is coupled into other neighboring OAM modes. However, as the turbulence strength increases, the power coupling into other OAM modes would become higher, which could induce severe signal fading and crosstalk.

12.2.4 Obstruction

FSO communication links rely on the LOS operation, and so obstructions in the beam path are another potential issue, particularly for OAM communications since completely or partially blocking an OAM beam will produce phase distortions in its phase front. This would reduce the orthogonality between the different OAM beams and cause power coupling from the desired OAM mode to neighboring OAM modes, thereby leading to signal fading and channel crosstalk, as shown in Figure 12.9 [20].

12.2.5 Summary

In summary, different potential challenges in designing OAM-multiplexed FSO communication systems have been discussed in this section. The critical issues that might induce power loss and crosstalk include: (i) beam divergence; (ii) misalignment between transmitter and receiver; (iii) atmospheric turbulence; and (iv) obstruction. Various trade-offs and compensation methods should be considered when designing real OAM multiplexed FSO links. For example, adaptive optics (AO) could be used to mitigate the turbulence effect. In addition, some special beams (e.g., Bessel beams) might help overcome the effects of beam obstruction.

12.3 Free-space Optical OAM Links

The first demonstration of using OAM modes in FSO communications was achieved in 2004 [11]. In this experiment, the OAM modes were utilized for data encoding by transmitting different OAM beams at a time, and eight different OAM modes represented eight different data symbols. Later, it was realized that OAM beams could also be used to carry different data streams [12, 28, 29]. This OAM-based MDM technique is compatible with PDM and WDM which could further increase the capacity and the spectral efficiency of a communication system [12]. In recent years, the use of OAM multiplexing for MDM communication links has been experimentally demonstrated under laboratory conditions and in the field [12, 30]. Through combining OAM-based MDM, PDM, and

WDM, the aggregate data rate of an FSO communication link which was demonstrated in the laboratory experiments has reached 100 Tbit/s and beyond [15, 16]. Meanwhile, significant efforts have also been made to achieve practical OAM FSO links over a longer distance in the field. Based on the OAM-based encoding scheme, the transmission of the OAM FSO link was demonstrated over a distance up to 143 km between two islands [31].

12.3.1 High-capacity OAM Multiplexed Communication Link Under Laboratory Conditions

Initial high-capacity MDM communication links based on OAM multiplexing were demonstrated in the laboratory. In 2012, a terabit per second FSO link was demonstrated by combining OAM multiplexing with polarization multiplexing. Figure 12.10a and b present the basic concept of generation and recovery of an information carrying OAM beam in their scheme, respectively. Specifically, a spiral phase plate (e.g., $\ell = +4$) converts a planar phase front into a helical one, resulting in the generation of an information-carrying OAM beam from an information-carrying Gaussian beam. Conversely, an inverse spiral phase mask (i.e., $\ell = -4$) recovers an information-carrying beam with a planar phase front from an information-carrying OAM beam with a helical phase front. In their experiment, specific spiral phase plates were achieved by adding different spiral phase masks using reflective liquid-crystal-based spatial light modulators (SLMs).

In order to achieve a terabit per second FSO link, four different OAM beams ($\ell = +4, +8, -8$, and $+16$) and two polarization states (in total eight channels) were used in their scheme. This could result in multiplexing of eight independent channels of data information on the same wavelength, which provides an eightfold improvement in capacity and spectral efficiency. Figure 12.11a1–a4 shows the experimental intensity profiles of the four polarization-multiplexed (pol-muxed) OAM beams with 'doughnut' shapes and Figure 12.11a5 presents the intensity profile for four superposed pol-muxed OAM beams. Interferograms were measured to indicate the charge of the OAM beams (Figure 12.11b1–b4), from which it is verified that the four pol-muxed OAM beams for multiplexing are OAM +4, OAM +8, OAM −8, and OAM +16, respectively. Figure 12.11c presents the measured spectra for demultiplexing (OAM +4, OAM +16) from four pol-muxed OAM beams (eight channels in total), with each channel carrying a 42.8 × 4 Gbit/s 16-ary quadrature amplitude modulation (16-QAM) signal. A power suppression of 30 dB was achieved at a frequency offset of 50 GHz from the center. A spectral efficiency of 25.6 bit.s^{-1}.Hz^{-1} was therefore obtained for

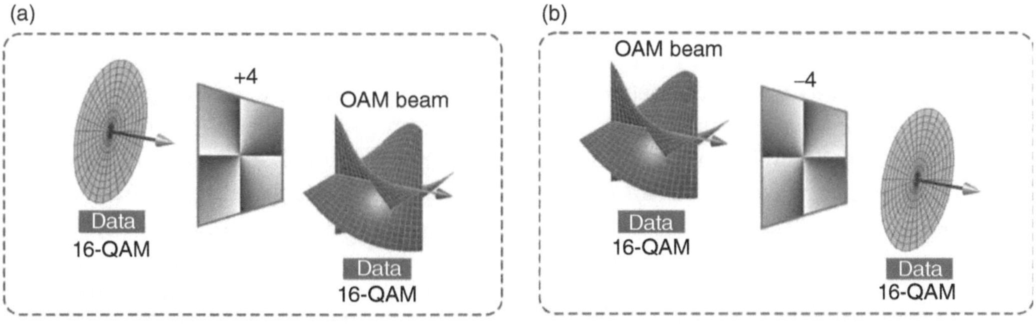

Figure 12.10 Concept of (a) generation of an information-carrying OAM beam with a helical phase front. (b) Recovery of an information-carrying beam with a planar phase front. *Source:* Wang et al. [12], © 2012 Macmillan Publishers.

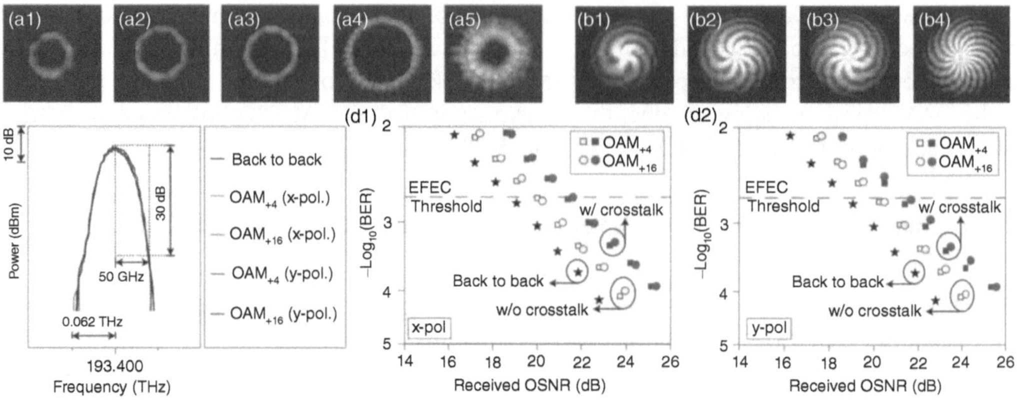

Figure 12.11 Experimental results of 16-QAM signals over pol-muxed OAM beams. (a1–a5) Observed intensity profiles of pol-muxed (a1) OAM +4, (a2) OAM −8, (a3) OAM +8, (a4) OAM +16 beams and (a5) their multiplexing; (b1–b4) measured interferograms corresponding to (b1) OAM +4, (b2) OAM +8, (b3) OAM −8 and (b4) OAM +16 beams; (c) measured spectra and (d1–d2) BER curves for demultiplexing of pol-muxed 42.8 × 4 Gbit/s 16-QAM-carrying OAM beams. *Source:* Wang et al. [12], © 2012 Macmillan Publishers.

a total capacity of 1369.6 ($42.8 \times 4 \times 4 \times 2$) Gbit/s (~1.4 Tbit/s). Figure 12.11d1–d2 plots the measured BER curves for the demultiplexing of pol-muxed OAM beams along the x- and y-polarizations without (x- or y-polarization of only one pol-muxed OAM beam is on) and with (all four pol-muxed OAM beams are on) crosstalk. OSNR penalties of < 1.5 dB without crosstalk and < 3.0 dB with crosstalk are observed at a BER of 2×10^{-3}.

The aforementioned experiment could be expanded by adding the wavelength dimension, resulting in a further increase of the capacity of the free-space data link. In 2014, a total capacity of 100 Tbit/s was achieved by combining OAM, polarization, and wavelength for multiplexing. Figure 12.12a presents the steps to generate 24 multiplexed OAM beams (12 OAM values and 2 polarizations). First, the SLM1 generated three OAM beams with $\ell = +4, +10, +16$, while the SLM2 generated another three OAM beams with $\ell = +7, +13, +19$. Then these two outputs were multiplexed together using a beam splitter (BS). Second, the multiplexed six OAM beams were split into two copies. One copy was reflected five times (three times by the mirrors and twice by the two BSs) to create another six OAM beams with inverse charges ($\ell = -4, -7, -10, -13, -16, -19$). These OAM beams were combined again to achieve the multiplexing of 12 OAM beams ($\ell = \pm 4, \pm 7, \pm 10, \pm 13, \pm 16, \pm 19$). Third, by using another BS, the superposition of the 12 OAM beams was split again into two copies. One copy was polarization-rotated by 90°, delayed by 33 symbols, and then recombined the other copy. Finally, the multiplexing of 24 OAM beams (with $\ell = \pm 4, \pm 7, \pm 10, \pm 13, \pm 16, \pm 19$ on two polarizations) was achieved. Each of the OAM beams carried a WDM signal comprising 100 GHz spaced 42 wavelengths (1536.34–1568.5 nm). Figure 12.12b presents the observed optical spectrum of a single OAM beam ($\ell = +10$) that carries the WDM signal. As a result, a total of 1008 data channels were carried by 12 OAM values, 2 polarizations, and 42 wavelengths. Each channel was encoded with a 100 Gbit/s quadrature phase-shift keying (QPSK) data, in which the information was encoded on the phase of an optical wave. In total, this system has an aggregate capacity of 100.8 Tbit/s. At the receiver side, the authors adjusted the OAM demultiplexer, the polarizer, and the filter to demultiplex each of 1008 channels and measured the BER and OSNR, as shown in Figure 12.12c2. As a result, each individual channel achieved a BER below 3.8×10^{-3}.

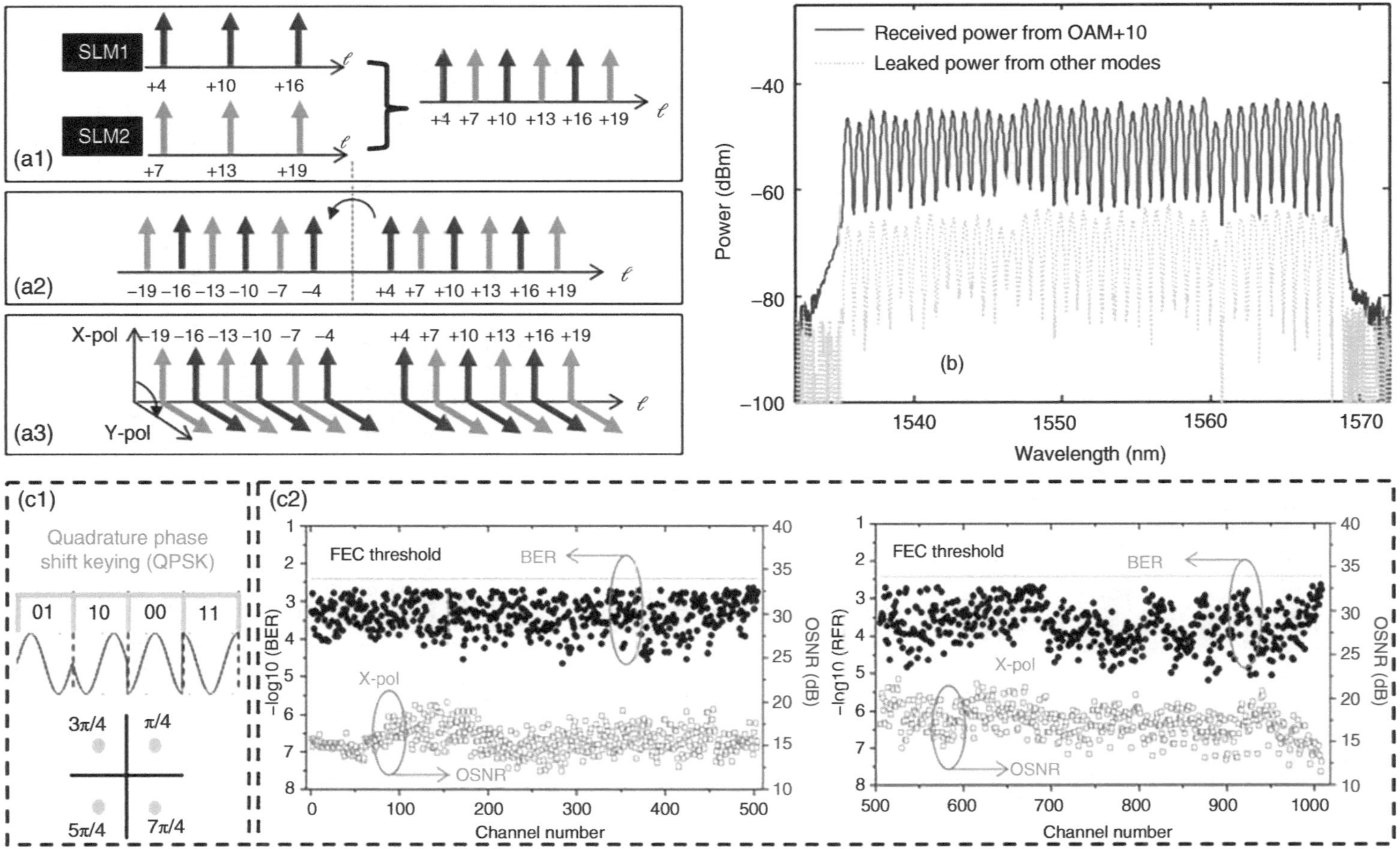

Figure 12.12 (a1–a3) Three steps to produce 24 multiplexed OAM beams using two SLMs, mirrors, beam splitters, a half-waveplate, and a PBS. *Source:* Willner et al. [32], ©2018 Optical Society of America; (b) the optical spectrum of the WDM signal carried on an OAM beam with $\ell = +10$; (c1) concept of quadrature phase-shift keying (QPSK) modulation; (c2) measured BER and OSNR for all 1008 channels (504 channels in x-pol and 504 channels in y-pol). *Source:* Willner et al. [32], © 2014 Optical Society of America.

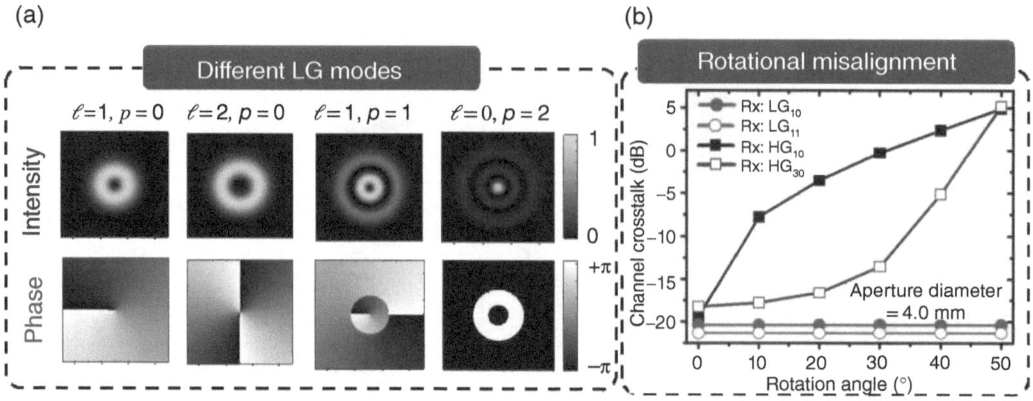

Figure 12.13 (a) The intensity profiles and phase fronts of LG beams with different ℓ values and p values; (b) the channel crosstalk of LG_{10}, LG_{11}, HG_{10}, and HG_{30} channels when all four LG modes or HG modes are transmitted with various rotational misalignments. *Source:* Pang et al. [35], ©2018 Optical Society of America.

One of the modal basis sets that carry OAM is the Laguerre–Gaussian (LG) modal group which can be characterized by (ℓ, p) referring to the azimuthal and radial indices, respectively [33, 34]. We note that OAM modes are considered as a subset of the full LG modal group. Figure 12.13a shows the intensity profiles and phase fronts of some LG modes with different azimuthal indices and radial indices. Many studies have focused on the multiplexing of data-carrying LG and LG-like beams with different ℓ values but the same p-value (p = 0) (i.e., OAM beams) in MDM links [12, 28, 29]. Besides the azimuthal index ℓ, the radial index p could also be utilized to provide more communication LG modes as well as form a complete orthogonal basis in an MDM link. For example, a 200-Gbit/s MDM system was demonstrated in 2016 that used two LG beams with equal ℓ but different p values [35, 36]. Due to the orthogonality of the LG beams with different p values, it was found that the channels crosstalk could be minimized with proper design of the demultiplexing patterns and both channels could achieve a bit error rate of 3.8×10^{-3}. In 2018, a fourfold multiplexing of LG modes was demonstrated to achieve a 400 Gbit/s MDM link. In this experiment, both modal indices (ℓ, p) were simultaneously utilized to achieve a larger potential orthogonal modal space. Moreover, the authors also investigated the effects of the rotational misalignment on such an MDM system. It was observed that due to the azimuthal symmetry of the LG modes, the rotational misalignment does not tend to affect the channel crosstalk when compared with the case of Hermite–Gaussian (HG) modes, as shown in Figure 12.13b.

12.3.2 OAM-based FSO Link Beyond Laboratory Distances

The experimental demonstrations of OAM multiplexed communication links in the laboratory were generally conducted over short distances of ~1 m. However, when OAM beams are used in a long-distance FSO link outside the laboratory, the system performance might be degraded due to several potential issues, such as beam divergence, system misalignment, and atmospheric turbulence effects [37].

In recent years, there are several experiments that investigated the potential of using OAM beams to achieve long-distance FSO links under the field environment [30, 31]. In 2016, a 260-m FSO data transmission link using OAM multiplexing and 16-QAM signals was demonstrated between two buildings. In addition, a 400 Gbit/s FSO communication link based on OAM multiplexing was performed over 120 m in Los Angeles. The optical setup containing the transmitter and the receiver was placed on the roof of a building, as shown in Figure 12.14a. Four OAM modes with $\ell = \pm 1$, ± 3, each carrying a 100-Gbit/s QPSK data channel were transmitted, thus allowing for a total capacity of 400 Gbit/s. The transmitted OAM beams were reflected twice by two flat mirrors placed 30 m away, achieving a 120 m propagation path. The beam tip/tilt aberrations of the link and its effects on system performance were measured and characterized. Due to beam jitter, the received power and crosstalk for each channel fluctuate. Figure 12.14b shows the instantaneous BERs of channel $\ell = 3$ in 180 s at a fixed OSNR of ∼21 dB. The recovered QPSK constellations of OAM channel $\ell = 3$ under three different conditions are also shown in the inset. It can be seen that the instantaneous BERs fluctuate temporally and link outage occurs (BERs larger than FEC threshold) under large jitter conditions. Moreover, the maximal beam jitter was also investigated by intentionally adjusting the displacement of the received beams with respect to the receiver axis. It is found that as the beam displacement increases, the BER and crosstalk increase rapidly. In addition, a maximum beam displacement of ~2 mm can be tolerated to achieve BERs below 3.8×10^{-3}.

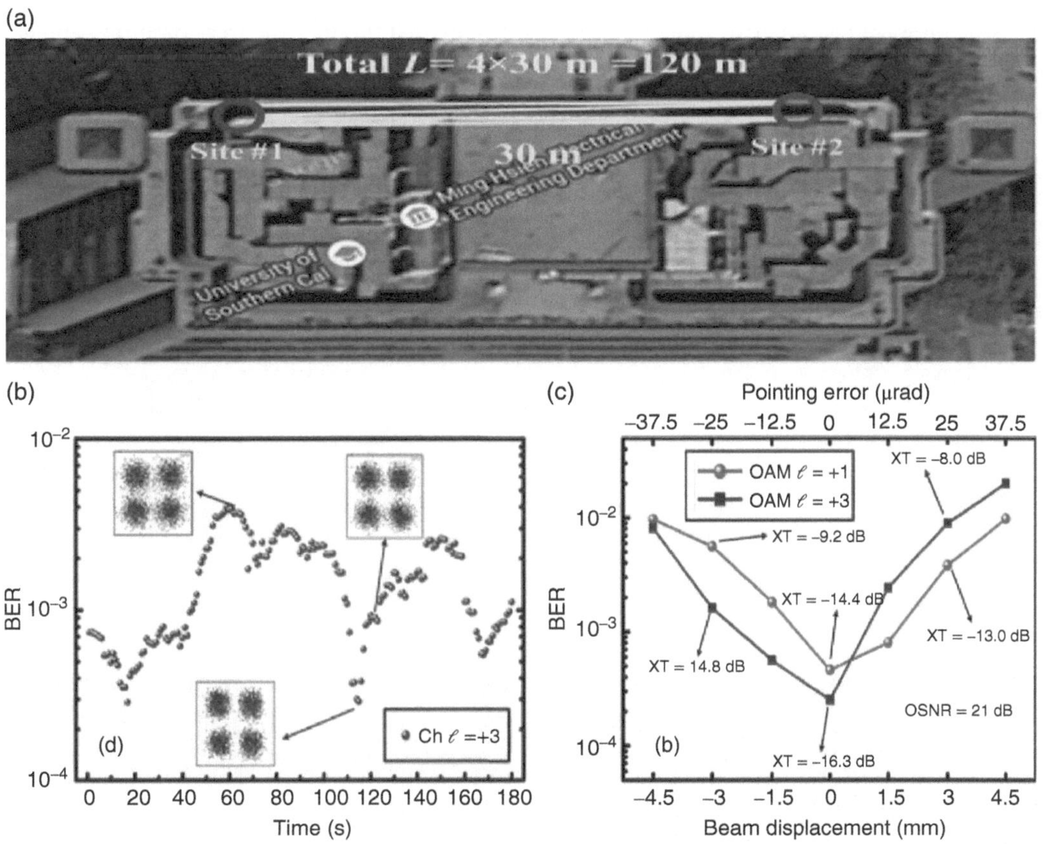

Figure 12.14 Experimental setup of a 120-m OAM-multiplexed link on a building roof. (a) Link layout on the building roof; (b) instantaneous BERs of channel $\ell = +3$ over 180 s at a fixed OSNR of 21 dB; (c) measured BERs and crosstalk for OAM channels $\ell = +1$ and $\ell = +3$ as a function of beam displacement; XT, crosstalk. *Source:* Ren et al. [30], ©2016 Optical Society of America.

Besides multiplexing techniques, data encoding is another way to use OAM beams for information transfer. In this approach, different OAM states can be encoded as different data symbols. At each time slot, only one of the OAM beams is transmitted, and a sequence of OAM states sent by the transmitter represents data information. At the receiver, in order to retrieve the information, the data can be decoded by checking the received OAM states. In 2016, such an OAM-based encoding link was demonstrated over a link distance of 143 km between the two Canary Islands of La Palma and Tenerife. Figure 12.15a shows the link layout and experimental setup for sending and receiving the spatial modes. At the sending station, a green laser with a wavelength of 532 nm and a power of 60 mW was used for data encoding. The output beam was modulated by a phase-only SLM with specific phase patterns and converted to the desired mode and the mode superposition. After being magnified by a telescope, the resulting beam propagated through the free space over 143 km. Finally, the received mode structure at the island of Tenerife was recorded by a regular camera. Examples of four OAM mode superpositions are shown in Figure 12.15b–e. Due to the relatively weak turbulence effects, the lobed modal structure was clearly visible for modes with $\ell = \pm 1$, ± 2, and ± 3. By introducing a relative phase of π in the $\ell = \pm 3$ mode superposition, the rotation of the mode structure was also clearly seen. To recover the encoded data, an artificial neural network-based pattern recognition algorithm was used to distinguish images of different twisted mode superpositions. The received mode superpositions could be identified with an accuracy of more than 80% up to the third mode order and the decoded message had an error rate of 8.33%. The results indicate that the free-space transmission of OAM modes over a kilometer-scale distance is feasible.

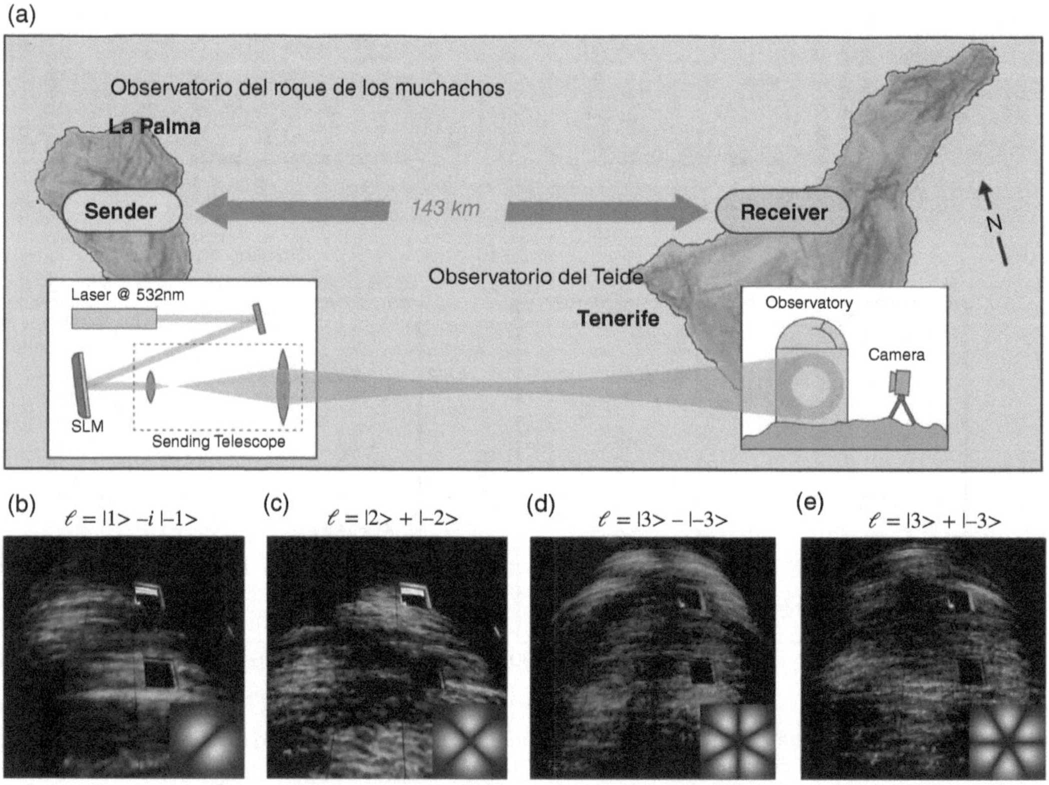

Figure 12.15 (a) Sketch of the experimental setup; (b–e) examples of superposed OAM (vortex) modes were observed at the receiver. *Source:* Krenn et al. [31], © 2016 United States National Academy of Sciences.

12.3.3 Summary

In this section, some experimental demonstrations of FSO OAM multiplexing and encoding links under or beyond laboratory conditions are presented. OAM multiplexing technology has been seen to improve the system capacity and spectral efficiency, which potentially help address the capacity demands in many applications. Up to now, the total data rate of an FSO communication link demonstrated in the laboratory experiments has reached 100 Tbit/s and beyond. In addition, the transmission of the OAM FSO link was demonstrated over a distance up to 143 km between two islands. In the future, it would be important to develop potential techniques for mitigating the degradation effects of harsh channel conditions. This would help push forward a high-capacity long-distance OAM-based FSO link outside the laboratory.

12.4 Inter-channel Crosstalk Mitigation Methods in OAM-multiplexed FSO Communications

In this section, we discuss various optical and electrical crosstalk mitigation methods for OAM-multiplexed FSO communications, including but not limited to the approaches shown in Figure 12.16.

12.4.1 Adaptive Optics for Crosstalk Mitigation

The inhomogeneity in temperature and pressure of the atmosphere can lead to variations of the refractive index in the FSO link path, i.e. atmospheric turbulence. This turbulence can introduce phase distortion to the helical phasefront of OAM beams, thereby increasing modal coupling and inter-modal crosstalk. AO is one of the turbulence mitigation approaches for OAM beams, and its concept is shown in Figure 12.17. In a typical AO system, the distorted beam profile is measured, and an error correction pattern can be derived and then sent to the beam correction module through a feedback loop. By compensating the distorted OAM beams, the AO approach can help to improve the performance of OAM-multiplexed FSO communication links. Several approaches for AO are discussed in the following subsections, including AO using (i) wavefront sensor, (ii) camera, and (iii) multiplane light converter (MPLC).

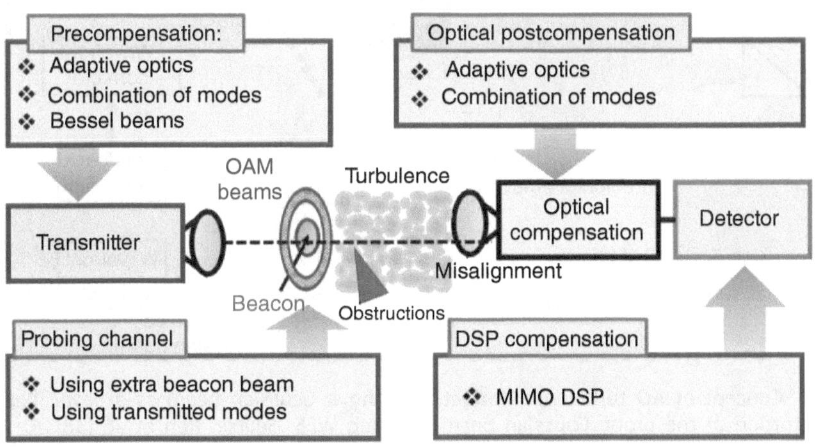

Figure 12.16 Different crosstalk mitigation approaches in an OAM-multiplexed FSO link.

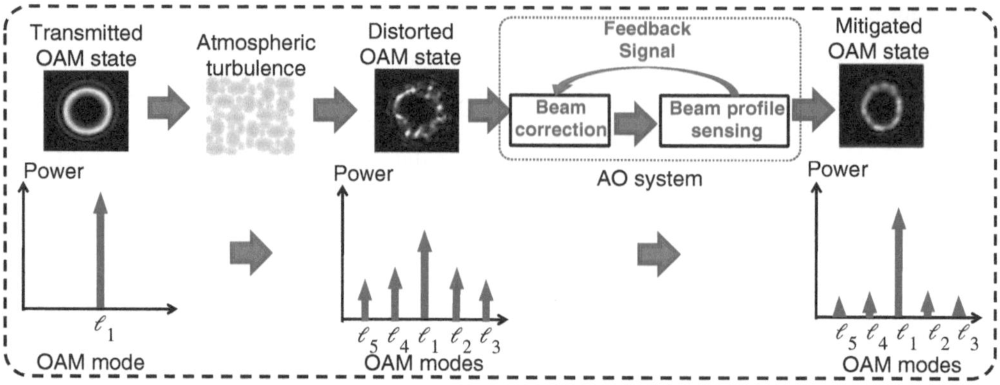

Figure 12.17 Concept of using the AO to mitigate turbulence-induced crosstalk for OAM beams. *Source:* Ren et al. [38], © 2014 Optical Society of America.

12.4.1.1 AO Using a Wavefront Sensor (WFS) and a Gaussian Probe Beam

A well-known method of AO turbulence mitigation is using WFS to detect the wavefront distortion of a beam. However, due to the phase structure of an OAM beam, it would be challenging to directly measure its phasefront. This is because there is a singular point at the center of the OAM beam, which could cause inaccuracy when measuring the phasefront using typical Shack–Hartmann WFSs. One approach could be using a Gaussian probe beam on different polarizations or wavelengths that can be measured by a commercially available WFS and be easily separated. The concept of this approach is shown in Figure 12.18.

This approach has been demonstrated in Ref. [38]. Figure 12.19 presents the intensity profiles of the Gaussian probe and OAM beams for a particular turbulence realization (a1–a6) without and (b1–b6) with compensation. It has been shown that by using the correction pattern obtained from the Gaussian probe beam in the AO system, the distorted OAM beams up to OAM $\ell = 9$ are efficiently compensated.

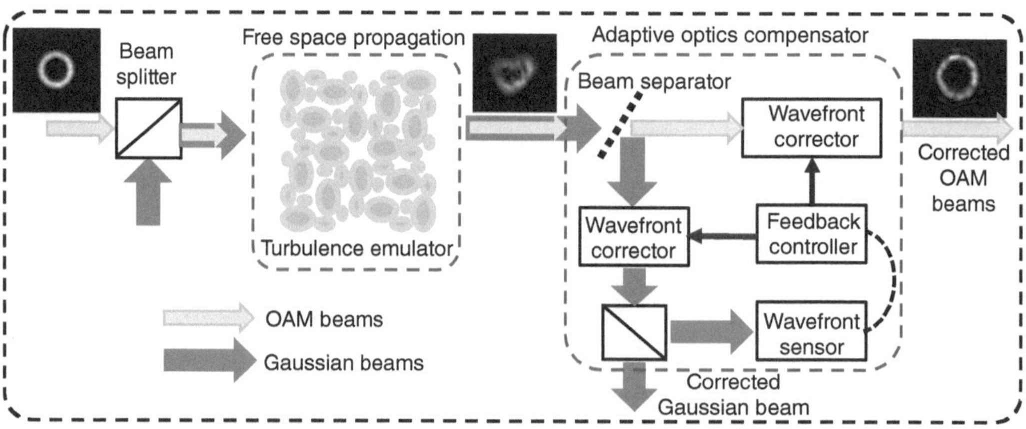

Figure 12.18 Concept of AO turbulence mitigation using a Gaussian beam as a probe beam to detect wavefront distortion of the probe Gaussian beam by using WFS. *Source:* Ren et al. [38], © 2014 Optical Society of America.

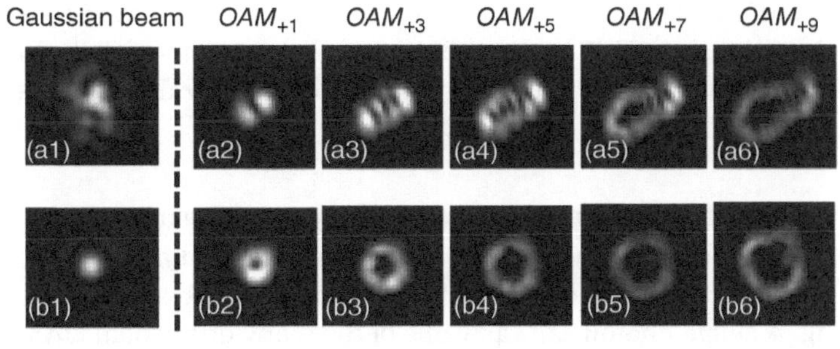

Figure 12.19 Far-field intensity images of the Gaussian probe beam and OAM beams (ℓ = 1, 3, 5, 7, and 9) before [upper, (a1–a6)] and after [lower, (b1–b6)] compensation. *Source:* Ren et al. [38], © 2014 Optical Society of America.

By compensating the distorted OAM beams, the AO approach can help to reduce the turbulence-induced crosstalk and improve the performance of OAM-multiplexed FSO communication links. This single-end AO module can be also used in simultaneously pre- and post-turbulence compensation in a bidirectional OAM link. Figure 12.18 shows the concept of postcompensation, where forward propagating OAM beams are first distorted by atmospheric turbulence, then compensated by the AO system. For the precompensation of turbulence, the backward propagating OAM first experiences compensation by the AO system, and then propagates through the same turbulence with the forward channel. Figure 12.20 shows the results from AO mitigation for the bidirectional OAM-multiplexed link in Ref. [39]. Three OAM-multiplexed channels each carrying 100 Gbit/s QPSK signal propagate through emulated weak-to-moderate atmospheric turbulence over 1 km. The QPSK modulation encodes 2 bits/symbol of information with four-phase points on the constellation as shown in Figure 12.20a0. The constellations of the channels with and without compensations at an OSNR of 23.5 dB are shown in Figure 12.20a1–a8 for the post- and precompensations (forward and backward links). Performance of different OAM mode spacings (Δ) are compared for $\Delta = 2$ and $\Delta = 4$, showing smaller Δ results in higher crosstalk and they can be both compensated by the AO approach. The BER curves are also presented as functions of OSNR, as shown in Figure 12.20b. It is shown that the BER curves for both forward and backward links are below

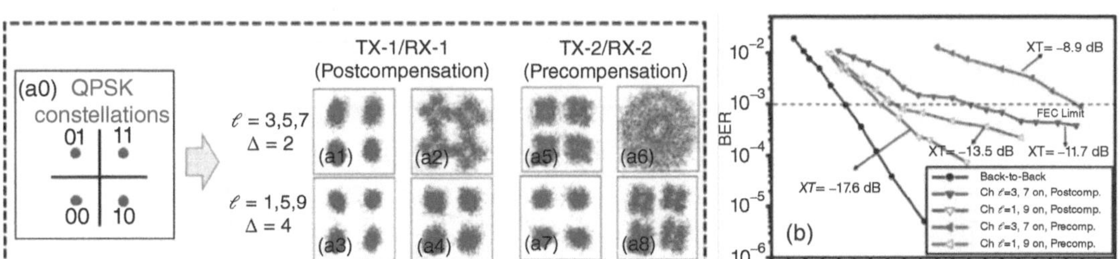

Figure 12.20 The performance of AO mitigation in an OAM-multiplexed FSO communication link. (a0) Concept of the constellation for QPSK signal; (a1–a8) recovered QPSK constellations of OAM channel with ℓ = 5, for forward (TX-1/RX-1) and backward (TX-2/RX-2) links and (b) BER for OAM channel ℓ = 5, when transmitting three multiplexed channels (ℓ = 3, 5, and 7 with Δ = 2, or ℓ =1, 5, and 9, with Δ = 4) under a random turbulence realization, with and without post- and precompensation. *Source:* Ren et al. [39], © 2014 Optical Society of America.

the FEC limit of 1×10^{-3} at OSNR values of 21 and 27.5 dB (for $\Delta = 2$), and 15.5 and 16.5 dB (for $\Delta = 4$), respectively. We note that the FEC limit is a BER upper limit under which the FEC method can be performed when coding the channel that can control the errors over unreliable or noisy communication channels.

12.4.1.2 AO Using WFS and Gaussian Probe Beam in a Quantum Communication Link

OAM modes can be used for encoding in quantum communication systems, thus providing potential advantages in system capacity and security [40]. Figure 12.21 shows the concept of OAM-based quantum encoding. A single photon can carry one of the many orthogonal OAM modes. By using OAM mode converter, within each symbol period, the coming single Gaussian photon is converted to the designed OAM photon occupies one of the N OAM states, resulting in information encoding of up to $\log_2 N$ bits/photon. The system performance degradation due to the atmospheric turbulence is also a key limitation of OAM-based quantum FSO links. Turbulence would affect the wavefront of a photon, such that the unique spatial phase profile that defines its OAM state would be distorted. Therefore, turbulence can decrease OAM modal purity and thereby increase intermodal crosstalk in the quantum communication systems.

The AO approach could also be applied to OAM-encoded quantum communication links by using a classical Gaussian (OAM $\ell = 0$) probe beam. To separate the probe beam from the quantum channel, one could transmit the probe beam with different polarizations, wavelengths, and OAM orders with the quantum channel. The single-end AO approach has been demonstrated for emulated turbulence compensation in an OAM-encoded, bidirectional FSO quantum communication link at 10 Mbit.s^{-1} per channel [41]. Figure 12.22 shows the channel transfer matrices (upper line), as well as the photon, counts ratio (lower line) on the received OAM modes when sending OAM modes $\{\ell = -3, ..., +3\}$ in different cases: (from left to right column) the back-to-back link (without the turbulence emulator in the optical path) with the probe being turned (i) off and (ii) on, under a random turbulence realization with the probe (iii) off and (iv) on without AO mitigation, and (v) with AO mitigation while the probe is on. The performance of the AO approach in quantum communications was also analyzed in Ref. [41] by measuring the quantum-symbol-error-rate (QSER) which is the ratio of the photons in the wrong OAM-states over the total detected photons

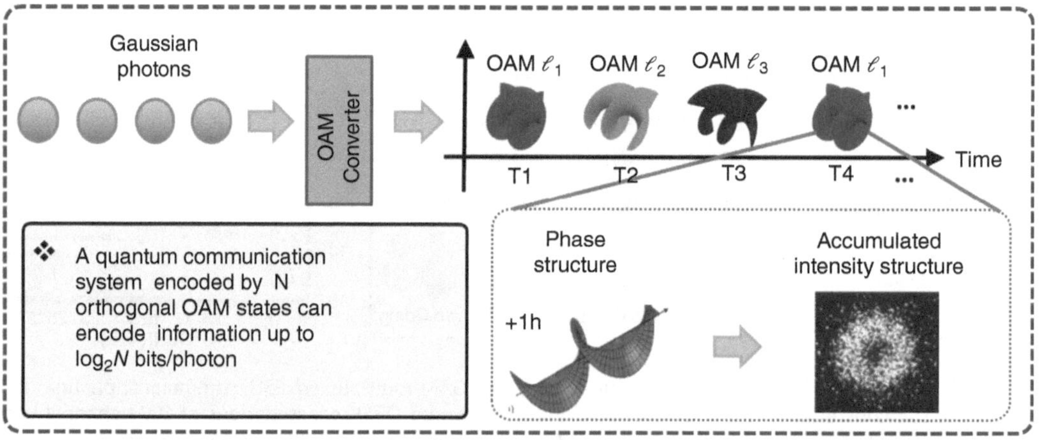

Figure 12.21 Concept of OAM-based quantum data encoding. Within each symbol period, the coming Gaussian photon is converted to an OAM photon occupying one of the N OAM states, results in information encoding of up to $\log_2 N$ bits/photon.

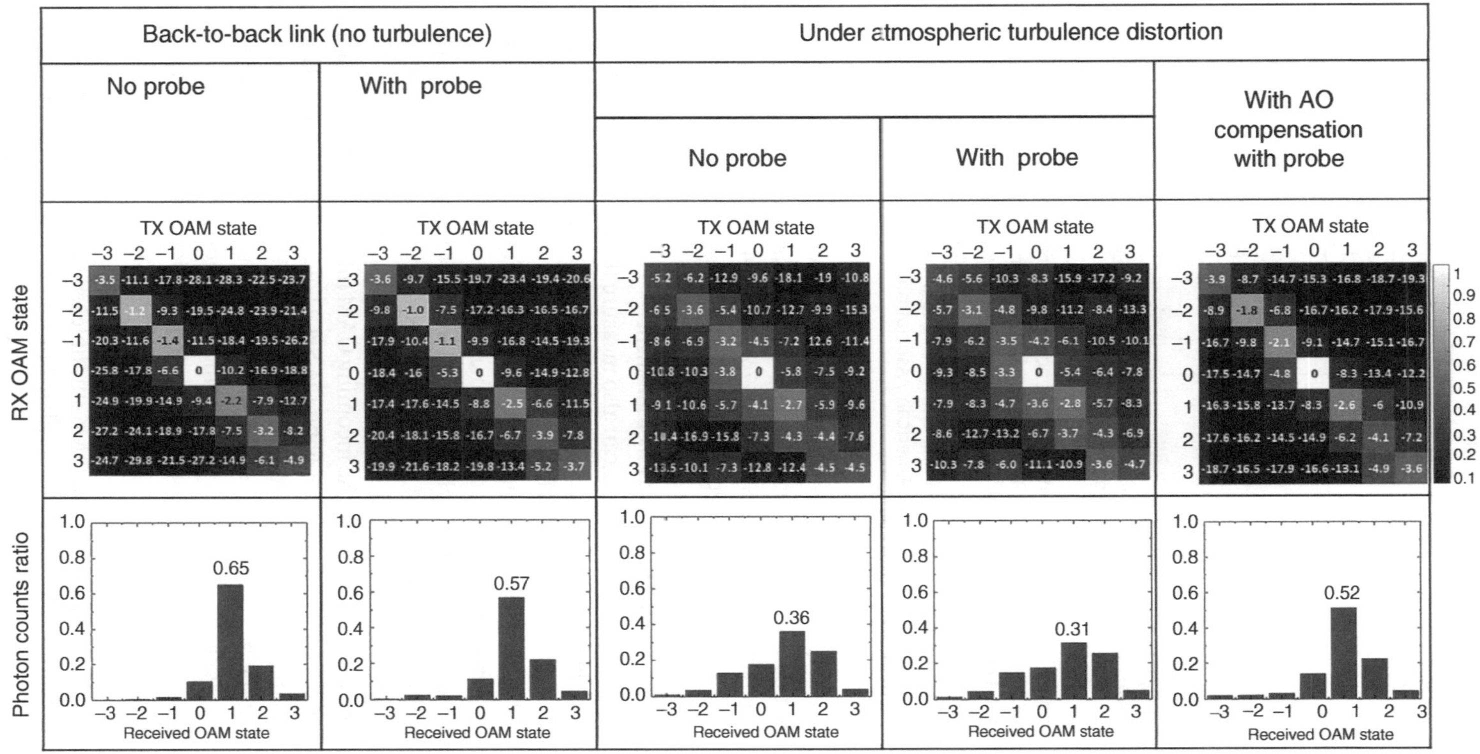

Figure 12.22 The performance of AO mitigation in OAM-based quantum link (Upper line) Channel transfer matrices of OAM-based quantum link when sending OAM modes $\{\ell = -3, \ldots, +3\}$, respectively, for different cases. The numbers are measured in the quantum domain as the ratio of the measured photon counts to the maximum photon counts in this matrix in a unit of dB. (Lower line) The photon counts ratio on received OAM modes $\{\ell = -3, \ldots, +3\}$ when sending only OAM $\ell = 1$ photons for different cases. TX: transmitter; RX: receiver. *Source:* Liu et al. [41]. Licenced under CC BY-4.0. © 2019 American Association for the Advancement of Science.

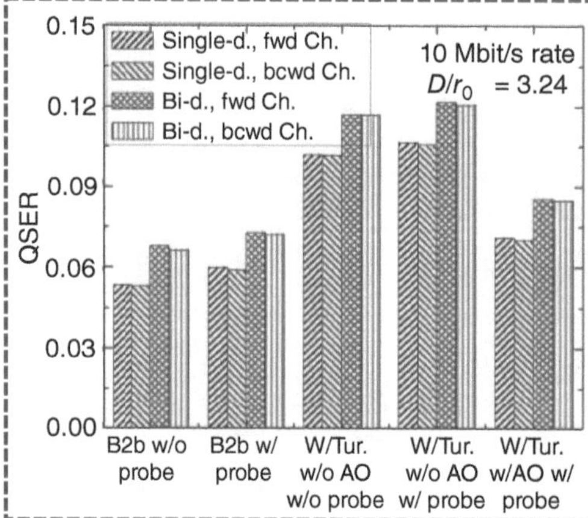

$$\text{QSER} = \frac{\text{Photons in the wrong states}}{\text{Total detected photons}}$$

Forward channel: OAM −1 & +2;
Backward channel: OAM −2 & +1;

Cases	Ch.	QSER improvement
Single-d	Fwd	~76.2%
	Bcwd	~75.7%
Bi-d	Fwd	~74.0%
	Bcwd	~74.1%

Figure 12.23 The performance of AO mitigation in an OAM-encoded, bidirectional quantum link: the QSER of an OAM-encoded quantum communication link for the single-directional and bidirectional cases under emulated turbulence effects. OAM $\ell = -1$ & $+2$ are used for forward channel encoding and OAM $\ell = -2$ & $+1$ are used for backward channel encoding. The turbulence strength of $\frac{D}{r_0} = 3.24$ is applied. The improvements of the AO mitigation are calculated and shown in the inset table; b2b: back-to-back. *Source:* Liu et al. [41]. Licenced under CC BY-4.0. © 2019 American Association for the Advancement of Science.

in an information sequence. Quantum information was encoded with OAM values of $\{\ell = -1, +2\}$ and $\{\ell = -2, +1\}$ for the forward and backward quantum channels, respectively. The AO system reduced the turbulence-effects-increased QSER by ~76 and ~74 %, for both channels in the uni-directional and bidirectional links, respectively, as shown in Figure 12.23.

12.4.1.3 AO Using a Camera for Beam Intensity Measurement

In an AO turbulence compensation system, one can also measure the incoming OAM intensity profile by using a camera and derive the phase correction pattern by applying the stochastic-parallel-gradient-descent (SPGD) algorithm [42, 43]. This approach does not need the Gaussian probe beam in an OAM-based link. The concept of this approach is shown in Figure 12.24. At the receiver, the intensity profile was measured by an infrared (IR) camera to provide the feedback signal to the SPGD algorithm. The algorithm generated a phase-correction pattern to correct the distorted wavefront. The information collected on one OAM order can be used to correct beams with other OAM orders as the beams pass through the same distorting transmission medium. In an AO turbulence compensation system, one can also measure the incoming OAM intensity profile by using a camera and derive the phase correction pattern by applying the SPGD algorithm [42, 43]. This approach does not need the Gaussian probe beam in an OAM-based link. The concept of this approach is shown in Figure 12.24. At the receiver, the intensity profile is measured by an infrared (IR) camera to provide the feedback signal to the SPGD algorithm. The algorithm will generate a phase-correction pattern to correct the distorted wavefront. The information collected on one OAM order can be used to correct beams with other OAM orders as the beams pass through the same distorting transmission medium.

This approach has been demonstrated for an OAM-multiplexed communication link in Ref. [44]. OAM $\ell = 3$ was firstly transmitted through the emulated turbulence and the SPGD algorithm was used to generate the phase-correction pattern. The turbulence emulation followed the Kolmogorov spectrum and the Fried parameter was held fixed to be $r_0 = 1$ mm. This correction pattern was then

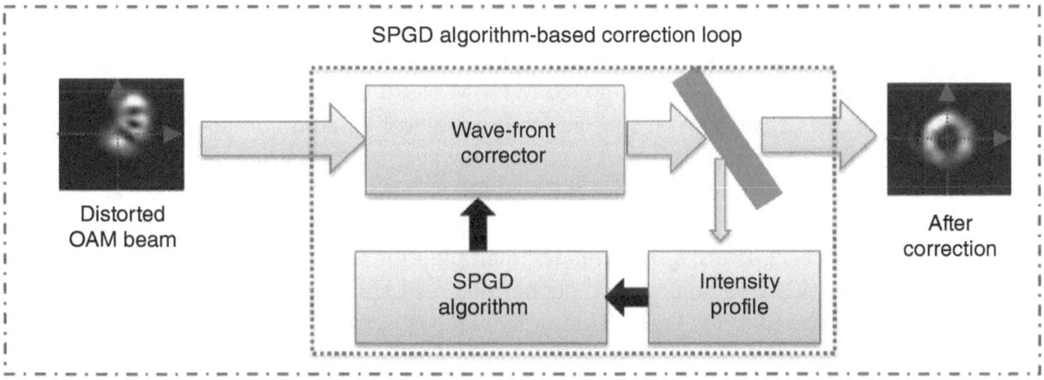

Figure 12.24 Concept of the phase correction for a distorted OAM beam using a camera for intensity profile detection and stochastic-parallel-gradient-descent (SPGD) algorithm. *Source:* Xie et al. [44], © 2015 Optical Society of America.

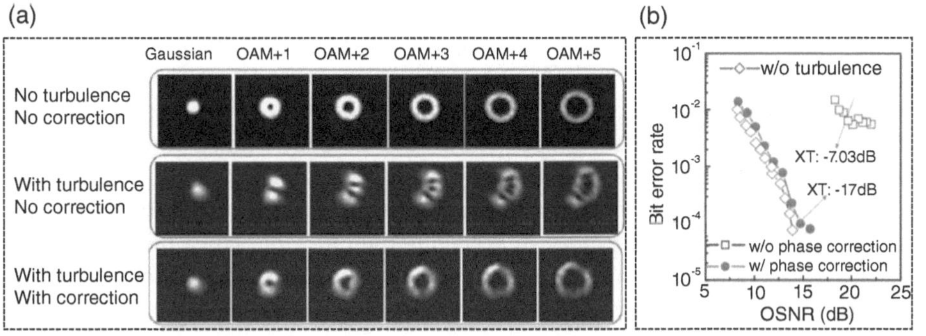

Figure 12.25 (a) Far-field intensity profiles for various OAM beams (OAM ℓ = 0, … , +5) before and after phase correction under a specific turbulence realization with r_0= 1 mm and (b) measured BERs as functions of OSNR when the phase pattern derived from OAM ℓ = 3 is used to correct the phases of three simultaneously transmitted OAM beams (OAM ℓ = +1, +3, and +5). *Source:* Xie et al. [44], © 2015 Optical Society of America.

used to correct the wavefront distortion of a Gaussian beam, as well as those of OAM $\ell = 1$ to OAM $\ell = 5$ beams. The far-field intensity profiles of each received OAM beam are shown in Figure 12.25a. The emulated turbulence distorted the intensity profiles of the OAM beams and after phase correction, the intensity profiles are compensated. A 50-Gbaud QPSK signal was encoded onto three multiplexed channels with OAM ℓ = +1, +3, and +5. The OAM-multiplexed channels propagate co-axially through emulated turbulence effects. The correction pattern derived from OAM $\ell = 3$ was applied to a multiplexed beam, comprising three channels (OAM ℓ = +1, +3, and +5), propagating through the same turbulence. Figure 12.25b shows the BERs of the OAM ℓ = 3 channel before and after phase correction for one turbulence realization. Without phase correction, the BER can barely reach the FEC limit of 3.8×10^{-3}, because of the large amount of crosstalk from the other two channels. With the phase correction, the BER could achieve the FEC limit.

12.4.2 Spatial Modes Manipulation for Crosstalk Mitigation

It would be valuable to develop an alternative optical approach for crosstalk mitigation without sensing the beam profile that could be applied to OAM-based FSO links. One potential approach would be to select and design the transmitted mode or the combination of modes, namely, spatial mode manipulation. This method is based on that, the wavefront of a structured beam (e.g. OAM) can be decomposed into a set of LG modes carrying OAM. The coefficient of each LG mode in the decomposition can be complexed, containing both amplitude and phase information, as shown in Figure 12.26. Approaches using spatial modes manipulations including (i) transmitting OAM modes combination, (ii) simultaneous orthogonalizing and shaping of multiple LG beams, and (iii) utilizing Bessel beams with nonzero OAM order will be discussed in this session for OAM beam crosstalk mitigation.

12.4.2.1 Turbulence Precompensation by OAM Mode Combination

The concept of turbulence compensation using mode combination is shown in Figure 12.27a. At the transmitter side, the inverse transmission matrix was applied using the compensation phase patterns at the transmitter for each channel, so that the signal from each transmitted channel was carried by the combination of multiple (e.g., 2) OAM modes with designed complex weights. The weights were calculated based on the inverse of the complex transmission matrix under the corresponding turbulence realization. After combining the beams from the two channels at the output port of the transmitter, the data-carrying beams from both channels become coaxial, and each of the two OAM modes in the combined beam carried the combination of signals from both channels. When the beams were transmitted through the turbulence, the signals on the two transmitted modes coupled to their neighboring modes and experience interference on those modes. Thus, the channels could have little power on the designated modes due to destructive interference and relatively high power on the others. By receiving the mode on which the undesired channel has little power, the desired channel could be recovered with little inter-channel crosstalk. The same concept can be applied to recover the second channel when receiving another mode. As an example, the compensation phase patterns used for beam generation in channel A and the intensity profiles of the generated beams are shown in Figure 12.27b.

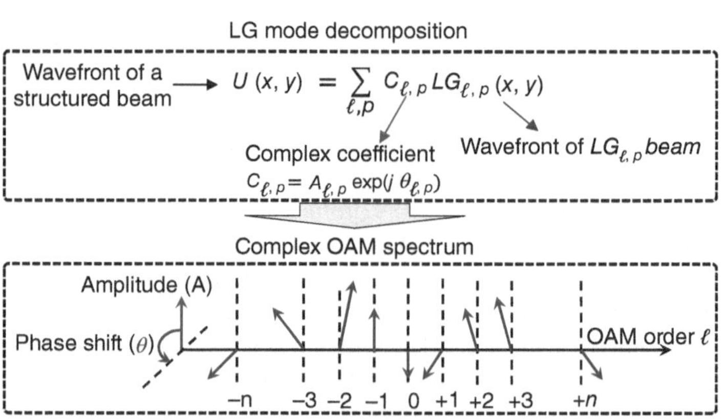

Figure 12.26 Concept of the LG mode decomposition and the complex OAM spectrum. The wavefront of a structured beam can be decomposed into a set of LG modes carrying OAM.

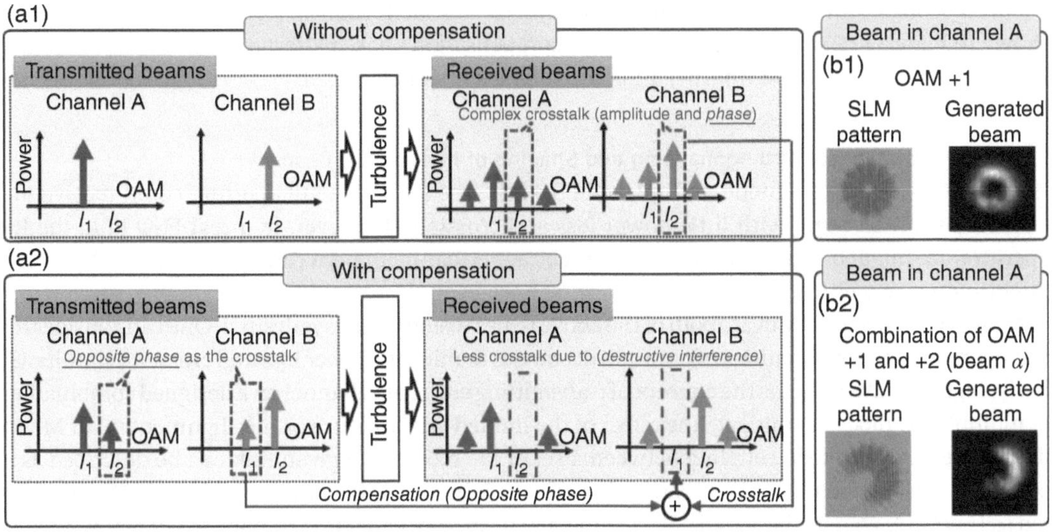

Figure 12.27 (a) The concept of the turbulence compensation utilizing inverse transmission matrix method: (a1) transmitting pure OAM modes, no compensation approach applied; and (a2) transmitting the combination of OAM modes to compensation the turbulence effects; (b) an example of transmitted beams in (b1) no compensation and (b2) with compensation cases, the SLM patterns, and the generated beams intensity profiles. *Source:* Song et al. [45], ©2020 Optical Society of America.

This approach is demonstrated in Ref. [45], two OAM-multiplexed channels each carrying a 100-Gbit/s QPSK signal were transmitted. Channels A and B transmitted OAM $\ell = +1$ and $\ell = +2$, respectively, in the link without the compensation. To apply the compensation approach, channels A and B transmitted beam α and β, which were combinations of OAM $\ell = +1$ and $\ell = +2$. The receivers of channels A and B always received and demultiplexed OAM $\ell = +1$ and $\ell = +2$, respectively. The results of this demonstration are shown in Figure 12.28. Figure 12.28a1 shows the back-to-back crosstalk. As can be seen from Figure 12.28a2, with the turbulence effect, the inter-channel crosstalk increases to −8.7 and −5.5 dB for channels A and B respectively in the absence of the compensation and this crosstalk decreases to −22.1 dB when receiving OAM +1, and −17.8 dB

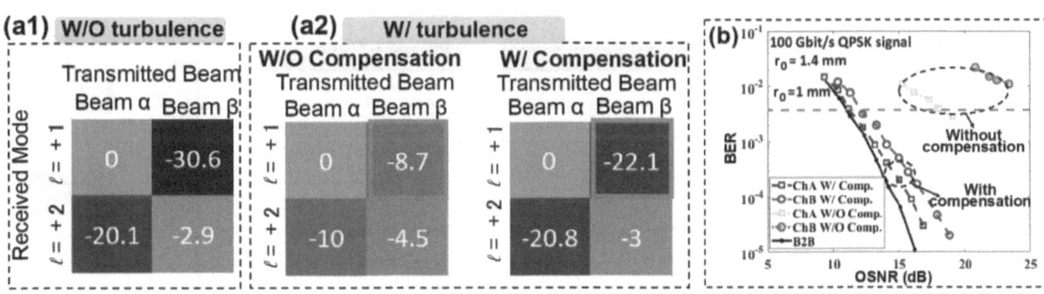

Figure 12.28 (a1–a2) The normalized transmission intensity matrices and (b) BER performance for the transmitted channels without and with compensation. In the demonstration, channel A receives OAM ℓ = +1, and channel B receives OAM ℓ = +2. In the link without compensation, channels A and B transmit OAM ℓ = +1, ℓ = +2, respectively. Beam α and β are the combinations of OAM +1 and +2 transmitted by channels A and B, respectively, when the compensations are applied. *Source:* Song et al. [45], © 2020 Optical Society of America.

when receiving OAM +2 with the compensation. The BER performance for the forward channels is shown in Figure 12.28b. Through applying precompensation phase patterns in the forward link, the BER performance could be improved for the forward channels.

12.4.2.2 Simultaneous Orthogonalizing and Shaping of Multiple LG Beams

In a perfectly aligned LG mode multiplexed FSO link, the Rx with a full aperture could receive the entire transmitted beam with little power loss and crosstalk. However, in a real FSO link, the Rx aperture size might be limited and there might be a misalignment between the Tx and the Rx, which might induce power loss and crosstalk.

An approach that can mitigate both of the above issues would be always desired. One can consider re-orthogonalize the combinations of multiple LG modes to mitigate power loss and crosstalk simultaneously. Figure 12.29a presents the concept of transmitting each data channel on a designed combination of multiple LG modes to mitigate the effect of the limited-size aperture or misalignment in an MDM link. Generally, the mode coupling between a set of LG modes in a given link can be described as a complex transmission matrix H. H can be factorized by singular value decomposition (SVD) and written as $H = U \cdot \Sigma \cdot V^*$. At the Tx side, the orthogonal beams are generated by complex combinations of multiple LG modes, of which the complex weights (amplitude and phase) are given by the column vectors of the V. After passing through the limited-size aperture or being affected by the misalignment, the transmitted beams on different channels are still orthogonal to each other. As a result, these independent beams could be demultiplexed with little crosstalk based on the row vectors of the inverse of the U matrix. When the beams are orthogonalized, the intensity profiles of the transmitted beams would also be shaped, which might simultaneously reduce the power loss. As an example, the intensity profiles of pure LG_{10}, pure LG_{11} beams, and the generated orthogonal beam are shown in Figure 12.29b.

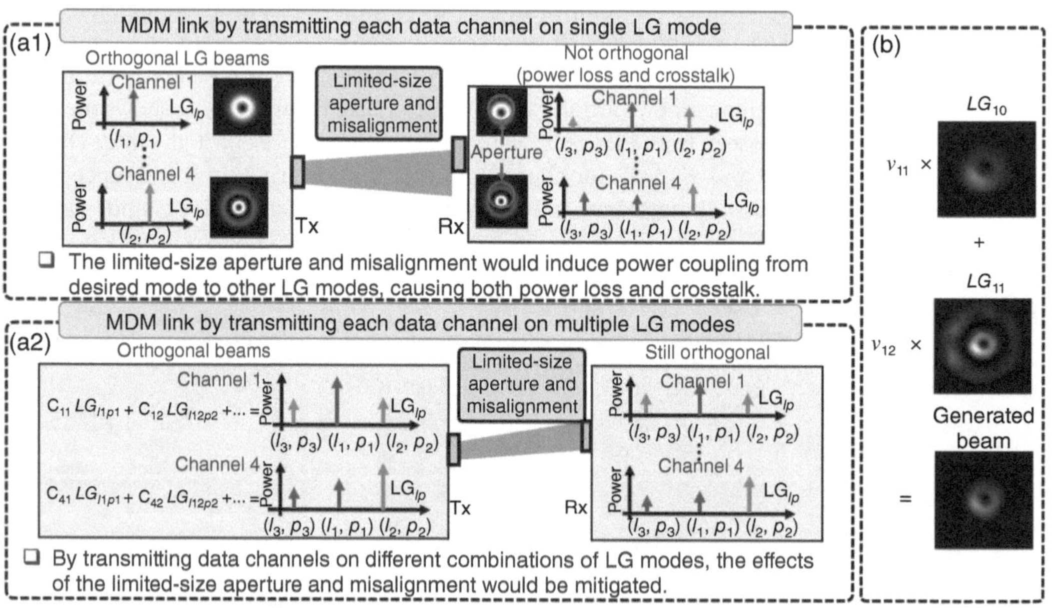

Figure 12.29 (a) Concept diagram of transmitting each data channel on a combination of multiple LG modes to mitigate the effect of the limited-size aperture and misalignment of an MDM link; (b) the intensity profiles of pure LG_{10}, pure LG_{11}, and the generated beam. Ch: Channel; Tx: Transmitter; Rx: Receiver. *Source:* Song et al. [45], © 2020 Optical Society of America.

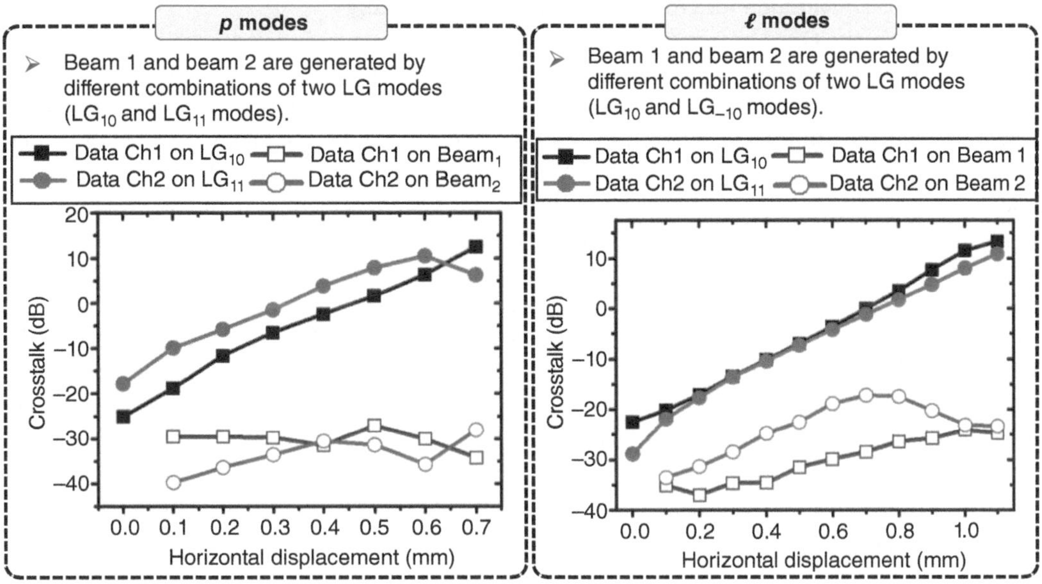

Figure 12.30 Displacement-induced channel crosstalk under various horizontal displacements when transmitting data channels on (left panel) pure LG modes with different p values (LG_{10}, LG_{11}) or designed orthogonal beams and (right panel) pure LG modes with different ℓ values (LG_{10}, LG_{-10}) or designed orthogonal beams. *Source:* Pang et al. [47], © 2020 Optical Society of America.

Ref. [46] demonstrates the above approach to a four-channel multiplexed link, each of which carries a 100-Gbit/s QPSK signal. The results are shown in Figure 4.15. The left panel of Figure 12.30 presents that when transmitting data channels on pure LG modes with different p values (LG_{10}, LG_{11}), the crosstalk would become larger with the horizontal displacement. However, when using the orthogonal beams (beams 1 and 2) which were generated by designed complex combinations of LG_{10}, LG_{11} modes, the crosstalk for both of the two generated beams could stay at a relatively low level (<−27 dB) with the displacement. The right panel of Figure 12.30 shows the effects of displacement on LG modes with different ℓ values (LG_{10}, LG_{-10}). The crosstalk of LG_{10} and LG_{-10} modes increase with the displacement and could stay at a relatively low level (<−17 dB) with the displacement for both channels.

12.4.3 Digital Signal Processing for Crosstalk Mitigation

An alternative option for crosstalk mitigation in OAM multiplexed links might be to shift the complexity of the optical setup to the digital domain, i.e. using digital signal processing (DSP). This section introduces digital approaches for the crosstalk mitigation in OAM multiplexed FSO links by (i) using multiple-input-multiple-output (MIMO) digital signal processing equalization in the laboratory environment, (ii) using MIMO DSP equalization in a link through a flying UAV, and (iii) using turbulence-resilient beam mixing.

12.4.3.1 MIMO Equalization for Crosstalk Mitigation in Laboratory

The wavefront of the OAM beams could be affected by various degradation effects. Therefore, after demultiplexing, the received signal carried on a particular OAM mode may include the signal leaked from other channels, as indicated in Figure 12.31. One potential crosstalk mitigation approach would be utilizing MIMO DSP-based equalization. Ref. [48] has demonstrated the implementation of a 4 × 4 adaptive MIMO equalizer in a four-channel OAM multiplexed link using heterodyne detection. Four OAM modes, each carrying 20 Gbit/s QPSK data, were collinearly propagated through a turbulence emulator to introduce crosstalk and are then demultiplexed. In this demonstration, the turbulence was emulated using a rotatable phase screen that obeys Kolmogorov spectrum statistics with a Fried parameter of ~5 mm.

Figure 12.32a and b illustrates the recovered constellations of all four 20 Gbit/s QPSK signals (with and without MIMO equalization) in each channel for a single turbulence realization. In this experiment, the channels 1–4, corresponds to OAM modes with $\ell = -1, -3, +1, +3$, respectively. With the assistance of MIMO DSP equalization, the EVM of four channels was improved from 0.24,

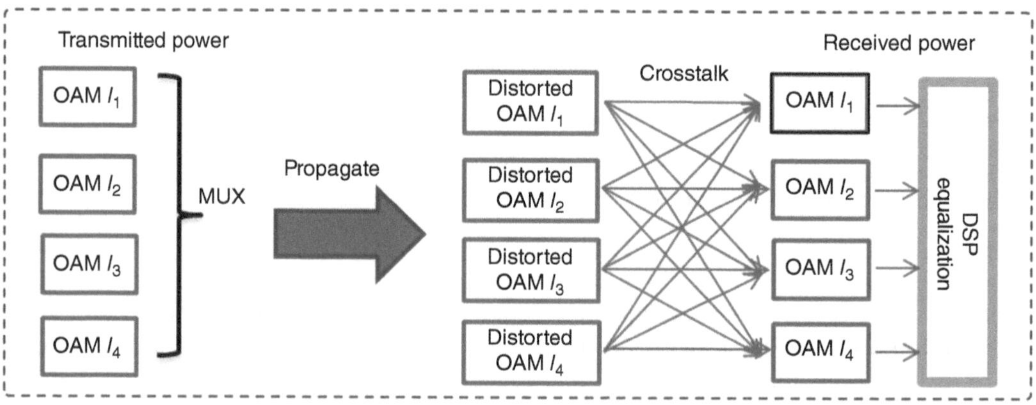

Figure 12.31 The concept of using MIMO DSP equalization for crosstalk mitigation . *Source:* Pang et al. [47], © 2014 Optical Society of America.

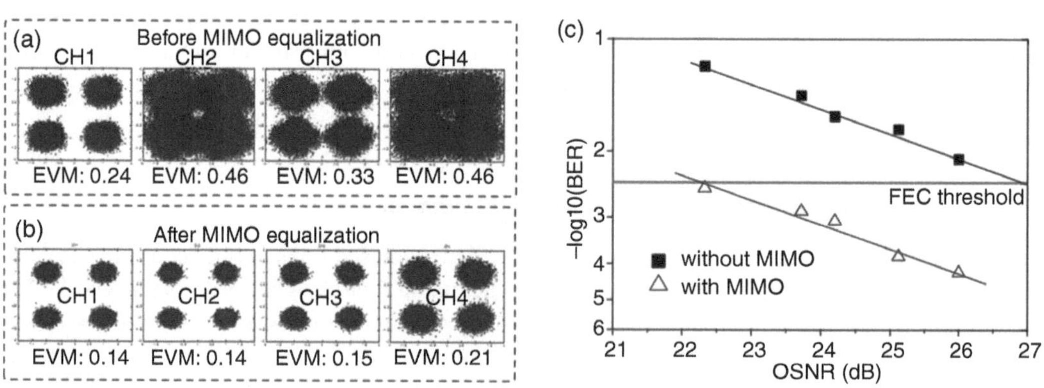

Figure 12.32 Recovered constellations of 20 Gbit/s QPSK signal in each of the four channels (a) without and (b) with MIMO equalization. The four channels in (a) and (b) are OAM modes with $\ell = -1, -3, +1, +3$; (c) measured BER as a function of OSNR for the 20 Gbit/s QPSK signal in channel 4 with $\ell = 8$. The four OAM modes in (c) are $\ell = +2, +4, +6$, and $+8$ [48]. *Source:* Huang et al. [47], © 2014 Optical Society of America.

0.46, 0.33, and 0.46 to 0.14, 0.14, 0.15, and 0.21, respectively. In the report, the MIMO equalization using another four OAM modes (ℓ =+2, +4, +6, and +8) was also demonstrated. Figure 12.32c shows the calculated BER as a function of OSNR for the channel carried on ℓ = +8 (the channel that experienced the most crosstalk). Without MIMO equalization, the measured BER for this channel was above 3.8×10^{-3}, i.e. the 7% FEC limit. The curve crosses the FEC threshold at the OSNR of ~26.8 dB. After MIMO equalization, the estimated required OSNR at the BER of 3.8×10^{-3} is ~22.3 dB.

12.4.3.2 Turbulence-Resilient Beam Mixing for Crosstalk Mitigation

Another approach for electrical crosstalk mitigation is to use pilot tones to mix beams at the receiver. The turbulence effects would distort the wavefront of the OAM beams, and such distortion could be potentially viewed as a unitary transform $U(x, y)$ [49–51]. If the entire distorted beams are received by a free-space photodetector, the output could be described by $|E_1U + E_2U|^2 = (E_1U \bullet U^*E_2^* + E_2U \bullet U^*E_1^*) + |E_1|^2 + |E_2|^2$, where E_1 and E_2 are the undistorted fields of the two beams (pilot and signal), respectively. In the signal-pilot beating term, the detector conjugates the distortion on the pilot tone, which in combination with the distorted signal, cancels the distortion on the signal. Therefore, when the same mode on two different frequencies is transmitted through the turbulence, the pilot tone and the signal are carried by the same OAM modes, a signal-pilot beating signal in the electrical domain will always be generated, as shown in Figure 12.33a. The signal carried by the OAM beam could then be recovered with little degradation since distortion transform U is canceled in the signal-pilot beating term [49–51]. When two different modes on the two frequencies are transmitted through the turbulence, the light is also distorted by the turbid medium. However, after the square-law detection, no beating term is generated because that the unitary distortion U is canceled by the detection, and the two modes are orthogonal with each other, as shown in Figure 12.33b. Therefore, although the turbulence induces the modal coupling, there is little coupling in the electrical domain after the square-law detection.

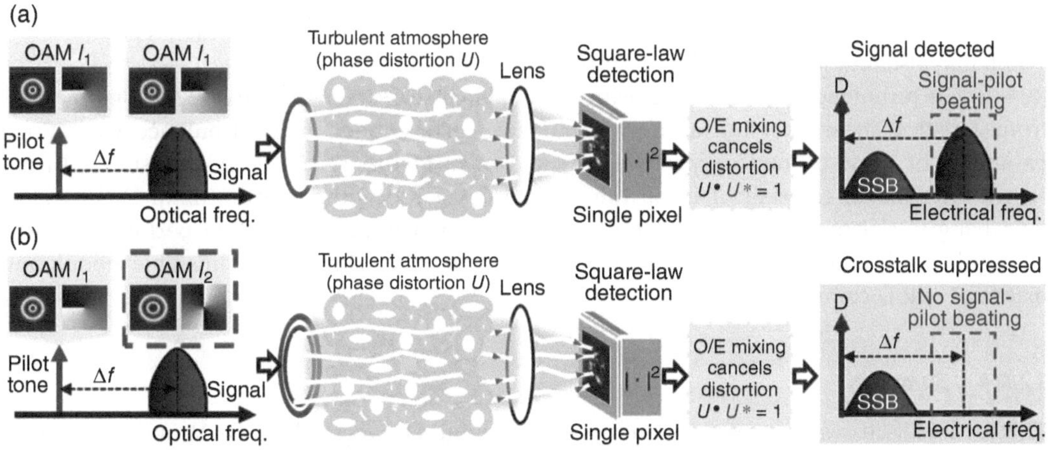

Figure 12.33 The concept of receiving signal and pilot tone using a free space PD; (a) when the turbulence-induced distortion is considered as a unitary transform and the distorted beam is received by a square-law photodetector, a beating signal could be generated in the electrical domain with turbulence effects automatically canceled; (b) when two beams are carried by different modes, the 'beating' will not be generated.

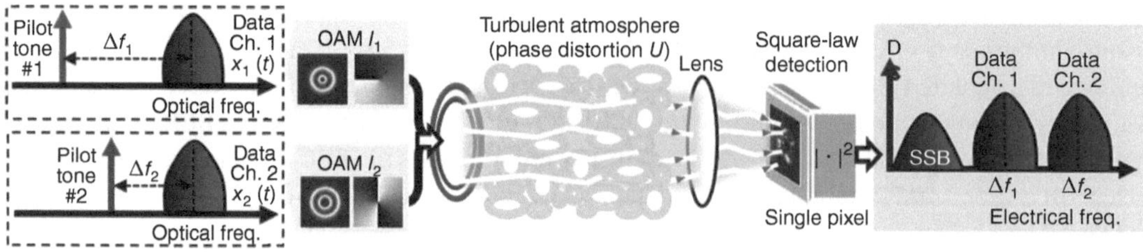

Figure 12.34 Concept of mixing OAM pilot tones and signal beams to achieve turbulence-resilient OAM-multiplexed FSO communications. The optoelectronic mixing would automatically cancel the turbulence distortion. *Source:* Zhang et al. [50], © 2020 Optical Society of America.

It was experimentally demonstrated in Ref. [50] that a 4-Gbit/s turbulence-resilient OAM-multiplexed FSO link using pilot tones to mix beams at the receiver. The concept is shown in Figure 12.34. Two OAM modes ℓ_1 and ℓ_2 are transmitted through emulated atmospheric turbulence, and each mode carries a pilot tone and an independent data channel. Two data channels are at the same wavelength and the frequency spacings between the pilot tones and the data channel are Δf_1 and Δf_2, respectively. The turbulence distortion is represented as a transformation U which distorts the wavefront of the OAM beams and causes inter-modal crosstalk if they are demultiplexed using their original waveshapes. However, by mixing pilot and data signals at the receiver and choosing the proper frequency spacing between the pilot tones and the signal, the turbulence distortion effects can be canceled and two data channels can be down-converted to different electrical frequencies. Therefore, these two data channels can be separated and recovered with little turbulence-induced crosstalk.

In this electrical approach, the multiplexed data channels are detected and recovered at different electrical intermediate frequencies. Therefore, the photodetector's bandwidth is divided for recovering different data channels. We note that this approach may be favorable in scenarios where: (i) the bandwidth of each generated data channel is limited at transmitter, and the detector has sufficient bandwidth for recovering multiple data channels simultaneously, or (ii) a sufficient number of detectors is not readily available, and a simplified receiver architecture is preferred.

Ref. [50] also investigates the link performance when using OAM modes $\{+1, -1\}$, $\{+1, -2\}$, or $\{0, -2\}$. The turbulence was emulated using a rotatable phase screen that obeys Kolmogorov spectrum statistics. The weaker and stronger turbulences corresponded to the turbulence with a Fried parameter of 1 and 0.4 mm, respectively. As shown in Figure 12.35, the system suffered from severe inter-modal crosstalk if using the normal OAM Rx (without beam mixing). However, by mixing the distorted OAM pilot tone with the signal beam, the system performance was not significantly affected by the turbulence-induced crosstalk (with residual unstable modulation bias for I and Q component).

12.4.4 Summary

This section discusses the causes and mitigation techniques for crosstalk in OAM multiplexed FSO communication links. Several related works are discussed for crosstalk mitigation within the topics of AO, spatial modes manipulation, and digital signal processing. There could also be other potential approaches that can mitigate the channel crosstalk in OAM multiplexed links. To further explore and develop the crosstalk mitigation techniques for OAM multiplexed links, one may need

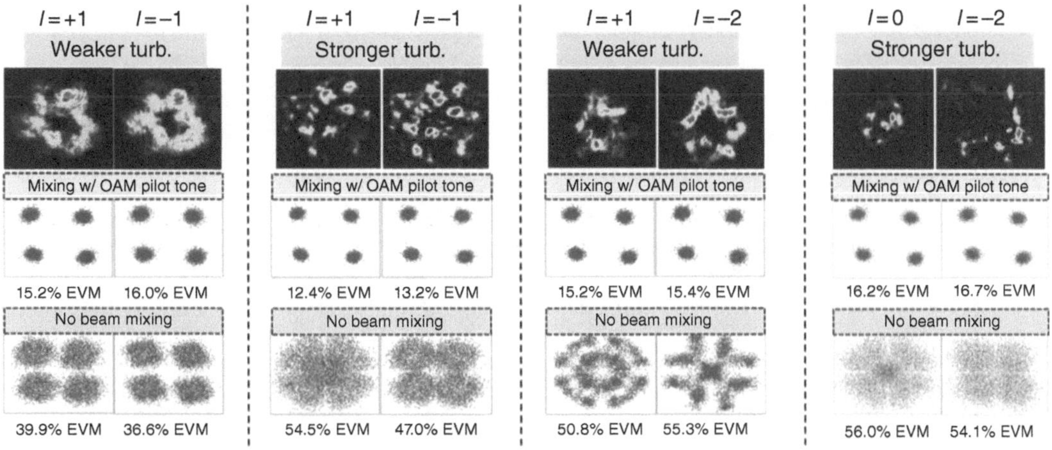

Figure 12.35 EVM performance for two multiplexed OAM channels under different turbulence strengths (r_0 = 0.4 and 1.0 mm): Constellation diagram for different modal sets $\{\ell_1, \ell_2\}$ = {+1, −1}, {+1, −2} or {0, −2} under turbulence effects [50]. The data rate for each channel is 4 Gbit/s. *Source:* Zhang et al. [50], ©2020 Optical Society of America.

to consider several issues, including but not limited to: (i) scalability: for scaling the compensation scheme to the system that multiplexes a larger number of modes; (ii) tuning speed: for real-time mitigation of the atmospheric turbulence effects; and (iii) complexity: for enabling the deploy of compensation scheme in a real OAM multiplexed link.

12.5 OAM Multiplexing for Unmanned Aerial Vehicle (UAV) Platforms

The data communications capacity needs of manned and unmanned aerial platforms have been increasing dramatically over the past several years, thereby driving the need for higher-capacity links between these platforms and their ground stations [2, 52–54]. OAM multiplexing techniques may be utilized to help increase the data capacity and spectral efficiency and reduce the probability of intercept in FSO airborne communications.

As shown in Figure 12.36, there are several scenarios in airborne FSO communications which require distinct applications and pose specific challenges including (i) ***Satellite-to-satellite communications*** usually require ultra-long-distance beam propagation (>1000 km) and careful control over the laser beam divergence. In such a scenario, the high sensitivity of the receiver detector is desirable since only a limited proportion of the transmitted beam can reach the receiver. (ii) ***Satellite-to-ground-station links*** generally utilize a laser beam to propagate through the Earth's atmosphere and the accumulated atmospheric turbulence effects would induce severe distortion of the wavefront of the optical beam. (iii) ***Airplane-to-ground communications*** involve a fast-moving airplane (>100 m/s) at a long distance of 1–10 km. Both the optical beam tracking and atmospheric turbulence effects are challenges in this scenario. And (iv) ***UAV-to-ground-station communications***: a relatively slow-moving (<10 m/s) UAV is hovering at a moderate distance of <100 m away from the ground station. In this scenario, pointing, acquisition, and tracking (PAT) of the data-carry optical beam is essential and the size, weight, and power (SWaP) of the

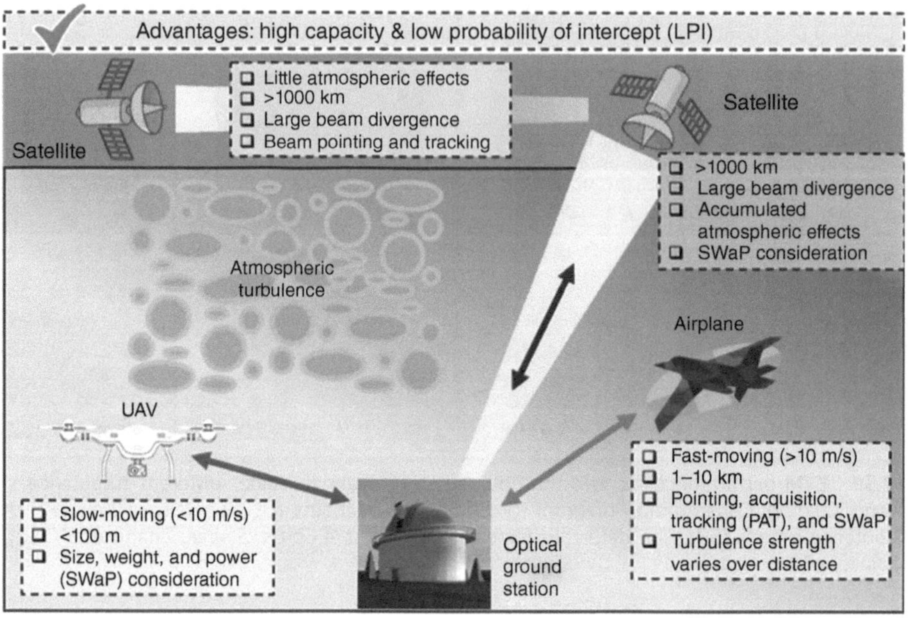

Figure 12.36 Challenges of different scenarios in OAM-multiplexed FSO airborne communications.

transmitter–receiver design is required. Drones, as an example of UAVs, have drawn a lot of attention over the recent years, due to their potential for proliferating numerous applications [6, 55–57].

In UAV platforms, such a scenario is likely to face a few major challenges arising from the special structured nature of the OAM beams themselves, including (i) **Alignment**: Low inherent crosstalk and power-coupling loss generally relies on accurate on-axis detection of the multiple OAM beams, thereby necessitating more tracking sophistication for an OAM-multiplexed link over a single conventional Gaussian-based link and (b) **Turbulence**: Turbulence resulting from the atmosphere or from a UAV's propellers could significantly distort the OAM beam's phase front, thus resulting in increased received power fluctuations and intermodal crosstalk, as compared to recovering a single conventional Gaussian beam.

12.5.1 OAM System Design and Demonstrations for UAV Platforms

To maintain the precise alignment between the Tx and the Rx, the pointing, acquisition, and tracking (PAT) system is typically used in UAV platforms [57]. Beam tracking can be achieved by using a probe beam which is either a fundamental Gaussian beam or an OAM beam [58, 59]. As shown in Figure 12.37a and b, an incoming beam with displacement is focused to a fast-steering mirror (FSM) and then equally split by a beam splitter (BS). One copy is sent to the receiver (i.e. signal decoding) and the other copy is detected by a position-sensitive detector (PSD). Such a PSD would determine the displacement of the light beam and feed this 'error signal' to the FSM controller. In such a feedback loop, the FSM is able to steer the received beams according to the feedback signal and correct the misalignment between the Tx and the Rx. Figure 12.37c1–c3 shows the statistics of displacement (after beam reduction) of an OAM $\ell = +1$ beam with respect to the center of the Rx aperture without tracking, with Gaussian beacon-based tracking, and with OAM beam-based tracking, respectively. Each marker represents the instantaneous location of the beam centroid. The

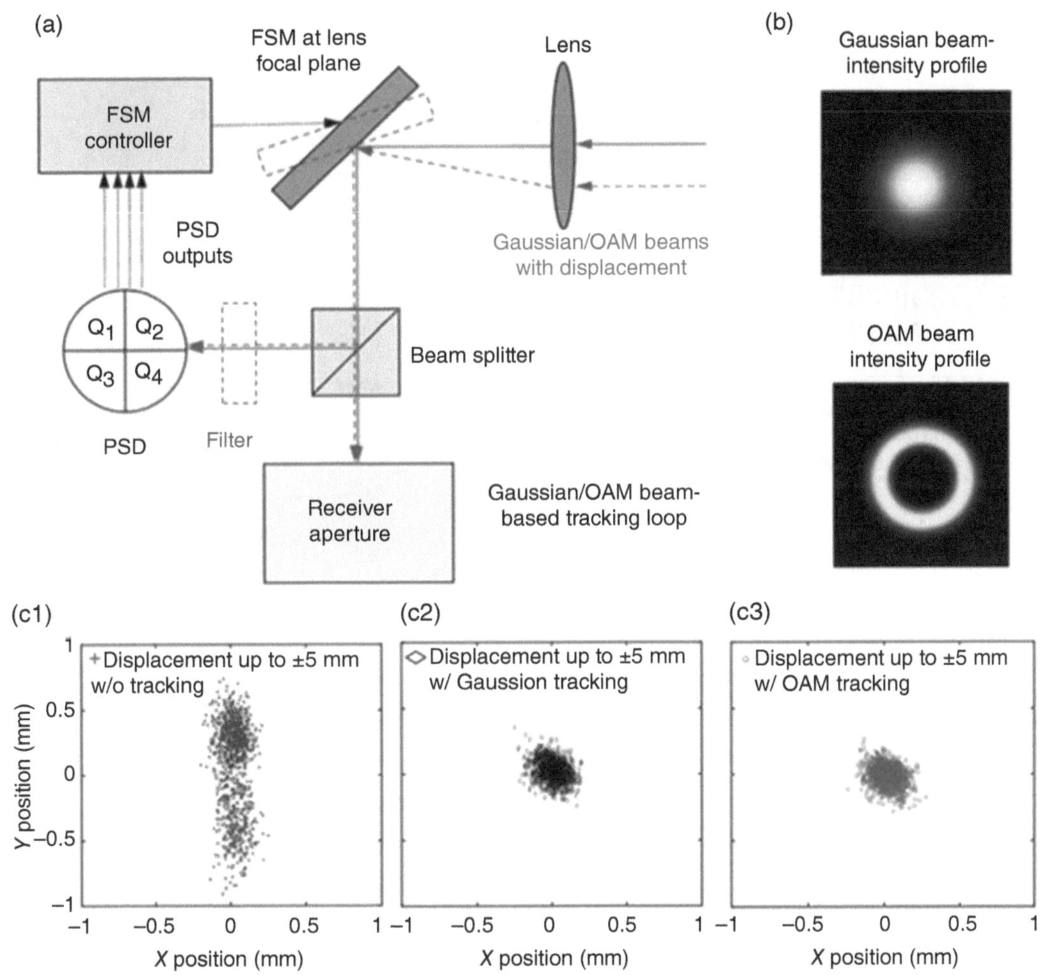

Figure 12.37 (a) Scheme of displacement tracking using Gaussian/OAM beam-based position detection; (b) intensity profiles of an OAM beam and a Gaussian beam. FSM, fast steering mirror; PSD, position sensitive detector; (c1–c3) measurement of beam centroid when displacement range is (c1) 5 mm, without tracking; (c2) 5 mm, with Gaussian beacon-based tracking; (c3) 5 mm, with OAM beam-based tracking when $\ell = +1$ is transmitted with a spot size of 1.2 mm on PSD. *Source:* Li et al. [59], ©2018 Optical Society of America.

misalignment was induced and emulated by a linear stage which randomly travels in $a \pm 5$ mm range. The results show that the FSM and PSD together could help the beam come closer to the center of the Rx, and OAM beam-based tracking and Gaussian beacon-based tracking show similar performance when the optical beam waist is 3 mm.

As shown in Figure 12.38, it has been demonstrated a high-capacity OAM multiplexed FSO link between a UAV and a ground station. The ground station contained an OAM Tx, an OAM Rx, and a beam tracking system. A retro-reflector carried by the UAV was flown up to ~50 m away (i.e. ~100 m round trip) from the ground station to efficiently reflect the OAM beams coming from the transmitter back to the receiver with little distortion. During the experiment, the octocopter UAV moved and hovered at different locations up to ~20 m above the ground and up to ~50 m away.

Figure 12.39a indicates the different locations of the UAV. In order to evaluate the effects of beam jitter, the statistics of the received beam centroid before beam reduction were measured.

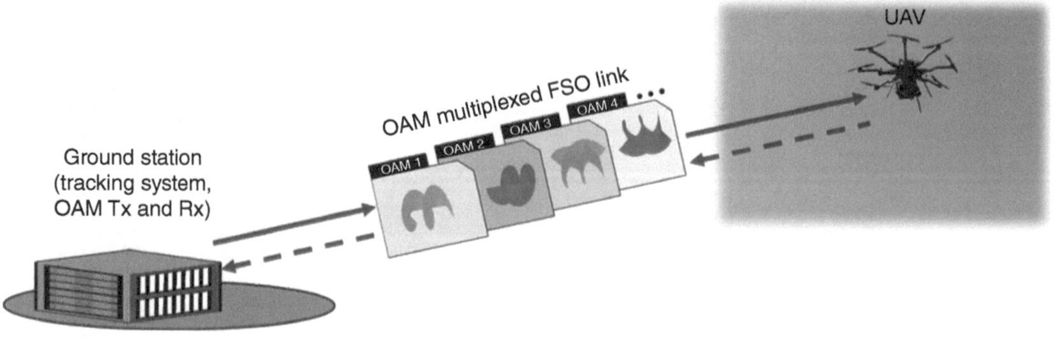

Figure 12.38 Concept of a free-space optical communication link between an unmanned-aerial-vehicle (UAV) and a ground station using OAM multiplexing. The ground station includes the tracking system as well as the OAM transmitter and receiver, and the UAV hovers in the air. *Source:* Li et al. [60], © 2017 Macmillan Publishers. Licenced under CC BY-4.0.

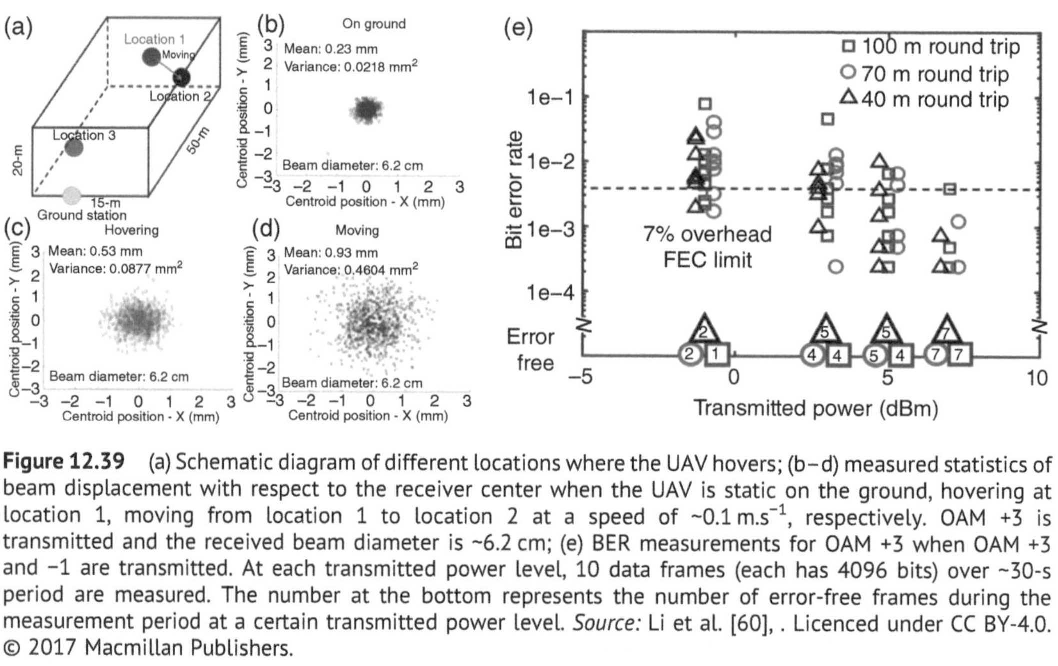

Figure 12.39 (a) Schematic diagram of different locations where the UAV hovers; (b–d) measured statistics of beam displacement with respect to the receiver center when the UAV is static on the ground, hovering at location 1, moving from location 1 to location 2 at a speed of ~0.1 m.s^{-1}, respectively. OAM +3 is transmitted and the received beam diameter is ~6.2 cm; (e) BER measurements for OAM +3 when OAM +3 and −1 are transmitted. At each transmitted power level, 10 data frames (each has 4096 bits) over ~30-s period are measured. The number at the bottom represents the number of error-free frames during the measurement period at a certain transmitted power level. *Source:* Li et al. [60], . Licenced under CC BY-4.0. © 2017 Macmillan Publishers.

Figure 12.39b–d shows the distributions of the relative position of beam centroid when the UAV was static on the ground with tracking system off, hovers in the air ~10 m above ground with tracking system on, moves horizontally in the air at a speed of ~0.1 m.s^{-1} with tracking system on, respectively. OAM $\ell = +3$ beam was transmitted during the measurements. The statistics of each scenario were obtained by continuously capture 1000 intensity profiles of the beam over a 120-s period using an infrared camera. The beam jitter variance was ~0.0218 mm^2 when the UAV was grounded, and it increased to ~0.0877 mm^2 and ~0.4604 mm^2 when the UAV was hovering and moving, respectively.

To verify link performance under flying conditions, Figure 12.39e shows the measurements of the BER for each 4096-symbol data frame when the OAM $\ell = +3$ and $\ell = -1$ beams were multiplexed and transmitted simultaneously for 40-m, 70-m, and 100-m round trip distance, each carrying a 40-Gbit/s QPSK signal. The transmitted power of each channel was fixed at 10 dBm. We observe that the measured BERs can be achieved under the 7% FEC limit most of the time and the OAM multiplexed link can achieve error-free transmission in this demonstration. A slight decrease in the number of error-free frames was observed when the transmission distance increases from 40 to 100 m, which may indicate a minor BER performance degradation.

12.5.2 Multiple-Input-Multiple-Output (MIMO) Mitigation for Atmospheric Turbulence in UAV Platforms

To mitigate atmospheric turbulence in UAV platforms, it has been demonstrated that the MIMO equalization algorithm can be utilized to compensate the turbulence-induced inter-channel cross-talk [61]. As shown in Figure 12.40, due to atmospheric turbulence, the OAM beams may be

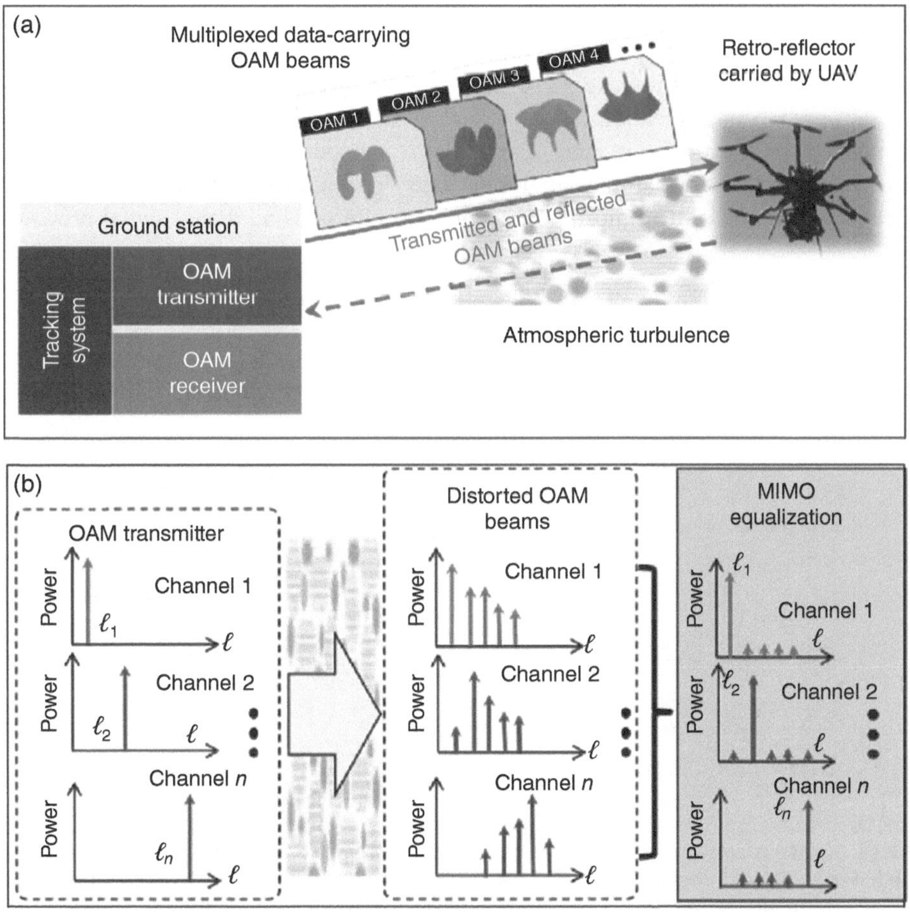

Figure 12.40 Concept of an OAM-multiplexed FSO link between a ground transmitter and a ground receiver via an unmanned-aerial-vehicle (UAV): (a) through atmospheric turbulence and (b) using MIMO equalization to mitigate turbulence. © 2019 Optical Society of America. *Source:* Li et al. [62]. (© [2018] IEEE)

distorted during their free-space propagation, such that the signal power on each particular OAM mode may be coupled to its neighboring modes. Therefore, the received signal at a particular mode may also contain crosstalk from its neighbors. MIMO equalization could help reduce crosstalk among channels by applying the inverse channel-coupling matrix to the received signals, thus mitigating performance degradation in a coherent optical communications link.

Figure 12.41a indicates the measured QPSK constellation diagrams and corresponding EVMs for OAM $\ell = +3$ and $\ell = -1$ beams with and without the MIMO equalization. The transmitted power for both channels was 10 dBm. We observe that the MIMO equalization can help reduce the EVM from 32 and 54 to 26 and 27% for OAM $\ell = +3$ and $\ell = -1$ beam, respectively. Figure 12.41b shows BERs for both channels as functions of transmitted power when the UAV was hovering ~50-m away with phase plate fixed at a random angle. We observe that the BERs dramatically decrease to below the 7% overhead FEC limit of 3.8×10^{-3} for both the OAM $\ell = +3$ and $\ell = -1$ beams after MIMO equalization.

12.5.3 Summary

This section discussed several demonstrations of OAM-multiplexed FSO links between a ground station and a UAV. OAM multiplexing has the potential to increase the data capacity and spectral efficiency of such aerial communications. To further employ OAM-based systems on different scenarios of airborne communications, several factors are likely to be taken into consideration: (i) Optimize OAM beams' size and divergence to maximize the received signal power; (ii) efficient PAT system to maintain alignment between the ground station and the UAV; and (iii) effective mitigation of the atmospheric turbulence effects using digital signal processing.

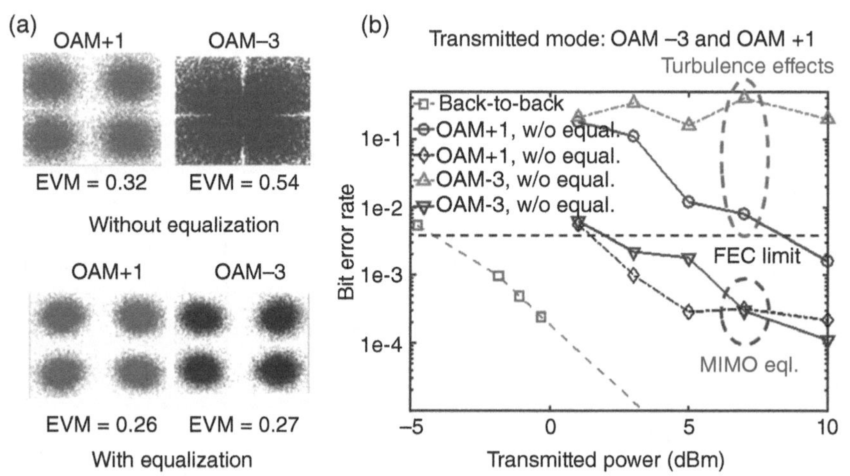

Figure 12.41 (a) Recovered QPSK constellations at transmitted power of 10 dBm for all channels with and without MIMO equalization when OAM +1 and −3 are transmitted under one turbulence realization; (b) BERs for both channels with and without MIMO equalization under one turbulence realization. The UAV is hovering ~50-m away. *Source:* Li et al. [62]. ()

12.6 OAM Multiplexing in Underwater Environments

There is growing interest in high-capacity underwater wireless communications in order to support the significant increase in demand for data, such as from sensor networks, unmanned vehicles, and submarines [63]. Traditionally, acoustic waves have been used for underwater communications, but this technique has quite limited bandwidth capacity [64]. Alternatively, optics in the low-attenuation blue-green region can enable higher-capacity underwater transmission links due to the much higher carrier-wave frequency [3]. In order to increase the capacity of underwater communications, one method would be to simultaneously transmit multiple independent data channels with each carried by a different OAM beam, i.e. OAM multiplexing.

As shown in Figure 12.42, maritime and underwater environments are sophisticated media because of various factors including water currents and oceanic turbulence, particle and aerosol scattering, the air–water interface, and water curvatures. In underwater FSO links, the oceanic turbulence and the particle scattering (e.g. $Mg(OH)_2$ and $Al(OH)_3$) are likely to distort the wavefront of the transmitted optical beams and induce dynamic beam wandering and possible inter-channel crosstalk. With respect to the FSO link across the air–water interface, it is desirable to manipulate the interaction between the light and water curvatures and enable the propagation of the light beams through that dynamic interface. One may also consider the effects of various aerosols on the beam's quality and on the link performance.

Moreover, the underwater environment presents several distinct challenges for a high-speed OAM link [32, 65]. For example, the OAM beam itself and the data it carries can be significantly degraded due to various widely varying effects, such as dynamic scattering/turbidity, water currents (i.e. underwater turbulence effects), and temperature gradients [66, 67]. Although these issues are challenging for nonOAM, single beam underwater links, the problem may escalate for systems using phase front-sensitive OAM beams.

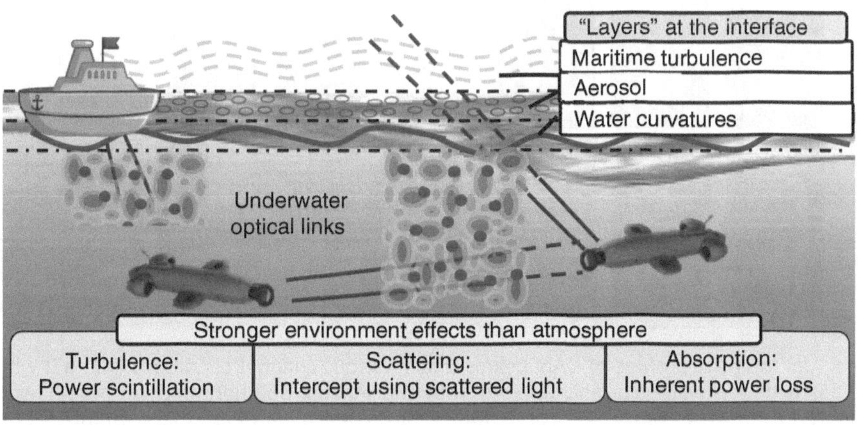

Figure 12.42 Challenges of different scenarios in underwater FSO communications.

12.6.1 Underwater Effects for OAM Beam Propagation

Generally, the underwater turbid environment includes various degrading effects for the OAM beam propagation, such as underwater turbulence induced by dynamic water currents [68], particle scattering effects consisting of suspended $Al(OH)_3$ and $Mg(OH)_2$ particles [4], and partial obstruction caused by floating air bubbles [69]. The spatial phase structure of the OAM beams is likely to be distorted by these environmental effects and an OAM-based underwater communication system may suffer from the distortion-induced inter-channel crosstalk.

There have been investigations in OAM beams' propagation through turbid media which includes water current and particle scattering effects. Figure 12.43 shows the distorted intensity profiles of OAM beams affected by different environmental effects. We observed that the ring-shaped intensity profiles of the OAM beams tend to be maintained after propagating through tap water and are slightly distorted by the water current. With respect to the Maalox solution, there was a small, time-varying change in the intensity profiles, which might be a result of the natural dynamic diffusive movement of $Al(OH)_3$ and $Mg(OH)_2$ particles in the water [66]. It has also been investigated that the OAM beams interact with the turbid medium in two scattering regimes including (i) ballistic (i.e. single-scattering events) regime: power loss was the main impairment in the ballistic scattering, while the mode purities of different OAM modes were not significantly affected and (ii) diffusive (i.e. multiple-scattering events) regime: the performance of an OAM-multiplexed link further suffered from the increased crosstalk between the transmitted OAM modes [71].

12.6.2 OAM Multiplexing Demonstrations in Underwater Environments

OAM multiplexing has the potential to increase the capacity (e.g., >10 Gbit/s) of optical wireless communications in the underwater environment. There have been various demonstrations of OAM multiplexed data transmission over a few meters in the emulated underwater environment

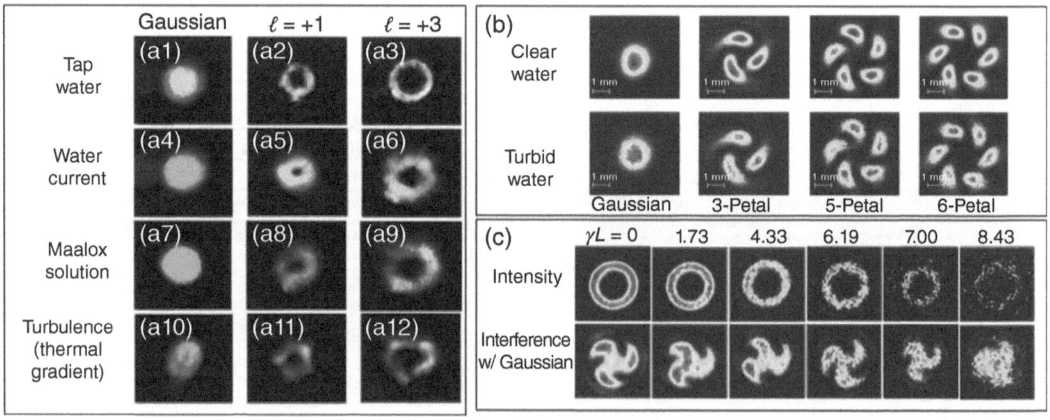

Figure 12.43 (a) Intensity profiles of OAM beams under various channel conditions: (a1–a3) with only tap water, (a4–a6) with the water current, (a7–a9) with the Maalox solution, and (a10–a12) with thermal gradient-induced turbulence. *Source:* Ren et al. [66], © 2017 Macmillan Publishers; (b) intensity profiles of Gaussian, 3-Petal (ℓ = +1, +2), 5-Petal (ℓ = +1, +4) and 6-Petal (ℓ = +2, +4) beams under clear water and turbid water condition. *Source:* Morgan et al. [70], © 2019 Optical Society of America.; (c) intensity and interferograms of an OAM ℓ = +3 beam after traversing the scattering medium for different optical depths γL. *Source:* Zhang et al. [71], © 2016 IOP Publishing Ltd.

[65, 66]. Figure 12.44 indicates the conceptual diagram of the high-capacity OAM multiplexed link in the underwater environment. A 1064-nm single-mode laser with a linewidth of less than 100 MHz was sent to a Lithium Niobate modulator to produce a 10 Gbit/s on–off keying (OOK) signal. The 10 Gbit/s OOK signal at 1064 nm was amplified by a ytterbium-doped fiber amplifier (YDFA) and fed into a collimator to convert the single-mode fiber output to a collimated Gaussian beam with a diameter of 2.6 mm. This Gaussian beam was then focused into the center of a periodically poled lithium niobate (PPLN) crystal for frequency doubling. A half-wave plate was inserted after the collimator to align the polarization of incident light with the polarization orientation of the PPLN crystal to maximize the conversion efficiency. As a result, a 532-nm green beam carrying a 10 Gbit/s OOK signal was produced. Furthermore, this frequency-doubling process was transparent to the intensity-based modulation format. After propagation through water, a phase mask with an inverse phase pattern of the desired OAM channel was used to convert the chosen OAM beam into a Gaussian-like beam. The other beams maintain their ring-shaped profiles and helical phases after passing through the phase mask. The Gaussian-like beam had a bright high intensity at the beam center and was thus separable from the other beams through spatial filtering (simply a pin hole). This beam was then focused onto a 9-GHz-bandwidth silicon APD, and the signal after the APD was amplified, filtered, and sent to a 10-Gbit/s OOK receiver for BER measurements. Given that four OAM beams each bearing a 10-Gbit/s data stream were transmitted, a total capacity of 40 Gbit/s was achieved.

Figure 12.45 presents the BER performance for an OAM-based underwater system with 4 multiplexed OAM beams each carrying 1-Gbit/s OOK signal under various conditions. Power penalties were measured to be 2.2 , 2.3 , and 2.7 dB in the cases of tap water, Maalox-induced scattering, and current, respectively. Due to the effects of thermal gradient-induced turbulence, the BERs were all above the 7% FEC limit, exhibiting a severe error-floor phenomenon, and power penalties were above 12 dB for all channels.

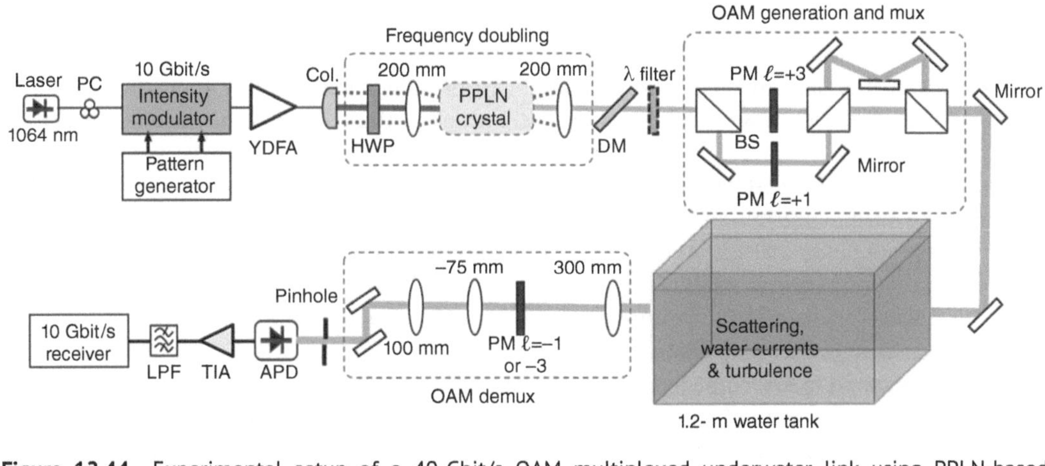

Figure 12.44 Experimental setup of a 40-Gbit/s OAM multiplexed underwater link using PPLN-based frequency doubling for signal generation. APD: avalanche photodetector. DM: dichroic mirror. HWP: half-wave plate. PM: metasurface phase mask. PPLN: periodically poled lithium niobate. YDFA: ytterbium-doped fiber amplifier. *Source:* Ren et al. [66]. Licenced under CC BY-4.0. © 2017 Macmillan Publishers

12.6.3 Multiple-Input-Multiple-Output (MIMO) Mitigation for Inter-Channel Crosstalk in Underwater Environments

MIMO electronic equalization has been widely employed in few-mode and multimode fiber-based mode division multiplexed systems to mitigate the mode coupling effects among multiple spatial modes [9, 72, 73]. In general, it is required that all the transmitted channels are simultaneously detected to enable multichannel equalization processing [9]. It has been demonstrated that the constant modulus algorithm (CMA)-based multichannel equalization can be employed to mitigate inter-channel crosstalk of OAM systems due to the thermal gradient-induced turbulence in the underwater environment [66].

As shown in Figure 12.46, a 2×2 CMA equalization algorithm was implemented in the DSP to recover two 10-Gbit/s OAM data channels with $\ell = +1$ and $\ell = +3$. The measured BERs were averaged over 1 min of data for both channels. Due to inter-channel crosstalk, the measured BER curves without 2×2 equalization reached the BER error floors. The power penalties at the FEC limit, compared to the B2B case, were below 2.0 dB for the two channels after equalization.

12.6.4 Summary

In this section, we discuss the utilization of OAM multiplexing in underwater communication systems. It has been demonstrated that OAM multiplexing can help increase the data capacity of FSO links in underwater environments. In the complicated turbid water environment, various complex physical phenomena including oceanic turbulence, particle scattering, and air bubbles, can significantly degrade the OAM beams' quality and induce inter-channel crosstalk. MIMO DSP has been applied to mitigate such channel interference and near error-free data transmission has been achieved.

12.7 Summary of this Chapter

OAM multiplexing is a potential technique to further increase the capacity and spectral efficiency of FSO links. Due to the unique spatial phase structure of the OAM beams, there are various challenges for OAM-based systems including (i) mode-dependent divergence; (ii) misalignment-induced power loss and inter-channel crosstalk; (iii) turbulence effects in either atmosphere or underwater environments; and (iv) possible obstructions in the light path of the co-propagating OAM beams.

To ensure high quality of data transmission in the OAM-multiplexed systems, various approaches to mitigating turbulence-induced inter-channel crosstalk have been demonstrated in different media. These methods include AO, optical beam combining, and MIMO electronic processing to reduce the coupling and crosstalk between the transmitted OAM beams. It has also been reported that a turbulence-resilient OAM multiplexing can be achieved by using OAM beam mixing at the Rx. With respect to the possible obstructions in the light path, the utilization of Bessel OAM beams is a potential method to overcome the effects of obstructions for limited-distance data transmission.

High-capacity OAM multiplexing systems have been demonstrated in different platforms and environments including the point-to-point fixed platform, the UAV platform, and the underwater environment. To employ an OAM multiplexing system in the field, one may consider several factors: (i) appropriate engineering of the OAM beams' divergence and the Rx aperture size; (ii) accurate beam tracking and misalignment correction; (iii) effective mitigation of turbulence-

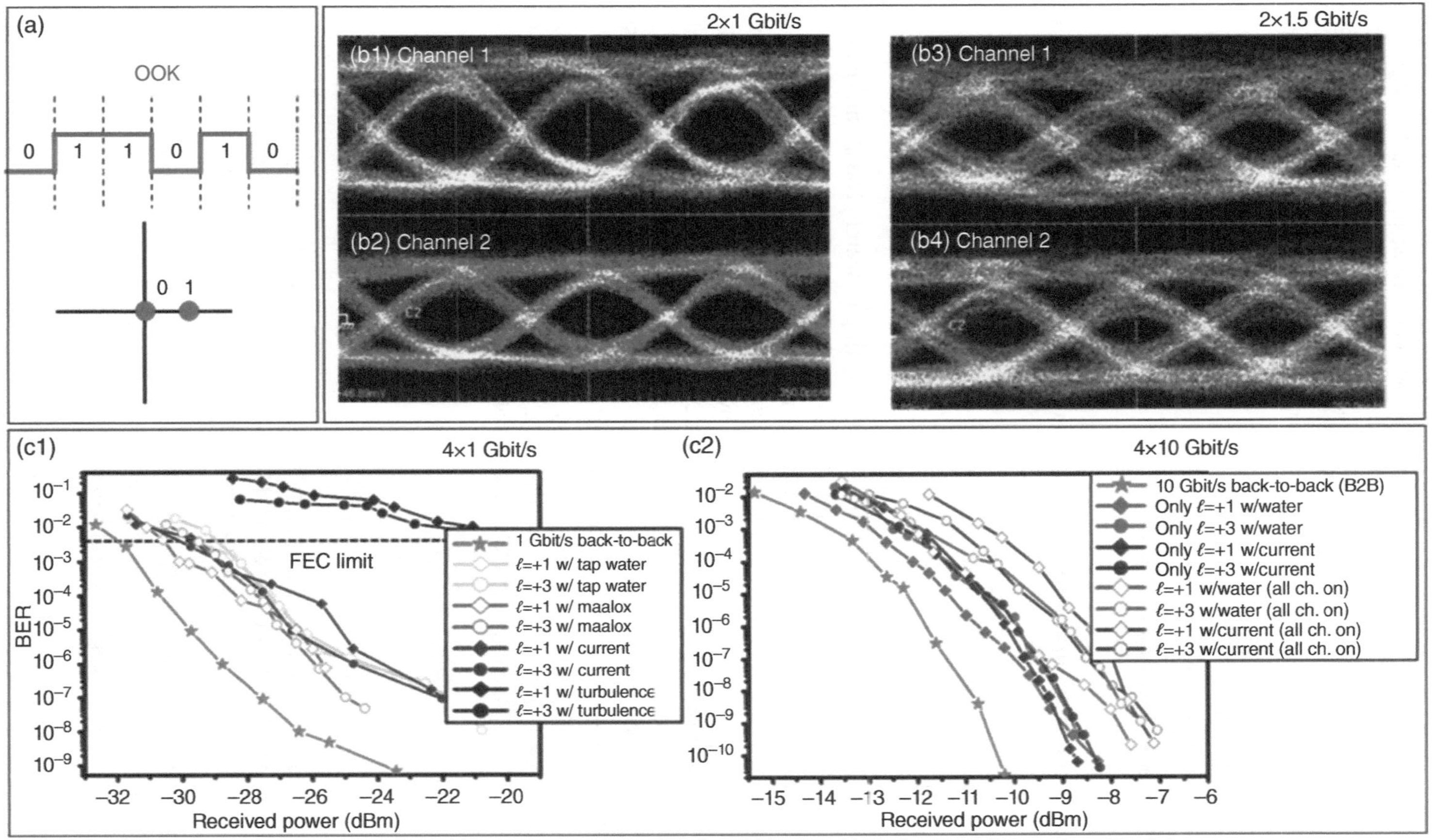

Figure 12.45 (a) Concept of on–off-keying (OOK) modulation. Eye diagram for *Source:* Modified from Baghdady et al. [65] (b1–b2) 1 Gbit/s channels and (b3–b4) 1.5 Gbit/s channels; (c1) BER performance with tap water and current for the 4-Gbit/s underwater link using directly modulated laser diodes with Maalox-induced scattering, current, and thermal gradient-induced turbulence; (c2) 40-Gbit/s underwater link using PPLN-based frequency doubling for the signal generation with tap water and current. *Source:* Ren et al. [66]. Licenced under CC BY-4.0. © 2016 Macmillan Publishers, and 2016 Optical Society of America.

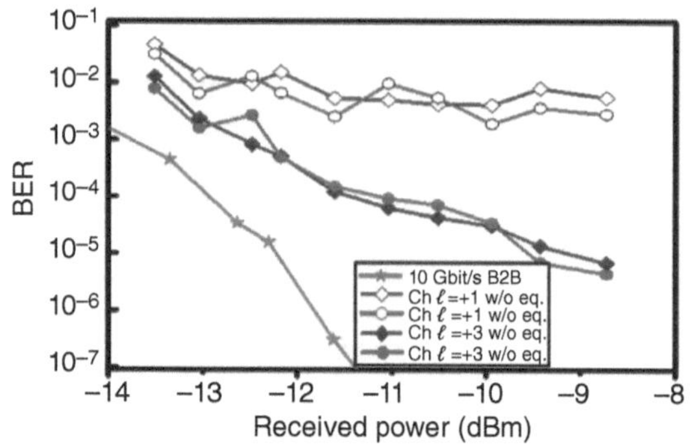

Figure 12.46 Measured BER curves of OAM channel $\ell = +1$ and $\ell = +3$ with and without CMA equalization. CMA: constant modulus algorithm. Ch.: channel. Eq.: equalization. *Source:* © 2017 Macmillan Publishers. *Source:* Ren et al. [66]. Licenced under CC BY-4.0.

induced inter-channel crosstalk; and (iv) low-loss (de)multiplexing of a large number (> 4) of OAM modes.

Acknowledgment

Qualcomm Innovation Fellowship; Airbus Institute for Engineering Research; Naval Information Warfare Center Pacific (N6600120C4704). We thank the support from Defense Advanced Research Projects Agency (DARPA) under the InPho (Information in a Photon) program; Vannevar Bush Faculty Fellowship sponsored by the Basic Research Office of the Assistant Secretary of Defense (ASD) for Research and Engineering (R&E) and funded by the Office of Naval Research (ONR) N00014-16-1-2813; National Science Foundation (NSF) ECCS-1509965 and IIP-1622777; Intel Labs Research Office; NSF MRI program; Air Force Office of Scientific Research (AFOSR) FA-9550–15–C-0024 and FA9550-16-C-0008; NxGen Partners; Air Force Research Laboratory (AFRL) FA8650-18-P-1699 and FA8650-20-C-1105; Office of Naval Research (ONR) N00014-15-1-2635; Office of Naval Research (ONR) N000141812352; Defense Security Cooperation Agency (DSCA) DSCA-4440646262 and DSCA-4441006051; Naval Air Warfare Center Aircraft Division N68335-18-C-0588; Nippon Telegraph and Telephone (NTT); Office of Naval Research through a MURI grant N00014-20-1-2558.

References

1 Ghassemlooy, Z., Arnon, S., Uysal, M. et al. (2015). Emerging optical wireless communications-advances and challenges. *IEEE Journal on Selected Areas in Communications* **33** (9): 1738–1749.

2 Majumdar, A.K. (2015). *Advanced Free Space Optics (FSO): A Systems Approach*. New York: Springer-Verlag.

3 Zhu, X. and Kahn, J.M. (2002). Free-space optical communication through atmospheric turbulence channels. *IEEE Transactions on Communications* **50** (8): 1293–1300.

4 Hanson, F. and Radic, S. (2008). High bandwidth underwater optical communication. *Applied Optics* **47** (2): 277–283.

5 Young, D.W., Sluz, J.E., Juarez, J.C., et al. (2007). Demonstration of high data rate wavelength division multiplexed transmission over a 150 km free space optical link. *MILCOM 2007 - IEEE Military Communications Conference*, 1–6.

6 Leitgeb, E., Zettl, K., Muhammad, S.S., et al. (2007). Investigation in free space optical communication links between unmanned aerial vehicles (UAVs). *2007 9th International Conference on Transparent Optical Networks.*

7 Ciaramella, E., Arimoto, Y., Contestabile, G., et al. (2009). 1.28 Terabit.s^{-1} (32×40 Gbit/s) WDM transmission over a double-pass free space optical link. *2009 Conference on Optical Fiber Communication*, San Diego, USA (22–26 March 2009) paper OMN7. IEEE.

8 Cvijetic, N., Qian, D., Yu, J. et al. (2010). Polarization-multiplexed optical wireless transmission with coherent detection. *Journal of Lightwave Technolology* **28** (8): 1218–1227.

9 Richardson, D.J., Fini, J.M., and Nelson, L.E. (2013). Space-division multiplexing in optical fibres. *Nature Photonics* **7** (5): 354–362.

10 Winzer, P.J. (2014). Making spatial multiplexing a reality. *Nature Photonics* **8** (5): 345–348.

11 Yan, Y., Xie, G., Lavery, M.P.J. et al. (2014). High-capacity millimetre-wave communications with orbital angular momentum multiplexing. *Nature Communications* **5** (1).

12 Gibson, G., Courtial, J., Padgett, M.J. et al. (2004). Free-space information transfer using light beams carrying orbital angular momentum. *Optics Express* **12** (22): 5448–5456.

13 Wang, J., Yang, J.-Y., Fazal, I.M. et al. (2012). Terabit free-space data transmission employing orbital angular momentum multiplexing. *Nature Photonics* **6** (7): 488–496.

14 Willner, A.E., Huang, H., Yan, Y. et al. (2015). Optical communications using orbital angular momentum beams. *Advances in Optics and Photonics* **7** (1): 66.

15 Huang, H., Xie, G., Yan, Y. et al. (2014). 100 Tbit/s free-space data link enabled by three-dimensional multiplexing of orbital angular momentum, polarization, and wavelength. *Optics Letters* **39** (2): 197.

16 Wang, J., Li, S., Luo, M. et al. (2014). N-dimentional multiplexing link with 1.036-Pbit.s^{-1} transmission capacity and 112.6-bit.s^{-1}.Hz^{-1} spectral efficiency using OFDM-8QAM signals over 368 WDM pol-muxed 26 OAM modes. *2014 The European Conference on Optical Communication (ECOC)*, Cannes, France (21–25 September 2014) paper Mo.4.5.1. IEEE.

17 Xie, G., Li, L., Ren, Y. et al. (2015). Performance metrics and design considerations for a free-space optical orbital-angular-momentum-multiplexed communication link. *Optica* **2** (4): 357–365.

18 Malik, M., O'Sullivan, M., Rodenburg, B. et al. (2012). Influence of atmospheric turbulence on optical communications using orbital angular momentum for encoding. *Optics Express* **20** (12): 13195–13200.

19 Ren, Y., Huang, H., Xie, G. et al. (2013). Atmospheric turbulence effects on the performance of a free space optical link employing orbital angular momentum multiplexing. *Optics Letters* **38** (20): 4062–4065.

20 Franke-Arnold, S., Barnett, S.M., Yao, E. et al. (2004). Uncertainty principle for angular position and angular momentum. *New Journal of Physics* **6**: 103–103.

21 Goodman, J.W. (2015). *Statistical Optics*. Hoboken, NJ, USA: John Wiley & Sons.

22 Willner, A.E., Ren, Y., Xie, G. et al. (2017). Recent advances in high-capacity free-space optical and radio-frequency communications using orbital angular momentum multiplexing. *Philosophical Transactions of the Royal Society A* **375**: 20150439.

23 Paterson, C. (2005). Atmospheric turbulence and orbital angular momentum of single photons for optical communication. *Physical Review Letters* **94** (15): 153901.

24 Tyler, G.A. and Boyd, R.W. (2009). Influence of atmospheric turbulence on the propagation of quantum states of light carrying orbital angular momentum. *Optics Letters* **34** (2): 142–144.

25 Anguita, J.A., Neifeld, M.A., and Vasic, B.V. (2008). Turbulence-induced channel crosstalk in an orbital angular momentum-multiplexed free-space optical link. *Applied Optics* **47** (13): 2414–2429.

26 Chandrasekaran, N. and Shapiro, J.H. (2014). Photon information efficient communication through atmospheric turbulence–part I: Channel model and propagation statistics. *Journal of Lightwave Technology* **32** (6): 1075–1087.

27 Djordjevic, I.B. (2011). Deep-space and near-Earth optical communications by coded orbital angular momentum (OAM) modulation. *Optics Express* **19** (15): 14277–14289.

28 Awaji, Y., Wada, N., and Toda, Y. (2010). Demonstration of spatial mode division multiplexing using Laguerre-Gaussian mode beam in telecom-wavelength. *2010 23rd Annual Meeting of the IEEE Photonics Society*, Denver, CO (7–11 November 2010), 551–552. IEEE.

29 Martelli, P., Gatto, A., Boffi, P., and Martinelli, M. (2011). Free-space optical transmission with orbital angular momentum division multiplexing. *Electronics Letters* **47** (17): 972–973.

30 Ren, Y., Wang, Z., Liao, P. et al. (2016). Experimental characterization of a 400 Gbit/s orbital angular momentum multiplexed free-space optical link over 120-meters. *Optics Letters* **41** (3): 622–625.

31 Krenn, M., Handsteiner, J., Fink, M. et al. (2016). Twisted light transmission over 143 km. *Proceedings of the National Academy of Sciences of the United States of America* **113** (48): 13648–13653.

32 Willner, A.E., Zhao, Z., Ren, Y. et al. (2018). Underwater optical communications using orbital angular momentum-based spatial division multiplexing. *Optics Communications* **408**: 21–25.

33 Yao, A.M. and Padgett, M.J. (2011). Orbital angular momentum: origins, behavior and applications. *Advances in Optics and Photonics* **3** (2): 161–204.

34 Allen, L., Beijersbergen, M.W., Spreeuw, R.J.C., and Woerdman, J.P. (1992). Orbital angular momentum of light and the transformation of Laguerre-Gaussian laser modes. *Physical Review A* **45** (11): 8185–8189.

35 Pang, K., Song, H., Zhao, Z. et al. (2018). 400-Gbit/s QPSK free-space optical communication link based on four-fold multiplexing of Hermite-Gaussian or Laguerre-Gaussian modes by varying both modal indices. *Optics Letters* **43** (16): 3889–3892.

36 Xie, G., Ren, Y., Yan, Y. et al. (2016). Experimental demonstration of a 200-Gbit/s free-space optical link by multiplexing Laguerre–Gaussian beams with different radial indices. *Optics Letters* **41** (15): 3447–3450.

37 Krenn, M., Handsteiner, J., Fink, M. et al. (2015). Twisted photon entanglement through turbulent air across Vienna. *Proceedings of the National Academy of Sciences of the United States of America* **112** (46): 14194–14201.

38 Ren, Y., Xie, G., Huang, H. et al. (2014). Adaptive optics compensation of multiple orbital angular momentum beams propagating through emulated atmospheric turbulence. *Optics Letters* **39** (10): 2845–2848.

39 Ren, Y., Xie, G., Huang, H. et al. (2014). Adaptive-optics-based simultaneous pre- and post-turbulence compensation of multiple orbital-angular-momentum beams in a bidirectional free-space optical link. *Optica* **1** (6): 376–382.

40 Mirhosseini, M., Magaña-Loaiza, O.S., O'Sullivan, M.N. et al. (2015). High-dimensional quantum cryptography with twisted light. *New Journal of Physics* **17** (3): 033033.

41 Liu, C., Pang, K., Zhao, Z. et al. (2019). Single-end adaptive optics compensation for emulated turbulence in a bi-directional 10-Mbit.s^{-1} per channel free-space quantum communication link using orbital-angular-momentum encoding. *Research* **2019**: 8326701.

42 Weyrauch, T. and Vorontsov, M.A. (2008). Free-space laser communications with adaptive optics: Atmospheric compensation experiments. In: *Free-Space Laser Communications: Principles and Advances* (eds. A.K. Majumdar and J.C. Ricklin), 247–271. New York, NY: Springer.

43 Weyrauch, T. and Vorontsov, M.A. (2005). Atmospheric compensation with a speckle beacon in strong scintillation conditions: directed energy and laser communication applications. *Applied Optics* **44** (30): 6388–6401.

44 Xie, G., Ren, Y., Huang, H. et al. (2015). Phase correction for a distorted orbital angular momentum beam using a Zernike polynomials-based stochastic-parallel-gradient-descent algorithm. *Optics Letters* **40** (7): 1197–1200.

45 Song, H., Song, H., Zhang, R. et al. (2020). Experimental mitigation of atmospheric turbulence effect using pre-signal combining for uni- and bi-directional free-space optical links with two 100-Gbit/s OAM-multiplexed channels. *Journal of Lightwave Technology* **38** (1): 82–89.

46 Pang, K., Song, H., Su, X., et al. (2020). Simultaneous orthogonalizing and shaping of multiple LG beams to mitigate crosstalk and power loss by transmitting each of four data channels on multiple modes in a 400-Gbit/s free-space link. *Optical Fiber Communication Conference (OFC) 2020*, paper W1G.2.

47 Pang, K., Song, H., Su, X. et al. (2020). Experimental mitigation of the effects of the limited size aperture or misalignment by singular-value-decomposition-based beam orthogonalization in a free-space optical link using Laguerre–Gaussian modes. *Optics Letters* **45** (22): 6310–6313.

48 Huang, H., Cao, Y., Xie, G. et al. (2014). Crosstalk mitigation in a free-space orbital angular momentum multiplexed communication link using 4×4 MIMO equalization. *Optics Letters* **39** (15): 4360–4363.

49 Yang, C., Xu, C., Ni, W. et al. (2017). Turbulence heterodyne coherent mitigation of orbital angular momentum multiplexing in a free space optical link by auxiliary light. *Optics Express* **25** (21): 25612–25624.

50 Zhang, R., Hu, N., Zou, K., et al. (2020). Experimental demonstration of crosstalk reduction to achieve turbulence-resilient multiple-OAM-beam free-space optical communications using pilot tones to mix beams at the receiver. *Conference on Lasers and Electro-Optics 2020*, Optical Society of America (10–15 May 2020), paper SW4L.4. Optical Society of America, Washington, DC, USA.

51 Zhou, H., Song, H., Zhang, R., et al., (2020). Demonstration of turbulence resiliency in a mode-, polarization-, and wavelength- multiplexed free-space optical link using pilot tones and optoelectronic wave mixing, *European Conference on Optical Communication (ECOC) 2020*, Belgium, Brussels, paper We2G-4.

52 Cunningham, J.S., Foulke, D., Goode, T., et al. (2009). Long range field testing of free space optical communications terminals on mobile platforms. *MILCOM 2009 – 2009 IEEE Military Communications Conference* Boston, USA (18–21 October 2009). IEEE.

53 Muhammad, S., Plank, T., Leitgeb, E., et al. (2008). Challenges in establishing free space optical communications between flying vehicles. *2008 6th International Symposium on Communication Systems, Networks and Digital Signal Processing*, IEEE, 82–86.

54 Fields, R., Lunde, C., Wong, R. et al. (2009). NFIRE-to-TerraSAR-X laser communication results: satellite pointing, disturbances, and other attributes consistent with successful performance. *SPIE Defense, Security, and Sensing* **2009**: 73300Q.

55 Zarganis, K.E. and Hatziefremidis, A. (2015). Performance analysis of coherent optical OFDM applied to UAV mobile FSO Systems. *Journal of Optics and Photonics* **3** (1): 5.

56 Kaadan, A., Refai, H., and Lopresti, P. (2016). Spherical FSO receivers for UAV communication: geometric coverage models. *IEEE Transactions on Aerospace and Electronic Systems* **52** (5): 2157–2167.

57 Kaadan, A., Refai, H.H., and LoPresti, P.G. (2014). Multielement FSO transceivers alignment for inter-UAV communications. *Journal of Lightwave Technology* **32** (24): 4183–4193.

58 Kazaura, K., Omae, K., Suzuki, T. et al. (2006). Enhancing performance of next generation FSO communication systems using soft computing based predictions. *Optics Express* **14** (12): 4958–4968.

59 Li, L., Zhang, R., Xie, G. et al. (2018). Experimental demonstration of beaconless beam displacement tracking for an orbital angular momentum multiplexed free-space optical link. *Optics Letters* **43** (10): 2392–2395.

60 Li, L., Zhang, R., Zhao, Z. et al. (2017). High-capacity free-space optical communications between a ground transmitter and a ground receiver via a UAV using multiplexing of multiple orbital-angular-momentum beams. *Scientific Reports* **7** (1).

61 Li, L., Zhang, R., Liao, P. et al. (2019). Mitigation for turbulence effects in a 40-Gbit/s orbital-angular-momentum-multiplexed free-space optical link between a ground station and a retro-reflecting UAV using MIMO equalization. *Optics Letters* **44** (21): 5181–5184.

62 Li, L., Zhang, R., Liao, P., et al., (2018). MIMO Equalization to Mitigate Turbulence in a 2-Channel 40-Gbit/s QPSK Free-Space Optical 100-m Round-Trip Orbital-Angular-Momentum-Multiplexed Link Between a Ground Station and a Retro-Reflecting UAV, 2018 European Conference on Optical Communication (ECOC). (© [2018] IEEE)

63 Friedman, N. and United States Naval Institute (1989). *The Naval Institute Guide to World Naval Weapons Systems*. Annapolis, Md: Naval Institute Press.

64 Stojanovic, M. (1996). Recent advances in high-speed underwater acoustic communications. *IEEE Journal of Oceanic Engineering* **21** (2): 125–136.

65 Baghdady, J., Miller, K., Morgan, K. et al. (2016). Multi-gigabit.s^{-1} underwater optical communication link using orbital angular momentum multiplexing. *Optics Express* **24** (9): 9794–9805.

66 Ren, Y., Li, L., Wang, Z. et al. (2016). Orbital angular momentum-based space division multiplexing for high-capacity underwater optical communications. *Scientific Reports* **6** (1): 33306.

67 Wang, W.B., Gozali, R., Shi, L. et al. (2016). Deep transmission of Laguerre–Gaussian vortex beams through turbid scattering media. *Optics Letters* **41** (9): 2069.

68 Johnson, L.J., Green, R.J., and Leeson, M.S. (2014). Underwater optical wireless communications: depth-dependent beam refraction. *Applied Optics* **53** (31): 7273–7277.

69 Zedini, E., Oubei, H., Kammoun, A. et al. (2019). Unified statistical channel model for turbulence-induced fading in underwater wireless optical communication systems. *IEEE Transactions on Communications* **67** (4): 2893–2907.

70 Morgan, K.S., Miller, J.K., Cochenour, B.M. et al. (2016). Free space propagation of concentric vortices through underwater turbid environments. *Journal of Optics* **18**: 104004.

71 Zhang, R., Li, L., Zhao, Z. et al. (2019). Coherent optical wireless communication link employing orbital angular momentum multiplexing in a ballistic and diffusive scattering medium. *Optics Letters* **44** (3): 691–694.

72 Randel, S., Ryf, R., Sierra, A. et al. (2011). 6×56-Gb.s^{-1} mode-division multiplexed transmission over 33-km few-mode fiber enabled by 6×6 MIMO equalization. *Optics Express* **19** (17): 16697–16707.

73 Ryf, R., Essiambre, R.-J., Randel, S. et al. (2011). MIMO-based crosstalk suppression in spatially multiplexed 3×56-Gb.s^{-1} PDM-QPSK signals for strongly coupled three-core fiber. *IEEE Photonics Technology Letters* **23** (20): 1469–1471.

Part IV

Multidisciplinary Explorations of Electromagnetic Vortices

13

Theory of Vector Beams for Chirality and Magnetism Detection of Subwavelength Particles

Mina Hanifeh and Filippo Capolino

Department of Electrical Engineering and Computer Science, University of California, Irvine, CA, USA

In this chapter, a theoretical treatment of two kinds of cylindrical vector beams and their composition is presented. A vector beam is an electromagnetic wave with a specific polarization and localized distribution in the transverse plane to the propagation axis [1]. Indeed, scalar electromagnetic beams [2, 3], which are a solution of the Helmholtz equation, are special cases of vector beams when a linear polarization for electric field is assumed. Cylindrical vector beams are a group of vector beams with polarization and field distribution in the transverse plane which respect cylindrical symmetry [4] and have been adopted for various applications, for example, in optical systems such as spectrometers [5–7], microscopes [8], and optical tweezers [9, 10]. Also, applications in radio frequency and antennas have been proposed [11–15].

Here, we focus our attention to practical cylindrical vector beams namely the radially and azimuthally polarized beams [16, 17] and their superposition [6, 18, 19]. Since applications of these beams rely on the specific ad hoc profiles and structures of their electric and magnetic field components, we start our discussions with the characterization of azimuthally (APB) and radially polarized beams (RPBs). These discussions are fundamental to the following sections where we briefly investigate how these beams are employed in magnetic and chirality force microscopy [6, 7, 20–23], and chirality detection of subwavelenth-size matter samples by using a generalization of circular dichroism. We conveniently use cylindrical coordinate system (ρ, φ, z) where the z axis is the optical axis of the beam, i.e. the longitudinal coordinate, for the characterization of cylindrical symmetric vector beams. Moreover, without loss of generality, time harmonic electromagnetic fields with time dependence $e^{-i\omega t}$ are assumed where ω is the angular frequency. Bold fonts denote phasor vector quantities and for the sake of simplicity, the time dependence factor $e^{-i\omega t}$ is dropped throughout this chapter. A hat '^' is used to denote unit vectors.

13.1 Characterization of Azimuthally and Radially Polarized Beams

In this section, we present an analytical treatment of APBs and RPBs as well as their superposition, which we call an azimuthally radially polarized beam (ARPB), to provide some insight about the potential applications of these vector beams discussed later in this chapter.

Electromagnetic Vortices: Wave Phenomena and Engineering Applications, First Edition.
Edited by Zhi Hao Jiang and Douglas H. Werner.

The electric component **E** of any free electromagnetic field with an arbitrary polarization is a solution of the differential vector wave equation [1]

$$\nabla \times \nabla \times \mathbf{E} - k^2\mathbf{E} = 0, \tag{13.1}$$

where k is the wave-number in the host medium. In an electromagnetic vector beam propagating along the z axis, the electric field takes the form $\mathbf{E} = \mathbf{E}_0 e^{ikz}$ and $\mathbf{E}_0$ is obtained by solving Eq. (13.1) while the paraxial approximation $\frac{\partial^2 \mathbf{E}_0}{\partial z^2} = 0$ guarantees localization of the beam in the transverse (ρ, φ) plane [1]. For example, we may assume that the electric field is polarized along the azimuthal direction, i.e. $\mathbf{E}^{\text{APB}} = E_0 e^{ikz}\hat{\varphi}$ and obtain the expression of E_0. This beam is called an azimuthally polarized vector beam [4, 16]. The electric field $\mathbf{E}^{\text{APB}}$ may also be obtained by superimposing Laguerre Gaussian (LG) beams carrying orbital angular momenta of orders +1 and −1 [16]

$$\mathbf{E}^{\text{APB}} = \frac{-i\sqrt{2}V_{\text{A}}}{2}(u_{-1,0}\,\hat{\mathbf{e}}_{\text{LH}} - u_{1,0}\,\hat{\mathbf{e}}_{\text{RH}})e^{ikz}. \tag{13.2}$$

Here, V_{A} is the complex amplitude of the beam with unit of Volt and $u_{l,\,p}$ is a normalized Laguerre–Gaussian (LG) mode with orbital l and radial p numbers given by [11, 14, 24]:

$$u_{l,p} = \sqrt{\frac{2p!}{\pi(p+|l|)!}}\frac{1}{w}\left[\frac{\rho\sqrt{2}}{w}\right]^{|l|} e^{-\frac{\rho^2}{w^2}\zeta} L_p^{(|l|)}\left(\frac{2\rho^2}{w^2}\right) e^{-i(2p+|l|+1)\tan^{-1}\left(\frac{z}{z_R}\right)} e^{il\varphi}. \tag{13.3}$$

This scalar field is a solutions of scalar Helmholtz equation in cylindrical coordinate system [14]. Unit vectors $\hat{\mathbf{e}}_{\text{RH}} = (\hat{\mathbf{x}} - i\hat{\mathbf{y}})/\sqrt{2}$ and $\hat{\mathbf{e}}_{\text{LH}} = (\hat{\mathbf{x}} + i\hat{\mathbf{y}})/\sqrt{2}$ are plane waves with right-handed and left-handed circular polarization (RCP and LCP). Please note that there are mainly two distinct definitions for circular polarization: one is commonly used in the optics community [25, section 2.4], while the other is frequently used by the classic electromagnetic community [26, p. 24]. In this paper we have adopted the first one. Also note that in the classical electromagnetic community, the time notation $e^{i\omega t}$ is often chosen, whereas here we have assumed that the time dependence is $e^{-i\omega t}$. To transform a field representation from one notation to the other, the complex conjugate operation shall be applied.

The parameters w, R, and ζ are defined as [11]

$$\begin{aligned} w &= w_0\sqrt{1+(z/z_R)^2} \\ R &= z\left(1+(z_R/z)^2\right) \\ \zeta &= 1 - iz/z_R, \end{aligned} \tag{13.4}$$

where w_0 is the beam parameter that corresponds to the one-half of the beam waist of the fundamental Gaussian beam (GB) with $(l, p) = (0, 0)$, $z_R = \pi w_0^2/\lambda$ is the Rayleigh range, and λ is the ambient wavelength. Moreover, $L_p^l(\cdot)$ is the associate Laguerre polynomial given as [11]

$$L_p^{(|l|)}\left(\frac{2\rho^2}{w^2}\right) = \sum_{m=0}^{p}\frac{(-1)^m(p+|l|)!}{(p-m)!(|l|+m)!m!}\left(\frac{2\rho^2}{w^2}\right)^m. \tag{13.5}$$

Let us consider an APB with radial order $p = 0$ and its electric field component simplifies to [16]

$$\mathbf{E}^{\text{APB}} = E_\varphi^{\text{APB}}\hat{\varphi} = \frac{2\rho V_{\text{A}}}{\sqrt{\pi}w^2}e^{-(\rho/w)^2\zeta}e^{-2i\tan^{-1}(z/z_R)}e^{ikz}\hat{\varphi}, \tag{13.6}$$

which shows that the electric field in an APB with radial number $p = 0$ is locally oriented along the azimuthal direction $\hat{\varphi}$ and also that the electric field $\mathbf{E}^{\mathrm{APB}}$ vanishes at the optical axis $\rho = 0$, at any given z. The field reaches the maximum of its magnitude [16]

$$\left|E_{\varphi}^{\mathrm{APB}}(\rho_M, z)\right| = |V_{\mathrm{A}}| \frac{2\rho}{\sqrt{\pi} w^2} e^{-\left(\frac{\rho}{w}\right)^2}\Bigg|_{\rho=\rho_M} = |V_{\mathrm{A}}| \frac{1}{w}\sqrt{\frac{2}{\pi e}} \tag{13.7}$$

at the radial distance $\rho = \rho_M$ where $\rho_M = w/\sqrt{2}$. The magnetic field of an APB is retrieved from the Maxwell equation $\mathbf{H} = (i\omega\mu_0)^{-1}\nabla \times \mathbf{E}$ that leads to the radial and longitudinal components [18]

$$\begin{aligned} H_{\rho}^{\mathrm{APB}} &= -\frac{E_{\varphi}^{\mathrm{APB}}}{\eta_0}\left(1 + \frac{1}{kz_R}\frac{\rho^2 - 2w_0^2}{w^2} - \frac{4i}{w^2}\frac{z}{z_R}\frac{1}{k^2}\left(1 - \frac{\rho^2\zeta}{w^2}\right)\right) \\ H_{z}^{\mathrm{APB}} &= \frac{2E_{\varphi}^{\mathrm{APB}}}{i\eta_0 k\rho}\left(1 - \frac{\rho^2}{w^2}\zeta\right), \end{aligned} \tag{13.8}$$

where μ_0 is the magnetic permeability of the vacuum host medium and η_0 is the free space wave impedance. The transverse component of the magnetic field H_{ρ}^{APB} is proportional to the azimuthal component of the electric field $E_{\varphi}^{\mathrm{APB}}$ and vanishes on the optical axis at $\rho = 0$. However, the longitudinal component of the magnetic field, H_{z}^{APB}, takes its maximum value on the beam axis and is equal to [16]

$$\left|H_{z}^{\mathrm{APB}}\right|_{\rho=0} = \frac{4|V_{\mathrm{A}}|}{w^2\omega\mu_0\sqrt{\pi}}. \tag{13.9}$$

Therefore, on the beam axis ($\rho = 0$) of an APB, for symmetry reasons, there are no transverse electric and transverse magnetic fields. This means that an APB provides a magnetic dominant region around its optical axis with purely longitudinal polarization, and this property can be used in a variety of applications [7, 16]. In Figure 13.1a, the components of electric and magnetic fields in an APB are shown normalized to the maximum value of the magnitude of the electric field.

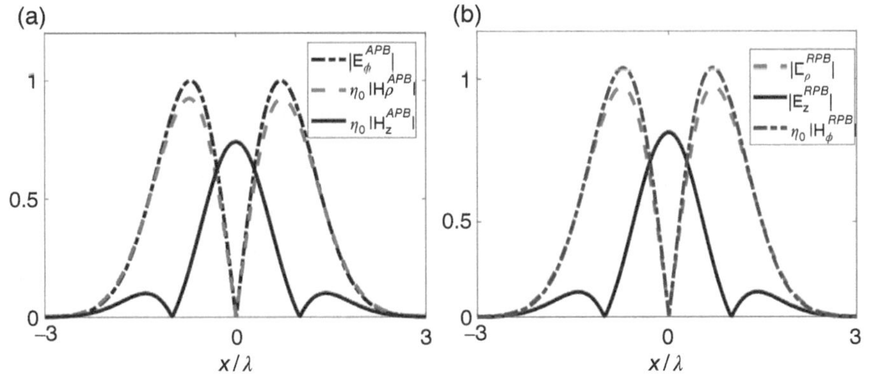

Figure 13.1 Electric and magnetic field magnitudes normalized to the maximum of electric field's magnitude, varying transverse coordinate x axis, evaluated at $z = 0$ plane (minimum waist plane) of an APB (a), and an RPB (b), with a beam waist parameter $w_0 = \lambda$.

The existence of a magnetic dominant region around the optical axis (black solid line) may have unique applications. For example, an APB enables the optical-magnetic interaction of light with matter that would be otherwise masked by the interaction of matter with the electric field. Indeed matter also expresses magnetic dipolar transitions [27, 28], besides the well-known electric dipolar transitions that are used in the majority of applications of light-matter interaction like solar cells, lasers, nonlinear optics, photodetectors, etc. This property of the APB empowers the development of future scanning probe microscopy and spectroscopy to study magnetism in matter at nanoscale [7, 20].

A similar approach to the analytical treatment of vector beams explained above is applied to the magnetic field by solving

$$\nabla \times \nabla \times \mathbf{H} - k^2\mathbf{H} = 0 \tag{13.10}$$

assuming the field to be represented as $\mathbf{H} = \mathbf{H}_0 e^{ikz}$ together with the paraxial approximation $\partial^2\mathbf{H}_0/\partial z^2 = 0$. For example, a particular solution is the RPB that has magnetic field polarized along the azimuthal direction $\mathbf{H}_0 = H_0 e^{ikz}\hat{\varphi}$ [4, 29]. Since Eqs. (13.10) and (13.11) are similar in form, one may readily obtain the solution

$$\mathbf{H}^{\mathrm{RPB}} = \frac{2\rho V_{\mathrm{R}}}{\eta_0\sqrt{\pi}w^2} e^{-(\rho/w)^2\zeta} e^{-2i\tan^{-1}(z/z_{\mathrm{R}})} e^{ikz}\hat{\varphi} \tag{13.11}$$

for the magnetic field of an RPB, where V_{R} is the complex amplitude with unit of Volt. The components of electric field [18]

$$\begin{aligned} E_\rho^{\mathrm{RPB}} &= \eta_0 H_\varphi^{\mathrm{RPB}}\left(1 + \frac{1}{kz_R}\frac{\rho^2 - 2w_0^2}{w^2} - \frac{4i}{w^2}\frac{z}{z_R}\frac{1}{k^2}\left(1 - \frac{\rho^2\zeta}{w^2}\right)\right) \\ E_z^{\mathrm{RPB}} &= -\frac{2\eta_0 H_\varphi^{\mathrm{RPB}}}{ik\rho}\left(1 - \frac{\rho^2}{w^2}\zeta\right) \end{aligned} \tag{13.12}$$

are obtained from Maxwell equation $\mathbf{E}^{\mathrm{RPB}} = (-i\omega\epsilon_0)^{-1}\nabla\times\mathbf{H}^{\mathrm{RPB}}$ where ϵ_0 is dielectric permittivity of the vacuum host medium. This beam is called an RPB since its transverse electric field component is polarized along the radial direction ρ. Often the electric field is the component chosen to describe the polarization of optical beams, a biased choice that also reveals that, in optics, the interaction with matter is historically dictated by the electric field, and that is why in this chapter we are focusing also on the magnetic field of light that paves the way to the exclusive interaction of magnetic field with matter. In an RPB, the longitudinal electric field is nonzero on its optical axis, where the other components of electric and magnetic fields vanish, which indicates the existence of an electric dominant region around the beam optical axis, a condition dual to the one for the APB. On the RPB optical axis, only the longitudinal component of the electric field vector exists, which makes it a good candidate for probing electric properties of matter at subwavelength scale. In Figure 13.1b, the components of electric and magnetic fields in an RPB, normalized to the maximum value of the magnitude of the electric field, are illustrated. An APB and an RPB with same beam parameters are electromagnetic dual and are linked via the relation

$$\begin{aligned} \mathbf{E}^{\mathrm{APB}} &= \eta_0\mathbf{H}^{\mathrm{RPB}}\left(\frac{V_{\mathrm{A}}}{V_{\mathrm{R}}}\right) \\ \mathbf{E}^{\mathrm{RPB}} &= -\eta_0\mathbf{H}^{\mathrm{APB}}\left(\frac{V_{\mathrm{R}}}{V_{\mathrm{A}}}\right). \end{aligned} \tag{13.13}$$

In addition to the features that each of APB and RPB offers, their combination, called ARPB in this chapter, provides a new interesting beam with unique properties [6, 18, 19]. As discussed earlier, APB and RPB, respectively, provide magnetic and electric dominant regions around their optical axes with their longitudinally polarized field components. The superposition of an APB and an RPB combines these features and a proper phase shift can also be added between the two longitudinal components. Let us assume we have a linear combination of an APB with complex amplitude V_A and an RPB with complex amplitude V_R as [18]

$$\begin{aligned} \mathbf{E}^{\mathrm{ARPB}} &= \mathbf{E}^{\mathrm{APB}} + \mathbf{E}^{\mathrm{RPB}}, \\ \mathbf{H}^{\mathrm{ARPB}} &= \mathbf{H}^{\mathrm{APB}} + \mathbf{H}^{\mathrm{RPB}}, \end{aligned} \tag{13.14}$$

where the superscript ARPB indicates 'azimuthally radially polarized beam.' The fascinating feature of this beam is that it possesses purely longitudinal electric and magnetic field vectors along its optical axis, where all transverse components of electric and magnetic field vectors are not present. The relative phases and magnitudes of the amplitudes of constituent APB and RPB, i.e. V_A and V_R, offer two degrees of freedom (their respective magnitudes and phases) in designing this beam for various applications. For example, colinear longitudinal electric and magnetic fields with controllable amplitudes may provide chiral light that is useful in chirality spectroscopy and microscopy of nanoscale matter [6, 18, 19]. As we discussed later in this chapter, it is possible to determine polarizabilities of a chiral sample of matter by using photoinduced force microscopy (PiFM) and a chiral ARPB. This is particularly important since the well-known chiral beam, i.e. an LG beam with circular polarization, only reveals chirality of a sample in transverse plane that could be confused with the anisotropy of the material sample under test [6, 22].

13.2 Circular Dichroism for a Particle of Subwavelength Size

In this section we briefly review the principles of chirality characterization at subwavelength scale, i.e. characterization of chirality of particle with subwavelength dimension. We assume an isotropic reciprocal nanoparticle with electric α_{ee}, magnetic α_{mm}, electromagnetic α_{em}, and magetoelectric α_{me} polarizabilities, that, upon external illumination, expresses induced electric $\mathbf{p}$ and magnetic $\mathbf{m}$ dipole moments [30]

$$\begin{aligned} \mathbf{p} &= \alpha_{\mathrm{ee}}\mathbf{E} + \alpha_{\mathrm{em}}\mathbf{H} \\ \mathbf{m} &= \alpha_{\mathrm{me}}\mathbf{E} + \alpha_{\mathrm{mm}}\mathbf{H}. \end{aligned} \tag{13.15}$$

Here, $\mathbf{E}$ and $\mathbf{H}$ are the local values of the electric and magnetic illuminating fields. The magnetic dipole moment is defined as $\mathbf{m} = \frac{1}{2}\int_V \mathbf{r} \times \mathbf{J}(\mathbf{r})\, dv$ where $\mathbf{r}$ and $\mathbf{J}$ are, respectively, the position vector and volume electric current density, and V is the volume occupied by the particle. Reciprocity rules that electromagnetic and magnetoelectric polarizabilities, which represent the chirality of the particle, are linked through relation $\alpha_{\mathrm{me}} = -\mu_0^{-1}\alpha_{\mathrm{em}}$. To determine chirality of the particle, we may excite the nanoparticle by two electromagnetic fields and employ the dissymmetry factor g, defined as [31]

$$g = \frac{\Delta P_{\mathrm{ext}}}{\overline{P}_{\mathrm{ext}}}, \tag{13.16}$$

where $\overline{P}_{\mathrm{ext}}$ and ΔP_{ext} are, respectively, the arithmetic average, denoted by a bar, and the difference, denoted by Δ, of the two extinction powers of the particle, i.e. $P_{\mathrm{ext}} = \omega\, \Im\{\mathbf{p}\cdot\mathbf{E}^* + \mu_0\mathbf{m}\cdot\mathbf{H}^*\}/2$, when using two kinds of illuminations, i.e. using two different fields in two different experiments. To investigate if the dissymmetry factor g reveals chirality, we substitute the extinction power expressions for the particle [19]

$$P_{\text{ext}} = 2\epsilon_0^{-1}\omega\Im\{\alpha_{\text{ee}}\}u_{\text{e}} + 2\omega\Im\{\alpha_{\text{mm}}\}u_{\text{m}} - 2c_0\omega^2\Re\{\alpha_{\text{em}}\}h \tag{13.17}$$

in Eq. (13.16). Here,$u_{\text{e}} = \epsilon_0|\mathbf{E}|^2/4$ and $u_{\text{m}} = \mu_0|\mathbf{H}|^2/4$ are, respectively, the time-averaged electric and magnetic energy densities of the local field and c_0 is the speed of light in vacuum. Symbols $\Im\{\cdot\}$ and $\Re\{\cdot\}$ denote the imaginary and real parts of a complex quantity, respectively. Moreover, the quantity h, which appears in the last term of Eq. (13.16) in the coefficient of the electromagnetic polarizability α_{em}, is the helicity density of the local field [32–34] defined as

$$h = \frac{\Im\{\mathbf{E}\cdot\mathbf{H}^*\}}{2\omega c_0} \tag{13.18}$$

in the International System of Units (SI) and is a measure of chirality of the electromagnetic field [35, 36]. The substitution of Eq. (13.17) in Eq. (13.16) leads to the detailed form [19]

$$g = \frac{P_{\text{ext}}^+ - P_{\text{ext}}^-}{\frac{(P_{\text{ext}}^- + P_{\text{ext}}^+)}{2}} = \frac{2\omega\epsilon_0^{-1}\Im\{\alpha_{\text{ee}}\}\Delta u_{\text{e}} + 2\omega\Im\{\alpha_{\text{mm}}\}\Delta u_{\text{m}} - 2c_0\omega^2\Re\{\alpha_{\text{em}}\}\Delta h}{2\omega\epsilon_0^{-1}\Im\{\alpha_{\text{ee}}\}\bar{u}_{\text{e}} + 2\omega\Im\{\alpha_{\text{mm}}\}\bar{u}_{\text{m}} + 2c_0\omega^2\Re\{\alpha_{\text{em}}\}\bar{h}}. \tag{13.19}$$

for the dissymmetry factor g, where the superscript + and − are used to distinguish between the two experiments, usually involving positive and negative values of helicity density. This equation involves the difference and average of various quantities, denoted by the Δ and bar symbols. Equation (13.19) shows that to enable the characterization of chirality of a nanoparticle α_{em} by using the dissymmetry factor g, two different experiments are required that possess different extinction powers. First, to remove the electric and magnetic polarizabilities α_{ee} and α_{mm} from the numerator, and the averaged helicity density $\bar{h}$ from the denominator, it is required that the two field excitations satisfy the following conditions [19]

$$\begin{aligned} u_{\text{m}}^+ &= u_{\text{m}}^- = u_{\text{m}} \\ u_{\text{e}}^+ &= u_{\text{e}}^- = u_{\text{e}} \\ h^+ &= -h^- = h, \end{aligned} \tag{13.20}$$

which means that $\Delta u_{\text{m}} = \Delta u_{\text{e}} = 0$ and $\bar{h} = 0$. This reduces the expression of the dissymmetry factor g to [19]

$$g = -\frac{2c_0\omega\Re\{\alpha_{\text{em}}\}h}{\epsilon_0^{-1}\Im\{\alpha_{\text{ee}}\}u_{\text{e}} + \Im\{\alpha_{\text{mm}}\}u_{\text{m}}}. \tag{13.21}$$

Now we need to remove energy and helicity densities u_{e}, u_{m}, and h from Eq. (13.21). It has been proved that when electromagnetic fields with *maximized helicity density* at a given energy density, called *optimally chiral* in Ref. [19], with electric and magnetic field components satisfying the relation

$$\mathbf{E}^\pm = \pm i\eta_0\mathbf{H}^\pm, \tag{13.22}$$

are used in each of the two experiments, the parameters h, u_{e}, and u_{m} that account for specific field values obey the relations [19]

$$\begin{aligned} h^\pm &= \pm\frac{u^\pm}{\omega} \\ u_{\text{e}}^\pm &= u_{\text{m}}^\pm = \frac{u^\pm}{2}. \end{aligned} \tag{13.23}$$

where $u = u_m + u_e$. This means that employing optimally chiral electromagnetic fields, that satisfy both conditions of Eqs. (13.20) and (13.22), reduces the dissymmetry factor g to [19]

$$g = -\frac{4c_0\Re\{\alpha_{em}\}}{\epsilon_0^{-1}\Im\{\alpha_{ee}\} + \Im\{\alpha_{mm}\}}. \tag{13.24}$$

In most of practical cases, particles have weak magnetic properties, i.e. $\epsilon_0^{-1}\Im\{\alpha_{ee}\} >> \Im\{\alpha_{mm}\}$, which further reduces the dissymmetry factor expression to

$$g = \frac{-4\eta_0^{-1}\Re\{\alpha_{em}\}}{\Im\{\alpha_{ee}\}}. \tag{13.25}$$

The framework for chirality characterization discussed in this chapter is schematically summarized in Figure 13.2. Equation (13.25) is quite remarkable since it says that the dissymmetry factor g is directly proportional to the chirality of the particle α_{em} normalized to the electric polarizability α_{ee} of the particle. Both depend on the size of the particle, and for subwavelength sizes, the ratio tends to become independent of the size of the particle. There are no other parameters that refer to the intensity and polarization of the fields used in the two experiments; the only important thing is that the two illuminating fields are *optimally chiral.*

13.2.1 Helicity of an Azimuthally Radially Polarized Vector Beam

There are various examples of optimally chiral fields. A trivial and notorious example consists of circularly polarized plane waves, and also LG beams which are not tightly focused, so they resemble circularly polarized plane waves. No wonder why beams of circularly polarized light have been used in circular dichroism till now, without knowing that they are just a single choice among all possible beams that exhibit optimal chirality. However, as described in this section, the response of a particle when using circularly polarized light also carries the information of the anisotropy of such particle, i.e. when the particle's electric polarizability α_{ee} is tensorial. A nontrivial example of an optimally

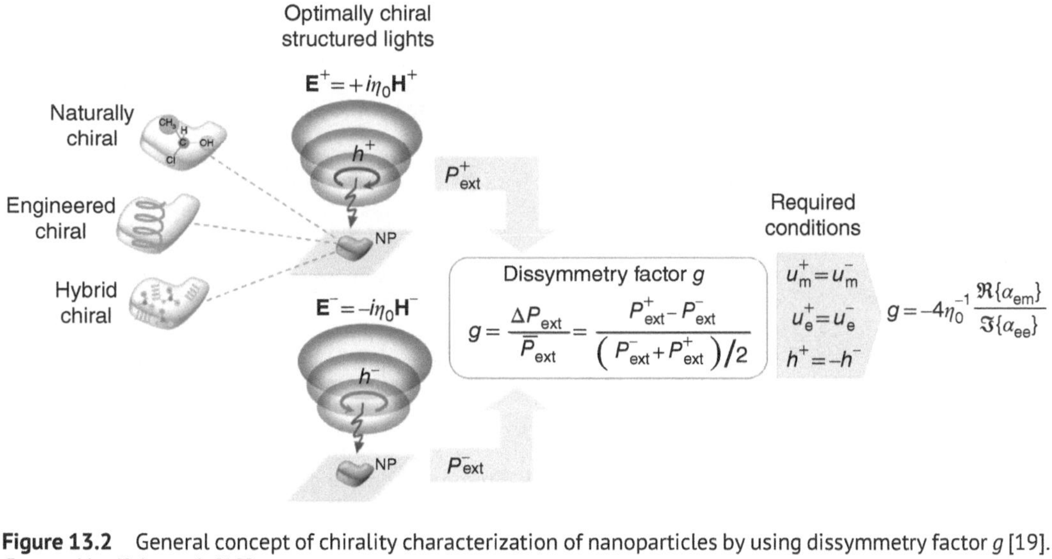

Figure 13.2 General concept of chirality characterization of nanoparticles by using dissymmetry factor *g* [19]. *Source:* Hanifeh et al. [19].

chiral vector beam is the ARPB with specific relative magnitude and phases of its constituent APB and RPB. Indeed substituting Eq. (13.13) in Eq. (13.41), we have

$$\begin{aligned}&\mathbf{E}^{\mathrm{ARPB}}=\mathbf{E}^{\mathrm{APB}}+\mathbf{E}^{\mathrm{RPB}}=\\&\eta_0\mathbf{H}^{\mathrm{RPB}}\left(\frac{V_{\mathrm{A}}}{V_{\mathrm{R}}}\right)-\eta_0\mathbf{H}^{\mathrm{APB}}\left(\frac{V_{\mathrm{R}}}{V_{\mathrm{A}}}\right)=\\&\eta_0\left(\frac{V_{\mathrm{A}}}{V_{\mathrm{R}}}\right)\left[\mathbf{H}^{\mathrm{RPB}}-\mathbf{H}^{\mathrm{APB}}\left(\frac{V_{\mathrm{R}}}{V_{\mathrm{A}}}\right)^2\right]\end{aligned} \tag{13.26}$$

which reduces to

$$\mathbf{E}^{\mathrm{ARPB}}=\pm i\eta_0\mathbf{H}^{\mathrm{ARPB}} \tag{13.27}$$

when $V_{\mathrm{A}}=\pm iV_{\mathrm{R}}$. The relation (13.26) guarantees satisfaction of conditions (13.23) in an ARPB when $V_{\mathrm{A}}=\pm iV_{\mathrm{R}}$. Moreover, two ARPBs with + and − signs in Eq. (13.27) have equal electric and magnetic energy densities and helicity densities with opposite signs, and therefore, two ARPB vector beams with $V_{\mathrm{A}}=iV_{\mathrm{R}}$ and $V_{\mathrm{A}}=-iV_{\mathrm{R}}$ that satisfy condition (13.20) are a convenient choice for chirality characterization of particles.

Helicity density of an ARPB is obtained from Eqs. (13.18) and (13.27) as

$$h=\frac{\mu_0}{2\omega}\left|H^{\mathrm{ARPB}}\right|^2\Im\{V_AV_R^{-1}\}=\frac{2u_m}{\omega}\Im\{V_AV_R^{-1}\} \tag{13.28}$$

or equivalently

$$h=-\frac{\epsilon_0}{2\omega}\left|E^{\mathrm{ARPB}}\right|^2\Im\{V_RV_A^{-1}\}=\frac{-2u_e}{\omega}\Im\{V_A^{-1}V_R\} \tag{13.29}$$

which shows that $h=\pm\frac{u}{\omega}$ when $V_{\mathrm{A}}=\pm iV_{\mathrm{R}}$. In Figure 13.3, helicity density of an ARPB is depicted. We observe that an ARPB attains its maximum helicity density on an annular ring around the beam axis. Note that although the intensity of helicity density of an ARPB varies, the relation $\mathbf{E}^{\mathrm{ARPB}}=\pm i\eta_0\mathbf{H}^{\mathrm{ARPB}}$ holds locally everywhere in space and an ARPB with $V_{\mathrm{A}}=\pm iV_{\mathrm{R}}$ is optimally

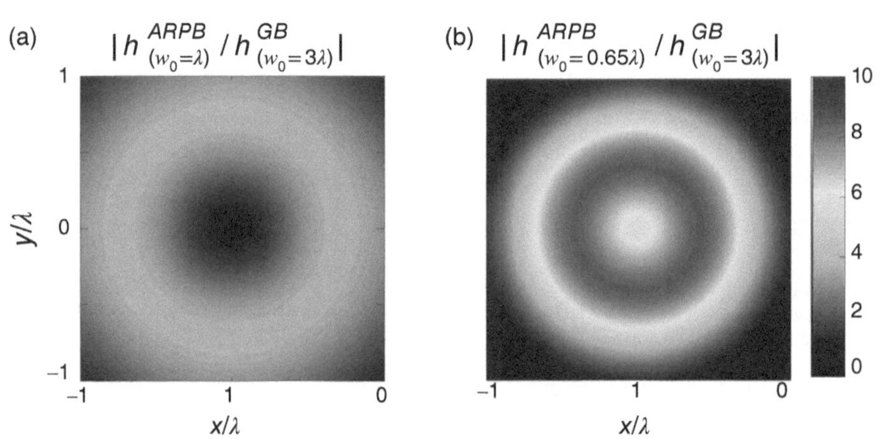

Figure 13.3 Helicity density at the $z=0$ plane of an ARPB propagating along the z direction with $V_{\mathrm{A}}=iV_{\mathrm{R}}$ and beam waist parameter $w_0=\lambda$(a) and $w_0=0.65\lambda$(b), and normalized to helicity density of a Gaussian beam with circular polarization and beam waist parameter $w_0=3\lambda$. The total powers of the GB and the ARPB beams are equal. Helicity at the center of the beam is mainly formed by only two field components: E_z^{ARPB} and H_z^{ARPB}.

chiral everywhere under paraxial approximation. One peculiarity of this beam is that helicity on the beam axis is formed by only two field components: E_z^{ARPB} and H_z^{ARPB}. Therefore this method constitutes a way to provide chiral light without resorting to circular polarization. Note that chirality of circularly polarized light involves two components of the electric field, along x and y axes, as well as two components of magnetic field along the same directions, therefore it cannot be used to determine chiral anisotropic properties. Furthermore, even more worrisome is that, as mentioned earlier, using circularly polarized light can bring us to a confusion between anisotropy and chirality when analyzing the scattering by a particle.

13.3 Photoinduced Force Microscopy at Nanoscale

In this section, we briefly revisit the optical force exerted on nanoparticles interacting with electromagnetic fields to explain how vector beams can be used in PiFM [6, 21, 37–39] to investigate material properties at nanoscale. Figure 13.4 depicts the schematic of a PiFM where the tip of a cantilever is located in proximity of sample nanoparticle and the system containing the sample and the probe tip is illuminated by an optical beam emerging from an objective lens from below. We show how the force exerted on the tip of a cantilever measured by the system of feedback laser and detector reveals some material properties of the sample.

To delve into the operation of a PiFM, the tip of the cantilever is generically modeled by photoinduced electric and magnetic dipole moments $\mathbf{p}_\text{t}$ and $\mathbf{m}_\text{t}$ with subscript t denoting the tip of the cantilever. The general expression for the time-averaged optical force $\mathbf{F}$ exerted on the tip, in International System of Units (SI), is obtained as [40–44]

$$\mathbf{F} = \frac{1}{2}\Re\left\{(\nabla\mathbf{E}(\mathbf{r}_\text{t}))^*\cdot\mathbf{p}_\text{t} + \mu_0(\nabla\mathbf{H}(\mathbf{r}_\text{t}))^*\cdot\mathbf{m}_\text{t} - \frac{\mu_0 c_0 k^4}{6\pi}\left(\mathbf{p}_\text{t}\times\mathbf{m}_\text{t}^*\right)\right\}, \tag{13.30}$$

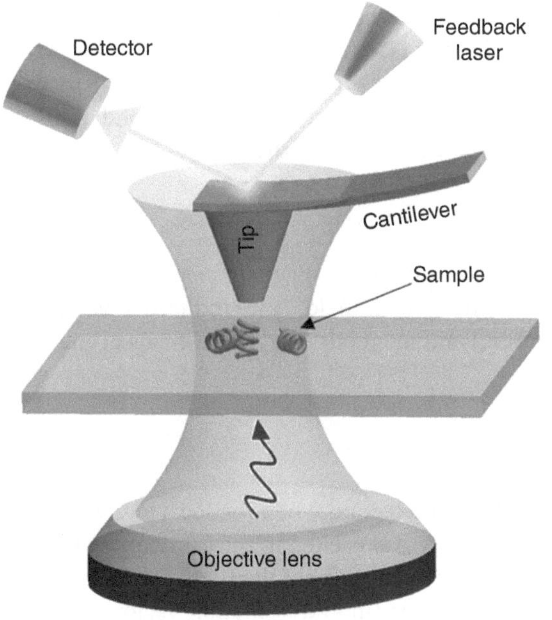

Figure 13.4 Schematic of photoinduced force microscopy [6]. *Source:* Kamandi et al. [6].

where $\mathbf{E}(\mathbf{r}_t)$ and $\mathbf{H}(\mathbf{r}_t)$ are the local electric and magnetic fields, respectively, at the location $\mathbf{r}_t$ of the tip. The gradient operator ∇ generates a second rank tensor whose ij component is equal to $\partial_i A_j$, where $i, j = x, y, z$ in Cartesian coordinates and ∂_i is partial derivative with respect to the i spatial coordinate when applied to a vector $\mathbf{A}$. Note that here the definition of magnetic dipole is slightly different from the one adopted in [6], therefore we shall include the magnetic permeability μ_0 in the expression in Eq. (13.30). For simplicity and without loss of generality, we assume that the tip is isotropic with electric and magnetic polarizabilities $\alpha_{\mathrm{ee,\,t}}$ and $\alpha_{\mathrm{mm,\,t}}$, which relate the local (illuminating) fields and the induced dipole moments of the tip by

$$\begin{aligned} \mathbf{p}_t &= \alpha_{\mathrm{ee,t}}\mathbf{E}(\mathbf{r}_t) \\ \mathbf{m}_t &= \alpha_{\mathrm{mm,t}}\mathbf{H}(\mathbf{r}_t). \end{aligned} \tag{13.31}$$

The optical force expression in Eq. (13.30) is simplified to [6]

$$\begin{aligned} \mathbf{F} = \frac{1}{2}\Re\{&\alpha_{\mathrm{ee,t}}(\nabla\mathbf{E}(\mathbf{r}_t))^*\cdot\mathbf{E}(\mathbf{r}_t) + \alpha_{\mathrm{mm,t}}\mu_0(\nabla\mathbf{H}(\mathbf{r}_t))^*\cdot\mathbf{H}(\mathbf{r}_t) \\ &- \alpha_{\mathrm{ee,t}}(\alpha_{\mathrm{mm,t}})^*\frac{\mu_0 c_0 k^4}{6\pi}\mathbf{E}(\mathbf{r}_t)\times(\mathbf{H}(\mathbf{r}_t))^*\}. \end{aligned} \tag{13.32}$$

The local fields $\mathbf{E}(\mathbf{r}_t)$ and $\mathbf{H}(\mathbf{r}_t)$ are made of two field species: the incident one, directly on the tip, plus the near field arising from the sample particle under test, just below the tip, shown in Figure 13.4. Since the latter one is a near field, its gradient is very intense and dominant. Therefore, the local fields $\mathbf{E}(\mathbf{r}_t)$ and $\mathbf{H}(\mathbf{r}_t)$ carry information about the material properties of the sample, and the optical force in Eq. (13.32) can be used to characterize the sample, if the tip and the illuminating vector beam are tailored properly. In what follows, we review some probe and beam concepts which may reveal magnetic and chiral properties of a particle with subwavelength scale.

13.3.1 Magnetic Photoinduced Force Microscopy by Using an APB

Let us investigate the possibility of revealing magnetic properties of the sample by measuring the optical force exerted on the tip of the cantilever presented in Eq. (13.32). This equation shows that when the electric local field $\mathbf{E}(\mathbf{r}_t)$ approximately vanishes and the magnetic field $\mathbf{H}(\mathbf{r}_t)$ dominates the region (e.g., when using an APB beam), only the second term in the right-hand side of Eq. (13.32) contributes to the force exerted on the tip:

$$\mathbf{F} = \frac{1}{2}\Re\{\alpha_{\mathrm{mm,t}}\mu_0(\nabla\mathbf{H}(\mathbf{r}_t))^*\cdot\mathbf{H}(\mathbf{r}_t)\}. \tag{13.33}$$

Local magnetic field $\mathbf{H}(\mathbf{r}_t)$ constitutes the magnetic near-field generated by the sample plus the magnetic field component of the incident field. Since induced dipole moments of an achiral sample are linked to the electric $\mathbf{E}(\mathbf{r}_s)$ and magnetic $\mathbf{H}(\mathbf{r}_s)$ local fields at the location of the sample $\mathbf{r}_s$ via relations

$$\begin{aligned} \mathbf{p}_s &= \alpha_{\mathrm{ee,s}}\mathbf{E}(\mathbf{r}_s) \\ \mathbf{m}_s &= \alpha_{\mathrm{mm,s}}\mathbf{H}(\mathbf{r}_s), \end{aligned} \tag{13.34}$$

where subscript 's' denotes the sample, and when the local electric field $\mathbf{E}(\mathbf{r}_s)$ vanishes, then only the induced magnetic dipole moment is excited in the sample. Local fields $\mathbf{E}(\mathbf{r}_s)$ and $\mathbf{H}(\mathbf{r}_s)$ include the excitation vector beam and the near-field of the tip at the location of the sample. Therefore, if there exists a combination of a probe-tip and excitation vector beam, which generates magnetic dominant region, the force exerted on the tip carries information about the magnetic polarizability

$\alpha_{mm,\,s}$ of the sample while its electric polarizability $\alpha_{ee,\,s}$ is excluded from the measurements. Based on Eq. (13.33), the measurement of the magnetic field generated by the sample is weighted by the magnetic polarizability of the probe-tip $\alpha_{mm,\,t}$, therefore to be able to make this measurement, it is necessary to have a probe-tip with a magnetic polarizability. A strong magnetic polarizability of the tip would enhance the response. For example, in Figures 13.5 and 13.6 we show that a probe with magnetic polarizability could be realized using a cluster of plasmonic nanoparticles. Another example of a probe with magnetic polarizability is provided by a Si truncated cone, shown in Figures 13.7, together with a schematic of a PiFM with gold-coated tip used to measure its magnetic resonance. The PiFM tip and Si truncated cone are illuminated by an APB and its measured force map is depicted in Fig. 13.8. Although it is very doable to design tips with strong magnetic polarizability at radio frequency, it is rather challenging at optical frequencies. Furthermore, a probe-tip with a magnetic polarizability shall be excited by an illuminating field that has a dominant magnetic field component.

As an illustrative example, it has been shown that a combination of an APB and a designed cluster of gold nanoparticles provides a magnetic dominant region for PiFM (Figures 13.5 and 13.6).

To explain the concept, we assume that the cluster, which consists of N plasmonic nanoparticles with induced electric dipole moments $\mathbf{p}_n$ located at $\mathbf{r}_n$ with $n = 1, \ldots, N$, is illuminated by an APB with axis coincident with the symmetry axis of the cluster. If the radius of the cluster is designed in a way that the nanoparticles are located close to the peak of the APB electric field to effectively excite electric dipole moments, the orientation of the electric dipole moments in nanoparticles forms an effective rotational current which can be effectively modeled as a magnetic dipole moment [45]

$$\mathbf{m} = \frac{i\omega}{2}\sum_{n=1}^{N} \mathbf{r}_n \times \mathbf{p}_n \tag{13.35}$$

at the center of the cluster which is responsible for the magnetic field enhancement at this location. Due to symmetry of the structure of the cluster and the excitation, electric field is not generated by the nanoparticles at the center of the cluster. Moreover, the electric field of the APB excitation vanishes on its optical axis which is aligned along the center of the cluster. Therefore a magnetic dominant region forms at the center of the cluster which is what we need for magnetic-PiFM. The use of an APB is important to excite the magnetic mode of the nanoprobe leading to enhanced

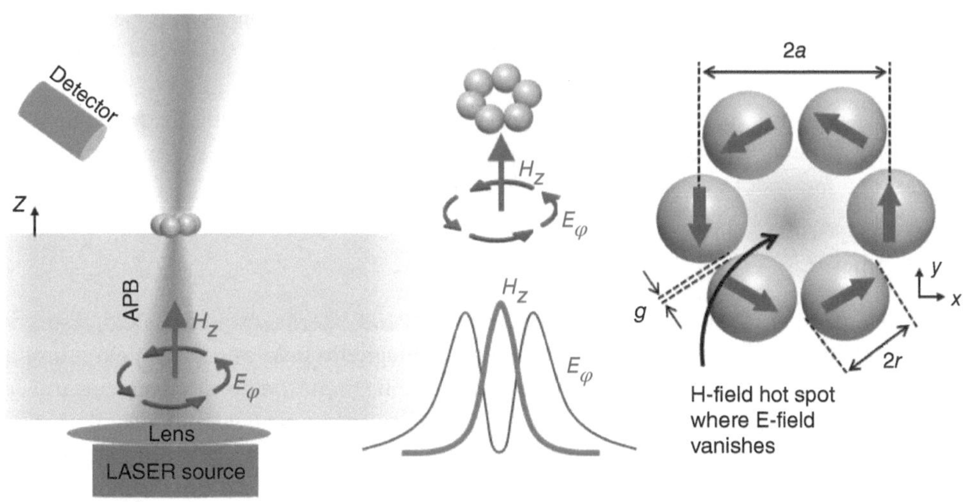

Figure 13.5 An example of setup for generation of a large magnetic to electric field contrast at the center of a nanoprobe made of a resonating circular cluster of plasmonic nanoparticles excited by an APB [45]. *Source:* Guclu et al. [45],

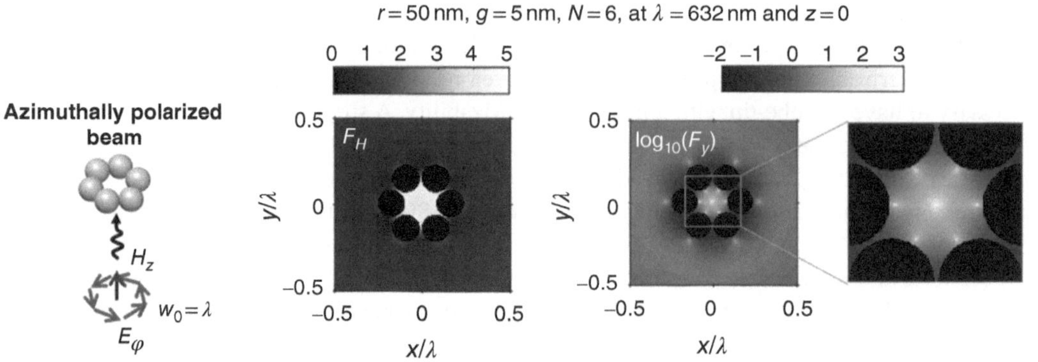

Figure 13.6 Local magnetic field enhancement F_H (first column of the plot) and normalized local field admittance F_Y (second column of the plot), both evaluated on the transverse symmetry plane of the cluster (i.e. $z = 0$ plane) excited by an APB with beam parameter $w_0 = \lambda$ whose beam axis coincides with the cluster axis (the z axis). The operating wavelength is $\lambda = 632$ nm, the radius of the nanoparticles is $r = 50$ nm, and the interparticle gap $g = 5$ nm [45]. *Source:* Guclu et al. [45],

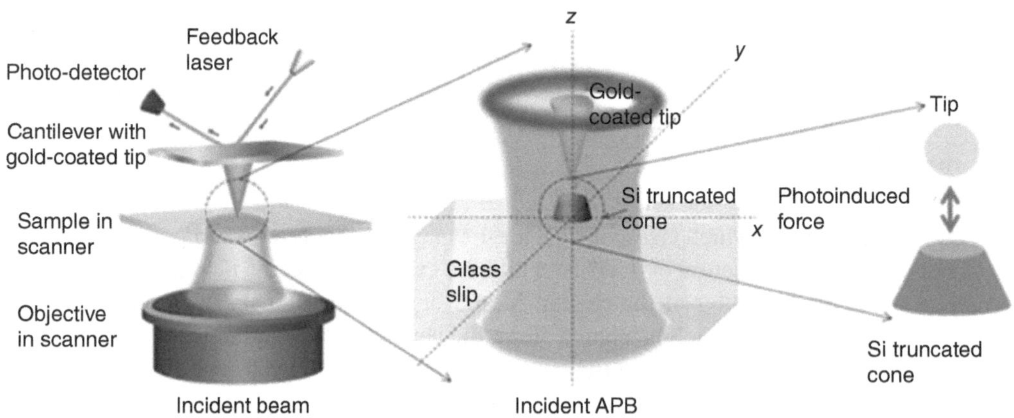

Figure 13.7 Left: Schematic of the PiFM that is used in our experiment; middle: the magnified region of the gold-plated tip and the sample on the glass slip; right: the model of the interactive gold tip-sample system. The tip and the sample are represented by a sphere and a Si truncated cone, respectively [7]. *Source:* Zeng et al. [7].

magnetic near-field and to suppress the electric near-field where magnetic field is enhanced. To quantify the magnetic field enhancement and magnetic to electric field ration at the center of the cluster, two figures of merit F_H and F_Y are defined as [45]

$$F_H = \frac{|\mathbf{H}^{\text{tot}}(\mathbf{r})|}{|\mathbf{H}^{\text{ext}}(\mathbf{r})|}, \qquad F_Y = \frac{\eta_0|\mathbf{H}^{\text{tot}}(\mathbf{r})|}{|\mathbf{E}^{\text{tot}}(\mathbf{r})|}, \tag{13.36}$$

where the superscripts 'tot' and 'ext' refer to the total field and external excitation field, respectively. The subscripts of figures of merit H and Y stand for magnetic field and local field admittance, respectively. Large values of F_H correspond to greater magnetic field enhancement and larger values of F_Y show exclusive existence of magnetic field. The magnetic field enhancement of a cluster depends on the nanosphere's radius r, intersphere gap g, and the number of nanoparticles N forming the cluster. Smaller values of N may result in higher magnetic field enhancement, however, they also reduce the accessible area in the center of the cluster. The choice of $N = 6$ seems preferable since it leads to a larger accessible area with enhanced magnetic field inside the cluster. In

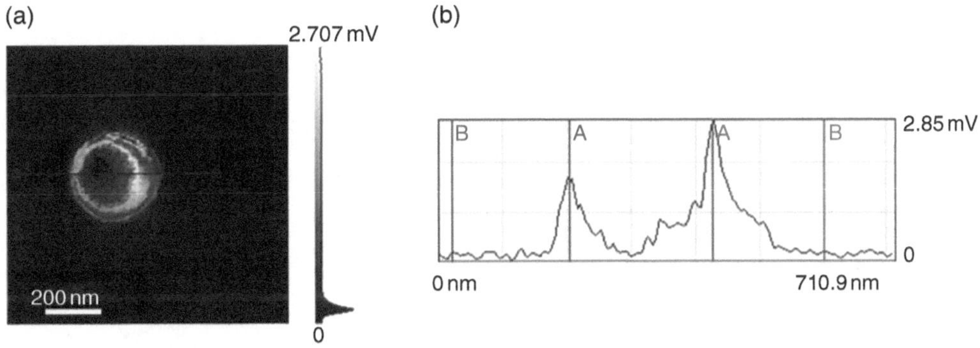

Figure 13.8 (a) Measured force map of the Si truncated cone under APB illumination with the solid line representing for the cross-line; (b) the cross-line of the measured force. In this measurement, tip and incident APB are aligned and fixed, while the stage with the Si truncated cone is moving [7]. *Source:* Zeng et al. [7].

Figure 13.6, figures of merit F_H and F_Y on the plane of a cluster with $N = 6$ nanoparticles with a radius $r = 50$ nm and interparticle gap $g = 5$ nm, which results is optimum magnetic field enhancement at operating wavelength $\lambda = 632$ nm. The cluster is placed at the APB minimum waist plane and nanosphere's silver relative permittivity is taken from Ref. [46]. Figure 13.6 shows that the magnetic nanocluster significantly boosts the total magnetic field and increases spatial magnetic resolution.

As another illustrative example, it has been shown that a dielectric disk illuminated by an APB exhibits a strong equivalent magnetic dipole if the disk has a 'magnetic' like resonance at the frequency of excitation and if its axis is coincident with the axis of the APB, shown in Fig. 13.8.

There are various possibilities to realize optical magnetic antennas, i.e. antennas that enhance the magnetic field component of light. See for example the studies in [28, 47–71]. However, we need to add that for a nanoantenna to be used as a magnetic probe-tip, it is important that the generated near-field of such probe has a large figure of merit F_Y, i.e. a very large local field admittance.

13.3.2 Chirality Photoinduced Force Microscopy

In this section, we show how the concept of differential force enables the characterization of chirality of a nanoparticle by using an optical vector beam in photoinduced force microscopy [6, 39]. Assume that the chiral sample is an anisotropic nanoparticle modeled by induced electric and magnetic dipole moments [30]

$$\begin{aligned} \mathbf{p}_s &= \underline{\alpha}_{ee,s}\mathbf{E}(\mathbf{r}_s) + \underline{\alpha}_{em,s}\mathbf{H}(\mathbf{r}_s) \\ \mathbf{m}_s &= \underline{\alpha}_{me,s}\mathbf{E}(\mathbf{r}_s) + \underline{\alpha}_{mm,s}\mathbf{H}(\mathbf{r}_s), \end{aligned} \tag{13.37}$$

where $\mathbf{E}(\mathbf{r}_s)$ and $\mathbf{H}(\mathbf{r}_s)$ are the local electric and magnetic fields, respectively, at the location $\mathbf{r}_s$ of the nanoparticle. Since the nanoparticle is anisotropic, the polarizabilities are second-rank tensors denoted by an underscore and their elements are indicated as superscripts ij with $i, j = x, y, z$. To simplify calculations and provide an analytical expression for the optical force exerted on the tip, we make the following assumptions: (i) The electric polarizability of the sample is isotropic, i.e. $\alpha_{ee,s}^{xx} = \alpha_{ee,s}^{yy} = \alpha_{ee,s}^{zz}$ but its electromagnetic polarizability is not; (ii) A plasmonic tip with electric dominant response is used, and the induced magnetic dipole moment and also the gradient of the magnetic field are neglected; (iii) Transverse components of the electromagnetic polarizability of

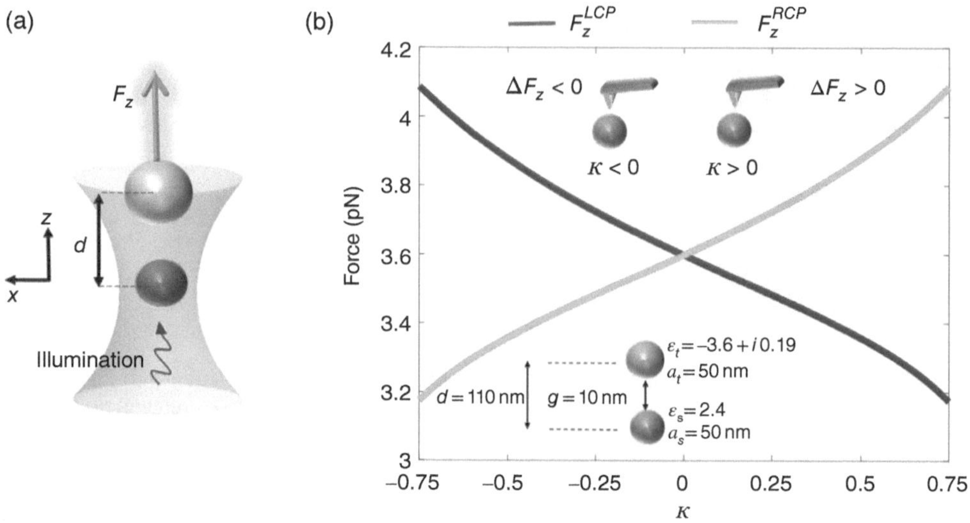

Figure 13.9 (a) Simplified model of a photoinduced force microscope, where the tip-apex and sample are represented by nanospheres with electric and magnetic dipole moments; (b) the exerted force on the tip-apex for two incident RCP and LCP illumination [6]. *Source:* Kamandi et al. [6].

the sample are equal $\alpha_{\mathrm{em,s}}^{\mathrm{xx}} = \alpha_{\mathrm{em,s}}^{\mathrm{yy}} = \alpha_{\mathrm{em,s}}^{\mathrm{tran}}$ while its longitudinal component $\alpha_{\mathrm{em,s}}^{\mathrm{zz}}$ is different; (iv) We also assume that the tip is located in the nearfield of the sample and use the dominant term of the scattered field by the sample; (v) The sample is reciprocal, meaning that $\alpha_{\mathrm{me}} = -\mu_0^{-1}\alpha_{\mathrm{em}}$, (vi) A simplified model, as shown in part (a) of Figure 13.9, is assumed for the tip-sample system.

Local fields at the location of the tip $\mathbf{r}_t$, i.e. $\mathbf{E}(\mathbf{r}_t)$ and $\mathbf{H}(\mathbf{r}_t)$, are superposition of the field generated by the sample's induced dipole moments $\mathbf{p}_s$ and $\mathbf{m}_s$, and the incident vector beam, and therefore the force exerted on the tip presented in Eq. (13.32) contains information about polarizabilities of the sample. Let us start with circularly polarized Gaussian beams excitation which are approximately modeled as plane waves when the beam is not tightly focused. The photoinduced force microscope measures the longitudinal force exerted on the tip which reads [6]

$$F_z = \frac{1}{2}\Re\left[\left(\frac{\partial E_x(z)}{\partial z}\right)^*_{z=z_t} p_{t,x} + \left(\frac{\partial E_y(z)}{\partial z}\right)^*_{z=z_t} p_{t,y}\right], \tag{13.38}$$

for circularly polarized plane waves. To remove electric and magnetic polarizabilities of the sample from the measured longitudinal force, we are required to employ two GB with RCP and LCP in two experiments and subtract the force exerted on the tip to obtain [6]

$$\Delta F_z = F_z^{RCP} - F_z^{LCP} \approx -\frac{3|\mathrm{E}_0|^2}{4\pi\sqrt{\varepsilon_0\mu_0}d^4}\Im\{\alpha_{\mathrm{ee,t}}(\alpha_{\mathrm{em,s}}^{\mathrm{tran}})^*\}, \tag{13.39}$$

which shows that the differential gradient force ΔF_z is proportional to the transverse component of the electromagnetic polarizability of the sample $\alpha_{\mathrm{em,s}}^{\mathrm{tran}}$ enhanced by electric polarizability $\alpha_{\mathrm{ee,t}}$ of the plasmonic tip. As an illustrative example, in Figure 13.9b the longitudinal force F_z for two GB excitations with RCP and LCP is depicted. The transverse component of the electromagnetic polarizability of the sample is assumed to be equal to that of a spherical nanoparticle with radius $a_s = 50$ nm made of an isotropic chiral material described by the constitutive relations

$\mathbf{D} = \epsilon_0\epsilon_s\mathbf{E} + i\kappa\sqrt{\epsilon_0\mu_0}\mathbf{H}$ and $\mathbf{B} = \mu_0\mathbf{H} - i\kappa\sqrt{\epsilon_0\mu_0}\mathbf{E}$, with relative permittivity $\epsilon_s = 2.4$ and chirality parameter κ [30, 72, 73]. The tip of the cantilever is assumed to be a spherical nanoparticle with radius $a_t = 50$ nm, relative permittivity $\epsilon_t = -3.6 + i0.19$, and the center-to-center distance between the sample and the tip is 110 nm.

The differential gradient force ΔF_z when the GBs are used reveals transverse components of electromagnetic polarizability of the sample $\alpha_{\text{em,s}}^{\text{tran}}$, however, probing the longitudinal component $\alpha_{\text{em,s}}^{zz}$ calls for utilizing alternative types of beam illuminations such as ARPB, which possess longitudinal electric and magnetic field components. Now let us investigate the exerted force on the tip when the system of chiral sample and plasmonic tip are illuminated by an ARPB beam with electric and magnetic components presented in Eq. (13.14). We place the tip-apex and chiral sample on the axis of the beam where the transverse components of the electric and magnetic fields vanish whereas the longitudinal components do not. We use two features of an ARPB as design parameters: (i) the phase difference between its APB and RPB constituent, i.e. the phase difference between V_A and V_R, denoted by parameter ψ, and (ii) the figure of merit F_Y defined in Eq. (13.36) at the focus of the beams (i.e. at the chiral sample location) which represents the magnitude of the longitudinal magnetic field divided by the longitudinal electric fields and reads $F_Y = |\frac{V_A}{V_R|}$. In order to detect chirality of the sample, we may use two ARPB illuminations with different phase-shift parameters ψ_1 and ψ_2 and similar figure of merit F_Y and calculate the difference between the exerted force on the tip-apex as [6]

$$\Delta F_z = F_z^{\text{ARPB}}\big|_{\psi_2} - F_z^{\text{ARPB}}\big|_{\psi_1} \approx -\frac{3F_Y|\mathrm{E}_0|^2}{4\pi\sqrt{\epsilon_0\mu_0}d^4}\Re\left\{\left(e^{-i\psi_2} - e^{-i\psi_1}\right)\alpha_{\text{ee,t}}\left(\alpha_{\text{em,s}}^{zz}\right)^*\right\}. \tag{13.40}$$

This equation shows that $|\psi_2 - \psi_1| = 180^o$ maximizes the differential force, and this choice leads us to [6]

$$\Delta F_z \approx -\frac{3F_Y|\mathrm{E}_0|^2}{2\pi\sqrt{\varepsilon_0\mu_0}d^4}\Re\{\alpha_{\text{ee,t}}\}\Im\left\{\left(\alpha_{\text{em,s}}^{zz}\right)^*\right\}. \tag{13.41}$$

Note that the difference $|\psi_2 - \psi_1| = 180^o$ is consistent with the earlier observation that the two optimally chiral fields used in the circular dichroism experiments have a phase difference of 180 degrees. By changing the figure of merit F_Y in an ARPB, when the total power of the beam is kept constant, the weight of the power of the APB and RPB constituents' changes. In Figure 13.10, the exerted force on the tip-apex versus logarithm values of F_Y^2 is depicted. The gradient force takes its

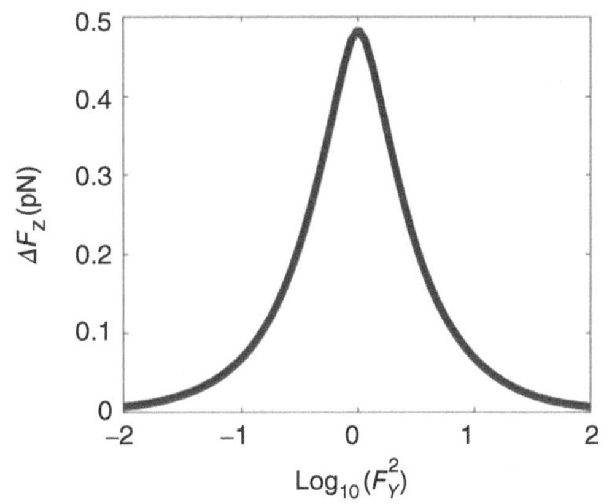

Figure 13.10 Differential force exerted on the tip of the probe in a PiFM when two ARPB beams are used sequentially. Total power of each of the beams is 2 mW and beam parameter of APB and RPB constituents is $w_0 = 0.7\lambda$ [6]. The maximum of the differential force is obtained when $F_Y^2 = 1$, i.e. when the weights of the APB and the RPB are equal. *Source:* Kamandi et al. [6].

maximum when the total power of each of the ARPB beams is divided equally between their APB and RPB constituents, i.e. $|V_A| = |V_R|$.

13.4 Conclusion

In this chapter, APBs and RPBs and their applications for spectroscopy and microscopy at nanoscale have been discussed. Vector beams offer their polarizations as a degree of freedom that can be tailored for specific applications. For example, an APB possesses an azimuthally polarized electric field vanishing on its optical axis where a dominant magnetic region is formed. This makes an APB a suitable candidate for studying magnetic properties of matter and enables building magnetic photoinduced force microscopes. An RPB is electromagnetic dual of an APB and has an electric dominant region in proximity of its optical axis. A superposition of an APB and an RPB, which has been referred to as ARPB in this chapter, with specific phase difference and relative amplitudes, provides an optimally chiral electromagnetic field with maximized helicity density. Optimally chiral fields enable chirality characterization of nanoparticles and has applications in chirality spectroscopy and microscopy at the nanoscale. We have shown how chirality at nanoscale can be detected using two experiments, both with optical chiral light with opposite helicity density, generalizing the concept of circular dichroism to any structured light.

References

1 Dennis, G. (1996). Hall. Vector-beam solutions of maxwell's wave equation. *Optics Letters* **21** (1): 9–11.

2 Siegman, A.E. (1973). Hermite–gaussian functions of complex argument as optical-beam eigenfunctions. *Journal of the Optical Society of America* **63** (9): 1093–1094.

3 Zauderer, E. (1986). Complex argument Hermite–Gaussian and Laguerre–Gaussian beams. *Journal of the Optical Society of America A* **3** (4): 465–469.

4 Zhan, Q. (2009). Cylindrical vector beams: from mathematical concepts to applications. *Advances in Optics and Photonics* **1** (1): 1–57.

5 Liu, M., Zhang, W., Lu, F. et al. (2019). Plasmonic tip internally excited via an azimuthal vector beam for surface enhanced Raman spectroscopy. *Photonics Research* **7** (5): 526–531.

6 Kamandi, M., Albooyeh, M., Veysi, M. et al. (2018). Unscrambling structured chirality with structured light at the nanoscale using photoinduced force. *ACS Photonics* **5** (11): 4360–4370.

7 Zeng, J., Darvishzadeh-Varcheie, M., Albooyeh, M. et al. (2018). Exclusive magnetic excitation enabled by structured light illumination in a nanoscale mie resonator. *ACS Nano* **12** (12): 12159–12168.

8 Xue, Y., Kuang, C., Li, S. et al. (2012). Sharper fluorescent super-resolution spot generated by azimuthally polarized beam in STED microscopy. *Optics Express* **20** (16): 17653–17666.

9 Nieminen, T.A., Heckenberg, N.R., and Rubinsztein-Dunlop, H. (2008). Forces in optical tweezers with radially and azimuthally polarized trapping beams. *Optics Letters* **33** (2): 122–124.

10 Kozawa, Y. and Sato, S. (2010). Optical trapping of micrometer-sized dielectric particles by cylindrical vector beams. *Optics Express* **18** (10): 10828–10833.

11 Veysi, M., Guclu, C., and Capolino, F. (2015). Vortex beams with strong longitudinally polarized magnetic field and their generation by using metasurfaces. *Journal of the Optical Society of America B* **32** (2): 345–354.

12 Wang, J., Yang, J.-Y., Fazal, I.M. et al. (2012). Terabit free-space data transmission employing orbital angular momentum multiplexing. *Nature Photonics* **6** (7): 488–496.

13 Djordjevic, I.B. (2011). Deep-space and near-earth optical communications by coded orbital angular momentum (OAM) modulation. *Optics Express* **19** (15): 14277–14289.

14 Allen, L., Beijersbergen, M.W., Spreeuw, R.J.C., and Woerdman, J.P. (1992). Orbital angular momentum of light and the transformation of Laguerre–Gaussian laser modes. *Physical Review A* **45**: 8185–8189.

15 Veysi, M., Guclu, C., Capolino, F., and Rahmat-Samii, Y. (2018). Revisiting orbital angular momentum beams: Fundamentals, reflectarray generation, and novel antenna applications. *IEEE Antennas and Propagation Magazine* **60** (2): 68–81.

16 Veysi, M., Guclu, C., and Capolino, F. (2016). Focused azimuthally polarized vector beam and spatial magnetic resolution below the diffraction limit. *Journal of the Optical Society of America B* **33** (11): 2265–2277.

17 Oron, R., Blit, S., Davidson, N. et al. (2000). The formation of laser beams with pure azimuthal or radial polarization. *Applied Physics Letters* **77** (21): 3322–3324.

18 Hanifeh, M., Albooyeh, M., and Capolino, F. (2020). Helicity maximization below the diffraction limit. *Physical Review B* **102**: 165419

19 Hanifeh, M., Albooyeh, M., and Capolino, F. (2020). Optimally chiral light: Upper bound of helicity density of structured light for chirality detection of matter at nanoscale. *ACS Photonics* **7** (10): 2682–2691.

20 Zeng, J., Kamandi, M., Darvishzadeh-Varcheie et al. (2018). In pursuit of photo-induced magnetic and chiral microscopy. *EPJ Applied Metamaterials* **5** (7): 2018002.

21 Rajaei, M., Almajhadi, M.A., Zeng, J., and Wickramasinghe, H.K. (2018). Near-field nanoprobing using Si tip-Au nanoparticle photoinduced force microscopy with 120:1 signal-to-noise ratio, sub-6-nm resolution. *Optics Express* **26** (20): 26365–26376.

22 Kamandi, M., Albooyeh, M., Guclu, C. et al. (2017). Enantiospecific detection of chiral nanosamples using photoinduced force. *Physical Review Applied* **8**: 064010.

23 Zeng, J., Huang, F., Guclu, C. et al. (2018). Sharply focused azimuthally polarized beams with magnetic dominance: Near-field characterization at nanoscale by photoinduced force microscopy. *ACS Photonics* **5** (2): 390–397.

24 Ivan, B., Djordjevic, M.A., Xu, L., and Wang, T. (2011). Spatial-domain-based multidimensional modulation for multi-Tb/s serial optical transmission. *Optics Express* **19** (7): 6845–6857.

25 Zvezdin, A.K. and Kotov, V.A. (1997). *Modern Magnetooptics and Magnetooptical Materials: Studies in Condensed Matter.* Institute of Physics Publishing Bristol and Philadelphia.

26 Pozar, D. (2005). *Microwave Engineering*, 4e. Hoboken, NJ: John Wiley & Sons.

27 Dodson, C.M. and Zia, R. (2012). Magnetic dipole and electric quadrupole transitions in the trivalent lanthanide series: calculated emission rates and oscillator strengths. *Physical Review B* **86**: 125102.

28 Sanz-Paz, M., Ernandes, C., Esparza, J.U. et al. (2018). Enhancing magnetic light emission with all-dielectric optical nanoantennas. *Nano Letters* **18** (6): 3481–3487.

29 Deng, D. and Guo, Q. (2007). Analytical vectorial structure of radially polarized light beams. *Optics Letters* **32** (18): 2711–2713.

30 Serdiukov, A., Semchenko, I., Tertyakov, S., and Sihvola, A. (2001). *Electromagnetics of Bi-anisotropic Materials—Theory and Application*, vol. **11**. Amsterdam, Netherlands: Gordon and Breach Science Publishers.

31 Kuhn, W. (1930). The physical significance of optical rotatory power. *Transactions of the Faraday Society* **26**: 293–308.

32 Cameron, R.P., Barnett, S.M., and Yao, A.M. (2012). Optical helicity, optical spin and related quantities in electromagnetic theory. *New Journal of Physics* **14** (5): 053050.

33 Arrayás, M. and Trueba, J.L. (2012). Exchange of helicity in a knotted electromagnetic field. *Annalen der Physik* **524** (2): 71–75.

34 Ranada, A.F. (1992). On the magnetic helicity. *European Journal of Physics* **13** (2): 70.

35 Tang, Y. and Cohen, A.E. (2011). Enhanced enantioselectivity in excitation of chiral molecules by superchiral light. *Science* **332** (6027): 333–336.

36 Tang, Y. and Cohen, A.E. (2010). Optical chirality and its interaction with matter. *Physical Review Letters* **104** (16): 163901.

37 Rajapaksa, I., Uenal, K., and Kumar Wickramasinghe, H. (2010). Image force microscopy of molecular resonance: A microscope principle. *Applied Physics Letters* **97** (7): 073121.

38 Rajapaksa, I. and Kumar Wickramasinghe, H. (2011). Raman spectroscopy and microscopy based on mechanical force detection. *Applied Physics Letters* **99** (16): 161103.

39 Zhao, Y., Saleh, A.A.E., van de Haar, M.A. et al. (2017). Nanoscopic control and quantification of enantioselective optical forces. *Nature Nanotechnology* **12** (11): 1055–1059.

40 Wang, S.B. and Chan, C.T. (2014) Lateral optical force on chiral particles near a surface. *Nature Communications* **5** (1): 3307.

41 Chaumet, P.C. and Nieto-Vesperinas, M. (2000). Time-averaged total force on a dipolar sphere in an electromagnetic field. *Optics Letters* **25** (15): 1065–1067.

42 A. D. Yaghjian (1999). Electromagnetic forces on point dipoles. *IEEE Antennas Propag. Soc. Int. Sympos.. 1999 Digest. Held in conjunction with: USNC/URSI National Radio Science Meeting (Cat. No. 99CH37010)*, vol. **4**, pp. 2868–2871.

43 Nieto-Vesperinas, M., Sáenz, J.J., Gómez-Medina, R., and Chantada, L. (2010). Optical forces on small magnetodielectric particles. *Optics Express* **18** (11): 11428–11443.

44 Alaee, R., Gurlek, B., Christensen, J., and Kadic, M. (2018). Optical force rectifiers based on PT-symmetric metasurfaces. *Physical Review B* **97**: 195420.

45 Guclu, C., Veysi, M., and Capolino, F. (2016). Photoinduced magnetic nanoprobe excited by an azimuthally polarized vector beam. *ACS Photonics* **3** (11): 2049–2058.

46 Aspnes, D.E. and Palik, E.D. (1985). *Handbook of Optical Constants of Solids*, vol. **I, II, and III**, Subject Index and Contributor Index. Orlando, FL: Elsevier Science & Tech.

47 Darvishzadeh-Varcheie, M., Guclu, C., and Capolino, F. (2017). Magnetic nanoantennas made of plasmonic nanoclusters for photoinduced magnetic field enhancement. *Physical Review Applied* **8**: 024033.

48 Darvishzadeh-Varcheie, M., Kamandi, M., Albooyeh, M., and Capolino, F. (2019). Optical magnetic field enhancement at nanoscale: a nanoantenna comparative study. *Optics Letters* **44** (20): 4957–4960.

49 Linden, S., Enkrich, C., Wegener, M. et al. (2004). Magnetic response of metamaterials at 100 terahertz. *Science* **306** (5700): 1351–1353.

50 Li, J., Verellen, N., and Van Dorpe, P. (2017). Enhancing magnetic dipole emission by a nano-doughnut-shaped silicon disk. *ACS Photonics* **4** (8): 1893–1898.

51 Hussain, R., Kruk, S.S., Bonner, C.E. et al. (2015). Enhancing Eu^{3+} magnetic dipole emission by resonant plasmonic nanostructures. *Optics Letters* **40** (8): 1659–1662.

52 Mivelle, M., Grosjean, T., Burr, G.W. et al. (2015). Strong modification of magnetic dipole emission through diabolo nanoantennas. *ACS Photonics* **2** (8): 1071–1076.

53 Chigrin, D.N., Kumar, D., Cuma, D., and von Plessen, G. (2016). Emission quenching of magnetic dipole transitions near a metal nanoparticle. *ACS Photonics* **3** (1): 27–34.

54 Choi, B., Iwanaga, M., Sugimoto, Y. et al. (2016). Selective plasmonic enhancement of electric- and magnetic-dipole radiations of Er ions. *Nano Letters* **16** (8): 5191–5196.

55 Baena, J.D., Marqués, R., Medina, F., and Martel, J. (2004). Artificial magnetic metamaterial design by using spiral resonators. *Physical Review B* **69**: 014402.

56 Linden, S., Enkrich, C., Dolling, G. et al. (2006). Photonic metamaterials: magnetism at optical frequencies. *IEEE Journal of Selected Topics in Quantum Electronics* **12** (6): 1097–1105.

57 Pendry, J.B., Holden, A.J., Robbins, D.J., and Stewart, W.J. (1999). Magnetism from conductors and enhanced nonlinear phenomena. *IEEE Transactions on Microwave Theory and Techniques* **47** (11): 2075–2084.

58 Zhou, J., Koschny, T., Kafesaki, M. et al. (2005). Saturation of the magnetic response of split-ring resonators at optical frequencies. *Physical Review Letters* **95**: 223902.

59 Katsarakis, N., Konstantinidis, G., Kostopoulos, A. et al. (2005). Magnetic response of split-ring resonators in the far-infrared frequency regime. *Optics Letters* **30** (11): 1348–1350.

60 Alu, A. and Engheta, N. (2009). The quest for magnetic plasmons at optical frequencies. *Optics Express* **17** (7): 5723–5730.

61 Corrigan, T.D., Kolb, P.W., Sushkov, A.B. et al. (2008). Optical plasmonic resonances in split-ring resonator structures: an improved LC model. *Optics Express* **16** (24): 19850–19864.

62 Krasnok, A.E., Miroshnichenko, A.E., Belov, P.A., and Kivshar, Y.S. (2012). All-dielectric optical nanoantennas. *Optics Express* **20** (18): 20599–20604.

63 Kuznetsov, A.I., Miroshnichenko, A.E., Brongersma, M.L. et al. (2016). Optically resonant dielectric nanostructures. *Science* **354** (6314).

64 Albella, P., Poyli, M.A., Schmidt, M.K. et al. (2013). Low-loss electric and magnetic field-enhanced spectroscopy with subwavelength silicon dimers. *The Journal of Physical Chemistry C* **117** (26): 13573–13584.

65 Zou, L., Withayachumnankul, W., Shah, C.M. et al. (2013). Dielectric resonator nanoantennas at visible frequencies. *Optics Express* **21** (1): 1344–1352.

66 Malheiros-Silveira, G.N., Wiederhecker, G.S., and Hernandez-Figueroa, H.E. (2013). Dielectric resonator antenna for applications in nanophotonics. *Optics Express* **21** (1): 1234–1239.

67 Verre, R., Yang, Z.J., Shegai, T., and Käll, M. (2015). Optical magnetism and plasmonic fano resonances in metal-insulator-metal oligomers. *Nano Letters* **15** (3): 1952–1958.

68 Savelev, R.S., Makarov, S.V., Krasnok, A.E., and Belov, P.A. (2015). From optical magnetic resonance to dielectric nanophotonics (A review). *Optics and Spectroscopy* **119** (4): 551–568.

69 Markovich, D., Baryshnikova, K., Shalin, A. et al. (2016). Enhancement of artificial magnetism via resonant bianisotropy. *Scientific Reports* **6** (1): 22546.

70 Yen, T.J., Padilla, W.J., Fang, N. et al. (2004). Terahertz magnetic response from artificial materials. *Science* **303** (5663): 1494–1496.

71 Dolling, G., Enkrich, C., Wegener, M. et al. (2005). Cut-wire pairs and plate pairs as magnetic atoms for optical metamaterials. *Optics Letters* **30** (23): 3198–3200.

72 Lindell, I.V., Sihvola, A.H., Tretyakov, S.A., and Viitanen, A.J. (1994). *Electromagnetic Waves in Chiral and Bi-isotropic Media*. Norwood, MA: Artech House Publishers.

73 Bohren C. F. (1975). Light scattering by optically active particles. Thesis. The University of Arizona.

14

Quantum Applications of Structured Photons

Alessio D'Errico and Ebrahim Karimi

Department of Physics, University of Ottawa, Ottawa, ON, Canada

14.1 Introduction

First observed by Johannes Kepler almost four centuries ago, it was understood that the radiation from the sun possesses linear momentum (solar wind) and pushes away the ion tail of comets orbiting in our solar system. At the beginning of the Twentieth century – after the unification of electric and magnetic fields by James Clerk Maxwell from 1861 to 1865 – John Henry Poynting found that circularly polarized electromagnetic waves can cause an object to 'spin.' Thus, light beams carry mechanical spin angular momentum (SAM) in addition to linear momentum. The radiation-induced spin motion on a doubly refracting plate due to a circularly polarized beam was experimentally observed by Richard Beth's students in 1936 [1]. However, a couple of years before in 1932, Charles Galton Darwin, in an attempt to find an analogy between photon (massless) and electron (massive) wavefunctions, showed that light beams possess both SAM, π, and orbital angular momentum (OAM), ℓ [2]. Thus, the total optical angular momentum, $\mathbf{j} = \pi + \ell$, can be split into two terms that are, respectively, coordinate-independent (SAM) and coordinate-dependent (OAM). Almost 60 years later, Les Allen and his coworkers experimentally generated microwave beams carrying well-defined OAM states [3]. In particular, an OAM-carrying beam has a point of phase dislocation (phase singularity) on axis, which makes them very interesting in the field of singular optics [4].

Albert Einstein's quantization of electromagnetic waves suggested that light beams are composed of quantized packets of energy, which we refer to as photons nowadays [5]. The quantization of energy ω suggested that photons carry quantized linear momentum $\mathbf{p} = \hbar\mathbf{k}$ and OAM $\ell = \hbar\ell$, where $\hbar$, ω, $\mathbf{k}$ and ℓ stand for the reduced Planck constant, photon frequency, photon wavevector and an integer number, respectively. Although it was known that a single photon carries a well-defined OAM state since the study by Darwin [2], its quantum traits were experimentally shown by Mair et al. [6] nearly 70 years later. Indeed, Mair et al. [6] showed that generated photon pairs in spontaneous parametric downconversion (SPDC) are entangled in the OAM degree of freedom. Since then, single- and multiphotons with engineered SAM and OAM have been used in quantum optics and, in particular, in exploring high-dimensional quantum information such as quantum communication, quantum sensing, and quantum simulation.

In the following sections, we present how photon pairs entangled in the optical angular momentum are generated, detected, and manipulated. In addition to various techniques to engineer

Electromagnetic Vortices: Wave Phenomena and Engineering Applications, First Edition.
Edited by Zhi Hao Jiang and Douglas H. Werner.

quantum states of photons, endeavoring to generate the desired structured photon state, the advantageous use of structured photons in quantum photonics will be discussed. Finally, we will provide a comprehensive review of the application of structured photons in quantum communication and quantum simulation.

14.2 Photonic Degrees of Freedom

The physical properties of photons, the quanta of light, can be described by means of several different entities, although quantum mechanics prohibits the simultaneous measurement of conjugate quantities deterministically and with certainty. A photon's state can be defined by a frequency ω, a polarization π, and temporal and spatial modes that define the photon energy, vectorial nature, and temporal and spatial amplitudes and phases, respectively. Frequency is a real (continuous) quantity, which can be controlled at the source or manipulated with optical elements, such as filters, and diffractive and refractive optics. The polarization, which determines how the tip and magnitude of the electric (and magnetic) field evolve in time and space, lies in a plane and, thus, will take only two independent values: $|L\rangle_\pi$ and $|R\rangle_\pi$ associated with left- and right-handed circular polarization, for example. Spatial modes determine the transverse quantization of the field and will also be represented by two independent parameters. Depending on the field quantization and the chosen coordinates, these parameters can be discrete or continuous values. For instance, the two parameters for a plane wave (in Cartesian coordinates) are associated with a tip-tilt along the x- and y-axes (assuming that the photon propagates along the z-axis). However, the quantization parameters (indices) for Hermite–Gauss modes in Cartesian coordinates are positive integers indicating how phase and amplitude of photon are distributed in the transverse plane along the x- and y-axes. Since optical beams and most of the optics used in the laboratory (lenses, mirrors, etc.) are circular in shape, i.e. cylindrically symmetric, we aim to quantize the field in the polar cylindrical coordinates ρ, ϕ, z. The photon 'wavefunction' $\psi_{\text{photon}}(\rho, \phi, z)$ must be physical, i.e. single-valued upon a 2π rotation of the coordinates, $\psi_{\text{photon}}(\rho, \phi, z) = \psi_{\text{photon}}(\rho, \phi + 2\pi, z)$; thus, the azimuthal index ℓ is an integer quantity. The radial index describing the radial structure depends on the boundary condition and normalization of the wavefunction. For instance, the radial parameter for (non-normalized) Bessel modes is a continuous real quantity, while for Laguerre–Gauss (LG) modes, the radial index p is a positive integer value. Figure 14.1 shows the transverse intensity profile and phase structure LG modes.

Therefore, the photon state in the LG basis is given by

$$|\psi\rangle_{\text{photon}} = \int d\omega \sum_{\pi,p,\ell} c_{p,\ell,\pi}(\omega) \mid \pi, \omega, p, \ell\rangle, \tag{14.1}$$

where $|\omega, \pi, p, \ell\rangle$ stands for a photon with a frequency of ω, a polarization of π, and a spatial mode of p, ℓ. LG modes are mutually orthogonal and form a complete basis, i.e.

$$\langle p', \ell' \mid p, \ell\rangle = \delta_{p,p'}\delta_{\ell,\ell'} \tag{14.2}$$

$$\sum_{p,\ell} \mid p, \ell\rangle\langle p, \ell \mid \; = \hat{1}, \tag{14.3}$$

where $\hat{1}$ and $\delta_{i,j}$ are the identity matrix and delta Kronecker, respectively. Their explicit expression in the position space $\langle \mathbf{r} \mid p, \ell\rangle$, here in the cylindrical polar coordinates ρ, ϕ, z, is given by

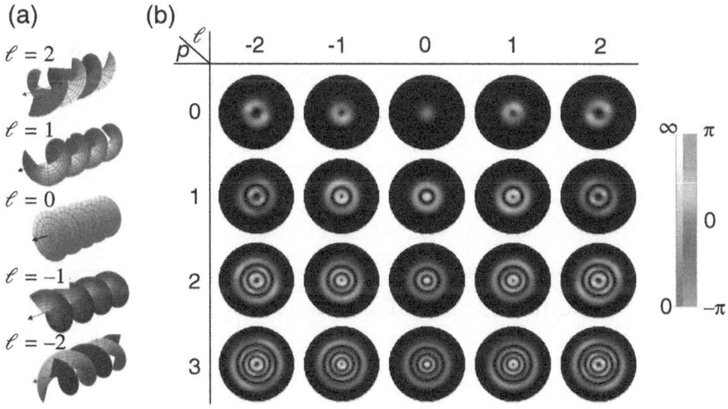

Figure 14.1 Phase and intensity profiles of Laguerre–Gauss modes. (a) Helical wavefronts are associated with beams with azimuthal indices of $\ell = \{-2, -1, 0, +1, +2\}$; the arrow indicates the propagation axis. (b) Intensity and phase profiles of Laguerre–Gauss beams in the transverse plane; p, ℓ are the azimuthal and radial indices, respectively.

$$
\begin{aligned}
\mathrm{LG}_{p,\ell}(\rho,\phi,z) &= \langle \mathbf{r} \mid p,\ell\rangle \\
&= \sqrt{\frac{2p!}{\pi(p+|\ell|)!}}\left(\frac{w_0}{w(z)}\right)\left(\frac{\sqrt{2}\rho}{w(z)}\right)^{|\ell|} L_p^{|\ell|}\left(\frac{2\rho^2}{w^2(z)}\right) e^{-\frac{\rho^2}{w^2(z)}} e^{i\ell\phi} \\
&\quad \times e^{-i\frac{k\rho^2}{z\left(1+\left(\frac{z_R}{z}\right)^2\right)} + i(2p+|\ell|+1)\tan^{-1}\left(\frac{z}{z_R}\right)},
\end{aligned}
\tag{14.4}
$$

where $w(z) = w_0(1 + (z/z_R)^2)^{1/2}$ is the waist of the Gaussian envelope at the plane z; $L_p^{|\ell|}(.)$ is the generalized Laguerre polynomials with azimuthal and radial indices of ℓ and p, respectively; $z_R = kw_0^2/2$ is the Rayleigh parameter; and k is the beam wavevector.

The representation in Eq. (14.1) is not a unique representation, and one can expand the photon state in different bases. In the Cartesian coordinates, i.e. in the plane-wave basis, the photon wavefunction can be expanded as

$$
\mid \psi\rangle_{\text{photon}} = \iiint d\omega\, dk_x\, dk_y \sum_{\pi} c_\pi\left(\omega, k_x, k_y\right) \mid \pi, \omega, k_x, k_y\rangle, \tag{14.5}
$$

where k_x, k_y are the transverse momenta along the x- and y-axes, respectively. Since both plane waves and LG modes form complete bases, one can find a transformation from one representation to the other.

Quantum information is based on the manipulation of qubits, the quantum analogue of classical bits, which consist of linear superpositions of two quantum states denoted as $|\,\text{`0'}\rangle$ and $|\,\text{`1'}\rangle$ [7]. In quantum optics, a qubit is often encoded in the polarization of a photon, e.g. through the identification of $|H\rangle \equiv |\,\text{`0'}\rangle$ and $|V\rangle \equiv |\,\text{`1'}\rangle$. While in the classical computation, binary encoding is used because it is simpler to find systems where one can have two clearly distinguishable states (e.g. voltage levels associated with saturation and cutoff states of a transistor), in quantum mechanics, there are many systems that can be in multiple discrete states (e.g. the energy levels of atoms). Hence, it is useful to consider quantum information and computational experiments based on *qudits*, i.e. linear superposition of d orthogonal states. The advantage in using qudits is that each particle can carry a larger amount of information. Moreover, qudit encoding allows for better security threshold quantum communication and for simulating processes in large Hilbert spaces without the need for a high number of particles, as we will discuss in this chapter. Structured photons are a promising candidate for encoding and manipulating qudits. Many applications are based on encoding information in OAM, in which case, a general $d + 1$-dimensional qudit will read

$$|\psi\rangle = \sum_{\ell=-d/2}^{d/2} c_\ell \,|\,\ell\rangle. \tag{14.6}$$

Here, without the loss of generality, we assumed d to be an even integer number. As we will see, other choices are possible, like the use of radial indices or even the coupling of OAM with the polarization.

14.3 Single Photon Source: SPDC

There are several different techniques to generate single photons as well as entangled photon pairs carrying well-defined optical angular momentum states. SPDC, although not deterministic, is a method that is now widely used in the quantum photonics laboratories. In the SPDC process, which is the inverse of the second-harmonic generation and occurs in nonlinear crystals, a photon with a high frequency (namely, the pump) is converted into a pair of photons (namely, the signal and idler) with lower frequencies. SPDC was experimentally observed by Burnham and Weinberg [8] followed by several theoretical studies. The Hamiltonian describing the interaction of electromagnetic waves with a nonlinear crystal is given by

$$\hat{\mathcal{H}}(t) = \epsilon_0 \int d^3r\,\hat{\mathbf{P}}(\mathbf{r};t)\cdot\hat{\mathbf{E}}(\mathbf{r};t). \tag{14.7}$$

Here, ϵ_0 is the vacuum permittivity, $\hat{\mathbf{E}}(\mathbf{r};t)$ is the quantized electric field of the electromagnetic wave, and $\hat{\mathbf{P}}(\mathbf{r};t)$ is the quantized polarization of the nonlinear crystal, which are themselves given by

$$\hat{\mathbf{E}}(\mathbf{r};t) = \frac{1}{\sqrt{V}}\sum_{\mathbf{k},\pi}\left(\frac{\hbar\omega}{2\epsilon_0 n^2}\right)^{1/2}\left(\mathbf{e}_\pi \hat{a}_{\mathbf{k},\pi} e^{-i(\omega t-\mathbf{k}\cdot\mathbf{r})} + \mathbf{e}_\pi^\star \hat{a}_{\mathbf{k},\pi}^\dagger e^{i(\omega t-\mathbf{k}\cdot\mathbf{r})}\right), \tag{14.8}$$

$$\hat{P}_i(\mathbf{r};t) = \int_0^\infty dt' \int_0^\infty dt''\chi_{ijk}^{(2)}\hat{E}_j(\mathbf{r};t')\hat{E}_k(\mathbf{r};t''), \tag{14.9}$$

where V is the quantization volume, $\hat{a}_{\mathbf{k},\pi}$ and $\hat{a}_{\mathbf{k},\pi}^\dagger$ are the field annihilation and creation operators of a mode with wavevector $\mathbf{k}$ and polarization state of $\mathbf{e}_\pi$, respectively; ω is the field frequency, and n is the crystal refractive index. The field quantum state $|\psi(t)\rangle$ after interacting with the nonlinear crystal is given by the Schrödinger picture as

$$\begin{aligned} |\,\psi(t)\rangle &= \exp\left(-\frac{i}{\hbar}\int_0^t dt'\hat{\mathcal{H}}(t')\right)|\,\psi(0)\rangle \\ &\simeq \left(|\,\psi(0)\rangle - \underbrace{\left(\frac{i}{\hbar}\int_0^t dt'\hat{\mathcal{H}}(t')\right)|\,\psi(0)\rangle}_{|\Psi\rangle} + \dots\right). \end{aligned} \tag{14.10}$$

where we have considered the interaction with the crystal to be very short compared to the time between downconversion photons, and thus have expanded the exponential term. Note that the higher terms (omitted in Eq. (14.10)) are associated with the creation of more than one photon pair, which have a much lower probability. The $|\Psi\rangle$ term, after simplifying, includes two types of terms, i.e. $\left(\hat{a}_{\mathbf{k}_1,\pi_1}\hat{a}^\dagger_{\mathbf{k}_2,\pi_2}\hat{a}^\dagger_{\mathbf{k}_3,\pi_3}\right)$ and $\left(\hat{a}^\dagger_{\mathbf{k}_1,\pi_1}\hat{a}_{\mathbf{k}_2,\pi_2}\hat{a}_{\mathbf{k}_3,\pi_3}\right)$, which express SPDC (one photon is absorbed and two photons are created) and second-harmonic generation (two photons are absorbed and one photon is created), respectively. Since this process is adiabatic and elastic, both energy and linear momentum are conserved during this process, and thus, $\hbar\omega_p = \hbar\omega_s + \hbar\omega_i$ and $\hbar\mathbf{k}_p = \hbar\mathbf{k}_s + \hbar\mathbf{k}_i$; these photons are named pump (subscript p), signal (subscript s) and idler (subscript i) photons.

The crystal type, which is defined based on the way that the crystal is cut, determines the quantum correlation among different degrees of freedom, e.g. frequency, spatial modes, and polarization degrees of freedom. For instance, for a crystal where the refractive index of frequency 2ω (pump) is equal to the refractive index of ω (signal and idler), i.e. $n_e(2\omega) = n_o(\omega)$, i.e. type-I, the created photon pair is correlated in position and anticorrelated in the momentum space (the correlation can also be seen in the spatial modes). For a crystal where $n(2\omega)$ is the average of the ordinary and extraordinary refractive indices at a frequency of ω, i.e. $n_e(2\omega) = (n_e(\omega) + n_o(\omega))/2$, the generated photon pair is correlated in polarization degrees of freedom; see Ref. [9] for more details on type-I and type-II crystals. These conditions will apply restrictions on the elements of the second-order susceptibility tensor.

For a thin and wide nonlinear crystal of either type-I or type-II, the above evolution operator acting on a coherent state $|\alpha\rangle$ yields

$$|\Psi\rangle_I \propto \iint d\mathbf{q}_i d\mathbf{q}_s \Phi(\mathbf{q}_i, \mathbf{q}_s)\mathcal{E}_{\text{pump}}(\mathbf{q}_i + \mathbf{q}_s) \mid 1\rangle_{\mathbf{q}_i,o} \mid 1\rangle_{\mathbf{q}_s,o}, \tag{14.11}$$

$$|\Psi\rangle_{\text{II}} \propto \iint d\mathbf{q}_i d\mathbf{q}_s \Phi(\mathbf{q}_i, \mathbf{q}_s)\mathcal{E}_{\text{pump}}(\mathbf{q}_i + \mathbf{q}_s) \times \left(\mid 1\rangle_{\mathbf{q}_i,e} \mid 1\rangle_{\mathbf{q}_s,o} + \mid 1\rangle_{\mathbf{q}_s,e} \mid 1\rangle_{\mathbf{q}_i,o}\right). \tag{14.12}$$

Here, $\mathcal{E}_{\text{pump}}$ is the pump amplitude, Φ is the phase matching function $\Phi(\mathbf{q}_i, \mathbf{q}_s) = \text{sinc}\left(4L|\mathbf{q}_i - \mathbf{q}_s|^2/K\right)e^{i4L|\mathbf{q}_i - \mathbf{q}_s|^2/K}$, L is the crystal thickness, and K is the pump wavenumber $K = n_e(2\omega)\omega/c$. The 'ket' $|1\rangle_{\mathbf{q},\ \sigma}$ indicates a one-photon state with polarization σ and transverse wavevector $\mathbf{q}$. Note that Eqs. (14.11) and (14.12) indicate that the generated photons in type-I and type-II crystals are nonseparable in different degrees of freedom. Figure 14.2 shows experimental results of generated photons in type-I and type-II crystals.

Let $|\Psi\rangle_\text{I}$ represent photon pairs that are anticorrelated in momentum space and correlated in position space, and let $|\Psi\rangle_\text{II}$ indicate that signal and idler photons are correlated in the polarization degree of freedom. In order to understand these physical features better, we can expand Eq. (14.11) in a desired transverse mode basis. Since our main interest in this chapter is transverse modes possessing OAM, we will use the LG modes $|p, \ell\rangle$ as the expansion basis. Therefore, expanding Eq. (14.11) in the LG basis yields

$$\begin{aligned} |\Psi\rangle_I &= \sum_{\substack{p_s,\ell_s \\ p_i,\ell_i}} (| p_s, \ell_s\rangle \otimes | p_i, \ell_i\rangle\langle p_s, \ell_s | \otimes \langle p_i, \ell_i |) \cdot |\Psi\rangle_I \\ &= \sum_{\substack{p_s,\ell_s \\ p_i,\ell_i}} c^{p_i,\ell_i}_{p_s,\ell_s} | p_s, \ell_s\rangle \otimes | p_i, \ell_i\rangle, \end{aligned} \tag{14.13}$$

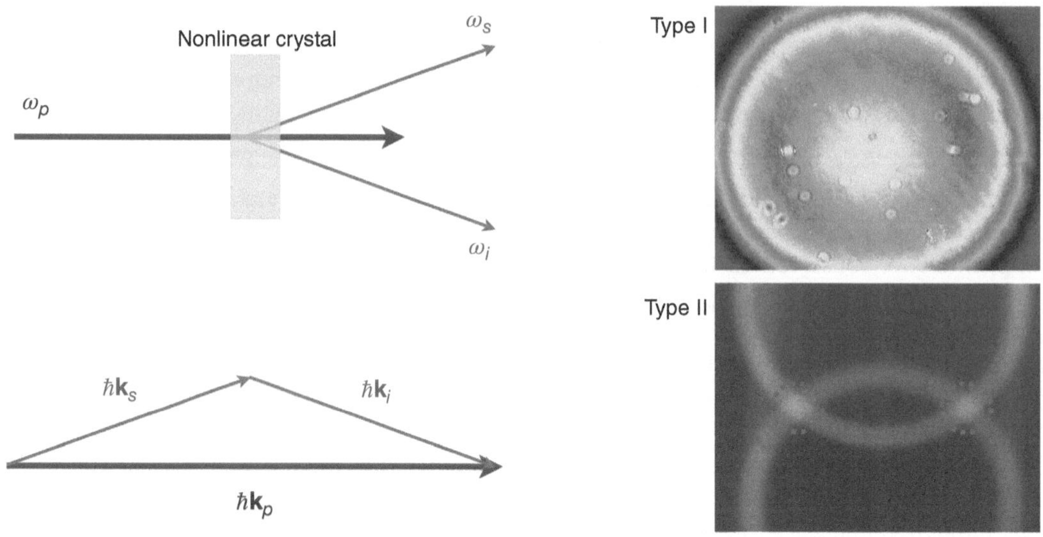

Figure 14.2 Illustration of spontaneous parametric downconversion in a nonlinear crystal. The generated photons must satisfy the conservation of energy $\hbar\omega_p = \hbar\omega_s + \hbar\omega_i$ as well as the linear momentum $\hbar\mathbf{k}_p = \hbar\mathbf{k}_s + \hbar\mathbf{k}_i$. The right insets show the transverse profiles of the generated photon pairs in type-I and type-II crystals that are taken with an intensified charge-coupled device (CCD) camera. The dotted circles indicate regions where the two photons are entangled in the polarization degree of freedom.

where

$$\begin{aligned} c_{p_s,\ell_s}^{p_i,\ell_i} &:= \langle \mathrm{p_s}, \ell_s \mid \otimes \langle \mathrm{p_i}, \ell_i \mid \cdot \mid \Psi \rangle_I \\ &\propto \iint d\mathbf{q}_i d\mathbf{q}_s\, \Phi(\mathbf{q}_i, \mathbf{q}_s) \mathcal{E}_{\mathrm{pump}}(\mathbf{q}_i + \mathbf{q}_s) \langle p_s, \ell_s \mid 1 \rangle_{\mathbf{q}_s, o} \langle p_i, \ell_i \mid 1 \rangle_{\mathbf{q}_i, o} \\ &\propto \iint d\mathbf{q}_i d\mathbf{q}_s\, \Phi(\mathbf{q}_i, \mathbf{q}_s) \mathcal{E}_{\mathrm{pump}}(\mathbf{q}_i + \mathbf{q}_s) \mathrm{LG}^*_{p_s,\ell_s}(\mathbf{q}_s)\, \mathrm{LG}^*_{p_i,\ell_i}(\mathbf{q}_i). \end{aligned} \tag{14.14}$$

Here, $\mathrm{LG}^*_{p,\ell}(\mathbf{q})$ is the complex conjugate of the LG modes in momentum space, i.e. Fourier transform of Eq. (14.4) at the $z = 0$ plane. For a very thin crystal, where $L/K \to 0$, the expansion coefficients in Eq. (14.14), referred to as coincidence amplitudes, can be further simplified to

$$c_{p_s,\ell_s}^{p_i,\ell_i} \propto \int_0^{2\pi} d\phi \int_0^{\infty} q\, dq\, \mathcal{E}_{\mathrm{pump}}\left(\sqrt{2}q, \phi\right) \mathrm{LG}^*_{p_s,\ell_s}(q, \phi)\, \mathrm{LG}^*_{p_i,\ell_i}(q, \phi). \tag{14.15}$$

Field continuity dictates that, for the Schmidt basis, the pump beam waist must be $\sqrt{2}$ of the idler and signal beam waists, i.e. $w_{\mathrm{pump}} = \sqrt{2}w$ and $w_s = w_i = w$; see Ref. [10]. The coefficients $c_{p_s,\ell_s}^{p_i,\ell_i}$ reveal that the correlation between the signal and idler photon spatial modes for a very thin crystal is given by the field continuity. For instance, consider a Gaussian pump beam, $\mathcal{E}_{\mathrm{pump}}\left(\sqrt{2}q, \phi\right) = \sqrt{\frac{2}{\pi w_p^2}} e^{-q^2/w_p^2}$. Only considering the azimuthal index, the correlation coefficients between signal and idler photons OAM states are given by

$$\begin{aligned} c_{\ell_s}^{\ell_i} &= \int_0^{2\pi} d\phi\, e^{-i(\ell_s + \ell_i)\phi} \\ &= c_{\ell_s} \delta_{\ell_s, -\ell_i}, \end{aligned} \tag{14.16}$$

where c_{ℓ_s} is a constant coefficient and is given by the integration over the radial part of the pump, signal and idler radial modes. Therefore, the generated SPDC state in Eq. (14.13) is

$$\begin{aligned} |\Psi\rangle_I &= \sum_{\ell=-\infty}^{+\infty} c_\ell \,|\ell\rangle_s\,|-\ell\rangle_i \\ &= c_0\,|0\rangle_s\,|0\rangle_i + c_1(|1\rangle_s\,|-1\rangle_i + |-1\rangle_s\,|1\rangle_i) \\ &\quad + c_2(|2\rangle_s\,|-2\rangle_i + |-2\rangle_s\,|2\rangle_i) + c_3(|3\rangle_s\,|-3\rangle_i + |-3\rangle_s\,|3\rangle_i) + \ldots, \end{aligned} \tag{14.17}$$

which indicates that the idler and signal photons possess opposite OAM states. In a similar fashion, it can be shown that if the pump carries the OAM value of m, then the symmetry is broken and the correlation becomes

$$|\Psi\rangle_I = \sum_{\ell=-\infty}^{+\infty} c_\ell\,|\ell+m\rangle_s\,|-\ell\rangle_i. \tag{14.18}$$

Figure 14.3 shows experimental results of correlation among different OAM states when the pump beam carries different OAM values. The result shows that OAM is conserved in the SPDC process, and thus the sum of the signal and idler OAM is exactly equivalent to the OAM of the pump, confirming Eq. (14.18). The value of coefficients c_ℓ, or $c^{p_i,\ell_i}_{p_s,\ell_s}$ in general, depends on the pump and the crystal geometry. Their values define the non-negligible term, i.e. the modal spiral bandwidth, in Eq. (14.18). Manipulating the expansion coefficients would provide access to high-dimensional Hilbert spaces, which will be needed in the following sections.

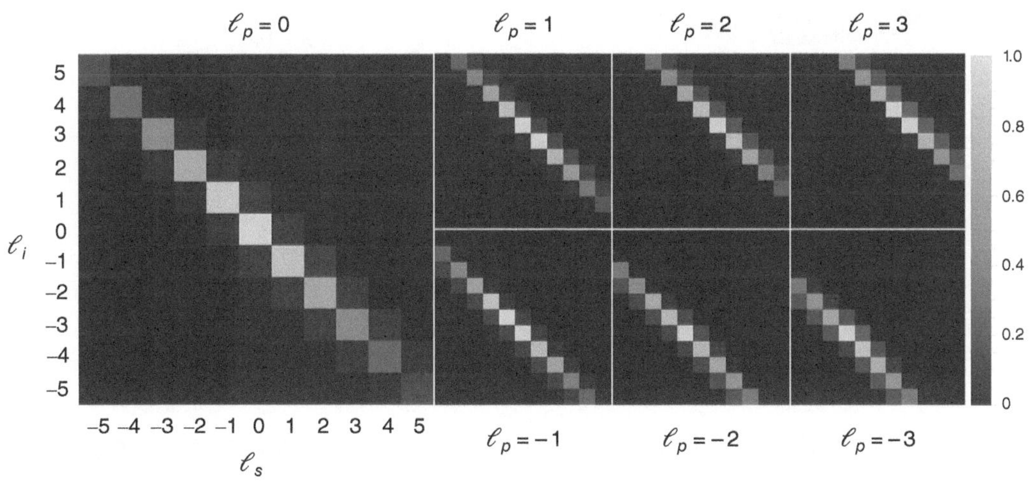

Figure 14.3 Detection matrix showing the correlation among different OAM states of signal and idler photons generated via SPDC in a type-I crystal. The pump carries $\ell_p = -3, -2, -1, 0, +1, +2, +3$ [11]. *Source:* Bouchard et al. [11]. Licenced under CC BY-4.0.

14.4 Generation and Detection of Structured Photon Quantum States

In this section, we will briefly review the main techniques for the generation and detection of structured light, focusing on those employed in quantum experiments.

14.4.1 Generation of Structured Photon States

The simplest way to generate a structured photon state is to start with a Gaussian beam and then use some optical element to shape its wavefront as desired. In particular, if one wants to generate a photon with OAM values of ℓ, then it suffices to use a phase element that multiplies the field of the input beam by a phase factor of $\exp(i\ell\phi)$. Such an element can be a refractive object whose thickness is proportional to the azimuthal angle ϕ. This device is termed the spiral phase plate (SPP) and was used to generate OAM beams for the first time by Beijersbergen et al. [12]. If h is the step height of the SPP and n_{SPP} is its refractive index, then the phase acquired by a plane wave crossing the plate is $\exp(i\Delta\ell\phi)$ with

$$\Delta\ell = \frac{h}{\lambda}(n_{\mathrm{SPP}} - 1). \tag{14.19}$$

where λ is the beam wavelength in vacuum. The working principle is illustrated in Figure 14.4a and b. A careful design of this device allowed the generation of high OAM beams ($\Delta\ell = 10\ 010$) [13].

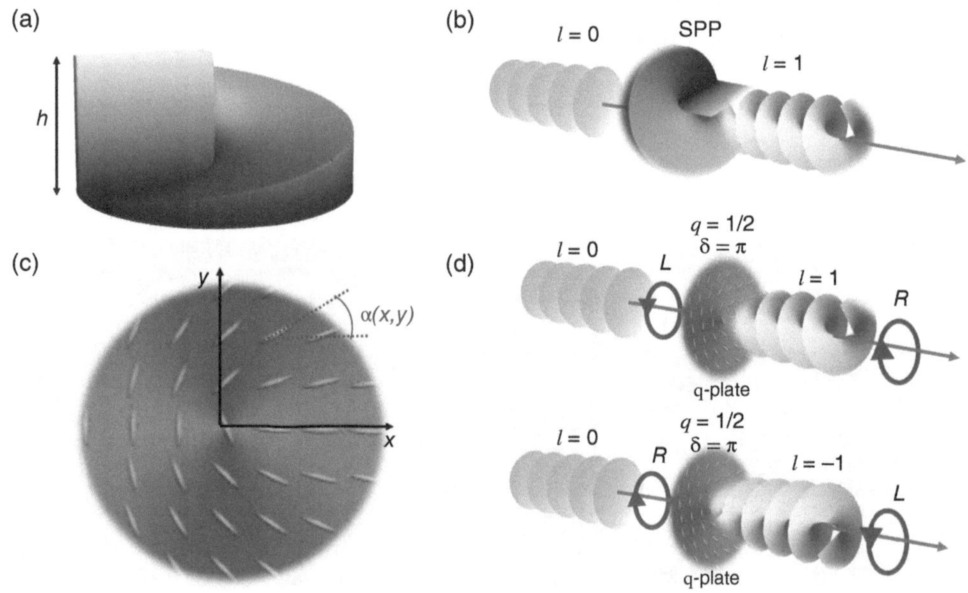

Figure 14.4 Generation of the helical beam with transmissive optics. (a) Profile of a spiral phase plate (SPP); (b) working principle of an SPP; (c) schematic of a q-plate, with a topological charge of $q = 1/2$; filled ellipses represent the orientation $\alpha(x, y)$ of the local optic axis. The background shows how the device looks like when observed between crossed polarizers under a white light illumination; (d) working principle of a $q = 1/2$-plate tuned at half-wave retardation, when it is illuminated with left- and right-handed circularly polarized Gaussian beams.

The wavefront of a photon can also be shaped by exploiting holographic techniques. This approach consists of recording the interference pattern generated by the superposition of a plane wave and the desired beam. In particular, this pattern can be calculated on a computer and displayed on a spatial light modulator (SLM). SLMs can be based on liquid-crystal displays (LCDs) or on digital micromirror devices (DMDs). In the former case (LCD SLM), each pixel implements a phase shift proportional to the amplitude in the calculated interference pattern, usually encrypted in eight bits (0–255). This shift is determined by the orientation of the liquid-crystal molecules, which is controlled by the voltage applied on local electrodes with an 8-bit resolution. In DMDs, each pixel corresponds to a micromirror whose orientation can be controlled independently. It is noteworthy that phase-only SLMs provide better efficiency than DMDs. Computer-generated SLMs, as well as DMDs, are reconfigurable; hence, they allow for the generation of a plethora of light modes. In principle, with the addition of amplitude shaping techniques, a single SLM provides a tool for the generation of arbitrary paraxial beams.

Liquid-crystal-based (phase-only) SLMs are the most widely used in quantum experiments with structured light. They can be modeled as an optical element that multiplies the electric field by a factor $\exp(ig(x,y))$, where (x,y) are the transverse coordinates. Usually, the phase $g(x,y)$ is the superposition of the desired phase $f(x,y)$ that we want to apply on the beam and a blazed (sawtooth-shaped) grating. The (blazed) grating gives the ability to spatially separate the residual light that is not transformed (due to the limited resolution of the SLM and an imperfect input polarization) and also maximizes the intensity of the light in the first-order of diffraction. Explicitly, a blazed hologram is given by the following mathematical expression:

$$g(x,y) = \text{Mod}\left(\frac{2\pi}{\Lambda}x + f(x,y), 2\pi\right), \tag{14.20}$$

where $\text{Mod}(x, 2\pi)$ and Λ are the reminder of $x/(2\pi)$ and the period of the grating, respectively. Theoretically, such a hologram modulates and deflects 100% of the input beam to the first-order of diffraction. In practice, due to a limited resolution of the SLM and experimental imperfections, the conversion efficiency is typically around 70% in the first-order of diffraction. Figure 14.5a shows an example of a typical pitch-fork blazed hologram used for inducing OAM on an input Gaussian beam.

Using a blazed hologram also allows for modulating not only the phase but also the amplitude of the input beam, resulting in the generation of a desired optical beam. This can be done at the expense of some losses by deviating only part of the input light to the first diffraction order. An exact method of doing this was devised in Bolduc et al. [14]. This method considers the beam obtained from selecting the first diffraction order of the phase grating described by the function

$$g(x,y) = M(x,y)\,\text{Mod}\left(\frac{2\pi}{\Lambda}x + F(x,y), 2\pi\right), \tag{14.21}$$

with

$$M(x,y) = 1 - \frac{1}{\pi}\text{sinc}^{-1}(A(x,y)) \tag{14.22}$$

$$F(x,y) = \Psi(x,y) - \pi M(x,y), \tag{14.23}$$

where $A(x, y)$ and $\Psi(x, y)$ are, respectively, the amplitude and phase of the beam that one wants to generate, i.e. $A(x, y)\, e^{i\Psi(x, y)}$.

The techniques discussed so far allow us to manipulate the scalar features of a light beam. In some applications, it is also important to have a means to couple the amplitude and phase with

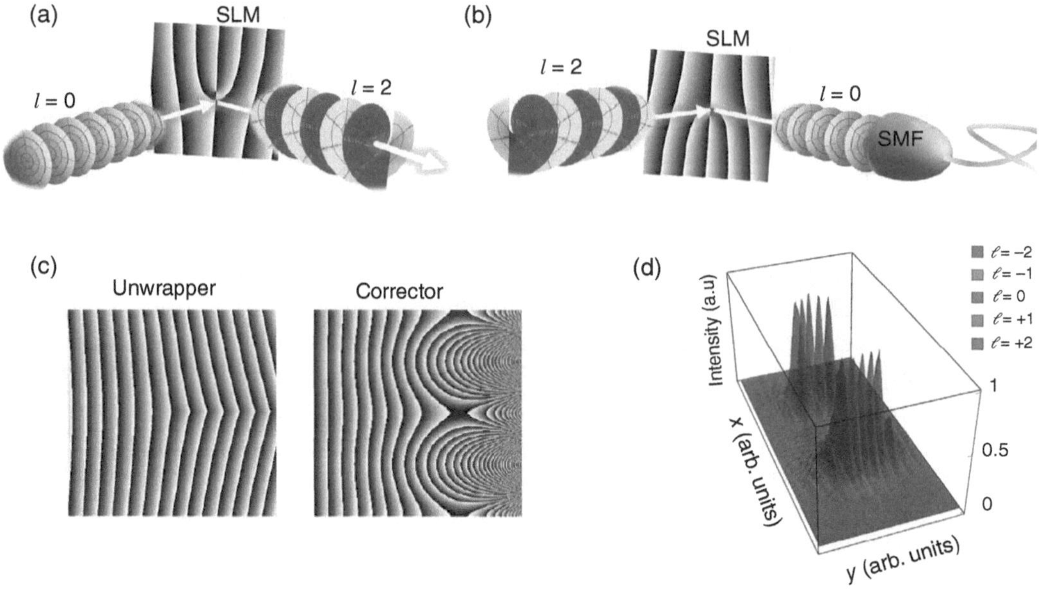

Figure 14.5 Generation and detection of structured light with SLMs. (a) Helical beams can be generated by shining a fork-shaped grating with a Gaussain beam; (b) the same kind of hologram can be used to convert an OAM beam into an $\ell = 0$ beam in order to measure its intensity after coupling with a single-mode fiber (SMF); (c) phase profiles of the unwrapper and corrector elements that constitute an OAM sorter; (d) experimental image of OAM modes transformed by a generalized sorter where the modes are deflected along the x coordinate depending on if their polarization is R or L (caption adapted from Larocque et al. [24]). *Source:* Adapted from Larocque et al. [24].

the polarization of a light beam. This gives the ability to generate beams with spatially varying polarization distributions or to apply transformations on an input beam that are conditioned by its polarization state. This task can be achieved with SLMs or SPPs by introducing these devices in properly designed interferometric setups. However, it is also possible to obtain the same results with more compact spatially varying waveplates. We next proceed to describe the principle of operation of these devices.

A waveplate is an optically anisotropic device that introduces a phase retardation between orthogonal polarizations parallel to the slow and fast axes of the medium. Its action can be described by a Jones matrix, on the basis of linear polarization, given by

$$W(\delta, \alpha) = R(\alpha)^{-1} \cdot \begin{pmatrix} 1 & 0 \\ 0 & e^{i\delta} \end{pmatrix} \cdot R(\alpha), \tag{14.24}$$

where $R(\alpha)$ is a rotation matrix that takes into account an inclination of the optic axis by an angle α in the reference system, and δ is the optical retardation. This object only transforms the polarization of the beam. However, one can simultaneously modify the polarization and the phase if the angle α is made to be spatially variant, i.e. $\alpha := \alpha(\mathbf{r})$. In this case, the resulting Jones matrix can be conveniently expressed in the circular polarization basis (where the left-circular polarization is given by the vector $|L\rangle = (1, 0)^{\top}$ and the right-circular polarization is given by $|R\rangle = (0, 1)^{\top}$):

$$W_c(\delta, \alpha) = \cos\left(\frac{\delta}{2}\right) 1 + i \sin\left(\frac{\delta}{2}\right) \begin{pmatrix} 0 & e^{i2\alpha(x,y)} \\ e^{-i2\alpha(x,y)} & 0 \end{pmatrix}. \tag{14.25}$$

In particular, at a half-wave retardation $\delta = \pi$, the wavefront is modified by a phase $\pm 2\alpha(x, y)$, where the sign depends on if the input polarization is $|L\rangle$ or $|R\rangle$. Remarkably, for an arbitrary input polarization, the beam will be in the coherent superposition of $\exp(i2\alpha(x, y))|R\rangle$ and $\exp(-i2\alpha(x, y))|L\rangle$, where both the spatial and polarization degrees of freedom become nonseparable – a feature that is often given the, sometimes confusing, name of *classical entanglement* [15]. Eq. (14.25) describes a class of optical elements known as the Pancharatnam–Berry phase optical elements (PBOE).

The most known PBOE is the q-plate, first introduced by Marrucci et al. [16], which couples the spin (polarization) and OAM of a photon. For this object, α is a linear function of the azimuthal angle ϕ:

$$\alpha(\phi) = q\phi + \alpha_0. \tag{14.26}$$

Typically, q-plates consist of liquid-crystal cells where the transverse projection of the molecular director (i.e. the vector describing the local optic axis orientation) is inclined by an angle $\alpha(\phi)$. As illustrated in Figure 14.4c and d, a light beam impinging on a q-plate acquires an OAM given by $\ell = \pm 2q$; recall that the sign is dictated by the input circular polarization. Since ℓ is an integer, q must be an integer or a semi-integer number (otherwise, the optical axis pattern would exhibit line dislocations) and is called the *topological charge* of the liquid-crystal pattern. Liquid-crystal-based q-plates have the remarkable property that their optical retardation can be tuned by temperature or electric voltage control [17]. Q-plates have been used in various quantum experiments, ranging from fundamental tests [18, 19] to quantum cloning [20], quantum communication (see the next section), and quantum simulation [21–23]. Some of these applications will be reviewed in the next sections.

14.4.2 Detection of Structured Photons

Many methods of structured light detection have been developed over the years. The first approaches were based on interferometric techniques. These allow for an easy way to identify phase singularities as dislocations in the interference pattern (see, e.g. [16]). Moreover, phase shifting holography can be used to fully characterize the amplitude and phase structure of a light beam [25]. However, since they are based on image recording, interferometric methods are not the most widely used in single-photon applications. In this field, the most common approach is based on performing appropriate beam shaping so that the desired mode is coupled into an SMF.

In particular, the detection of OAM states can be performed by means of a *phase-flattening* optical element, which applies a (conjugate) phase $\exp(-i\ell\phi)$ so that the unknown photon's OAM spectrum is shifted by $-\ell$, i.e. the state $|m\rangle$ is transformed into $|m - \ell\rangle$ (see Figure 14.5b). If the beam is then coupled to an SMF, whose fundamental mode carries no OAM $|0\rangle$, only the photon state (before the OAM shifting) $|m = \ell\rangle$ will be guided through the fiber to the detector. Hence, the combination of the phase-flattening element and the SMF/detector system corresponds to the optical implementation of the projector $|\ell\rangle\langle\ell|$. The phase-flattening element can be an SPP, as well as a q-plate, or an SLM. The latter is the most widely used since, being reconfigurable, it can be used to reconstruct the full OAM spectrum. This technique was first applied in a quantum optics framework by Mair et al. [6] to experimentally reconstruct the OAM-entangled state that arises from a type-I SPDC process. SLMs can also project onto states that are combinations of different OAM modes; thus, one can perform full quantum state tomography [26, 27].

The phase-flattening approach presents some issues when one wants to realize detection at a high speed and over a large number of modes. The detection of a single photon is limited by a success rate of $1/M$, where M is the number of modes. As well as for high values of ℓ, the detection efficiency is lower due to the fact that the radial shape is not controlled by the phase-flattening technique; hence, the resulting beam coupled to the SMF, even if the OAM is zero, does not have a perfect overlap with the fundamental mode [28]. The latter limitation implies that a calibration procedure is needed before the actual measurement to avoid an underestimation of higher-order mode contributions.

An alternative approach is based on using some optical element that spatially separate modes corresponding to different OAM values. Such an OAM sorter relies on implementing a map from OAM eigenstates to plane waves with different wavevectors. Separation by spatial propagation of each wave will allow for the simultaneous detection of the relative intensities of each mode using a fiber array or a camera. The sorting can be achieved by means of a single optical element, called the 'unwrapper,' if the coordinate transformation is a conformal mapping from Cartesian to logpolar coordinates, $(x, y) \rightarrow (u, v)$, where $u = a\tan^{-1}(y/x)$ and $v = a\log\left(\sqrt{x^2+y^2}/b\right)$, a and b are independent parameters. The unwrapper is then implemented by an optical element with phase profile given by

$$\phi_1(x,y) = \frac{2\pi a}{\lambda f}\left(y\arctan(y/x) - x\log\left(\sqrt{x^2+y^2}/b\right) + x - (x^2+y^2)\right), \tag{14.27}$$

where we added a paraboloidal phase contribution that incorporated the effect of a converging lens with focal length f. Such a transformation will introduce a phase distorsion due to the different path lengths. A second element, to be placed in the Fourier plane of the unwrapper, is thus needed to compensate for this effect. The phase profile of this 'phase corrector' is given by

$$\phi_2(u,v) = -\frac{2\pi ab}{\lambda f}e^{-u/a}\cos(v/a). \tag{14.28}$$

Examples of unwrapper and corrector holograms are shown in Figure 14.5c. Figure 14.5d shows an example of the resulting intensity distribution. The OAM sorter usually leads to a relevant cross-talk between different detected modes (around 20%). However, this issue can be addressed with an additional 'fan-out' phase element, introduced by Mirhosseini et al. [29], that allows for a separation efficiency of the various spots higher than 92%. OAM sorters have also been realized with PBOEs allowed to simultaneously sort OAM and polarization states, allowing for an efficient reconstruction of beams with space variant polarization [24].

In recent years, followed by the theoretical study of Karimi and Santamato [30], more efforts have been put into place to exploit and characterize the radial degree of freedom in quantum applications. New ways of sorting and measuring LG modes have been developed [31–35].

14.5 Quantum Key Distribution

One of the applied fields where quantum mechanics is expected to be an important contributing factor in the near future is cryptography, more specifically the possibility of sharing cryptographic keys in a secure way, where the security will be guaranteed by the laws of Nature. Current classical cryptosystems, e.g. the RSA protocol [36], are under the potential threat by advances in number theory [37] and will become unreliable when quantum computers, capable of efficiently factoring high prime numbers, will be developed [7]. In principle, one could follow the approach of developing novel classical cryptosystems, known as postquantum cryptography, which cannot be broken by known quantum algorithms. However, new quantum algorithms may then be discovered that could break these cryptosystems. The ultimate solution to secret key sharing is instead quantum key distribution (QKD). In QKD, there are two parties, called Alice and Bob, who wish to privately share a message, and a third party, Eve, trying to eavesdrop on their communication. The basic idea behind quantum communication is that any attempt from Eve to read part of the message Alice is sending to Bob will partially modify that message. Indeed, the message (the cryptographic key in the case of QKD) is encoded in a quantum state, and eavesdropping will consist of a measurement (either strong or weak), which will modify the quantum state, at least for some qubits of information. The latter statement holds true if Alice and Bob agree to encode the bits of information in randomly chosen and nonorthogonal quantum states. Hence, if Alice and Bob have at their disposal a protocol that estimates the rate of qubits that have been modified, the presence of Eve can be revealed and the shared message can be classified as insecure. More formally, one may think of Eve as being able to take the quantum state sent by Alice, create a copy of it, and send the original one to Bob. In this way, Eve can measure the cloned state without changing the original one. However, this action turns out to be impossible in quantum mechanics. This result is known as the *no-cloning theorem* [7]. The proof of this theorem is straightforward: suppose a cloning machine is represented by a unitary operator U_{clone} whose action is

$$U_{clone}(\mid \psi\rangle_A \otimes \mid 0\rangle_E) = \mid \psi\rangle_A \otimes \mid \psi\rangle_E \tag{14.29}$$

which should hold true for any ψ. The subscripts A and E refer to the states held by Alice and Eve, respectively. $|0\rangle_E$ is an ancilla state that Eve uses as a resource to clone Alice's state. We notice that Eq. (14.29) cannot hold simultaneously for two nonorthogonal states $|\psi_1\rangle$ and $|\psi_2\rangle$. Indeed, onehas $\langle\psi_1 \mid \psi_2\rangle = \langle\psi_1 \mid U_{\text{clone}}^{-1} U_{\text{clone}} \mid \psi_2\rangle = (\langle\psi_1|_A \otimes \langle\psi_1|_E) \cdot (\mid \psi_2\rangle_A \otimes \mid \psi_2\rangle_E) = (\langle\psi_1 \mid \psi_2\rangle)^2$. However, this result is possible if and only if $|\psi_1\rangle$ and $|\psi_2\rangle$ are either equal or orthogonal; hence, the theorem is proven.

In practical implementations of QKD, however, there will always be errors caused not by eavesdroppers but rather by environmental noise and system imperfections – for example, turbulence and misalignment in free-space links, or birefringence and dispersion effects in optical fiber communications. Eve may thus exploit these imperfections to build an optimal (instead of perfect) cloning machine whose effects may be indistinguishable from environmental noise. Hence, practical QKD must establish an upper limit for the acceptable qubit error rate (QBER), below which it is impossible for Eve to retrieve enough information. Addressing this issue while guaranteeing a high secret key rate is the fundamental challenge of practical QKD. Generally speaking, this will require the following technological developments: (i) sources with a high transmission rate, (ii) reduction of environmental noise or compensation for its effects, and (iii) minimization of detector imperfections (or implementation of device-independent QKD protocols).

In this section, we will review the simplest and most widely used protocol for QKD, the BB84 protocol [38], and its generalization to qudits. Then, we will show how structured light can be useful to surpass errors due to misalignment between Alice and Bob's reference frames (alignment-free QKD) and to access higher-dimensional QKD that allows for higher QBER thresholds. We will also mention how the communication distance, which for structured light is limited by the mode divergence, can, in principle, be improved by using quantum repeaters based on entanglement swapping. A recent and more detailed review on the field of QKD can be found in Ref. [39].

14.5.1 BB84 Protocol

In the BB84 protocol, Alice wants to send Bob the key string needed to decipher her message. The key is encoded in a sequence of qubits as follows: Alice and Bob agree on using two nonorthogonal and mutually unbiased bases (MUB), e.g. the bases given by the eigenstates of, respectively, the σ_x and σ_y Pauli matrices. This choice is necessary for preventing eavesdropping with a cloning machine. Often, the two bases are given by the horizontal/vertical polarization basis, $X := \{|H\rangle, |V\rangle\}$, and diagonal/antidiagonal polarization basis, $Y := \{|D\rangle, |A\rangle\}$. The two parties agree that the bit '0' corresponds to $|H\rangle$ and $|D\rangle$, while '1' is encoded in $|V\rangle$ and $|A\rangle$.

The protocol works as follows. Alice first randomly chooses in which basis to encode each bit. For example, she may encode the key '10011011011000...' as 'VDDAVHAAHVAHDD....' Bob then randomly chooses in which basis to measure each qubit, e.g. he chooses the sequence: 'XXYYXXXYX-XYYXX....' When the chosen basis corresponds to Alice's for a given qubit, then the outcome of the measurement will give the correct bit. When the basis is different, the outcome will be random. In our example, Bob will obtain '1R0110R1R11R 00...,' where 'R' stands for a random outcome. Using a public channel, Alice shares with Bob the sequence of bases she chose so that Bob can recognize which one of his bits are random. Those bits are discarded and the remaining string ('1011011100...') will be the new key. How does Bob knows if some of his measurement outcomes corresponding to the right basis choice have not been flipped by Eve? Alice and Bob can agree to test if there are errors by sharing some randomly chosen bits of the new key and check if they are equal. If some of them turn out to be different, then the presence of the eavesdropper is detected. In practice, due to the unavoidable presence of noise in the channel, they check if the percentage of errors is lower than the maximum acceptable QBER. If such a condition is achieved, Alice and Bob need to correct the remaining errors in order to share identical keys. This is done via *information reconciliation*: an error-correcting procedure conducted via the public channel. Different algorithms can be used; a common one is the *cascade protocol* that consists of recursively dividing the two keys in blocks and doing parity checks for each corresponding block (see Ref. [40] for details). After the cascade protocol is stopped, Alice and Bob reorder their keys in the same random way. During information reconciliation, Eve may gain additional information from the public results of the parity checks. A further process, called *privacy amplification* (PA), can be applied to reduce Eve's knowledge about the key. PA consists of reducing the key size, by an amount dependent on the QBER, using a universal hash function chosen randomly from a publicly known set (see Ref. [7]). The resulting key that is ultimately used by Alice and Bob is called the *sifted key*.

A quantitative estimate of the QBER threshold is thus crucial for ascertaining the security of a QKD scheme. Security proofs are based on estimating the relation between the information gained by Eve and the corresponding error rate. This relation depends on the kind of attack performed by Eve. Attacks can be individual (Eve performs individual measurements on her ancillary qubits), collective (all the output ancillary qubits are stored and measured collectively after the information reconciliation), or general-coherent (which where shown to be equivalent to collective attacks in the asymptotic scenario, where a large number of signals are exchanged; see Ref. [39]).

For the two-basis BB84 protocol, it has been shown (see Ref. [41]) that the QBER below which it is still possible for Alice and Bob to obtain a positive secure key is 11%. This limit was slightly improved to 12.6% by a generalization of the BB84, where Alice and Bob can choose all three different mutually unbiased bases to encode information (*six-state protocol*, [42]).

Cerf et al. (2002) [43] showed that high-dimensional QKD protocols, where information is encoded in qudits, have higher QBER thresholds. Hence, using structured light to encode qudits is a promising route for practical QKD, as we will show in the following discussion.

14.5.2 Alignment-free QKD

Before investigating applications of structured light in high-dimensional QKD, we first show how qubit encoding in specific structured light states can be advantageous to get rid of the effect of misalignment in the measuring apparatuses between Alice and Bob. If the qubits are encoded in the polarization of single photons, measurement errors can be caused by a rotation of Bob's reference frame with respect to Alice's (see Figure 14.6a). Establishing a shared reference frame can be demanding in terms of the required amount of classical communication, especially in network systems involving a high number of parties. This issue can be surpassed by using rotationally invariant qubits. This approach was first proposed by D'Ambrosio et al. [44]. The basic idea is to exploit optical modes with zero *total angular momentum*. The simplest modes satisfying this relationship are circularly polarized eigenstates with an OAM opposite to the SAM corresponding to the circular polarization handedness. To these modes, we can associate the logical basis $\{|\text{'0'}\rangle, |\text{'1'}\rangle\}$, with

$$\begin{aligned} |\text{'1'}\rangle &:= |\text{R}\rangle_\pi \otimes |1\rangle_o =: |\text{R}, 1\rangle \\ |\text{'0'}\rangle &:= |\text{L}\rangle_\pi \otimes |-1\rangle_o =: |\text{L}, -1\rangle, \end{aligned} \tag{14.30}$$

where the subscripts π and o refer, respectively, to the SAM and OAM Hilbert spaces. It is worth noticing that if these modes were not to be combined together, they would not be rotationally invariant, since they pick up a phase $\exp(\pm i\theta)$ under a physical rotation of θ around the propagation axis.

In order to perform alignment-free QKD, at least two rotationally invariant (mutually unbiased) bases are needed. This requirement is automatically satisfied since any linear combination of the modes $|\text{'0'}\rangle$ and $|\text{'1'}\rangle$ also carries zero angular momentum. As the second basis to encode Alice's bits, one can choose:

$$\begin{aligned} |\tilde{1}\rangle &= \frac{|\text{'0'}\rangle + |\text{'1'}\rangle}{\sqrt{2}} = \frac{|L\rangle_\pi \otimes |-1\rangle_o + |R\rangle_\pi \otimes |1\rangle_o}{\sqrt{2}} \\ |\tilde{0}\rangle &= \frac{|\text{'0'}\rangle - |\text{'1'}\rangle}{\sqrt{2}} = \frac{|L\rangle_\pi \otimes |-1\rangle_o - |R\rangle_\pi \otimes |1\rangle_o}{\sqrt{2}}, \end{aligned} \tag{14.31}$$

which corresponds to the well-known radially and azimuthally polarized vector vortex modes; Figure 14.6 shows the four states used in this experiment. All four states needed for alignment-free QKD are efficiently obtained by means of a q-plate with topological charge $q = 1/2$, tuned at half-wave retardation $\delta = \pi$ [45]. Indeed, given a Gaussian mode prepared in an arbitrary polarization state $|\pi\rangle = \alpha|L\rangle + \beta|R\rangle$, the action of the q-plate will be:

$$|\pi\rangle_\pi \otimes |0\rangle_o \xrightarrow{q-\text{plate}} \alpha|\text{'0'}\rangle + \beta|\text{'1'}\rangle. \tag{14.32}$$

Hence, the q-plate can be viewed as a tool to automatically switch from encoding information with polarization to that of rotationally invariant modes. Moreover, the same device, coupled with quarter- and half-wave plates followed by a polarizing beam splitter (needed for polarization projection), can be used for the detection of these modes via phase-flattening measurements.

D'Ambrosio et al. [44] verified the feasibility of this scheme implementing a BB84 protocol and obtaining a QBER that corresponded to a secret key rate of $r = (70 \pm 1)\%$. Moreover, the possibility to distribute entangled states and violate the CHSH inequality with this scheme was successfully proven. Remarkably, the use of rotationally invariant qubits is robust to a large set of spatial mode transformations that may severely affect the efficiency of the key distribution. Indeed, such a transformation will transform the state $\alpha|R\rangle_\pi \otimes |1\rangle_o + \beta|L\rangle_\pi \otimes |-1\rangle_o$ into

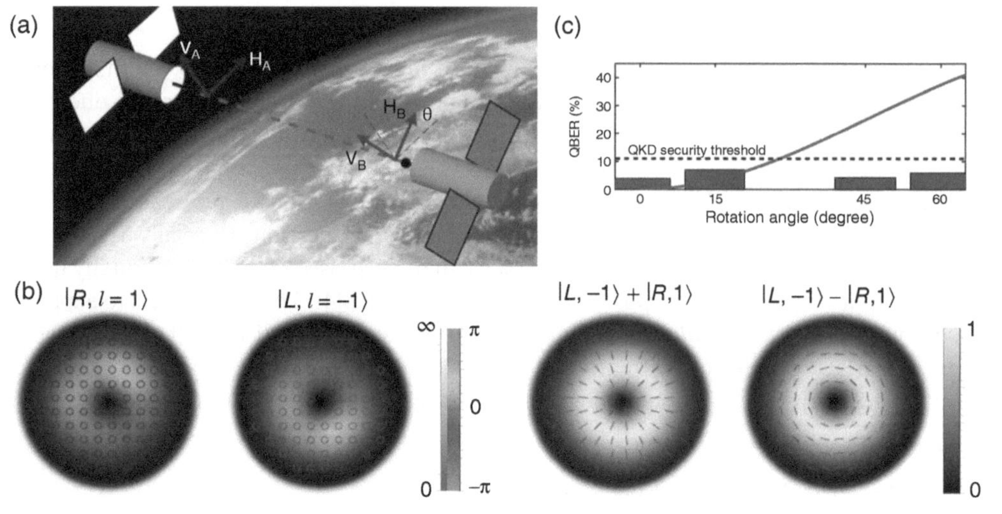

Figure 14.6 Concept of alignment-free QKD. (a) Alice and Bob (depicted as two satellites) may not share the same convention for *H* and *V* polarization states, due to relative rotations of their respective frames [44]. *Source:* D'Ambrosio et al. [44]. (b) Plots of phase, amplitude, and polarization distribution of the rotationally invariant optical modes used in Ref. [44]. *Source:* Based on D'Ambrosio et al. [44]. (c) QBER obtained by Vallone et al. [46] as a function of the rotation angle of Bob's frame. The continuous line indicates the corresponding theoretical QBER in the case of polarization-based BB84. *Source:* Vallone et al. [46].

$$\sum_{\ell}(C_{1,\ell}\alpha \mid R\rangle_{\pi}\otimes \mid \ell\rangle_o + C_{-1,\ell}\beta \mid L\rangle_{\pi}\otimes \mid \ell\rangle_o), \tag{14.33}$$

which after the detection q-plate will become

$$\sum_{\ell}(C_{1,\ell}\alpha \mid L\rangle_{\pi}\otimes \mid \ell-1\rangle_o + C_{-1,\ell}\beta \mid R\rangle_{\pi}\otimes \mid \ell+1\rangle_o). \tag{14.34}$$

The SMF in the detection stage will project (postselect) this state into

$$C_{1,1}\alpha \mid L\rangle + C_{-1,-1}\beta \mid R\rangle. \tag{14.35}$$

If the perturbation is such that $C_{1,\,1} = C_{-1,\,-1}$, the qubit detected by Bob will be identical to that prepared by Alice. This condition is satisfied by mirror-symmetrical transformations with respect to a plane containing the propagation axis. Such transformations are satisfied by a large class of perturbations: beam parallel displacements, elliptical deformations, finite apertures (on- and off-axis), planar obstacles, and also turbulence-induced effects, such as beam wandering and spreading.

In Ref. [46], the alignment-free QKD scheme was tested in an indoor 210-m free-space link in Padua, achieving a QBER of around 5% and estimated key rates up to 20%. The QBER was shown to be essentially independent of the relative rotation between the two reference frames (Figure 14.6c).

14.5.3 High-dimensional QKD

In recent years, there has been an increasing interest in high-dimensional QKD, where each signal sent by Alice consists of a d-dimensional qudit, i.e. a quantum state $|i\rangle$, with $i = 1, \ldots, d$. The first advantage is that each qudit carries a larger amount of information, given by $\log_2 d$ (although one must consider that the information density per mode scales as $\log_2(d)/d$). Due to the increase in information per qudit, the key rate can be increased beyond the limits of the photon flux imposed by the detector dead-time. Moreover, it has been shown that high-dimensional QKD allows for higher QBER thresholds. The first security proof was given by Ref. [43], where the effects of Eve's individual attacks carried out with an optimal cloning machine were considered in order to estimate the mutual information between Eve and Alice (I_{EA}), and Alice and Bob (I_{AB}). The difference $I_{AB} - I_{EA}$ gives the amount of the secured key rate under individual attacks. The effect of coherent attacks actually gives more stringent bounds on the key rate. These bounds are obtained from the condition $I_{AB} + I_{EA} \leq \log_2 d$, which translates to the following upper bound for the error rate D [43]:

$$(1-D)\log_2(1-D) + D\log_2\left(\frac{d-1}{D}\right) \leq \frac{1}{2}\log_2 d. \tag{14.36}$$

Remarkably, the upper bound for D increases with the dimensionality d; hence, high-dimensional protocols are more tolerant to external disturbances. For example, while for the two-basis BB84 with $d = 2$ the maximum QBER is $D_{max} = 11\%$, for $d = 4$ one has $D_{max} = 18.93\%$ and $D_{max} = 24.7\%$ for $d = 8$. Similar bounds were found for $d + 1$ state protocols. These bounds have been confirmed by Sheridan and Scarani [47] for coherent attacks.

The OAM of light provides an ideal degree of freedom for realizing high-dimensional QKD. One of the first proofs of principle realizations was reported by Mirhosseini et al. [48]. The experiment consisted of a two-basis BB84 protocol with seven-dimensional qudits. The two bases chosen for the encoding were given by the OAM basis $\{|\ell\rangle\}_{\ell = -3, \dots, 3}$, and the corresponding MUB of angular modes,

$$|\psi^k_{\text{ANG}}\rangle = \frac{1}{\sqrt{d}}\sum_{\ell=-3}^{3} e^{ik\left(\frac{2\pi\ell}{d}\right)}|\ell\rangle. \tag{14.37}$$

Both bases can be easily generated by sending each pulse to an SLM. Each signal was then imaged to Bob and analyzed using a mode sorter and a beam splitter (the latter needed to randomly select one of the two bases). A QBER of $D \approx 10\%$ was achieved for the above proof-of-principle experiment.

In Ref. [49], different high-dimensional QKD protocols were compared and realized experimentally with a fixed optical setup (see Figure 14.7a, b). The qudit states were encoded and detected using SLMs. It was thus possible to explore the relative advantages and drawbacks of the considered protocols. In particular, it was found that the Chau15 protocol [50] is more tolerant in a high-noise environment while offering lower performances in terms of key rates in low-noise scenarios. Moreover, tomographic protocols, such as the high-dimensional extension of the six-state BB84 and the Singapore protocol [51], allow for a complete characterization of the quantum channel, thereby introducing the possibility of distinguishing between environment-induced and eavesdropper-induced noise (see also Ref. [52]). The latter analysis was made possible by mimicking different quantum hacking scenarios with optical setups, e.g. inserting an optimal cloning machine in the communication link [53].

The last few years have seen considerable improvements in QKD with structured light outside the lab environment. QKD protocols have been successfully implemented in free-space [54], optical fiber [55, 56], and underwater [57] links. These schemes used specific spin–orbit states that are shape-invariant during propagation. In the following, we briefly describe these implementations.

High-dimensional free space QKD in urban environments. Sit et al. [54] successfully realized a 300-m turbulent free-space link in Ottawa, Canada (Figure 14.7d) for two- (2D) and four-dimensional

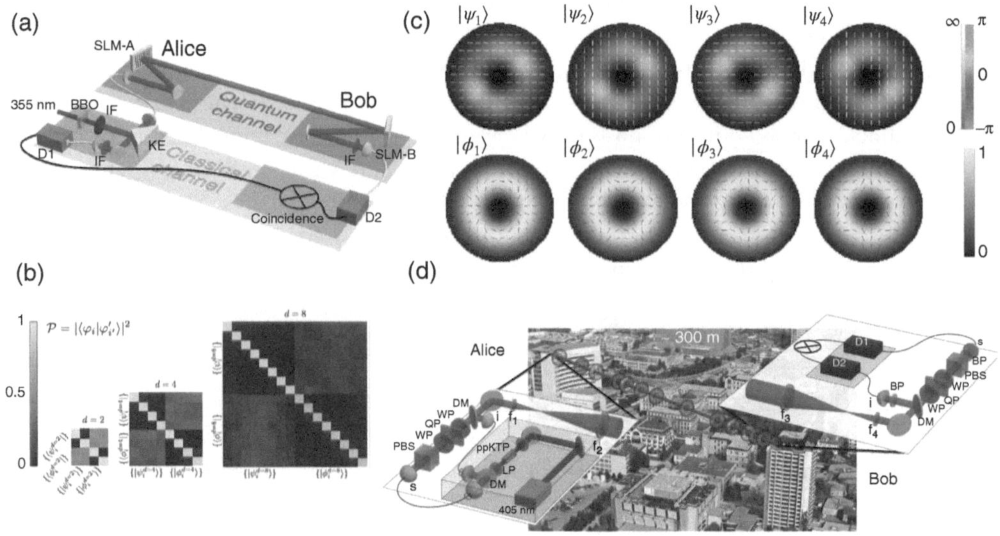

Figure 14.7 Experimental high-dimensional QKD. (a) Setup used by Bouchard et al. [49] to test various QKD protocols for different Hilbert spaces; (b) corresponding experimental detection matrices, showing the probability that Bob detects a state $|\phi'\rangle$, assuming that Alice sent the state $|\phi\rangle$ [49]. *Source:* Bouchard et al. [49]. ; (c) spatial modes used to encode the two four-dimensional MUB in Ottawa's free space link, shown in panel (d) [54]. *Source:* Sit et al. [54].

QKDs. A secret key rate of 0.65 bits per sifted photon was attained for the four-dimensional protocol, demonstrating the superiority over the 2D protocol where the secret key rate was 0.43 bits per sifted photon. The 2D protocol was a BB84 encoded in vector vortex modes as in Ref. [44]. Such a protocol was generalized to four dimensions using the following two sets of mutually unbiased bases (see Figure 14.7c):

$$\begin{aligned}
&\{|\psi_i\rangle\}_{i=1,\dots,4} = \{|H,\ell\rangle, |H,-\ell\rangle, |V,\ell\rangle, |V,-\ell\rangle\}\\
&\{|\phi_i\rangle\}_{i=1,\dots,4} = \{(|L,\ell\rangle + |R,-\ell\rangle)/\sqrt{2}, (|L,\ell\rangle - |R,-\ell\rangle)/\sqrt{2},\\
&(|L,-\ell\rangle + |R,\ell\rangle)/\sqrt{2}, (|L,-\ell\rangle - |R,+\ell\rangle)/\sqrt{2}\}.
\end{aligned} \tag{14.38}$$

These modes can be efficiently generated by sandwiching a q-plate between two pairs of half-wave and quarter-wave plates.

The main source of noise in free space quantum links is turbulence, which alters the spatial structure of the optical mode and its alignment with respect to the receiver. While the structure of vector modes is considered to be robust in turbulent scenarios, the displacement of the beam center with respect to the receiver axis is a major drawback; we remark that, while the 2D protocol is alignment-free, this is not anymore true for the higher-dimensional ones based on vector modes. This issue was surpassed by an appropriate postselection procedure. This experiment exploited for the first time a heralded single photon source in a free-space quantum communication link; previous experiments were based on attenuated laser sources. The heralded (idler) photon was selected as a Gaussian mode and sent to Bob through the same optical path of the signal photon, which was the information carrier, so that both photons experience the same noise effects (mainly the same turbulence). In this way, measurements, where the idler count rates were smaller than a certain threshold due to beam misalignment, could be discarded.

The turbulence in the channel was characterized by observing beam-wandering effects. The relative average displacement s with respect to the beam axis could be measured by multiple beam images taken with a short exposure time. This parameter is directly related to the light deflection angle β induced by turbulent cells as $\beta = s/L$, where L is the length of the link. The deflection angle can be used to evaluate the Fried parameter [58] $r_0 = 0.98\,\lambda/\beta$ - λ is the beam wavelength. This parameter estimates the average fluctuations in the refractive index given by the atmospheric structure constant [59], i.e.

$$C_n^2 = |\mathbf{r}|^{-2/3}\left\langle [n(\mathbf{r}_1 + \mathbf{r}) - n(\mathbf{r}_1)]^2 \right\rangle = \frac{r_0^{3/5}}{0.423 k^2 L}, \tag{14.39}$$

where k is the wavenumber and $\mathbf{r}_1$ is a reference point. The structure constant was estimated by Sit et al. [50] for their experiment to be of the order $C_n^2 \approx 10^{-15} - 10^{-16}$.

QKD through vortex fiber links. In Ref. [60], an optical fiber designed for sustaining propagation over long distances of vector vortex modes was demonstrated. This kind of *vortex fiber* is a step-index fiber where the core section is shaped in such a way to mimic the intensity distribution of a doughnut beam. This lifts the degeneracy between radially and azimuthally polarized modes, thus reducing their cross-talk. Sit et al. [56] demonstrated a 2D BB84 protocol through a 60-m vortex fiber link, exploiting the protocol introduced by D'Ambrosio et al. [44], reaching a QBER=8.6%.

Vortex fibers can sustain higher-order modes, thus allowing for high-dimensional QKD. A four-dimensional BB84 protocol through a 1.2-km fiber link was demonstrated by Cozzolino et al. [54]. In this implementation, superpositions of $|\pm 6\rangle_o$ and $|\pm 7\rangle_o$ OAM eigenstates were used to select the set of MUB to be used for the information encoding. An interferometric OAM sorter allowed for a high transmission rate (37.85 kbit.s^{-1}), which was improved to 42.3 kbit.s^{-1} using a 2D×2D BB84 protocol, where bits were encoded in four different 2D bases.

Developments in underwater quantum communication. Underwater quantum communication is under investigation for secure communication between submersibles. Photon transmission through water is challenging due to the high absorption coefficient at infrared wavelengths and relevant turbulence-induced aberration. Underwater high-dimensional QKD was explored by Bouchard et al. [57] for the first time in an outdoor swimming pool. The turbulence was characterized through a 3-m link in terms of induced aberration on the beam profile through phase-retrieval algorithms and decomposition in Zernike polynomials. The measured QBER was below the threshold only for 2D and three-dimensional (3D) BB84 protocols. Hufnagel et al. [61] performed a similar characterization of a 5.5-m link in the Ottawa river. It was found that, despite the low absorption of light at visible wavelengths, scattering and turbulence may severely affect the communication efficiency. However, promising progress has been recently made: at the time of writing, a preprint by Hufnagel et al. [62] shows secure quantum communication with vector vortex modes in a underwater link up to 10.5-m in a flume tank (with key rate 0.84). The same key rate was achieved in a 30-m link for polarization-based BB84 QKD.

We conclude this section by mentioning that distribution over long distances of structured photons may be improved with quantum repeaters. The latter may function based on entanglement swapping protocols, which can introduce correlations between particles that have never interacted. In entanglement swapping, one measures the Hong–Ou–Mandel interference of two photons coming from two independent entangled photons. The two photons will be entangled after the measurement process. A first step toward this direction was achieved by Zhang et al. [63], where entanglement swapping of multiple-entangled OAM states was demonstrated. Another useful element in a quantum network will be the possibility to store the information in a quantum memory. This has been achieved in vector beams by Parigi et al. [64] where vector beams were stored in an ensemble of cesium atoms with the additional possibility of on-demand retrieval.

14.6 Quantum Simulation with Quantum Walks

As we have shown in the beginning of this chapter, structured photons can be described in terms of modes labeled by integer indices. These indices are unbounded; hence, the associated states span an Hilbert space that is (in principle) infinite-dimensional. We can thus think of exploiting these modes to investigate physics in high-dimensional spaces. For instance, we can imagine to associate the index ℓ of the photon's OAM with the position in a one-dimensional crystal. Similarly, the radial index, $p > 0$, of $\mathrm{LG}_{p,\ell}$ modes can be considered as labeling energy levels of a harmonic oscillator. Then, we can imagine to implement an optical experiment where the indices ℓ or p are modified following some general rules. Such an experiment will then mimic the evolution of a particle on a lattice (e.g. the electron dynamics in solids) or transitions between different energy levels. This is the basic idea behind quantum simulation (first proposed by Feynman [65]): to use a *simple* and *easy to control* quantum system to simulate the behavior of other physical systems whose properties are less accessible experimentally.

In this section, we will focus on applications where the structured photon was exploited to investigate condensed matter physics. This was done with experiments that implemented specific unitary processes known as *quantum walks* (QWs). The idea of a QW was introduced by Aharonov et al. [66] as a quantum analogue of the random walk. In a classical random walk, a particle moves on a lattice, or a graph, in a way that is conditioned by a random outcome (often depicted as the tossing of a coin). In the simplest case, if the outcome is 'heads' ('tails'), then the particle, usually named as the *walker*, will move to the left (right) site. Quantum-mechanically, the coin tossing can be translated into a rotation operator on a qubit, whose space of possible states is spanned by the set $\{|\mathrm{tails}\rangle, |\mathrm{heads}\rangle\}$. In this case, the outcome can be $|\mathrm{tails}\rangle$, $|\mathrm{heads}\rangle$ or a superposition of the two: $a\,|\mathrm{heads}\rangle + b\,|\mathrm{tails}\rangle$. The latter possibility gives rise to a new feature, absent in the classical case, that is, the quantum superposition of coin and position states. As a consequence, the dynamics of a QW is completely different from the classical counterpart: the probability distribution evolves in a ballistic rather than diffusive fashion [67] and its shape is dependent on the initial coin state. Since the original proposal of Yakir Aharonov, QW has found applications in quantum computing [68, 69], quantum search algorithms [70], and in the investigation of condensed matter physics, in particular, topological insulators [71], fermionic and bosonic statistics [72], Anderson localization [73], Floquet systems, and nonhermitian processes [74].

QWs have been implemented in many different physical platforms: optical lattices [75, 76], cavity QED [77], Bose–Einstein condensates [78], and photonic platforms. In the majority of photonic experiments, the coin has been encoded in the polarization state, which implements rotation operators by means of standard birefringent waveplates. Apart from the structured light implementations, the walker space has been encoded in the optical path [72, 79] and in the time-bin domain [80].

In this section, we will show how structuring light can implement QWs within a single beam, thus realizing platforms that are versatile (in the sense of disposing of different tunable parameters) and scalable (that is, there is a subexponential growth of the number of optical elements with respect to the number of steps).

14.6.1 Quantum Walks in the OAM Space

As a first example of a QW, we consider a platform based on the OAM of light [81]. The OAM eigenstates, $\{|\ell\rangle\}_{\ell \in \mathbb{Z}}$, span the walker space: the index ℓ thus labels the position on a one-dimensional infinite lattice. As in the majority of the photonic implementations, we consider the polarization as the coin degree of freedom, e.g. we identify the left- and right-circular polarization states, $|L\rangle$ and $|R\rangle$, with, respectively, $|\text{heads}\rangle$ and $|\text{tails}\rangle$. A QW will be realized as a unitary evolution in the spin–orbit space.

The coin operator is implemented by any birefringent waveplate. We consider the case of a quarter-wave plate, which, if properly oriented, can be described by the operator (written in the circular polarization basis):

$$\hat{W} = \left(\frac{\sigma_0 + i\sigma_1}{\sqrt{2}}\right)_c \otimes 1_w, \tag{14.40}$$

where $\{\sigma_i\}_{i=0,\dots 3}$ are the Pauli matrices (with the identity matrix σ_0 included) and w is the identity in the walker space. The subscripts 'c' and 'w' indicate operators acting on the coin and walker space, respectively.

For the coin-dependent translation, we need an optical device that couples the SAM and OAM degrees of freedom, e.g. a q-plate. The q-plate action, for half-wave retardation, can be described by the unitary operator:

$$\hat{Q} = \sum_{\ell} (|\ell + 1\rangle_w \langle \ell| \otimes |R\rangle_c \langle L| + |\ell - 1\rangle_w \langle \ell| \otimes |L\rangle_c \langle R|), \tag{14.41}$$

from which we see that, using the QW nomenclature, if the coin is $|L\rangle$, the walker position is increased by one lattice site (and also the coin state is flipped). Conversely, if the coin is $|R\rangle$, the walker position is decreased by one lattice site. Notice that we are considering a q-plate with $q = 1/2$; however, the same holds for arbitrary q, with the only difference that the translation in the walker space will be $\pm 2q$. We also remark that the q-plate optical retardation, δ, can be electrically tuned in such a way to realize the more general operator:

$$\hat{Q}(\delta) = \cos\left(\frac{\delta}{2}\right)1 + i\sin\left(\frac{\delta}{2}\right)\sum_{\ell} (|\ell + 1\rangle_w \langle \ell| \otimes |R\rangle_c \langle L| + |\ell - 1\rangle_w \langle \ell| \otimes |L\rangle_c \langle R|), \tag{14.42}$$

which introduces a new feature: the possibility that, between one step and the other, the particle stays in its original position, with probability $\cos(\delta/2)^2$. This feature can be applied to a variety of different protocols by changing the parameter δ and, as we will see, it can be used to explore different topological phases. The QW evolution is obtained by a repeated application of the operator $\hat{U}_{1D}(\delta) = \hat{W}\cdot\hat{Q}(\delta)$; i.e. given an initial state in the spin–orbit space $|\psi_0\rangle$, the state after t-steps will be $|\psi_t\rangle = (\hat{U}_{1D}(\delta))^t |\psi_0\rangle$.

This platform was first realized in Ref. [81] by cascading alternatively q-plates and quarter waveplates. The final probability distribution can be measured with the standard techniques, e.g. phase flattening and mode sorting (the latter has been used in Ref. [82]). In Figure 14.8, we show the layout of the experimental setup (panel a) with an example of an experimentally reconstructed QW evolution (panel b).

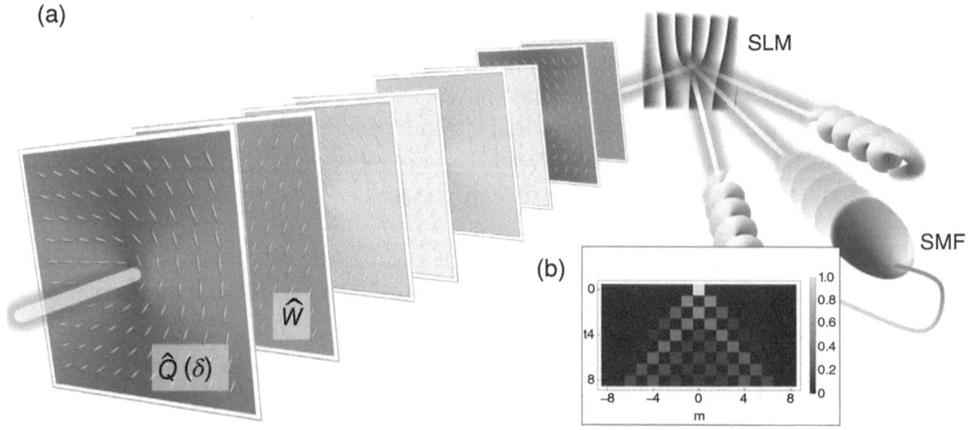

Figure 14.8 Quantum walk in the spin–orbit space. (a) Schematic representation of a setup implementing a quantum walk in the polarization–OAM space. The process is obtained by alternating q-plates $\hat{Q}(\delta)$ with quarter waveplates $\hat{W}$. The final probability distribution is obtained by phase-flattening measurements achieved with an SLM whose first diffraction order is coupled with a single-mode optical fiber (SMF). (b) Example of the probability distribution evolution in a QW with eight-steps ($\delta = \pi$).

14.6.2 Shaping the Walker Space: Cyclic Walks and Walks on 2D Lattices

The OAM space is isomorphic to a one-dimensional lattice; thus, it is not ideal to study phenomena on other kinds of graphs. Different implementations are needed to exploit structured light in these more general settings. Here, we will consider QWs on cyclic and on 2D lattices.

Cyclic QWs can be of interest for exploring the physics of condensed matter under periodic boundary conditions and electron dynamics in cyclic molecules. One way to implement a QW on a circular lattice is to encode the walker position in some spatial degree of freedom that is intrinsically periodic [83]. This can be an angular mode whose intensity distribution is confined within some values of the azimuthal angle ϕ (defined assuming a cylindrical coordinate system (r, ϕ, z)). The intensity distribution of this mode is represented in Figure 14.9a.

A translation operator in this space will correspond to a rotation of the image by an angle $\alpha = \pm 2\pi/N$, where N is the number of sites in the cyclic lattice. This task is achieved using a Dove prism oriented at an angle $\alpha/2$. Depending on the entrance facet of the prism, the rotation will be either clockwise or counterclockwise. To make this transformation polarization-dependent, it is sufficient to place the prism into a polarizing Sagnac interferometer (see Figure 14.9b). Horizontal polarization, $|H\rangle$, will experience clockwise rotations, while vertical polarization, $|V\rangle$, will be subject to a counterclockwise rotation. This interferometer thus represents the coin-dependent translation on a cyclic lattice:

$$\hat{S} = \sum_{i=1,\dots,N} (|\,\mathrm{mod}_N(i+1)\rangle_w\langle i\,| \otimes |\,H\rangle_c\langle H\,| \;+\; |\,\mathrm{mod}_N(i-1)\rangle_w\langle i\,| \otimes |\,V\rangle_c\langle V\,|), \tag{14.43}$$

where $\mathrm{mod}_N(a)$ indicates that a should be evaluated 'modulo N.' As in other examples, the coin of this setup will correspond to the polarization. In this platform, the operator $\hat{W}$, defined in Eq. (14.40), will be implemented by a half-wave plate instead of a quarter-wave plate; notice that in this case, we are indeed considering the linear polarization basis. The unitary evolution after t steps is obtained by applying the operator $\hat{U}^t_{\text{cyclic}} = \left(\hat{S}\cdot\hat{W}\right)^t$. Figure 14.9c gives an example of the

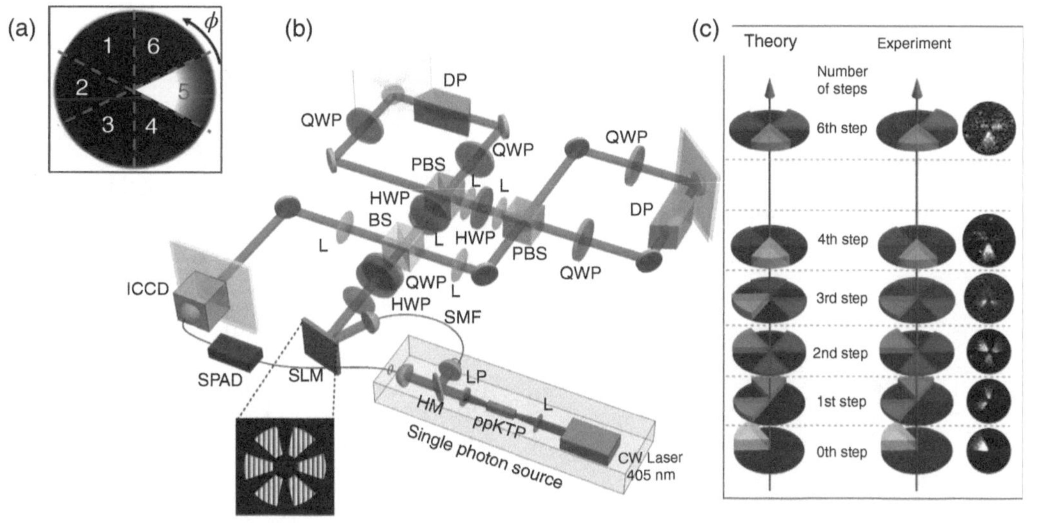

Figure 14.9 Quantum walk on a cyclic graph. (a) Intensity plot of an optical mode associated with a particle localized on site $i = 5$, for a cyclic lattice with six sites; (b) experimental setup; (c) theoretical and experimental distributions. Panels (b) and (c) are adapted from Ref. [83]. *Source:* Adapted from Nejadsattari et al. [83].

probability distribution dynamics obtained with this setup. Cyclic QWs have been recently realized also in the time-bin domain, exploiting a four-dimensional coin space [84] and in the phase space of optical modes coupled with superconducting qubits [77].

Let us now consider the possibility to implement QWs on a 2D lattice. The realization of this kind of platform has strong interest since it can explore the condensed matter physics of 2D materials, e.g. graphene or quantum Hall and spin-Hall insulators. Also, some quantum search algorithms are based on performing QWs on square lattices [85]. In D'Errico et al. [86], the lattice position was encoded in the transverse wavevector of light. Consider the following optical mode:

$$\mathbf{E}_i(x, y, z) = \exp\left(i\Delta k\left(m_x x + m_y y\right)\right)\exp\left(ik_z z\right)\mathbf{e}_i, \tag{14.44}$$

where $m_{x,\,y}$ are integer numbers, Δk is a constant with the dimensions of an inverse length, k_z is the longitudinal component of the wavevector, and $\mathbf{e}_i$ is the complex polarization vector. This mode represents a plane wave with a transverse component of the wavevector $k_\perp = \Delta k\,(m_x,\, m_y)$. In the following, we will assume that $\Delta k \ll k_z$, so that $k_z \approx 2\pi/\lambda$. Thus, we are considering plane waves propagating approximately along the z-direction but with a slightly tilted wavefront. The amount of 'tilting' will encode the position on a 2D space labeled by the pair $(m_x,\, m_y)$ (see Figure 14.10a). The requirement $\Delta k \ll k_z$ is necessary to ensure that, during the QW process, the different optical modes do not separate spatially leading to the loss of interference effects. Notice that in practice, Eq. (14.44) is modulated by a Gaussian amplitude; hence, during propagation, the overlap between different modes decreases.

For a 2D lattice, we need to implement two translation operators that shift the position along independent directions. Polarization-dependent changes in wavevector can be implemented by a PBOE whose optic axis angle α is given by $\alpha(x) = (\pi/\Lambda)x + \alpha_0$, where x is the direction along which the beam is deflected, Λ is a constant determining the periodicity of the optic axis pattern and α_0 is an offset angle. This device was termed g-plate in Ref. [86]. Its action is mathematically equivalent a q-plate. On the (discretized) wavevector space, a g-plate can be described by the operator,

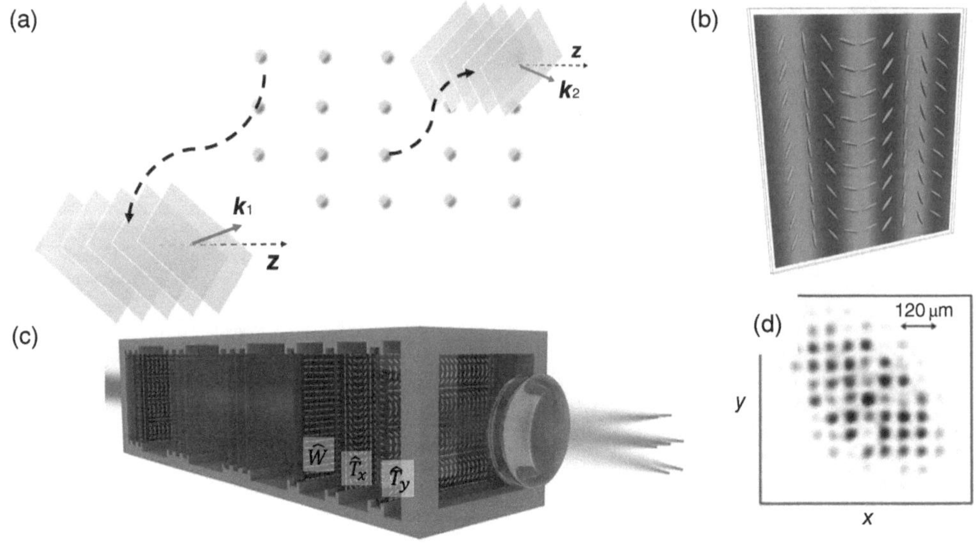

Figure 14.10 Quantum walk on a square lattice. (a) Position eigenstates are encoded in tilted plane waves, with the wavevector being associated with the lattice site; (b) schematic of the optic axis pattern of a *g*-plate. The background shows how this device appears when placed between crossed polarizers and illuminated with a white light; (c) sketch of the experimental setup. The probability distributions are retrieved by looking at the intensity distribution on the focal plane of the output lens. An example is reported in panel (d). *Source:* Adapted from D'Errico et al. [86].

$$\hat{T}_x(\delta) = \cos\left(\frac{\delta}{2}\right) + i\sin\left(\frac{\delta}{2}\right)\sum_{m_x,m_y}\left(e^{i2\alpha_0} \mid m_x+1, m_y\rangle\langle m_x, m_y \mid \otimes \mid R\rangle\langle L \mid \; + h.c.\right), \tag{14.45}$$

and a similar formula holds for $\hat{T}_y$, which describes g-plates patterned along y. The optic axis pattern of the g-plate is shown in Figure 14.10b.

The building blocks of 2D QWs are g-plates, oriented along the directions that correspond to the lattice translation vectors, and waveplates. In Ref. [84], the latter were made of tunable liquid-crystal slabs, fabricated with the same technology used for q-plates and g-plates. Figure 14.10c and d shows a sketch of the experimental platform and the experimentally recorded probability distribution after five steps of a QW. The single-step operator was given by $\hat{U}_{2D} = \hat{W}\cdot\hat{T}_x(\pi/2)\cdot\hat{T}_y(\pi/2)$. This protocol, as we will see in the next section, exhibits interesting physical properties; in particular, it can be shown to simulate a periodically driven Chern insulator.

14.6.3 Applications: Wavepacket Dynamics and Detection of Topological Phases

Important features of a QW can be found by looking at the spectral properties of the single-step operator U. The translational symmetry of the processes considered so far means we can take Fourier transform of the position eigenfunctions $|\mathbf{m}\rangle$:

$$\mid \mathbf{q}\rangle = \sum_{\mathbf{m}} \exp(i\mathbf{q}\cdot\mathbf{m}) \mid \mathbf{m}\rangle. \tag{14.46}$$

These states are known as *quasi-momentum states*, where the quasi-momentum $\mathbf{q}$ is a variable defined within a finite interval (or region in the 2D case) known as the Brillouin zone. For a one-dimensional infinite lattice, $\mathbf{q} = q \in [-\pi, \pi]$. The same relation holds for the cyclic lattice, but this time, q takes discrete values in between the Brillouin zone: $q = -\pi, -(N-1)\pi/N, \ldots(N-1)\pi/N$. For a 2D square lattice, $\mathbf{q} = (q_x, q_y) \in [-\pi, \pi) \times [-\pi, \pi)$. When decomposed in the quasi-momentum eigenstate basis, the single-step operator takes the block-diagonal form:

$$\hat{U} = \int_{BZ} \frac{d^s\mathbf{q}}{(2\pi)^s} \mathcal{U}(\mathbf{q}) \otimes |\,\mathbf{q}\rangle\langle\mathbf{q}\,|, \tag{14.47}$$

where s is the lattice dimensionality, and BZ indicates that the integral has to be evaluated within the Brillouin zone; in the case of the cyclic walk, the integral must be replaced by a sum over the allowed quasi-momenta. $\mathcal{U}(\mathbf{q})$ is a 2×2 unitary matrix acting on the coin space. In general, it can be written in terms of the Pauli matrices:

$$\mathcal{U}(\mathbf{q}) = \exp(-iE(\mathbf{q})\hat{n}(\mathbf{q})\cdot\boldsymbol{\sigma}), \tag{14.48}$$

where $\boldsymbol{\sigma} = (\sigma_1, \sigma_2, \sigma_3)$, $\hat{n}(\mathbf{q})$ is a unit vector, and $\pm E(\mathbf{q})$ are the phases defining the two eigenvalues of $\mathcal{U}(\mathbf{q})$. Since these are phases of an evolution operator, they can be interpreted as the energy eigenvalues of the system. However, since the QW takes place in discrete steps, i.e. time-translational symmetry is discretized, these energies are defined modulo 2π and thus called *quasi-energies* (by an equivalent line of reasoning that leads to the definition of the quasi-momentum). In analogy with solid-state systems, the quasi-energies are functions of the quasi-momentum, i.e. they form energy bands. This simple result shows that we can use QWs to explore the electron dynamics in solids, e.g. semiconductors, whose energy band structure is analogous to the quasi-energy spectrum of the QW. The vectors $\pm\,\hat{n}(\mathbf{q})$ can be interpreted as the coin part of the system eigenstates, as represented on the Bloch sphere. This can be immediately seen recalling that Eq. (14.48) represents a rotation on the Bloch sphere by an angle $2E(\mathbf{q})$ around $\hat{n}(\mathbf{q})$. The full eigenstates of the system can then be written as $|\,\psi(\mathbf{q})\rangle = |\,\hat{n}(\mathbf{q})\rangle_c \otimes |\,\mathbf{q}\rangle_w$. These states are plane waves with wavevector $\mathbf{q}$, equally distributed over all the lattice sites. Their evolution is trivial since, after each step, their wavefunction changes only by a global phase factor $\exp(iE(\mathbf{q}))$. On infinite lattices, these states cannot be generated experimentally. However, this is possible in cyclic QWs, as shown in Ref. [83], where stationary states were observed in a six-step cyclic walk.

More features manifest in the propagation of initial states that form wavepackets, with a quasi-momentum distribution centered around $\mathbf{q}_0$ and the coin equal to the corresponding eigenstate $|\,\hat{n}(\mathbf{q}_0)\rangle$:

$$|\,G(\mathbf{q}_0)\rangle \propto \int_{BZ} \frac{d^s\mathbf{q}}{(2\pi)^s} \exp\left(-(\mathbf{q}-\mathbf{q}_0)^2/\omega_q^2\right) \exp(i\mathbf{q}_0\cdot\mathbf{m})\,|\,\hat{n}(\mathbf{q}_0)\rangle_c \otimes |\,\mathbf{q}\rangle_w. \tag{14.49}$$

These states approximate an eigenstate, and, by having a well-defined center of mass, it is possible to investigate their trajectory on the lattice. Their center of mass will evolve with a speed given by the *group velocity* $\mathbf{v}_g = \nabla_\mathbf{q} E(\mathbf{q})\big|_{\mathbf{q}_0}$. In addition to this, due to the dispersion in the energy band structure, the wavepacket's width will experience a spreading proportional to the *group velocity dispersion* $\mathbf{GVD} = \nabla_\mathbf{q}\mathbf{v}_g(\mathbf{q})\big|_{\mathbf{q}_0}$. Wavepacket dynamics have been observed in both the 1D and 2D infinite lattice QW platforms presented in the previous sections (see Refs. [81, 86]).

The distribution of the eigenvectors $\hat{n}(\mathbf{q})$ in the Brillouin zone may carry information about the topology of the system under question. This observation has lead, in the last decades, to the understanding of phenomena like the quantum Hall effect and the discovery of new phases of matter known as *topological insulators* [87]. These systems are characterized by robust and dissipationless edge modes manifesting in soliton states in 1D polyacetilene chains [88] and quantized electric and spin currents in 2D semiconductors (quantum Hall effect and spin Hall effect) [89]. These phenomena observed at the boundary of these materials are directly connected to bulk properties, i.e. the nontrivial topology of the eigenstates of the bulk system. This connection is known as the *bulk boundary correspondence*. It is indeed possible to associate a *topological invariant* with the bulk Hamiltonian, i.e. a quantity that is *global* and *invariant under smooth transformations*. This fascinating feature has found a plethora of potential applications, from metrology [90] to spintronics and quantum computation [91].

QWs are ideal platforms to study these systems due to the possibility to engineer the energy eigenvalues and eigenvectors [71, 92]. Interestingly, the QW platforms presented in this chapter have nontrivial topological phases. Although implementing edges in the OAM or the transverse wavevector space is experimentally challenging, it is still possible to look for topological effects on the single particle bulk dynamics.

Let us start with the protocol U_{1D}. This system exhibits an important discrete symmetry, called *chiral* or *sublattice symmetry*, for which the vectors $\hat{n}(\mathbf{q})$ are all constrained to lie on the same plane, all being perpendicular to a constant vector $\mathbf{v}_\Gamma$. This defines a topological invariant called the winding number. This can be visualized by considering the vector $\mathbf{h}(q) = E(q)\hat{n}(q)\cdot\boldsymbol{\sigma}$, which describes a curve $(h_1(q), h_2(q))$ in the 2D plane. In this plane, the origin represents a point with zero energy (i.e. a degeneracy point). The winding number is given by the number of times this curve encloses the origin. Obviously, this quantity is ill-defined if there is a gap closing, since in this case the origin is part of the curve. The protocol $U_{1D}(\delta)$ has winding number $\nu = 1$ for $\pi/2 < \delta < 3\pi/2$ and $\nu = 0$ for the other values. In Ref. [22], the behavior of the momenta of the probability distribution as a function of δ was investigated. It was found that slope singularities in these functions appear (in the long step-number limit) in correspondence with the topological phase transition. However, these may be effects of only the gap closing and not of the change in topological invariant. A more direct probe of the topological phase is, instead, the measurement of a bulk property, proposed in Ref. [23], called the *mean chiral displacement*. This is given by the operator $\exp(i\mathbf{v}_\Gamma\cdot\sigma)\hat{m}$ averaged over the state obtained by the QW evolution of an initially localized particle:

$$\mathcal{C} = \left\langle m = 0, \pi \mid \hat{U}^{-t} \exp(i\mathbf{v}_\Gamma\cdot\sigma)\hat{m}\hat{U}^{t} \mid m = 0, \pi \right\rangle \tag{14.50}$$

where $|\pi\rangle$ is an arbitrary coin state. It can be shown that, in the limit $t \to \infty$:

$$\mathcal{C} = \nu/2 + O(1/\sqrt{t}). \tag{14.51}$$

Hence, the mean chiral displacement can directly measure the winding number. This task was achieved in Ref. [23], where the whole topological phase diagram of the protocol U_{1D} was experimentally reconstructed. The result in Eq. (14.50) was later generalized to higher-dimensional coin spaces [93] and to disordered lattice system, showing the first experimental proof of the existence of an Anderson topological insulator [78]. Moreover, the constraint on the initial localized state was later relaxed by D'Errico et al. [94], where it was shown both theoretically and experimentally that the mean chiral displacement is able to track the winding number during dynamical topological phase transitions.

In 2D spaces, there is a larger variety of possible topological phases. Anomalous quantum Hall insulators are semiconductors that exhibit a quantized edge transverse conductivity, which is proportional to a topological invariant called the Chern number. This invariant is a generalization of the winding number concept in two dimensions. In this case, the vector $\mathbf{h}(\mathbf{q}) = E(q)\hat{n}(\mathbf{q})\cdot\boldsymbol{\sigma}$ describes a 2D closed surface embedded in the 3D space. The Chern number is given by the number of times the origin is enclosed into this surface.

Although no observable similar to the mean chiral displacement is known to allow a bulk measurement of this invariant, it is possible to infer the value of this quantity from bulk transport measurements when an external constant (electric or magnetic) field is applied in order to induce an *anomalous displacement* proportional to the Chern number. Remarkably, an external field can be easily simulated in the platform implementing the protocol $\hat{U}_{2D}$ by introducing a transverse shift of the g-plates that is proportional to the step number [86]. This approach was able to measure the effect of the topological phase on the anomalous displacement, as shown in Figure 14.11b.

We conclude this section by mentioning how structured light QWs are also of interest in quantum-computational experiments. Recent works have exploited OAM-based QWs to generate specific high-dimensional states by means of machine learning techniques aimed to optimize the different tunable parameters available in the platform [95]. Moreover, QW setups may be employed in multiphoton experiments to investigate particle statistics and correlations or to perform Boson sampling [96].

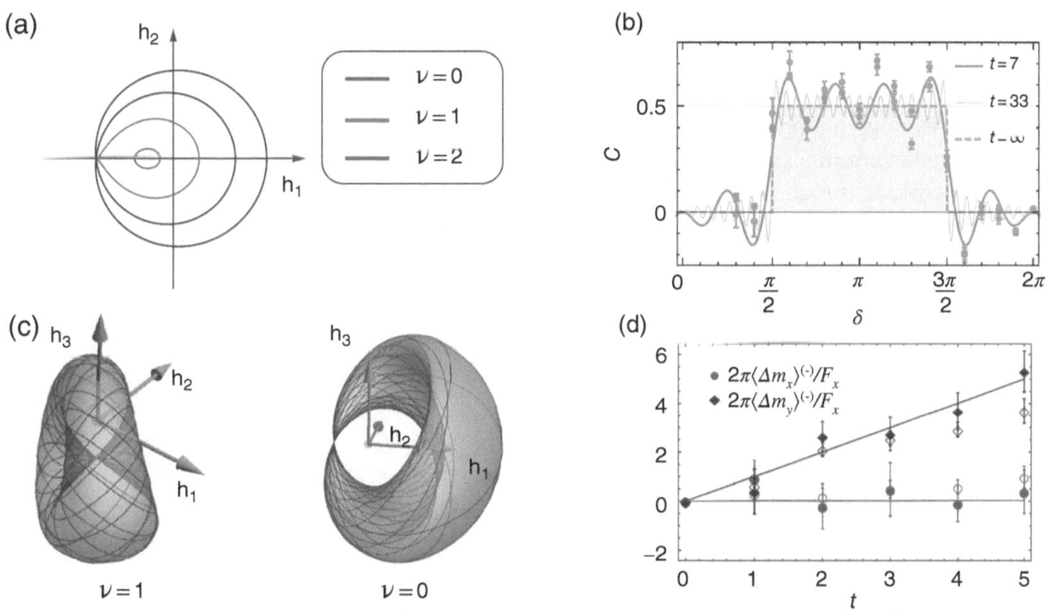

Figure 14.11 Bulk measurement of topological invariants. (a) Examples of curves with different winding numbers; (b) experimental reconstruction of the topological phase diagram of the protocol $\hat{U}_{1D}$ from measurements of the mean chiral displacement (continuous lines show theoretical results for different time steps). Results are from Ref. [23]. *Source:* Cardano et al. [23]. Licenced under CC BY-4.0.; (c) examples of surfaces with different values of the Chern number. This number is different from zero in the origin if enclosed by the surface; (d) experimental detection of the anomalous displacement for the protocol $\hat{U}_{2D}(\delta = \pi/2)$ with Chern number $\nu = 1$. *Source:* Adapted from D'Errico et al. [86].

14.7 Outlook

In this chapter, we showed how structured light plays a vital role in quantum optics and quantum information. The most common method for generating photon pairs, the SPDC process, naturally gives rise to entangled quantum states in the OAM degree of freedom. The OAM, eventually coupled with the polarization, can play an essential role in quantum communication, by increasing the security threshold in QKD as well as the amount of information that is shared, and allowing for alignment-free communications. We have reviewed some recent experiments where the possibility of exploiting high-dimensional structured light for QKD through free space and fiber links was successfully demonstrated. Moreover, exciting progress is currently being made toward underwater QKD. Future improvements toward practical QKD links will require speeding up the transmission rate (by, e.g. developing fast switching q-plates and exploiting fast electronics) and increasing the link length (where turbulence effects will have to be addressed by exploiting adaptive optics methods). Moreover, further research in entanglement swapping, high-dimensional quantum teleportation protocols, and storage/retrieval of photonic qudits is needed to develop the essential ingredients for a structured light-based quantum network.

We have also seen how structured light allows accessing high-dimensional Hilbert spaces for quantum simulation, giving examples of simulators of electron dynamics in crystals. QWs based on structured light have allowed for the experimental unveiling of wavepacket dynamics and for testing the relationship between bulk evolution and topological phases. Multiparticle experiments in the described platforms may lead to interesting developments in Boson sampling and simulation of molecular vibrionic spectra [97]. The main challenges to deal with would be the increase of the achievable number of steps, which may be attained through the development of loop architectures, and the reduction of photon losses (which are relevant in liquid-crystal devices but may be overcome by applying properly designed antireflection coatings).

Structured quantum light is currently the subject of intense research in many other areas that we could not cover in this chapter. We mention developments in quantum imaging [98, 99], metrology [100], quantum cloning [19, 53], quantum interference [101–103], and fundamental tests of quantum mechanics [18, 104].

References

1 Beth, R.A. (1936). Mechanical detection and measurement of the angular momentum of light. *Physical Review* **50**: 115–125.

2 Darwin, C.G. (1932). Notes on the theory of radiation. *Proceedings of the Royal Society A* **136** (829): 36–52.

3 Allen, L., Beijersbergen, M.W., Spreeuw, R.J.C., and Woerdman, J.P. (1992). Orbital angular momentum of light and the transformation of laguerre-gaussian laser modes. *Physical Review A* **45** (11): 8185.

4 Nye, J.F. and Berry, M.V. (1974). Dislocations in wave trains. *Proceedings of the Royal Society of London* **336** (1605): 165–190.

5 Einstein, A. (1905). Über einen die erzeugung und verwandlung des lichtes betreffenden heuristischen gesichtspunkt. *Annalen der Physik* **322** (6): 132–148.

6 Mair, A., Vaziri, A., Weihs, G., and Zeilinger, A. (2001). Entanglement of the orbital angular momentum states of photons. *Nature* **412** (6844): 313–316.

7 Nielsen, M. A. and Chuang, I. (2002). *Quantum Computation and Quantum Information*. Cambridge University Press.

8 Burnham, D.C. and Weinberg, D.L. (1970). Observation of simultaneity in parametric production of optical photon pairs. *Physical Review Letters* **25**: 84–87.

9 Boyd, R.W. (2019). *Nonlinear Optics*. London, UK: Academic Press.

10 Miatto, F.M., Yao, A.M., and Barnett, S.M. (2011). Full characterization of the quantum spiral bandwidth of entangled biphotons. *Physical Review A* **83** (3): 033816.

11 Bouchard, F., Harris, J., Mand, H. et al. (2015). Observation of quantum recoherence of photons by spatial propagation. *Scientific Reports* **5** (15330).

12 Beijersbergen, M.W., Coerwinkel, R.P.C., Kristensen, M., and Woerdman, J.P. (1994). Helical-wavefront laser beams produced with a spiral phaseplate. *Optics Communications* **112** (5-6): 321–327.

13 Fickler, R., Campbell, G., Buchler, B. et al. (2016). Quantum entanglement of angular momentum states with quantum numbers up to 10,010. *Proceedings of the National Academy of Sciences of the United States of America* **113** (48): 13642–13647.

14 Bolduc, E., Bent, N., Santamato, E. et al. (2013). Exact solution to simultaneous intensity and phase encryption with a single phase-only hologram. *Optics Letters* **38** (18): 3546–3549.

15 Karimi, E. and Boyd, R.W. (2015). Classical entanglement? *Science* **350** (6265): 1172–1173.

16 Marrucci, L., Manzo, C., and Paparo, D. (2006). Optical spin-to-orbital angular momentum conversion in inhomogeneous anisotropic media. *Physical Review Letters* **96** (16): 163905.

17 Piccirillo, B., D'Ambrosio, V., Slussarenko, S. et al. (2010). Photon spin-to-orbital angular momentum conversion via an electrically tunable q-plate. *Applied Physics Letters* **97** (24): 4085–4090.

18 Karimi, E., Cardano, F., Maffei, M. et al. (2014). Hardy's paradox tested in the spin-orbit hilbert space of single photons. *Physical Review A* **89**: 032122.

19 Nagali, E., Sciarrino, F., De Martini, F. et al. (2009). Quantum information transfer from spin to orbital angular momentum of photons. *Physical Review Letters* **103**: 013601

20 Nagali, E., Giovannini, D., Marrucci, L. et al. (2010). Experimental optimal cloning of four-dimensional quantum states of photons. *Physical Review Letters* **105** (7): 073602.

21 Cardano, F., Massa, F., Qassim, H. et al. (2015). Quantum walks and wavepacket dynamics on a lattice with twisted photons. *Science Advances* **1**: e1500087.

22 Cardano, F., Maffei, M., Massa, F. et al. (2016). Statistical moments of quantum-walk dynamics reveal topological quantum transitions. *Nature Communications* **7**: 11439.

23 Cardano, F., D'Errico, A., Dauphin, A. et al. (2017). Detection of zak phases and topological invariants in a chiral quantum walk of twisted photons. *Nature Communications* **8**: 15516.

24 Larocque, H., Gagnon-Bischoff, J., Mortimer, D. et al. (2017). Generalized optical angular momentum sorter and its application to high-dimensional quantum cryptography. *Optics Express* **25** (17): 19832–19843.

25 D'Errico, A., D'Amelio, R., Piccirillo, B. et al. (2017). Measuring the complex orbital angular momentum spectrum and spatial mode decomposition of structured light beams. *Optica* **4** (11): 1350–1357.

26 Agnew, M., Leach, J., McLaren, M. et al. (2011). Tomography of the quantum state of photons entangled in high dimensions. *Physical Review A* **84** (062101)

27 Langford, N.K., Dalton, R.B., Harvey, M.D. et al. (2004). Measuring entangled qutrits and their use for quantum bit commitment. *Physical Review Letters* **93**: 053601.

28 Qassim, H., Miatto, F.M., Torres, J.P. et al. (2014). Limitations to the determination of a laguerre–gauss spectrum via projective, phase-flattening measurement. *Journal of the Optical Society of America B* **31** (6): A20–A23.

29 Mirhosseini, M., Malik, M., Shi, Z., and Boyd, R.W. (2013). Efficient separation of the orbital angular momentum eigenstates of light. *Nature Communications* **4**: 2781.

30 Karimi, E. and Santamato, E. (2012). Radial coherent and intelligent states of paraxial wave equation. *Optics Letters* **37** (13): 2484–2486.

31 Bouchard, F., Valencia, N.H., Brandt, F. et al. (2018). Measuring azimuthal and radial modes of photons. *Optics Express* **26** (24): 31925–31941.

32 Fickler, R., Bouchard, F., Giese, E. et al. (2020). Full-field mode sorter using two optimized phase transformations for high-dimensional quantum cryptography. *Journal of Optics* **22** (2) ISSN 20408986.

33 Nicolas, K., Fontaine, R.R., Chen, H. et al. (2019). Laguerre-Gaussian mode sorter. *Nature Communications* **10** (1): 1865.

34 Fu, D., Zhou, Y., Qi, R. et al. (2018). Realization of a scalable Laguerre–Gaussian mode sorter based on a robust radial mode sorter. *Optics Express* **26** (25): 33057.

35 Gu, M., Krenn, M., Erhard, M., and Zeilinger, A. (2018). Gouy phase radial mode sorter for light: Concepts and experiments. *Physical Review Letters* **120**: 103601.

36 Mukesh Kumar Rana and Shaina Garg. A review on rsa encryption algorithm. *International Journal of Engineering and Computer Science* **5**(7), 2017.

37 Agrawal, M., Kayal, N., and Saxena, N. (2004). Primes is in p. *Annals of Mathematics* **160** (2): 781–793.

38 Bennett, C. H. and Brassard, G. (1984). *Proceedings of the IEEE International Conference on Computers, Systems, and Signal Processing, Bangalore, India*, (10 December 1984).

39 Pirandola, S., Andersen, U.L., Banchi, L. et al. (2019). Advances in quantum cryptography. *Advances in Optics and Photonics* **12** (4): 1012–1236.

40 Brassard, G. and Salvail, L. (1994). *Secret-key Reconciliation by Public Discussion*, 410–423. Springer-Verlag.

41 Shor, P.W. and Preskill, J. (2000). Simple proof of security of the bb84 quantum key distribution protocol. *Physical Review Letters* **85** (2): 441.

42 Bruß, D. (1998). Optimal eavesdropping in quantum cryptography with six states. *Physical Review Letters* **81** (14): 3018.

43 Cerf, N.J., Bourennane, M., Karlsson, A., and Gisin, N. (2002). Security of quantum key distribution using d-level systems. *Physical Review Letters* **88** (12): 127902.

44 D'Ambrosio, V., Nagali, E., Walborn, S.P. et al. (2012). Complete experimental toolbox for alignment-free quantum communication. *Nature Communications* **3**: 961.

45 Cardano, F., Karimi, E., Slussarenko, S. et al. (2012). Polarization pattern of vector vortex beams generated by q-plates with different topological charges. *Applied Optics* **51** (10): C1–C6.

46 Vallone, G., D'Ambrosio, V., Sponselli, A. et al. (2014). Free-space quantum key distribution by rotation-invariant twisted photons. *Physical Review Letters* **113** (6): 060503.

47 Sheridan, L. and Scarani, V. (2010). Security proof for quantum key distribution using qudit systems. *Physical Review A* **82** (3): 030301.

48 Mirhosseini, M., Magaña-Loaiza, O.S., O'Sullivan, M.N. et al. (2015). High-dimensional quantum cryptography with twisted light. *New Journal of Physics* **17** (3): 033033.

49 Bouchard, F., Heshami, K., England, D. et al. (2018). Experimental investigation of quantum key distribution protocols with twisted photons. *Quantum* **2**: 111.

50 Chau, H.F. (2015). Quantum key distribution using qudits that each encode one bit of raw key. *Physical Review A* **92** (6): 062324.

51 Berthold-Georg, E., Chua, W.K., Anders, J. et al. (2004). Highly efficient quantum key distribution with minimal state tomography. https://arxiv.org/abs/quant-ph/0412075v2.

52 Bouchard, F., Hufnagel, F., Koutný, D. et al. (2019). Quantum process tomography of a high-dimensional quantum communication channel. *Quantum* **3**: 138.

53 Bouchard, F., Fickler, R., Boyd, R.W., and Karimi, E. (2017). High-dimensional quantum cloning and applications to quantum hacking. *Science Advances* **3** (2): e1601915.

54 Sit, A., Bouchard, F., Fickler, R. et al. (2017). High-dimensional intracity quantum cryptography with structured photons. *Optica* **4** (9): 1006–1010.

55 Cozzolino, D., Bacco, D., Da Lio, B. et al. (2019). Orbital angular momentum states enabling fiber-based high-dimensional quantum communication. *Physical Review Applied* **11**: 064058.

56 Sit, A., Fickler, R., Alsaiari, F. et al. (2018). Quantum cryptography with structured photons through a vortex fiber. *Optics Letters* **43** (17): 4108–4111.

57 Bouchard, F., Sit, A., Hufnagel, F., et al.. (2018). Quantum cryptography with twisted photons through an outdoor underwater channel. *Optics Express* **26** (17): 22563–22573.

58 Fried, D.L. (1966). Optical resolution through a randomly inhomogeneous medium for very long and very short exposures. *Journal of the Optical Society of America* **56** (10): 1372–1379.

59 Gang, S., Ning-Quan, W., Li-Ming, X., and Wu, Y. (2012). Profile and character of atmospheric structure constant of refractive index cn2. *Atmospheric and Oceanic Science Letters* **5** (3): 270–272.

60 Ramachandran, S., Kristensen, P., and Yan, M.F. (2009). Generation and propagation of radially polarized beams in optical fibers. *Optics Letters* **34** (16): 2525–2527.

61 Hufnagel, F., Sit, A., Grenapin, F. et al. (2019). Characterization of an underwater channel for quantum communications in the ottawa river. *Optics Express* **27** (19): 26346–26354.

62 Hufnagel, F., Sit, A., Bouchard, F. et al. (2020). Investigation of underwater quantum channels in a 30 meter flume tank using structured photons. *New Journal of Physics* **22** (9): 093074.

63 Zhang, Y., Agnew, M., Roger, T. et al. (2017). Simultaneous entanglement swapping of multiple orbital angular momentum states of light. *Nature Communications* **8** (1): 632.

64 Parigi, V., D'Ambrosio, V., Arnold, C. et al. (2015). Storage and retrieval of vector beams of light in a multiple-degree-of-freedom quantum memory. *Nature Communications* **6** (1): 7706.

65 Feynman, R.P. (1982). Simulating physics with computers. *International Journal of Theoretical Physics* **21** (6–7): 467–488.

66 Aharonov, Y., Davidovich, L., and Zagury, N. (1993). Quantum random walks. *Physical Review A* **48**: 1687–1690.

67 Kempe, J. (2003). Quantum random walks: An introductory overview. *Contemporary Physics* **44** (4): 307–327.

68 Childs, A.M. (2009). Universal computation by quantum walk. *Physical Review Letters* **102** (18): 180501.

69 Lovett, N.B., Cooper, S., Everitt, M. et al. (2010). Universal quantum computation using the discrete-time quantum walk. *Physical Review A* **81** (4): 042330.

70 Childs, A.M. and Goldstone, J. (2004). Spatial search by quantum walk. *Physical Review A* **70** (2): 022314.

71 Kitagawa, T., Rudner, M.S., Berg, E., and Demler, E. (2010). Exploring topological phases with quantum walks. *Physical Review A* **82** (3): 1–12.

72 Sansoni, L., Sciarrino, F., Vallone, G. et al. (2012). Two-Particle Bosonic-Fermionic quantum walk via integrated photonics. *Physical Review Letters* **108** (1): 010502.

73 Crespi, A., Osellame, R., Ramponi, R. et al. (2013). Anderson localization of entangled photons in an integrated quantum walk. *Nature Photonics* **7**: 322–328.

74 Xiao, L., Deng, T., Wang, K. et al. (2020). Non-Hermitian bulk–boundary correspondence in quantum dynamics. *Nature Physics* **16**: 761–766.

75 Genske, M., Alt, W., Steffen, A. et al. (2013). Electric quantum walks with individual atoms. *Physical Review Letters* **110** (19): 190601.

76 Karski, M., Förster, L., Choi, J.M. et al. (2009). Quantum walk in position space with single optically trapped atoms. *Science* **325** (5937): 174–177.

77 Ramasesh, V.V., Flurin, E., Rudner, M.S. et al. (2016). Direct probe of topological invariants using bloch oscillating quantum walks. http://arxiv.org/abs/1609.09504.

78 Meier, E.J., An, F.A., Dauphin, A. et al. (2018). Observation of the topological Anderson insulator in disordered atomic wires. *Science* **362**(6417): 929–933.

79 Broome, M.A., Fedrizzi, A., Lanyon, B.P. et al. (2010). Discrete single-photon quantum walks with tunable decoherence. *Physical Review Letters* **104** (15): 1–5.

80 Schreiber, K., Cassemiro, N., Potoček, V. et al. (2010). Photons walking the line: A quantum walk with adjustable coin operations. *Physical Review Letters* **104** (5): 3–6.

81 Cardano, F., Massa, F., Qassim, H. et al. (2015). Quantum walks and wavepacket dynamics on a lattice with twisted photons. *Science Advances* **1** (2): e1500087.

82 Sephton, B., Dudley, A., Ruffato, G. et al. (2019). A versatile quantum walk resonator with bright classical light. *PLOS ONE* **14** (4): e0214891.

83 Nejadsattari, F., Zhang, Y., Bouchard, F. et al. (2019). Experimental realization of wave-packet dynamics in cyclic quantum walks. *Optica* **6** (2): 174–180.

84 Lorz, L., Meyer-Scott, E., Nitsche, T. et al. (2019). Photonic quantum walks with four-dimensional coins. *Physical Review Research* **1**: 033036.

85 Jeong, Y.C., Di Franco, C., Lim, H.T. et al. (2013). Experimental realization of a delayed-choice quantum walk. *Nature Communications* **4**: 3471.

86 D'Errico, A., Cardano, F., Maffei, M. et al. (2020). Two-dimensional topological quantum walks in the momentum space of structured light. *Optica* **7** (2): 108–114.

87 Andrei Bernevig, B. (2013). *Topological Insulators and Topological Superconductors*. Princeton: Princeton University Press.

88 Su, W.P., Schrieffer, J.R., and Heeger, A.J. (1979). Solitons in Polyacetylene. *Physical Review Letters* **42** (25): 1698–1701.

89 Hasan, M.Z. and Kane, C.L. (2010). Colloquium: Topological insulators. *Reviews of Modern Physics* **82** (4): 3045–3067.

90 von Klitzing, K. (1986). The quantized Hall effect. *Reviews of Modern Physics* **58** (3): 519–531.

91 He, M., Sun, H., and He, Q.L. (2019). Topological insulator: Spintronics and quantum computations. *Frontiers in Physics* **14** (4): 43401.

92 Kitagawa, T. (2012). Topological phenomena in quantum walks: Elementary introduction to the physics of topological phases. *Quantum Information Processing* **11** (5): 1107–1148.

93 Maffei, M., Dauphin, A., Cardano, F. et al. (2018). Topological characterization of chiral models through their long time dynamics. *New Journal of Physics* **20** (1): 1–17.

94 D'Errico, A., Di Colandrea, F., Barboza, R. et al. (2020). Bulk detection of time-dependent topological transitions in quenched chiral models. *Physical Review Research* **2**: 023119.

95 Giordani, T., Polino, E., Emiliani, S. et al. (2019). Experimental engineering of arbitrary qudit states with discrete-time quantum walks. *Physical Review Letters* **122** (2): 20503.

96 Bentivegna, M., Spagnolo, N., Vitelli, C. et al. (2015). Experimental scattershot boson sampling. *Science Advances* **1** (3): e1400255.

97 Huh, J., Guerreschi, G.G., Peropadre, B. et al. (2015). Boson sampling for molecular vibronic spectra. *Nature Photonics* **9** (9): 615–620.

98 Jack, B., Leach, J., Romero, J. et al. (2009). Holographic ghost imaging and the violation of a bell inequality. *Physical Review Letters* **103**: 083602.

99 Chen, L., Lei, J., and Romero, J. (2014). Quantum digital spiral imaging. *Light: Science and Applications* **3**: e153.

100 D'Ambrosio, V., Spagnolo, N., Del Re, L. et al. (2013). Photonic polarization gears for ultra-sensitive angular measurements. *Nature Communications* **4**: 2432.

101 Nagali, E., Sansoni, L., Sciarrino, F. et al. (2009). Optimal quantum cloning of orbital angular momentum photon qubits through Hong-Ou-Mandel coalescence. *Nature Photonics* **3** (12): 720–723.

102 Karimi, E., Boyd, R.W., De La Hoz, P. et al. (2014). Radial quantum number of laguerre-gauss modes. *Physical Review A* **89** (6): 063813.

103 Zhang, Y., Roux, F.S., Konrad, T. et al. (2016). Engineering two-photon high-dimensional states through quantum interference. *Science Advances* **2** (2): e1501165.

104 Cardano, F., Karimi, E., Marrucci, L. et al. (2013). Violation of leggett-type inequalities in the spin-orbit degrees of freedom of a single photon. *Physical Review A* **88**: 032101.

Index

Note: Page numbers in *italics* denote figures and page numbers in **bold** denote tables.

Electromagnetic Vortices: Wave Phenomena and Engineering Applications, First Edition.
Edited by Zhi Hao Jiang and Douglas H. Werner.

n

o

q

r

s

t

IEEE PRESS SERIES ON ELECTROMAGNETIC WAVE THEORY

Field Theory of Guided Waves, Second Edition
Robert E. Collin

Field Computation by Moment Methods
Roger F. Harrington

Radiation and Scattering of Waves
Leopold B. Felsen, Nathan Marcuvitz

Methods in Electromagnetic Wave Propagation, Second Edition
D. S. J. Jones

Mathematical Foundations for Electromagnetic Theory
Donald G. Dudley

The Transmission-Line Modeling Method: TLM
Christos Christopoulos

Methods for Electromagnetic Field Analysis
Ismo V. Lindell

Singular Electromagnetic Fields and Sources
Jean G. Van Bladel

The Plane Wave Spectrum Representation of Electromagnetic Fields, Revised Edition
P. C. Clemmow

General Vector and Dyadic Analysis: Applied Mathematics in Field Theory
Chen-To Tai

Computational Methods for Electromagnetics
Andrew F. Peterson

Finite Element Method Electromagnetics: Antennas, Microwave Circuits, and Scattering Applications
John L. Volakis, Arindam Chatterjee, Leo C. Kempel

Waves and Fields in Inhomogenous Media
Weng Cho Chew

Electromagnetics: History, Theory, and Applications
Robert S. Elliott

Plane-Wave Theory of Time-Domain Fields: Near-Field Scanning Applications
Thorkild B. Hansen

Foundations for Microwave Engineering, Second Edition
Robert Collin

Time-Harmonic Electromagnetic Fields
Robert F. Harrington

Antenna Theory & Design, Revised Edition
Robert S. Elliott

Differential Forms in Electromagnetics
Ismo V. Lindell

Conformal Array Antenna Theory and Design
Lars Josefsson, Patrik Persson

Multigrid Finite Element Methods for Electromagnetic Field Modeling
Yu Zhu, Andreas C. Cangellaris

Electromagnetic Vortices: Wave Phenomena and Engineering Applications, First Edition.
Edited by Zhi Hao Jiang and Douglas H. Werner.

Electromagnetic Theory
Julius Adams Stratton

Electromagnetic Fields, Second Edition
Jean G. Van Bladel

Electromagnetic Fields in Cavities: Deterministic and Statistical Theories
David A. Hill

Discontinuities in the Electromagnetic Field
M. Mithat Idemen

Understanding Geometric Algebra for Electromagnetic Theory
John W. Arthur

The Power and Beauty of Electromagnetic Theory
Frederic R. Morgenthaler

Electromagnetic Modeling and Simulation
Levent Sevgi

Multiforms, Dyadics, and Electromagnetic Media
Ismo V. Lindell

Low-Profile Natural and Metamaterial Antennas: Analysis Methods and Applications
Hisamatsu Nakano

From ER to E.T.: How Electromagnetic Technologies Are Changing Our Lives
Rajeev Bansal

Electromagnetic Wave Propagation, Radiation, and Scattering: From Fundamentals to Applications, Second Edition
Akira Ishimaru

Time-Domain Electromagnetic Reciprocity in Antenna Modeling
Martin Štumpf

Boundary Conditions in Electromagnetics
Ismo V. Lindell, Ari Sihvola

Substrate-Integrated Millimeter-Wave Antennas for Next-Generation Communication and Radar Systems
Zhi Ning Chen, Xianming Qing

Electromagnetic Radiation, Scattering, and Diffraction
Prabhakar H. Pathak, Robert J. Burkholder

Electromagnetic Vortices: Wave Phenomena and Engineering Applications
Zhi Hao Jiang, Douglas H. Werner